Edited by
Paolo Samorì and
Franco Cacialli

Functional Supramolecular Architectures

Related Titles

Siebbeles, L. D. A., Grozema, F. C. (eds.)

Charge and Exciton Transport through Molecular Wires

2011

ISBN: 978-3-527-32501-6

Leclerc, M., Morin, J.-F. (eds.)

Design and Synthesis of Conjugated Polymers

2010

ISBN: 978-3-527-32474-3

Cosnier, S., Karyakin, A. (eds.)

Electropolymerization

Concepts, Materials and Applications

2010

ISBN: 978-3-527-32414-9

Guldi, D. M., Martín, N. (eds.)

Carbon Nanotubes and Related Structures

Synthesis, Characterization, Functionalization, and Applications

2010

ISBN: 978-3-527-32406-4

Brabec, C., Scherf, U., Dyakonov, V. (eds.)

Organic Photovoltaics

Materials, Device Physics, and Manufacturing Technologies

2008

ISBN: 978-3-527-31675-5

Atwood, J. L., Steed, J. W. (eds.)

Organic Nanostructures

2008

ISBN: 978-3-527-31836-0

Diederich, F., Stang, P. J., Tykwinski, R. R. (eds.)

Modern Supramolecular Chemistry

Strategies for Macrocycle Synthesis

2008

ISBN: 978-3-527-31826-1

Hadziioannou, G., Malliaras, G. G. (eds.)

Semiconducting Polymers

Chemistry, Physics and Engineering

2007

ISBN: 978-3-527-31271-9

Samori, P. (ed.)

Scanning Probe Microscopies Beyond Imaging

Manipulation of Molecules and Nanostructures

2006

ISBN: 978-3-527-31269-6

Edited by
Paolo Samorì and Franco Cacialli

Functional Supramolecular Architectures

for Organic Electronics and Nanotechnology

Volume 1

WILEY-VCH Verlag GmbH & Co. KGaA

The Editors

Prof. Dr. Paolo Samorì
Université de Strasbourg and CNRS (UMR 7006)
Nanochemistry Laboratory, ISIS
8, allée Gaspard Monge
67083 Strasbourg Cedex
France

Prof. Dr. Franco Cacialli
Department of Physics and Astronomy
University College London
Gower Street
London WC1E 6BT
United Kingdom

All books published by **Wiley-VCH** are carefully produced. Nevertheless, authors, editors, and publisher do not warrant the information contained in these books, including this book, to be free of errors. Readers are advised to keep in mind that statements, data, illustrations, procedural details or other items may inadvertently be inaccurate.

Library of Congress Card No.: applied for

British Library Cataloguing-in-Publication Data
A catalogue record for this book is available from the British Library.

Bibliographic information published by the Deutsche Nationalbibliothek
The Deutsche Nationalbibliothek lists this publication in the Deutsche Nationalbibliografie; detailed bibliographic data are available on the Internet at http://dnb.d-nb.de.

© 2011 WILEY-VCH Verlag & Co. KGaA, Boschstr. 12, 69469 Weinheim, Germany

All rights reserved (including those of translation into other languages). No part of this book may be reproduced in any form – by photoprinting, microfilm, or any other means – nor transmitted or translated into a machine language without written permission from the publishers. Registered names, trademarks, etc. used in this book, even when not specifically marked as such, are not to be considered unprotected by law.

Composition   Thomson Digital, Noida, India
Printing and Binding   betz-druck GmbH, Darmstadt
Cover Design   Adam Design, Weinheim

Printed in the Federal Republic of Germany
Printed on acid-free paper

ISBN: 978-3-527-32611-2

To Cristiana, Emma and Anna
and
To Anna, Chiara and Marco

Contents

Preface *XXIII*
List of Contributors *XXV*

Volume 1

Part One Modeling and Theory *1*

1 **Charge Transport in Organic Semiconductors: A Multiscale Modeling** *3*
   Nicolas Martinelli, Yoann Olivier, Luca Muccioli, Andrea Minoia, Patrick Brocorens, Mari-Carmen Ruiz Delgado, Claudio Zannoni, David Beljonne, Roberto Lazzaroni, Jean-Luc Brédas, and Jérôme Cornil
1.1 Introduction *3*
1.2 Organic Single Crystals *6*
1.2.1 Molecular Parameters for Charge Transport *6*
1.2.2 Influence of Intermolecular Vibrations *9*
1.2.3 Charge Mobility Values *13*
1.3 Tetrathiafulvalene Derivatives *15*
1.4 Polythiophene Derivatives *18*
1.5 Phthalocyanine Stacks *22*
1.5.1 Structural Properties *22*
1.5.2 Charge Transport Properties *25*
1.6 Polymer Dielectrics *27*
1.6.1 Electrostatic Disorder *29*
1.6.2 Charge Mobility Values *29*
1.7 Outlook *31*
   References *33*

2 **Monte Carlo Studies of Phase Transitions and Cooperative Motion in Langmuir Monolayers with Internal Dipoles** *39*
   Christopher B. George, Mark A. Ratner, and Igal Szleifer
2.1 Introduction *39*
2.2 Computational Details *40*

Functional Supramolecular Architectures. Edited by Paolo Samorì and Franco Cacialli
Copyright © 2011 WILEY-VCH Verlag GmbH & Co. KGaA, Weinheim
ISBN: 978-3-527-32611-2

2.3	Results and Discussion 42	
2.3.1	Uncharged System 42	
2.3.2	Charged System 45	
2.4	Summary 49	
	References 51	

3 **Molecules on Gold Surfaces: What They Do and How They Go Around to Do It** 55
Nadja Sändig and Francesco Zerbetto

3.1	Introduction 55
3.2	A Simple Description of the Geometrical Structure of Metals 57
3.3	A Simple Description of the Geometrical Structure of Molecules 58
3.4	Electronegativity Governs Chemical Interactions: Charge Equilibration, QEq, Models 58
3.5	A Simple Description of the Interaction between Metal Surfaces and Molecules 59
3.6	Presence of an External Electric Potential or Field 60
3.7	Generality of the Model and Its Transferability 60
3.8	Thiolates on Gold 61
3.9	Adsorption of a Large Molecule: C_{60} 65
3.10	Simple Packing Problems 67
3.11	The Presence of an Electrostatic Potential 69
3.12	Challenges and Conclusion 71
	References 72

Part Two **Supramolecular Synthetic Chemistry** 79

4 **Conjugated Polymer Sensors: Design, Principles, and Biological Applications** 81
Mindy Levine and Timothy M. Swager

4.1	Introduction 81
4.1.1	Amplification 81
4.1.2	Role of Dimensionality and Sensory Mechanism 82
4.2	Water Solubility 83
4.2.1	Conjugated Polyelectrolytes 83
4.2.2	Nonspecific Binding 84
4.2.3	Nonionic Water-Soluble Polymers 86
4.3	Protein Detection 89
4.3.1	Introduction 89
4.3.2	Protease Detection 89
4.3.3	Kinase and Phosphatase Detection 93
4.3.4	Lectin Detection 94
4.3.5	Other Protein Detection 95
4.3.6	Protein Conformation Detection 96
4.4	DNA Detection 97

4.4.1	Polythiophene-Based DNA Detection	97
4.4.2	FRET-Based DNA Detection	98
4.5	Bacteria Detection	101
4.5.1	Introduction	101
4.5.2	Mannose-Containing Polymers	102
4.5.3	Detection of Bacterial Excretion Products	103
4.6	Electron-Deficient Polymers	104
4.6.1	Fluorinated PPEs	104
4.7	Aggregation-Based Detection	106
4.7.1	Introduction	106
4.7.2	Detection via Polymer Quenching	106
4.7.3	Detection via Low-Energy Emission	108
4.8	Temperature-Responsive Fluorescent Polymers	112
4.8.1	Introduction	112
4.8.2	Examples	112
4.9	Nonhomogeneous Detection Schemes	115
4.9.1	Introduction	115
4.9.2	CPs Adsorbed on Surfaces	116
4.9.3	CP Particles from Collapsed Polymer Chains	119
4.10	Mechanism of Energy Transfer	122
4.10.1	Theory	122
4.10.2	Modifications to Förster Theory	123
4.10.3	Evidence for Dexter-Type Energy Migration	123
4.11	Conclusions and Future Directions	127
	References	128
5	**Chromophoric Polyisocyanide Materials**	**135**
	Bram Keereweer, Erik Schwartz, Stéphane Le Gac, Roeland J.M. Nolte, Paul H. J. Kouwer, and Alan E. Rowan	
5.1	Introduction	135
5.2	Polyisocyanide Materials	137
5.3	Perylene Polyisocyanides in Devices	143
5.4	TFT Devices	147
5.5	Morphology Control in Perylene/Crystal Systems	147
5.6	Postmodification of Polyisocyanopeptides	149
5.7	Toward Larger Length Scales: Polyisocyanopeptide Brushes	151
5.8	Summary and Outlook	152
	References	152
6	**Functional Polyphenylenes for Supramolecular Ordering and Application in Organic Electronics**	**155**
	Martin Baumgarten and Klaus Müllen	
6.1	Introduction	155
6.2	Conjugated Polymers	156
6.2.1	Linear and Ladderized Polyphenylenes	156

6.2.2	Polycyclic Aromatic Hydrocarbon-Based Conjugated Polymers	160
6.2.3	Polyphenylene Macrocycles	164
6.2.4	Donors and Donor–Acceptor Approaches	168
6.2.4.1	Defined Oligomers of Heteroacenes	168
6.2.4.2	Poly(benzodithiophene)s	169
6.2.4.3	Poly(cyclopentadithiophene-benzothiadiazole)	171
6.3	Graphene Molecules and Their Alignment	173
6.3.1	Extended Size Nanographene Molecules	173
6.3.2	Polymeric Extensions to Graphene Layers	176
6.3.3	Variation of the Aromatic Core and Its Symmetry	177
6.3.4	Influence of the Side Chains and Their Polarities	179
6.4	Rylenes Dyes	183
6.4.1	Core-Extended Rylenes	183
6.4.2	Rylenes for Ambipolar Field-Effect Transistors	185
6.4.3	Rylenes for Solar Cell Applications	186
6.5	Dendritic Polyphenylenes: The Three-Dimensional Case	187
6.6	Conclusion and Outlook	189
	References	190

7 **Molecular Tectonics: Design of Hybrid Networks and Crystals Based on Charge-Assisted Hydrogen Bonds**
Sylvie Ferlay and Mir Wais Hosseini

7.1	Introduction	195
7.2	Examples of Robust Charge-Assisted H-Bonded (CAHB) Networks	197
7.2.1	Organic Networks	197
7.2.2	Inorganic Networks	198
7.3	Charge-Assisted H-Bonded Networks Based on Amidinium Tectons	199
7.3.1	Amidinium-Based Tectons	199
7.3.2	Organic Networks Based on Amidinium Tectons	200
7.3.3	Inorganic Networks Based on Amidinium Tectons	201
7.3.3.1	Metallatectons Bearing a Carboxylate at Their Periphery	201
7.3.3.2	Oxalatometallate as H-Bond Acceptor	202
7.3.3.3	Thiocyanatometallate as H-Bond Acceptor	203
7.3.3.4	Polycyanometallate as H-Bond Acceptor	204
7.4	Charge-Assisted H-Bonded Networks Based on Amidinium and Polycyanometallate Tectons	205
7.4.1	Octahedral Cyanometallate Anions	205
7.4.1.1	$[M^{III}(CN)_6]^{3-}$ and 2-2H$^+$ or 3-2H$^+$ (M = Fe, Co, and Cr)	205
7.4.1.2	$[M^{III}(CN)_6]^{3-}$ and 4-2H$^+$ (M = Fe, Co, and Cr)	207
7.4.1.3	$[M^{II}(CN)_6]^{4-}$ with 2-2H$^+$ or 4-2H$^+$ (M = Fe and Ru)	208
7.4.1.4	$[M^{II}(CN)_6]^{4-}$ with 3-2H$^+$ (M = Fe and Ru)	208
7.4.1.5	$[M^{II}(CN)_6]^{4-}$ with 4-2H$^+$ (M = Fe and Ru)	209
7.4.2	Pentadentate Cyanometallate Anions, $[Fe^{II}(CN)_5NO]^{2-}$ and 2-2H$^+$	210
7.4.3	Square Planar Cyanometallate Anions	212

7.4.3.1	$[M^{II}(CN)_4]^{2-}$ and 2-2H$^+$ or 4-2H$^+$ (M = Ni, Pd, and Pt)	212
7.4.3.2	$[M^{II}(CN)_4]^{2-}$ and 3-2H$^+$ (M = Ni, Pd, and Pt)	213
7.4.4	Linear Cyanometallate Anions $[M^I(CN)_2]^-$ (M = Ag and Au)	214
7.5	Properties of Charge-Assisted H-Bonded Networks Based on Amidinium Tectons	215
7.5.1	Luminescence	215
7.5.2	Porosity	215
7.5.3	Redox Properties in the Solid State	217
7.5.3.1	Reduction in the Solid State	218
7.5.3.2	Formation of Mixed Valence Compounds	218
7.6	Design of Crystals Based on CAHB Networks	218
7.6.1	Solid Solutions	219
7.6.1.1	Solid Solutions Using (2-2H$^+$)$_2$([M^{II}(CN)$_6$])(M = Fe and Ru)	219
7.6.1.2	Solid Solutions Using (X-2H$^+$)$_3$([M^{III}(CN)$_6$])$_2$ (X = 2 or 3, M = Fe, Cr, or Co)	219
7.6.2	Epitaxial Growth of Isostructural Crystals	220
7.6.2.1	Principles and Examples	220
7.6.2.2	Study of the Interface	222
7.6.2.3	*In Situ* Study of the Growth	223
7.7	Conclusion	224
	References	225
8	**Synthesis and Design of π-Conjugated Organic Architectures Doped with Heteroatoms**	**233**
	Simon Kervyn, Claudia Aurisicchio, and Davide Bonifazi	
8.1	Introduction	233
8.2	Boron	234
8.2.1	Boron Connected to Aryl Unit	235
8.2.1.1	Boron Derivatives Used as Chemical Sensors	235
8.2.1.2	Covalent Organic Frameworks	242
8.2.1.3	Boron Connected to Electron-Rich Aryls Groups	243
8.2.1.4	Vinyl Borane	249
8.2.2	Boron Inserted into Aryl Unit	251
8.2.2.1	Borabenzene	251
8.2.2.2	Boranaphathalene and Higher Acenes	252
8.2.2.3	Borepin	255
8.3	Sulfur, Selenium, and Tellurium	257
8.3.1	Sulfur	258
8.3.1.1	Oligothiophene-Based Structures	258
8.3.2	Selenium and Tellurium	267
8.3.2.1	Selenium-Doped Macrocyclic Structures	267
8.3.2.2	Selenium- and Tellurium-Annulated Aromatic Structures	269
8.4	Miscellaneous	271
8.5	Conclusions	276
	References	277

Part Three Nanopatterning and Processing *281*

9 Functionalization and Assembling of Inorganic Nanocontainers for Optical and Biomedical Applications *283*
André Devaux, Fabio Cucinotta, Seda Kehr, and Luisa De Cola
9.1 Introduction *283*
9.2 Zeolite L as Inorganic Nanocontainers *285*
9.2.1 General Concept for Nanocontainers *285*
9.2.1.1 The Internal Cavity System *287*
9.2.1.2 The Pore Openings *287*
9.2.1.3 The External Surface *288*
9.2.2 A Practical Sample Material: Zeolite L *288*
9.3 Functionalization of Zeolites: Host–Guest Chemistry and Surface and Channel Functionalizations *290*
9.3.1 Inorganic–Organic Host–Guest Systems *290*
9.3.2 Dye Loading of Zeolites *291*
9.3.2.1 Inserting Molecules *291*
9.3.2.2 Doping of Zeolites with Rare Earth Ions *293*
9.3.3 Functionalization of the External Surface *294*
9.3.3.1 Reactions with the Whole External Surface of Zeolite L *294*
9.3.3.2 Selective Functionalization of the Channel Entrances *294*
9.3.4 More Complex Functionalizations *295*
9.3.5 Metal–Organic Frameworks as Host Materials *296*
9.4 Photoinduced Processes in Zeolites *297*
9.4.1 Förster Resonance Energy Transfer *298*
9.4.2 Electron Transfer *300*
9.4.3 Aggregates Formation in Nanochannels *303*
9.5 Self-Assembly in Solution and on Surfaces *305*
9.5.1 Self-Assembly in Solution *305*
9.5.2 Self-Assembly on Surfaces *309*
9.5.2.1 Zeolite L Monolayers: Synthesis and Characterization *309*
9.6 Possible Optical and Biomedical Applications of Nanocontainers *312*
9.6.1 Optical Applications *312*
9.6.1.1 Hybrid Solar Cells *313*
9.6.1.2 Hybrid Light-Emitting Devices (LEECs and LEDs) *314*
9.6.2 Biomedical Applications *316*
References *323*

10 Soft Lithography for Patterning Self-Assembling Systems *343*
Xuexin Duan, David N. Reinhoudt, and Jurriaan Huskens
10.1 Introduction *343*
10.2 Self-Assembling Systems *344*
10.3 Soft Lithography *345*
10.3.1 Microcontact Printing *345*
10.3.2 Micromolding Injection in Capillaries *346*

10.3.3	Main Limitations of Soft Lithography	*347*
10.4	Contact Printing of SAMs with High Resolution	*349*
10.4.1	Improvements of the Stamp	*350*
10.4.1.1	Hard PDMS and Composite Stamps	*350*
10.4.2	Other Stamp Materials	*351*
10.4.3	Flat Stamps	*351*
10.4.4	New Ink Materials	*353*
10.4.5	Alternative μCP Strategies	*354*
10.4.5.1	High-Speed μCP	*354*
10.4.6	Catalytic Printing	*355*
10.5	Soft Lithography to Pattern Assemblies of Nanoparticles	*355*
10.5.1	Patterning by Contact Printing	*356*
10.5.1.1	Nanoparticles as Inks	*356*
10.5.1.2	Convective Assembly	*357*
10.5.1.3	Templated Assembly	*359*
10.5.2	Patterning by Micromolding in Capillaries	*359*
10.5.3	Patterning by Soft Lithography with Solvent Mediation	*360*
10.6	Soft Lithography Pattern Supramolecular Assembly	*361*
10.6.1	Affinity Contact Printing	*361*
10.6.2	Supramolecular Nanostamping	*362*
10.6.3	Molecular Printboards	*363*
10.7	Concluding Remarks	*364*
	References	*365*
11	**Colloidal Self-Assembly of Semiconducting Polymer Nanospheres: A Novel Route to Functional Architectures for Organic Electronic Devices** *371*	
	Evelin Fisslthaler and Emil J. W. List	
11.1	Introduction	*371*
11.2	Formation of Semiconducting Polymer Nanospheres	*372*
11.3	Driving Forces behind Nanoparticle Self-Assembly Processes	*373*
11.4	Deposition Methods for Aqueous Dispersions of Semiconducting Polymer Nanospheres	*377*
11.5	Organic Electronic Devices from Semiconducting Polymer Nanospheres	*380*
11.5.1	Organic Electronic Devices from SPNs with Homogeneous Functional Layers	*382*
11.5.2	Functional Micro- and Nanopatterns from SPNs for Organic Electronic Devices	*382*
11.5.2.1	Functional Micro- and Nanopatterns from SPNs Formed by Ink-Jet Printing and Template-Assisted Self-Assembly	*385*
11.5.2.2	Functional Micro- and Nanopatterns from SPNs Formed by Micromolding in Capillaries and Template-Assisted Fluidic Self-Assembly	*389*
11.6	Conclusions	*393*
	References	*395*

12	**Photolithographic Patterning of Organic Electronic Materials** *399*	
	John DeFranco, Alex Zakhidov, Jin-Kyun Lee, Priscilla Taylor,	
	Hon Hang Fong, Margarita Chatzichristidi, Ha Soo Hwang,	
	Christopher Ober, and George Malliaras	
12.1	Introduction *399*	
12.1.1	Patterning Requirements in Organic Electronics *399*	
12.1.2	Common Methods of Patterning Organic Electronic Materials *401*	
12.1.2.1	Solution-Deposited Materials *401*	
12.1.2.2	Vapor-Deposited Materials *403*	
12.1.3	Photolithographic Patterning Techniques *403*	
12.2	Photolithographic Methods for Patterning Organic Materials *406*	
12.2.1	Hybrid Techniques *406*	
12.2.2	Organic Materials Directly Processed with UV Light *406*	
12.2.3	Materials Choice or Modification *408*	
12.2.4	Patterning with an Interlayer *409*	
12.2.5	Modification of the Resist Chemistry *412*	
12.3	Conclusions and General Considerations *415*	
	References *415*	
Part Four	**Scanning Probe Microscopies** *421*	
13	**Toward Supramolecular Engineering of Functional Nanomaterials: Preprogramming Multicomponent 2D Self-Assembly at Solid–Liquid Interfaces** *423*	
	Carlos-Andres Palma, Artur Ciesielski, Massimo Bonini, and Paolo Samorì	
13.1	Introduction *423*	
13.2	van der Waals Interactions *425*	
13.2.1	Adsorption *425*	
13.3	Hydrogen–bonding Interactions *433*	
13.3.1	Weak and Single Hydrogen Bonds *433*	
13.3.2	Carboxylic Acids as Dihapto Hydrogen–bonding Moieties: Chirality and Polymorphism *434*	
13.3.3	Combined Dihapto Hydrogen–bond and van der Waals Interactions *437*	
13.3.4	Trihapto Hydrogen–bonding Moieties *439*	
13.3.5	Tetrahapto Hydrogen–bonding Moieties *441*	
13.4	Metal–Ligand Interactions *441*	
13.5	Conclusions and Outlook *445*	
	References *446*	
14	**STM Characterization of Supramolecular Materials with Potential for Organic Electronics and Nanotechnology** *457*	
	Kevin R. Moonoosawmy, Jennifer M. MacLeod, and Federico Rosei	
14.1	Introduction *457*	

14.2	Characterization Using STM and Related Technologies	458
14.2.1	The Capabilities of STM	458
14.2.1.1	High-Resolution Imaging	458
14.2.1.2	Electronic Characterization	459
14.2.1.3	Manipulation and Synthesis with the STM	459
14.2.2	Modified STM Systems for Specialized Probing of Electronic, Magnetic, and Photonic Properties	460
14.2.2.1	Two- and Four-Probe STM Systems	460
14.2.2.2	Spin-Polarized STM	461
14.2.2.3	Light-Emission STM	461
14.3	Molecular Systems with Applications in Electronics	462
14.3.1	Thiophenes	462
14.3.1.1	Thiophene Layers	462
14.3.1.2	Host–Guest Systems Containing Thiophenes and Fullerenes	464
14.3.2	Influence of Symmetry on Adsorption of Rubrene	464
14.4	Optically Active Molecules	467
14.4.1	Obliqueness of Chiral Unit Cell and its Angular Mismatch	469
14.4.2	Height Difference Revealed by Contrast	470
14.4.3	STM Reveals the Nature of First Layer Growth	471
14.5	Magnetic Systems	472
14.5.1	The Spin–Electron Interaction	472
14.5.2	Metal–Organic Coordination Networks	473
14.5.3	Controlling Magnetic Anisotropy of Supramolecular MOCN	474
14.6	STM Characterization and Biological Surface Science	475
14.6.1	Probing the Fundamentals of Life	475
14.6.2	Bionanotechnological Applications	477
14.7	Using the STM to Initiate Chemical Reactions	478
14.7.1	Molecular Dissociation	478
14.7.2	Molecular Synthesis	479
14.7.3	Inducing Reactions Over a Larger Spatial Extent	480
14.8	Conclusions and Perspective	481
	References	483
15	**Scanning Probe Microscopy Insights into Supramolecular π-Conjugated Nanostructures for Optoelectronic Devices**	**491**
	Mathieu Surin, Gwennaëlle Derue, Simon Desbief, Olivier Douhéret, Pascal Viville, Roberto Lazzaroni, and Philippe Leclère	
15.1	Introduction: SPM Techniques for the Nanoscale Characterization of Organic Thin Films	491
15.2	Controlling the Supramolecular Assembly and Nanoscale Morphology of π-Conjugated (Macro)Molecules	495
15.2.1	Biotemplates for Assembling Conjugated Molecules into Supramolecular Nanowires	495
15.2.2	Nanophase Separation in Rod–Coil Block Copolymers	499

15.2.3	Nanorubbing: a Tool for Orienting π-Conjugated Chains in Thin Films *501*	
15.3	Effect of the Nanoscale Morphology on the Optoelectronic Properties and Device Performances *504*	
15.3.1	Luminescence Properties of Supramolecular π-Conjugated Structures and Applications for White Light OLED *504*	
15.3.2	Polythiophene Nanostructures for Charge Transport Applications: Rubbing, Annealing, and Doping *507*	
15.3.2.1	Nanorubbing P3HT Thin Films in OFET Channels *507*	
15.3.2.2	Influence of the Film Morphology on Field-Effect Transistor's Performances *509*	
15.3.2.3	Doping of P3HT Nanofibrils *511*	
15.3.3	SPM Characterization of Organic Bulk Heterojunction Solar Cells *514*	
15.3.3.1	A Brief Introduction to Organic Photovoltaics *514*	
15.3.3.2	Morphological Analysis of Bulk Heterojunction OPVs *516*	
15.3.3.3	Electrical Characterization of Bulk Heterojunction OPVs by Conductive-AFM *518*	
15.4	Conclusions and Perspectives *520*	
	References *521*	
16	**Single-Molecule Organic Electronics: Toward Functional Structures** *527*	
	Simon J. Higgins and Richard J. Nichols	
16.1	Introduction *527*	
16.2	Techniques *527*	
16.2.1	DNA *531*	
16.2.2	Base Pair Junctions *535*	
16.2.3	Porphyrins *536*	
16.2.4	Environmental Effects on Junctions *539*	
16.2.5	Influence of π-Stacking on Metal|Molecule|Metal Junctions *543*	
16.3	Summary and Outlook *546*	
	References *546*	

Volume 2

Part Five Electronic and Optical Properties *551*

17 Charge Transfer Excitons in Supramolecular Semiconductor Nanostructures *553*
Jean-François Glowe, Mathieu Perrin, David Beljonne, Ludovic Karsenti, Sophia C. Hayes, Fabrice Gardebien, and Carlos Silva

17.1	Introduction *553*
17.2	Experimental Methodologies *554*
17.3	Delayed PL Decay Dynamics: Evidence of Charge Transfer Exciton Recombination in T6 *555*
17.4	Distribution of Charge Transfer Exciton Radii *561*

17.5	Conclusions *563*	
	References *564*	

18	**Optical Properties and Electronic States in Anisotropic Conjugated Polymers: Intra- and Interchain Effects** *567*	
	Davide Comoretto, Valentina Morandi, Matteo Galli, Franco Marabelli, and Cesare Soci	
18.1	Introduction *567*	
18.2	Polymer Properties and Orientation *568*	
18.3	Intrachain Effects *571*	
18.4	Interchain Effects *575*	
18.4.1	Polarized Photoluminescence Spectra *575*	
18.4.1.1	Pump Polarization Anisotropy *577*	
18.4.2	Emission Anisotropy *577*	
18.4.3	Polarized Optical Spectroscopy under Hydrostatic Pressure *578*	
18.4.3.1	Raman Spectroscopy *579*	
18.4.3.2	Reflectance Spectroscopy *580*	
18.4.3.3	Oscillator Strength *582*	
18.5	Conclusions *583*	
	References *584*	

19	**Nanoscale Shape of Conjugated Polymer Chains Revealed by Single-Molecule Spectroscopy** *589*	
	Enrico Da Como and John M. Lupton	
19.1	Introduction *589*	
19.2	The Single-Molecule Approach *591*	
19.3	Chain Shape *591*	
19.3.1	Extracting Chain Shape by Polarization Anisotropy *591*	
19.3.2	Difference between Polarization Anisotropy in Excitation and Emission *595*	
19.3.3	Material Examples *596*	
19.3.3.1	Polyfluorenes: Chain Bending versus Twisting *596*	
19.3.3.2	Phenylene Vinylenes *601*	
19.4	Conclusions *606*	
	References *607*	

20	**Electronic Structure Engineering Through Intramolecular Polar Bonds** *611*	
	Georg Heimel and Norbert Koch	
20.1	Introduction *611*	
20.2	Electrostatic Considerations *612*	
20.3	Introducing Energy Levels *618*	
20.3.1	Isolated Molecules *618*	
20.3.2	Orientation-Dependent Energy Levels in Ordered Molecular Layers *620*	

20.3.3	Implications for Experiments	625
20.4	Intrinsic Intramolecular Surface Dipoles	627
20.4.1	Surface Dipole from the π-Electron System	627
20.4.2	Intramolecular Polar Bonds	630
20.4.2.1	The Concept	630
20.4.2.2	Materials Design from First Principles	632
20.4.2.3	Experimental Evidence	634
20.5	Implications for Materials and Devices	635
20.5.1	Polarization Energy in Organic Solids	636
20.5.2	Level Alignment in Organic Electronic Devices	637
20.5.2.1	Organic Heterojunction Solar Cells	637
20.5.2.2	Organic Field-Effect Transistors	639
20.6	Conclusions	642
	References	643

Part Six   Field-Effect Transistors 649

21   Crystal Structure Performance Relationship in OFETs 651
Marta Mas-Torrent and Concepció Rovira

21.1	Introduction	651
21.2	Single-Crystal OFETs	655
21.3	Crystal Packing Motifs	660
21.4	Polymorphism	663
21.4.1	Pentacene and Related Acenes	664
21.4.2	Oligothiophenes	667
21.4.3	Tetrathiafulvalenes	670
21.5	Summary	674
	References	674

22   Bioactive Supramolecular Architectures in Electronic Sensing Devices 683
Luisa Torsi, Gerardo Palazzo, Antonia Mallardi, Maria D. Angione, and Serafina Cotrone

22.1	Introduction	683
22.2	Supramolecular Architectures for Organic Thin-Film Field-Effect Sensing Transistors	685
22.3	Bioactive Sensing Layer	691
22.3.1	Membrane Proteins	691
22.3.1.1	Bacteriorhodopsin	693
22.3.1.2	Photosynthetic Proteins	693
22.3.1.3	Receptors	694
22.3.1.4	Biological Nanopores	696
22.3.2	Embedding Proteins	696
22.3.2.1	Solid-Supported Phospholipid Bilayers	697
22.3.2.2	Polyelectrolyte Multilayers	699

22.4	Sensing Devices with Polyelectrolyte Multilayer Architectures	*701*
22.5	Electronic Sensing Devices with Phospholipid Layer Architectures	*704*
22.6	Conclusions and Perspectives	*711*
	References	*712*

23 Field-Effect Devices Based on Organic Semiconductor Heterojunctions *719*
Annalisa Bonfiglio and Piero Cosseddu

23.1	Introduction	*719*
23.2	Field-Effect Devices	*720*
23.2.1	Ambipolar Field-Effect Transistors	*721*
23.2.1.1	Organic Planar Heterojunction	*723*
23.2.1.2	Organic Bulk Heterojunctions	*726*
23.2.1.3	Organic Polymer Blends	*728*
23.2.2	Organic Bulk Heterojunctions for Threshold Voltage Tuning in OFETs	*732*
23.3	Conclusions	*735*
	References	*736*

24 Functional Semiconducting Blends *741*
Natalie Stingelin

24.1	Introduction	*741*
24.2	Processing Aids	*742*
24.2.1	Polymeric Binders	*742*
24.2.2	Use of Vitrifiers	*750*
24.2.3	Epitaxial Growth	*754*
24.3	Mechanically Tough Semiconducting Blends	*755*
24.4	Ferroelectric Semiconducting Blends	*756*
24.5	Photovoltaic Blends	*761*
24.6	Conclusions	*763*
	References	*763*

Part Seven Solar Cells *767*

25 Hybrid Organic–Inorganic Photovoltaic Diodes: Photoaction at the Heterojunction and Charge Collection Through Mesostructured Composites *769*
Henry J. Snaith

25.1	Introduction	*769*
25.2	Basic Operating Principles of Hybrid Solar Cells	*770*
25.2.1	Solid-State Dye-Sensitized Solar Cells	*770*
25.2.2	Polymer Metal Oxide Devices	*772*
25.3	Photoaction at the Heterojunction: Light Harvesting, Charge Generation, and Recombination	*773*
25.3.1	Electron and Hole Transfer Phenomena	*773*

25.3.1.1	Solid-State DSSCs 773
25.3.1.2	Polymer Metal Oxide Solar Cells 775
25.3.2	Energy Transfer from Light-Harvesting Antenna to Charge Generation Centers 777
25.3.2.1	Dye-Sensitized Solar Cells 777
25.3.2.2	Polymer Metal Oxide Solar Cells 780
25.3.3	The Influence of and Controlling Electron–Hole Recombination 783
25.3.3.1	The Influence of Recombination on Solar Cell Operation 783
25.3.3.2	The Mechanisms of Electron–Hole Recombination 786
25.3.3.3	Controlling the Spatial Separation of the Electron and Hole 789
25.3.3.4	Influence of Metal Salts on Recombination 791
25.4	Pore Filling and Current Collection in Hybrid Solar Cells 793
25.5	Summary and Outlook 796
	References 797

26 Nanostructured Hybrid Solar Cells 801
Lukas Schmidt-Mende

26.1	Introduction 801
26.2	Motivation 803
26.2.1	Morphology Control 803
26.2.2	Ideal Morphology 803
26.2.3	Interface Engineering 805
26.2.4	Stability/Lifetime 806
26.2.5	Plasmonic Effects 806
26.3	Materials 807
26.3.1	TiO_2 807
26.3.2	ZnO 807
26.3.3	SnO_2 and Other Materials 809
26.4	Nanostructures 810
26.4.1	Nanotubes 810
26.4.2	Nanowires 812
26.4.3	Nanoporous Metal Oxide Layers 816
26.4.4	Diblock Copolymer Morphologies 817
26.4.5	Bulk Heterojunction Solar Cells 817
26.4.6	Interface Modifications 819
26.4.7	Conclusion 821
26.5	Summary and Outlook 822
	References 823

27 Determination and Control of Microstructure in Organic Photovoltaic Devices 827
Christoph J. Brabec, Iain McCulloch, and Jenny Nelson

27.1	Introduction 827
27.2	Measurement of Microstructure in Polymer:Fullerene Blend Films 829
27.3	Control of the Structure of Organic Photovoltaic Materials through Chemical Design 830

27.3.1	Factors Influencing Microstructure of Conjugated Polymer Films	831
27.3.1.1	Thiophene Polymers	831
27.3.1.2	Molecular Weight	832
27.3.1.3	Regioregularity	833
27.3.1.4	Structural Control of Thiophene Polymers through Design	834
27.3.2	Control of Polymer Optical Gap through Chain Conformation	835
27.3.3	Factors Controlling the Structure of Fullerene Films	838
27.3.4	Influence of Chemical Structure on the Microstructure of Polymer: Fullerene Blend Films	839
27.3.4.1	Polymer Molecular Weight and Regioregularity	839
27.3.4.2	Side Chain Density/Control of Intercalation	840
27.3.5	Alternative Strategies to Control the Microstructure of Donor–Acceptor Composite Films	842
27.4	Control of Microstructure in Polymer:Fullerene Blend Devices via Processing	843
27.4.1	Solvent	843
27.4.2	Thermal Annealing	844
27.4.3	Blend Composition	847
27.4.4	Other Approaches	848
27.4.5	Substrate Control	850
27.5	Numerical Simulations of Microstructure	851
27.6	Conclusions	852
	References	852

28   Morphology and Photovoltaic Properties of Polymer–Polymer Blends   *861*
Dieter Neher

28.1	Introduction	861
28.2	Neutral Excitations at Polymer–Polymer Heterojunctions	862
28.2.1	Energetics at the DA Heterojunction	866
28.2.2	Blend Morphology and Phase Composition	867
28.3	Polymer–Polymer Blends Formed in Nanoparticles	869
28.4	Excited State and Photovoltaic Properties of Blends of M3EH-PPV with CN-Ether-PPV	874
28.5	Correlation Between Heterojunction Topology and Fill Factor	882
28.6	Concluding Remarks	887
	References	888

Part Eight   LEDs/LECs   *893*

29   The Light-Emitting Electrochemical Cell: Utilizing Ions for Self-Assembly and Improved Device Operation   *895*
Ludvig Edman

29.1	Introduction	895
29.2	Historical Background and Liquid-Containing LECs	896
29.3	The Solid-State LEC	898

29.3.1	Pros and Cons with the LEC Concept 899
29.4	The Controversial Operational Mechanism 900
29.4.1	Doping of Organic Semiconductors 901
29.4.2	Electrochemical Doping or Not... 902
29.5	The Self-Assembled and Dynamic p–n Junction 905
29.6	Device Performance 908
29.6.1	The Turn-On Time 908
29.6.2	Toward Long-Term and Efficient Operation 910
29.7	Concluding Remarks 913
	References 914

30 **Optical and Electroluminescent Properties of Conjugated Polyrotaxanes** 919
Sergio Brovelli and Franco Cacialli

30.1	Introduction 919
30.2	Conjugated Polyrotaxanes as Insulated Molecular Wires and Organic Nanostructures 921
30.3	Solution Optical Properties of Conjugated Polyrotaxanes: Control and Tuning of Intermolecular Interactions 922
30.3.1	Photophysics of Polyrotaxanes in Diluted Aqueous Solutions 923
30.3.2	Time Evolution of the Average Energy of Photoluminescence 928
30.3.3	Quantitative Analysis of the Luminescence Dynamics of Unthreaded PDV.Li in Aqueous Solution 929
30.4	Role of Progressive Encapsulation in the Control of the Photophysics of Conjugated Polyrotaxanes 931
30.5	Role of Cyclodextrin Size on the Photophysics and Resistance to Quenching of Conjugated Polyrotaxanes 933
30.6	Ionic Interactions in the Solid State and Solutions with Poly(ethylene Oxide) (PEO) 939
30.7	Solid-State Optical and Electroluminescent Properties of Conjugated Polyrotaxanes 942
30.7.1	Optical Properties in the Solid State 943
30.7.2	PEO Solid-State Solutions 944
30.7.3	Ultrafast Time-Resolved Photoluminescence Studies of Polyrotaxanes in the Solid State 946
30.8	Electroluminescence of Conjugated Polyrotaxanes 949
30.8.1	Neat Polyrotaxanes Active Layers 949
30.8.2	PEO:Polyrotaxanes Blends 950
30.8.3	Effect of Threading Ratio on Electroluminescence Properties 951
30.8.4	Influence of the Countercations on the Electrical and Luminescence Properties of Conjugated Electrolytes and Related Polyrotaxanes 952
30.9	Conclusions 955
	References 957

Index 961

Preface

Is it possible to understand how order emerges from chaos from a molecular perspective, and how? How can information encoded in a single molecule be transferred and exploited to perform a useful function in an ensemble of the same molecules? How can one tailor a complex chemical system featuring several sophisticated functions? What's the technological relevance of the supramolecular approach? These are only a few examples of questions that require a deep understanding of how molecules interact both between themselves and with a variety of functional surfaces. More often than we perhaps realise, such interactions are mediated by so-called non-covalent interactions, which define the domain of supramolecular chemistry.

The supramolecular approach makes it possible to manipulate the local molecular environment with elegance, precision, and endless versatility, thus providing access to fundamentally new classes of organic functional materials with unprecedented properties and performance. The influence of such interactions is wide ranging, affecting properties as diverse as luminescence, electrical transport, as well as chemical and mechanical stability. And indeed, accurate control of such interactions is needed to allow optimum exploitation of molecular materials, not only in today's most common optoelectronic devices such as light-emitting diodes (LEDs), field-effect transistors (FETs), and solar cells, but also in emerging nanoscale technology applications. Despite the progress of the last ten years in the area of organic semiconductors, much remains to be done to achieve control, at the nanoscale, of the local environment of functional macromolecules, both regarding interaction with other molecular units or species, and with the electrodes.

By bringing together the contribution of world-leading scientists in the fields of supramolecular functional systems and materials, nanoscience, and conjugated semiconductors, this interdisciplinary book, at the cross-road between chemistry, physics, and engineering, aims at giving a wide overview of different supramolecular strategies to organic electronics and photonics.

Attention is paid to theoretical approaches and modelling of charge transport at different length scales, as well as to the science and engineering of surface dipoles and the role of the substrate for both self-assembly and electronic properties. Novel synthetic methodologies and the formation of complex architectures are discussed as a route to achieve full control over properties of potential interest, both at the single molecule level, and as a relevant step towards predictive self-assembly capacities in 2 and 3D. Further emphasis is placed on nanoscale-controlled patterning, as well as multiscale physico-chemical characterization of various properties. These span from optical to electrical ones, with a particular focus on those that can be investigated by the use of scanning probe methods, a variety of different spectroscopies and surface sensitive techniques. Various examples of the three most relevant classes of organic (opto)electronic devices, i.e. FETs, solar cells and light-emitting devices are also highlighted to offer an all-encompassing view of the field.

We hope that through the series of outstanding contributions covering such a broad variety of domains, this book will offer valuable highlights on the design, fabrication, characterization, and exploitation of supramolecular, and at the same time complex and functional architectures. The contributions we have selected should especially provide unambiguous evidence of the high potential for breakthroughs and innovation that characterizes this R&D area. Indeed, we would like this book to be a stimulating playground for further elaboration and development. This might appear as an ambitious goal, but as the book amply demonstrates, the relevant knowledge already gathered is impressive and constitutes a worthwhile intellectual capital to build on, and invest in, to enable future radical advances in functional materials and their applications. The fact that this diverse knowledge is not yet fully "interwoven" and exploited only means this is right time to look in this area for crucial breakthroughs to be made.

We would like to extend our special thanks to all the colleagues who enthusiastically contributed to this book, and thus made it possible. We are also grateful to Martin Ottmar and Martin Preuss for the invitation to edit this book, and to Lesley Belfit who has been working closely with us to make it a reality.

October 2010

Paolo Samorì, Strasbourg
Franco Cacialli, London

List of Contributors

Maria D. Angione
Università degli Studi di Bari
Dipartimento di Chimica
via Orabona 4
70126 Bari
Italy

Claudia Aurisicchio
University of Namur
FUNDP
Department of Chemistry
Rue de Bruxelles 61
5000 Namur
Belgium

Martin Baumgarten
Max-Planck-Institut für
Polymerforschung
Ackermannweg 10
55128 Mainz
Germany

David Beljonne
University of Mons/Materia Nova
Laboratory for Chemistry of Novel
Materials
Place du Parc 20
7000 Mons
Belgium

and

Georgia Institute of Technology
School of Chemistry and Biochemistry
901 Atlantic Drive NW
Atlanta, GA 30332-0400
USA

Annalisa Bonfiglio
University of Cagliari
Department of Electrical and Electronic
Engineering
Piazza d'Armi
09123 Cagliari
Italy

List of Contributors

Davide Bonifazi
University of Namur
FUNDP
Department of Chemistry
Rue de Bruxelles 61
5000 Namur
Belgium

and

Università degli Studi di Trieste
Dipartimento di Scienze Farmaceutiche
Piazzale Europa 1
34127 Trieste
Italy

Massimo Bonini
BASF SE
67056 Ludwigshafen
Germany

Christoph J. Brabec
Friedrich-Alexander-Universität
Erlangen-Nürnberg and Bayerisches
Zentrum für Angewandte
Energieforschung (ZAE Bayern)
Martensstrasse 7
91058 Erlangen
Germany

Jean-Luc Brédas
University of Mons/Materia Nova
Laboratory for Chemistry of Novel
Materials
Place du Parc 20
7000 Mons
Belgium

and

Georgia Institute of Technology
School of Chemistry and Biochemistry
901 Atlantic Drive NW
Atlanta, GA 30332-0400
USA

Patrick Brocorens
University of Mons/Materia Nova
Laboratory for Chemistry of Novel
Materials
Place du Parc 20
7000 Mons
Belgium

Sergio Brovelli
University College London
Department of Physics and Astronomy
Gower Street
London WC1E 6BT
UK

and

University College London
London Centre for Nanotechnology
Gower Street
London WC1E 6BT
UK

Franco Cacialli
University College London
Department of Physics and Astronomy
Gower Street
London WC1E 6BT
UK

and

University College London
London Centre for Nanotechnology
Gower Street
London WC1E 6BT
UK

Margarita Chatzichristidi
Cornell University
Department of Materials Science and
Engineering
Bard Hall 327
Ithaca, NY 14853-1501
USA

Artur Ciesielski
Université de Strasbourg and CNRS
(UMR 7006)
Nanochemistry Laboratory, ISIS
Allée Gaspard Monge 8
67083 Strasbourg Cedex
France

Davide Comoretto
Università degli Studi di Genova
Dipartimento di Chimica e Chimica
Industriale
via Dodecaneso 31
16146 Genova
Italy

Jérôme Cornil
University of Mons/Materia Nova
Laboratory for Chemistry of Novel
Materials
Place du Parc 20
7000 Mons
Belgium

and

Georgia Institute of Technology
School of Chemistry and Biochemistry
901 Atlantic Drive NW
Atlanta, GA 30332-0400
USA

Piero Cosseddu
University of Cagliari
Department of Electrical and Electronic
Engineering
Piazza d'Armi
09123 Cagliari
Italy

Serafina Cotrone
Università degli Studi di Bari
Dipartimento di Chimica
via Orabona 4
70126 Bari
Italy

Fabio Cucinotta
Westfälische Wilhelms-Universität
Münster
Phyiskalisches Institut
Mendelstrasse 7
48149 Münster
Germany

Enrico Da Como
Ludwig-Maximilians-Universität
München
Department of Physics
Center for NanoScience (CeNS)
Photonics and Optoelectronics Group
80799 Munich
Germany

Luisa De Cola
Westfälische Wilhelms-Universität
Münster
Phyiskalisches Institut
Mendelstrasse 7
48149 Münster
Germany

John DeFranco
Cornell University
Department of Materials Science and
Engineering
Bard Hall 327
Ithaca, NY 14853-1501
USA

Mari-Carmen Ruiz Delgado
Georgia Institute of Technology
School of Chemistry and Biochemistry
901 Atlantic Drive NW
Atlanta, GA 30332-0400
USA

Gwennaëlle Derue
University of Mons (UMONS)
Centre of Innovation and Research in Materials and Polymers (CIRMAP)
Laboratory for Chemistry of Novel Materials/Materia Nova
20 Place du Parc
7000 Mons
Belgium

Simon Desbief
University of Mons (UMONS)
Centre of Innovation and Research in Materials and Polymers (CIRMAP)
Laboratory for Chemistry of Novel Materials/Materia Nova
20 Place du Parc
7000 Mons
Belgium

André Devaux
Westfälische Wilhelms-Universität Münster
Phyiskalisches Institut
Mendelstrasse 7
48149 Münster
Germany

Olivier Douhéret
University of Mons (UMONS)
Centre of Innovation and Research in Materials and Polymers (CIRMAP)
Laboratory for Chemistry of Novel Materials/Materia Nova
20 Place du Parc
7000 Mons
Belgium

Xuexin Duan
University of Twente
MESA+ Institute for Nanotechnology
Molecular Nanofabrication Group
P.O. Box 217
7500 AE Enschede
The Netherlands

Ludvig Edman
Umeå University
Department of Physics
The Organic Photonics and Electronics Group
901 87 Umeå
Sweden

Sylvie Ferlay
Université de Strasbourg
Institut Le Bel
Laboratoire de Chimie de Coordination Organique (UMR 7140)
4 rue Blaise Pascal
67000 Strasbourg
France

Evelin Fisslthaler
Graz Centre for Electron Microscopy
Steyrergasse 17/III
8010 Graz
Austria

and

NanoTecCenter Weiz Forschungsgesellschaft mbH
Franz-Pichler-Strasse 32
8160 Weiz
Austria

List of Contributors

Hon Hang Fong
Cornell University
Department of Materials Science and Engineering
Bard Hall 327
Ithaca, NY 14853-1501
USA

Matteo Galli
Università degli Studi di Pavia
Dipartimento di Fisica "A. Volta"
via Bassi 7
27100 Pavia
Italy

Fabrice Gardebien
Université de Mons-Hainaut
Service de Chimie des Matériaux Nouveaux
Place du Parc 20
7000 Mons
Belgium

Christopher B. George
Northwestern University
Department of Chemistry
2145 Sheridan Road
Evanston, IL 60208-3113
USA

Jean-François Glowe
Université de Montréal
Département de physique & Regroupement québécois sur les matériaux de pointe
C.P. 6128, Succursale centre-ville
Montréal (Québec) H3C 3J7
Canada

Sophia C. Hayes
University of Cyprus
Department of Chemistry
P.O. Box 20537
1678 Nicosia
Cyprus

Georg Heimel
Humboldt-Universität zu Berlin
Institut für Physik
Brook-Taylor-Strasse 6
12489 Berlin
Germany

Simon J. Higgins
University of Liverpool
Department of Chemistry
Crown Street
Liverpool L69 7ZD
UK

Mir Wais Hosseini
Université de Strasbourg
Institut Le Bel
Laboratoire de Chimie de Coordination Organique (UMR 7140)
4 rue Blaise Pascal
67000 Strasbourg
France

Jurriaan Huskens
University of Twente
MESA+ Institute for Nanotechnology
Molecular Nanofabrication Group
P.O. Box 217
7500 AE Enschede
The Netherlands

Ha Soo Hwang
Cornell University
Department of Materials Science and Engineering
Bard Hall 327
Ithaca, NY 14853-1501
USA

Ludovic Karsenti
Université de Montréal
Département de physique &
Regroupement québécois sur les
matériaux de pointe
C.P. 6128, Succursale centre-ville
Montréal (Québec) H3C 3J7
Canada

Bram Keereweer
Radboud University Nijmegen
Institute for Molecules and Materials
Heijendaalseweg 135
6525 AJ Nijmegen
The Netherlands

Seda Kehr
Westfälische Wilhelms-Universität
Münster
Phyiskalisches Institut
Mendelstrasse 7
48149 Münster
Germany

Simon Kervyn
University of Namur
FUNDP
Department of Chemistry
Rue de Bruxelles 61
5000 Namur
Belgium

Norbert Koch
Humboldt-Universität zu Berlin
Institut für Physik
Brook-Taylor-Strasse 6
12489 Berlin
Germany

Paul H. J. Kouwer
Radboud University Nijmegen
Institute for Molecules and Materials
Heijendaalseweg 135
6525 AJ Nijmegen
The Netherlands

Roberto Lazzaroni
University of Mons (UMONS)
Centre of Innovation and Research in
Materials and Polymers (CIRMAP)
Laboratory for Chemistry of Novel
Materials/Materia Nova
20 Place du Parc
7000 Mons
Belgium

Stéphane Le Gac
Radboud University Nijmegen
Institute for Molecules and Materials
Heijendaalseweg 135
6525 AJ Nijmegen
The Netherlands

Philippe Leclère
University of Mons (UMONS)
Centre of Innovation and Research in
Materials and Polymers (CIRMAP)
Laboratory for Chemistry of Novel
Materials/Materia Nova
20 Place du Parc
7000 Mons
Belgium

Jin-Kyun Lee
Cornell University
Department of Materials Science and
Engineering
Bard Hall 327
Ithaca, NY 14853-1501
USA

Mindy Levine
Massachusetts Institute of Technology
Department of Chemistry
77 Massachusetts Avenue, Bldg. 18-597
Cambridge, MA 02139
USA

List of Contributors

Emil J.W. List
NanoTecCenter Weiz
Forschungsgesellschaft mbH
Franz-Pichler-Strasse 32
8160 Weiz
Austria

and

Graz University of Technology
Institute of Solid State Physics
Petersgasse 16
8010 Graz
Austria

John M. Lupton
University of Utah
Department of Physics and Astronomy
115 South 1400 East
Salt Lake City, UT 84112
USA

Jennifer M. MacLeod
University of Quebec
INRS-EMT
1650 Boul. Lionel Boulet
J3X 1S2 Varennes
Quebec
Canada

Antonia Mallardi
Istituto per i Processi Chimico-Fisici (IPCF)
CNR
via Orabona 4
Bari 70126
Italy

George Malliaras
Cornell University
Department of Materials Science and Engineering
Bard Hall 327
Ithaca, NY 14853-1501
USA

Franco Marabelli
Università degli Studi di Pavia
Dipartimento di Fisica "A. Volta"
via Bassi 7
27100 Pavia
Italy

Nicolas Martinelli
University of Mons/Materia Nova
Laboratory for Chemistry of Novel Materials
Place du Parc 20
7000 Mons
Belgium

Marta Mas-Torrent
Institut de Ciència de Materials de Barcelona (CSIC)
and Networking Research Center on Bioengineering, Biomaterials and Nanomedicine (CIBER-BBN)
Campus UAB
08193 Bellaterra
Spain

Iain McCulloch
Imperial College London
Department of Chemistry
London SW7 2AZ
UK

Andrea Minoia
University of Mons/Materia Nova
Laboratory for Chemistry of Novel Materials
Place du Parc 20
7000 Mons
Belgium

Kevin R. Moonoosawmy
University of Quebec
INRS-EMT
1650 Boul. Lionel Boulet
J3X 1S2 Varennes
Quebec
Canada

Valentina Morandi
Università degli Studi di Pavia
Dipartimento di Fisica "A. Volta"
via Bassi 7
27100 Pavia
Italy

Luca Muccioli
Università di Bologna
Dipartimento di Chimica Fisica e
Inorganica and INSTM
Viale Risorgimento 4
40136 Bologna
Italy

Klaus Müllen
Max-Planck-Institut für
Polymerforschung
Ackermannweg 10
55128 Mainz
Germany

Dieter Neher
University of Potsdam
Institute of Physics and Astronomy
Karl-Liebknecht-Str.24/25
14476 Potsdam
Germany

Jenny Nelson
Imperial College London
Department of Physics
London SW7 2AZ
UK

Richard J. Nichols
University of Liverpool
Department of Chemistry
Crown Street
Liverpool L69 7ZD
UK

Roeland J.M. Nolte
Radboud University Nijmegen
Institute for Molecules and Materials
Heijendaalseweg 135
6525 AJ Nijmegen
The Netherlands

Christopher Ober
Cornell University
Department of Materials Science and
Engineering
Bard Hall 327
Ithaca, NY 14853-1501
USA

Yoann Olivier
University of Mons/Materia Nova
Laboratory for Chemistry of Novel
Materials
Place du Parc 20
7000 Mons
Belgium

Gerardo Palazzo
Università degli Studi di Bari
Dipartimento di Chimica
via Orabona 4
70126 Bari
Italy

Carlos-Andres Palma
Université de Strasbourg and CNRS
(UMR 7006)
Nanochemistry Laboratory, ISIS
Allée Gaspard Monge 8
67083 Strasbourg Cedex
France

Mathieu Perrin
Université de Montréal
Département de physique &
Regroupement québécois sur les
matériaux de pointe
C.P. 6128, Succursale centre-ville
Montréal (Québec) H3C 3J7
Canada

Mark A. Ratner
Northwestern University
Department of Chemistry
2145 Sheridan Road
Evanston, IL 60208-3113
USA

David N. Reinhoudt
University of Twente
MESA+ Institute for Nanotechnology
Molecular Nanofabrication Group
P.O. Box 217
7500 AE Enschede
The Netherlands

Federico Rosei
University of Quebec
INRS-EMT
1650 Boul. Lionel Boulet
J3X 1S2 Varennes
Quebec
Canada

Concepció Rovira
Institut de Ciència de Materials de
Barcelona (CSIC)
and Networking Research Center on
Bioengineering, Biomaterials and
Nanomedicine (CIBER-BBN)
Campus UAB
08193 Bellaterra
Spain

Alan E. Rowan
Radboud University Nijmegen
Institute for Molecules and Materials
Heijendaalseweg 135
6525 AJ Nijmegen
The Netherlands

Paolo Samorì
Université de Strasbourg and CNRS
(UMR 7006)
Nanochemistry Laboratory, ISIS
Allée Gaspard Monge 8
67083 Strasbourg Cedex
France

Nadja Sändig
Università di Bologna
Dipartimento di Chimica
"G. Ciamician"
via F. Selmi 2
40126 Bologna
Italy

Lukas Schmidt-Mende
Ludwig-Maximilians-Universität
München
Department of Physics
Center for NanoScience (CeNS)
Amalienstr., 54
80799 Munich
Germany

Erik Schwartz
Radboud University Nijmegen
Institute for Molecules and Materials
Heijendaalseweg 135
6525 AJ Nijmegen
The Netherlands

Carlos Silva
Université de Montréal
Département de physique &
Regroupement québécois sur les
matériaux de pointe
C.P. 6128, Succursale centre-ville
Montréal (Québec) H3C 3J7
Canada

Henry J. Snaith
University of Oxford
Department of Physics
Clarendon Laboratory
Parks Road
Oxford OX1 3PU
UK

Cesare Soci
Nanyang Technological University
School of Physical and Mathematical
Sciences
Division of Physics and Applied Physics
Singapore 637371
Singapore

and

Nanyang Technological University
School of Electrical and Electronic
Engineering
Division of Microelectronics
Singapore 637371
Singapore

Natalie Stingelin
Imperial College London
Department of Materials
Exhibition Road
London SW7 2AZ
UK

Mathieu Surin
University of Mons (UMONS)
Centre of Innovation and Research in
Materials and Polymers (CIRMAP)
Laboratory for Chemistry of Novel
Materials/Materia Nova
20 Place du Parc
7000 Mons
Belgium

Timothy M. Swager
Massachusetts Institute of Technology
Department of Chemistry
77 Massachusetts Avenue, Bldg. 18-597
Cambridge, MA 02139
USA

Igal Szleifer
Northwestern University
Department of Chemistry
2145 Sheridan Road
Evanston, IL 60208-3113
USA

and

Northwestern University
Department of Biomedical Engineering
2145 Sheridan Road
Evanston, IL 60208-3113
USA

Priscilla Taylor
Cornell University
Department of Materials Science and
Engineering
Bard Hall 327
Ithaca, NY 14853-1501
USA

Luisa Torsi
Università degli Studi di Bari
Dipartimento di Chimica
via Orabona 4
Bari 70126
Italy

Pascal Viville
University of Mons (UMONS)
Centre of Innovation and Research in Materials and Polymers (CIRMAP)
Laboratory for Chemistry of Novel Materials/Materia Nova
20 Place du Parc
7000 Mons
Belgium

Alex Zakhidov
Cornell University
Department of Materials Science and Engineering
Bard Hall 327
Ithaca, NY 14853-1501
USA

Claudio Zannoni
Università di Bologna
Dipartimento di Chimica Fisica e Inorganica and INSTM
Viale Risorgimento 4
40136 Bologna
Italy

Francesco Zerbetto
Università di Bologna
Dipartimento di Chimica
"G. Ciamician"
via F. Selmi 2
40126 Bologna
Italy

Part One
Modeling and Theory

1
Charge Transport in Organic Semiconductors: A Multiscale Modeling

Nicolas Martinelli, Yoann Olivier, Luca Muccioli, Andrea Minoia, Patrick Brocorens, Mari-Carmen Ruiz Delgado, Claudio Zannoni, David Beljonne, Roberto Lazzaroni, Jean-Luc Brédas, and Jérôme Cornil

1.1
Introduction

The concept of fabricating electronic devices based on organic conjugated materials dates back to mid-1980s, with the exploitation of the remarkable semiconducting properties of π-conjugated small molecules and polymers [1–6]. Considerable efforts have been made since then in both industry and academia to significantly improve the lifetime and efficiency of such organic-based devices [7]; they have resulted in the appearance on the marketplace of the first commercial applications, in particular light-emitting displays [8]. However, major efforts are still needed to understand and optimize all electronic and optical processes taking place in devices and ensure a continuous increase of their performance. One key process in the operation of most devices, which will be the focus of this chapter, is *charge transport* [9].

The transport of electrical charges is indeed at the heart of the working principle of the devices in the field of organic electronics. In organic light-emitting diodes (OLEDs) [10], good charge transport properties (and hence high charge mobilities) are required to reduce the impact of image effects upon charge injection, to confine light emission into the bulk of the organic layers, and to limit exciton-polaron quenching processes occurring at high polaron concentrations. In organic solar cells [11, 12], high charge carrier mobilities are required to facilitate the dissociation of the generated electron–hole pairs in their hot state (i.e., prior to their full nuclear and electronic relaxation) [13, 14] and to limit the efficiency of recombination of the generated free carriers along the way to the electrodes. In field-effect transistors (FETs), optimal mobility values are desirable to yield short switching times between on and off states and, by extension, to build organic-based electronic circuits with high-frequency operation [15, 16].

At the experimental level, charge transport is quantified via the charge carrier mobility μ that reflects the ease for holes or electrons to travel in a conducting medium and is defined by

$$\mu = \frac{v_D}{F} = \frac{d_{tot}}{t_{tot} \cdot F} \qquad (1.1)$$

where F is the amplitude of the electric field inducing charge migration, v_D is the drift velocity in the field direction, and d_{tot} is the distance traveled by the charge in the field direction during time t_{tot}. The mobility is known to be governed by a large number of parameters such as (i) the chemical structure and molecular packing of the organic semiconductor [17–21]; (ii) the presence of impurities or traps [22, 23]; (iii) the presence of *static* energetic disorder (also referred to as diagonal disorder) introducing a distribution in energy of the electronic transport levels (HOMO – highest occupied molecular orbital – for holes and LUMO – lowest unoccupied molecular orbital – for electrons) [24–27]. The complex interplay between all these parameters generally makes the comparison among experimental measurements rather difficult. Thus, theoretical modeling has proven over the years to be a useful tool to shed light on the mechanism of charge transport at the microscopic level and on the molecular parameters controlling the charge mobility values.

The transport mechanisms of charge carriers in organic semiconductors are still under debate. Depending on the nature of the materials and the degree of spatial or energetic disorder, two extreme models are generally considered [9]. In highly purified molecular single crystals, transport at low temperature operates in a band model [28] similar to that prevailing in inorganic semiconductors, in which the charge carriers are delocalized over a large number of molecular units. In this regime, the charge carrier mobility is very high (up to a few hundred cm^2 V^{-1} s^{-1}) at low temperature and is reduced with increasing temperature due to scattering phenomena induced by the thermal activation of lattice phonons [29]. When the temperature is further increased, the *dynamical* energetic and positional disorder induced by the lattice vibrations strongly reduces the width of the bands (formed by the interaction among the transport levels of the individual molecule) and tends to localize the charge carriers. We then enter into a hopping regime in which polarons (i.e., charges coupled to a local geometric distortion of the molecules) jump from one unit to another to migrate across the organic layer [27, 30–36]. This hopping picture is usually adopted to describe charge transport in systems characterized by a significant static energetic and/or spatial disorder, such as amorphous materials or many polymer thin films. Since all simulations are performed at room temperature to describe transport in devices in standard operating conditions, charge hopping thus appears to be a good model to provide a reliable description of charge migration.

A hopping process can be conveniently described within the electron transfer theories developed by Marcus and others; for a reaction such as $M_a^{\pm} - M_b \rightarrow M_a - M_b^{\pm}$, a semiclassical description of the electron transfer process leads to an expression of the transfer rate between an initial state (charge localized on molecule M_a) and a final state (charge localized on molecule M_b) given by [37]

$$k_{ET} = \frac{2\pi}{\hbar} \frac{|J|^2}{\sqrt{4\pi k_B T \lambda}} \exp\left[-\frac{(\Delta G^0 + \lambda)^2}{4\lambda k_B T}\right] \quad (1.2)$$

where

- T is the temperature and k_B is the Boltzmann constant.
- J is the transfer integral that reflects the strength of the electronic coupling between the HOMO (LUMO) orbitals of the two molecules involved in the hole (electron) transfer.
- λ is the total reorganization energy. It measures the strength of the so-called local electron–phonon coupling, which arises from the modulation of the site energies by vibrations.
- ΔG^0 is the Gibbs energy associated with the hole (electron) transfer process.

This represents the energy difference between the site energies (or in a standard one-electron picture between the HOMO (LUMO) levels) of the two interacting molecules. ΔG^0 is induced by the presence of energetic disorder and by the external electric field, which promotes a preferential direction for charge transport by creating an energy gradient of the electronic levels [30]:

$$\Delta G^0 = (E_f - E_i) - e \cdot \vec{F} \cdot \vec{d} \quad (1.3)$$

where E_i and E_f are the energies of the transport levels on the initial and final sites (often referred to as site energies), respectively, $\vec{F}$ is the electric field vector, and $\vec{d}$ is the vector connecting the centroids of the electronic distribution associated with the electronic level of the two individual molecules. Interestingly, the charge mobility can be inferred from such calculated transfer rates by injecting them into kinetic Monte Carlo (KMC) algorithms [38, 39]. Note that the semiclassical Marcus theory assumes a weak electronic coupling (nonadiabatic regime) and that all vibrations can be treated at a classical level [9]. This simple formalism is generally adopted to provide useful trends, whereas more quantitative values are accessed by treating high-energy modes at a quantum mechanical (QM) level to account for tunneling effects; this is done in Sections 1.5 and 1.6 using the Marcus–Levich–Jortner formalism.

Over the years, a large number of quantum chemical studies have been performed to characterize the amplitude of the different parameters appearing into Marcus charge transfer rates [40–53]. Most of these results have been obtained so far by considering model dimers where parameters such as the intermolecular distance or rotational angles are systematically varied or taken from experimental structures determined by X-ray diffraction. However, a severe drawback is the lack of direct comparison with many experimental measurements for which the exact molecular packing is generally not known. This has recently motivated the coupling of force field calculations, which can provide structural parameters, with a quantum chemistry approach, which allows an estimation of the electronic parameters. Force field calculations make possible the search of structures minimizing the total energy via molecular mechanics (MM) approaches and the generation of trajectories by injecting the force field into Newton's equations in molecular dynamics (MD) simulations [54, 55]. We stress that it is still a formidable task for quantum chemistry to predict the molecular packing of organic semiconductors, primarily due to the poor description of the van der Waals terms, though interesting developments are currently made in conjunction with density functional theory (DFT) [56–61].

In this context, the main goal of this chapter is to review some recent works showing the benefit of coupling force field and quantum chemical calculations to shed light on the transport properties of organic semiconductors. The selected studies focus on organic conjugated materials widely investigated at the experimental level to ease comparison between experimental and theoretical data and aim at demonstrating that our modeling tools can be applied to both small molecules and polymer chains. Section 1.2 deals with molecular crystals that are the best defined systems and hence very attractive to understand intrinsic charge transport properties. In many studies, the transfer integrals in molecular crystals have been computed on the basis of the frozen crystal geometry, thus neglecting the impact of thermal fluctuations. In this context, the work reported here addresses the impact of intermolecular vibrations in single crystals of anthracene (ANT) and perfluoropentacene on the electronic couplings and mobility values [62, 63]. In the previous case, the force field calculations were exploited to depict the lattice dynamics; however, they also proved useful in modeling the packing of organic semiconductors into organized nanostructures, as illustrated in Section 1.3 focusing on the molecular packing of tetrathiafulvalene (TTF) derivatives and the resulting charge transport properties [64]. Since it is highly desirable to validate the structures provided by force field calculations, we introduce in Section 1.4 an original approach where X-ray diffraction spectra are generated on the basis of the calculated structures to be compared to corresponding experimental spectra; this is illustrated here for polythiophene chains incorporating thienothiophene units in order to discuss their hole transport properties [65]. Section 1.5 further illustrates through a study of charge transport along one-dimensional stacks made of phthalocyanine (PC) derivatives that lattice dynamics can generate structural defects that are dynamic in nature [66]. Finally, Section 1.6 shows that layers adjacent to the transporting layer can also impact the charge mobility values; in particular, we demonstrate the influence of the chemical structure of the polymer dielectrics used in field-effect transistors on the charge mobility in layers of pentacene [67].

1.2
Organic Single Crystals

1.2.1
Molecular Parameters for Charge Transport

In this section, we first illustrate the sensitivity of (i) the internal reorganization energy to the nature of the molecular compounds, and (ii) the transfer integral to

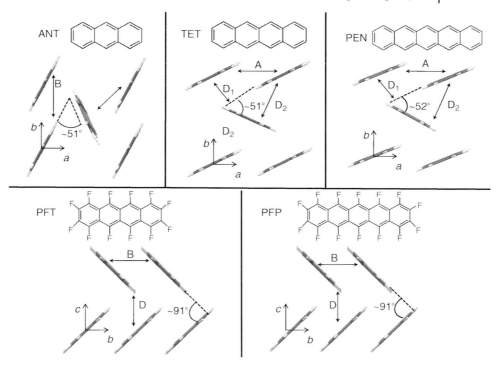

Figure 1.1 Chemical structures of the oligoacenes examined in this study with the notation used throughout the text. ANT: anthracene; TET: tetracene; PEN: pentacene; PFT: perfluorotetracene; PFP: perfluoropentacene. Representation of the crystal unit cell of the five molecules; the labeling of the different dimers is also displayed. Adapted from Refs [62, 63].

the molecular packing, by focusing on the single crystals of tetracene (TET), pentacene (PEN), and perfluorinated derivatives (perfluorotetracene (PFT) and perfluoropentacene (PFP) (see Refs [62, 63] for original references). Crystal and chemical structures are presented in Figure 1.1. Oligoacenes form an important class of hole-transporting semiconductors for organic field-effect transistors (OFETs); however, it has turned out to be more difficult to achieve electron transport in these materials. For instance, pentacene OTFTs exhibit electron mobilities of about 0.04 cm^2 V^{-1} s^{-1} [68]. Perfluorination is a successful strategy to convert a p-type organic semiconductor into n-type because it increases the electron affinity (in absolute terms) with respect to the parent molecule without strongly affecting the molecular structure and hence favors electron injection in devices [69–77]. Transistors based on perfluoropentacene exhibit electron mobilities as high as 0.22 cm^2 V^{-1} s^{-1} and the combination of perfluoropentacene (n-channel) and pentacene (p-channel) offers the possibility to fabricate bipolar transistors and complementary circuits [73, 74, 78]; this justifies the choice of the four compounds under study. In a second stage, we will describe the way the electronic coupling between adjacent molecules is affected by intermolecular vibrations in the crystal of anthracene and perfluoropentacene and the impact on the charge mobility values. To do so, a force field approach is required to depict the fluctuations in molecular packing induced by the intermolecular vibrational modes.

As illustrated in Figure 1.1, both perfluorinated and unsubstituted acenes present a herringbone motif in the b–c (or a–b) planes; however, there are significant differences that are induced by the introduction of the fluorine atoms:

(i) The molecular planes of adjacent molecules along the herringbone diagonal axis form an angle of ∼52° in PEN (TET), while these molecules are nearly perpendicular (∼91°) in the perfluorinated crystals.
(ii) Although the interplanar distance within the π-stacks in PFT and PFP (∼3.25 Å) is larger than that found in PEN (∼2.55 Å), the displacements along the short molecular axis are much less pronounced and no long-axis sliding is observed upon perfluorination (see Figure 1.1).

As a consequence of the differences in the mode of packing upon perfluorination, the electronic couplings for nearest-neighbor pairs of molecules along the various crystal directions are notably different. The transfer integrals between adjacent monomers were evaluated using the PW91 functional and the TZP basis set as implemented in the ADF package [79]. While in the perfluoroacene crystals, the largest electronic couplings are found only for the π-stacked dimers along the b-axis (i.e., for dimer B in PFP, $J_h = 132$ meV for holes and $J_e = 73$ meV for electrons), in the unsubstituted oligoacenes, large couplings are also present along the diagonal directions within the a–b plane (i.e., for dimer D_1 (D_2) in PEN, $J_h = 51$ (85) meV and $J_e = 82$ (81) meV) [9, 42]. For comparison, the electronic coupling between π-stacked dimers (dimer A for PEN and dimer B for PFP) (Figure 1.1) increases by a factor of ∼4 for holes (electrons) in PFP versus that of the parent PEN, whereas values up to 40 times smaller are found for the face-to-edge dimers along the diagonal directions (i.e., for dimer D in PFP, $J_h = 2$ meV and $J_e = 3$ meV). This is not surprising since the electronic coupling is driven by wavefunction overlap, which is expected to be much higher in a 51° tilted PEN dimer than in the perpendicular (∼91°) PFP dimer.

The internal reorganization energy λ_i can be estimated as the sum of two components: (i) the difference between the energy of the radical cation in its equilibrium geometry and that in the geometry characteristic of the ground state, and (ii) the difference between the energy of the neutral molecule in its equilibrium geometry and that in the geometry characteristic of the charged state. Upon oxidation and reduction, the major geometrical changes in the perfluoroacenes occur not only on the C—C bonds as found for their unsubstituted counterparts but also on the C—F bonds since a small electron density is observed on the peripheral fluorine atoms in the HOMO and LUMO wavefunctions [62]. The DFT estimates of the internal reorganization energies are found to be twice as large in the fluorinated systems for both holes and electrons (see Table 1.1). As the size of the system increases, that is, when going from PFT to PFP, λ_i decreases as expected (by ∼0.035 eV in the case of holes and ∼0.060 eV in the case of electrons). It is also possible via a normal mode

Table 1.1 Reorganization energies λ_i (in eV), associated with the hole and electron vibrational couplings, calculated at the B3LYP/6-31G** level.

	λ_i (Holes)		λ_i (Electrons)	
	AP[a]	NM[b]	AP	NM
PFT	0.258	0.259	0.289	0.286
TET	0.112	0.108	0.160	0.156
PFP	0.222	0.222	0.224	0.225
PEN	0.092	0.092	0.129	0.127

Adapted from Ref. [62].
a) Values calculated from the adiabatic potential (AP) surfaces for the neutral and charged species.
b) Values obtained from a normal mode (NM) analysis.

analysis to decompose the intramolecular reorganization energies into individual contributions from the intramolecular vibrational modes (see Figure 1.2). This analysis shows that the main contributions to λ_i in perfluoroacenes come from C—C/C—F stretching modes in the 1200–1600 cm^{-1} range.

1.2.2
Influence of Intermolecular Vibrations

We now turn to the description of the influence of the intermolecular vibrations on the transfer integrals (also referred to as the nonlocal electron–phonon couplings) [36, 80–83]. For the sake of illustration, we will show how this can be assessed by relying on molecular dynamics simulations in the case of the anthracene and perfluoropentacene crystals. This choice is motivated by the fact that these two crystals provide a large variety of transfer integral distributions at room temperature (Figure 1.3).

In the equilibrium crystal geometry optimized with the COMPASS [84] force field (see Table 1.2 for lattice parameters comparison to experiment), significant transfer

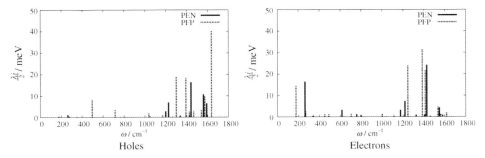

Figure 1.2 Contributions of the intramolecular vibrational modes to the hole and electron relaxation energy in PEN and PFP. Adapted from Ref. [62].

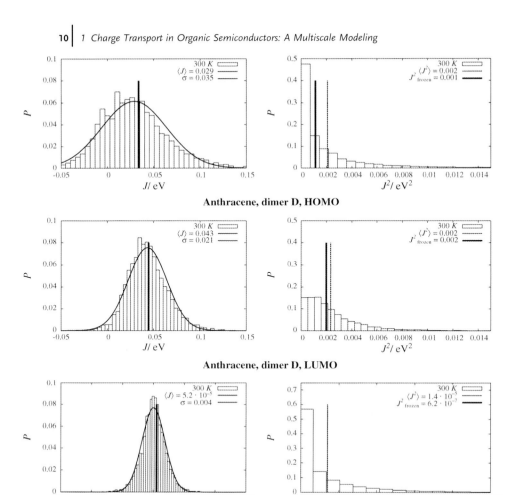

Figure 1.3 Probability distribution in arbitrary units of the transfer integrals for the three considered cases, as extracted from 5000 snapshots generated with the COMPASS force field. The transfer integrals calculated at the INDO level are reported both with their proper sign (left) and with their square value (right). When justified, the distribution has been fitted with a Gaussian function; the average value $\langle J \rangle$ and the standard deviation σ are also reported. Adapted from Ref. [63].

integrals are calculated at the semiempirical Hartree–Fock INDO (intermediate neglect of differential overlap) level only along the b-axis ($J_h = 44$ meV, $J_e = 26$ meV) and along the diagonal d-axis ($J_h = 34$ meV, $J_e = 45$ meV); the transfer integrals are vanishingly small between molecules located in adjacent layers, which suggests that transport in this material is anisotropic. Note that the INDO method has been preferred to DFT for the evaluation of the transfer integrals in this part of the work;

Table 1.2 Lattice parameters of the anthracene and perfluoropentacene unit cell, as provided by the COMPASS force field and experimental X-ray diffraction data. (Cell lengths are in Å.).

	a	b	c	α	β	γ
Anthracene						
COMPASS	8.30	6.01	11.07	90.0°	125.6°	90.0°
Experimental [93]	8.55	6.02	11.17	90.0°	124.6°	90.0°
Perfluoropentacene						
COMPASS	14.89	4.87	10.58	90.0°	92.2°	90.0°
Experimental [74]	15.51	4.49	11.45	90.0°	91.6°	90.0°

Adapted from Ref. [63].

indeed, it would be prohibitive to compute at the *ab initio* level the transfer integrals for thousands of snapshots extracted from MD runs, especially with the goal of extending this approach to more amorphous materials. The approach used to evaluate transfer integrals at the INDO level is detailed in Refs [85, 86]. In the equilibrium PFP crystal geometry, significant transfer integrals are calculated at the INDO level only along the *b*-axis ($J_h = 156$ meV, $J_e = 127$ meV), while very small values are obtained along the *c*- and diagonal axes, in good qualitative agreement with the DFT results. This implies quasi-one-dimensional transport when neglecting the impact of lattice dynamics.

Figure 1.3 portrays the distribution of transfer integrals for the HOMO and LUMO levels in dimer D of ANT (along the herringbone direction). This distribution has been calculated at the INDO level from 5000 snapshots generated with COMPASS and separated by 30 fs. In each case, we display the transfer integrals with their actual signs as well as the square values since the transfer integrals enter in the Marcus expression of transfer rates as squares; note that a positive sign implies that the negative combination of the two individual levels corresponds to the most stable level (i.e., HOMO-1 level) in the dimer and vice versa. Similarly, we report in Figure 1.3 the distribution of the transfer integrals estimated at the INDO level for the LUMO of dimer D of PFP, as extracted from 5000 MD snapshots. Periodic boundary conditions are used in all MD simulations and the size of the supercell is chosen such as the dimers of interest are surrounded by a full shell of neighboring molecules in order to prevent artificial symmetry effects ($4 \times 4 \times 3$ molecules for ANT and $4 \times 6 \times 3$ for PFP).

The selected distributions of transfer integrals reported in Figure 1.3 allow us to identify different degrees of impact of the lattice vibrations on the charge transport properties when compared to the equilibrium geometry. Hereafter, we distinguish among various cases by considering the ratio $\eta = |\langle J \rangle / \sigma|$, with $\langle J \rangle$ being the average value of the transfer integral in the distribution (i.e., corresponding to the center of the Gaussian distribution) and σ the standard deviation of the Gaussian distribution, as well as the inverse of the coherence parameter; the latter has been defined as $\langle J \rangle^2 / \langle J^2 \rangle$ in previous studies [87]. When the frequency of the intermolecular modes ($\sim 10^{12}\,\text{s}^{-1}$ for vibrational energies of 5–20 meV) is larger than the hopping frequency, the role of the lattice vibrations can be accounted for by injecting the corresponding $\langle J^2 \rangle$ value into the Marcus rate associated with a given jump, as generally done in the description of biological systems [87]. In contrast, when the frequency of the modes is smaller than the hopping frequency, each jump occurs at a rate extracted from the $\langle J \rangle^2$ distribution. These two cases can be referred to as "thermalized" and "static" limits, respectively. The actual situation often lies in between these two extreme cases in organic semiconductors. Since $\sigma^2 = \langle J^2 \rangle - \langle J \rangle^2$ for a Gaussian function, we can write

$$\frac{\langle J^2 \rangle}{\langle J \rangle^2} = \left(\frac{1}{\eta^2} + 1\right) \tag{1.4}$$

When η is large (small width σ), $\langle J^2 \rangle \sim \langle J \rangle^2$ and the impact of the lattice vibrations is expected to be weak. In contrast, smaller η values imply that $\langle J^2 \rangle$ becomes significantly larger than $\langle J \rangle^2$ and that the Marcus transfer rates in the thermalized limit are globally increased by the lattice vibrations compared to the equilibrium crystal geometry. In this framework, the impact of the lattice vibrations in a given crystal can be described from the trends observed for all inequivalent dimers found in the periodic structure of the molecular crystal.

A first scenario is obtained when the distribution is found to have almost entirely positive *or* negative transfer integral values and can be fitted with a Gaussian function centered close to the value for the equilibrium crystal structure; this is the situation for the LUMO of ANT (Figure 1.3) [88]. In this case, $\eta = 2.1$ and $\langle J^2 \rangle / \langle J \rangle^2 = 1.2$. The impact of the lattice vibrations is thus moderate owing to partial compensation between the slower and faster jumps compared to the average $\langle J \rangle$ value. In the case of ANT, a similar distribution is obtained for all possible inequivalent dimers for electron transport, thus suggesting that globally the lattice dynamics only slightly perturbs the charge mobility values. Another case occurs when the transfer integral calculated for the equilibrium geometry is vanishingly small and yields a Gaussian distribution centered around zero when the lattice dynamics is included; this is the situation for the LUMO of PFP for transport along c (Figure 1.3). Here, $\eta = 0.01$ and translates into a huge value for the ratio $\langle J^2 \rangle / \langle J \rangle^2 = 5200$. The intermolecular vibrations thus open new hopping pathways along c, thereby increasing the dimensionality of the charge transport in the crystal, which might prove important in the presence of structural defects. Finally, an intermediate situation is observed when the distribution of the transfer integrals has a Gaussian shape and is found to have both positive *and* negative values, with the average of the Gaussian distribution matching closely the transfer integral value characteristic of the equilibrium crystal structure. This is the case for the HOMO in dimer D of ANT (Figure 1.3) that exhibits a much larger broadening than the LUMO, despite the fact that the transfer integral values are similar for the two electronic levels in the equilibrium geometry. Here, $\eta = 0.8$ and $\langle J^2 \rangle / \langle J \rangle^2 = 2.5$, thus leading to an enhancement of the transport properties in the thermalized limit compared to the mobility values computed for the equilibrium crystal geometry.

1.2.3
Charge Mobility Values

The influence of lattice dynamics has been further assessed by performing kinetic Monte Carlo (MC) simulations of electron transport in the ANT and PFP crystals, with structural parameters extracted from the MD simulations and microscopic charge transport parameters calculated at the quantum chemical level [63]. Since our goal here is not to provide absolute values of the charge carrier mobilities, we have considered a pure hopping regime for charge transport with the transfer rates between two neighboring molecules i and j calculated according to the simple semiclassical Marcus expression (see Equation 1.2). We have used an internal reorganization energy for electrons of 196 meV for ANT [49] and 224 meV for PFP [62], as calculated at the DFT/B3LYP level from the adiabatic potential energy surfaces of the neutral and charged states; the external part has been neglected here.

The charge mobility values have been evaluated using a single-particle biased Monte Carlo algorithm in which a charge initially positioned at a random starting site performs a biased random walk under the influence of the electric field [89]. At each MC step, the time for the charge at site i to hop to any of the six neighboring sites j is calculated from an exponential distribution as

$$\tau_{ij} = -\frac{1}{k_{ij}} \ln(X) \tag{1.5}$$

with X being a random number uniformly distributed between 0 and 1 and k_{ij} the Marcus transfer rate. The hop requiring the smallest time is selected and executed; the position of the charge and the simulation time are then updated accordingly. A transit time t_{tot} averaged over 10 000 simulations is calculated for the charge migration over a distance d_{tot} along the field direction. The charge carrier mobility is ultimately evaluated from Equation 1.1 for an applied electric field F in a given direction within the molecular layer. The charge samples a periodically repeated cell consisting of a single layer of 32 molecules.

The impact of lattice dynamics on the charge mobility values has been assessed by using Marcus transfer rates involving (i) the square of the electronic coupling averaged over the MD snapshots $\langle J^2 \rangle$ for all inequivalent dimers (thermalized limit), (ii) the square of the average coupling $\langle J \rangle^2$ of the dimers corresponding to the center of the Gaussian distribution and hence to the electronic coupling characteristic of the equilibrium geometry at room temperature, and (iii) instantaneous J_{ij} and molecular positions for selected MD configurations (static limit). In this case, the mobility is then averaged over all molecular frames (1000 MD frames separated by 30 fs).

The results are summarized in Figure 1.4, which displays a polar plot of the electron mobility for the ANT and PFP crystals at $T = 300$ K for an applied electric field of 250 kV cm^{-1} (such a field is in the typical range of time-of-flight (TOF) experiments). For the ANT crystal, we have calculated the two-dimensional electron mobility plot within the a–b plane. For small values of the electric field, the transport is nearly isotropic, with a small enhancement along the a-axis direction observed for

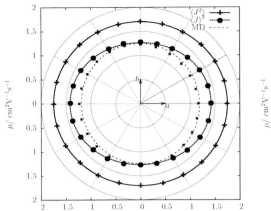
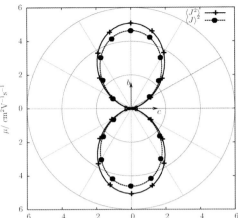

Figure 1.4 Polar plot of μ_e within the a–b plane for anthracene (left) and within the b–c plane for perfluoropentacene (right) crystal at $T = 300$ K and for an applied electric field of 250 kV cm^{-1}. Solid lines refer to MC simulations using the values of $\langle J^2 \rangle$ characteristic of the equilibrium structure (thermalized limit) and bold dotted lines refer to MC simulations using the values of $\langle J \rangle^2$. In the left panel, dashed lines are associated with mobility values averaged over all MD snapshots within a static limit. Adapted from Ref. [63].

cases (i) and (ii). The plot shows that the charge carrier mobility is enhanced in the thermalized limit when replacing $\langle J \rangle^2$ by $\langle J^2 \rangle$ in Equation 1.2, in agreement with the calculated η ratios equal to 1.2 and 1.5 for dimers D and B, respectively. There is an increase of ∼23% along a and of ∼6% along b, leading to calculated mobility values of $\mu_e = 1.74$ and 1.70 cm^2 V^{-1} s^{-1}, respectively, in the hopping regime. In contrast, the charge mobility calculated in the static limit by averaging over a large number of frozen geometries in the course of the trajectory is lower than the value computed from the equilibrium structure. The mobility values obtained using the equilibrium structure are within the range obtained from the various individual snapshots. We note that although our approach is not expected to provide absolute values of the mobility in these crystals, the magnitude of the calculated room-temperature electron mobility in ANT is in good agreement with experimental values [90–92].

In the case of the perfluoropentacene crystal, we observe a pronounced anisotropy in the mobility within the b–c plane that reflects the large variations in the amplitude of the electronic coupling along the different directions (see Figure 1.4). Electron mobility values are very low for transport along the c direction, while the highest values are obtained for transport along the b-axis. As in the case of anthracene, an increase in the mobility is found in the thermalized limit compared to the value characteristic of the equilibrium structure when injecting $\langle J^2 \rangle$ values in the transport simulations. This enhancement is observed for all directions within the b–c plane. In particular, we find an increase of ∼45% for a deviation of 5° with respect to the c direction, which is further substantially amplified when approaching a 0° deviation. Though there is clearly the opening of a new conducting pathway along the c-axis induced by lattice dynamics, transport along this direction remains strongly limited due to the two-dimensional character of the system that favors electron migration in the directions exhibiting the largest electronic couplings. The electron mobility increases by ~10% in the thermalized limit along the *b*-axis to reach a value of 5.1 cm^2 V^{-1} s^{-1}.

1.3
Tetrathiafulvalene Derivatives

The creation of circuits at the nanoscale is important for miniaturization of electronic devices. Candidate components for the wires in the circuits are generally considered to be conventional metals [94] or carbon nanotubes [95]. However, several reports have shown the possibility to form supramolecular wires [96, 97] by a bottom-up approach [98], using organic molecules that self-assemble [99] at a surface under ambient conditions from a solution. Recently, interest has focused on tetrathiafulvalene as functional component, since compounds of that family can behave as conductors and superconductors in a crystalline environment [100]. Intrinsically, these π-electron-rich units tend to adsorb flat on a graphite surface, which precludes the formation of nanowires [101, 102]. In order to overcome this problem, the TTF molecules have been functionalized by introducing amide groups in the structure to generate hydrogen bonds [103, 104], which are expected to modify the molecular orientation of the TTF units and create one-dimensional assemblies. We will refer to this compound as TTF-1; its molecular structure is shown in Figure 1.5a.

Scanning tunneling microscopy (STM) investigation of the self-assembled monolayer of TTF-1 at the solvent (octanoic acid)/HOPG graphite interface shows equally spaced continuous lines of high tunneling current, indicative of the formation of supramolecular rod-like fibers at the surface (Figure 1.5b and c). The fibers are separated by 4.65 ± 0.15 nm, that is, approximately the length of one extended molecule of TTF-1. Assuming that the bright spots in the images correspond to the molecular TTF moieties, the distance between TTF units within a fiber is found to be approximately 0.44 ± 0.03 nm. These observations indicate that the planes of the TTF moieties are not parallel to the HOPG surface (the spots are too close), but are tilted with a high angle with respect to the surface plane. In such conformation, the TTF moieties are expected to interact with each other rather than with the graphite, leading to the formation of a delocalized π-system that extends along the columnar structures.

In this context, force field calculations have been performed to describe at the atomistic level the organization of the TTF molecules in the nanowires (see Ref. [64] for the original publication). The goal of such calculations is twofold: (i) to explore in detail the energetics and the role of the different and competitive interactions acting in the assembly, and (ii) to provide a reliable geometry to calculate electronic and optical properties for the system. Based on the structural information obtained at the force field level, quantum chemical calculations have been performed at the DFT level (using the B3LYP functional and a TZP basis set) to evaluate transfer integral values along the stacks.

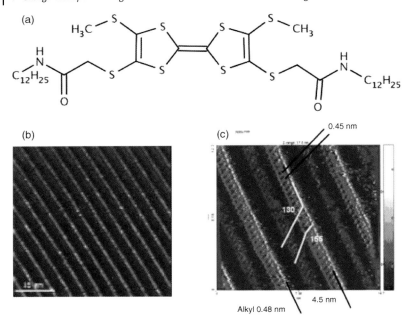

Figure 1.5 (a) Molecular structure of the diamide tetrathiafulvalene derivative TTF-1. (b and c) STM images of the TTF-1 monolayer. In (c), some structural parameters are shown (left: $I_{set} = 0.45$ nA, $V_{bias} = -0.463$ V; the bar length is 15 nm; right: 12.3 × 14.7 nm). Images obtained at the KUL University, Leuven, Belgium.

Looking at the molecular structure of TTF-1, different and competitive interactions can take place in the assembly since the molecules of TTF-1 are able to interact through π–π stacking, involving the conjugated molecular cores, and through H-bonding involving the amide groups, as shown in Figure 1.6a. These intermolecular interactions are not the only ones acting in the system: the molecules also interact with the solvent and with the surface. π–π stacking can also take place between the conjugated molecular cores (TTF moieties) and the π-system of the graphite, driving the molecules to physisorb flat on the surface. In view of all these interactions and the huge number of degrees of freedom (molecular configurations and orientations), the system is highly complex from the theoretical point of view and the most suitable approach to model it relies on force field calculations.

Based on the information from the STM images, reasonable starting geometries can be built for modeling the columnar stacks. The force field (here MM3) is selected by considering the type of molecules to model, as well as the ability of the force field to describe the interactions that are likely to occur in the assembly. In order to avoid edge effects, an infinite HOPG graphite is constructed by applying periodic boundary conditions (PBCs) to the system. The unit cell is cubic and its size in the plane of the graphite (XY) is tailored on the system. A slab of vacuum 50 Å thick along the Z-direction is used as a spacer between the unit cell and its periodic images to avoid

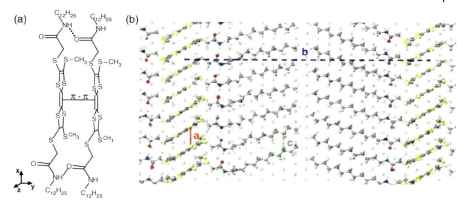

Figure 1.6 (a) Schematic representation of a dimer of TTF-1: the two conjugated cores are cofacial and can interact via π–π stacking, while amide groups can form H-bonds to link the molecules together into 1D assemblies. In this scheme, the long side chains are considered to be parallel to the plane of the graphite surface (*XY* plane), while the TTF moieties are orthogonal to this plane (edge-on geometry). (b) A model of two parallel stacks of TTF-1 on HOPG, after periodic MD/MM simulations. In blue is indicated the interstack distance and in red the intermolecular distance within a single stack. The green dotted line indicates the distance between two adjacent alkyl chains. Adapted from Ref. [64].

interactions along the third dimension. The effect of solvent has been to some extent implicitly included using its dielectric constant to screen long-range electrostatic interactions. The resulting model not only allows reproducing the experimental geometry of the monolayer but also provides new useful data. Figure 1.6b shows part of the structure obtained with the model, in which well-ordered stacks are clearly visible. Molecules have their alkyl chains fully adsorbed on graphite and parallel to each other along the stacks, forming compact alkyl rows. There is no interdigitation of the alkyl groups. The TTF moieties are edge-on with respect to the surface with a tilt angle of about 80° and are all aligned to form π-stacking. The amide groups are oriented in such a way that H-bond patterns are formed on both sides of the TTF units to bind the molecules together and hence to maintain the molecular alignment along the stack. Statistics of some meaningful geometric parameters, such as the interstack distance and the separation along the main axis of the stack between two adjacent molecules, have been collected along the molecular dynamics trajectories and compared with the information extracted from the STM images. The results are collected in Table 1.3, revealing that the experimental morphology is fully consistent with that predicted by the model.

The distance between the planes of the TTF conjugated core of adjacent molecules is about 3.5 Å, that is, well in the range of action of intermolecular π–π interactions between neighboring molecules. This has strong implications regarding the charge transport capabilities of the stacks. Quantum chemical calculations were performed to quantify the electronic coupling between two adjacent molecules in the stack. The transfer integral is computed to be 134 meV between their HOMOs, and 111 meV between the LUMOs. This indicates a strong coupling between the π-systems of the

Table 1.3 Comparison between theoretical and experimental meaningful geometric parameters.

Parameter	Experiment	Theory
Interstack distance (nm)	4.5	4.4
Intermolecular distance a (nm)	0.46	0.46
Intermolecular distance b (nm)	—	0.35

The molecular distance a is the distance between two adjacent molecules along the main axis of the stack. The molecular distance b is the distance between the two π-systems of adjacent molecules (these data have not been provided by the analysis of the STM images). The two distances are not the same since molecules are tilted and shifted with respect to the main axis of the stacks.

molecules. According to Hückel theory, for an infinite one-dimensional stack, the width of the valence band and conduction band is four times the transfer integral associated with the HOMO and LUMO levels, respectively, that is, 0.54 and 0.44 eV, respectively. These large bandwidths show that those TTF stacks can act as molecular wires for both hole and electron transport [94].

1.4
Polythiophene Derivatives

Force field calculations can also be used to predict the packing of conjugated chains and describe on that basis their charge transport properties. An original approach validating the use of force field calculations via the simulation of corresponding X-ray diffraction patterns is also introduced here (see Ref. [65] for the original reference).

Conjugated polymers are generally made of a conjugated backbone along which alkyl groups are grafted to improve solubility. During film formation, these two components self-segregate, often giving rise to a lamellar structure with stacks of conjugated backbones separated by layers of alkyl groups, which act as insulators. The lamellae can orient differently – parallel or normal – to the substrate, dramatically changing the mobility in the plane of the film (by more than a factor of 100 for poly(hexyl thiophene), P3HT) [105]. Such a high anisotropy reflects a transport of the charge carriers that is much more efficient along the π-stacking direction and along the backbones than through the layers of packed alkyl groups. The mobility is thus maximized when the π-stacking direction or the long-chain axes are aligned along the flow of current, that is, when the lamellae are parallel to the substrate.

Poly(2,5-bis(3-alkylthiophen-2-yl)thieno[3,2-b]thiophene) (PBTTT) (see chemical structure in Figure 1.7, top) has been recently shown to have improved stability to air and light and higher mobilities (up to 0.6 cm^2 V^{-1} s^{-1} in long-channel and 1 cm^2 V^{-1} s^{-1} in short-channel FETs) relative to P3HT, so far one of the most promising semiconducting polymers [106, 107]. The structural ordering in the films shows large lateral terraces extending over several hundreds of nanometers [106, 108]. Calculations of the transport properties carried out on model stacks of PBTTT and P3HT oligomers indicate that the improved mobility of PBTTT is not related to the nature of

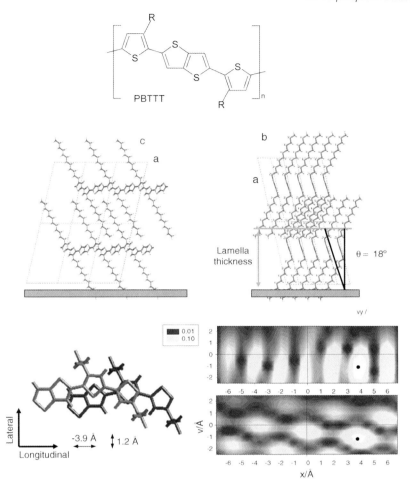

Figure 1.7 *Top*: Chemical structure of PBTTT. *Middle*: Interdigitated polymorph of PBTTT viewed along *b* (left) and *c* (right) and represented in its preferential orientation with respect to the substrate. *Bottom left*: Monomer units of adjacent chains viewed orthogonally to the backbone planes and with their relative displacements indicated. *Bottom right*: ZINDO-calculated transfer integrals *J* (in eV) between HOMOs (top) and LUMOs (bottom) for two adjacent PBTTT dimers separated by an intermolecular distance of 3.6 Å. The transfer integrals have been calculated for the simulated geometry (in 0,0) and for different relative positions of the oligomers obtained by sliding one oligomer laterally (*y*-axis) and longitudinally (*x*-axis); the dot represents the position of the perfectly cofacial dimer. Adapted from Ref. [65].

the repeat unit, but instead could partly be linked to the relative positioning of the chains [86].

So far, only few detailed packing structures have been reported for semiconducting polymers in films since X-ray diffraction (XRD), the preferential structural analysis method, generally gives too few well-resolved diffraction peaks, thus leaving uncertainties about the structures. In contrast, PBTTT films have well-defined XRD patterns due to the unprecedentedly high crystallinity and orientation of the polymer chains, thus giving the opportunity to obtain an accurate evaluation of the film structure by combining theoretical and experimental approaches. The systems considered here are thin films (70–100 nm) of PBTTT substituted by dodecyl groups (PBTTT-C12), which were spin coated on silicon oxide surfaces modified by octyltrichlorosilane, followed by annealing at 180 °C [106, 109]. The structure of the crystalline domains of the films was inferred by simulation techniques, following a procedure in four steps to funnel the search toward the equilibrium structure:

(i) The torsion potentials around the bonds connecting the thiophene units of the PBTTT backbone were first evaluated to determine whether the backbone adopts a planar conformation upon packing of the molecules. This information is necessary as the next step consists in modeling the PBTTT chain conformation in crystalline domains from a conformational search performed on isolated molecules, that is, where packing effects are absent. Data in the literature [110, 111] and our DFT calculations show that both junctions can planarize easily upon packing. As a result, the junctions in PBTTT were set planar.

(ii) A conformational search was performed with the PCFF force field on a long isolated oligomer. A lamellar organization of the polymers implies that backbone layers and alkyl layers alternate, with a dense packing expected in both types of layers. Few conformations of a polymer chain are compatible with such an organization, thus requiring the use of both geometry and energy criteria to select the best candidates. Six PBTTT conformers were generated, all having alkyl groups that are out of plane with respect to the conjugated backbone. The most stable conformer corresponds to an all-*anti* conjugated backbone and has the best geometrical characteristics: the backbone is straight (with bent backbones, helical or disordered structures occur), and the alkyl groups are oriented in the same direction (in the solid state, this structure favors a dense packing in the alkyl layer via tilting, interdigitation, and nesting).

(iii) A crystal cell was built containing one monomer unit (i.e., the substituted thiophen-2-ylthieno[3,2-*b*]thiophene motif) repeated along the backbone direction. The monomer unit was linked to its images in the neighboring cells to produce an all-*anti* polymer chain. The initial parameters of the cell were adjusted to reproduce an infinite stack of infinite polymer chains, with either interdigitated (I) or noninterdigitated (NI) alkyl side groups; those systems were optimized with the PCFF force field. The very good packing of the alkyl groups in both polymorphs is related to their aptitude to tilt and orient differently with respect to the backbone so as to maintain an optimal density in the crystal. This explains why the density does not change much from NI to I (1.2 versus 1.3). NI is much less stable than I, by 14.7 kcal mol^{-1} per monomer unit [65]. PBTTT is therefore expected to be interdigitated (see middle of Figure 1.7).

(iv) Interdigitation has been further confirmed by simulating the 2D XRD patterns of I and NI and comparing them to experimental patterns: the pattern of I fits the experiment much better than that of NI. Still, the matching can be improved if we get rid of the small differences between the simulated and real cells inherent to the approximations of the simulation methods. To do so, an indexation of selected experimental spots was proposed from the simulated pattern; the positions of these spots were then calculated varying systematically the six cell parameters, and were compared to the experimental positions using a rms deviation criterion. Sets of parameters corresponding to the smallest rms were applied to crystal cells, which were then optimized with these new cell parameters set fixed. The 2D patterns were finally simulated and compared to experiment. Figure 1.8 compares the experimental 2D pattern to that of the simulated cell with refined parameters, showing a remarkable agreement between theory and experiment. A detailed assignment of the experimental spots is provided in Ref. [65].

Finally, the relationship between the supramolecular organization and the charge transport properties was analyzed by computing the transfer integrals J between PBTTT dimers at the INDO level. The initial relative positions and geometric structures of the oligomers were extracted from the proposed cell. In order to assess the influence of deviations from that packing geometry on charge transport, J was calculated for different relative positions of the oligomers, as obtained by sliding one oligomer laterally and longitudinally by steps of 0.5 Å, with the intermolecular distance fixed at 3.6 Å. That distance corresponds to the interbackbone spacing obtained in the simulated cell. The evolution of J for the HOMO (hole transport) and LUMO (electron transport) levels is illustrated in the bottom part of Figure 1.7. The transfer integrals are among the highest for the proposed structure (i.e., the 0,0 coordinate in Figure 1.7) for both holes and electrons. This result further supports our analysis that the proposed structure is close to the PBTTT equilibrium packing

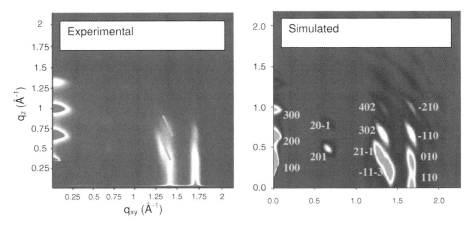

Figure 1.8 Experimental and simulated 2D XRD patterns of PBTTT for the interdigitated configuration: a low level of disorder of the crystallites in the film (with a standard deviation $\sigma = 4°$) and a small peak broadening (0.5°) are considered. Adapted from Ref. [65].

geometry since it is consistent with the high mobility values measured for PBTTT and points to the ambipolar transport properties of the material (provided that the charge injection is optimized for both carriers, as observed recently for other polymers) [112]. Note that the transfer integrals can drop by at least one order of magnitude upon small relative displacements (less than 1 Å) of the chains, thus suggesting that the conformational dynamics of the chains might affect transport properties.

1.5
Phthalocyanine Stacks

Liquid crystals (LCs) have attracted a lot of fundamental and technological interest owing to their remarkable properties of self-assembly over large areas [113]. This feature of LCs is of particular interest for the field of organic electronics that requires well-ordered active materials to ensure efficient charge and energy transport. Another advantage is the possibility to modulate the nature of the side chains attached to the conjugated core to achieve a LC character at room temperature and hence avoid the formation of grain boundaries due to the self-healing properties. We focus hereafter on phthalocyanine molecules substituted by four alkoxy chains that self-organize into one-dimensional columns [114]. Recently, PR-TRMC (pulse-radiolysis time-resolved microwave conductivity) measurements probing charge transport at a very local scale have pointed to hole mobility values around $0.2\,\mathrm{cm^2\,V^{-1}\,s^{-1}}$ [115]. On the other hand, time-of-flight measurements yield charge mobility values on the order of only $10^{-3}\,\mathrm{cm^2\,V^{-1}\,s^{-1}}$ [116], thus suggesting that the presence of structural defects strongly affects the charge transport properties. This apparent discrepancy is rationalized here by combining molecular dynamics simulations to characterize structural properties and kinetic Monte Carlo approaches to access the charge transport properties (see Ref. [66] for the original reference).

1.5.1
Structural Properties

Atomistic MD simulations have been applied to a sample of 80 tetraalkoxy phthalocyanine molecules (see Figure 1.9 for chemical structure) described at the united atom level (with hydrogen atoms condensed with the closest heavier atom) by using the AMBER force field with periodic boundary conditions; the simulation time is about 100 ns.

By progressively equilibrating within a cooling sequence at atmospheric pressure the initial configuration of the system consisting of four columns of 20 regularly stacked molecules arranged on a regular square lattice, a hexagonal lattice was obtained in the temperature range 450–350 K and a phase transition to the rectangular lattice between 325 and 300 K. This is in agreement with experimental data showing a hexagonal to rectangular transition at 340 K [114]; the rectangular phase is characterized by a tilt of about 15° of the molecules with respect to the columnar axis. The phase transition was confirmed by Arrhenius plots of the characteristic time of

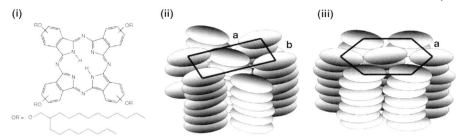

Figure 1.9 (i) Chemical structure of the tetraalkoxy-substituted phthalocyanine molecules under study featuring branched C_{12}–C_8 side chains. Schematic representation of a rectangular (i) and hexagonal (ii) mesophase. Adapted from Ref. [66].

rotation of the x-axis (see Figure 1.10a), which allows singling out two different regimes: at high temperature (columnar hexagonal phase), the alkyl chains display a liquid-like behavior, so the columns can translate with respect to the other and molecules can rotate around the columnar axis; at low temperature (columnar rectangular phase), these motions become frozen.

The agreement with the experimental morphologies was further confirmed by confronting the density values (MD: 1.08, 1.01, 0.99 g cm^{-3} at 300, 400, and 425 K, respectively; X-ray: 1.09–1.1 and 1.00 g cm^{-3} at 300 and 383 K, respectively [114, 117])

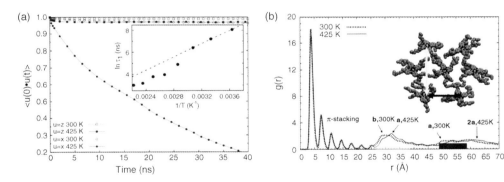

Figure 1.10 (a) Autocorrelation function of the molecular axes x and z in the rectangular (empty symbols) and hexagonal (filled symbols) phases. In the inset, an Arrhenius plot of the rotational time τ_1 associated with the x-axis based on data at all simulated temperatures shows the presence of two different regimes, corresponding to hexagonal (high temperature, thin dashed line) and rectangular (low temperature, thick dashed line) phases. The lines represent fits of τ_1 with the equation $\tau_1(T) = \tau^* \exp(-E_A/kT)$ for $T > 325$ K and $T \leq 325$ K, respectively. In-plane rotational times τ_1 were obtained at all simulated temperatures by fitting in the 0–40 ns range the autocorrelation functions of the main plot with triexponential functions and integrating them from $t=0$ to $t=\infty$. (b) Radial distribution of the core hydrogens in the rectangular (dotted line) and hexagonal (full line) phases, with explicit assignment of the main peaks. In the inset, a snapshot of seven molecules belonging to different columns evidences the hexagonal packing and the interdigitation of alkyl chains. Adapted from Ref. [66].

and by calculating the radial distributions of the inner hydrogens at 300 and 425 K on a larger sample made of 1440 molecules; the latter was built by replicating the 80-molecule cell and subjected to a simulation run of 5 ns. At both temperatures, the distributions (see Figure 1.10b) are dominated at short distances (<25 Å) by a sequence of peaks centered at multiples of the intermolecular stacking distance within a column. Their intensity is rapidly attenuated with increasing distance, in agreement with the short correlation length (around 3.7 nm) revealed by atomic force microscopy measurements on spin-coated phthalocyanines of similar nature [117]. At larger distances, a broader peak associated with intercolumnar separations appears at about 30 Å and its multiple at 60 Å, in full consistency with the X-ray values of $b = 29.8$ Å (at 300 K) for the rectangular phase and $a = 30.0$ Å (at 383 K) for the hexagonal phase [114].

Since a hopping process involves a charge transfer between two adjacent molecules, the characterization of the geometry of the dimers in the columns is of key importance. To do so, each dimer was characterized by the relative translation (Δx, Δy, Δz) between the centers of mass and the rotation angle θ. The distribution of (Δx, Δy) presents a cylindrical symmetry in the hexagonal phase that is lost at 300 K (Figure 1.11a), thus confirming the rotational freezing in the rectangular phase. The configuration displaying two neighboring molecules with exactly superimposed mass centers ($\Delta x = 0$, $\Delta y = 0$) clearly becomes disfavored at 300 K following the onset of a tilt angle in the columns.

The distributions of θ (see Figure 1.11b) in the interval 0–180° show in both phases a broad peak extending between 60° and 120°, with a maximum around 90°. The latter geometry prevents the inner hydrogen atoms of the PC cores to get superimposed; this feature has also been predicted by DFT-D calculations for a dimer of porphine [118] sharing structural similarities with PC.

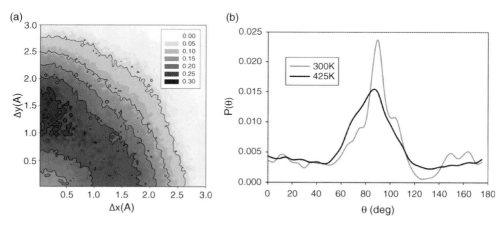

Figure 1.11 (a) Two-dimensional probability maps of the horizontal displacements of the intermolecular vector (Δx, Δy) in the rectangular phase. (b) Probability distribution of the rotational angle θ at 300 and 425 K. Adapted from Ref. [66].

1.5.2
Charge Transport Properties

Since the transfer integrals are highly sensitive to the relative positions of the interacting molecules [44], the differences in the distribution of geometrical parameters in the LC phases should translate into significant variations of the transfer integrals along the molecular dynamics trajectory. To verify this hypothesis, we have calculated all transfer integrals in the unit cell on the basis of 650 configurations separated by 100 ps, which were extracted from the MD simulations. In both phases, the transfer integrals are negligible between molecules located in adjacent columns, so the charge transport is strictly one dimensional. Unexpectedly, in spite of the different structural properties of the two mesophases, we obtain similar distributions and similar average transfer integral values (60 and 58 meV, respectively) for the rectangular and hexagonal mesophases. Plotting the distributions in a log–log scale clearly shows a tail associated with geometric configurations promoting very small transfer integrals, which might thus act as traps.

Fast fluctuations of the electronic coupling are evidenced in Figure 1.12 showing the evolution over 5 ps of the transfer rate from a given molecule to the nearest neighbors (i.e., in a direction parallel and opposite to the applied electric field) in the rectangular phase. The rates have been estimated here using the Marcus–Levich–Jortner formalism [119] treating at the quantum mechanical level a single effective intramolecular mode assisting the transfer, with an internal reorganization energy of 114 meV calculated at the DFT level for holes, an external reorganization energy set at

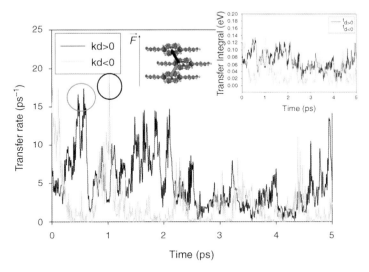

Figure 1.12 Time evolution of the transfer rate along the electric field direction (d > 0, dark curve) and in the direction opposite the field (d < 0, light curve) in a trimer of PC molecules extracted from a column. The left (right) circle represents a situation when charge transfer is more likely to occur along the electric field direction (in the direction opposite to the electric field). The inset shows the corresponding time evolution of the transfer integrals.

200 meV for all hopping events, and an external electric field of 1000 V cm^{-1}. Interestingly, situations where the charge transfer in the direction opposite to the electric field is more favorable do happen regularly and are not attenuated by the presence of high electric fields (up to 10^6 V cm^{-1}). In these instances, the charge is thus frozen out for a finite time over a part of the column that acts as a dynamic defect. Such a defect is temporarily created by specific translations and/or rotations yielding a geometry characterized by a very small transfer integral.

In order to get a deeper insight into the influence of such dynamical defects on the transport properties, we have calculated the charge mobility along the stacks in two different ways, namely, in a purely static limit and with an approach taking into account lattice dynamics. In the static picture, we have propagated via Monte Carlo simulations a hole in 650 snapshots separated by 100 ps extracted from the MD simulation and have calculated the average hole mobility; in this case, since the molecules are frozen during the KMC evolution, the lifetime of a defect is infinite. Doing so, we obtain similarly low mobility values of 3.12×10^{-3} and 2.92×10^{-3} cm^2 V^{-1} s^{-1} in the rectangular and hexagonal mesophases, respectively. Interestingly, the calculated mobility has the same order of magnitude as the value provided by TOF measurements ($\mu = 10^{-3}$ cm^2 V^{-1} s^{-1} [116]) for a very similar PC derivative; this is consistent with the fact that the TOF mobility values are known to be limited by the presence of static defects. In the dynamic approach, in view of the fast thermal fluctuations, we have performed Monte Carlo simulations based on hopping rates averaged for each dimer over the 650 snapshots. In this approach, the system is defect-free and the transfer rate in the forward direction (i.e., parallel to the electric field) is always larger than that in the backward direction. The mobility here is limited only by the hopping events in the direction opposite to the electric field. In this framework, the hole mobility is estimated to be two orders of magnitude larger (around 0.1 cm^2 V^{-1} s^{-1}) at both temperatures. Interestingly, this value is very close to that given by the PR-TRMC technique ($\mu = 0.08-0.37$ cm^2 V^{-1} s^{-1} [115]) that probes the transport at the nanoscale in defect-free regions.

The huge difference between the mobility values provided by the two approaches suggests the presence of many dynamic structural defects along the columns. This has been quantified by spotting every single defect in the 650 configurations of the system and by classifying them according to the transfer probability $p_\rightarrow$ to cross the defect (namely, the ratio of the transfer rate along the electric field direction over the sum of the forward and backward transfer rates). Defects characterized by $p_\rightarrow$ lower than 10%, which are likely to influence charge transport the most, are by far the most numerous (see Table 1.4). We have also calculated the time required to cross the defect; since the distance between the molecules is always approximately the same (3.6 Å), this parameter directly reflects the impact of molecular misalignment on the charge mobility. The results reveal that the transit time is reduced by two orders of magnitude in the presence of defects with the lowest transfer probability ($p_\rightarrow = 10\%$), thus demonstrating that the static mobility values are predominantly limited by the presence of such defects. We stress that the impact of static defects is amplified in phthalocyanine derivatives due to the one-dimensional character of charge transport; indeed, there is no other solution than to keep

Table 1.4 Percentage of defects as a function of the transfer probability along the electric field direction in the rectangular and hexagonal phases, as calculated for a field $F = 1000\,\text{V}\,\text{cm}^{-1}$.

Probability	Percentage of defects	
	300 K	425 K
0–10%	23.0	22.4
10–20%	8.8	8.9
20–30%	6.8	6.8
30–40%	5.9	6.1
40–50%	5.4	5.6

Adapted from Ref. [66].

propagating the charge along a given column, in comparison to two- or three-dimensional transport encountered in molecular crystals. The situation is also different for one-dimensional transport involving compounds with degenerate electronic transport levels (such as triphenylene or hexabenzocoronene derivatives); in this case, when the efficiency of a conduction pathway involving a given level is poor, the problem is generally compensated by another pathway relying on another degenerate level [39, 120].

1.6
Polymer Dielectrics

We show in this section how the presence of an interface between two different materials, whose morphology has been simulated by force field calculations, can affect charge transport properties (see Ref. [67] for the original reference). The previous sections dealt so far with charge transport in the bulk of organic semiconductors. However, charge transport does also occur in the vicinity of interfaces; this is especially the case in OFETs where the charges are confined within a few nanometers from the surface of the dielectrics at low charge density, so the transport mostly takes place within the first molecular layer [121, 122]. The transport properties, and hence the charge mobility values, are thus expected to be further affected in the conducting channel by the following:

(i) The electrical properties of the insulator layer. A significant drop of the mobility by up to one order of magnitude was reported in OFETs based on polytriarylamine chains when replacing low-k polymer dielectrics by polymethylmethacrylate (PMMA) [123]. Similar observations were made in the case of pentacene layers [124, 125]. This deterioration of the mobility was attributed to an increase in the energetic disorder promoted by the polar carbonyl bonds of the PMMA chains [123].

(ii) The morphology of the organic layer. The formation of grain boundaries is one of the major structural limitations to high charge mobility values [126–128].

Recent studies on OFETs based on pentacene have shown that the grain size varies as a function of the nature of the polymer dielectrics and that the carrier mobility is very sensitive to the grain size below a root mean square value of about 0.8 μm [124, 129].

(iii) The nature of the electronic states at the surface. Organic semiconductors typically display a p-type behavior when using SiO_2 as the dielectric layer in OFETs due to the presence of electron traps on the surface [112, 125].

We focus here on pentacene molecules (i.e., one of the most studied and efficient organic semiconductors) deposited on top of polystyrene (PS) versus PMMA chains, widely used as polymer dielectrics. This choice is motivated by the fact that PS is nonpolar, while PMMA features polar bonds associated with the carbonyl groups; moreover, the performance of OFETs involving these interfaces have been assessed [124, 125]. In a way similar to the experimental fabrication of an OFET [130], we obtained with atomistic molecular dynamics simulations 3D periodic cells consisting of a polymer (PS or PMMA)/pentacene system equilibrated at 300 K (see Ref. [67] for details). Each slab is about 60 Å thick, allowing the extraction of two independent dielectrics/pentacene interfaces, with the organic semiconductor forming four crystalline layers homeotropically oriented (Figure 1.13). Note that the finite size of the supercell (60 × 60 Å^2) does not allow us to describe morphological defects such as grain boundaries.

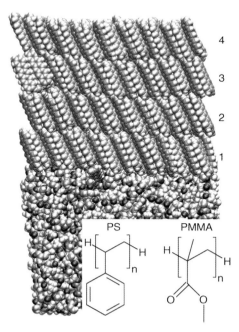

Figure 1.13 Atomistic description of the pentacene/polymethylmethacrylate interface. The inset shows the chemical structures of the two polymer dielectrics under study. Adapted from Ref. [67].

1.6.1
Electrostatic Disorder

In order to characterize the impact of energetic disorder in the different pentacene layers on the charge transport properties, we have defined the parameter ΔE_{ij} associated with a charge hopping process:

$$\Delta E_{\text{diel}} = E_i^0 + E_j^+ - \left(E_i^+ + E_j^0 \right) \quad (1.6)$$

where E is the electrostatic interaction between a pentacene molecule and the PMMA or PS chains. $0\,(+)$ denotes a neutral (positively charged) molecule and i and j the initial and final sites involved in the hopping process, respectively. The parameter ΔE_{ij} only represents the energetic difference between the initial and final states induced by the polymer dielectrics; the influence of an applied electric field is not considered in this parameter and will be considered in the next step. The Coulomb energies were calculated from atomic point charges obtained with DFT calculations using the B3LYP functional and the aug-cc-pVDZ basis set. Static electronic polarization effects due to the polymer dielectrics are found to be limited due to the low polarizability of the saturated polymer chains in the dielectric layer, as supported by additional quantum chemical calculations [67].

Figure 1.14 displays the distribution of ΔE_{ij} for all pairs of pentacene molecules in the different layers on top of PMMA versus PS, as averaged over 100 snapshots extracted every 100 ps from a MD run of 10 ns. The distributions are symmetric with respect to zero since each molecule is considered at the same time as a possible initial or final site; all distributions can be fitted with a Gaussian distribution. For each polymer dielectrics, the broadening is most pronounced in the layer in direct contact with the polymer chains. The evolution of the standard deviation σ of the Gaussian distribution as a function of the distance z from the interface can be fitted by a $1/z$ function, as expected from the Coulomb law. When comparing the two polymers, the main difference is observed in the surface layer where $\sigma_{\text{PMMA}} \sim 2\sigma_{\text{PS}}$. From a simple qualitative reasoning, we thus expect that the charge transport properties within the pentacene surface layer (i.e., in contact with the polymer) should be significantly affected, with a lower mobility predicted for PMMA.

1.6.2
Charge Mobility Values

The impact of the polymer dielectrics on the charge transport properties has been assessed quantitatively by propagating charge carriers within a hopping regime in the different pentacene layers by means of the same Monte Carlo algorithm as that described in Ref. [38]. These simulations explicitly account for the electrostatic interactions between the pentacene molecules and the polymer chains by introducing the energetic difference between the initial and final states in the ΔG^0 parameter of the transfer rate. The transfer rate has been expressed within the Marcus–Levich–Jortner formalism [119], with the electronic couplings estimated in a direct

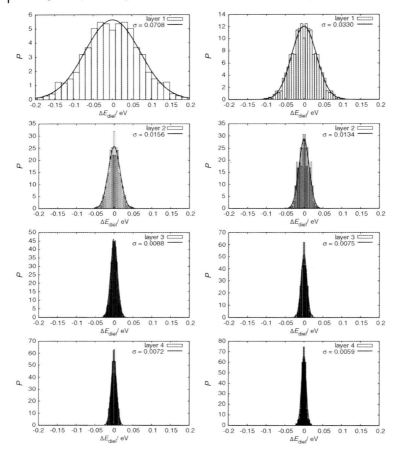

Figure 1.14 Distribution of the site energy difference ΔE_{diel} in the four pentacene layers in the presence of PMMA (left) and PS (right). A fit with a Gaussian distribution is done in each case (the corresponding standard deviation is given). The labeling of the layers is given in Figure 1.1. Adapted from Ref. [67].

way at the INDO level. λ_i has been estimated in a previous study to be 97 meV for holes in pentacene [49] and λ_s has been set equal to a reasonable value of 0.2 eV in both cases. Figure 1.15 shows the polar plot of the mobility values obtained for the different pentacene layers (with the charge constrained to remain in the same layer) in the presence of PMMA versus PS chains. These plots are generated by rotating the direction of the external electric field (with a typical amplitude of 10^4 V cm^{-1}) within the layer in order to explore the anisotropy of charge transport. The trends observed for the mobility reflect the distributions obtained for the ΔE values. For layers 2–4, the influence of the dielectrics on the mobility values is very small, as supported by the narrow energetic distributions. On the contrary, the mobility is significantly lowered in the surface layer 1 due to the increased energetic disorder. The mobility is found to

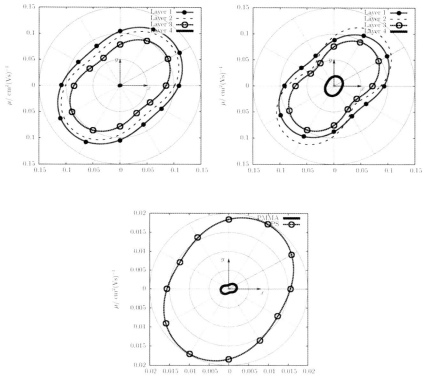

Figure 1.15 Polar plots of the mobility calculated for the four pentacene layers in the presence of PMMA (top left) and PS (top right); zoom on layer 1 for both polymers (bottom). In each plot, the radius corresponds to the mobility value and the angle to the electric field orientation. Adapted from Ref. [67].

be reduced by a factor of 60 for PMMA and 5 for PS compared to bulk pentacene. The calculated ratio of 12 between PMMA and PS is larger than the reported experimental values on the order of 4 [124, 125]. However, it is worth recalling that our simulations do not take into account any macroscopic morphological defects such as grain boundaries that are likely to affect the experimental measurements; it has been shown, for instance, that using PS chains of different molecular weight yields various grain sizes and different mobility values [124].

1.7 Outlook

We have illustrated through several examples that force field calculations prove extremely useful in the field of organic electronics by giving insight at the atomistic level into the supramolecular organization of materials, both in the bulk and at interfaces. Fragments of the generated structures can then be injected into quantum chemical calculations to depict key electronic processes, for instance, charge transport for the studies reviewed here. This can also be applied to other processes such as energy transfer, photoinduced charge transfer, or charge recombination processes. The use of classical force field simulations has nevertheless some limitations. The description of vibrational modes at the classical level is valid only for low-frequency modes with $\hbar\omega \leq kT$; this is the case for the lattice modes modulating the transfer integrals in Section 1.2, even though the temperature dependence of the broadening of the distributions is not correctly described when neglecting that any chemical system does vibrate at 0 K (i.e., the zero-point energy) [131, 132]. On the other hand, the normal modes of the system can be described via MM calculations, generally with an explicit account of anharmonic effects, in contrast to many corresponding quantum chemical calculations relying on the harmonic approximation. Before running MD simulations, it is a prerequisite to validate the force field for the study of the systems under consideration since the transferability of force field parameters is not always fully ensured. For instance, many force fields tend to poorly describe conjugation effects and lead to torsion potentials that are extremely different from those obtained with state-of-the-art quantum chemical calculations [133, 134]. Another drawback is the size of the unit cell used in MD simulations, which is generally too small due to computational limits to depict macroscopic phenomena, for instance, the formation of grain boundaries in organic thin films or the roughness of interfaces.

Interesting current developments in the field aim at increasing the size of the systems subjected to the MD simulations via the use of coarse-grained approaches where atoms are condensed into molecular objects or beads and parameters are defined to depict the interactions between the beads [135, 136]. Another important contribution is expected from increasing the size of the fragments injected into the quantum chemical calculations by exploiting hybrid QM/MM schemes where a central core is treated at a quantum mechanical level with an explicit account of the electrostatic interactions induced by a neighboring medium described at the MM level. Polarizable force fields or force fields with charge equilibration schemes are required here to allow the description of back-polarization effects, that is, changes in the charge distribution of the medium triggered by the polarization of the central core. Such approaches appear to be very attractive to provide a mapping of the polarization energies associated with individual units in supramolecular structures and thus the energetic landscape defining the conducting pathways within the system [137].

Acknowledgments

The authors acknowledge stimulating discussions and collaborations with David Amabilino, Stavros Athanasopoulos, Frédéric Castet, Michael Chabinyc, Veaceslav Coropceanu, Demetrio da Silva Filho, Steven De Feyter, Raffaele Della Valle, Yves Geerts, Martin Heeney, Vincent Lemaur, Iain McCulloch, Joseph Norton, Kathryn

Pigg, Josep Puigmarti-Luis, Concepcio Rovira, Roel Sigifredo Sanchez-Carrera, Maxim Shkunov, and Elisabetta Venuti. The work in Mons is partly supported by the Interuniversity Attraction Pole IAP 6/27 of the Belgian Federal Government, the European projects MODECOM (NMP3-CT-2006-016434), ONE-P (NMP3-LA-2008-212311), and MINOTOR (FP7-NMP-228424), and the Belgian National Fund for Scientific Research (FNRS/FRFC). The work at Georgia Tech has been primarily supported by STC Program under Award DMR-0120967 as well as by the MRSEC Program of the National Science Foundation under Award Number DMR-0819885. The work in Bologna is partly supported by the European projects MODECOM, ONE-P, and MINOTOR and also by the Emilia-Romagna regional project PRITT "NANOFABER." J.C., D.B., and Y.O. are FNRS Research Fellows; N.M. acknowledges a grant from Fonds pour la Formation à la Recherche dans l'Industrie et dans l'Agriculture (FRIA). M.C.R.D. is grateful to the MEC/Fulbright for her Postdoctoral Fellowship at the Georgia Institute of Technology.

References

1 Friend, R.H., Gymer, R.W., Holmes, A.B., Burroughes, J.H., Marks, R.N., Taliani, C., Bradley, D.D.C., dos Santos, D.A., Lögdlund, M., and Salaneck, W.R. (1999) *Nature*, **397**, 121.
2 Tang, C.W. and Vanslyke, S.A. (1987) *Appl. Phys. Lett.*, **51**, 913.
3 Sheats, J.R., Antoniadis, H., Hueschen, M., Leonard, W., Miller, J., Moon, R., Roitman, D., and Stocking, A. (1996) *Science*, **273**, 884.
4 Sariciftci, N.S., Smilowitz, L., Heeger, A.J., and Wudl, F. (1992) *Science*, **258**, 1474.
5 Garnier, F., Hajlaoui, R., Yassar, A., and Srivastava, P. (1994) *Science*, **265**, 1684.
6 Baldo, M.A., O'Brien, D.F., You, Y., Shoustikov, A., Sibley, S., Thompson, M.E., and Forrest, S.R. (1998) *Nature*, **395**, 151.
7 Skotheim, T.A. and Reynold, J.R. (2007) *Handbook of Conducting Polymers*, 3rd edn, CRC Press, Boca Raton, FL.
8 http://www.oled-display.net/.
9 Coropceanu, V., Cornil, J., da Silva Filho, D.A., Olivier, Y., Silbey, R., and Brédas, J.-L. (2007) *Chem. Rev.*, **107**, 926.
10 Yersin, H. (2007) *Highly Efficient OLEDs with Phosphorescent Materials*, Wiley-VCH Verlag GmbH, Weinheim.
11 Lenes, M., Morana, M., Brabec, C.J., and Blom, P.W.M. (2009) *Adv. Funct. Mater.*, **19**, 1106.
12 Sun, S. and Sariciftci, N.S. (2005) *Organic Photovoltaics: Mechanisms, Materials, Devices*, Marcel Dekker, New York.
13 Peumans, P. and Forrest, S.R. (2004) *Chem. Phys. Lett.*, **398**, 27.
14 ersson, L.M. and Inganas, O. (2009) *Chem. Phys.*, **357**, 120.
15 Braga, D. and Horowitz, G. (2009) *Adv. Mater.*, **21**, 1473.
16 Zaumseil, J. and Sirringhaus, H. (2007) *Chem. Rev.*, **107**, 1296.
17 de Boer, R.W.I., Gershenson, M.E., Morpurgo, A.F., and Podzorov, V. (2004) *Physica Status Solidi A*, **201**, 1302.
18 Chang, P.C., Lee, J., Huang, D., Subramanian, V., Murphy, A.R., and Frechet, J.M.J. (2004) *Chem. Mater.*, **16**, 4783.
19 Adam, D., Schuhmacher, P., Simmerer, J., Haussling, L., Siemensmeyer, K., Etzbach, K.H., Ringsdorf, H., and Haarer, D. (1994) *Nature*, **371**, 141.
20 Sheraw, C.D., Jackson, T.N., Eaton, D.L., and Anthony, J.E. (2003) *Adv. Mater.*, **15**, 2009.
21 Mas-Torrent, M., Hadley, P., Bromley, S.T., Ribas, X., Tarres, J., Mas, M., Molins, E., Veciana, J., and Rovira, C. (2004) *J. Am. Chem. Soc.*, **126**, 8546.
22 Jurchescu, O.D., Baas, J., and Palstra, T.T.M. (2004) *Appl. Phys. Lett.*, **84**, 3061.

23 Zen, A., Pflaum, J., Hirschmann, S., Zhuang, W., Jaiser, F., Asawapirom, U., Rabe, J.P., Scherf, U., and Neher, D. (2004) *Adv. Funct. Mater.*, **14**, 757.
24 Dieckmann, A., Bässler, H., and Borsenberger, P.M. (1993) *J. Chem. Phys.*, **99**, 8136.
25 Borsenberger, P.M. and Fitzgerald, J.J. (1993) *J. Phys. Chem.*, **97**, 4815.
26 Dunlap, D.H., Parris, P.E., and Kenkre, V.M. (1996) *Phys. Rev. Lett.*, **77**, 542.
27 Bässler, H. (1993) *Physica Status Solidi B*, **175**, 15.
28 Glaeser, R.M. and Berry, R.S. (1966) *J. Chem. Phys.*, **44**, 3797.
29 Warta, W. and Karl, N. (1985) *Phys. Rev. B*, **32**, 1172.
30 Miller, A. and Abrahams, E. (1960) *Phys. Rev.*, **120**, 745.
31 Holstein, T. (1959) *Ann. Phys.*, **8**, 325.
32 Holstein, T. (1959) *Ann. Phys.*, **8**, 343.
33 Hannewald, K., Stojanovic, V.M., Schellekens, J.M.T., Bobbert, P.A., Kresse, G., and Hafner, J. (2004) *Phys. Rev. B*, **69**, 075211.
34 Hannewald, K. and Bobbert, P.A. (2004) *Appl. Phys. Lett.*, **85**, 1535.
35 Hultell, M. and Stafstrom, S. (2006) *Chem. Phys. Lett.*, **428**, 446.
36 Troisi, A. and Orlandi, G. (2006) *J. Phys. Chem. A*, **110**, 4065.
37 Marcus, R.A. (1993) *Rev. Mod. Phys.*, **65**, 599.
38 Olivier, Y., Lemaur, V., Brédas, J.-L., and Cornil, J. (2006) *J. Phys. Chem. A*, **110**, 6356.
39 Kirkpatrick, J., Marcon, V., Nelson, J., Kremer, K., and Andrienko, D. (2007) *Phys. Rev. Lett.*, **98**, 227402.
40 Valeev, E.F., Coropceanu, V., da Silva, D.A., Salman, S., and Brédas, J.-L. (2006) *J. Am. Chem. Soc.*, **128**, 9882.
41 Binstead, R.A., Reimers, J.R., and Hush, N.S. (2003) *Chem. Phys. Lett.*, **378**, 654.
42 Cornil, J., Calbert, J.-P., and Brédas, J.-L. (2001) *J. Am. Chem. Soc.*, **123**, 1250.
43 Cheng, Y.C., Silbey, R.J., da Silva, D.A., Calbert, J.-P., Cornil, J., and Brédas, J.-L. (2003) *J. Chem. Phys.*, **118**, 3764.
44 Brédas, J.-L., Beljonne, D., Coropceanu, V., and Cornil, J. (2004) *Chem. Rev.*, **104**, 4971.
45 Brédas, J.-L., Calbert, J.-P., da Silva, D.A., and Cornil, J. (2002) *Proc. Natl. Acad. Sci. USA*, **99**, 5804.
46 Lemaur, V., da Silva Filho, D.A., Coropceanu, V., Lehmann, M., Geerts, Y., Piris, J., Debije, M.G., Van de Craats, A.M., Senthilkumar, K., Siebbeles, L.D.A., Warman, J.M., Brédas, J.-L., and Cornil, J. (2004) *J. Am. Chem. Soc.*, **126**, 3271.
47 Kwon, O., Coropceanu, V., Gruhn, N.E., Durivage, J.C., Laquindanum, J.G., Katz, H.E., Cornil, J., and Brédas, J.-L. (2004) *J. Chem. Phys.*, **120**, 8186.
48 Gruhn, N.E., da Silva, D.A., Bill, T.G., Malagoli, M., Coropceanu, V., Kahn, A., and Brédas, J.-L. (2002) *J. Am. Chem. Soc.*, **124**, 7918.
49 Coropceanu, V., Malagoli, M., da Silva Filho, D.A., Gruhn, N.E., Bill, T.G., and Brédas, J.-L. (2002) *Phys. Rev. Lett.*, **89**, 275503.
50 Kato, T. and Yamabe, T. (2001) *J. Chem. Phys.*, **115**, 8592.
51 Kato, T., Yoshizawa, K., and Yamabe, T. (2001) *Chem. Phys. Lett.*, **345**, 125.
52 Kato, T. and Yamabe, T. (2003) *J. Chem. Phys.*, **119**, 11318.
53 Sánchez-Carrera, R.S., Coropceanu, V., da Silva, D.A., Friedlein, R., Osikowicz, W., Murdey, R., Suess, C., Salaneck, W.R., and Brédas, J.-L. (2006) *J. Phys. Chem. B*, **110**, 18904.
54 Frenkel, D. and Smit, B. (1996) *Understanding Molecular Simulations: From Algorithms to Applications*, Academic Press, San Diego.
55 Leach, A. (2001) *Molecular Modelling: Principles and Applications*, Prentice Hall, New Jersey.
56 Lynch, B.J. and Truhlar, D.G. (2001) *J. Phys. Chem. A*, **105**, 2936.
57 Cramer, C.J. and Truhlar, D.G. (1999) *Chem. Rev.*, **99**, 2161.
58 Albu, T.V., Espinosa-Garcia, J., and Truhlar, D.G. (2007) *Chem. Rev.*, **107**, 5101.
59 Grimme, S. (2004) *J. Comput. Chem.*, **25**, 1463.
60 Grimme, S. (2003) *J. Chem. Phys.*, **118**, 9095.
61 Silvestrelli, P.L. (2008) *Phys. Rev. Lett.*, **100**, 053002.

62 Delgado, M.-C.R., Pigg, K.R., Filho, D.A.D.S., Gruhn, N.E., Sakamoto, Y., Suzuki, T., Osuna, R.M., Casado, J., Hernandez, V., Navarrete, J.T.L., Martinelli, N.G., Cornil, J., Sanchez-Carrera, R.S., Coropceanu, V., and Brédas, J.-L. (2009) *J. Am. Chem. Soc.*, **131**, 1502.

63 Martinelli, N.G., Olivier, Y., Athanasopoulos, S., Ruiz Delgado, M.C., Pigg, K.R., da Silva Filho, D.A., Sánchez-Carrera, R.S., Venuti, E., Della Valle, R.G., Brédas, J.-L., Beljonne, D., and Cornil, J. (2009) *ChemPhysChem.*, **10**, 2265.

64 Puigmarti-Luis, J., Minoia, A., Uji-I, H., Rovira, C., Cornil, J., De Feyter, S., Lazzaroni, R., and Amabilino, D.B. (2006) *J. Am. Chem. Soc.*, **128**, 12602.

65 Brocorens, P., Van Vooren, A., Chabinyc, M.L., Toney, M.F., Shkunov, M., Heeney, M., McCulloch, I., Cornil, J., and Lazzaroni, R. (2009) *Adv. Mater.*, **21**, 1193.

66 Olivier, Y., Muccioli, L., Lemaur, V., Geerts, Y.H., Zannoni, C., and Cornil, J. (2009) *J. Phys. Chem. B*, **113**, 14102.

67 Martinelli, N.G., Savini, M., Muccioli, L., Olivier, Y., Castet, F., Zannoni, C., Beljonne, D., and Cornil (2009) J., *Adv. Funct. Mater.*, **19**, 3254.

68 Singh, T.B., Senkarabacak, P., Sariciftci, N.S., Tanda, A., Lackner, C., Hagelauer, R., and Horowitz, G. (2006) *Appl. Phys. Lett.*, **89**, 033512.

69 Bao, Z.A., Lovinger, A.J., and Brown, J. (1998) *J. Am. Chem. Soc.*, **120**, 207.

70 Facchetti, A., Mushrush, M., Katz, H.E., and Marks, T.J. (2003) *Adv. Mater.*, **15**, 33.

71 Facchetti, A., Yoon, M.H., Stern, C.L., Katz, H.E., and Marks, T.J. (2003) *Angew. Chem., Int. Ed.*, **42**, 3900.

72 Heidenhain, S.B., Sakamoto, Y., Suzuki, T., Miura, A., Fujikawa, H., Mori, T., Tokito, S., and Taga, Y. (2000) *J. Am. Chem. Soc.*, **122**, 10240.

73 Inoue, Y., Sakamoto, Y., Suzuki, T., Kobayashi, M., Gao, Y., and Tokito, S. (2005) *Jpn. J. Appl. Phys., Part 1*, **44**, 3663.

74 Sakamoto, Y., Suzuki, T., Kobayashi, M., Gao, Y., Fukai, Y., Inoue, Y., Sato, F., and Tokito, S. (2004) *J. Am. Chem. Soc.*, **126**, 8138.

75 Sakamoto, Y., Suzuki, T., Miura, A., Fujikawa, H., Tokito, S., and Taga, Y. (2000) *J. Am. Chem. Soc.*, **122**, 1832.

76 Yoon, M.H., Facchetti, A., Stern, C.E., and Marks, T.J. (2006) *J. Am. Chem. Soc.*, **128**, 5792.

77 Jones, B.A., Facchetti, A., Wasielewski, M.R., and Marks, T.J. (2007) *J. Am. Chem. Soc.*, **129**, 15259.

78 Sakamoto, Y., Suzuki, T., Kobayashi, M., Gao, Y., Inoue, Y., and Tokito, S. (2006) *Mol. Cryst. Liq. Cryst.*, **444**, 225.

79 Te Velde, G., Bickelhaupt, F.M., Baerends, E.J., Fonseca Guerra, C., Van Gisbergen, S.J.A., Snijders, J.G., and Ziegler, T. (2001) *J. Comput. Chem.*, **22**, 931.

80 Munn, R.W. (1997) *Chem. Phys.*, **215**, 301.

81 Palenberg, M.A., Silbey, R.J., and Pfluegl, W. (2000) *Phys. Rev. B*, **62**, 3744.

82 Troisi, A. (2007) *Adv. Mater.*, **19**, 2000.

83 Troisi, A., Orlandi, G., and Anthony, J.E. (2005) *Chem. Mater.*, **17**, 5024.

84 Bunte, S.W. and Sun, H. (2000) *J. Phys. Chem. B*, **104**, 2477.

85 Van Vooren, A., Lemaur, V., Ye, A., Beljonne, D., and Cornil, J. (2007) *ChemPhysChem*, **8**, 1240.

86 Milián Medina, B., Van Vooren, A., Brocorens, P., Gierschner, J., Shkunov, M., Heeney, M., McCulloch, I., Lazzaroni, R., and Cornil, J. (2007) *Chem. Mater.*, **19**, 4946.

87 Balabin, I.A. and Onuchic, J.N. (2000) *Science.*, **290**, 114.

88 A chi square test has been performed on each plot to validate the use of a Gaussian fit: Pisani, R. and Purves, R. (eds) (2007) *Statistics*, 4th edn, W.W. Norton & Company.

89 Athanasopoulos, S., Kirkpatrick, J., Martinez, D., Frost, J.M., Foden, C.M., Walker, A.B., and Nelson, J. (2007) *Nano Lett.*, **7**, 1785.

90 Kajiwara, T., Inokuchi, H., and Minomura, S. (1967) *Bull. Chem. Soc. Jpn.*, **40**, 1055.

91 Karl, N. and Marktanner, J. (2001) *Mol. Cryst. Liq. Cryst.*, **355**, 149.

92 Kepler, R.G. (1960) *Phys. Rev.*, **119**, 1226.

93 Brock, C.P. and Dunitz, J.D. (1990) *Acta Crystallogr. B*, **46**, 795.

94 Chen, J., Hsu, J.H., and Lin, H.N. (2005) *Nanotechnology*, **16**, 1112.

95 Mihara, T., Miyamoto, K., Kida, M., Sasaki, T., Aoki, N., and Ochiai, Y. (2003) *Superlattices Microstruct.*, **34**, 383.
96 Schenning, A.P.H.J. and Meijer, E.W. (2005) *Chem. Commun.*, 3245.
97 Ashkenasy, N., Horne, W.S., and Ghadiri, M.R. (2006) *Small*, **2**, 99.
98 Hogg, T., Chen, Y., and Kuekes, P.J. (2006) *IEEE Trans. Nanotechnol.*, **5**, 110.
99 Philip, D. and Stoddart, J.F. (1996) *Angew. Chem., Int. Ed. Engl.*, **35**, 1154.
100 Williams, J.M., Ferrar, J.R., Thorn, R.J., Carlson, K.D., Ans Geiser, U., Wang, H.H., Kini, A.M., and Whangbo, M. (1992) *Organic Semiconductors (Including Fullerenes)*, Prentice Hall, New Jersey.
101 Abdel-Mottaleb, M.M.S., Gomar-Nadal, E., Surin, M., Uji-i, H., Mamdouh, W., Veciana, J., Lemaur, V., Rovira, C., Cornil, J., Lazzaroni, R., Amabilino, D.B., De Feyter, S., and De Schryver, F.C. (2005) *J. Mater. Chem.*, **15**, 4601.
102 Lu, J., Zeng, Q.D., Wan, L.J., and Bai, C.L. (2003) *Chem. Lett.*, **32**, 856.
103 van Gorp, J.J., Vekemans, J.A.J.M., and Meijer, E.W. (2002) *J. Am. Chem. Soc.*, **124**, 14759.
104 Shirakawa, M., Kawano, S., Fujita, N., Sada, K., and Shinkai, S. (2003) *J. Org. Chem.*, **68**, 5037.
105 Sirringhaus, H., Brown, P.J., Friend, R.H., Nielsen, M.M., Bechgaard, K., Langeveld-Voss, B.M.W., Spiering, A.J.H., Janssen, R.A.J., Meijer, E.W., Herwig, P., and de Leeuw, D.M. (1999) *Nature*, **401**, 685.
106 McCulloch, I., Heeney, M., Bailey, C., Genevicius, K., MacDonald, I., Shkunov, M., Sparrowe, D., Tierney, S., Wagner, R., Zhang, W., Chabinyc, M.L., Kline, R.J., McGehee, M.D., and Toney, M.F. (2006) *Nat. Mater.*, **5**, 328.
107 Hamadani, B.H., Gundlach, D.J., McCulloch, I., and Heeney, M. (2007) *Appl. Phys. Lett.*, **91**, 243512.
108 Kline, R.J., DeLongchamp, D.M., Fischer, D.A., Lin, E.K., Heeney, M., McCulloch, I., and Toney, M.F. (2007) *Appl. Phys. Lett.*, **90**, 062117.
109 Chabinyc, M.L., Toney, M.F., Kline, R.J., McCulloch, I., and Heeney, M. (2007) *J. Am. Chem. Soc.*, **129**, 3226.
110 Takayanagi, M., Gejo, T., and Hanozaki, I. (1994) *J. Phys. Chem. A*, **98**, 12893.
111 Sancho-García, J.C. and Cornil, J. (2004) *J. Chem. Phys.*, **121**, 3096.
112 Chua, L.L., Zaumseil, J., Chang, J.F., Ou, E.C.W., Ho, P.K.H., Sirringhaus, H., and Friend, R.H. (2005) *Nature*, **434**, 194.
113 Tracz, A., Makowski, T., Masirek, S., Pisula, W., and Geerts, Y.H. (2007) *Nanotechnology*, **18**, 485303.
114 114.Tant, J., Geerts, Y.H., Lehmann, M., De Cupere, V., Zucchi, G., Laursen, B.W., Bjørnholm, T., Lemaur, V., Marcq, V., Burquel, A., Hennebicq, E., Gardebien, F., Viville, P., Beljonne, D., Lazzaroni, R., and Cornil, J. (2005) *J. Phys. Chem. B*, **109**, 20315;Additions, corrections. *J. Phys. Chem. B*, 2006, **110**, 3449.
115 van de Craats, A.M., Schouten, P.G., and Warman, J.M. (1997) *J. Jpn. Liq. Cryst. Soc.*, **2**, 12.
116 Deibel, C., Janssen, D., Heremans, P., De Cupere, V., Geerts, Y., Benkhedir, M.L., and Adriaenssens, G.J. (2006) *Org. Electron.*, **7**, 495.
117 Gearba, R.I., Bondar, A.I., Goderis, B., Bras, W., and Ivanov, D.A. (2005) *Chem. Mater.*, **17**, 2825.
118 Mück-Lichtenfeld, C. and Grimme, S. (2007) *Mol. Phys.*, **105**, 2793.
119 Jortner, J. (1976) *J. Chem. Phys.*, **64**, 4860.
120 Kirkpatrick, J., Marcon, V., Kremer, K., Nelson, J., and Andrienko, D. (2008) *J. Chem. Phys.*, **129**, 094506.
121 Dinelli, F., Murgia, M., Levy, P., Cavallini, M., Biscarini, F., and de Leeuw, D.M. (2004) *Phys. Rev. Lett.*, **92**, 116802.
122 Allard, S., Forster, M., Souharce, B., Thiem, H., and Scherf, U. (2008) *Angew. Chem., Int. Ed.*, **47**, 4070.
123 Veres, J., Ogier, S.D., Leeming, W., Cupertino, D.C., and Khaffaf, S.M. (2003) *Adv. Funct. Mater.*, **13**, 199.
124 Kim, C., Facchetti, A., and Marks, T.J. (2007) *Science*, **318**, 76.
125 Benson, N., Melzer, C., Schmechel, R., and von Seggern, H. (2008) *Physica Status Solidi A*, **205**, 475.
126 Horowitz, G. (1998) *Adv. Mater.*, **10**, 365.
127 Verlaak, S. and Heremans, P. (2007) *Phys. Rev. B*, **75**, 115127.

128 Kim, D.H., Lee, H.S., Yang, H., Yang, L., and Cho, K. (2008) *Adv. Funct. Mater.*, **18**, 1363.

129 Kim, C., Facchetti, A., and Marks, T.J. (2007) *Adv. Mater.*, **19**, 256.

130 Gershenson, M.E., Podzorov, V., and Morpurgo, A.F. (2006) *Rev. Mod. Phys.*, **78**, 973.

131 Coropceanu, V., Sánchez-Carrera, R.S., Paramonov, P., Day, G.M., and Brédas, J.-L. (2009) *J. Phys. Chem. C*, **113**, 4679.

132 Kwiatkowski, J.J., Frost, J.M., Kirkpatrick, J., and Nelson, J. (2008) *J. Phys. Chem. A*, **112**, 9113.

133 Marcon, V. and Raos, G. (2004) *J. Phys. Chem. B*, **108**, 18053.

134 Berardi, R., Cainelli, G., Galletti, P., Giacomini, D., Gualandi, A., Muccioli, L., and Zannoni, C. (2005) *J. Am. Chem. Soc.*, **127**, 10699.

135 Wang, H., Junghans, C., and Kremer, K. (2009) *Eur. Phys. J. E*, **28**, 221.

136 Loison, C., Mareschal, M., Kremer, K., and Schmid, F. (2003) *J. Chem. Phys.*, **119**, 13138.

137 Norton, J.E. and Brédas, J.-L. (2008) *J. Am. Chem. Soc.*, **130**, 12377.

2
Monte Carlo Studies of Phase Transitions and Cooperative Motion in Langmuir Monolayers with Internal Dipoles

Christopher B. George, Mark A. Ratner, and Igal Szleifer

2.1
Introduction

The increasing long-term trend in the chemical sciences, away from an overwhelming focus on covalent systems toward a more catholic interest in bonding interactions of all types, is particularly noticeable in two areas where interactions much weaker than covalent bonds dominate the interesting behavior [1]. The first focuses on single entities and includes crown ethers and cryptand conjugates, rotaxanes, catenanes, calixarenes, and similar species [2–8]. The second deals with collections of molecules, and therefore includes situations of self-assembled monolayers, membranes, and block copolymers [9–11].

The structure and phase behavior of monolayers assembled on an aqueous subphase play a pivotal role in controlling the functionality of devices such as electro-optic and biomedical sensors based on Langmuir–Blodgett films [12, 13] and molecular electronic junctions consisting of monolayers chemisorbed to liquid mercury electrodes [14, 15]. Understanding and controlling the structural properties of the surfactant monolayers used in the creation of these technological applications is a necessary step for fine-tuning device performance and longevity; structural defects in the monolayer can lead to performance inefficiencies and potentially to device failure.

Similarly, in applications that depend on the dynamic behavior of molecules within a Langmuir monolayer, knowledge of the monolayer's internal organization and dynamics is requisite for understanding the system's functionality. Lateral electron transfer in two-dimensional (2D) arrays of electroactive molecules, a topic of interest in the fields of solar energy conversion [16] and chemical sensing [17], has been shown to depend on the diffusion rates of molecules within a monolayer [18]. Local crowding and confinement effects can have a dramatic effect on the mobility of such molecules.

Several theoretical and experimental studies have been conducted with the goal of elucidating the rich phase behavior of Langmuir monolayers [19]. Tilting transitions, backbone ordering transitions, and fluid–fluid transitions have all been examined,

Functional Supramolecular Architectures. Edited by Paolo Samorì and Franco Cacialli
Copyright © 2011 WILEY-VCH Verlag GmbH & Co. KGaA, Weinheim
ISBN: 978-3-527-32611-2

typically with studies focusing on monolayers composed of amphiphilic molecules. In monolayers composed of amphiphiles, dipole–dipole interactions between hydrophilic head groups and van der Waals interactions between hydrophobic tail groups often dictate packing structures [20–24]. Recently, however, Langmuir monolayers have been formed using surfactants in which an internal dipole exists [25–30]. In these systems, packing structures and phase transitions are expected to depend on the magnitude of the internal dipole.

Of interest in this work is the interplay between short-range steric interactions due to van der Waals forces and long-range interactions due to internal dipoles. Following the example of previous simulation studies of Langmuir monolayers [22, 31–37], we use a coarse-grained model that includes only the degrees of freedom necessary for understanding the phase behavior in question. High-density configurations are studied, and both tilting and melting transitions are characterized using a range of order parameters.

Results indicate that the system exists in a frustrated state; the potential energy due to dipole–dipole interactions is minimized by increasing collective tilt, but the degree of tilting is limited by steric interactions. This results in dipole-dependent tilting transitions and melting transitions. The two transitions are found to couple at high dipole strengths, and nanodomains with highly ordered head groups and increased molecular tilt are found to form near the transition density. Lateral diffusion within the monolayers at densities near the melting transition is examined as well, and both heterogeneous dynamics and cooperative motion are observed. Systems with stronger internal dipoles exhibit pronounced heterogeneous motion near the melting transition due to the formation of nanodomains. Results from the canonical ensemble have been reported previously [38].

2.2
Computational Details

Surfactant molecules are modeled by four Lennard-Jones centers fixed along a rigid rod, with 0.67σ between monomers (Figure 2.1). Each Lennard-Jones potential is truncated and shifted to be purely repulsive. Equal positive and negative point charges are located at the centers of the second and third monomers, respectively, with the first and fourth monomers uncharged. Monomers on different chains interact via a modified Lennard-Jones potential defined by

$$U_{LJ}(r_{ij}) = \begin{cases} 4\varepsilon_{LJ}\left[\left(\dfrac{\sigma}{r_{ij}}\right)^{12} - \left(\dfrac{\sigma}{r_{ij}}\right)^{6}\right] - 4\varepsilon_{LJ}\left[\left(\dfrac{\sigma}{r_c}\right)^{12} - \left(\dfrac{\sigma}{r_c}\right)^{6}\right] & \text{for } r_{ij} \leq r_c \\ 0 & \text{for } r_{ij} > r_c \end{cases}$$
(2.1)

Here the cutoff radius of the Lennard-Jones potential is $r_c = 2^{1/6}\sigma$. The range of the potential is given by σ and ε_{LJ} is a measure of the "hardness" of the repulsive potential. The distance between monomers is r_{ij}. Electrostatic interactions between

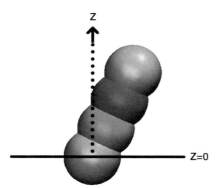

Figure 2.1 A model surfactant composed of four monomers fixed along a rigid rod. The second and third monomers have equal positive and negative charge. The head monomer is constrained to translate in the $z=0$ plane.

monomers with point charges are described by

$$U_C(r_{ij}) = \frac{q_i q_j e^2}{4\pi\varepsilon_s r_{ij}} \tag{2.2}$$

where q_i and q_j are the charges on monomers i and j (in units of electrons), e is the charge of an electron, and ε_s is the static permittivity of the system. The first monomer in each rod, defined as the head monomer, is fixed to lie in the $z=0$ plane but allowed to translate in the x and y directions. The rest of the rod is allowed to rotate such that all other monomers remain above the xy plane ($z > 0$). The $z=0$ plane is flat and electrically insulating.

Off-lattice Monte Carlo simulations are conducted following a Metropolis scheme in both the canonical (NAT) and isobaric–isothermal (NΠT) ensembles. In both cases, monolayers consist of 1024 rods packed in a rectangular simulation box with an aspect ratio of $L_y/L_x = \sqrt{3}/2$. Periodic boundary conditions are implemented in the x and y directions, and electrostatic interactions are truncated at $L_y/2$.

In the canonical ensemble, a rod is selected at random and a rotation or translation move is attempted; both types of moves are attempted with equal probability. The move is accepted if

$$e^{-\frac{\Delta U}{kT}} > \gamma \tag{2.3}$$

where γ is a random number between zero and one and ΔU is the difference in energy between the new and old configurations [39].

In the isobaric–isothermal ensemble, both rotation and translation moves occur as before, but changes in box size are attempted as well. Moves to change the simulation area are attempted on average once every tenth Monte Carlo step (MCS), where one Monte Carlo step is defined as 1024 attempted rotation/translation moves. Changes in the area maintain the $\sqrt{3} : 2$ aspect ratio of the simulation cell. Moves are accepted if

$$e^{-\frac{1}{kT}\left\{\Delta U + \Pi \Delta A - NkT \log\left(\frac{A_n}{A_o}\right)\right\}} > \gamma \tag{2.4}$$

where γ is a random number between zero and one, ΔU is the difference in energy between the configurations, Π is the pressure, ΔA is the change in area, and $A_{n,o}$ is the area of the new/old configuration [39].

In both ensembles, the maximum size of each trial move is adjusted throughout the simulations to maintain an acceptance ratio of 50%. An initial 500 000 MCS are performed to equilibrate the systems followed by another 500 000 for data collection. One Monte Carlo step defines the timescale used in the study of particle dynamics.

Reduced units are chosen with $\sigma = 1$ and $(e^2/4\pi\varepsilon_s) = 1$. In these units, the reduced temperature is defined as

$$T^* = \frac{4\pi\varepsilon_s \sigma kT}{e^2} = \frac{\sigma}{l_b} = \frac{1}{l_b^*} \tag{2.5}$$

where l_b is the Bjerrum length. The strength of the steric repulsion between monomers is set to be $\varepsilon_{LJ} = kTl_b^*$. Simulations in the canonical ensemble are performed for configurations with densities ranging $\varrho^* = 0.7$–0.95 ($\varrho^* = \varrho\sigma^2$) with charges between $q = 0$ and $q = 1.5$. Simulations in the isobaric–isothermal ensemble are performed with the same densities, but only for $q = 0$. Results from the NΠT ensemble are used to verify data obtained from the NAT ensemble, particularly near the transition density. All simulations are conducted at $T^* = l_b^* = 1$.

The coarse-grained model used in this study is comparable to models used in previous works examining the phase behavior [31–36] and translational diffusion [40] of Langmuir monolayers. The difference in this work is the inclusion of an internal dipole. Many molecules that comprise adlayer films, both biological and synthetic, have such dipoles (such as Si–O or C–O bonds). In addition, the models of earlier studies often included a degree of flexibility in the chain of monomers while our rods are entirely rigid. This will preclude the observation of certain fluid–fluid phase transitions [34, 41], but is not expected to significantly affect the tilt behavior, melting transitions, or translational diffusion due to the high densities being inspected.

2.3
Results and Discussion

2.3.1
Uncharged System

Surface pressure–density isotherms calculated from both NAT and NΠT simulations are shown in Figure 2.2 with error bars representing standard deviations. In the canonical ensemble, pressures are calculated using the virial theorem [39]:

$$\Pi = \frac{NkT}{A} + \frac{1}{2A}\left\langle \sum_{i=1}^{N}\sum_{j>i} f(r_{ij}) \cdot r_{ij} \right\rangle \tag{2.6}$$

Here N is the number of rods, $f(r_{ij})$ is the force between two rods, and r_{ij} is the distance between the centers of mass of the rods. Results from the two ensembles are

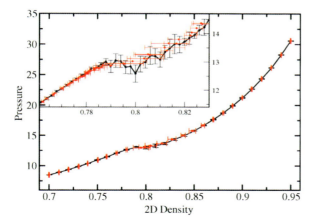

Figure 2.2 Surface pressure–density isotherms for uncharged systems from NAT (black) and NΠT (red) ensembles. Error bars represent the standard deviation of each value. An order–disorder phase transition is responsible for the dips in the isotherms. Inset shows enlarged view of transition region. Units of pressure and density are (kTl_b/σ^2) and σ^{-2}, respectively.

in excellent agreement. Large fluctuations are present in values obtained from both the NAT and NΠT ensembles near the transition region with large variations in the densities persisting until $\varrho^* = 0.85-0.86$. The fluctuations highlight the difficulty in identifying an exact transition density.

The phase transition responsible for the dip in the surface pressure–density isotherm is similar to melting transitions seen in two-dimensional systems [42–45]. As the density increases at constant temperature, particles are forced to pack closer together until they begin to pack with well-defined hexagonal order (Figure 2.3). The onset of this hexagonal packing is apparent in surface pressure–density isotherms (as in Figure 2.2) as well as a range of order parameters. Because the rods in this study are essentially two-dimensional particles with an additional degree of freedom (rotation of the tail), the hexagonal order of the head monomers may be characterized using order parameters employed in two-dimensional melting studies [42].

A local bond orientation correlation order parameter may be defined by

$$\psi_{6,i} = \frac{1}{n_i} \sum_{j=1}^{n_i} e^{i6\theta_{ij}} \quad (2.7)$$

where n_i is the number of nearest neighbors of particle i and θ_{ij} is the angle between the line connecting head monomers i and j and an arbitrary axis. Using this order parameter, a bond orientation correlation function may be written:

$$g_6(r) = \left\langle \frac{\psi_6(r)\psi_6^*(0)}{\sum_i \sum_{j \neq i} \delta(r - r_{ij})} \right\rangle \quad (2.8)$$

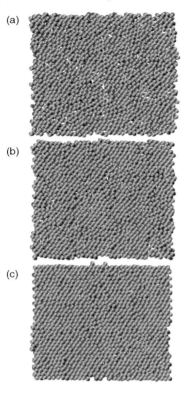

Figure 2.3 Snapshots of simulation boxes for uncharged systems at (a) $\varrho^* = 0.77$, (b) $\varrho^* = 0.81$, and (c) $\varrho^* = 0.85$ densities below, at, and above the order–disorder. Hexagonal packing increases with increasing density. Results are from the NAT ensemble.

with

$$\psi_6(r) = \sum_i^N \delta(r-r_i)\psi_{6,i} \qquad (2.9)$$

Here N is the number of rods, r_i is the position of head monomer i, and δ is the Kronecker delta function. Bond orientation correlation functions (Figure 2.4) show a slight rise in the $\varrho^* = 0.79$ curve with the $\varrho^* = 0.80$ line falling back to zero. This is unsurprising given the large fluctuations near the transition density in the surface pressure–density isotherms seen in Figure 2.2; the exact transition density is difficult to locate. A more pronounced jump in the correlation function occurs at $\varrho^* = 0.81$, and this increase is sustained at higher densities.

Density-dependent translational diffusion in two-dimensional liquids has been studied in detail with the intent of understanding cooperative dynamics that occur near the melting transition [46–52]. It has been found that kinetic inhomogeneities exist in which some particles are restricted to oscillate within a "cage" defined by nearest neighbors, while other particles travel with string-like cooperative motion

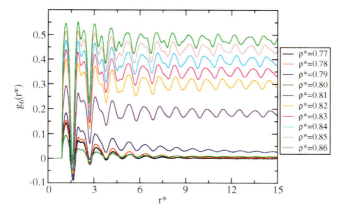

Figure 2.4 Bond orientation correlation functions $g_6(r^*)$ of uncharged systems in the canonical ensemble with densities near the melting transition. The jump in $g_6(r^*)$ near $\varrho^* = 0.81$ indicates an order–disorder transition. r^* is in units of σ.

through the dense liquid. Given the existence of a similar order–disorder transition in the pseudo-2D system studied here, analogous heterogeneous dynamics may be expected.

Trajectories of particles in configurations near the melting transition are shown in Figure 2.5. In the $\varrho^* = 0.80$ system, areas of both hexagonal order and disorder are seen. In the enlarged region, evidence of cooperative jump dynamics is visible: a rod "jumps" to the position that a neighboring rod had previously occupied shortly after the neighbor moves. The $\varrho^* = 0.82$ system shows increased hexagonal order with particles moving with string-like cooperative motion along directions with strong bond orientation correlation. Enlarged regions indicate that jump behavior is present as well. As in purely 2D systems [50, 51], "ordered" domains exist in which particle motion is completely restricted by neighboring molecules. Surrounding these ordered areas, domains of "disorder" exist in which particles concertedly diffuse in one dimension.

2.3.2
Charged System

Introducing an internal dipole affects both the internal organization and the lateral diffusion of the surfactant system. At high densities, both the repulsive and attractive dipole–dipole interactions will have a significant effect on the packing structure. Ordering of the head monomers will increase for rods with stronger dipoles due to the increased intermolecular repulsions, while greater dipole–dipole attractions will cause the rods to tilt more. The amount of tilt will be constrained by steric repulsions.

To quantitatively describe the effect that internal charges have on the monolayer's phase behavior, several additional order parameters are calculated. As a measure of the hexagonal order, a global bond orientation order may be defined as [42]

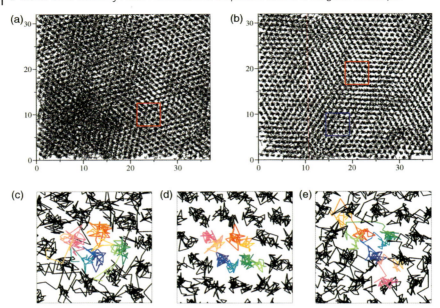

Figure 2.5 Sections of simulation boxes of uncharged rods in the canonical ensemble with (a) $\varrho^* = 0.80$ and (b) $\varrho^* = 0.82$ showing rods' center of mass trajectories. A total of 50 frames are shown with each frame separated by 5000 Monte Carlo steps. Regions of jump dynamics and string-like motion are visible in addition to sections where particles are completely caged by neighbors. Enlarged areas for (c) $\varrho^* = 0.80$, (d) $\varrho^* = 0.82$ (red box), and (e) $\varrho^* = 0.82$ (blue box) provide a view of the time dependence of the correlated motion. Trajectories change shade (orange–yellow, pink–yellow, green–yellow, blue–green) as a function of time. Larger simulation box lengths are in units of σ, and enlarged areas have sides of length 5σ.

$$\langle |\Psi_6| \rangle = \left\langle \left| \frac{1}{N} \sum_i^N \psi_{6,i} \right| \right\rangle \tag{2.10}$$

where N is the number of rods. Values of $\langle |\Psi_6| \rangle$ for different charge strengths are shown in Figure 2.6 and vary from 0 to 1, with 1 indicating perfect hexagonal order and 0 complete hexagonal disorder. Charge-dependent jumps in the global bond orientation order parameter indicate that the order–disorder transition drops to lower densities for systems composed of rods with stronger internal dipoles. This is to be expected given the increased dipole–dipole repulsion in systems with larger internal charges.

Interestingly, the order–disorder transition density remains relatively constant until $q \geq 1.2$ at which point the transition shifts monotonically to lower densities as a function of charge. This may be understood by examining the tilt behavior of the rods. The tilting transitions discussed here are due to dipole–dipole attractions as opposed to van der Waals attractions [31–33, 35, 36, 53, 54].

The tilt order of the rods may be described by two order parameters, $\langle \cos \theta \rangle$, where θ is the tilt angle of the rods (measured from vertical), and R_{xy}, a measure of collective

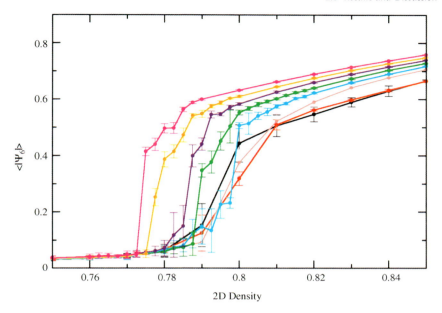

Figure 2.6 Global bond orientation order parameters $\langle|\Psi_6|\rangle$ versus density in units of σ^{-2} for $q=0$ (black), $q=0.5$ (red), $q=1.0$ (brown), $q=1.1$ (cyan), $q=1.2$ (green), $q=1.3$ (violet), $q=1.4$ (orange), and $q=1.5$ (magenta), all in the canonical ensemble. A jump in $\langle|\Psi_6|\rangle$ indicates the presence of an order–disorder transition. In systems with $q \geq 1.2$, the phase transition shows a charge dependence in which transition densities monotonically decrease as internal dipole strength increases. For lower charges, minimal charge dependence is seen. Image published with permission from the *Journal of Chemical Physics*.

tilt given by [55]

$$R_{xy} = \frac{\sqrt{\langle [x]^2 + [y]^2 \rangle}}{l} \quad (2.11)$$

Here $[x]$ and $[y]$ are the average x and y projections for all rods in a configuration and l is the length of the rods. The value of R_{xy} may vary from 0 to 1 with 0 meaning configurations have no collective tilt and 1 indicating that all rods are lying flat on the surface in the same direction. The charge dependence of the tilt order is evident from both order parameters (Figure 2.7). From the $\langle \cos\theta \rangle$ values, it is clear that stronger internal charges lead to greater tilt. Dipole–dipole attractions compete with steric interactions to determine the rods' average tilt with stronger dipole–dipole attractions permitting larger average tilt.

Jumps in R_{xy} and dips in $\langle \cos\theta \rangle$ indicate the presence of a tilt transition. In contrast to the order–disorder transition, the tilting transition densities show a monotonic dependence on the dipole strength for the entire range of charges studied. Also of note is the sharpness of the transition as a function of charge; systems with larger internal dipoles exhibit more pronounced peaks in $\langle \cos\theta \rangle$ as well as sharper jumps in R_{xy}.

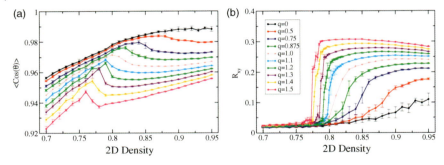

Figure 2.7 Tilt order parameters (a) $\langle \cos\theta \rangle$ and (b) R_{xy}. Tilt transitions are charge dependent for the entire range of dipole strengths studied, as is evident from the locations of jumps in R_{xy} and dips in $\langle \cos\theta \rangle$. Results are from simulations at constant NAT. Image published with permission from the *Journal of Chemical Physics*.

Comparing these results to $\langle |\Psi_6| \rangle$ values in Figure 2.6 suggests a possible coupling between the two transitions at higher charge strengths ($q \geq 1.2$). The tilt transitions and order–disorder transitions occur at nearly the same densities in systems with stronger dipoles, while at lower charge strengths the two occur separately.

To understand this apparent coupling and search for domains with higher local collective tilt or head group order, local order parameters within the simulation box are calculated. Each simulation box is discretized into a grid, and individual local order parameters are calculated within each bin. The local order parameter measuring hexagonal order is defined as

$$\langle |\Psi_6| \rangle_a = \left\langle \left| \frac{1}{M} \sum_i^M \frac{1}{n_i} \sum_{j=1}^{n_i} e^{i6\theta_{ij}} \right| \right\rangle_a \tag{2.12}$$

where i runs over all M monomers found within a particular bin a for a single configuration. The sums are then averaged over all values in a single bin, yielding one order parameter per bin. The collective tilt within each bin is measured by

$$(R_{xy})_a = \frac{\sqrt{[x]_a^2 + [y]_a^2}}{l} \tag{2.13}$$

where $[x]_a$ and $[y]_a$ are the averages over all configurations of the x and y projections of rods within a particular bin a. Color maps of these order parameters (Figure 2.8) for a high charge and low charge case near their respective order–disorder transitions ($q = 1.4$, $\varrho^* = 0.7825$ and $q = 0.5$, $\varrho^* = 0.80$) reveal distinctly different mechanisms for the formation of hexagonal ordering. In the high charge case, inhomogeneous melting occurs in which a domain with a high degree of head group order and collective tilt exists. For the low charge case, no such domain appears. The color maps support the hypothesis that at high charge, the melting and tilting transitions couple, while at low charge they occur independently.

The high charge systems exhibit heterogeneous dynamics in addition to heterogeneous melting. Trajectories of rods within the well-ordered nanodomain show

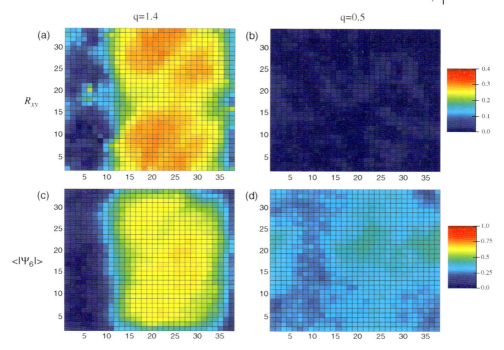

Figure 2.8 Color maps of $\langle R_{xy}\rangle_a$ for (a) $q=1.4$, (b) $q=0.5$ and $\langle|\Psi_6|\rangle_a$ for (c) $q=1.4$ and (d) $q=0.5$ in the canonical ensemble. Densities shown correspond to the relative melting transitions ($\varrho^* = 0.7825$ for $q=1.4$ and $\varrho^* = 0.8000$ for $q=0.5$) of the systems. The melting mechanism is dramatically different for the high charge and low charge cases. In the high charge case, a domain with increased tilt and head group order forms, while in the low charge case, nearly homogeneous melting occurs.

collective motion with rods moving in conjunction with nearest neighbors along the directions with strong bond orientation correlation (Figure 2.9). The type of motion within the domain is similar to that seen in the $q=0$, $\varrho^* = 0.82$ system, except with much greater uniformity in the direction of the rods' motions. Outside the ordered domain, particles exhibit no sign of collective motion, and near the domain boundary, jump dynamics appear to exist. The inhomogeneity in diffusive motion within the system is to be expected given the relationship between head group ordering and dynamics seen in the uncharged system (collective motion is more prevalent in systems with greater hexagonal ordering).

2.4 Summary

The phase behavior and lateral diffusion of surfactant molecules tethered to a flat interface have been explored using Monte Carlo simulations. In systems with no internal dipoles, a melting transition is observed that has a dramatic effect on the

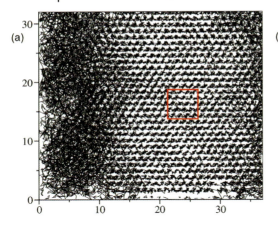
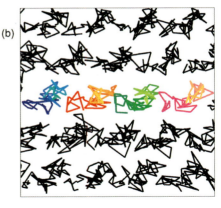

Figure 2.9 Section of a simulation box at constant NAT with charged rods at the melting transition ($\varrho^* = 0.7825$, $q = 1.4$). The larger simulation box (a) shows extremely heterogeneous motion with particles in the ordered domain moving with string-like cooperative motion, while rods outside the domain move with disordered dynamics. The concerted motion within the domain is highlighted by an enlarged view (b) of the area within the red box. The enlarged box in (b) has side lengths of 5σ, and the units of length in (a) are in σ. As in Figure 2.5, trajectories change shade (orange–yellow, pink–yellow, green–yellow, blue–green) as a function of time. Trajectory timescales are identical to those in Figure 2.5.

dynamics of particles within the monolayer. Collective motion is seen in which rods move in a string-like manner in the direction of strong bond orientation correlation. Jump dynamics are also observed in which particles hop to fill in holes left by the movement of nearest neighbors.

In charged systems, a tilting transition is found in addition to a melting transition. Due to the high-density configurations being studied, monolayers with internal dipoles exist in a frustrated state in which dipole–dipole interactions contribute to increased head group ordering (due to dipole–dipole repulsions) as well as collective tilting (dipole–dipole attractions). As the internal dipole strength increases, the average molecular tilt angle increases and the tilt transition occurs at progressively lower densities until the tilting transition and melting transition couple. After coupling, melting occurs via the formation of nanodomains with increased head group order and collective tilt. In systems with coupled transitions, heterogeneous dynamics are observed in which rods within the well-ordered domains move uniformly with collective motion, while particles outside the nanodomains diffuse with disordered motion.

The size and shape of the domains observed in the charged systems likely depend on the size and shape of the simulation box. Also, the exact densities of the melting transitions are known to be size dependent in purely 2D studies [42–44], and quantitative results are expected to be size dependent in this study as well. Qualitatively, the results are expected to remain the same for larger system sizes.

Wikipedia states, "Supramolecular chemistry refers to the area of chemistry beyond the molecules and focuses on the chemical systems made up of a discrete number of assembled molecular subunits or components" [56]. In this sense, the adlayer systems discussed here are exemplary of such species. Because the individual molecular components (the four-center monomers) are not allowed to vary, the supramolecular adlayer has motions and configurations that change only with the intermolecular interactions – the extensive set of geometries and dynamics that can be accessed even with these restrictions is indicative of the richness of supramolecular structure/behavior. In more common usage, the term describes finite molecular adducts (rotaxanes, catenanes, foldamers, etc.), while these also show fascinating dynamical and structural variation, the adlayers define a quasi-two-dimensional supramolecular phase, whose properties follow from the components and their density. The diffusive behavior of particles within dense monolayers observed in this study highlights the close relationship between structure and motion in crowded systems. Both experimental and theoretical studies of two-dimensional liquids report similar findings. Cui *et al.* [46] observe regions of ordered and disordered domains near the freezing transition of a quasi-two-dimensional colloidal liquid. The dynamical behavior of particles in the ordered domains show caging effects with minimal displacement, while particles in the disordered regions are much more mobile. Molecular dynamics studies have confirmed the correlation between increased particle mobility and decreased order [50, 52].

More complex systems, such as those found in cellular membranes, also exhibit diffusive behavior that depends on crowding effects [57]. Recent work has shown that collective motion plays a critical role in the diffusion of quasi-two-dimensional biological structures such as lipid membranes [58]. The results presented in this work indicate that modifying the internal organization of similar two-dimensional supramolecular structures via electrostatics may provide a means to control dynamic behavior and ultimately the functionality of the system. Applications can be envisioned in which external electric fields are used to modulate the size and shape of nanodomains within high-density monolayers, potentially facilitating the creation of functional organic thin films in which diffusive motion may be controlled.

Acknowledgments

This material is based upon work supported as part of the NERC (Non-Equilibrium Research Center), an Energy Frontier Research Center funded by the U.S. Department of Energy, Office of Science, Office of Basic Energy Sciences under Award Number DE-SC0000989. C.B.G is supported by a Graduate Research Fellowship from the NSF.

References

1 Steed, J.W., Turner, D.R., and Wallace, K.J. (2007) *Core Concepts in Supramolecular Chemistry and Nanochemistry*, John Wiley & Sons, Inc., Chichester.

2 Raymo, F.M. and Stoddart, J.F. (1999) *Chem. Rev.*, **99**, 1643.

3 Rowan, S.J., Cantrill, S.J., Cousins, G.R.L., Sanders, J.K.M., and Stoddart, J.F. (2002) *Angew. Chem., Int. Ed.*, **41**, 898.

4 Saha, S. and Stoddart, J.F. (2007) *Chem. Soc. Rev.*, **36**, 77.
5 Gokel, G.W., Leevy, W.M., and Weber, M.E. (2004) *Chem. Rev.*, **104**, 2723.
6 Harada, A. (2001) *Acc. Chem. Res.*, **34**, 456.
7 Ritch, J.S. and Chivers, T. (2007) *Angew. Chem., Int. Ed.*, **46**, 4610.
8 Gutsche, C.D. (2008) *Calixarenes: An Introduction*, 2nd edn, Royal Society of Chemistry, Cambridge.
9 Fasolka, M.J. and Mayes, A.M. (2001) *Annu. Rev. Mater. Res.*, **31**, 323.
10 Love, J.C., Estroff, L.A., Kriebel, J.K., Nuzzo, R.G., and Whitesides, G.M. (2005) *Chem. Rev.*, **105**, 1103.
11 Simon, J. and Bassoul, P. (2000) *Design of Molecular Materials: Supramolecular Engineering*, John Wiley & Sons, Inc., Chichester.
12 Ulman, A. (1991) *An Introduction to Ultrathin Organic Films*, Academic Press, Boston.
13 Talham, D.R. (2004) *Chem. Rev.*, **104**, 5479.
14 Rampi, M.A., Schueller, O.J.A., and Whitesides, G.M. (1998) *Appl. Phys. Lett.*, **72**, 1781.
15 Rampi, M.A. and Whitesides, G.M. (2002) *Chem. Phys.*, **281**, 373.
16 O'Regan, B. and Gratzel, M. (1991) *Nature*, **353**, 737.
17 Bardea, A., Patolsky, F., Dagan, A., and Willner, I. (1999) *Chem. Commun.*, 21.
18 Forster, R.J., Keyes, T.E., and Majda, R. (2000) *J. Phys. Chem. B*, **104**, 4425.
19 Kaganer, V.M., Mohwald, H., and Dutta, P. (1999) *Rev. Mod. Phys.*, **71**, 779.
20 McConnell, H.M. and Moy, V.T. (1988) *J. Phys. Chem.*, **92**, 4520.
21 Schmid, F. and Lange, H. (1997) *J. Chem. Phys.*, **106**, 3757.
22 Duchs, D. and Schmid, F. (2001) *J. Phys. Condens. Matter*, **13**, 4853.
23 Thirumoorthy, K., Nandi, N., and Vollhardt, D. (2005) *J. Phys. Chem. B*, **109**, 10820.
24 Thirumoorthy, K., Nandi, N., and Vollhardt, D. (2007) *Langmuir*, **23**, 6991.
25 Krafft, M.P. and Riess, J.G. (2009) *Chem. Rev.*, **109**, 1714.
26 Tamam, L., Kraack, H., Sloutskin, E., Ocko, B.M., Pershan, P.S., Ulman, A., and Deutsch, M. (2005) *J. Phys. Chem. B*, **109**, 12534.
27 Broniatowski, M., Macho, I.S., Minones, J., and Dynarowicz-Latka, P. (2004) *J. Phys. Chem. B*, **108**, 13403.
28 Schneider, M.F., Andelman, D., and Tanaka, M. (2005) *J. Chem. Phys.*, **122**, 94717.
29 Trabelsi, S., Zhang, S., Lee, T.R., and Schwartz, D.K. (2008) *Phys. Rev. Lett.*, **100**, 037802.
30 Escobedo, F.A. and Chen, Z. (2004) *J. Chem. Phys.*, **121**, 11463.
31 Haas, F.M. and Hilfer, R. (1996) *J. Chem. Phys.*, **105**, 3859.
32 Haas, F.M., Hilfer, R., and Binder, K. (1995) *J. Chem. Phys.*, **102**, 2960.
33 Haas, F.M., Hilfer, R., and Binder, K. (1996) *J. Phys. Chem.*, **100**, 15290.
34 Schmid, F. and Schick, M. (1995) *J. Chem. Phys.*, **102**, 2080.
35 Stadler, C., Lange, H., and Schmid, F. (1999) *Phys. Rev. E*, **59**, 4248.
36 Stadler, C. and Schmid, F. (1999) *J. Chem. Phys.*, **110**, 9697.
37 Schmid, F. (2009) *Macromol. Rapid Commun.*, **30**, 741.
38 George, C.B., Ratner, M.A., and Szleifer, I. (2010) *J. Chem. Phys*, **132**, 014703.
39 Frenkel, D. and Smit, B. (1996) *Understanding Molecular Simulation*, Academic Press, San Diego.
40 Sintes, T., Baumgaertner, A., and Levine, Y.K. (1999) *Phys. Rev. E*, **60**, 814.
41 Barton, S.W., Goudot, A., Bouloussa, O., Rondelez, F., Lin, B.H., Novak, F., and Acero, A. et al. (1992) *J. Chem. Phys.*, **96**, 1343.
42 Strandburg, K.J. (1988) *Rev. Mod. Phys.*, **60**, 161.
43 Jaster, A. (1999) *Phys. Rev. E*, **59**, 2594.
44 Mak, C.H. (2006) *Phys. Rev. E*, **73**, 065104.
45 Xu, X.L. and Rice, S.A. (2008) *Phys. Rev. E*, **78**, 011602.
46 Cui, B.X., Lin, B.H., and Rice, S.A. (2001) *J. Chem. Phys.*, **114**, 9142.
47 Hurley, M.M. and Harrowell, P. (1995) *Phys. Rev. E*, **52**, 1694.
48 Hurley, M.M. and Harrowell, P. (1996) *J. Chem. Phys.*, **105**, 10521.
49 Marcus, A.H., Schofield, J., and Rice, S.A. (1999) *Phys. Rev. E*, **60**, 5725.
50 Zangi, R. and Rice, S.A. (2003) *Phys. Rev. E*, **68**, 061508.

51 Zangi, R. and Rice, S.A. (2004) *Phys. Rev. Lett.*, **92**, 035502.
52 Shiba, H., Onuki, A., and Araki, T. (2009) *Europhys. Lett.*, **86**, 66004.
53 Schmid, F. (1997) *Phys. Rev. E*, **55**, 5774.
54 Swanson, D.R., Hardy, R.J., and Eckhardt, C.J. (1996) *J. Chem. Phys.*, **105**, 673.
55 Scheringer, M., Hilfer, R., and Binder, K. (1992) *J. Chem. Phys.*, **96**, 2269.
56 Wikipedia (2009) Supramolecular chemistry). Retrieved December 18, 2009, from http://en.wikipedia.org/w/index.php?title=Supramolecular_chemistry&oldid=331799436.
57 Dix, J.A. and Verkman, A.S. (2008) *Annu. Rev. Biophys.*, **37**, 247.
58 Falck, E., Rog, T., Karttunen, M., and Vattulainen, I. (2008) *J. Am. Chem. Soc.*, **130**, 44.

3
Molecules on Gold Surfaces: What They Do and How They Go Around to Do It

Nadja Sändig and Francesco Zerbetto

3.1
Introduction

The geometrical and structural control of molecular assembly processes and the accurate positioning of individual molecules in 2D and 3D supramolecular structures is one of the ambitious objectives of nanoscience [1, 2]. This effort has generated a rich complexity of nano-sized objects that include clusters, wires, and extended networks [3–8]. Under a variety of different conditions also, spheres [9], stripes [10], rods [11], toroids [12], dendrimers [13], and dots [14] have been prepared. Traditionally, hydrogen bonds and π-stacking interactions are the structural motifs most widely exploited to create new systems [15]. More recently, long-range repulsive intermolecular interactions have come to play a significant role in supramolecular assembly in the form of surface-enhanced dipole–dipole interactions and substrate-mediated repulsion between rows of molecules [16, 17]. The dipole interactions between adsorbates can be tuned via a surface, which, in turn, is perturbed by the presence of adsorbates [18–22]. An interesting example is the adsorption of styrene molecules on a gold surface, where the orientations of the molecular dipoles have a local ferroelectric ordering and a long-range antiferroelectric one [23].

Modifications of molecular dipoles influence not only the structure but also the other physical quantities such as the work function of a metallic surface, which can be appraised by Kelvin probe experiments [24–26]. A classical example is the change of the work function due to the presence of self-assembled monolayers (SAMs) of thiolates [27, 28]. Dipolar interactions are crucial to understand the controversial [29–32] structure of the SAM terminated by COOH groups, where atomistic calculations of the dipole–dipole interactions showed that chains equal, or longer, than 12 carbons simultaneously have a crystalline-like order – with respect to the chain backbones – and a high degree of disorder – with respect to the COOH-groups [33].

Functional Supramolecular Architectures. Edited by Paolo Samorì and Franco Cacialli
Copyright © 2011 WILEY-VCH Verlag GmbH & Co. KGaA, Weinheim
ISBN: 978-3-527-32611-2

Understanding the intermolecular and molecule/substrate interactions is also critical in the fabrication of devices. Molecule/substrate interactions can (i) decrease the mobility of the adsorbing molecules and cause the formation of defects and/or adsorption in random conformations, (ii) modify the charges of the molecules or the substrate, and (iii) ultimately dictate the morphology of the adsorbed layer(s).

The high stability of gold to reactivity has made its surface very attractive as a possible substrate [34] and has been our workhorse for understanding the interaction between molecules and metals over the last few years. In particular, the Au(111) surface is highly stable and its reconstruction is rather small in geometrical terms, especially when compared to other metals. In general, molecular adsorption is driven by van der Waals and Coulomb interactions that can at times be qualitatively understood in terms of the "image charges" model, a basic tool in electrostatics [35]. Image charges occur because charges external to the metal, such as the local atomic charges of a molecule, can polarize the conducting metal. The new charge distribution in the metal creates an additional electrostatic potential/force on the molecules. Explicit treatments of image charges to describe the chemical physisorption of molecules are not abundant [36] and the effect is often ignored altogether even in high-level quantum chemical calculations [37]. A second Coulomb contribution arises from the fact that metals can also donate/accept charges, which can be obtained by quantum chemical calculations at any of the many density functional theory levels available. Less computationally demanding, and at a lower state of sophistication, charge equilibration models [38–42] are a possible alternative to determine the charge flow between molecules and metals. These models derive from the intuitive concept that electronegativity is equalized in chemical systems. Their rational foundation has recently been examined in terms of grand canonical ensemble and valence bond theories [43, 44].

To further complicate things, in the interaction with molecules, metals are hardly passive spectators when it comes to their geometrical structure. Indeed, their behavior is drastically different from molecules because their atoms can be displaced, removed altogether, or even added, often without affecting the chemical nature of the system. The geometrical description of a metal at the atomic level, or better of its geometrical structure, therefore requires methods that do not freeze atom–atom connectivity, as is often done, either implicitly or explicitly, with molecules. In other words, the model should allow detaching and rebinding of an atom to the surface, although this can hardly happen at room temperature. Embedded atom models (EAMs) have been developed for the purpose. They are a set of potential energy functions that describe bonding in many (transition) metals. Embedded atom models have been developed for Cu, Ag, Au, Ni, Pd, Pt, and alloys of these metals [45–56].

In the following, we examine in some detail a computational model able to address the issues mentioned above and provide an overview of the measure of success that is reached. Specific examples are provided for a variety of rather large molecular systems. A number of challenges and future goals will also be discussed.

3.2
A Simple Description of the Geometrical Structure of Metals

Embedded atom models are often used to describe the structure and dynamics of metals in the bulk. They share some of the features of molecular mechanics models that are typically used to describe molecules at the simplest possible level where the presence of atoms is preserved as opposed to mesoscopic models [57] where the smallest unit is made by several atoms. In a standard molecular mechanics picture, bonds are assigned at the beginning of the simulation. In EAM, such assignment is not necessary and actually is not sought.

The potential energy in EAMs contains two terms, the first is a two body, bond-like, term, and the second term depends on the atomic density around each one of the atoms of the metal. The binding contribution of this term changes with atomic valence or, in other words, the number of nearest-neighbor atoms modifies the contribution to the stability of the system. To clarify the concept, one could consider the well-known case of carbon, where the number of nearest-neighbor atoms generates sp, sp^2, and sp^3 hybridizations.

The potential energy of an embedded atom model has the form given in Equation 3.1:

$$V = \frac{1}{2} \sum_{\substack{i,j=1 \\ i \neq j}}^{N} \phi(r_{ij}) + \sum_{i}^{N} U(n_i) \tag{3.1}$$

where the first term on the right-hand side of the equation is the two-body contribution, with r_{ij} being the distance between atoms i and j, N the number of atoms, n_i the coordination of atom i, $\phi(r_{ij})$ a standard two-body potential that can be harmonic or anharmonic, and $U(n_i)$ is the energy associated with the coordination or valence of a metal atom.

In the glue model formulation of the potential that was originally developed for gold [58], $\phi(r_{ij})$, $U(n_i)$, and $\varrho(r_{ij})$ were used in and determined by a fitting procedure. They are expressed as polynomials that are connected adiabatically and have a different analytical representation in different regions. $\phi(r_{ij})$ is a polynomial in that r_{ij} contains terms up to an order of 6, while $U(n_i)$ contains terms up to an order of 4 and $\varrho(r_{ij})$ contains terms up to an order of 3.

The glue model is often employed by us to describe gold in the presence of physisorption processes. The model does not describe the adsorption (see below), but the interactions between gold atoms. This glue model reproduced with high accuracy the lattice parameters of the metal, its cohesive and surface energies, bulk modulus, melting temperature, melting entropy, and latent melting heat. It underestimated, partly, vacancy and vacancy migration energies. It overestimated the thermal expansion coefficient. The model also reproduced the (34 × 5) reconstruction of Au(100), preferred a (1 × 2) reconstruction for Au(110), found (1 × 3) and (1 × 4) energetically close, in agreement with the experiments, and gave an (11 × sqrt(3)) reconstruction of Au(111), instead of the experimental (23 × sqrt(3)).

3.3
A Simple Description of the Geometrical Structure of Molecules

In 1970s and perhaps even later, molecular mechanics was a form of art where even the analytical expression of the potential energy functions was a matter of discussion and the parameters were difficult to evaluate. Molecules are now described by a variety of techniques and models that accurately describe bond stretchings, bendings, torsions, and pyramidalizations, together with the nonbonding, that is, Coulomb and van der Waals, interactions. The force fields, that is, the set of functions and their parameters, are either aimed at classes of molecules such as proteins [59–61] or at providing a general description of a large set of molecules such as the organic molecules [62–66]. Trade-offs between these two extreme cases are now well understood. For the sake of generality, to describe organic molecules, we usually select the MM3 force field of Allinger and coworkers [67–69].

This force field describes stretchings and bendings with a potential energy function with quadratic, cubic, and quartic terms; a three-term Fourier series expansion describes the torsions for four atoms; coupled stretch–bend, stretch–torsion, and bend–bend interactions are also included. The function of the van der Waals repulsive component is exponential, while its attractive term is proportional to r^{-6}, that is, the van der Waals interaction is described by a Buckingham potential energy function.

In MM3, the Coulomb interactions were cleverly parameterized in terms of bond dipoles, although the authors made allowance for charge–charge and charge–dipole interactions and claimed that no particular advantage was gained by the use of bond dipoles. In the parameterization, most CH bonds and CC bonds do not carry any dipole moment, exceptions are CC bonds with atoms of different valence, such as sp^2–sp^3 bonds. This aspect must be considered carefully when dealing with adsorption on metal surfaces.

3.4
Electronegativity Governs Chemical Interactions: Charge Equilibration, QEq, Models

As atoms or molecules approach each other to produce a chemical process, be it the formation of a bond or physical vicinity, the electronegativity of each atom change until they become equal. Atoms accomplish such equilibration by transferring electron density. For different spatial arrangements of atoms, different amounts of transfer take place.

The starting point of charge equilibration methods, QEq, is the potential energy V_Q of a system/atom, which depends on the effective nearby charges (Equation 3.2)

$$V_Q = \sum_i \left(\frac{1}{2} \varkappa_i Q_i^2 + (\chi_i + F_i^{ext}) Q_i \right) + \sum_{i,j>i} \frac{Q_i Q_j}{4\pi \varepsilon_0 r_{ij}} \quad (3.2)$$

where $\varkappa_i$ is a restoring force, or electronic hardness, for the charge of the system, $\chi_i = 1/2\,(IP_i + EA_i)$ is its electronegativity, with IP_i being the ionization potential of atom i and EA_i its electron affinity, F_i^{ext} is a possible external electrostatic potential, and r_{ij} are the distances from external charges Q_j. The existence of a charge-restoring term can be justified from first principles arguments [70, 71]. During molecular optimizations and dynamics, charges can be re-equilibrated at every step or set of steps.

Minimization of the energy in Equation 3.2 by deriving with respect to the atomic charges, under the constraint of fixed total charge, leads to a set of equations whose solutions are the partial atomic charges of the atoms that equilibrate electronegativity. The approach is equivalent to the application of Sanderson's electronegativity equalization principle [72]. The set forms a linear system that can be solved even for large chemical systems of the order of several thousands of atoms.

One of the main failures of QEq models is that they predict that two unlike molecules in vacuum carry opposite nonzero charges, even if there is a large separation between them. This result contradicts chemical intuition and quantum chemical calculations. To ensure that molecules with closed electron shells remain neutral in the QEq description, one can simply impose it [73].

Another crucial point is the evaluation of the interactions between the charges of the atoms. At long distances, it is represented by a classical Coulomb term, while at short distances, where quantum effects are obviously important, to the best of our knowledge, there is no simple and unique way to represent it. In the formulation by Goddard and Rappé [74], it is calculated as an integral over Slater functions. This is an empirical approach that will probably require in the future the development of a deeper insight.

Although a great degree of success can be met with electronegativity equilibration models and a number of observables can be determined, there remain some intrinsically quantum mechanical quantities that require a wavefunction. An example is the short-range repulsion arising from the Pauli exclusion principle. QEq does not account for it and charges can eventually interact up to an effective "fusion."

With all these caveats in mind, the charges determined by QEq – in the presence of rationally physical situations – are reasonable and can be used to calculate the Coulomb interaction between atoms in semiempirical approaches.

3.5
A Simple Description of the Interaction between Metal Surfaces and Molecules

Molecular adsorption is driven by van der Waals and Coulomb interactions. In the model that we use, Coulomb interactions play a dominant role. They are calculated from the atomic charges obtained by the QEq electronegativity equilibration approach. The charges change as a function of the interatomic distances, a feature that, in practice, introduces polarization effects in the description of the interaction. The atoms of the metal, that is, gold, can acquire or donate electron density with amounts that vary during simulations of molecular dynamics. Charges of opposite signs obviously attract each other and the simple use of QEq charges can ultimately result in nuclear fusion. Van der Waals interactions between gold atoms and the atoms of the molecules are present in the model with the double intent of tuning the energy of adsorption so as to reproduce the experimental adsorption energy and of avoiding nuclear fusion.

To describe the van der Waals interactions between metal atoms and the atoms of a molecule, we retain the potential energy form of MM3. In practice, Equation 3.3 is used:

$$V = A \exp\left(-\frac{r}{\varrho}\right) - Cr^{-6} \tag{3.3}$$

where ϱ has a constant value set to 0.296 Å, A and C are developed for pairs of atom types, one of which is gold. In general, in the calculations, after an initial equilibration of up 0.2 ns, the production runs in the molecular dynamics simulations are performed for times of up to a few nanoseconds. The typical time step varies from 0.1 to 0.5 fs. Geometries are optimized to root mean square (RMS) values in the gradient components smaller than 0.1 kcal mol^{-1} Å^{-1}. Details of the procedure may vary from case to case and are always defined in each cited work.

3.6
Presence of an External Electric Potential or Field

In practical devises or in several types of experimental measurements, the presence of a metal surface where molecules are deposited is accompanied by an electric potential that goes to zero at a certain distance from the surface, thereby creating an electric field. The approach we advocate can incorporate the presence of an electric potential and the associated field in a more complete way than usually done in standard molecular dynamics calculations. In a classical molecular dynamics simulation, the field acts on the partial atomic charges either attracting or repelling them. In practice, the presence of the field entails an additional term in the energy gradient. The procedure is extremely efficient and has been recently used in our laboratory to simulate the growth of a water pillar from a drop [75]. However, the presence of an electrostatic potential also modifies the ionization potential and the electronic affinity of a molecule. Modification of electronic properties, such as IPs and EAs, is usually not accounted for in classical molecular mechanics/dynamics approaches, unless they contain a quantum chemical part. QEq is founded on the atomic ionization potentials and the electronic affinities that are combined to give the electronegativity χ_i and the electronic hardness $\varkappa_i$. The electrostatic potential shifts the former by an amount equal to its value, while the latter is unaffected.

In the simulations [76], two issues are considered: (i) in keeping with Gauss theorem, the electronegativity and the electronic hardness of the metal atoms are not affected by the electric potential, and (ii) the potential V_0 decays linearly as a function of the height z. Such variation of the potential is a gross simplification, but more accurate models of the decay can be introduced quite easily.

3.7
Generality of the Model and Its Transferability

An important issue is whether the model can be extended to metal other than gold. To date, only silver has been examined by us. A variety of applications have been discussed in a rather comprehensive paper [77]. They include the adsorption of C_{60}, benzene, ethylene, formaldehyde, acetone, several *n*-alkanes and aniline on Ag(111). Also, adsorption of a chemically complicated system such as 3,4,9,10-perylene tetracarboxylic acid dianhydride (PTCDA) on the Ag(110) surface has been examined. At present, we are seeking to extend the model to Hg.

Van der Waals and electrostatic interactions usually have limited or no coupling to other internal degrees of freedom, such as stretching and bendings. This, in turn, should make the approach and the parameters transferable to other force fields. Care should be taken in the transfer to verify the accuracy of the prediction of the molecular torsions, which may be effectively coupled to van der Waals and electrostatic interactions.

There is an important exception where strong coupling between internal degrees of freedom and charge transfer and the Coulomb interactions between metals and molecules occur. This is the case of catalysis, which is at present not described by the model. An interesting discussion of QEq to describe reactivity, and therefore possibly catalysis, has appeared [78].

The model relies on the two parameters of Equation 3.3 for each atom, which must be determined from experimental adsorption energies for individual molecules. The use of high-level *ab initio* or density functional theory based calculations was avoided because of the large variation of the energies that appear in the literature. To date, we have determined the *A* and *C* parameters for H, C, N, O, and S. The training set for their tuning included several hydrocarbons, both saturated and unsaturated, the DNA bases, ammonia, benzene, formaldehyde, thiophene, and thiolates. Comparison with experimental data showed that the adsorption energies were reproduced within $1 \, \text{kcal mol}^{-1}$ [79].

The merits of the results of the binding energy ought not to be exaggerated. At odds with other semiempirical procedures that are characterized by a host of functions and parameters and that describe completely a chemical system, here only the metal/molecule interactions are accounted for by the two fundamental types of interactions, van der Waals and Coulomb, whose shape has long been known. If emphasis should be given, the results vouch for the brilliant performance in the calculation of the charges of the QEq model developed and implemented by Goddard and Rappé.

3.8
Thiolates on Gold

The experimental investigation of the photophysical properties of a classic fluorophore, namely, pyrene, attached to gold nanoparticles via thiolate chains gave counterintuitive results [80]. If pyrene is linked via "short" alkyl chains of four carbon atoms, pyrene emits as it does when it is free. If pyrene is attached via "long" alkyl chains of 11 carbon atoms, pyrene does not emit. The quenching of the metal core is, therefore, more efficient when the fluorophore is more distant from it!

In the calculations [81], the pyrene derivatives were attached to the Au(111) surface via alkyl chains of length of 4 and 11 carbon atoms, C_4 and C_{11}, and the surface was fully packed. In the optimized geometries, the tilt angle of the short chains, namely, C_4, with respect to the normal to the metal surface is ~20° (see Figure 3.1, top). This value is about half as large as that calculated for the more flexible and longer chains, where it increases to 45–48°.

The origin of the effect observed experimentally is readily explained. When the chains are short, the dominant energy interaction that governs the packing of the

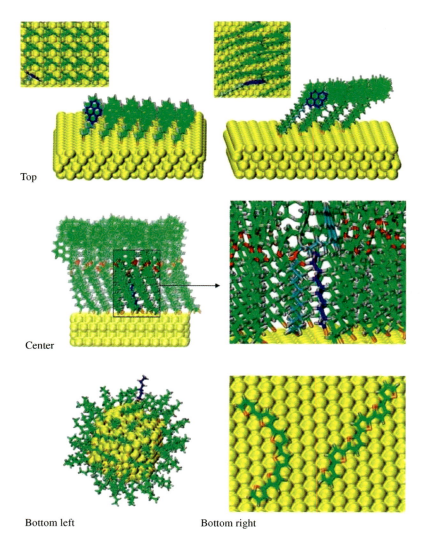

Figure 3.1 Molecules adsorbed on gold surfaces. *Top*: Pyrene linked via "short" and "long" alkyl chains. *Center*: An early stage of the interactions between thiols and a single decanethiol (blue). *Bottom left*: "Hairy ball" nanoparticle with one outstanding thiol. *Bottom right*: All-*cis* and all-*trans* sexithiophene molecules.

organic layer is between the pyrene moieties. Because of the short chain length, the chain/chain interaction is small and the molecules arrange themselves to maximize pyrene/pyrene interactions [82]. When the chains elongate, the chain/chain interactions become more important and the molecules arrange themselves accordingly.

The difference in tilt angle of the two types of chains has an important consequence: for the short C_4 chains, the pyrene fluorophore is almost perpendicular to the gold surface, while for the longer C_{11} chains, pyrene lies flatter with respect to the metal surface.

A flatter orientation of the molecule, as in the case of C_{11}, increases the probability of charge transfer to gold, which is zero for a perpendicularly standing pyrene.

The origin of the quenching of the fluorescence is geometrical in nature, but is different from the celebrated odd–even effect well known for thiols attached to gold [83] and is mainly caused by the large van der Waals interaction energy between pyrenes that competes with the chain/chain interaction energy to create structures with different local morphologies. The calculations show that the unexpected photophysical behavior is ultimately the result of the flexible chains that effectively modulate the fluorescence properties of the pyrene-functionalized nanoparticles via the modification of the tilt angle and the subsequent change of the interactions between metal and organic moiety.

This promising mechanism has also been discussed by McGuiness *et al.* [84] and highlighted in a "perspective" contribution [85]. In our opinion, calculations can drive the appropriate selection of design motifs that may be tuned to obtain a variety of desired fluorescence properties.

The thiolate chains on a nanoparticle undergo dynamical movements and molecular dynamics simulations provided information on the structure and dynamics that takes place on the surface. We considered an average-sized particle of 309 Au atoms whose shape is a cube octahedron with a diameter of $\sim$2.3 nm with 80 chains (see Figure 3.1, bottom left) [86].

The molecular dynamics of the chains were analyzed with the intent of identifying the less crowded terminal atoms. This situation should favor reactivity of the top atom of a chain, if properly functionalized. In practice, we sought the ligands whose terminal (methyl) groups were furthest away from their four closest neighbors, and the ligands whose chains were more perpendicular from the core of the nanoparticle.

If the system had the most symmetric configuration, only five types of atoms/distances would be present, with relative populations of 8:24:12:24:12. However, in the timescale of $\sim$1 ns, a handful of ligands are more "separate" from and taller than the others. Generally, the more exposed chains are at the antipodal edges of the nanoparticle.

Analysis of the distances of all the carbon atoms from the metal surface shows the presence of s-*cis* conformations in about one third of the chains. Experimentally, it has been determined [87] that the average distance from the Au surface of the fifth and sixth carbon atoms of the thiolate chains are 6.35 and 7.35 Å, respectively. These values agree well with the present average distances, over slightly less than a nanosecond of dynamics, of 6.54 and 7.40 Å. It appears that the timescale of the simulation captures the dynamics of the thiolates.

It is known that it is impossible to comb the hair of a ball smoothly so as to ensure that there is no bald spot. The reason is the same for which it is not possible to cover a sphere with square magnets so that they do not repel each other somewhere and is formally expressed by the "hairy ball theorem" [88]. In practice, as the thiolate chains tilt to maximize their mutual interactions, the effects of the theorem kick in and there must be (at least) one chain that stands apart from the others. In time, each chain/hair may play this role, but the simulations show that the chain(s) that stand(s) apart is/are the same for at least 1 ns.

The existence of terminal atoms more accessible from the external environment suggests that when properly functionalized, they could react more easily and the decoration of the nanoparticle should therefore occur at its antipodes. The more accessible thiolates could also be involved in the fast exchange dynamics of ligands [89].

The thiolate chains of the nanoparticle can be exchanged with thiolates present in the solution. The mechanical nature of the modifications caused by insertion of a single decanethiol in the middle of pyrene-functionalized thiols attached to a stable Au surface was also calculated (see Figure 3.1 center) [90]. The process may represent the early stage of the interactions between the thiols to displace one or more chains or set a thiol free to the solution. This is a drastic scenario that exaggerates what must occur in reality where the incoming thiol inserts itself in a less crowded region such as a terrace or a generic defect. The pristine system was optimized with the sulfur atoms sitting at the threefold hollow sites of the gold surface. Then an intruder chain was introduced. The four thiols nearest to the intruder were then allowed to relax. The resulting stabilization was 41 kcal mol^{-1} and was accompanied by the displacement of two thiols. All the thiols were then relaxed with a further stabilization of 28 kcal mol^{-1} and the shift of another pyrene-functionalized chain. The energy gained during the geometry optimization is the balance between a number of interactions that are either established anew or removed by the addition of the decanethiol chain. An important effect present in the final structure is the partial absence of the initial π-stacking of the pyrene moieties. This complicated scenario agrees with the kinetic analysis, which shows the presence of two regimes and suggests that for each decanethiol that becomes attached to the surface, more than one of the pyrene-labeled thiols is involved in the rate-determining step of the reaction. The model shows that the exchange process of an aliphatic thiol bound to a gold surface occurs through the interaction between several chains. Modifications of the thiol chains can therefore yield substantially different reaction patterns, which make these systems even more versatile than previously reported.

Sulfur interacts with gold surfaces not only in the form of thiolates but also in the form of thiophene chains (see Figure 3.1, bottom right). Oligomers of thiophene adopt in the vacuum the energetically most stable all-*trans* conformation. Adsorption on metal surfaces, such as gold, may modify the picture, as is now being shown by a number of experiments. In the scanning tunneling microscopy (STM) investigation of the conformations of a 10 nm long oligothiophene wire on Au(111), Yokoyama and coworkers [91] found that its tetrathiophene fragments often adopt s-*cis* linkages at the terminal bonds. Glowatzki *et al.* [92] found that T6 on Au(111) gives bright features, separated by dark ones, along substrate step edges and observed that the T6 molecules follow the shape of the step and deviate from the straight conformation of the gas phase and the bulk.

We investigated the stability of the 20 rotamers of sexithiophene on Au [93]. As expected, the stability of the isolated rotamers does not correlate with the molecular dipole moment. The most stable adsorbed system is formed by the all-*cis* rotamer. The energy penalty due to the isomerization is more than that overcome by the energy gain of the molecular dipole moment that interacts with the metal.

The energy of adsorption correlates directly with the purely electrostatic contribution to the total energy of the entire system (T6 plus gold), which implies that the van der Waals contribution is not the driving force for the adsorption of T6 rotamers. The sulfur atoms of the 20 rotamers interact with 46 fcc sites, 41 hcp sites, and 24 bridge sites (Au–Au bonds) and only in 9 cases are located on the top of a gold atom. The dienes interact directly on the top of a gold atom in 72 cases, while in 32 cases are located on hcp sites, in 14 cases on fcc sites, and only in 2 cases are located on the top of a bridge. Qualitatively, one can conclude that the electron-rich sulfur atoms tend to be located at the center of the electron-poor triangular cavity formed by three gold atoms, the fcc or hcp sites, while the metal atoms, instead, attempt to complex the dienes present in the rings.

3.9
Adsorption of a Large Molecule: C_{60}

C_{60} has become one of the most investigated molecules. The interaction of C_{60} with gold surfaces has attracted much interest and the structural properties of the C_{60}/metal interface have been studied for Au(111) [94, 95], Au(110) [96, 97], Au(001) [98], and also for polycrystalline Au substrates [99–101]. The low surface stability of Au(110) missing row (1 × 2) reconstructed surface makes it quite versatile, and two entirely independent patterns of C_{60} deposition were observed. Gimzewski and coworkers [102] showed that C_{60} adsorption induces a (1 × 5) interfacial reconstruction, where the maximum number of C_{60} molecules bonded to the (110) surface is reached through the formation of a distorted (6 × 5) overlayer (Figure 3.2a). Pedio *et al.* [103] found a more complex C_{60}–Au interface with a large number of gold surface atoms displaced from the original position to form calyxes or cups that accommodate the fullerenes (Figure 3.2b).

The existence of the two interfacial patterns poses the question of their relative stability and provides a test for any theory of adsorption that requires the careful balance of C_{60}–C_{60} van der Waals interactions with the C_{60}–Au ionic binding and the Au metal state. A computational difficulty is the different number of Au atoms in the two cells since the structure proposed by Pedio *et al.* has six less Au atoms. Even assuming that the six Au atoms have the energy of the bulk, that is, the maximum stability, the cell proposed by Gimzewski *et al.* is 27.7 kcal mol^{-1} lower in energy.

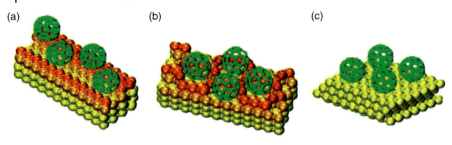

Figure 3.2 (a–c) Fullerenes adsorbed on gold surfaces, (a) Au(110) structure proposed by Gimzewski et al. [102] and (b) by Pedio et al. [103], and (c) on the Au(111) surface.

Interestingly, however, both the interaction and the binding of the interface of the cell of Pedio et al. are much larger than that for the structure of Gimzewski et al.

If the growth conditions favor metal–molecule interactions, for instance when free Au atoms stick to C_{60} rather than to each other, the structure of Pedio et al. becomes important.

The model also provides the amount of charge transfer from the Au surface to the C_{60} molecules and the relevant gold–carbon distances. The fullerene charge is higher for the structure of Pedio et al. than for the other structure. The amount of charge transfer is below one atomic unit (1 electron), in agreement with that deduced from the vibrational spectra of this interface [104]. It is also consistent with a charge transfer lower than $1.0 + 0.2$ atomic units, deduced from the comparison of valence band photoemission intensity near the Fermi edge of C_{60} on Au(110) [105] and C_{60} on polycrystalline Au [106].

While the structure and stability of C_{60} adsorbed on Au(110) have been discussed above, the flatter less-corrugated Au(111) surface could serve as an ideal substrate to test the lubrication by C_{60} (Figure 3.2c) [107, 108]. Indeed, it was long thought that C_{60} may represent an ideal system for lubrication since it is the molecular system with the closest resemblance to a ball bearing [109]. The strong van der Waals interactions established by the cage with neighboring atoms make friction far too high for lubrication. However, it is the possibility of modifying friction that is often of interest in many practical applications. In this respect, temperature can be considered the simplest external stimulus capable of tuning the balance between molecule–molecule and molecule–substrate interactions and consequently of affecting the tribological properties of a decorated surface [110].

The calculated binding energy of a single C_{60} adsorbed in a monolayer of molecules on Au(111) is $-61\,\text{kcal mol}^{-1}$ and includes both cage–metal and cage–cage interactions. The pure C_{60}–Au interaction energy is $-48\,\text{kcal mol}^{-1}$. In the optimized structure, the carbon cages alternate their positions on the top and bridge sites both along the $[0\bar{1}1]$ and the $[1\bar{2}1]$ directions of the Au surface. During the dynamics, the ideal separation in two different types of adsorption sites ceases to exist. The molecules wobble about and explore the vicinity of the adsorption minimum.

The cage movements occur in two characteristic ways that can loosely be described as bouncing and spinning. The timescales of these motions range from a few to several picoseconds. A more quantitative analysis can be obtained by Fourier transform techniques of the molecular dynamics runs and reveal the presence of a third type of motion similar to that of billiard balls shaken in a partly filled roll-a-rack triangle. Its timescale is intermediate between the other two.

The spinning motion hardly depends on the temperature and occurs between 3.5 and 4.5 cm^{-1}. Larger frequency shifts, as a function of temperature, are observed for the bouncing motion, which occurs from 23 to 28 cm^{-1}. This value agrees with the lowest experimental data of 24 cm^{-1} that was measured by sophisticated techniques in single-molecule transistors [111] and supports the accuracy of the approach to model molecular–metal interactions. Higher temperatures increase the average distance of the cage from the surface and reduce the binding strength. This results in a lower frequency of bouncing. The picosecond timescale of the movements of the C_{60} cages is similar to that of the reorientational dynamics of liquids such as water or acetonitrile, where the relaxation times are 13 and 9.6 ps [112]. Indeed, the C_{60} molecular adlayer on Au(111) is known to be highly mobile and the C_{60} molecules quickly diffuse on the surface [113, 114].

Evaluation of even the simplest quantity related to friction, namely, the shear viscosity η, demands long simulation times and large numbers of molecules to achieve an adequate statistical description of the dissipation of the instantaneous velocity gradients built by temperature. Integration of the stress correlation function gives the shear viscosity. Fourier transform of the functions shows that the dominant contributions are due to bouncing and rattling of the cages on the surface. These two motions, together with the oscillation in time of the charges, are therefore responsible for the dissipative properties of the C_{60} layer.

3.10
Simple Packing Problems

A comprehensive experimental study of alkanes adsorbed on Au(111) with chain lengths n ranging from 10 to 38 carbon atoms showed the absence of ordered monolayers for $18 \leq n \leq 26$ [115]. The appearance of disorder was not ascribed to a decrease of binding energy with n. This possible hypothesis was deemed unsatisfactory because it was impossible to identify the origin of the decrease in energy and it was, instead, proposed that for certain lengths the molecules do not properly lock on the gold lattice. The incommensurability is easily proved on geometrical grounds. However, neither the Au surface nor the alkane chains are rigid and they can and do adjust with each other. In the computational investigation [79], six fully packed surfaces of alkanes on Au(111) were studied with molecules initially oriented parallel to the $\langle 110 \rangle$ direction of Au(111). The adsorption energies of C_{14}, C_{16}, C_{22}, C_{24}, C_{30}, and C_{32} scale linearly with the number of atoms of the alkyl chain ($r^2 > 0.9999$) confirming that the adsorption energy cannot be advocated as the origin of disorder in the deposition of the intermediate length chains.

Two simple geometrical features indicate that intermediate length chains, C_{22} and C_{24}, differ from the others. The first feature is that the C_{methyl}–Au distance of the two terminal carbon atoms is not the same for a given molecule, which in turn produces instability. In practice, the alkane attempts to rearrange its interactions with gold to switch the long and the short C_{methyl}–Au distances, whose differences in Å are 0.032 (for C_{14}), 0.011 (for C_{16}), 0.058 (for C_{22}), 0.117 (for C_{24}), 0.004 (for C_{30}), and 0.046 (for C_{32}).

The second geometrical feature is the number of anchor points along the chains on the Au surface. A short C–Au distance entails a large local binding interaction. The presence of several anchor points entails a greater rigidity of the systems, with the molecules what can behave as blocks. If one sets at 2.5 Å, the maximum distance for points of docking, from C_{14} to C_{32}, in the six systems investigated there are 3, 3, 1, 1, 5, and 5 docking points. Short and long chains are able to nail themselves down on the Au(111) surface at more points than intermediate length chains, which once absorbed should display a higher local mobility. The fundamental reason for this behavior is the (in)commensurability of the Au and CH_2 lattices, which is here found to persist after full geometry optimization and manifests itself as a variety of Au–C distances.

Fumaramide [2]rotaxane, fumaramide thread, and benzylic amide macrocycle are shown schematically in Figure 3.3a–d. They only contain amide, phenyl, and methylene groups that give rather complicated infrared (IR) spectra [116].

The detailed assignment of all the infrared peaks was not attempted, apart from the identification of the main vibrational bands that served as a reference for the

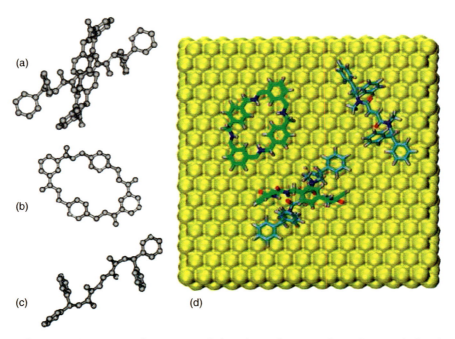

Figure 3.3 (a) Fumaramide [2]rotaxane, (b) benzylic amide macrocycle, (c) fumaramide thread, and (d) the rotaxane and its components adsorbed on Au(111).

subsequent high-resolution electron energy loss spectroscopy (HREELS) characterization, which, in turn, provided the molecular orientations of these species on the Au(111) surface. Further insight into film formation was given by the analysis of the elastic peak of HREELS. In short, the growth of the first layer is ordered for the macrocycle all the way to full coverage, ordered only up to 50% for the rotaxane and disordered for the thread. In the calculations, the macrocyclic ring adopts a planar conformation on Au(111) that readily explains the tendency to order. Less trivial is to explain the origin of the partial order observed for the largest interlocked system.

The dynamics shows that the four phenyls of the macrocycle undergo very small oscillations, and the entire molecule is lying flat on the surface. In the case of the thread, only two phenyls interact strongly with the surface. They belong to different stoppers and are in a *trans* arrangement with each other. One of these two phenyls is not consistently lying flat and temporarily leaves the surface. For the rotaxane, there are three phenyls lying flat on the surface, two of them belong to the macrocycle and are linked to each other, and one of them belongs to the stopper further away from the adsorbed part of the macrocycle.

The picture provided by the simulations is consistent with the HREELS spectra where the out-of-plane phenyl vibrations are more intense for the macrocycle because of its four phenyls lying flat on the surface, followed by the rotaxane with its three flat-lying phenyls adsorbed, and finally by the thread.

Inspection of the simulations also shows the origin of the greater degree of disorder observed for the thread. Both macrocycle and rotaxane are firmly anchored on the Au(111) surface and undergo relatively small displacements. The thread, however, shifts along the metal surface. Mobility was estimated as the largest displacement, in Angstroms, from the initial position for any phenyl group. The ratios are 4.8:23.0:7.5 for macrocycle/thread/rotaxane. The molecular dynamics simulations therefore imply that the disorder observed up to monolayer coverage for the thread is due to its high mobility on the metal surface, a feature not shared by the rotaxane and the macrocycle.

3.11
The Presence of an Electrostatic Potential

Citric acid (1,2,3-tricarboxy-2-hydroxypropane) and its anion citrate are effective stabilizers of various nanomaterials [117], are important participants in many biochemical processes [118], and are also used as anticlotting reagents in health care. Cyclic voltammetry of citrate on the Au(111) electrode surface showed for a well-defined electrode surface a pair of small peaks at ∼0.8 V. They were assigned to the reversible adsorption of citrate [119, 120]. Large-scale scanning tunneling microscopy images of an adlayer of citrate were acquired at 0.5 V, and the adlayer was ordered also at potentials negative with respect to 0.8 V. More positive potentials produced clusters of citrates and a disordered adlayer. *In situ* infrared spectroscopy experiments found the maximum of intensity of the symmetric carboxylate stretch of the adsorbed citrate for a potential of 0.8 V.

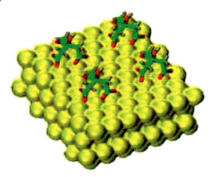

Figure 3.4 Citric acid molecules adsorbed on Au(111).

Calculations (Figure 3.4) [121] show that the interaction energy with gold steadily destabilizes as a function of voltage. This apparently suggests that the adsorption is hindered by a voltage increase. However, the energy of the entire molecular adlayer is stabilized by the increase of the electrostatic potential. The origin of the two effects is readily explained by the analysis of the calculated charges: as voltage increases, the molecules become less negative. The calculated molecular charge is −1.62 at 0 V and becomes −1.5 at 1 V. The net result is that as the potential becomes more positive, the adsorbed molecules repel each other to a lesser extent. The charge variation is linear with voltage. The balance between the two effects gives a total energy of the cell, which is minimal between 0.5 and 0.6 V and is still quite stable up to 0.8 V. The range of values where the calculations give enhanced cell stability agrees with the experimental observations of ordered adsorption.

The stabilization of the molecular adlayer is due to a reduced charge transfer from the metal and not an increase of interaction between the molecules and the metal that drives the adsorption in the presence of the external voltage. Note that the O–Au adsorption distance is 2.7 Å at 0 V. This distance increases ∼0.1 Å over the voltage range examined here. This confirms that molecule–Au interactions are not the driving force for adsorption and further emphasizes the role of the intermolecular interactions within the adlayer.

Thiolates show voltage-dependent desorption rates from Au(111) that were estimated experimentally by Sumi and Uosaki [122], Vericat et al. [123], and Vinokurov et al. [124]. Assuming an Arrhenius form for the desorption rate constants and that the frequency factor does not depend on voltage (as the experimental linear trend suggests), one can obtain $\partial E_{ads}(V)/\partial V = -10.8 \, \text{kcal mol}^{-1} \text{V}^{-1}$. Thus, as the voltage increases, the adsorption (desorption) energy linearly decreases (increases), and it is more difficult to remove a thiolate from the adlayer.

As in the other case, for thiolates on Au(111), the intermolecular interactions of the adlayer become more stable as voltage increases. This is due to a reduced electron transfer from the metallic substrate. The calculated molecular charges scale linearly with voltage for the singly adsorbed chain and the half-packed and the fully packed monolayers. For these three cases, the calculated slopes $\partial E_{ads}/\partial V$ are 7.81, −8.75, and

$-12.09\,\text{kcal}\,\text{mol}^{-1}\,\text{V}^{-1}$. The change of sign is remarkable and can be understood considering that for the isolated molecule, there is no intermolecular energy that contributes to the adsorption energy. Consequently, a diminished charge transfer cannot stabilize the intermolecular interaction but can only make easier the desorption of the isolated thiolate chain. The two negative values for the complete or semicomplete adlayers bracket the experimental data of $-10.8\,\text{kcal}\,\text{mol}^{-1}\,\text{V}^{-1}$ and agree with the intuitive notion that desorption takes place from defect sites where the chain does not have a fully packed environment.

Consistently, in all cases examined, the adsorption of the systems is driven by the stabilization of the organic adlayer through reduced charge transfer between metal and molecules.

3.12 Challenges and Conclusion

Decoration of surfaces by molecules has become an important area of chemistry because of both the beautiful structures that can be prepared and the practical applications. A host of new 2D structures are produced daily, but the predicting power of the computational tools available to date is limited, and we are at present unable to predict what can be formed and many of the details of how the systems behave. Gold is the paradigm of a well-behaved metal and its most stable surface has been used for a myriad of applications. And yet, even recently, it has become clear that our understanding of its properties is not as deep as one would like it to be. It is now clear that even its most stable surface, Au(111), when it interacts with molecules, can strongly reconstruct and connect sulfur atoms via the presence of a Au adatom [125], which also characterize nanoparticles such as that based on the 102 magic number [126]. There are, however, other important challenges that remain open. Foremost is probably that of catalysis. In a time of ever-increasing computer power and wide availability of quantum chemical approaches, one might wonder whether there is need for a (semi-)empirical approach as that presented here. The accuracy of the model is, at present, for the few systems it can treat higher than standard quantum chemical approaches. This is no wonder since it has tuning parameters that improve its accuracy. The model is mainly founded on ionization potentials, electron affinities, and a representation of the overlap of atomic electronic clouds (admittedly calculated rather crudely). Relying on these three quantities only guarantees that the physical insight of the results is straightforward. Failure of the model entails situations where the physics and chemistry require more than electronegativity and electronic hardness. More than 20 years ago, Shankar and Parr [127] concluded for inorganic structures that these parameters suffice to describe the nature of bonding in solids. The implication is that this applies also to one of the best behaved metals. In the future, it will be interesting to try and pursue these concepts to develop a qualitative theory of adsorption on metals.

Acknowledgments

FZ would like to thank Drs. R. Baxter, G. Teobaldi, S. Rapino, J.-P. Jalkanen, and T. Cramer for the long hours spent on what are unfairly called "technical details," but in reality form the working heart of the model. Collaborations and discussions with the experimental groups of Profs. P. Rudolf, F. Biscarini, F. Kajzar, D.A. Leigh, and all their coworkers have been of the utmost importance in the development of the model.

References

1 Joachim, C., Gimzewski, J.K., and Aviram, A. (2000) Electronics using hybrid-molecular and mono-molecular devices. *Nature*, **408**, 541–548.
2 Barth, J.V., Costantini, G., and Kern, K. (2005) Engineering atomic and molecular nanostructures at surfaces. *Nature*, **437**, 671–679.
3 Bühringer, M., Morgenstern, K., Schneider, W.-D., Berndt, R., Mauri, F., DeVita, A., and Car, R. (1999) Two-dimensional self-assembly of supramolecular clusters and chains. *Phys. Rev. Lett.*, **83** (2), 324–327.
4 Weckesser, J., DeVita, A., Barth, J.V., Cai, C., and Kern, K. (2001) Mesoscopic correlation of supramolecular chirality in one-dimensional hydrogen-bonded assemblies. *Phys. Rev. Lett.*, **87** (9), 096101–096105.
5 Yokoyama, T., Yokoyama, S., Kamikado, T., Okuno, Y., and Mashiko, S. (2001) Selective assembly on a surface of supramolecular aggregates with controlled size and shape. *Nature*, **413**, 619–621.
6 Yokoyama, T., Kamikado, T., Yokoyama, S., and Mashiko, S. (2004) Conformation selective assembly of carboxyphenyl substituted porphyrins on Au (111). *J. Chem. Phys.*, **121** (23), 11993–11997.
7 Palma, C.-A., Bjork, J., Bonini, M., Dyer, M.S., Llanes-Pallas, A., Bonifazi, D., Persson, M., and Samorì, P. (2009) Tailoring bicomponent supramolecular nanoporous networks: phase segregation, polymorphism, and glasses at the solid–liquid interface. *J. Am. Chem. Soc.*, **131** (36), 13062–13071.

8 Palma, C.-A., Bonini, M., Llanes-Pallas, A., Breiner, T., Maurizio Prato, M., Bonifazi, D., and Samorì, P. (2008) Pre-programmed bicomponent porous networks at the solid–liquid interface: the low concentration regime. *Chem. Commun.*, **42**, 5289–5291.
9 Zhou, S.Q., Burger, C., Chu, B., Sawamura, M., Nagahama, N., Toganoh, M., Hackler, U.E., Isobe, H., and Nakamura, E. (2001) Spherical bilayer vesicles of fullerene-based surfactants in water: a laser light scattering study. *Science*, **291** (5510), 1944–1947.
10 Leclère, P., Calderone, A., Marsitzky, D., Francke, V., Geerts, Y., Müllen, K., Brédas, J.L., and Lazzaroni, R. (2000) Highly regular organization of conjugated polymer chains via block copolymer self-assembly. *Adv. Mater.*, **12** (14), 1042–1046.
11 Georgakilas, V., Pellarini, F., Prato, M., Guldi, D.M., Melle-Franco, M., and Zerbetto, F. (2002) Supramolecular self-assembled fullerene nanostructures. *Proc. Natl. Acad. Sci. USA*, **99** (8), 5075–5080.
12 Zubarev, E.R., Pralle, M.U., Li, L.M., and Stupp, S.I. (1999) Conversion of supramolecular clusters to macromolecular objects. *Science*, **283** (5401), 523–526.
13 Ungar, G., Liu, Y.S., Zeng, X.B., Percec, V., and Cho, W.D. (2003) Giant supramolecular liquid crystal lattice. *Science*, **299** (5610), 1208–1211.
14 Biscarini, F., Cavallini, M., Kshirsagar, R., Bottari, G., Leigh, D.A., Leon, S., and Zerbetto, F. (2006) Self-organization of

15 Lehn, J.-M. (1995) *Supramolecular Chemistry: Concept and Perspectives*, Wiley-VCH Verlag GmbH, Weinheim.
16 Yokoyama, T., Takahashi, T., and Shinozaki, K. (2007) Quantitative analysis of long-range interactions between adsorbed dipolar molecules on Cu(111). *Phys. Rev. Lett.*, **98** (20), 206102–206106.
17 Lukas, S., Witte, G., and Wöll, Ch. (2001) Novel mechanism for molecular self-assembly on metal substrates: unidirectional rows of pentacene on Cu (110) produced by a substrate-mediated repulsion. *Phys. Rev. Lett.*, **88** (2), 028301–028305.
18 Kamna, M.M., Stranick, S.J., and Weiss, P.S. (1996) Imaging substrate-mediated interactions. *Science*, **274** (5284), 118–119.
19 Koutecký, J. (1958) A contribution to the molecular-orbital theory of chemisorption. *Trans. Faraday Soc.*, **54**, 1038–1052.
20 Merrick, M.L., Luo, W.W., and Fichthorn, K.A. (2003) Substrate-mediated interactions on solid surfaces: theory, experiment, and consequences for thin-film morphology. *Prog. Surf. Sci.*, **72** (5–8), 117–134.
21 Stranick, S.J., Kamna, M.M., and Weiss, P.S. (1995) Interactions and dynamics of benzene on Cu{111} at low temperature. *Surf. Sci.*, **338** (1–3), 41–59.
22 Sykes, E.C.H., Han, P., Kandel, S.A., Kelly, K.F., McCarty, G.S., and Weiss, P.S. (2003) Substrate-mediated interactions and intermolecular forces between molecules adsorbed on surfaces. *Acc. Chem. Res.*, **36** (12), 945–953.
23 Baber, A.E., Jensen, S.C., and Sykes, E.C.H. (2007) Dipole-driven ferroelectric assembly of styrene on Au{111}. *J. Am. Chem. Soc.*, **129** (20), 6368–6369.
24 Evans, S.D. and Ulman, A. (1990) Surface potential studies of alkyl-thiol monolayers adsorbed on gold. *Chem. Phys. Lett.*, **170** (5–6), 462–466.
25 Lu, J., Delamarche, E., Eng, L., Bennewitz, R., Meyer, E., and Guntherodt, H.-J. (1999) Kelvin probe force microscopy on surfaces: investigation of the surface potential of self-assembled monolayers on gold. *Langmuir*, **15** (23), 8184–8188.
26 Cui, X.D., Freitag, M., Martel, R., Brus, L., and Avouris, P. (2003) Controlling energy-level alignments at carbon nanotube/Au contacts. *Nano Lett.*, **3** (6), 783–787.
27 Taylor, D.M. (2000) Developments in the theoretical modelling and experimental measurement of the surface potential of condensed monolayers. *Adv. Colloid Interface Sci.*, **87** (2–3), 183–203.
28 Wang, R.L.C., Krenzer, H.J., Grunze, M., and Pertsin, A.J. (2000) The effect of electrostatic fields on an oligo(ethylene glycol) molecule: dipole moments, polarizabilities and field dissociation. *Phys. Chem. Chem. Phys.*, **2**, 1721–1727.
29 Himmel, H.-J., Weiss, K., Jäger, B., Dannenberger, O., Grunze, M., and Wöll, C. (1997) Ultrahigh vacuum study on the reactivity of organic surfaces terminated by OH and COOH groups prepared by self-assembly of functionalized alkanethiols on Au substrates. *Langmuir*, **13** (19), 4943–4947.
30 Dannenberger, O., Weiss, K., Himmel, H.-J., Jäger, B., Buck, M., and Wöll, C. (1997) An orientation analysis of differently endgroup-functionalised alkanethiols adsorbed on Au substrates. *Thin Solid Films*, **307** (1–2), 183–197.
31 Nuzzo, R.G., Dubois, L.H., and Allara, D.L. (1990) Fundamental studies of microscopic wetting on organic surfaces. 1. Formation and structural characterization of a self-consistent series of polyfunctional organic monolayers. *J. Am. Chem. Soc.*, **112** (2), 558–569.
32 Ito, E., Konno, K., Noh, J., Kanai, K., Ouchi, Y., Seki, K., and Hara, M. (2005) Chain length dependence of adsorption structure of COOH-terminated alkanethiol SAMs on Au(1 1 1). *Appl. Surf. Sci.*, **244** (1–4), 584–587.
33 Sushko, M.L. and Shluger, A.L. (2007) Dipole–dipole interactions and the structure of self-assembled monolayers. *J. Phys. Chem. B*, **111** (16), 4019–4025.

34 Love, J.C., Estroff, L.A., Kriebel, J.K., Nuzzo, R.G., and Whitesides, G.M. (2005) Self-assembled monolayers of thiolates on metals as a form of nanotechnology. *Chem. Rev.*, **105** (4), 1103–1170.

35 Smyth, W.R. (1968) *Static and Dynamic Electricity*, 3rd edn, Chapter 3, especially Eq. 3.08, McGraw-Hill, New York.

36 Bakalis, E. and Zerbetto, F. (2007) Charge–metal interaction of a carbon nanotube. *Chem. Phys. Chem.*, **8** (7), 1005–1008.

37 Mujica, V. and Ratner, M.A. (2003) Molecular conductance junctions: a theory and modeling progress report, in *Handbook of Nanoscience, Engineering, and Technology* (eds D. Brenner, S. Lyshevski, G. Iafrate and W.A. Goddard), 2nd edn, CRC Press, Boca Raton, FL.

38 Rick, S.W. and Stuart, S.J. (2002) Potentials and algorithms for incorporating polarizability in computer simulations. *Rev. Comput. Chem.*, **18**, 89–146.

39 Mortier, W.J., Van Genechten, K., and Gasteiger, J. (1985) Electronegativity equalization: application and parametrization. *J. Am. Chem. Soc.*, **107** (4), 829–835.

40 Rappé, A.K. and Goddard, W.A., III (1991) Charge equilibration for molecular dynamics simulations. *J. Phys. Chem.*, **95** (8), 3358–3363.

41 Rick, S.W., Stuart, S.J., and Berne, B.J. (1994) Dynamical fluctuating charge force fields: application to liquid water. *J. Chem. Phys.*, **101** (7), 6141–6156.

42 MacKerell, A.D., Jr. (2005) *Handbook of Materials Modeling* (ed. S. Yip), Springer, Dordrecht, p. 509.

43 Morales, J. and Martinez, T.J. (2001) Classical fluctuating charge theories: the maximum entropy valence bond formalism and relationships to previous models. *J. Phys. Chem. A*, **105** (12), 2842–2850.

44 Morales, J. and Martinez, T.J. (2004) A new approach to reactive potentials with fluctuating charges: quadratic valence-bond model. *J. Phys. Chem. A*, **108** (15), 3076–3084.

45 Finnis, M.W. and Sinclair, J.E. (1984) A simple empirical n-body potential for transition-metals. *Philos. Mag. A*, **50**, 45–55.

46 Ercolessi, F., Parrinello, M., and Tosatti, E. (1988) Simulation of gold in the glue model. *Philos. Mag. A*, **58** (1), 213–226.

47 Sutton, A.P. and Chen, J. (1990) Long-range Finnis–Sinclair potentials. *Philos. Mag. Lett.*, **61**, 139–164.

48 Qi, Y., .6pÇ51ptagin, T., Kimura, Y., and Goddard, W.A., III (1999) Molecular dynamics simulations of glass formation and crystallization in binary liquid metals: Cu–Ag and Cu–Ni. *Phys. Rev. B*, **59** (5), 3527–3533.

49 Tartaglino, U., Tosatti, E., Passerone, D., and Ercolessi, F. (2002) Bending strain-driven modification of surface reconstructions: Au(111). *Phys. Rev. B*, **65** (24), 241406–241410.

50 Daw, M.S. and Baskes, M.I. (1984) Embedded-atom method: derivation and application to impurities, surfaces, and other defects in metals. *Phys. Rev. B*, **29** (12), 6443–6453.

51 Foiles, S.M., Baskes, M.I., and Daw, M.S. (1986) Embedded-atom-method functions for the fcc metals Cu, Ag, Au, Ni, Pd, Pt, and their alloys. *Phys. Rev. B*, **33** (12), 7983–7991.

52 Johnson, R.A. (1989) Alloy models with the embedded-atom method. *Phys. Rev. B*, **39** (17), 12554–12559.

53 Lu, J. and Szpunar, J.A. (1997) Applications of the embedded-atom method to glass formation and crystallization of liquid and glass transition-metal nickel. *Philos. Mag. A*, **75** (4), 1057–1066.

54 Voter, A.F. (1995) *Intermetallic Compounds: Principles and Practice* (eds J.H. Westbrook and R.L. Fleischer), John Wiley & Sons, Inc., New York, p. 77.

55 Daw, M.S. (1989) Model of metallic cohesion: the embedded-atom method. *Phys. Rev. B*, **39** (11), 7441–7452.

56 Voter, A. and Chen, S. (1987) Accurate interatomic potentials for Ni, Al and Ni$_3$Al. *Mater. Res. Soc. Symp. Proc.*, **82**, 175–180.

57 Zannoni, C. (2001) Molecular design and computer simulations of novel

58 Ercolessi, F., Bartolini, A., Garofalo, M., Parrinello, M., and Tosatti, E. (1987) Au surface reconstructions in the glue model. *Surf. Sci.*, **189–190**, 636–640.

59 Neria, E., Fischer, S., and Karplus, M. (1996) Simulation of activation free energies in molecular systems. *J. Chem. Phys.*, **105** (5), 1902–1921.

60 Wang, J., Cieplak, P., and Kollman, P.A. (2000) How well does a restrained electrostatic potential (RESP) model perform in calculating conformational energies of organic and biological molecules? *J. Comput. Chem.*, **21** (12), 1049–1074.

61 Kaminski, G. and Jorgensen, W.L. (1996) Performance of the AMBER94, MMFF94, and OPLS-AA force fields for modeling organic liquids. *J. Phys. Chem.*, **100** (46), 18010–18013.

62 Halgren, T.A. (1996) Merck molecular force field. I. Basis, form, scope, parameterization, and performance of MMFF94. *J. Comput. Chem.*, **17** (5), 490–519.

63 Halgren, T.A. (1996) Merck molecular force field. II. MMFF94 van der Waals and electrostatic parameters for intermolecular interactions. *J. Comput. Chem.*, **17** (5), 520–552.

64 Halgren, T.A. (1996) Merck molecular force field. III. Molecular geometries and vibrational frequencies for MMFF94. *J. Comput. Chem.*, **17** (5), 553–586.

65 Halgren, T.A. (1996) Merck molecular force field. IV. Conformational energies and geometries for MMFF94. *J. Comput. Chem.*, **17** (5), 587–615.

66 Halgren, T.A. (1996) Merck molecular force field. V. Extension of MMFF94 using experimental data, additional computational data, and empirical rules. *J. Comput. Chem.*, **17** (5), 616–641.

67 Allinger, N.L., Yuh, Y.H., and Lii, J.-H. (1989) Molecular mechanics. The MM3 force field for hydrocarbons. 1. *J. Am. Chem. Soc.*, **111** (23), 8551–8566.

68 Lii, J.-H. and Allinger, N.L. (1989) Molecular mechanics. The MM3 force field for hydrocarbons. 2. Vibrational frequencies and thermodynamics. *J. Am. Chem. Soc.*, **111** (23), 8566–8575.

69 Lii, J.-H. and Allinger, N.L. (1989) Molecular mechanics. The MM3 force field for hydrocarbons. 3. The van der Waals' potentials and crystal data for aliphatic and aromatic hydrocarbons. *J. Am. Chem. Soc.*, **111** (23), 8576–8582.

70 Elstner, M., Porezag, D., Jungnickel, G., Elsner, J., Haugk, M., Frauenheim, Th., Suhai, S., and Seifert, G. (1998) Self-consistent-charge density-functional tight-binding method for simulations of complex materials properties. *Phys. Rev. B*, **58** (11), 7260–7268.

71 Tabacchi, G., Mundy, C.J., Hutter, J., and Parrinello, M. (2002) Classical polarizable force fields parametrized from *ab initio* calculations. *J. Chem. Phys.*, **117** (4), 1416–1433.

72 Sanderson, R.T. (1951) An interpretation of bond lengths and a classification of bonds. *Science*, **144** (2973), 670–672.

73 Nistor, R.A., Polihronov, J.G., Müser, M.H., and Mosey, N.J. (2006) A generalization of the charge equilibration method for non-metallic materials. *J. Chem. Phys.*, **125** (9), 094108-1–094108-10.

74 Rappé, A.K. and Goddard, W.A., III (1991) Charge equilibration for molecular dynamics simulations. *J. Phys. Chem.*, **95** (8), 3358–3363.

75 Cramer, T., Zerbetto, F., and García, R. (2008) Molecular mechanism of water bridge buildup: field-induced formation of nanoscale menisci. *Langmuir*, **24** (12), 6116–6120.

76 Teobaldi, G. and Zerbetto, F. (2007) Adsorption of organic molecules on gold electrodes. *J. Phys. Chem. C*, **111** (37), 13879–13885.

77 Jalkanen, J.-P. and Zerbetto, F. (2006) Interaction model for adsorption of organic molecules on the silver surface. *J. Phys Chem. B*, **110** (11), 5595–5601.

78 Chen, J., Hundertmark, D., and Martinez, T.J. (2008) A unified framework for fluctuating-charge models in atom-space and in bond-space. *J. Chem. Phys.*, **129** (21), 214113-1–214113-11.

79 Baxter, R.J., Teobaldi, G., and Zerbetto, F. (2003) Modeling the adsorption of alkanes on an Au(111) surface. *Langmuir*, **19** (18), 7335–7340.

80 Tovmachenko, O.G., Graf, C., van den Heuvel, D.J., van Blaaderen, A., and Gerritsen, H.C. (2006) Fluorescence enhancement by metal-core/silica-shell nanoparticles. *Adv. Mater.*, **18** (1), 91–95.

81 Battistini, G., Cozzi, P.G., Jalkanen, J.-P., Montalti, M., Prodi, L., Zaccheroni, N., and Zerbetto, F. (2008) The erratic emission of pyrene on gold nanoparticles. *ACS Nano*, **2** (1), 77–84.

82 Tam, F., Goodrich, G.P., Johnson, B.R., and Halas, N.J. (2007) Plasmonic enhancement of molecular fluorescence. *Nano Lett.*, **7** (2), 496–501.

83 Cheng, D.M. and Xu, Q.H. (2007) Separation distance dependent fluorescence enhancement of fluorescein isothiocyanate by silver nanoparticles. *Chem. Commun.*, 248–250.

84 McGuiness, C.L., Blasini, D., Masejewski, J.P., Uppili, S., Cabarcos, O.M., Smilgies, D., and Allara, D.L. (2007) Molecular self-assembly at bare semiconductor surfaces: characterization of a homologous series of n-alkanethiolate monolayers on GaAs(001). *ACS Nano*, **1** (1), 30–49.

85 Bent, S.F. (2007) Heads or tails: Which is more important in molecular self-assembly? *ACS Nano*, **1** (1), 10–12.

86 Rapino, S. and Zerbetto, F. (2007) Dynamics of thiolate chains on a gold nanoparticle. *Small*, **3** (3), 386–388.

87 Terril, H.R., Postlethwaite, T.A., Chen, C.-H., Poon, C.-D., Terzis, A., Chen, A., Hutchison, J.E., Clark, M.R., Wignall, G., Londono, J.D., Superfine, R., Falvo, M., Johnson, C.S., Jr., Samulski, E.T., and Murray, R.W. (1995) Monolayers in three dimensions: NMR, SAXS, thermal, and electron hopping studies of alkanethiol stabilized gold clusters. *J. Am. Chem. Soc.*, **117** (50), 12537–12548.

88 http://en.wikipedia.org/wiki/Hairy_ball_theorem.

89 Hostetler, M.J., Templeton, A.C., and Murray, R.W. (1999) Dynamics of place-exchange reactions on monolayer-protected gold cluster molecules. *Langmuir*, **15** (11), 3782–3789.

90 Montalti, M., Prodi, L., Zaccheroni, N., Baxter, R., Teobaldi, G., and Zerbetto, F. (2003) Kinetics of place-exchange reactions of thiols on gold nanoparticles. *Langmuir*, **19** (12), 5172–5174.

91 Nishiyama, F., Ogawa, K., Tanaka, S., and Yokoyama, T. (2008) Direct conformational analysis of a 10nm long oligothiophene wire. *J. Phys. Chem. B*, **112** (17), 5272.

92 Glowatzki, H., Duhm, S., Braun, K.-F., Rabe, J.P., and Koch, N. (2007) Molecular chains and carpets of sexithiophenes on Au(111). *Phys. Rev. B*, **76** (12), 125425–125437.

93 Sändig, N., Biscarini, F., and Zerbetto, F. (2008) Driving force for the adsorption of sexithiophene on gold. *Phys. Chem. C*, **112** (49), 19516–19520.

94 Gallagher, P.M., Coffey, M.P., Krukonis, V.J., and Klasutis, N. (1989) Gas anti-solvent recrystallization: new process to recrystallize compounds insoluble in supercritical fluids. *ACS Symp. Ser.*, **406**, 334–354.

95 Wei, M., Musie, G.T., Busch, D.H., and Subramaniam, B. (2002) CO_2-expanded solvents: unique and versatile media for performing homogeneous catalytic oxidations. *J. Am. Chem. Soc.*, **124** (11), 2513–2517.

96 Ventosa, N., Sala, S., Torres, J., Llibre, J., and Veciana, J. (2001) Depressurization of an expanded liquid organic solution (DELOS): a new procedure for obtaining submicron- or micron-sized crystalline particles. *Cryst. Growth Des.*, **1** (4), 299–303.

97 Gallagher, P.M., Krukonis, V., and Botsaris, G.D. (1991) Gas antisolvent recrystallization: application to particle design. *AIChE Symp. Ser.*, **87** (284), 96–112.

98 Eckert, C.A., Bush, D., Brown, J.S., and Liotta, C.L. (2000) Tuning solvents for sustainable technology. *Ind. Eng. Chem. Res.*, **39** (12), 4615–4621.

99 De la Fuente Badilla, J.C., Peters, C.J., and de Swaan Arons, J. (2000) Volume expansion in relation to the gas-

100. Giacobbe, F.W. (1992) Thermodynamic solubility behavior of carbon dioxide in acetone. *Fluid Phase Equilib.*, **72**, 277–297.
101. Zhang, X., Han, B., Hou, Z., Zhang, J., Liu, Z., Jiang, T., He, J., and Li, H. (2002) Why do co-solvents enhance the solubility of solutes in supercritical fluids? New evidence and opinion. *Chem. Eur. J.*, **8** (22), 5107–5111.
102. Gimzewski, J.K., Modesti, S., and Schlittler, R.R. (1994) Cooperative self-assembly of Au atoms and C_{60} on Au(110) surfaces. *Phys. Rev. Lett.*, **72**, 1036.
103. Pedio, M., Felici, R., Torrelles, X., Rudolf, P., Capozi, M., Rius, J., and Ferrer, S. (2000) Study of C_{60}/Au(110)-$p(6\times5)$ reconstruction from in-plane X-ray diffraction data. *Phys. Rev. Lett.*, **85** (5), 1040–1043.
104. Modesti, S., Cerasari, S., and Rudolf, P. (1993) Determination of charge states of C_{60} adsorbed on metal surfaces. *Phys. Rev. Lett.*, **71** (15), 2469–2472.
105. Maxwell, A.J., Brühwiler, P.A., Nilsson, A., Mårtensson, N., and Rudolf, P. (1994) Photoemission, autoionization, and X-ray-absorption spectroscopy of ultrathin-film C_{60} on Au(110). *Phys. Rev. B*, **49** (15), 10717–10725.
106. Hoogenboom, B.W., Hesper, R., Tjeng, L.H., and Sawatzky, G.A. (1998) Charge transfer and doping-dependent hybridization of C_{60} on noble metals. *Phys. Rev. B*, **57** (19), 11939–11942.
107. Cao, T., Wei, F., Yang, Y., Huang, L., Zhao, X., and Cao, W. (2002) Microtribologic properties of a covalently attached nanostructured self-assembly film fabricated from fullerene carboxylic acid and diazoresin. *Langmuir*, **18** (13), 5186–5189.
108. Luengo, G., Campbell, S.E., Srdanov, V.I., Wudl, F., and Israelachvili, J.N. (1997) Direct measurement of the adhesion and friction of smooth C_{60} surfaces. *Chem. Mater.*, **9** (5), 1166–1171.
109. Urbakh, M., Klafter, J., Gourdon, D., and Israelachvili, J. (2004) The nonlinear nature of friction. *Nature*, **430**, 525–528.
110. Braun, O.M., Palily, M., and Consta, S. (2004) Ordering of a thin lubricant film due to sliding. *Phys. Rev. Lett.*, **92** (25), 256103-1–256103-4.
111. Park, H., Park, J., Lim, A.K.L., Anderson, E.H., Alivisatos, A.P., and McEuen, P.L. (2000) Nanomechanical oscillations in a single-C_{60} transistor. *Nature*, **407**, 57–60.
112. Cook, D.J., Chen, J.X., Morlino, E.A., and Hochstrasser, R.M. (1999) Terahertz-field-induced second-harmonic generation measurements of liquid dynamics. *Chem. Phys. Lett.*, **309** (3–4), 221–228.
113. Marchenko, A. and Cousty, J. (2002) C_{60} self-organization at the interface between a liquid C_{60} solution and a Au(111) surface. *Surf. Sci.*, **513** (1), 233–237.
114. Yoshimoto, S., Narita, R., Tsutsumi, E., Matsumoto, M., Itaya, K., Ito, O., Fujiwara, K., Murata, Y., and Komatsu, K. (2002) Adlayers of fullerene monomer and $[2 + 2]$-type dimer on Au(111) in aqueous solution studied by *in situ* STM. *Langmuir*, **18** (22), 8518–8522.
115. Wetterer, S.M., Lavrich, D.J., Cummings, T., Bernasek, S.L., and Scoles, G. (1998) Energetics and kinetics of the physisorption of hydrocarbons on Au (111). *J. Phys. Chem. B*, **102** (46), 9266–9275.
116. Whelan, C.M., Gatti, F., Leigh, D.A., Rapino, S., Zerbetto, F., and Rudolf, P. (2006) Adsorption of fumaramide [2] rotaxane and its components on a solid substrate: a coverage-dependent study. *J. Phys. Chem. B*, **110** (34), 17076–17081.
117. Özkar, S. and Finke, R.G. (2002) Nanocluster formation and stabilization fundamental studies: ranking commonly employed anionic stabilizers via the development, then application, of five comparative. *J. Am. Chem. Soc.*, **124** (20), 5796–5810.
118. Glusker, J.P. (1980) Citrate conformation and chelation: enzymic implications. *Acc. Chem. Res.*, **13** (10), 345–352.
119. Lin, Y., Pan, G.-B., Su, G.J., Fang, X.H., Wan, L.J., and Bai, C.L. (2003) Study of citrate adsorbed on the Au(111) surface by scanning probe microscopy. *Langmuir*, **19** (24), 10000–10003.
120. Floate, S., Hosseini, M., Arshadi, M.R., Ritson, D., Young, K.L., and Nichols, R.J. (2003) An *in-situ* infrared spectroscopic

study of the adsorption of citrate on Au (111) electrodes. *J. Electroanal. Chem.*, **542**, 67–74.

121 Teobaldi, G. and Zerbetto, F. (2007) Adsorption of organic molecules on gold electrodes. *J. Phys. Chem. C*, **111** (37), 13879–13885.

122 Sumi, T. and Uosaki, K. (2004) Electrochemical oxidative formation and reductive desorption of a self-assembled monolayer of decanethiol on a Au(111) surface in KOH ethanol solution. *J. Phys. Chem. B*, **108** (20), 6422–6428.

123 Vericat, C., Andreasen, G., Vela, M.E., Martin, H., and Salvarezza, R.C. (2001) Following transformation in self-assembled alkanethiol monolayers on Au(111) by *in situ* scanning tunneling microscopy. *J. Chem. Phys.*, **115** (14), 6672–6678.

124 Vinokurov, I.A., Morin, M., and Kankare, J. (2000) Mechanism of reductive desorption of self-assembled monolayers on the basis of Avrami theorem and diffusion. *J. Phys. Chem. B*, **104** (24), 5790–5796.

125 Yu, M., Bovet, N., Satterley, C.J., Bengio, S., Lovelock, K.R.J., Milligan, P.K., Jones, R.G., Woodruff, D.P., and Dhanak, V. (2006) True nature of an archetypal self-assembly system: mobile Au–thiolate species on Au(111). *Phys. Rev. Lett.*, **97** (16), 166102–166106.

126 Jadzinsky, P.D., Calero, G., Ackerson, C.J., Bushnell, D.A., and Kornberg, R.D. (2007) Structure of a thiol monolayer-protected gold nanoparticle at 1.1 Å resolution. *Science*, **318** (5849), 430–433.

127 Shankar, S. and Parr, R.G. (1985) Electronegativity and hardness as coordinates in structure stability diagrams. *Proc. Natl. Acad. Sci. USA*, **82** (2), 264–266.

Part Two
Supramolecular Synthetic Chemistry

4
Conjugated Polymer Sensors: Design, Principles, and Biological Applications
Mindy Levine and Timothy M. Swager

4.1
Introduction

4.1.1
Amplification

Conjugated polymers (CPs) are an extremely important class of materials for sensory applications, as they can transform a binding event into an easily detectable electrical or optical readout signal. The widespread interest in CPs is primarily due to their ability to "amplify" molecular level events, that is, their increased sensitivity compared to monomeric analogues [1]. Swager's research group originally elucidated the mechanistic underpinnings of this increased sensitivity [2]. In this work, poly (phenylene ethynylene) (PPE) **1** was synthesized. This polymer contains cyclophane receptors that serve as well-defined binding sites for paraquat (compound **2**), a well-known fluorescence quencher. The polymer demonstrated substantially enhanced fluorescence quenching in the presence of paraquat compared to the quenching displayed by monomeric cyclophane receptor **3**.

Figure 4.1 Demonstration of amplified quenching in a conjugated polymer (reprinted from Ref. [1a]).

This amplification is due to the mobility of the generated exciton (excited state) in the CP to encounter multiple cyclophane receptor sites prior to relaxation, thereby increasing the likelihood of an encounter with a paraquat-containing receptor, leading to fluorescence quenching. In contrast, the quenching of a monomeric receptor by paraquat depends solely on the binding affinity of the paraquat–cyclophane complex (Figure 4.1). This method also allows a direct measurement of the exciton diffusion length.

4.1.2
Role of Dimensionality and Sensory Mechanism

The sensitivity of CP-based sensors depends on a number of factors. In dilute solutions, CPs behave as isolated rigid-rod chromophores, and exciton migration occurs via a one-dimensional random walk. This random walk is relatively inefficient for fluorescence amplification, as the exciton will sample the same receptor sites multiple times. In contrast, aggregated CPs or CPs in thin films can exhibit enhanced rates of energy transfer due to interchain exciton migration. The primary drawback associated with polymer aggregates is their tendency to form excimer-like species with greatly attenuated fluorescence. Nonetheless, sensors have been developed that rely on the aggregation/deaggregation of the CP.

A second factor that influences the sensitivity of CP sensors is whether they operate via a "turn-off" (quenching) or "turn-on" (unquenching) mechanism. The quenching mechanism, which relies on the binding of a small-molecule quencher to turn off the polymer's fluorescence, has historically been the predominant mechanism used in CP sensors. Turn-on fluorescence detection has the advantage of being potentially more sensitive and selective, if the new signal can be generated on a completely dark background.

In this chapter, we review the use of CPs for the detection of biologically relevant analytes. As this is an active field of research, this chapter highlights important contributions, but does not function as a comprehensive review. To the extent that it is relevant, we also discuss the theory and mechanism of energy migration within these polymeric sensors.

4.2 Water Solubility

4.2.1 Conjugated Polyelectrolytes

A primary concern in utilizing CPs as biological sensors is their solubility in aqueous media. Historically, this issue has been solved by using ionic CPs. One of the first examples of conjugated polyelectrolyte sensors was reported by Whitten and coworkers [3]. Avidin, a biotin binding protein, was added to a solution of anionic polymer **4** and biotinylated paraquat **5**, resulting in a partial restoration of the polymer's fluorescence. The claims of greater than 10^6-fold amplification in this system are inappropriate and additional experiments conducted by the group of Bazan revealed major problems with this methodology (see below). However, this report is nonetheless one of the first to propose the use of conjugated polyelectrolytes in biological sensing.

Conjugated polyelectrolytes have also been used as biosensors based on analyte-induced polymer aggregation. For example, cationic polymer **6** [4] aggregates in the presence of DNA, leading to enhanced energy transfer to the benzothiadiazole (BT) moieties [5].

A biosensor based on analyte-induced conformational changes was described by Leclerc and coworkers, who reported the use of cationic polythiophene **7** for DNA detection [6]. In the presence of single-stranded DNA (ssDNA), polymer **7** adopts a planar conformation, resulting in a red shift in its absorption maximum. The addition of a complementary DNA strand results in the formation of a double-stranded DNA (dsDNA) helix, which weakened the planar conformation of the polythiophene, leading to a blue shift in the polymer's absorbance and an increase in its fluorescence intensity.

An interesting backbone architecture for conjugated polyelectrolytes was developed by Bazan and coworkers (compound **8**) [7]. This polymer contains varying percentages of *meta*- and *para*-linkages between the fluorene and phenyl moieties. The authors demonstrated that increasing the proportion of the *meta*-linkages in the polymer backbone resulted in improved energy transfer efficiency to a fluorescein-labeled dsDNA, as well as to ethidium bromide intercalated within the dsDNA.

4.2.2
Nonspecific Binding

Unfortunately, the problem of nonspecific binding severely limits the utility of conjugated polyelectrolytes in real-world applications [8]. Biological samples contain a myriad of charged species. Thus, attempts to utilize a conjugated polyelectrolyte *in vivo* invariably result in electrostatically driven binding of nonspecific analytes that hinders the selectivity and sensitivity of the system.

The literature is replete with cautionary tales of nonspecific binding in conjugated polyelectrolyte sensory schemes. For example, Heeger and coworkers reported the detection of cytochrome *c* using a conjugated polyelectrolyte, supposedly via electron transfer quenching [9]. However, a subsequent investigation [10] found that both the ferric and ferrous forms of the protein efficiently quenched the fluorescence of a similar polymer (**9**). Moreover, researchers found in this investigation that the emission of polymer **9** is quenched more efficiently by polyamidoamine (PAMAM) dendrimers [11] than by cytochrome *c*. They proposed that the likely quenching mechanism is conformational changes in the polymer upon binding to the electrostatically complementary protein or dendrimer.

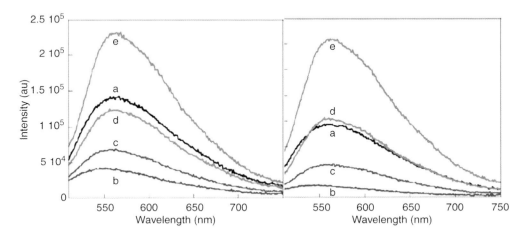

Another cautionary tale can be found in the work of Whitten and coworkers as alluded to earlier, who reported that the addition of avidin to a solution of polymer **4** and biotinylated paraquat **5** resulted in a restoration of the polymer's fluorescence [3]. They attributed this to avidin's binding to the biotinylated quencher, which removes the quencher from the polymer's vicinity.

Bazan and coworkers conducted a thorough reinvestigation of Whitten's work [12], and their key findings are summarized below:

1) The addition of avidin to compound **4** in water led to an about two-fold fluorescence enhancement. This result is likely due to the role of cationic avidin [13] in disrupting the anionic polymer aggregates.
2) Both biotinylated and unfunctionalized paraquat behaved identically, in that the addition of avidin to the polymer/paraquat solution led to a recovery of the polymer's fluorescence (Figure 4.2). This result demonstrated conclusively that specific biotin–avidin binding was not responsible for Whitten's observations.

Figure 4.2 Fluorescence emission (500 nm excitation) of polymer **4** and paraquat **5** with the addition of avidin (left), and polymer **4** and unfunctionalized paraquat **2** with the addition of avidin (right). (a) Polymer **4** (1.0×10^{-5} M); (b) polymer **4** + paraquat (3.2×10^{-7} M); (c) polymer **4** + paraquat + avidin (3.0×10^{-8} M); (d) polymer **4** + paraquat + avidin (8.0×10^{-8} M); and (e) polymer **4** + paraquat + avidin (2.3×10^{-7} M) (reprinted from Ref. [12]).

3) The addition of various anionic and cationic proteins (that do not bind biotin) to an aqueous solution of polymer **4** and paraquat **5** caused a recovery of the polymer's fluorescence.

The problem of nonspecific interactions can be avoided by forming a charge neutral complex (CNC), for example, between anionic conjugated polyelectrolyte **10** and saturated cationic polyelectrolyte **11** [14]. The CNC eliminated the problem of nonspecific binding of cationic antibodies to polymer **10**. Selective antibody detection was possible using this complex; however, the detection limit was modest (300 nM).

4.2.3
Nonionic Water-Soluble Polymers

The problem of nonspecific electrostatic binding can be avoided by using nonionic water-soluble polymers. Kuroda and Swager reported that arborol-type substituents [15] conferred water solubility to polymer **12**, which was synthesized via the Sonogashira coupling of monomers **13** and **14** (Scheme 4.1) [16].

Oligoethylene glycol substituents also confer water solubility to CPs. In addition, their aprotic nature minimizes undesired hydrogen bonding. For example, water-soluble PPE **15** was synthesized [17] via an $A_2 + BB'$ polycondensation (Scheme 4.2) [18].

This polymerization strategy used the aryl bis-iodide **16**, with two equivalent functional groups, as the "A_2" monomer. Commercially available trimethylsilylacetylene **17**, which contains both a protected and unprotected alkyne terminus, functioned as the BB' monomer. The unprotected alkyne coupled with the aryl iodide, followed by *in situ* TMS deprotection, to yield AB'-type monomer **18**. Subsequent polymerization yielded defect-free polymer **15**. This polymer had a high aqueous quantum yield (0.43). Similar poly(*para*-phenylene)s (PPPs) have also been developed [19].

Oligoethylene glycol linkers have also been utilized in water-soluble saccharide-substituted CPs [20], for example, in polythiophenes (**19**) [21], polyfluorenes (**20**) [22], and PPPs (**21**) [23]. The use of an alkyl linker instead of oligoethylene glycol leads to water-insoluble polymers such as polymer **22** [21].

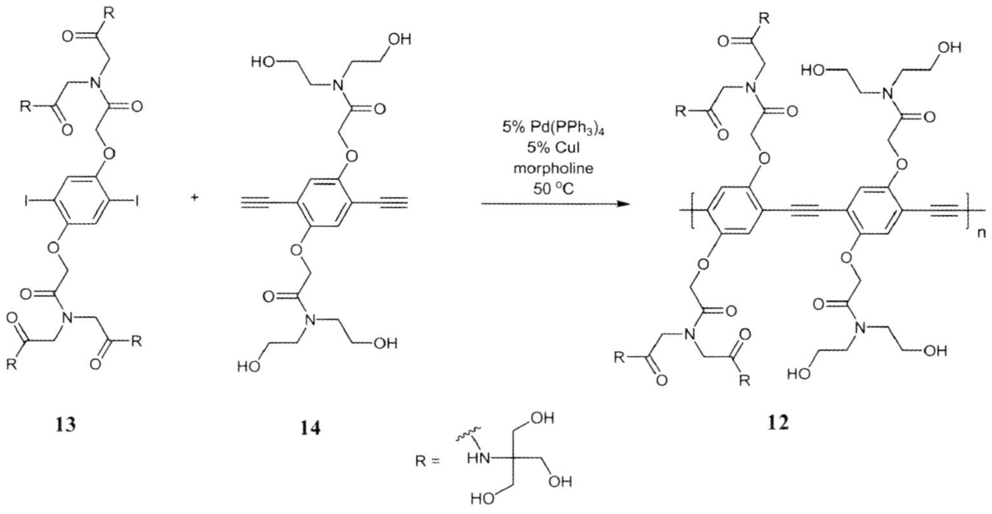

Scheme 4.1 Synthesis of nonionic water-soluble polymer **12**.

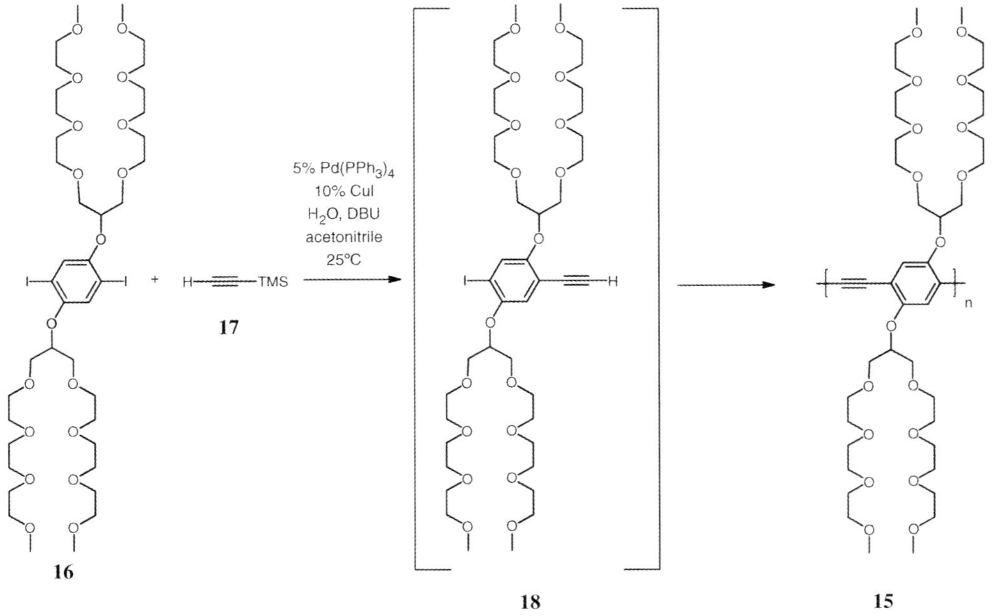

Scheme 4.2 Synthesis of PPE **15**.

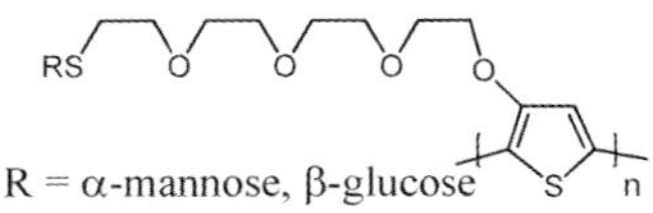

R = α-mannose, β-glucose

19

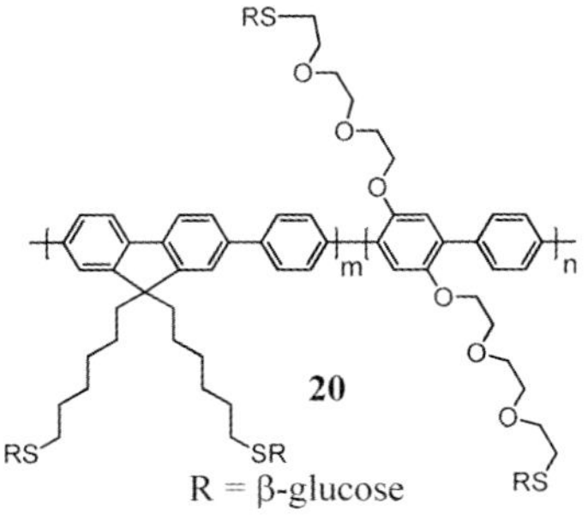

20 R = β-glucose

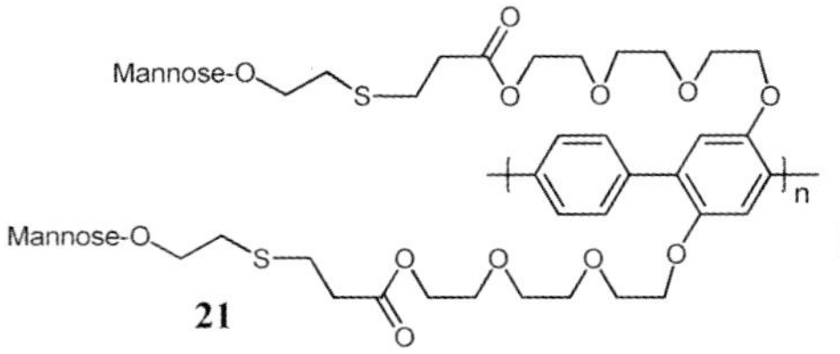

21 Mannose-O-...

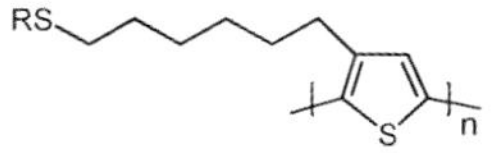

22 R = α-mannose, β-glucose

The mannose-substituted CPs were utilized to detect concanavalin A (ConA), a known mannose binding protein [24]. The glucose-substituted polymers demonstrated no interaction with ConA, because ConA binds only weakly to glucose moieties.

An interesting example of a nonionic water-soluble CP was reported by Ogoshi et al. who synthesized PPE **23** with β-cyclodextrin substituents [25]. The addition of 1-adamantanecarboxylic acid **24** to an aqueous solution of the polymer resulted in the polymer's deaggregation, because the adamantane moiety binds in the cyclodextrin cavity. Interestingly, adamantane-modified paraquat **25** quenched the polymer's fluorescence, whereas unmodified paraquat and adamantane-bearing pyridinium **26** had no effect on fluorescence of polymer **23**.

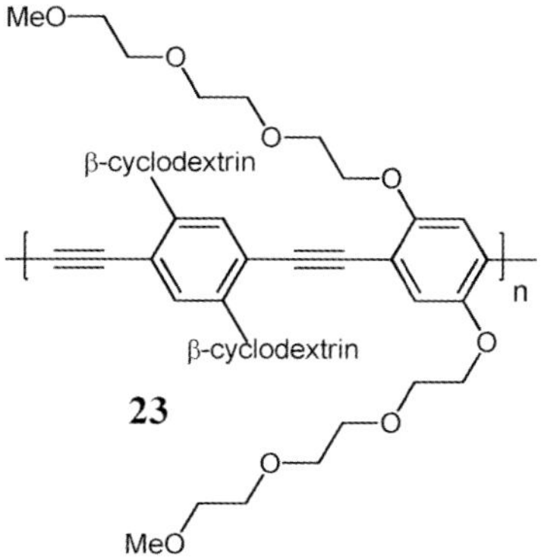

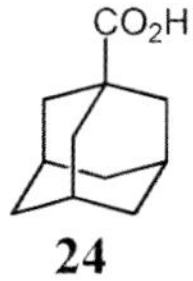

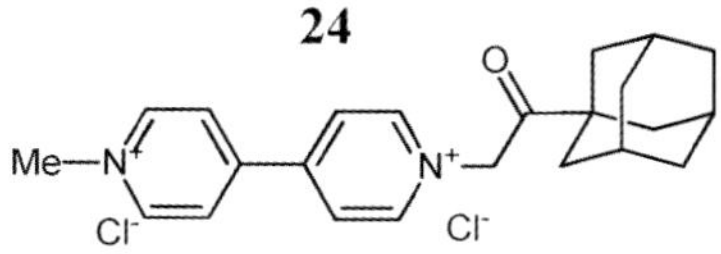

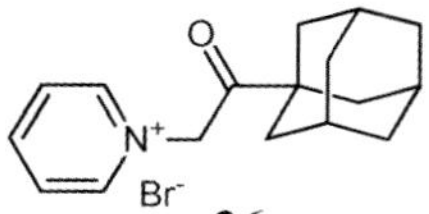

The authors explained that binding of the adamantane-modified paraquat in β-cyclodextrin brought the quencher into proximity with the polymer backbone, leading to fluorescence quenching. Unmodified paraquat does not bind in β-cyclodextrin and therefore did not act as an efficient quencher in this case. The adamantane-bearing pyridinium, which binds in the cyclodextrin cavity, does not quench the fluorescence because the pyridinium cannot act as a suitable electron acceptor.

Water-soluble CPs (both ionic and nonionic) have been utilized extensively for the detection of biologically relevant analytes, including proteins (Section 4.3), DNA (Section 4.4), and bacteria (Section 4.5).

4.3 Protein Detection

4.3.1 Introduction

The pathophysiology of several diseases is caused by malfunctioning enzymes such as proteases, kinases, and phosphatases [26]. Other proteins such as lectins (sugar binding proteins) can be potentially toxic and their detection is, therefore, critical in combating bioterrorism. For example, the highly toxic heterodimer ricin [27] is a lectin [28]. CPs, in particular conjugated polyelectrolytes, have been utilized extensively for protein detection, both *in vivo* and *in vitro* [29].

4.3.2 Protease Detection

Although there are a variety of methods that are currently utilized for protease detection, many of them suffer from drawbacks. The classical method for protease detection is the enzyme-linked immunosorbent assay (ELISA) [30]. However, developing an ELISA assay is very time-consuming, and ELISA assays do not provide quantitative data. Recently, proteolytic beacons have been developed for protease detection [31]. These beacons are very sensitive, however utilizing CPs for protease detection can provide even more sensitive assays.

Small-molecule fluorogenic probes have also been used for protease detection [33]. These probes typically consist of a fluorescent donor and a nonfluorescent acceptor, separated by a peptide tether. In the absence of the protease, the donor-tether-acceptor is non-fluorescent due to internal energy transfer. Cleavage of the peptide tether by the target protease leads to a restoration of the donor's fluorescence.

CP-based protease detection was accomplished using PPE **27** [33], which was conjugated to a peptide that contained a dinitrophenyl moiety as a fluorescence quencher. This peptide is a known substrate for matrix metalloproteinase 13 (MMP-13), which has been implicated in both tumor progression and arthritis [34].

The peptidic linker can also be cleaved by the serine protease trypsin, which hydrolyzes proteins on the carboxyl side of lysine and arginine residues [35]. Cleavage of the peptide by trypsin led to an increase of the polymer's fluorescence by one order of magnitude (Figure 4.3). A small-molecule mimic, compound **28**, demonstrated a

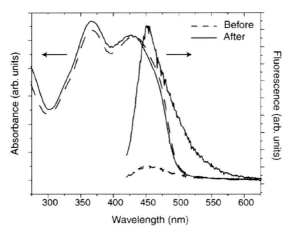

Figure 4.3 Absorbance and fluorescence (405 nm excitation) of polymer **27** (1.1 μM in repeat units) with high peptide loading (1.7 peptides/repeat unit) before (dashed line) and after (solid line) treatment with trypsin (3 μg ml^{-1}) (reprinted from Ref. [33]).

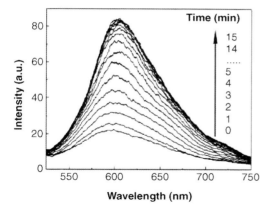

Figure 4.4 Emission spectra of polymer **29**/Cu^{2+}/BSA as a function of trypsin digestion time in PBS buffer solution (5 mM, pH 7.8). Polymer **29** = 2.0 × 10^{-5} M (repeat units); Cu^{2+} = 1.2 × 10^{-5} M; BSA = 0.25 mg ml^{-1}; trypsin = 3.0 μg ml^{-1} (reprinted from Ref. [38]).

smaller increase in fluorescence after trypsin hydrolysis, illustrating the enhanced sensory response of CPs.

Other assays for trypsin activity rely on the hydrolysis of bovine serum albumin (BSA) by trypsin. BSA causes enhanced fluorescence emission of CPs such as polymer **9** [36] due to its surfactant-like effect in enhancing polymer deaggregation [37]. Trypsin-mediated cleavage of BSA caused the emission of polymer **9** to decrease, because the BSA peptide fragments are ineffective in promoting polymer deaggregation.

Another example of using a CP to monitor trypsin-mediated BSA hydrolysis utilizes the zwitterionic polythiophene **29** and Cu^{2+} (Figure 4.4) [38].

Cu^{2+} efficiently quenched the fluorescence of the polythiophene by binding to the amino acid side chains [39]. Trypsin-mediated cleavage of BSA led to the release of short peptide fragments that bind Cu^{2+} efficiently (whereas the intact protein did not bind Cu^{2+}). The polymer's fluorescence was restored to ~70% of its Cu^{2+}-free levels, with some fraction of the Cu^{2+} remaining bound to the polymer and preventing complete fluorescence restoration.

An interesting three-component system was designed for the simultaneous detection of protease and nuclease activity (Figure 4.5) [40]. The system consisted of anionic CP **30**, a cationic fluorescein-labeled peptide, and Texas Red-labeled

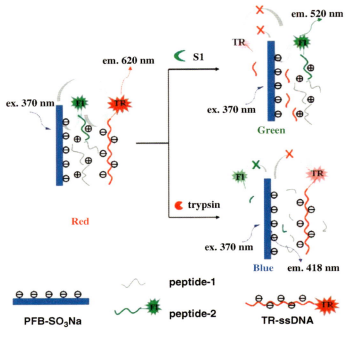

Figure 4.5 A three-component detection scheme for nuclease and protease activity (reprinted from Ref. [40]). PFB-SO$_3$Na = polymer **30**, S1 = nuclease, trypsin = protease, and TR = Texas Red dye.

ssDNA. In the absence of enzymatic activity, excitation of the polymer results in energy transfer to Texas Red via a relay mechanism involving the fluorescein-labeled peptide. In the presence of trypsin, the cationic peptide was hydrolyzed, the complex dissociated, and polymer emission occurred. In the presence of nuclease, the ssDNA was hydrolyzed and energy transfer from the polymer to fluorescein occurred. Thus, this system has three potential "readout" signals: (a) without enzymatic activity, Texas Red emission occurred (620 nm); (b) with trypsin activity, polymer emission occurred (418 nm); and (c) with nuclease activity, fluorescein emission occurred (520 nm).

Finally, trypsin activity can be detected via polymer de-aggregation [41]. For example, cationic hexapeptide arginine-6 caused anionic polysulfonate **31** to aggregate. Excitation of the aggregated polymer's fluorene moieties resulted in energy transfer to the BT moieties, and BT emission at 540 nm was observed. Trypsin-induced cleavage of the hexapeptide led to de-aggregation of the polymer. In its de-aggregated state, excitation of the polymer's fluorene subunits resulted in fluorene emission (and weak BT emission). The same principle was utilized to develop a sensor for adenosine triphosphate (ATP) using polymer **6** with 5% BT subunits.

4.3.3
Kinase and Phosphatase Detection

Kinases [42] and phosphatases catalyze the phosphorylation and dephosphorylation of proteins, respectively [43], and aberrant expression of these enzymes has been linked to various diseases [44]. The detection of these enzymes has been accomplished via CPs. Polyelectrolyte/Ga^{3+} hybrid microspheres, which exploit the affinity of Ga^{3+} for phosphate ions [45], were utilized as sensors for these enzymes [46]. When a rhodamine-labeled peptide was phosphorylated, the binding of phosphate to Ga^{3+} brought the peptide close to the fluorescent microsphere, efficiently quenching the polymer's fluorescence (Figure 4.6), presumably via energy transfer to the rhodamine fluorophore (although the effect on the rhodamine's emission was not reported). When phosphatase catalyzed the dephosphorylation of the peptide, the peptide did not bind to the microsphere, and the polymer's fluorescence was restored.

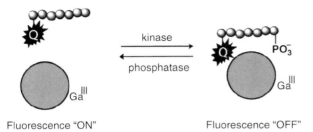

Figure 4.6 Microsphere-based assay for kinases and phosphatases using phosphate–Ga^{3+} interaction (reprinted from Ref. [46]).

4.3.4
Lectin Detection

Sugar-functionalized CPs have been utilized for lectin detection. Lectins are involved in crucial biological events [47], and some lectins such as ricin [28] are highly toxic. Mannose-containing polymer **32** [48] was efficiently quenched in the presence of the mannose-binding lectin concanavalin A (ConA) [24].

32

Similarly, mannose-containing polymer **33**, synthesized by Swager and co-workers, interacted strongly with ConA [49], as did mannose-functionalized PPE **34**, which formed microparticles in solution [50]. The galactose-functionalized analogue **35** demonstrated nonspecific binding to ConA, even though it lacks a substrate for ConA. This binding may be due to van der Waals interactions, hydrogen bonding, or protein adsorption on the polymer bead.

33

34: R = Mannose
35: R = Galactose

4.3.5
Other Protein Detection

Even in the absence of specific protein substrates, CPs interact efficiently with a variety of proteins. In many cases, proteins demonstrate nonspecific quenching of the polymer's fluorescence [9, 29, 51]. Array-based detection with a variety of CPs can circumvent the problem of nonspecific binding. For example, Bunz and coworkers utilized six CPs (**36–41**) to develop a sensor for 17 different protein classes [52]. This array was used to identify 68 unknown proteins with 97% accuracy.

40: n = 12
41: n = 21

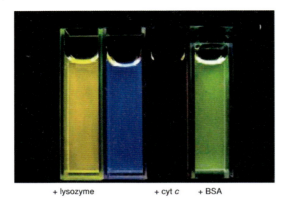

Figure 4.7 Photographs for solutions of polymer **42** (repeat unit = 1.0×10^{-6} M in Na_2CO_3/$NaHCO_3$ buffer (pH = 9)) in the presence of lysozyme, cyt c, and BSA under UV excitation (365 nm) (reprinted from Ref. [53]).

A colorimetric detection scheme was also developed for protein detection [53] based on the aggregation of BT-containing polymer **42** [54].

This polymer displayed unique responses to three different proteins:

1) BSA increased the emission of both the fluorene and the BT segments, consistent with its surfactant-like behavior [37].
2) Metalloporphyrin-containing cytochrome c quenched the polymer's emission (porphyrin emission was not observed).
3) Positively charge lysozyme caused aggregation of the anionic polymer to occur, leading to efficient interchain energy transfer from the fluorene to the BT moieties and enhanced BT emission (Figure 4.7).

4.3.6
Protein Conformation Detection

Polythiophenes can detect changes in protein conformation, which can have marked effects on disease expression and morphology. For example, amyloid plaque formation is associated with Alzheimer's disease [55] and spongiform encephalopathy [56]. Anionic polymer **43** [57] displayed a substantial bathochromic shift in its absorption and emission spectra in the presence of amyloid fibrils of bovine insulin, which cause the polymer to adopt a more planar conformation.

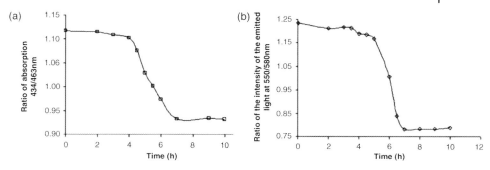

Figure 4.8 Kinetics of amyloid fibril formation of bovine insulin (320 μM) monitored by (a) absorption of polymer **43**; (b) fluorescence of polymer **43** (400 nm excitation). The fibrillation occurred at pH 1.6 and at 65 °C (reprinted from Ref. [57]).

43

The real-time formation of amyloid fibrils can be measured by monitoring the red-shift in either the absorption or the emission spectra (Figure 4.8). These results were identical to those obtained using Congo Red, a known indicator of amyloid fibril formation [58].

4.4
DNA Detection

4.4.1
Polythiophene-Based DNA Detection

CPs that have been utilized for DNA detection include both polythiophenes and polyfluorenes [59]. Both ssDNA and dsDNA induce conformational changes in polythiophenes that alter the polymer's photophysical properties. For example, Leclerc and coworkers described the detection of DNA at zeptomole concentrations [60] using polymer **7** [6]. In addition, polythiophene **29** can detect DNA via hydrogen bonding-induced conformational changes [38a]. ssDNA disrupted the internal hydrogen bonding of the polymer, causing the polymer to adopt an extended conformation with a red-shifted emission (595 nm). The DNA bases then formed hydrogen bonds with the amino acid side chains, causing the formation of defined, emissive polymer aggregates over time (as opposed to nonemissive aggregates discussed in other sections) and a further red-shifted emission (670 nm). Adding the complementary DNA strand to the mixture disrupted the aggregates, causing a blue shift in the polymer's emission (595 nm).

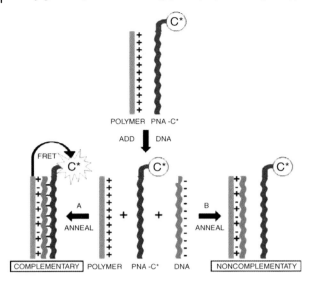

Figure 4.9 Schematic representation for the use of a water-soluble CP with a specific PNA-C* optical reporter probe to detect a complementary ssDNA sequence (option A) resulting in FRET, whereas a noncomplementary DNA strand (option B) will not lead to FRET. C* = fluorescent energy acceptor (reprinted from Ref. [62]).

4.4.2
FRET-Based DNA Detection

DNA detection using polyfluorenes generally occurs via fluorescence resonance energy transfer (FRET). This detection relies on Förster energy transfer [61], which occurs from an excited donor chromophore to an acceptor chromophore through the coupling of two molecular dipoles. In the work pioneered by Bazan and coworkers [62], FRET-based DNA detection was accomplished via a peptide nucleic acid (PNA) probe that was covalently linked to fluorescein. PNA is a DNA mimic in which the negatively charged phosphate linkers have been replaced by neutral amide linkers [63], and it is therefore resistant to nuclease degradation [64]. In addition, PNA/DNA duplexes do not suffer from electrostatic repulsion and are more stable than the analogous all-DNA duplexes [65]. When a complementary DNA strand hybridized to the PNA probe, the negatively charged duplex bound to the cationic CP 44, causing efficient FRET from the polymer to the fluorescein-containing duplex (Figure 4.9). Some nonspecific binding of PNA to the polymer occurred, likely as a result of favorable hydrophobic interactions [66].

44 $R = -(CH_2)_6 NMe_3^+ I^-$

The PNA probe can be replaced with an ssDNA probe [67]. The detection of dsDNA duplexes over the unbound ssDNA probe relies on the higher charge density of the dsDNA complex. In the event, a threefold increase in FRET efficiency was observed in the presence of dsDNA compared to ssDNA.

PNA-based detection has been adopted for use in solid-state microarrays [68] using polymer **45** [69], which has red-shifted absorption and emission spectra relative to polymer **44**. As a result of the red-shifted polymer spectra, the solid-state microarrays could be excited at 488 nm, which is a wavelength utilized in many commercial instruments. In the microarrays, various PNA probes were bound to a solid support and washed with a solution containing ssDNA. Only complementary ssDNA bound to the PNA probe, resulting in a site-isolated negatively charged complex. Cationic polymer **45** bound to the negatively charged complexes, causing a site-isolated fluorescence signal.

$R = -(CH_2)_6NMe_3^+Br^-$

45

Many variables can affect the FRET efficiency between conjugated polyelectrolytes and probe-labeled DNA [70]. Polymers **44** and **46–48** have similar emission spectra, but different frontier orbital energy levels [71]. Only polymer **47** functions as an efficient energy donor with fluorescein-labeled DNA, as fluorine substitution lowers the orbital energy levels by ∼0.2 eV. This example highlights the effect of orbital energy levels on the efficiency of FRET.

44: R' = H
46: R' = OMe
47: R' = F
48: R' = OR

$R = -(CH_2)_6NMe_3^+Br^-$

The efficiency of FRET also depends on the orientation factor (κ) between the polymer donor and dye acceptor. Tetrahedral molecule **49** has a large cross-sectional area with multiple transition dipole moment orientations that facilitate FRET [72]. A complex with compound **49** and fluorescein-labeled DNA showed 5.5-fold fluorescence amplification from excitation of compound **49** compared to direct fluorescein excitation. This amplification is greater than that observed with analogous linear polymer **44**.

49

R = (CH$_2$)$_6$NMe$_3^+$Br$^-$

CP-based DNA detection has been utilized to detect single nucleotide polymorphisms (SNPs) [73], which are small genetic variations that have been implicated in disease morphology and expression [74]. A three-component FRET system composed of a CP, fluorescein-labeled ssDNA, and ethidium bromide (EB) was utilized. Binding of a complementary DNA strand resulted in tandem FRET: the excited polymer transferred its energy to fluorescein, which in turn transferred its energy to EB, causing EB emission at 617 nm. Without the complementary DNA strand, EB did not intercalate and fluorescein emission at 521 nm was observed.

At room temperature, the complementary DNA strand must contain a minimum of eight incorrect base pairs (BPs) to sufficiently destabilize the duplex and prevent

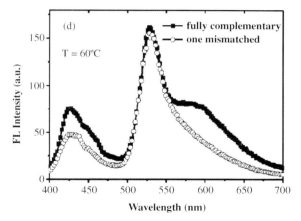

Figure 4.10 Comparison of fluorescence spectra (380 nm excitation) of complex for fully complementary and one base pair-mismatched dsDNA at 60 °C. ssDNA = 5 × 10^{-8} M; EB = 3.67 × 10^{-6} M; conjugated polymer = 1.5 × 10^{-6} M in repeat units. (Reprinted from Ref. [73].)

EB intercalation. Varying the temperature of the assay allowed the detection of single BP mismatches, as a duplex with a single BP mismatch melted 5 °C lower than the fully complementary case (60 versus 65 °C). At 60 °C, therefore, there was no EB emission at 617 nm in the mismatched duplex, whereas the fully complementary case retained some EB emission (Figure 4.10).

ethidium bromide

4.5 Bacteria Detection

4.5.1 Introduction

Methods of rapidly detecting pathogenic bacteria using CPs have been reported. Numerous other methods for bacterial detection have been developed, including the use of fluorescently labeled antibodies [75], DNA probes [76], or bacteriophages [77]. CPs can provide many advantages over other methods for pathogen detection, including the ability to provide multivalent binding [78], which mimics the actual multivalent pathogen–cell interactions.

4.5.2
Mannose-Containing Polymers

Swager, Seeberger, and coworkers utilized post-polymerization reactions to synthesize polymer **33** with mannose side chains [49]. These mannose ligands were utilized to detect *E. coli*, which binds to mannose groups on the surfaces of cells they infect [79]. Incubation of **33** with *E. coli* resulted in the formation of large fluorescent clusters (Figure 4.11), containing between 30 and several thousand bacterial cells. This aggregation was a necessary prerequisite for bacterial detection, as a fluorescein-labeled mannose monomer did not display any observable fluorescence in the presence of *E. coli*.

Other mannose-containing CPs have also been utilized for *E. coli* detection, including PPP **21** [23], polyfluorene **50** [80], and polythiophene **51** [81]. The red-shifted emission that was observed when polymer **51** binds to *E. coli* is due to conformational changes in the polymer, consistent with other polythiophene-based detection schemes [6].

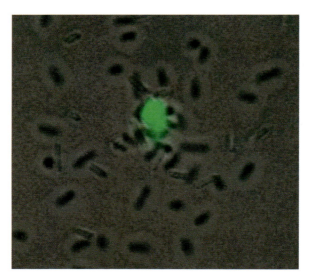

Figure 4.11 Laser scanning confocal microscopy image of a fluorescent bacterial aggregate due to multivalent interactions between the mannose binding bacterial pili and polymer **33** (superimposed fluorescence and transmitted light images) (reprinted from Ref. [49]).

4.5.3
Detection of Bacterial Excretion Products

A different approach to pathogen detection relies on the detection of bacterial excretion products [82]. Many bacterial strains, including *E. coli*, excrete membrane-active components [83], which interfere with the host's membrane function. These membrane-active components can be detected using polydiacetylene (PDA)-containing phospholipid vesicles. The color of PDA (blue or red) varies depending on conformational changes in the polymer backbone [84]. In addition, the red-phase PDA is strongly fluorescent, whereas the blue-phase PDA is not. The use of PDA-containing vesicles in pathogen detection is well precedented [85].

In this case, a bacterial sensor was developed by embedding PDA-phospholipid vesicles in an agar medium, which facilitates bacterial growth. Addition of bacteria to the sensor resulted in agar-promoted bacterial proliferation. The bacteria excreted membrane-active components, which diffused through the semipermeable medium and penetrated the PDA-containing vesicles. PDA's conformation changed, causing a blue-to-red color change accompanied by a turn-on fluorescent signal (Figure 4.12). In principle, any quantity of bacteria can be detected via this methodology, which includes agar-promoted bacterial growth. A potential downside of this sensor is that it cannot distinguish between different types of bacteria; however, there are cases in which the presence of any type of bacteria is harmful, such as in food contamination and poisoning.

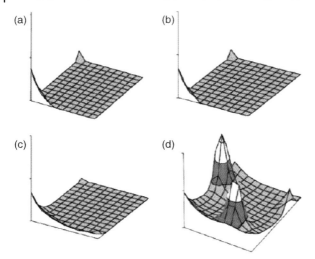

Figure 4.12 Fluorescence induced by bacterial colonies. Three-dimensional fluorescence intensity distributions, recorded using a multiwell fluorescence reader, in conventional bacterial growth agar plates used as a control (a and b) and phospholipid–PDA agar plates (c and d) are shown. The z-axes depict the fluorescence intensity (arbitrary units), while the x- and y-planes correspond to the well surface area. (a and c) Initial measurements (time zero). (b and d) Measurements 6 h after bacterial colonies were streaked on the agar surface (reprinted from Ref. [82]).

4.6
Electron-Deficient Polymers

4.6.1
Fluorinated PPEs

Most CPs are electron-rich, and are thus well suited for the detection of electron-deficient analytes such as trinitrotoluene (TNT) and 2,4-dinitrotoluene (DNT). However, the detection of biologically relevant electron-rich analytes requires electron-deficient CPs, such as PPEs synthesized by Swager and coworkers (i.e., polymer **52**) [86]. This perfluoroalkyl-containing PPE is highly resistant to photobleaching [87], as 85% of the fluorescence intensity of polymer **52** was retained after 30 minutes of irradiation (compared to 50% of nonfluorinated polymer **53**). As expected, polymer **52**, which has an extremely high ionization potential (6.75 eV), displayed excellent quenching efficiency in the presence of indole vapor, a typical electron-rich biological analyte (90% quenching after 5 minutes). Conversely, polymer **53** responded to DNT vapor (Figure 4.13), demonstrating the differential sensing abilities of these two polymers.

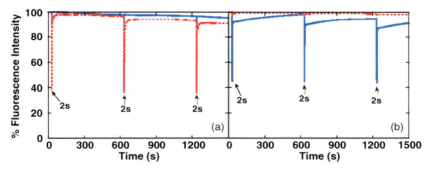

Figure 4.13 Plot of changes in fluorescence intensity (380 nm excitation) of (a) **52** and (b) **53** film exposed to indole (blue line) and DNT (red dotted line) vapors for the indicated time (reprinted from Ref. [86]).

An analogous water-soluble fluorinated PPE **54** was synthesized [88]. This anionic polymer displayed highly efficient quenching in the presence of tryptophan ($K_{SV} = 1400\,M^{-1}$). This efficient quenching occurred despite possible electrostatic repulsion between the anionic polymer and tryptophan (which is anionic under the experimental conditions). Neurotransmitters serotonin and dopamine displayed even larger quenching responses, consistent with their superior electron-donating abilities. Moreover, lysozyme, which contains six tryptophan residues, also efficiently quenched the polymer's fluorescence.

4.7
Aggregation-Based Detection

4.7.1
Introduction

As mentioned in the introductory remarks, CPs have a tendency to aggregate in thin films or in solvents in which they are not fully soluble ("poor" solvents). Aggregation generally results in severely attenuated broad, red-shifted emission from the polymer aggregates [89], although it also can result in enhanced exciton migration to well-defined low-energy "traps" and emission from those traps [90].

4.7.2
Detection via Polymer Quenching

Both fluorescence quenching due to aggregation and fluorescence emission from low-energy traps in aggregates have been applied in sensory schemes. For example, aggregation-induced quenching was utilized to detect K^+ with polymer **55** [89]. As K^+ forms 2:1 sandwich complexes with the 15-crown-5 substituents [91], the presence of K^+ induced polymer aggregation (Figure 4.14), which resulted in a red-shifted absorption spectrum and quenching of the polymer's emission. In contrast, smaller cations Li^+ and Na^+ did not cause polymer aggregation, because they formed 1:1 complexes with the crown ether moieties.

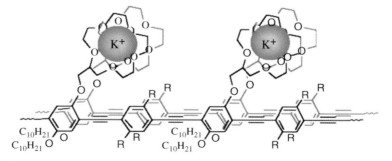

Figure 4.14 K^+-induced aggregation of polymer **55**.

Similarly, Pb^{2+} detection was accomplished via polymer aggregation. There is widespread interest in methods for lead detection, as lead has been implicated in a variety of undesired physiological effects [92]. Specifically, Pb^{2+} caused polymer **36** to aggregate [93], leading to a quenching of the polymer's emission that is >10^3-fold more efficient than the quenching of model compound **56**. Pb^{2+} also induced aggregation of polymer **57** [94] via formation of 2 : 1 sandwich complexes with the 18-crown-6 substituents [95].

Copper [96] and mercury [97] cations have also been detected via polymer aggregation. The sensitive and selective detection of Hg^{2+} is particularly relevant, as mercury is highly toxic both to humans and to the environment [98] and has been implicated in a variety of diseases [99].

In addition to cations, biologically relevant anions can also be detected by aggregation-induced fluorescence quenching. For example, the biologically relevant [100] oxalate anion caused the aggregation and fluorescence quenching of polymer **58** [101]. This polymer can discriminate between oxalate and longer chain dicarboxylates (Figure 4.15) because the short oxalate induced the formation of tight aggregates with complete fluorescence quenching. Longer chain dicarboxylates, by contrast, caused the formation of looser aggregates with minimal fluorescence quenching.

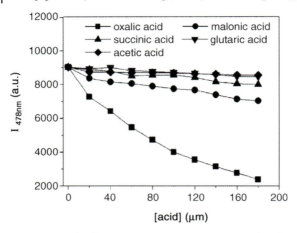

Figure 4.15 The fluorescence intensity (excitation wavelength 431 nm) of polymer **58** (4.0×10^{-6} M) as a function of concentrations of dicarboxylic acids (reprinted from Ref. [101]).

4.7.3
Detection via Low-Energy Emission

Fluorescence emission from low-energy traps in polymer aggregates has also been utilized to detect analytes. Specifically, Bazan and coworkers synthesized BT-containing polymers that aggregate in the presence of DNA or protein, leading to enhanced BT emission from the aggregates [4, 5, 54]. An analogous pH-sensitive BT-containing polymer **59** was synthesized [102]. At pH values above 4, electrostatic repulsion between the negatively charged carboxylates prevented polymer aggregation. At pH values less than 3, complete neutralization of the side chains caused the polymer to aggregate, leading to enhanced energy transfer to the BT moieties (Figure 4.16).

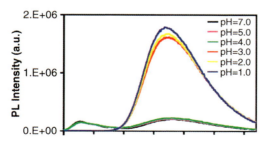

Figure 4.16 Fluorescence spectra (370 nm excitation) of polymer **59** (4.3×10^{-5} M in repeat units) in water as a function of pH (reprinted from Ref. [102]).

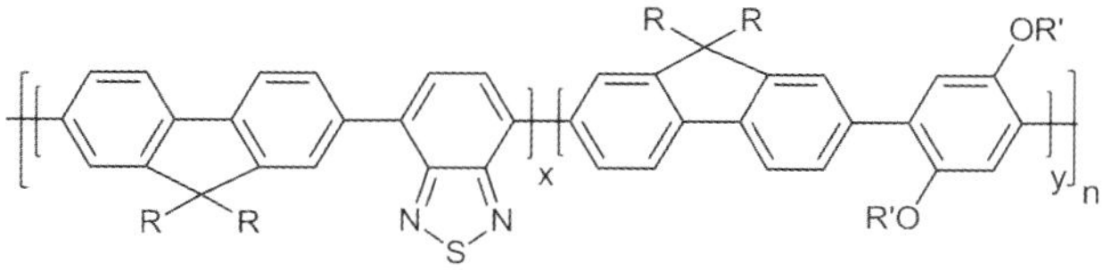

4.7 Aggregation-Based Detection | 109

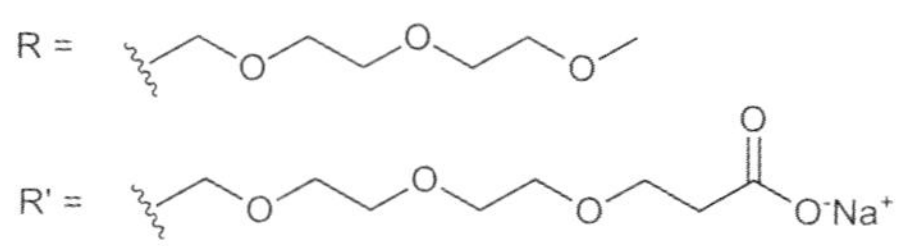

Similarly, polymers **60–63** displayed enhanced low-energy emission in thin films compared to dilute solutions [103]. This effect was initially attributed to the presence of highly emissive excimers. The crystal structure of monomer **64** provided tentative support for the presence of excimers, as the phenylenes of two molecules in the crystal structure were sufficiently close to allow electronic coupling (Figure 4.17) [104].

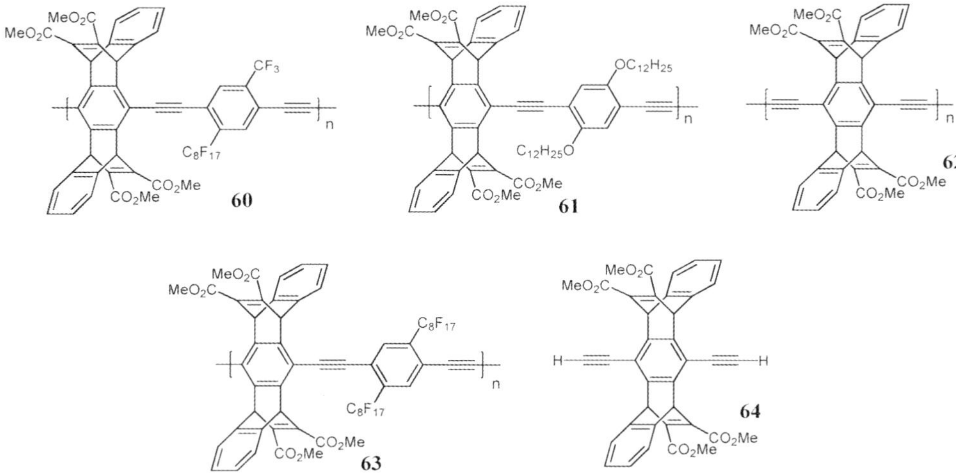

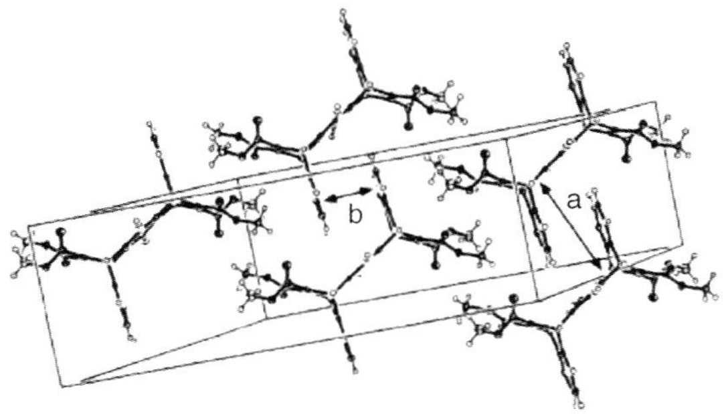

Figure 4.17 Perspective view of crystal packing of monomer **64** (reprinted from Ref. [103]).

Further investigation [90] indicated that the low-energy emission occurred from anthrynyl defects, which arise from a retro-Diels–Alder reaction of the [2.2.2] bicyclic ring system (promoted by irradiation of the thin films). In this case, the fluorescence spectra of irradiated PPEs were compared to those of PPE **65** that was intentionally doped with anthrynyl defects. The similar emission spectra (Figure 4.18) indicated that the green emission in thin films is likely due to aggregation-induced energy transfer to anthrynyl defects. The effects of incorporating anthracene into a CP backbone had been investigated previously [105].

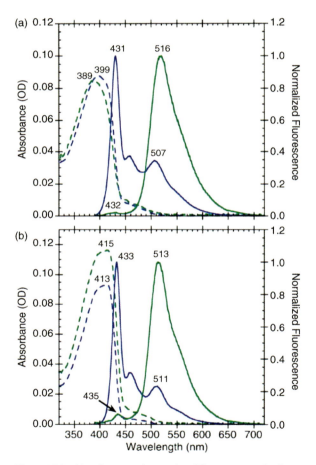

Figure 4.18 Absorbance and normalized fluorescence of polymer **65** (a) and thermally degraded polymer **60** (b) in chloroform solution (blue) and as spin-cast films (green). Fluorescence spectra were obtained using an excitation wavelength of 375 nm (reprinted from Ref. [90]).

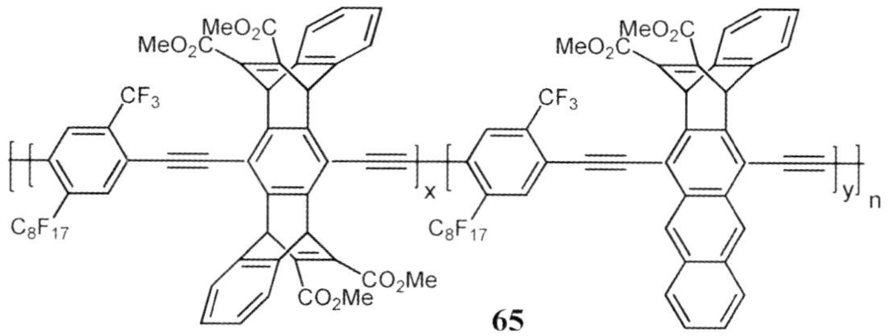

[Structure of polymer 65]

Anionic polymer **66**, obtained from the hydrolysis of polymer **65**, demonstrated enhanced anthracene emission upon aggregation [106].

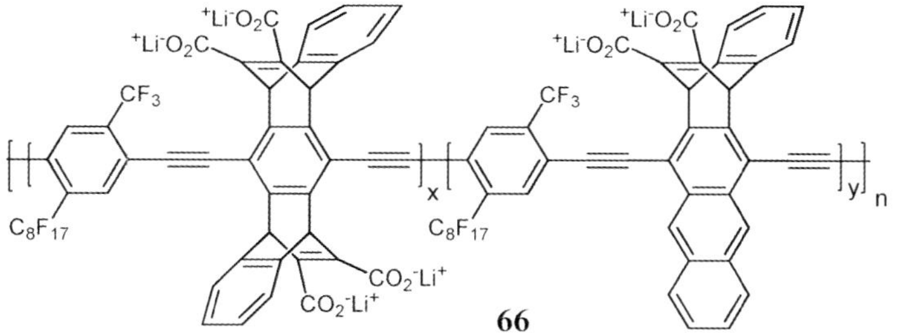

[Structure of polymer 66]

The aggregation of polymer **66** was used for the detection of multicationic polyamines spermine and spermidine. These polyamines play a critical role in cellular growth and differentiation [107], and aberrant polyamine expression may be correlated with a variety of cancers [108]. Because current methods for detection of these polyamines suffer from a variety of drawbacks [109], CP-based detection of these polyamines fills an unmet need.

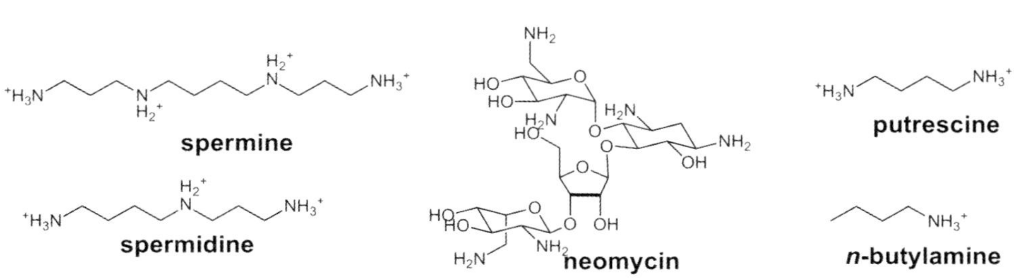

[Structures of spermine, spermidine, neomycin, putrescine, n-butylamine]

Polymer **66** could detect as little as 0.69 μM spermine via aggregation-enhanced anthracene emission. Spermidine, with one fewer positive charge, had a somewhat higher detection limit (1.6 μM). Neither the presence of putrescine (+2 charge) nor n-butylamine (+1 charge) resulted in aggregation-enhanced anthracene emission. Finally, this system proved suitable for the detection of the antibiotic neomycin, which contains six primary amine groups.

4.8
Temperature-Responsive Fluorescent Polymers

4.8.1
Introduction

As discussed in previous sections, exciton migration is enhanced in polymer aggregates due to the exciton's ability to migrate in three dimensions rather than along a single polymer chain. In addition to the previously discussed methods for inducing aggregation, CP aggregation can also occur via the incorporation of CPs into temperature-sensitive poly(N-isopropylacrylamide) (polyNIPA, **67**) copolymers [110]. PolyNIPA undergoes a phase transition at 32 °C, termed its lower critical solution temperature (LCST) [111]. Below the LCST, polyNIPA exists in a chain-extended, highly hydrated conformation. Above the LCST, the polymer collapses into a hydrophobic, globular-like structure [112], and thus CP-polyNIPA copolymers should demonstrate enhanced exciton migration in this temperature range.

67

4.8.2
Examples

As an example, thermally induced precipitation in fluorescent polyNIPA copolymers was utilized to achieve enhanced FRET from polymer **68** to rhodamine-containing polymer **69** [113]. A co-precipitate of **68** and **69** demonstrated energy transfer from donor polymer **68** to the acceptor polymer **69**, resulting in a five-fold increase in fluorescence from exciting **68** compared to exciting rhodamine directly (Figure 4.19). Although this system demonstrated only modest fluorescence amplification, it nonetheless showed that thermally induced precipitation is a viable method for enhancing the efficiency of FRET.

4.8 Temperature-Responsive Fluorescent Polymers

68

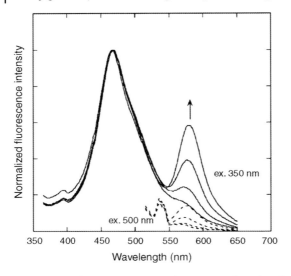

Figure 4.19 Fluorescence spectra of a mixture of polymers **68** and **69** on filter paper. Polymer **68** = 0.12 mg ml^{-1}; polymer **69** = 0.05, 0.025, 0.0125, 0.00 625 mg ml^{-1} in solutions. Dashed line represents fluorescence spectra from directly exciting the rhodamine at 500 nm; solid line is the fluorescence spectra from exciting polymer **68** at 350 nm (which then transfers its energy to the rhodamine fluorophore) (reprinted from Ref. [113].)

Similarly, block copolymer **70** formed aggregates above the LCST that displayed an enhanced quenching response to sodium 3-(4-nitrophenyl)propane-1-sulfonate [114].

This improved sensory response is due to the ability of a single quencher to quench multiple polymer chains in the aggregated state. Analogous polyfluorene-polyNIPA copolymers have also been synthesized (**71**, **72**) [115].

4.9
Nonhomogeneous Detection Schemes

4.9.1
Introduction

Particle-based platforms for biological detection have numerous advantages compared to homogeneous analogues. First, CPs adsorbed on particles do not need to be fully water soluble in order to be utilized in aqueous media. Second, adsorption of a CP on a solid support leads to enhanced exciton migration. Third, the resulting fluorescent particles can be injected in or absorbed by living systems, providing *in vivo* fluorescent probes whose properties equal or surpass those of conventional dyes [116] and quantum dots [117].

4.9.2
CPs Adsorbed on Surfaces

CP particles can be divided into two classes: (i) particles formed by adsorbing the polymer on a preexisting (inorganic) particle surface, and (ii) particles formed by collapse of the polymer chains into spherical aggregates. In the first class of particles, polymers **73–75** were adsorbed on europium-functionalized silica microspheres [118]. Polymer **73** had been previously shown to display a sensitive quenching response when coated on microspheres [119], and polymers **74** and **75** were utilized to study the ability of cyclophane receptors to bind a quencher while adsorbed on a solid support.

The polymer-coated microspheres were efficiently quenched by both paraquat **2** and naphthyl-functionalized viologen **76**. Moreover, normalization of the spectra to the unperturbed Eu^{3+} fluorescence signal allowed for quantitative determination of fluorescence quenching. Interestingly, whereas cyclophane-containing **74** displayed better sensitivity to quenchers in solution studies compared to polymer **75**, the quenching responses of both polymers on the microspheres were equivalent. This disparity may be due to the geometric constraints of the cyclophane in the solid state that prevent it from forming the desired sandwich complex with the quencher.

Whitten and coworkers demonstrated that a lipid bilayer coating protected CP microspheres from analyte-induced quenching [120]. In the event, polymer **77** was adsorbed on borosilicate particles and coated with a lipid bilayer. Minute amounts of melittin, a lipase that disrupts lipid bilayers [121], destroyed the protective bilayer and rendered polymer **77** susceptible to quencher **78**. Detergent Triton-X also caused polymer quenching via membrane rupture. This melittin sensor is substantially more sensitive than previously developed systems for melittin detection [122].

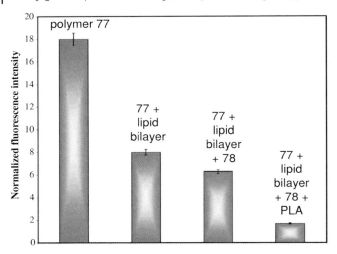

Figure 4.20 Fluorescence "turn off" of physisorbed **77** particles coated with lipid bilayer (1×10^5 lipobeads in 200 μl PBS). Lipobeads were examined in the presence and absence of PLA after the addition of compound **78** (25 μM) (reprinted from Ref. [123].)

Similar lipid-coated polymer beads were utilized to detect phospholipase-A2 (PLA) activity [123], which is an enzyme that is a potential biomarker for atherosclerosis [124]. PLA caused degradation of the lipid bilayer surrounding polymers **79** and **80**, leading to quenching of the polymers by compound **78** (Figure 4.20).

4.9.3
CP Particles from Collapsed Polymer Chains

CP particles have also been fabricated by the collapse of polymer chains into spherical aggregates [125]. This collapse is due to the sudden decrease in solubility that occurs from adding a solution of the polymer in a "good" solvent (one in which it is highly soluble) to water [126]. Nanoparticles fabricated from polymer **81** [127] demonstrated excellent *in vivo* photostability (Figure 4.21) and no obvious cytotoxicity (Figure 4.22).

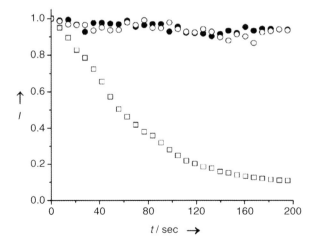

Similarly, McNeill and coworkers synthesized a series of CP particles from polymers **82–86** [128]. AFM and TEM studies confirmed the spherical shape and narrow size distribution of the particles. These particles also demonstrated excellent photostability and no obvious cytotoxicity, as well as high two-photon absorption

Figure 4.21 Photostability comparison of particles fabricated from polymer **81** (circles) compared to fluorescein dye (squares) (reprinted from Ref. [127]). Open circles indicate that the particles were excited at 80% of 488 nm laser power; closed circles indicate particle excitation at 4% laser power; and open squares indicate fluorescein excitation at 4% laser power.

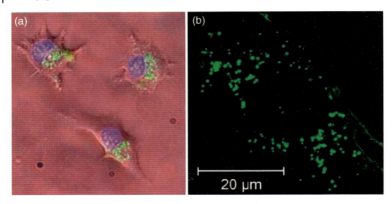

Figure 4.22 Fluorescence image of live (a) and fixed (b) cells that were incubated with particles of polymer **81** (reprinted from Ref. [127]).

(2PA) cross sections [129]. Two-photon spectroscopy allows for the use of near-infrared excitation wavelengths that limit the damage to biological tissue [130]. The reported 2PA cross sections of the particles are higher than those of conventional fluorescent dyes [131] and quantum dots [132]; however, these values were not normalized to monomer molecular weight, thus artificially inflating the values. The ability to utilize multiphoton spectroscopy, combined with the particles' well-defined size and facile cellular uptake, renders these particles highly promising candidates for biological detection.

McNeill and coworkers also studied intra-particle energy transfer in blended nanoparticles. Polyfluorene **87** was utilized as the energy donor with a variety of energy acceptors, including both polymers (**86**, **88**, **89**) [133] and small-molecule fluorophores (**90–93**) [134]. These blended particles demonstrated facile intra-particle energy transfer, measured by the quenching of the donor's fluorescence. In particular, blended particles with 6 wt % dopant polymer or 2 wt % small-molecule fluorophore showed near-complete quenching of the donor's fluorescence (Figure 4.23).

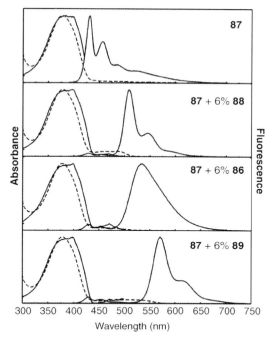

Figure 4.23 Normalized absorption (dashed line) and fluorescence excitation (375 nm excitation) and emission spectra (solid line) of pure **87** and polymer blend nanoparticles (reprinted from Ref. [133]).

4.10
Mechanism of Energy Transfer

4.10.1
Theory

The mechanism of energy transfer in the preceding examples can occur either via dipole–dipole coupling (Förster energy transfer) [61] or via electronic exchange (Dexter energy transfer) [135]. In the former Förster mechanism, commonly referred to as fluorescence resonance energy transfer, energy transfer is described by a point-dipole (Coulombic) interaction between the excited state of a donor and the ground state of an acceptor. The classical Förster energy transfer rate (k_{ET}) and related terms are described by Equations 4.1–4.3:

$$k_{ET} = \frac{1}{\tau_D} \cdot \left(\frac{R_0}{R}\right)^6 \tag{4.1}$$

$$R_0^6 = \frac{9000 \ln 10}{128 \pi^5 N_A} \frac{\kappa^2 Q_D}{n^4} J \tag{4.2}$$

$$J = \int_0^\infty F_D(\lambda) \varepsilon_A(\lambda) \lambda^4 d\lambda \tag{4.3}$$

where τ_D is the lifetime of the donor; R_0 is the Förster radius, defined as the distance at which the transfer efficiency is 50%, typically in the range of 20–90 Å; R is the distance between a donor and an acceptor; κ^2 accounts for the relative orientation of the donor and acceptor transition dipole moments, with a value of 2/3 which represents a random donor–acceptor orientation; Q_D is the quantum yield of the donor without an acceptor; n is the refractive index of the medium; N_A is Avogadro's number; $F_D(\lambda)$ is the normalized fluorescence intensity of the donor; $\varepsilon_A(\lambda)$ is the molar extinction coefficient of the acceptor; and J is the spectral overlap integral, calculated over the whole spectrum with respect to wavelength λ. These equations reveal that the rate of the energy transfer is proportional to the spectral overlap and inversely proportional to R^6.

In contrast, Dexter energy transfer occurs from direct electron exchange between the excited state of a donor and the ground state of an acceptor through orbital overlap. The Dexter energy transfer rate (k_{ET}) is described by Equation 4.4:

$$k_{ET} = KJ' \exp\left(-\frac{2R}{L}\right) \tag{4.4}$$

where K is related to the specific orbital interaction, J' is the spectral overlap integral normalized by the extinction coefficient of the acceptor, R is the distance between a donor and an acceptor, and L is their van der Waals radii. Unlike Förster energy transfer, where the spectral overlap integral (J) has a great influence on rate, the rate of Dexter energy transfer is comparatively unaffected by the normalized spectral overlap integral (J'). The rate of electron exchange that requires strong orbital integration decreases exponentially in terms of R. It becomes insignificant beyond 5–10 Å, a distance past which strong orbital interactions cannot occur.

4.10.2
Modifications to Förster Theory

In CPs, where the size of the excited state can be of similar dimensions to the distance between the donor and acceptor, modifications to the Förster theory have been proposed [136]. These modifications view CPs as multichromophoric systems, composed of conjugated oligomers of varying lengths linked by "kinks" or "defects" in the polymer chain [137]. This multichromophoric view of CPs is supported by the similar photophysical properties of a CP and its corresponding oligomers.

Förster theory assumes that the excitonic coupling is accurately represented by the dipole–dipole interaction between the transition dipoles, each located at the center of the respective molecule, commonly referred to as a "point-dipole" approximation [138]. This assumption is invalid for multichromophoric CPs, and calculations based on this approximation have been shown to differ greatly from experimental results [139]. Improved Förster models have been devised based on a "line-dipole" approximation [140], where fractional transition dipoles along the polymer chain produce results that more closely match experimental data.

Although FRET is the most commonly used theory in energy transfer, the Dexter energy transfer mechanism has garnered recent interest in the field of CPs [141]. In polymer–fluorophore blend films, where the distance between the donor and the acceptor is shorter than the size of the molecules themselves, orbital coupling is feasible [142]. It is possible that both Förster- and Dexter-like energy transfer mechanisms may exist in a given system, even though either one can predominate for a specific case.

4.10.3
Evidence for Dexter-Type Energy Migration

Swager and coworkers have demonstrated evidence for Dexter-type energy migration in systems with triphenylene-containing PPEs (**94–97**) [143]. These polymers showed approximately 30% longer fluorescence lifetimes compared to polymers that lacked the triphenylene moieties, due to the symmetry-forbidden S_0–S_1 transition of triphenylene chromophores [144]. The energy migration in these polymers was measured by the depolarization of the polymer (Figure 4.24) [145], which occurs predominantly via energy migration (as the polymers are rotationally static over their fluorescence lifetime). If energy migration occurred predominantly through a Förster-type mechanism, the depolarization of triphenylene-containing polymers should be slower than the analogous triphenylene-free polymers, because Förster theory predicts that decreasing the rate of energy migration correlates with an increase in the fluorescence lifetime. In fact, the opposite result was obtained, with triphenylene polymers undergoing faster depolarization under all excitation wavelengths measured.

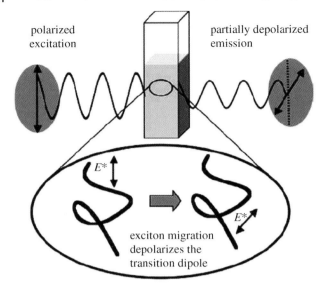

Figure 4.24 A simplified depiction of depolarization due to energy migration in CPs. The excitation beam is vertically polarized and therefore only vertical transition dipoles are initially excited on the CP chain. Vertically polarized excitons on the polymer chain migrate. As they move over a disordered polymer chain, they lose their initial polarization. The emission of polymer is depolarized relative to the excitation beam. This amount of measured depolarization indicates the extent of energy migration in the conjugated polymer (reprinted from Ref. [145]).

Faster depolarization in longer lifetime polymers was also observed for a series of chrysene-based PPEs (**98–100**) [145]. These results provide evidence that Dexter energy migration may be operative in intrachain energy migration.

In addition, Dexter energy migration may also occur in intermolecular energy transfer. In particular, biotin-labeled polymer **101** was incubated with a series of streptavidin-functionalized dyes [146]. The dye with the least spectral overlap between the donor emission and acceptor absorption spectra, Texas Red, demonstrated the greatest degree of fluorescence amplification from polymer excitation compared to direct dye excitation. This result is opposite what would be expected based on Förster theory, in which the efficiency of energy transfer depends strongly on the spectral overlap integral J. In this system, the streptavidin–biotin interaction brought Texas Red in proximity to the polymer backbone, at which point the planar conformation and hydrophobic character of the dye determined its interaction with the polymer. This subtle interplay between the dye's properties and the efficiency of energy transfer points to the likelihood of energy transfer based on orbital overlap mechanisms.

Further, PPE **53** was utilized as an energy donor in thin film blends with a series of fluorophores, compounds **93** and **102–113** as energy acceptors [147]. Squaraine [148] and terrylene [149] fluorophores were found to be excellent energy acceptors, with up to 99-fold amplification of fluorescence observed from exciting the polymer compared to exciting the fluorophore directly. This highly efficient energy transfer occurred despite the minimal spectral overlap observed between the emission spectrum of the donor and the absorption spectra of the near-infrared acceptors.

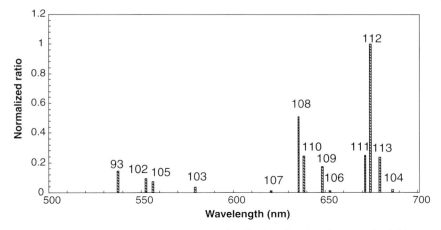

Figure 4.25 Fluorophore emission intensity in thin film blends with polymer **53** divided by spectral overlap integral *J* and plotted versus the absorption maxima of the fluorophores in PMMA thin films.

To probe the mechanism of energy transfer in these thin films, the spectral overlap integral *J* was calculated for each donor–acceptor pair. The ratio of fluorophore emission intensity divided by its spectral overlap integral was calculated for all fluorophores and is summarized in Figure 4.25.

Assuming that Förster-type energy transfer is occurring, this ratio should be constant because energy transfer is proportional to the spectral overlap integral. Of course, however, the intrinsic properties of fluorophores such as quantum yields, lifetimes, aggregation properties, orientations and strengths of transition dipole moment, or subtle compositional variations may induce discrepancies. Nonetheless, the significantly different ratios observed suggest that the observed energy transfer and the polymer–fluorophore interactions fall outside the scope of what can be accounted for with Förster energy transfer alone.

These results offer substantial possibilities for designing turn-on fluorescent sensors. In such sensory schemes, based on Dexter energy transfer, the polymer emission does not overlap the acceptor's emission. Moreover, both squaraines and terrylenes fluoresce in the near-infrared region, an optimal spectral area for biological imaging. Thus, the new emission will occur on a completely dark background (free of both polymer emission and interfering biological analytes), leading to even greater sensitivity in turn-on sensors.

4.11
Conclusions and Future Directions

From this review, we hope that the reader has an appreciation for the state of the field of conjugated polymer sensors and the physical principles that govern their responses. This area of research remains rich with opportunities for molecular design, novel transduction, and applications. Although many applications have been investigated, few meet the robustness and sensitivities needed to compete against conventional technologies. However, given that it has been less than 15 years since the initial demonstrations of fluorescence-based conjugated polymer sensors and their amplifying abilities [2], the field is obviously still in its early stages of development. The ability to create systems with unique functionality and spectral properties holds great promise for chemical sensing, inexpensive assays, and *in vivo* imaging. We hope that this chapter serves as a primer to encourage new researchers to join in this exciting area of research.

References

1. (a) Thomas, S.W., Joly, G.D., and Swager, T.M. (2007) *Chem. Rev.*, **107**, 1339; (b) McQuade, D.T., Pullen, A.E., and Swager, T.M. (2000) *Chem. Rev.*, **100**, 2537.
2. (a) Zhou, Q. and Swager, T.M. (1995) *J. Am. Chem. Soc.*, **117**, 7017; (b) Zhou, Q. and Swager, T.M. (1995) *J. Am. Chem. Soc.*, **117**, 12593.
3. Chen, L., McBranch, D.W., Wang, H.-L., Helgeson, R., Wudl, F., and Whitten, D.G. (1999) *Proc. Natl. Acad. Sci. USA*, **96**, 12287.
4. Liu, B. and Bazan, G.C. (2004) *J. Am. Chem. Soc.*, **126**, 1942.
5. Hong, J.W., Hemme, W.L., Keller, G.E., Rinke, M.T., and Bazan, G.C. (2006) *Adv. Mater.*, **18**, 878.
6. (a) Aberem, M.B., Najari, A., Ho, H.-A., Gravel, J.-F., Nobert, P., Boudreau, D., and Leclerc, M. (2006) *Adv. Mater.*, **18**, 2703; (b) Ho, H.-A., Boissinot, M., Bergeron, M.G., Corbeil, G., Dore, K., Boudreau, D., and Leclerc, M. (2002) *Angew. Chem., Int. Ed.*, **41**, 1548; (c) Leclerc, M. (1999) *Adv. Mater.*, **11**, 1491.
7. Xue, C., Liu, B., Wang, S., Bazan, G.C., and Mikhailovsky, A. (2003) *J. Am. Chem. Soc.*, **125**, 13306.
8. Heeger, P.S. and Heeger, A.J. (1999) *Proc. Natl. Acad. Sci. USA*, **96**, 12219.
9. Fan, C., Plaxco, K.W., and Heeger, A.J. (2002) *J. Am. Chem. Soc.*, **124**, 5642.
10. Liu, M., Kaur, P., Waldeck, D.H., Xue, C., and Liu, H. (2005) *Langmuir*, **21**, 1687.
11. Tomalia, D.A., Baker, H., Dewald, J., Hall, M., Kallos, G., Martin, S., Roeck, J., Ryder, J., and Smith, P. (1986) *Macromolecules*, **19**, 2466.
12. Dwight, S.J., Gaylord, B.S., Hong, J.W., and Bazan, G.C. (2004) *J. Am. Chem. Soc.*, **126**, 16850.
13. Woolley, D.W. and Longsworth, L.G. (1942) *J. Biol. Chem.*, **142**, 285.
14. Wang, D., Gong, X., Heeger, P.S., Rininsland, F., Bazan, G.C., and Heeger, A.J. (2002) *Proc. Natl. Acad. Sci. USA*, **99**, 49.
15. Newkome, G.R., Lin, X., Yaxiong, C., and Escamilla, G.H. (1993) *J. Org. Chem.*, **58**, 3123.
16. Kuroda, K. and Swager, T.M. (2003) *Chem. Commun.*, 26.
17. Khan, A., Muller, S., and Hecht, S. (2005) *Chem. Commun.*, 584.
18. Khan, A. and Hecht, S. (2004) *Chem. Commun.*, 300.
19. Lauter, U., Meyer, W.H., Enkelmann, V., and Wegner, G. (1998) *Macromol. Chem. Phys.*, **199**, 2129.
20. Kim, I.-B., Erdogan, B., Wilson, J.N., and Bunz, U.H.F. (2004) *Chem. Eur. J.*, **10**, 6247.
21. Xue, C., Luo, F.-T., and Liu, H. (2007) *Macromolecules*, **40**, 6863.
22. Xue, C., Donuru, V.R.R., and Liu, H. (2006) *Macromolecules*, **39**, 5747.
23. Xue, C., Jog, S.P., Murthy, P., and Liu, H. (2006) *Biomacromolecules*, **7**, 2470.
24. Simanek, E.E., McGarvey, G.J., Jablonowski, J.A., and Wong, C.-H. (1998) *Chem. Rev.*, **98**, 833.

25 Ogoshi, T., Takashima, Y., Yamaguchi, H., and Harada, A. (2006) *Chem. Commun.*, 3702.
26 (a) Roose, J.P. and Van Noorden, C.J.F. (1995) *J. Lab. Clin. Med.*, **125**, 433; (b) Hardy, J. and Selkoe, D.J. (2002) *Science*, **297**, 353.
27 Wedin, G.P., Neal, J.S., Everson, G.W., and Krenzelok, E.P. (1986) *Am. J. Emerg. Med.*, **4**, 259.
28 Wales, R., Richardson, P.T., Roberts, L.M., Woodland, H.R., and Lord, J.M. (1991) *J. Biol. Chem.*, **266**, 19172.
29 Ambade, A.V., Sandanaraj, B.S., Klaikherd, A., and Thayumanavan, S. (2007) *Polym. Int.*, **56**, 474.
30 Zangar, R.C., Daly, D.S., and White, A.M. (2006) *Expert Rev. Proteomics*, **3**, 37.
31 (a) Liu, G., Wang, J., Wunschel, D.S., and Lin, Y. (2006) *J. Am. Chem. Soc.*, **128**, 12382; (b) McIntyre, J.O., Fingleton, B., Wells, K.S., Piston, D.W., Lynch, C.C., Gautam, S., and Matrisian, L.M. (2004) *Biochem. J.*, **377**, 617; (c) Thurley, S., Roglin, L., and Seitz, O. (2007) *J. Am. Chem. Soc.*, **129**, 12693.
32 (a) Yan, Z.-H., Ren, K.-J., Wang, Y., Chen, S., Brock, T.A., and Rege, A.A. (2003) *Anal. Biochem.*, **312**, 141; (b) Rehault, S., Brillard-Bourdet, M., Bourgeois, L., Frenette, G., Juliano, L., Gauthier, F., and Moreau, T. (2002) *Biochim. Biophys. Acta*, **1596**, 55; (c) Muller, J.C.D., Ottl, J., and Moroder, L. (2000) *Biochemistry*, **39**, 5111; (d) Yang, C.F., Porter, E.S., Boths, J., Kanyi, D., Hsieh, M., and Cooperman, B.S. (1999) *J. Pept. Res.*, **54**, 444.
33 Wosnick, J.H., Mello, C.M., and Swager, T.M. (2005) *J. Am. Chem. Soc.*, **127**, 3400.
34 (a) Curran, S. and Murray, G.I. (2000) *Eur. J. Cancer*, **36**, 1621; (b) Deng, S.J., Bickett, D.M., Mitchell, J.L., Lambert, M.H., Blackburn, R.K., Carter, H.L., III, Neugebauer, J., Pahel, G., Weiner, M.P., and Moss, M.L. (2000) *J. Biol. Chem.*, **275**, 31422; (c) Lovejoy, B., Welch, A.R., Carr, S., Luong, C., Broka, C., Hendricks, R.T., Campbell, J.A., Walker, K.A.M., Martin, R., Van Wart, H., and Browner, M.F. (1999) *Nat. Struct. Biol.*, **6**, 217; (d) Knauper, V., Cowell, S., Smith, B., Lopez-Otin, C., O'Shea, M., Morris, H., Zardi, L., and Murphy, G. (1997) *J. Biol. Chem.*, **272**, 7608; (e) Knauper, V., Lopez-Otin, C., Smith, B., Knight, G., and Murphy, G. (1996) *J. Biol. Chem.*, **271**, 1544; (f) Freije, J.M.P., Diez-Itza, I., Balbin, M., Sanchez, L.M., Blasco, R., Tolivia, J., and Lopez-Otin, C. (1994) *J. Biol. Chem.*, **269**, 16766.
35 Ionescu, R.E., Cosnier, S., and Marks, R.S. (2006) *Anal. Chem.*, **78**, 6327.
36 Tan, C., Atas, E., Muller, J.G., Pinto, M.R., Kleiman, V.D., and Schanze, K.S. (2004) *J. Am. Chem. Soc.*, **126**, 13685.
37 Zhang, T., Fan, H., Zhou, J., Liu, G., Feng, G., and Jin, Q. (2006) *Macromolecules*, **39**, 7839.
38 (a) An, L., Liu, L., and Wang, S. (2009) *Biomacromolecules*, **10**, 454; (b) Nilsson, K.P.R. and Inganas, O. (2004) *Macromolecules*, **37**, 9109.
39 (a) Zhao, X., Liu, Y., and Schanze, K.S. (2007) *Chem. Commun.*, 2914; (b) Liu, Y. and Schanze, K.S. (2008) *Anal. Chem.*, **80**, 8605.
40 Zhang, Y., Wang, Y., and Liu, B. (2009) *Anal. Chem.*, **81**, 3731.
41 An, L., Tang, Y., Feng, F., He, F., and Wang, S. (2007) *J. Mater. Chem.*, **17**, 4147.
42 Manning, G., Whyte, D.B., Martinez, R., Hunter, T., and Sudarsanam, S. (2002) *Science*, **298**, 1912.
43 (a) Hunter, T. (1995) *Cell*, **80**, 225; (b) Franklin, R.A. and McCubrey, J.A. (2000) *Leukemia*, **14**, 2019; (c) Hunter, T. (2000) *Cell*, **100**, 113.
44 (a) Garrett, M.D. and Workman, P. (1999) *Eur. J. Cancer*, **35**, 2010; (b) Bhagwat, S.S., Manning, A.M., Hoekstra, M.F., and Lewis, A. (1999) *Drug Discov. Today*, **4**, 472.
45 Andersson, L. and Porath, J. (1986) *Anal. Biochem.*, **154**, 250.
46 Rininsland, F., Xia, W., Wittenburg, S., Shi, X., Stankewicz, C., Achyuthan, K., McBranch, D., and Whitten, D. (2004) *Proc. Natl. Acad. Sci. USA*, **101**, 15295.
47 Lis, H. and Sharon, N. (1998) *Chem. Rev.*, **98**, 637.
48 Kim, I.-B., Wilson, J.N., and Bunz, U.H.F. (2005) *Chem. Commun.*, 1273.

49 Disney, M.D., Zheng, J., Swager, T.M., and Seeberger, P.H. (2004) *J. Am. Chem. Soc.*, **126**, 13343.
50 Kelly, T.L., Lam, M.C.W., and Wolf, M.O. (2006) *Bioconjugate Chem.*, **17**, 575.
51 (a) Kim, I.-B., Dunkhorst, A., and Bunz, U.H.F. (2005) *Langmuir*, **21**, 7985; (b) Herland, A. and Inganas, O. (2007) *Macromol. Rapid Commun.*, **28**, 1703.
52 Miranda, O.R., You, C.-C., Phillips, R., Kim, I.-B., Ghosh, P.S., Bunz, U.H.F., and Rotello, V.M. (2007) *J. Am. Chem. Soc.*, **129**, 9856.
53 Yu, D., Zhang, Y., and Liu, B. (2008) *Macromolecules*, **41**, 4003.
54 Zhang, Y., Liu, B., and Cao, Y. (2008) *Chem. Asian J.*, **3**, 739.
55 Terzi, E., Holzemann, G., and Seelig, J. (1997) *Biochemistry*, **36**, 14845.
56 Caughey, B. (2000) *Nat. Med.*, **6**, 751.
57 Nilsson, K.P.R., Herland, A., Hammarstrom, P., and Inganas, O. (2005) *Biochemistry*, **44**, 3718.
58 (a) Glenner, G.G. (1981) *Prog. Histochem. Cytochem.*, **13**, 1; (b) Klunk, W.E., Pettegrew, J.W., and Abraham, D.J. (1989) *J. Histochem. Cytochem.*, **37**, 1273; (c) Klunk, W.E., Pettegrew, J.W., and Abraham, D.J. (1989) *J. Histochem. Cytochem.*, **37**, 1293.
59 Liu, B. and Bazan, G.C. (2004) *Chem. Mater.*, **16**, 4467.
60 Dore, K., Dubus, S., Ho, H.-A., Levesque, I., Brunette, M., Corbeil, G., Boissinot, M., Boivin, G., Bergeron, M.G., Boudreau, D., and Leclerc, M. (2004) *J. Am. Chem. Soc.*, **126**, 4240.
61 Förster, T. (1959) *Discuss. Faraday Soc.*, **27**, 7.
62 Gaylord, B.S., Heeger, A.J., and Bazan, G.C. (2002) *Proc. Natl. Acad. Sci. USA*, **99**, 10954.
63 Stender, H., Fiandaca, M., Hyldig-Nielsen, J.J., and Coull, J. (2002) *J. Microbiol. Methods*, **48**, 1.
64 Nielsen, P.E. (1999) *Curr. Opin. Biotechnol.*, **10**, 71.
65 Egholm, M., Buchardt, O., Christensen, L., Behrens, C., Freier, S.M., Driver, D.A., Berg, R.H., Kim, S.K., Norden, B., and Nielsen, P.E. (1993) *Nature*, **365**, 566.
66 (a) Ganachaud, F., Elaissari, A., Pichot, C., Laayoun, A., and Cros, P. (1997) *Langmuir*, **13**, 701; (b) Smith, J.O., Olson, D.A., and Armitage, B.A. (1999) *J. Am. Chem. Soc.*, **121**, 2686; (c) Gaylord, B.S., Wang, S., Heeger, A.J., and Bazan, G.C. (2001) *J. Am. Chem. Soc.*, **123**, 6417.
67 Gaylord, B.S., Heeger, A.J., and Bazan, G.C. (2003) *J. Am. Chem. Soc.*, **125**, 896.
68 (a) Liu, B. and Bazan, G.C. (2005) *Proc. Natl. Acad. Sci. USA*, **102**, 589; (b) Sun, C., Gaylord, B.S., Hong, J.W., Liu, B., and Bazan, G.C. (2007) *Nat. Protoc.*, **2**, 2148; (c) Raymond, F.R., Ho, H.-A., Peytavi, R., Bissonnette, L., Boissinot, M., Picard, F.J., Leclerc, M., and Bergeron, M.G. (2005) *BMC Biotech.*, **5**, 10.
69 Liu, B. and Bazan, G.C. (2006) *Nat. Protoc.*, **1**, 1698.
70 Pu, K.-Y. and Liu, B. (2009) *Biosens. Bioelectron.*, **24**, 1067.
71 Liu, B. and Bazan, G.C. (2006) *J. Am. Chem. Soc.*, **128**, 1188.
72 Liu, B., Dan, T.T.T., and Bazan, G.C. (2007) *Adv. Funct. Mater.*, **17**, 2432.
73 Tian, N., Tang, Y., Xu, Q.-H., and Wang, S. (2007) *Macromol. Rapid Commun.*, **28**, 729.
74 Suh, Y. and Cantor, C. (2005) *Mutat. Res. Fund. Mol. Mech. Mutag.*, **573**, 1.
75 (a) Yu, L.S.L., Reed, S.A., and Golden, M.H. (2002) *J. Microbiol. Methods*, **49**, 63; (b) Nakamura, N., Burgess, J.G., Yagiuda, K., Kudo, S., Sakaguchi, T., and Matsunaga, T. (1993) *Anal. Chem.*, **65**, 2036; (c) Yamaguchi, N., Sasada, M., Yamanaka, M., and Nasu, M. (2003) *Cytometry*, **54**, 27.
76 (a) Jung, W.-S., Kim, S., Hong, S.-I., Min, N.-K., Lee, C.-W., and Paek, S.-H. (2004) *Mater. Sci. Eng. C*, **24**, 47; (b) Stender, H., Oliveira, K., Rigby, S., Bargoot, F., and Coull, J. (2001) *J. Microbiol. Methods*, **45**, 31.
77 Goodridge, L., Chen, J., and Griffiths, M. (1999) *Int. J. Food Microbiol.*, **47**, 43.
78 Mammen, M., Choi, S.-K., and Whitesides, G.M. (1998) *Angew. Chem., Int. Ed.*, **37**, 2754.

79 (a) Karlsson, K.-A. (1999) *Biochem. Soc. Trans.*, **27**, 471; (b) Karlsson, K.-A. (2001) *Adv. Exp. Med. Biol.*, **491**, 431.

80 Xue, C., Velayudham, S., Johnson, S., Saha, R., Smith, A., Brewer, W., Murthy, P., Bagley, S.T., and Liu, H. (2009) *Chem. Eur. J.*, **15**, 2289.

81 Baek, M.-G., Stevens, R.C., and Charych, D.H. (2000) *Bioconjug. Chem.*, **11**, 777.

82 Silbert, L., Ben Shlush, I., Israel, E., Porgador, A., Kolusheva, S., and Jelinek, R. (2006) *Appl. Environ. Microbiol.*, **72**, 7339.

83 (a) Bendtsen, J.D., Kiemer, L., Fausboll, A., and Brunak, S. (2005) *BMC Microbiol.*, **5**, 58; (b) Thanassi, D.G. and Hultgren, S.J. (2000) *Curr. Opin. Cell Biol.*, **12**, 420.

84 Kobayashi, T., Yasuda, M., Okada, S., Matsuda, H., and Nakanishi, H. (1997) *Chem. Phys. Lett.*, **267**, 472.

85 (a) Jelinek, R. and Kolusheva, S. (2001) *Biotechnol. Adv.*, **19**, 109; (b) Ma, G. and Cheng, Q. (2005) *Langmuir*, **21**, 6123; (c) Ma, Z., Li, J., Liu, M., Cao, J., Zou, Z., Tu, J., and Jiang, L. (1998) *J. Am. Chem. Soc.*, **120**, 12678; (d) Ma, Z., Li, J., Jiang, L., Cao, J., and Boullanger, P. (2000) *Langmuir*, **16**, 7801; (e) Okada, S., Peng, S., Spevak, W., and Charych, D. (1998) *Acc. Chem. Res.*, **31**, 229.

86 Kim, Y., Whitten, J.E., and Swager, T.M. (2005) *J. Am. Chem. Soc.*, **127**, 12122.

87 Kim, Y. and Swager, T.M. (2005) *Chem. Commun.*, 372.

88 Kim, Y. and Swager, T.M. (2006) *Macromolecules*, **39**, 5177.

89 Kim, J., McQuade, D.T., McHugh, S.K., and Swager, T.M. (2000) *Angew. Chem., Int. Ed.*, **39**, 3868.

90 Satrijo, A., Kooi, S.E., and Swager, T.M. (2007) *Macromolecules*, **40**, 8833.

91 Gokel, G.W. (1991) *Crown Ethers and Cryptands*, The Royal Society of Chemistry, Cambridge, UK, pp. 99–128.

92 Clarkson, T.W. (1997) *Crit. Rev. Clin. Lab. Sci.*, **34**, 369.

93 Kim, I.-B., Dunkhorst, A., Gilbert, J., and Bunz, U.H.F. (2005) *Macromolecules*, **38**, 4560.

94 Yu, M., He, F., Tang, Y., Wang, S., Li, Y., and Zhu, D. (2007) *Macromol. Rapid Commun.*, **28**, 1333.

95 (a) Xia, W.-S., Schmehl, R.H., Li, C.-J., Mague, J.T., Luo, C.-P., and Guldi, D.M. (2002) *J. Phys. Chem. B*, **106**, 833; (b) Ito, T., Hioki, T., Yamaguchi, T., Shinbo, T., Nakao, S., and Kimura, S. (2002) *J. Am. Chem. Soc.*, **124**, 7840.

96 Zeng, D., Cheng, J., Ren, S., Sun, J., Zhong, H., Xu, E., Du, J., and Fang, Q. (2008) *React. Funct. Polym.*, **68**, 1715.

97 Tang, Y., He, F., Yu, M., Feng, F., An, L., Sun, H., Wang, S., Li, Y., and Zhu, D. (2006) *Macromol. Rapid Commun.*, **27**, 389.

98 (a) Descalzo, A.B., Martinez-Manez, R., Radeglia, R., Rurack, K., and Soto, J. (2003) *J. Am. Chem. Soc.*, **125**, 3418; (b) Zhang, X.-B., Guo, C.-C., Li, Z.-Z., Shen, G.-L., and Yu, R.-Q. (2002) *Anal. Chem.*, **74**, 821; (c) Valeur, B. and Leray, I. (2000) *Coord. Chem. Rev.*, **205**, 3.

99 Harada, M. (1995) *Crit. Rev. Toxicol.*, **25**, 1.

100 (a) Egashira, N., Kumasako, H., Kurauchi, Y., and Ohga, K. (1994) *Anal. Sci.*, **10**, 405; (b) Nelson, B.C., Uden, P.C., Rockwell, G.F., Gorski, K.M., and Aguilera, Z.B. (1997) *J. Chromatogr. A*, **771**, 285.

101 Sun, H., Feng, F., Yu, M., and Wang, S. (2007) *Macromol. Rapid Commun.*, **28**, 1905.

102 Wang, F. and Bazan, G.C. (2006) *J. Am. Chem. Soc.*, **128**, 15786.

103 Kim, Y., Bouffard, J., Kooi, S.E., and Swager, T.M. (2005) *J. Am. Chem. Soc.*, **127**, 13726.

104 (a) Cornil, J., dos Santos, D.A., Crispin, X., Silbey, R., and Bredas, J.-L. (1998) *J. Am. Chem. Soc.*, **120**, 1289; (b) Blatchford, J.W., Jessen, S.W., Lin, L.-B., Gustafson, T.L., Fu, D.-K., Wang, H.-L., Swager, T.M., MacDiarmid, A.G., and Epstein, A.J. (1996) *Phys. Rev. B*, **54**, 9180.

105 Swager, T.M., Gil, C.J., and Wrighton, M.S. (1995) *J. Phys. Chem.*, **99**, 4886.

106 Satrijo, A. and Swager, T.M. (2007) *J. Am. Chem. Soc.*, **129**, 16020.

107 Gaugas, J.M. (ed.) (1980) *Polyamines in Biomedical Research*, John Wiley & Sons, Inc., New York.

108 (a) Russell, D.H. (1971) *Nat. New Biol.*, **233**, 144; (b) Fujita, K., Nagatsu, T., Shinpo, K., Maruta, K., Teradaira, R., and Nakamura, M. (1980) *Clin. Chem.*, **26**, 1577; (c) Bachrach, U. (2004) *Amino Acids*, **26**, 307.

109 (a) Teti, D., Visalli, M., and McNair, H. (2002) *J. Chromatogr. B*, **781**, 107; (b) Jin, Y., Jang, J.-W., Lee, M.-H., and Han, C.-H. (2006) *Clin. Chim. Acta*, **364**, 260.

110 (a) Magoshi, T., Ziani-Cherif, H., Ohya, S., Nakayama, Y., and Matsuda, T. (2002) *Langmuir*, **18**, 4862; (b) Rao, G.V.R., Krug, M.E., Balamurugan, S., Xu, H., Xu, Q., and Lopez, G.P. (2002) *Chem. Mater.*, **14**, 5075; (c) Neradovic, D., van Nostrum, C.F., and Hennink, W.E. (2001) *Macromolecules*, **34**, 7589.

111 Dautzenberg, H., Gao, Y., and Hahn, M. (2000) *Langmuir*, **16**, 9070.

112 (a) Sun, T., Wang, G., Feng, L., Liu, B., Ma, Y., Jiang, L., and Zhu, D. (2004) *Angew. Chem., Int. Ed.*, **43**, 357; (b) Sun, T., Song, W., and Jiang, L. (2005) *Chem. Commun.*, 1723.

113 Kuroda, K. and Swager, T.M. (2004) *Macromolecules*, **37**, 716.

114 Wang, W., Wang, R., Zhang, C., Lu, S., and Liu, T. (2009) *Polyhedron*, **50**, 1236.

115 (a) Xiao, X., Fu, Y., Zhou, J., Bo, Z., Li, L., and Chan, C.-M. (2007) *Macromol. Rapid Commun.*, **28**, 1003; (b) Ma, Z., Qiang, L., Zheng, Z., Wang, Y., Zhang, Z., and Huang, W. (2008) *J. Appl. Poly. Sci.*, **110**, 18.

116 Giepmans, B.N.G., Adams, S.R., Ellisman, M.H., and Tsien, R.Y. (2006) *Science*, **312**, 217.

117 (a) Michalet, X., Pinaud, F.F., Bentolila, L.A., Tsay, J.M., Doose, S., Li, J.J., Sundaresan, G., Wu, A.M., Gambhir, S.S., and Weiss, S. (2005) *Science*, **307**, 538; (b) Derfus, A.M., Chan, W.C.W., and Bhatia, S.N. (2004) *Nano Lett.*, **4**, 11; (c) Cho, S.J., Maysinger, D., Jain, M., Roder, B., Hackbarth, S., and Winnik, F.M. (2007) *Langmuir*, **23**, 1974.

118 Liao, J.H. and Swager, T.M. (2007) *Langmuir*, **23**, 112.

119 Wosnick, J.H., Liao, J.H., and Swager, T.M. (2005) *Macromolecules*, **38**, 9287.

120 (a) Zeineldin, R., Piyasena, M.E., Bergstedt, T.S., Sklar, L.A., Whitten, D., and Lopez, G.P. (2006) *Cytometry A*, **69**, 335; (b) Zeineldin, R., Piyasena, M.E., Sklar, L.A., Whitten, D., and Lopez, G.P. (2008) *Langmuir*, **24**, 4125.

121 (a) Bechinger, B. (2004) *Crit. Rev. Plant Sci.*, **23**, 271; (b) Ladokhin, A.S., Selsted, M.E., and White, S.H. (1997) *Biophys. J.*, **72**, 1762; (c) Papo, N. and Shai, Y. (2003) *Biochemistry*, **42**, 458; (d) Ladokhin, A.S. and White, S.H. (2001) *Biochim. Biophys. Acta*, **1514**, 253.

122 (a) Dempsey, C.E. (1990) *Biochim. Biophys. Acta*, **1031**, 143; (b) Constantinescu, I. and Lafleur, M. (2004) *Biochim. Biophys. Acta*, **1667**, 26.

123 Chemburu, S., Ji, E., Casana, Y., Wu, Y., Buranda, T., Schanze, K.S., Lopez, G.P., and Whitten, D.G. (2008) *J. Phys. Chem. B*, **112**, 14492.

124 (a) Sudhir, K. (2005) *J. Clin. Endocrinol. Metab.*, **90**, 3100; (b) Ballantyne, C.M., Hoogeveen, R.C., Bang, H., Coresh, J., Folsom, A.R., Heiss, G., and Sharrett, A.R. (2004) *Circulation*, **109**, 837.

125 Moon, J.H., Deans, R., Krueger, E., and Hancock, L.F. (2003) *Chem. Commun.*, 104.

126 Kurokawa, N., Yoshikawa, H., Hirota, N., Hyodo, K., and Masuhara, H. (2004) *ChemPhysChem*, **5**, 1609.

127 Moon, J.H., McDaniel, W., MacLean, P., and Hancock, L.F. (2007) *Angew. Chem., Int. Ed.*, **46**, 8223.

128 (a) Wu, C., Szymanski, C., and McNeill, J. (2006) *Langmuir*, **22**, 2956; (b) Szymanski, C., Wu, C., Hooper, J., Salazar, M.A., Perdomo, A., Dukes, A., and McNeill, J. (2005) *J. Phys. Chem. B*, **109**, 8543; (c) Wu, C., Bull, B., Szymanski, C., Christensen, K., and McNeill, J. (2008) *ACS Nano*, **2**, 2415.

129 Wu, C., Szymanski, C., Cain, Z., and McNeill, J. (2007) *J. Am. Chem. Soc.*, **129**, 12904.

130 (a) Denk, W., Strickler, J.H., and Webb, W.W. (1990) *Science*, **248**, 73; (b) Xu, C., Zipfel, W., Shear, J.B.,

Williams, R.M., and Webb, W.W. (1996) *Proc. Natl. Acad. Sci. USA*, **93**, 10763.

131 Xu, C. and Webb, W.W. (1996) *J. Opt. Soc. Am. B*, **13**, 481.

132 Larson, D.R., Zipfel, W.R., Williams, R.M., Clark, S.W., Bruchez, M.P., Wise, F.W., and Webb, W.W. (2003) *Science*, **300**, 1434.

133 Wu, C., Peng, H., Jiang, Y., and McNeill, J. (2006) *J. Phys. Chem. B*, **110**, 14148.

134 Wu, C., Zheng, Y., Szymanski, C., and McNeill, J. (2008) *J. Phys. Chem. C*, **112**, 1772.

135 (a) Dexter, D.L. (1953) *J. Chem. Phys.*, **21**, 836; (b) Turro, N.J. (1991) *Modern Molecular Photochemistry*, University Science Books, Sausalito, CA, Chapter 9; (c) Kavarnos, G.J. and Turro, N.J. (1986) *Chem. Rev.*, **86**, 401.

136 (a) Jang, S., Newton, M.D., and Silbey, R.J. (2004) *Phys. Rev. Lett.*, **92**, 218301; (b) Murphy, C.B., Zhang, Y., Troxler, T., Ferry, V., Martin, J.J., and Jones, W.E. (2004) *J. Phys. Chem. B*, **108**, 1537.

137 Kersting, R., Lemmer, U., Mahrt, R.F., Leo, K., Kurz, H., Baessler, H., and Goebel, E.O. (1993) *Phys. Rev. Lett.*, **70**, 3820.

138 Beenken, W.J.D. and Pullerits, T. (2004) *J. Chem. Phys.*, **120**, 2490.

139 Wiesenhofer, H., Beljonne, D., Scholes, G.D., Hennebicq, E., Bredas, J.-L., and Zojer, E. (2005) *Adv. Funct. Mater.*, **15**, 155.

140 Beljonne, D., Cornil, J., Silbey, R., Millie, P., and Bredas, J.-L. (2000) *J. Chem. Phys.*, **112**, 4749.

141 (a) Zhang, Y. and Xu, Z. (2008) *Appl. Phys. Lett.*, **93**, 083106; (b) Dias, F.B., Knaapila, M., Monkman, A.P., and Burrows, H.D. (2006) *Macromolecules*, **39**, 1598; (c) Thompson, A.L., Gaab, K.M., Xu, J., Bardeen, C.J., and Martinez, T.J. (2004) *J. Phys. Chem. A*, **108**, 671.

142 (a) Faure, S., Stern, C., Guilard, R., and Harvey, P.D. (2004) *J. Am. Chem. Soc.*, **126**, 1253; (b) Didraga, C., Malyshev, V.A., and Knoester, J. (2006) *J. Phys. Chem. B*, **110**, 18818.

143 Rose, A., Lugmair, C.G., and Swager, T.M. (2001) *J. Am. Chem. Soc.*, **123**, 11298.

144 Markovitsi, D., Germain, A., Millie, P., Lecuyer, P., Gallos, L., Argyrakis, P., Bengs, H., and Ringsdorf, H. (1995) *J. Phys. Chem.*, **99**, 1005.

145 Rose, A., Tovar, J.D., Yamaguchi, S., Nesterov, E.E., Zhu, Z., and Swager, T.M. (2007) *Philos. Trans. R. Soc. Lond. A*, **365**, 1589.

146 Zheng, J. and Swager, T.M. (2004) *Chem. Commun.*, 2798.

147 Levine, M., Song, I., Andrew, T.L., Kooi, S.E., and Swager, T.M. (2010) *J. Polym. Sci. A Polym. Chem.*, **48**, 3382.

148 (a) Block, M.A.B. and Hecht, S. (2004) *Macromolecules*, **37**, 4761; (b) Snee, P.T., Somers, R.C., Nair, G., Zimmer, J.P., Bawendi, M.G., and Nocera, D.G. (2006) *J. Am. Chem. Soc.*, **128**, 13320; (c) Isgor, Y.G. and Akkaya, E.U. (1997) *Tetrahedron Lett.*, **38**, 7417.

149 Nolde, F., Qu, J., Kohl, C., Pschirer, N.G., Reuther, E., and Müllen, K. (2005) *Chem. Eur. J.*, **11**, 3959.

5
Chromophoric Polyisocyanide Materials

Bram Keereweer, Erik Schwartz, Stéphane Le Gac, Roeland J.M. Nolte, Paul H. J. Kouwer, and Alan E. Rowan

5.1
Introduction

The relationship between architecture and functionality lies at the heart of nature. One of the most striking examples being the chromophoric arrays found in the chlorophyll and bacteriochlorophyll photosynthetic systems that are organized in such a precise manner that the transfer of excitation energy and electrons occurs with a phenomenally high efficiency over large distances. The purple bacteria *Rhodopseudomonas acidophila* that contains a circular antenna complex, named LH2 [1], is a perfect example of a system with a clear architecture–function relationship. In the LH2, the peptidic alpha helices fix the chlorophyll chromophores in a precise orientation that allows an efficient energy transfer process [2]. This natural use of a biomolecular scaffold to organize functional units has inspired many researchers in mimicking the positioning of chromophores on artificial scaffold systems, such as dendrimers [3], supramolecular systems [4], or covalent systems [5]. Recently, the application of *rigid* scaffolds, such as carbon nanotubes (CNTs), DNA, or helical polymers, has extended this approach to give excellent spatial organization of chromophores in functional arrays.

Carbon nanotubes, which can be envisaged as polymeric carbon arrays, have been used for immobilization of chromophores by grafting them on the nanotube, directly resulting in a system that combines the favorable properties of the conducting CNTs with the optical properties of the chromophores [6]. The disadvantage of this approach is that the scaffold although rigid has no defined structural information built-in and hence the resulting arrays are more randomly decorated, thereby having a less controlled architecture. In contrast, DNA has a very high level of structural information encoded in the polymer, being built up from two programmable self-assembled strands based on four basic building blocks, and can be readily prepared using automated synthesis procedures. The distance between two base pairs is 3.3 Å that agrees well with the typical π–π distance in chromophoric arrays. A clear advantage of such systems is the controlled [7] or (purposefully) random [8] introduction of a variety of chromophoric moieties on the same scaffold. In addition, many systems that employ the DNA scaffold are water soluble, which allows their application as sensors in biological systems. Although various aspects of using DNA as a rigid scaffold for chromophore organization have been touched upon [9], the field is still in its early stages. When its main advantage, that is, the exact programmability of the building blocks, can be used to its full extent, many exciting applications will certainly emerge.

Numerous examples in literature show that synthetic polymers can be easily functionalized with chromophores. The dynamic character of the polymer main chain, however, often causes severe disorder in the chromophore architecture, which renders such systems unsuitable for many optoelectronic applications, for which structural definition is essential. In synthetic "helical polymers," the chain dynamics are restricted by the helical structural motive that result in well-defined rigid scaffolds similar in persistence length to that found for DNA and α-helices in proteins. Such synthetic polymers are therefore ideal candidates for a controlled organization of chromophores in side chain arrays. The helical polymers of choice are polyisocyanates, polysilanes, polyguanidines, and polyisocyanides.

Polyisocyanates [10] are characterized by an N-substituted amide repeating unit and fold in a 8_3 helical conformations (Figure 5.1). These resultant helical polymers are dynamic in nature, owing to the low inversion barrier. Due to the harsh conditions for the preparation of the monomer and anionic nature of the polymerization, there are not many examples of functionalized polyisocyanates. An azobenzene chromophore could be introduced in a postfunctionalization approach by the *trans*-esterification of the methyl ester resulting in a 30% degree of functionalization. Despite the incomplete substitution reaction, the optical rotation value showed a 10-fold increase, which indicated that the chromophores experienced the helical environment of the polyisocyanate.

Another class of dynamic helical polymers are polysilanes [11]. Similar to polyisocyanates, the rigorous reaction conditions in the polysilane synthesis impede the introduction of various functionalities. Postfunctionalization of the polymer allowed the introduction of azobenzene-based chromophores. The chromophore-loaded polymer showed a three-order increase in photoconductivity compared to simple polysilanes [12].

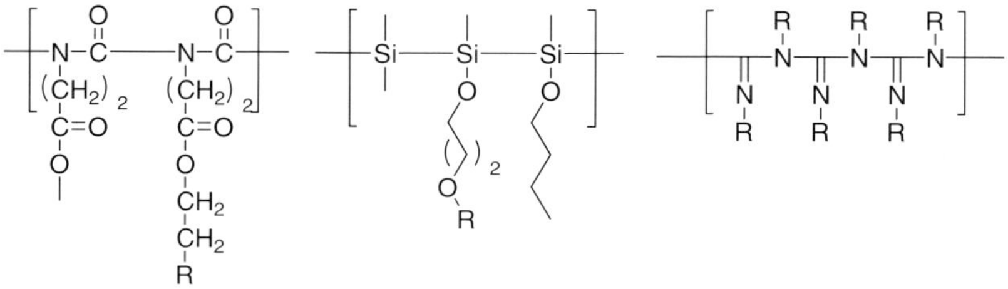

Figure 5.1 Examples of polymers that form helical structures: polyisocyanates, polysilanes, and polyguanidines.

Polyguanidines [13] also form very stable helical structures with high persistence length [14]. The polymer is obtained by a controlled polymerization of carbodiimides in the presence of a titanium(IV) or a copper(II) catalyst [15], the copper catalyst having a higher tolerance to impurities such as water and air than the titanium catalyst. An advantage of these polymers is that the functional groups are allowed in the polymerization, resulting in a fully covered polymer. For instance, anthracene-decorated polyguanidines have been used as a chiroptical switch in display technology [16].

5.2
Polyisocyanide Materials

An alternative class of stable helical polymers is polyisocyanides, which allow a facile introduction of chromophores as a result of the mild polymerization conditions. This chapter focuses on the use of polyisocyanides as scaffolds for chromophore organization. Polyisocyanides were first polymerized in 1965 in a catalytic system by Millich [17], starting from α-phenylethyl isocyanide. Nolte [18] discovered that isocyanides could also be polymerized with simple nickel(II) salts, such as nickel (II) perchlorate, that polymerize the isocyanides in high yields at room temperature. The polymerization is believed to proceed via the "merry-go-round" mechanism, in which the first step is the formation of a square–planar nickel complex, as can be seen in Scheme 5.1. Deming and Novak [19] suggested that the mechanism is slightly more complex and probably involves Ni(I).

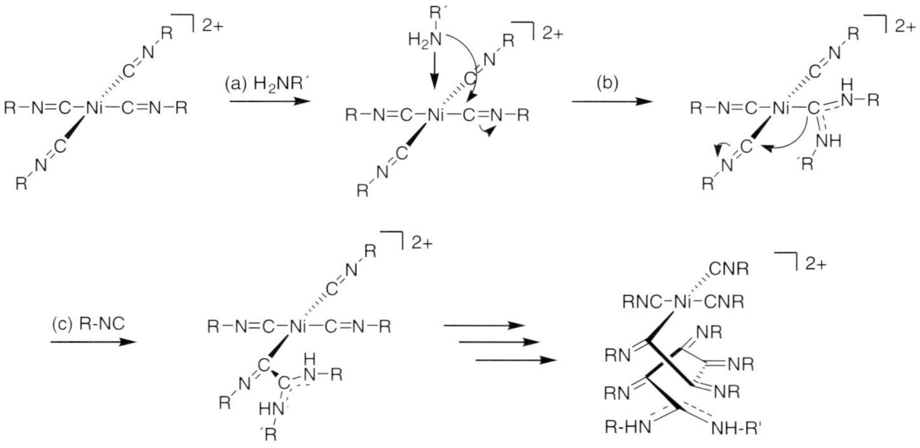

Scheme 5.1 Merry-go-round mechanism.

The polymerization is initiated when a nucleophile, commonly an amine, attacks one of the isocyanides coordinated to the nickel center. This nucleophile coordinates to the nickel center (a) and migrates into one of the isocyanides forming a carbene complex (b). The carbene ligand attacks a neighboring nickel(II)-coordinated

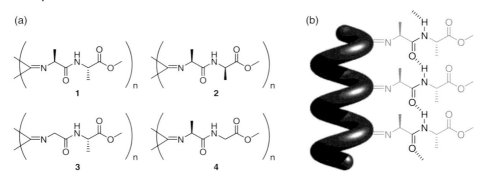

Figure 5.2 (a) Dipeptide-derived polyisocyanides LL-PIAAOMe (**1**), LD-PIAAOMe (**2**), L-PIGAOMe (**3**), and L-PIAGOMe (**4**). (b) Hydrogen bonding interactions between side chains in L-PIAAOMe.

isocyanide, forming the first C—C bond. The resulting vacant site is reoccupied by an isocyanide monomer from the solution (c). With a chiral bias in the system (from the monomer, the nucleophile or the solvent), there is a preference for one of the two neighboring isocyanides, which results in helices with only the preferred handedness. The helicity results in part from the repulsive interaction between the nitrogen atoms and the steric repulsion between the substituents. Therefore, the helices of the polyisocyanides are stable in solution only when the side groups are sufficiently bulky. For less sterically demanding side chains, such as polyphenylisocyanides, the initially formed helical backbone slowly uncoils after standing in solution [20].

Cornelissen *et al.* found that when optically active polyisocyanides derived from dipeptides (alanine–alanine and alanine–glycine) are used (Figure 5.2a), the helical form is stabilized as a result of hydrogen bonding arrays between the peptide side groups [21].

In case of PIGAOMe (**3**) and PIAGOMe (**4**), an achiral methylene group is present, instead of a stereocenter. This methylene group leads to a reduction of the steric interactions and thereby to a smaller preference for a specific conformation of the side chain.

In LL-PIAAOMe (**1**) and LD-PIAAOMe (**2**), the steric demands of the alanine methyl group stabilize the helix. Because of the hydrogen bonding array, the distance between the amide groups is 4.7 Å in the peptide polyisocyanides, instead of the 4.2 Å postulated for the conventional polyisocyanides.

Initial work on the application of stiff polyisocyanides as a chromophoric templates used the grafting strategy. The commonly incomplete grafting reaction, however, resulted in inhomogeneous polymer stacks [22]. In addition, the chromophore-functionalized polyisocyanides lacked structural stability. The discovery of the peptide-based polyisocyanides offered a significant improvement in the application of the polymers as macromolecular scaffolds. An additional benefit of these very rigid polymers (persistence length of ∼76 nm) is that they can be synthesized with lengths of hundreds of nanometers [23].

Scheme 5.2 shows the Ni(II)-catalyzed polymerization of a porphyrin isocyanide [24]. Atomic force microscopy (AFM) measurements showed porphyrin poly-

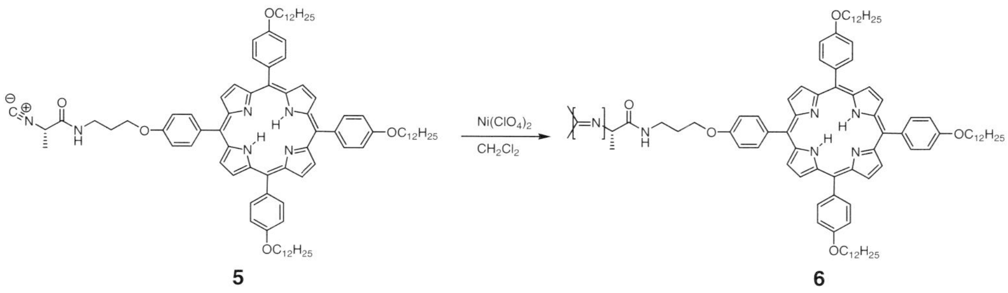

Scheme 5.2 Polymerization of a porphyrin-functionalized isocyanide with Ni(ClO₄)₂.

mers with average lengths of approximately 100 nm. This corresponds to a degree of polymerization of ∼830. The UV/Vis absorption spectrum of the polymer changed dramatically in comparison to that of the monomer (**5**). The latter displays a Soret band at 421 nm, whereas in the polymer (**6**), this band has disappeared and two new Soret bands are present at 413 and 437 nm. The absorption at 437 nm is indicative of porphyrin molecules that are arranged as J-aggregates and is attributed to an offset stacking of the first and the fifth porphyrins. The blueshifted peak is attributed to a combination of interactions between the first and the fourth porphyrins and the first and the second porphyrins.

The circular dichroism (CD) spectrum of the porphyrin polyisocyanide in CHCl₃ showed a strong bisignate Cotton effect at 437 nm that decreases reversibly upon heating. The spectrum signals the presence of chiral interactions between nth and $(n + 4)$th monomers in a slipped conformation that closely resembles the organization of porphyrins in natural antenna systems. Their strong CD spectra are associated with the energy transfer processes and are ascribed to exciton delocalization over large distances. Resonance light scattering (RLS) experiments confirm that upon excitation, the excited state is delocalized over at least 25 porphyrin molecules in one stack, the delocalization distance being ∼100 Å. Depolarized RLS studies revealed that the slip angle between the first and the fifth porphyrins amounted to 30° (Figure 5.3), which results in a helical twist angle of 22° and an overall helical pitch of ∼68–71 Å.

Fluorescence anisotropy studies yielded additional information on the orientation of the porphyrins along the polymer scaffold. Subsequent calculations showed that the porphyrin moieties are tilted by approximately 25° with respect to the helical axis of the polyisocyanide. The anisotropy measurements, in combination with the RLS studies, showed a model architecture in which the chromophores form a fourfold helter-skelter arrangement along the polymer scaffold (Figure 5.3).

Incorporation of metals in the porphyrin was employed to modify the organization and optoelectronic properties of the polymer. Zinc insertion required forceful conditions as a result of the well-defined organization of the free base porphyrin in the polymer stack. UV-Vis experiments indicated a near complete insertion of the metal [25]. When this polymer, decorated with Zn(II) metal ions, is visualized by

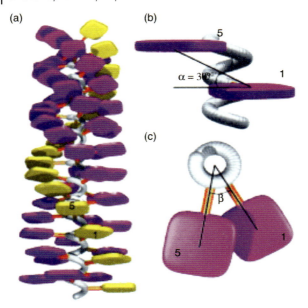

Figure 5.3 Schematic drawings of a porphyrin polyisocyanide and the proposed orientation of the chromophores obtained from DLS and fluorescence anisotropy measurements.

transmission electron microscopy (TEM), a network of fibers can be seen without any staining technique required.

These metal–porphyrin arrays are interesting as supramolecular building block since they can bind axial ligands, such as 1,4-diazabicyclo[2.2.2]octane (DABCO). Upon addition of 0.1–0.2 equivalents of the bifunctional ligand to this zinc-decorated polymer, the CD spectrum changed dramatically. The large effect with even small amounts of DABCO can be explained by the fact that the porphyrins are covalently bound to the polymer backbone. In order to bind the DABCO, one of the porphyrins needs to rotate, which has an influence on the neighboring porphyrins that need to rotate as well. The change in CD signal can be explained by the porphyrins having a left-handed helical organization, which becomes right-handed upon addition of DABCO [25]. This is in agreement with the infrared and CD spectra of the polyisocyanide backbone that remained unaltered by the addition of DABCO.

Amabilino and coworkers developed (chiral) electroactive polyisocyanides bearing TTF (tetrathiafulvalene) derivatives [26, 27]. The polymer has three extreme univalent states (UVSs) and two very wide mixed valence states (MVSs), which are fully interconvertible due to fully reversible redox processes (Figure 5.4). It was shown using CD spectroscopy that the polymer has, in contrast to the monomer, very surprisingly different chiroptical properties that are induced by the stereocenter located at ~18 Å from the isocyanide backbone. The different redox states of the polymer were investigated by cyclic voltammetry (CV) and UV-Vis spectroscopy. These findings showed that by using a rigid scaffold, redox systems can be

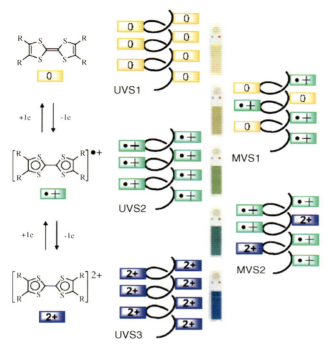

Figure 5.4 Schematic representation of the redox properties of the TTF unit showing the three univalent states (UVSs) and the two mixed valence states (MVSs) and the corresponding color of the polymer solutions.

incorporated in a well-defined manner and this allows to use these polymers as a multistate redox-switchable organic system in molecular devices [27]. More recently, the same group synthesized a related TTF-derivatized polyisocyanide and based on various spectroscopic techniques concluded that due to the precise arrangement of the π-electron-rich units, a charge transfer between the TTF moieties is observed upon oxidation [26].

Perylene diimides (PDIs) have unique properties, such as strong absorptivity in the visible part of the spectrum, high thermal and photochemical stability, and high electron affinity [28], which make them promising components as n-type semiconductors in organic photovoltaic cells. To further explore the concept of organizing chromophores on rigid polymer scaffolds, polyisocyanides 7–9 with pendant PDI groups (Figure 5.5) were explored [29–33]. The combination of hydrogen bonding arrays together with the additional π–π stacking interactions of the perylene side groups resulted in a very stable helical polymers. Even when heated, the arrangement of the perylenes remained intact.

Three polymers were studied that differed in the size of lateral and terminal substituents. Compared to the PDI monomer, the absorption spectra of the polymers **7** and **8** were broadened and redshifted and the relative intensities had changed, all indicative of strong chromophore–chromophore interactions. Fluorescence

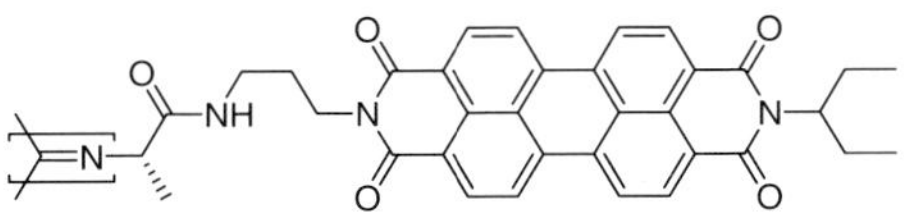

Figure 5.5 Chemical structures of polymers **7–9**.

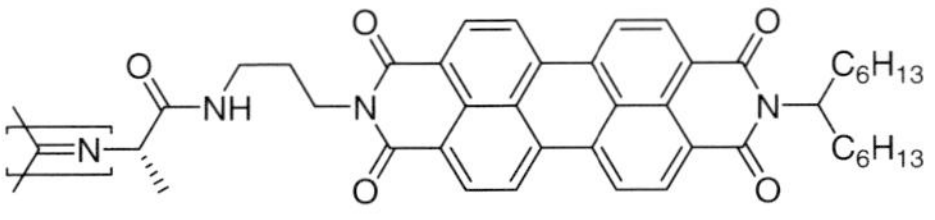

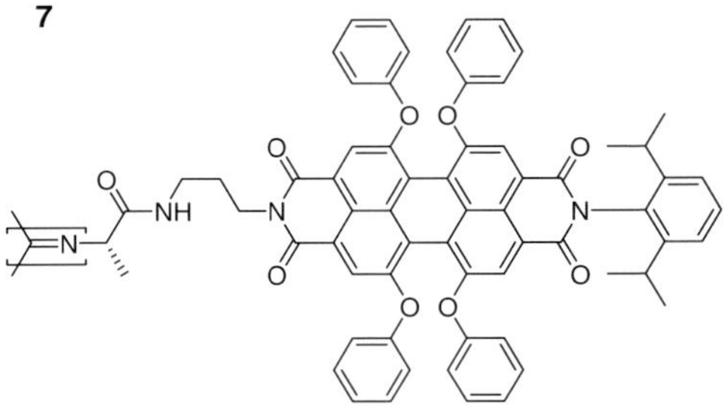

spectroscopy studies also showed broad, structureless, and redshifted bands in the emission spectra confirming aggregation. The emission spectra were concentration independent pointing solely toward intramolecular stacking.

Polymer **9** has bulky phenoxy substituents in the perylene bay area, which prevents good π–π stacking of the PDIs. In contrast to **7** and **8**, the absorption spectrum of **9** closely resembles that of its monomer. Also, the emission of **9** gave no evidence for the formation of excimers, which is consistent with the conclusion that by the introduction of steric bulk, the degree of chromophore assembly can be fine-tuned.

The CD spectrum of **7** showed several positive Cotton effects in the absorption region of the perylene π–π* transitions, besides the Cotton effect around 310 nm, which is typically assigned to the imines connected to the backbone.

Fluorescence decay measurements revealed the presence of species with a long lifetime of 19.9 ns – a monomeric perylene is typically around 4 ns – which further supports the idea that emission from this polymer is through an excimer species.

A combination of confocal fluorescence microscopy and AFM made it possible to distinguish two species obtained from the polymerization reaction: ill-defined oligomeric species and well-defined polymeric species (Figure 5.6) [29]. The oligomeric species (green arrow) could not be visualized by AFM since they are too short. They display monomer-like fluorescence. The polymer species (red arrow) have a well-defined rigid structure and could be visualized as independent fibers with AFM. These longer fibers were found to primarily exhibit excimer emission. It could be concluded from these experiments that upon excitation of the longer rods, the emission is moved along the isoelectronic perylene array and is quenched at perylene dimer sites, which results in excimer-like emission. Dimer sites can be fixed defects in the polymer or dynamic excimers that are formed throughout the polymer.

The absence of a bulky tail (polymer **7**) renders the polymers insoluble, however polymer **8** showed the combination of good processability and desired electronic

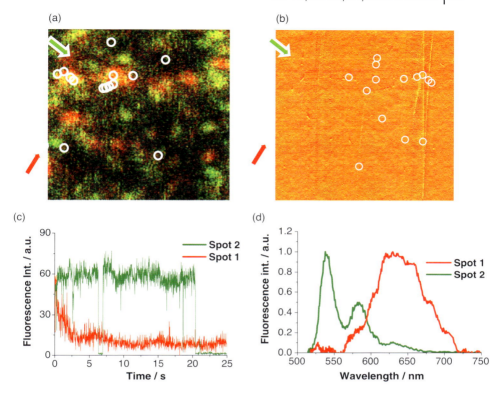

Figure 5.6 (a) Confocal and (b) AFM image from the same area of a solution of perylene polyisocyanide **7** on glass, the white circles show the position of the polymer species. (c) Corresponding fluorescence intensity trajectory for an emitting species, with (red) and without (green) topographic features in AFM. (d) Idem for emission spectra.

properties. Extensive dynamic, molecular modeling studies [33] suggest that the most probable conformation for the helical arrays in solution and the solid state is the architecture, which is presented in Figure 5.7a. This model combined with the observed spectroscopic properties indicated that polymer **8** is an ideal system for electron or energy transfer. This hypothesis was further strengthened by transient absorption spectroscopy studies, which indicated extremely rapid exciton migration rates and high charge densities. The modeled architecture as depicted in Figure 5.7b shows the calculated PDI ordering, which accounts for all the physical observations (UV-Vis and CD) observed.

5.3
Perylene Polyisocyanides in Devices

Perylene polyisocyanide **8** has shown to be a macromolecular system with high rigidity, good processability, and the capacity to generate and transport charges and

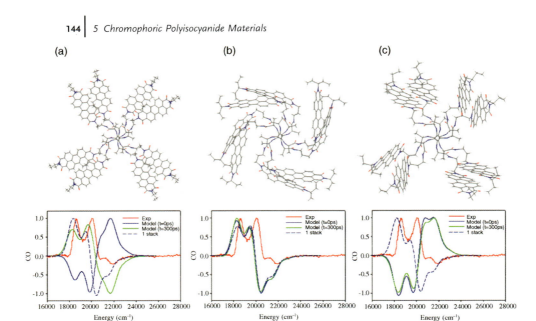

Figure 5.7 Top view of the low-energy conformations of the perylene polyisocyanide **8** with the relative orientation of the perylene molecules. In (b), the calculated spectra overlap best with the experimental spectra.

thus is potentially ideal candidate for photovoltaic devices. These devices were prepared from a mixture of perylene polyisocyanide (**8**) as the n-type material and poly(3-hexylthiophene) (P3HT) as the p-type material. Solutions of different donor–acceptor ratios were spin coated on an indium tin oxide (ITO)/polyethylenedioxythiophene: polystyrenesulfonate (PEDOT:PSS) electrode, after which an electrode of lithium fluoride (1 nm) and aluminum (100 nm) was deposited on top. The best performance was obtained for the thin film with a 1:1 ratio [31]. This device gave an external quantum efficiency (EQE) of 8.2% at 500 nm. Although the overall efficiency of ~0.2% is far from the state of the art for organic solar cells, there was a 20-fold efficiency improvement for the thin film when the polymer was used instead of the monomeric species. This improvement in efficiency is probably because the polymeric backbone hinders the formation of crystals, which increases the interpenetration of P-PDI and P3HT, favorable for efficient charge separation between the donor and acceptor. Also, the linear structure of the P-PDI wires is expected to preserve the percolation continuity of the electron accepting phase, forming a continuous network and thus decreasing the presence of dead ends and bottlenecks.

In order to visualize the relationship between the architecture and the photovoltaic efficiency, AFM and Kelvin probe force microscopy (KPFM)[1]) measurements were

1) KPFM is a contactless technique that allows the quantitative measurement of the electric surface potential (SP) with nanoscale resolution without significantly disturbing the potential of the system under study, thereby making it possible to perform an *in situ* exploration of the operation of electronic devices.

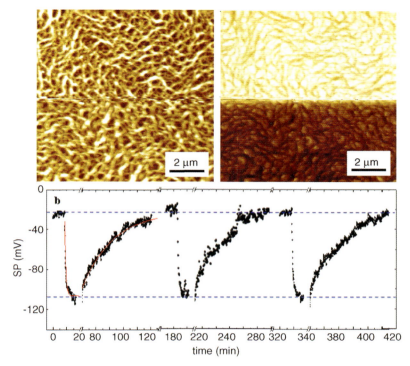

Figure 5.8 AFM and KPFM images of perylene polyisocyanide 8/P3HT films spin coated on PEDOT:PSS.

carried out on polymer films (Figure 5.8). This allowed the direct visualization of the photovoltaic activity occurring in polymeric bundles of electron-accepting PDI wires and bundles of electron-donating P3HT chains with true nanoscale spatial resolution.

Thick layers of P-PDI/P3HT were spin coated on PEDOT:PSS and were measured with AFM revealing a surface consisting of intertwined yet elongated bundles of fibers with cross sections of ~200 nm. The corresponding KPFM image showed that when the sample was illuminated with white light, the blends had a significant photovoltaic activity, with the average surface potential (SP) becoming more negative. The change in the average SP as the illumination was turned on and off is shown in Figure 5.9b.

Submonolayer thick films of donor/acceptor blends were made to gain more insight into the process of charge separation at the nanoscale level. Figure 5.9 shows the results for ultrathin blends of P-PDI with P3HT. The thick fibers have an average width of 190 ± 50 nm and a height of 17 ± 4 nm and the thin fibers have an average width of 70 ± 24 nm and a height of 3.5 ± 1.0 nm, both being much larger than either of the polymers diameter suggesting that they are composed of bundled polymer chains. On measuring these bundles with KPFM, a clear difference in contrast can be

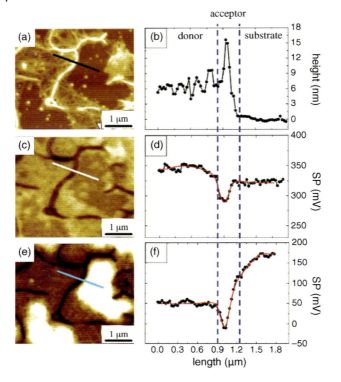

Figure 5.9 (a) AFM image of an ultrathin blend of **8** and P3HT deposited on silicon. (c and e) Surface potential images of the same area as in (a), under (c) no illumination and (e) illumination with white light (∼60 mW cm^{-2}). (b, d, and f) Measured (black lines and circles) and simulated (red lines) profiles obtained by tracing the arbitrary lines in the corresponding images (a), (c), and (e).

seen. The SP of the thick bundles appears much more negative (darker in the image) and the thin bundles appear more positive than the silicon substrate. These differences indicate that the thick fibers are composed of electron-accepting **8** and the thin fibers of electron-donating P3HT. The results also show that **8** and P3HT form effectively phase-separated yet interdigitated architectures on a hundred nanometer scale, resulting in a high contact area between the materials and defined percolation paths for the charges.

Results from monomeric PDI/P3HT blends showed that the difference between the two phases with the light off was much smaller than when **8** was used. The KPFM images of these blends also showed that a part of the monomeric PDI cluster that interfaces the P3HT has a higher negative charge, while the other part of the same cluster has a lower charge, indicating that the charges generated at the M-PDI/P3HT interface do not diffuse very far from the accepter–donor interface.

5.4
TFT Devices

Polyisocyanides **8** (perylene), **9** (bulky perylene) and **1** (non-chromophoric) were used as the active layer of a transistor [32]. Thin film transistor (TFT) measurements with the polyisocyanides with the perylene functionality showed that these films had a much better transconductance than **1**. Working transistors with very low n-type mobilities were obtained for **9**, because the sterically hindered perylenes prevented the required chromophore–chromophore interactions. As expected, **8** gave the best results with mobilities between 10^{-6} and $10^{-5}\,\mathrm{cm^2\,V^{-1}\,s^{-1}}$.

The charge can be transported by two mechanisms: intrachain hopping (along the backbone) and interchain hopping (to adjacent polymers). To understand the relative contributions of these processes, the carrier mobility as a function of temperature was studied. Significant differences were observed in the Arrhenius plots of the field-effect mobilities for **8** and **9**. In the case of **8**, two distinct regimes of temperature dependence are observed. At low temperature, activation energy of ~200 meV is observed, whereas at higher temperature, a rapid increase in mobility with temperature was measured (approximately two orders of magnitude between room temperature and $T = 360$ K), albeit with some hysteresis between the upward and downward sweep. In contrast, in the case of **9**, only a single nonhysteretic activation process with a characteristic energy of around 100 meV could be observed. Qualitatively, the temperature dependence in the mobility of **8** at low temperature is broadly consistent with variable-range transport phenomena [34], where long-range tunneling over small energy barriers (intrachain) predominates. An activated hopping mechanism over larger energy barriers (interchain) is significant at higher temperatures. It has been speculated that the hysteresis may be due to the partial unwinding and rewinding of the polyisocyanide helix with temperature, leading to some small changes in the conformal arrangement of adjacent chains. A measured mobility of order $10^{-3}\,\mathrm{cm^2\,V^{-1}\,s^{-1}}$ is reached at 360 K, which indicates that the mobilities for purely intrachain transport will be significantly larger than those in the present case where interchain hopping is a limiting factor.

5.5
Morphology Control in Perylene/Crystal Systems

Although the well-defined polyisocyanide chromophoric systems have promising characteristics when arranged in their helical environments, absolute value for their mobilities remain small compared to the single-crystal devices. Typical devices, however, are based on nanocrystalline or microcrystalline layers. In such systems, the interfaces between different crystals act as bottlenecks for charge transport. To improve contact between nanocrystals of monomeric PDI, the perylene polyisocyanide could be used as a kind of bridge [35]. When **8** and monomeric PDI are used together in a blend, three different outcomes are anticipated (Figure 5.10).

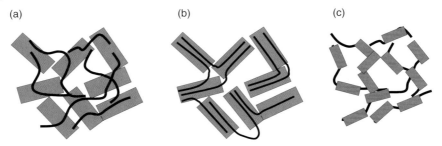

Figure 5.10 Possible outcomes of **8**/monomeric PDI blends.

When the monomer and polymer do not interact with each other, a simple superimposition of the individual deposition patterns will be visible (a). Alternatively, the monomer self-assembles in large-crystal domains and the polymer adapts its conformation to this morphology, connecting to one or more monomer crystals (b). The last possible outcome is that the monomers are forced to crystallize on the polymer network (c). These different morphologies can be obtained by changing the deposition conditions, such as the solvent and the substrate (Figure 5.11). When $CHCl_3$ is used as solvent and mica as substrate, the monomeric PDI and **8** do not influence each other's morphology. The nanocrystals rule the morphology when THF is used as the solvent and Si/SiO_x as the substrate. And when THF and mica are used, the polymer is first absorbed on the surface with the PDI crystallizing on these chains. This is also the case when the polymer is deposited from THF on Si/SiO_x, followed by the deposition of the monomer.

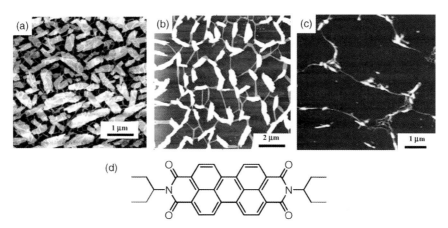

Figure 5.11 AFM images of **8** and monomeric PDI deposited (a) from $CHCl_3$ on mica, (b) from THF on Si/SiO_x, and (c) **8** from THF, followed by monomeric PDI from MeOH on Si/SiO_x. In (d), the molecular structure of the monomeric PDI is shown.

5.6
Postmodification of Polyisocyanopeptides

Although promising candidates for electron transfer, the syntheses of **7–9** are demanding and laborious. A modular postmodification method to the polyisocyanide backbone would be favorable, if only the grafting reactions could be guaranteed to proceed completely. An efficient method based on click chemistry [36] was recently developed.

A peptide polymer scaffold containing a terminal acetylene functionality (**10**, Scheme 5.3) was coupled to dodecyl azide to demonstrate the potential of the postmodification approach [37]. Analysis by NMR and IR spectroscopy as well as AFM showed complete conversion of the acetylene moieties. The wide variety of azides available could give rise to the accessibility of a vast array of functionalized polymers with varying properties.

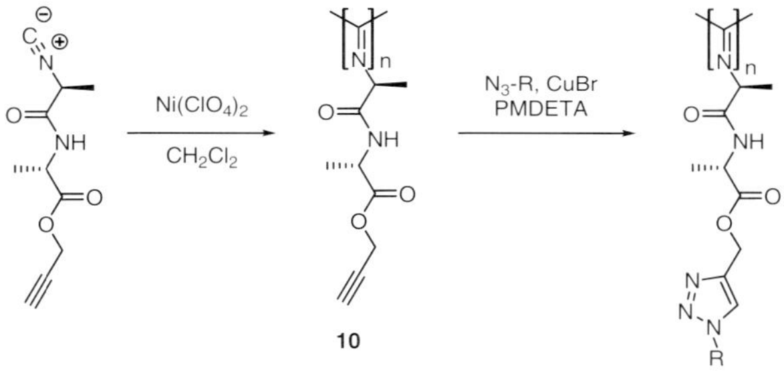

Scheme 5.3 Polyisocyanide **10** containing an acetylene functionality.

When an azide-functionalized perylene was grafted onto the **10**, its absorption spectrum showed the same features as the perylene polyisocyanide **8** [38]. Also, the fluorescence showed the redshifted peak, consistent with the formation of an excimer-like species. However, this peak has a much lower intensity compared to **8** and also monomer emission is observed. These differences with **8** are likely due to defects in the grafting reaction of the azide onto the polymers. The CD spectrum of the clicked polymer shows the typical peak for the imines around 310 nm, but does not exhibit any effect reminiscent of that observed for **8** in the region of 450–550 nm. This absence of signal may be due to increased distance between the chiral backbone and the chromophore. This reduced definition in ordering of the perylenes may also be responsible for the difference in fluorescence.

The click approach is suitable to introduce different functional groups on the polymer scaffolds. This was demonstrated when random copolymers were prepared by coclicking perylene azides and ethylene glycol azides to **10**. This approach yielded, for the first time, chromophoric water-soluble peptide polyisocyanides, which exhibited the same CD spectrum in water as in dichloromethane. For a sample containing minimal amounts of perylene, only monomer-like emission was observed in fluorescence experiments. Increasing the perylene fraction first increases the

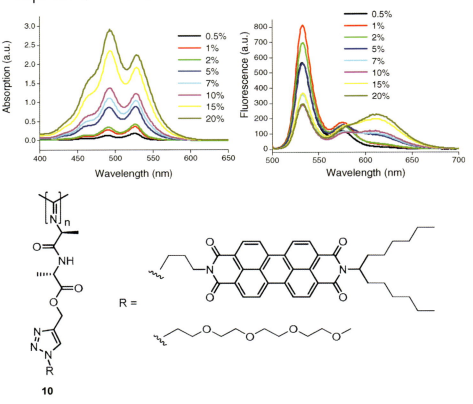

Figure 5.12 UV-Vis and fluorescence spectra of randomly clicked polymers obtained after reaction of **10** with ethylene glycol azides and perylene azides (with an increasing percentage of perylene) in CH_2Cl_2.

intensity of the monomer emission. However, after a further increase in the perylene content, excimer emission starts to dominate (Figure 5.12).

Even more complex systems are feasible when multiple chromophores are grafted simultaneously. For instance, copolymers based on a coumarin dye and a perylene dye were formed in which the absorption and emission from both chromophores could be seen. Interactions between the chromophores in proximity were observed, evidenced by a quenched and blueshifted emission of the coumarin dyes. This shows that two different chromophores can now be readily incorporated into the polymers and opens the way to polymeric materials with tunable properties.

The approach, presented above, in which rigid polymer chains are used as scaffolds in order to control the position of dyes within a macromolecular structure allows the construction of a whole class of stiff, chromophore-appended nanowires. The rigidity of the wire on the nanometer scale, the linker flexibility, and the stacking of the dye molecules on the molecular level can be adjusted by chemical modification, thereby allowing a readily accessible preparation route toward novel materials.

5.7
Toward Larger Length Scales: Polyisocyanopeptide Brushes

For application of the helical polymers in actual devices, the precise organization of the chromophores in arrays is a prerequisite but not sufficient. Also, their organization at much larger length scales, typically micrometers, needs to be controlled. A way to control the morphology on a macroscopic scale is to grow polyisocyanides from a surface. These brushes have several advantages over conventionally processed polymers. By tethering the polymers to the surface at high grafting densities, the chains are forced to stretch away from the surface resulting in unidirectional alignment. Besides the higher grafting densities that are obtained, the alignment of the polymers will result in improved charge or energy transport properties. For responsive polymers sensitive to changes in pH, salt concentration, temperature, and solvent conditions, this will give rise to a unidirectional response. Moreover, conventional patterning techniques allow more complex spatially controlled surface morphologies.

Polyisocyanide brushes can be grown from the surface by anchoring the catalyst to the surface [39]. Two strategies can be followed: first, immobilize $(tBuNC)Ni(ClO_4)_2$ on an amino-terminated monolayer. Second, use presynthesized triethoxysilane Ni (II) complex to form the activated Ni-functionalized monolayers in a single step (Scheme 5.4).

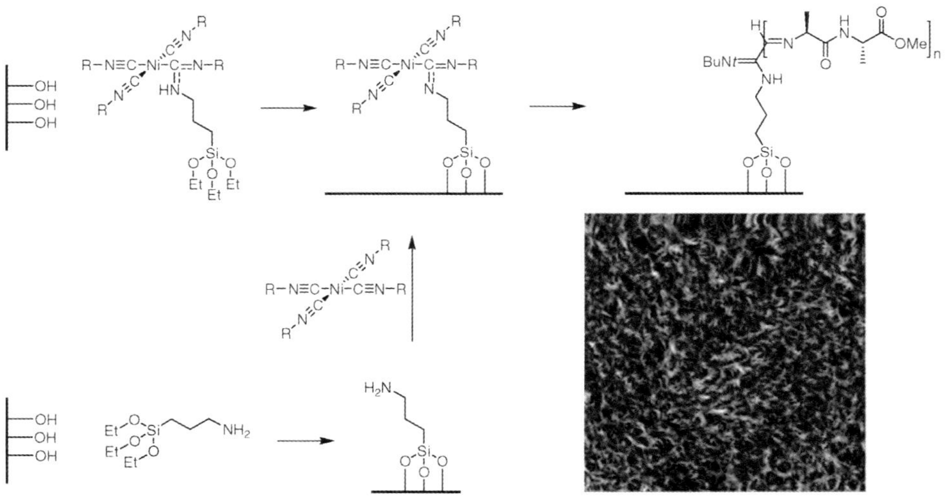

Scheme 5.4 Functionalization of the surface with a nickel catalyst and an AFM image of the obtained surface morphology.

The first method results in the formation of rather thin brushes. This, tentatively, is attributed to the challenging formation of high-quality 3-aminopropyltrimethoxysilane (APTS) self-assembled monolayers, causing an insufficient surface density of the Ni(II) catalyst to provide efficient brush growth. When the second method was used, brushes up to 200 nm could be grown in 3 h. The surface growth could readily be controlled by changing the polymerization time as well as the monomer concentration. The high tolerance to impurities, oxygen, and water is an additional advantage of the procedure.

Analyzing these brushes with CD spectroscopy showed the characteristic strong positive Cotton effect centered at 315 nm and IR spectroscopy showed the same peaks as for the corresponding bulk polymer, implying that the brushes-stabilized well-defined helices are formed.

The polyisocyanide brushes could serve as a rigid and well-aligned backbone for functional polymer thin films. Isocyanides containing electroactive groups, such as carbazole, have been prepared for the use in field-effect transistors (FETs) and photovoltaic devices. Alternatively, the modular concept of click chemistry can be applied: in the first step, the (patterned) surface is functionalized with the catalyst, which is used to grow the azide or acetylene-functionalized polyisocyanide. In the final step, one or more desired chromophores can be attached to the rigid helical scaffold.

5.8
Summary and Outlook

The ability to orient chromophores in space into a more efficient energy or electron transfer geometry, which is difficult to achieve by self-assembly, offers considerable opportunities for materials to have much improved mechanical and electronic properties. Polyisocyanides have shown to be ideal candidates since they offer a well-defined and highly stable scaffold that can be grown up to micrometers in length. The functional side groups can be introduced either at the monomer level or using the postmodification approach. Various studies have shown that the polymer backbone controls the organization and orientation of the chromophoric side groups, which renders them suitable materials for many electronic or optical applications. The challenges to incorporate these materials in real devices will be to not only control the organization at the nanometer scale but also at device dimensions.

References

1 Mcdermott, G., Prince, S.M., Freer, A.A., Hawthornthwaitelawless, A.M., Papiz, M.Z., Cogdell, R.J., and Isaacs, N.W. (1995) *Nature*, **374**, 517.
2 Bahatyrova, S., Frese, R.N., van der Werf, K.O., Otto, C., Hunter, C.N., and Olsen, J.D. (2004) *J. Biol. Chem.*, **279**, 21327; Fleming, G.R. and vanGrondelle, R. (1997) *Curr. Opin. Struct. Biol.*, **7**, 738; Linnanto, J. and Korppi-Tommola, J.E.I. (2002) *Phys. Chem. Chem. Phys.*, **4**, 3453; Pullerits, T. and Sundstrom, V. (1996) *Acc. Chem. Res.*, **29**, 381; Scheuring, S., Seguin, J., Marco, S., Levy, D., Robert, B., and Rigaud, J.L. (2003) *Proc. Natl. Acad. Sci. USA*, **100**, 1690.
3 Bosman, A.W., Janssen, H.M., and Meijer, E.W. (1999) *Chem. Rev.*, **99**, 1665;De Schryver, F.C., Vosch, T., Cotlet, M., Van der Auweraer, M., Mullen, K., and Hofkens, J. (2005) *Acc. Chem. Res.*, **38**, 514.

4 Schenning, A.P.H.J. and Meijer, E.W. (2005) *Chem. Commun.*, 3245.
5 Hoeben, F.J.M., Jonkheijm, P., Meijer, E.W., and Schenning, A. (2005) *Chem. Rev.*, **105**, 1491.
6 Li, H.P., Martin, R.B., Harruff, B.A., Carino, R.A., Allard, L.F., and Sun, Y.P. (2004) *Adv. Mater.*, **16**, 896.
7 Tinnefeld, P., Heilemann, M., and Sauer, M. (2005) *ChemPhysChem*, **6**, 217; Heilemann, M., Kasper, R., Tinnefeld, P., and Sauer, M. (2006) *J. Am. Chem. Soc.*, **128**, 16864; Sanchez-Mosteiro, G., van Dijk, E.M.H.P., Hernando, J., Heilemann, M., Tinnefeld, P., Sauer, M., Koberlin, F., Patting, M., Wahl, M., Erdmann, R., van Hulst, N.F., and Garcia-Parajo, M.F. (2006) *J. Phys. Chem. B*, **110**, 26349.
8 Teo, Y.N., Wilson, J.N., and Kool, E.T. (2009) *J. Am. Chem. Soc.*, **131**, 3923.
9 Wagenknecht, H.-A. (2009) *Angew. Chem., Int. Ed.*, **48**, 2838; Varghese, R. and Wagenknecht, H.-A. (2009) *Chem. Commun.*, 2615.
10 Yu, H., Bur, A.J., and Fetters, L.J. (1966) *J. Chem. Phys.*, **44**, 2568; Troxell, T.C. and Scheraga, H.A. (1971) *Macromolecules*, **4**, 528; Itou, T., Chikiri, H., Teramoto, A., and Aharoni, S.M. (1988) *Polym. J.*, **20**, 143.
11 Fujiki, M. (2001) *Macromol. Rapid. Commun.*, **22**, 539.
12 Tang, H.D., Liu, Y.Y., Huang, B., Qin, J.G., Fuentes-Hernandez, C., Kippelen, B., Li, S.J., and Ye, C. (2005) *J. Mater. Chem.*, **15**, 778.
13 Goodwin, A. and Novak, B.M. (1994) *Macromolecules*, **27**, 5520.
14 Deming, T.J. and Novak, B.M. (1992) *J. Am. Chem. Soc.*, **114**, 7926.
15 Shibayama, K., Seidel, S.W., and Novak, B.M. (1997) *Macromolecules*, **30**, 3159; Kim, J., Novak, B.M., and Waddon, A.J. (2004) *Macromolecules*, **37**, 8286.
16 Tang, H.Z., Novak, B.M., He, J.T., and Polavarapu, P.L. (2005) *Angew. Chem., Int. Ed.*, **44**, 7298; Tang, H.Z., Lu, Y.J., Tian, G.L., Capracotta, M.D., and Novak, B.M. (2004) *J. Am. Chem. Soc.*, **126**, 3722.
17 Millich, F. (1972) *Chem. Rev.*, **72**, 101.
18 Nolte, R.J.M. (1994) *Chem. Soc. Rev.*, **23**, 11; Yashima, E., Maeda, K., Iida, H., Furusho, Y., and Nagai, K. (2009) *Chem. Rev.*, **109**, 6102; Kumaki, J., Sakurai, S., and Yashima, E. (2009) *Chem. Soc. Rev.*, **38**, 737.
19 Deming, T.J. and Novak, B.M. (1993) *J. Am. Chem. Soc.*, **115**, 9101.
20 Huang, J.T., Sun, J.X., Euler, W.B., and Rosen, W. (1997) *J. Polym. Sci. A Polym. Chem.*, **35**, 439.
21 Cornelissen, J.J.L.M., Donners, J.J.J.M., de Gelder, R., Graswinckel, W.S., Metselaar, G.A., Rowan, A.E., Sommerdijk, N.A.J.M., and Nolte, R.J.M. (2001) *Science*, **293**, 676; Cornelissen, J.J.L.M., Graswinckel, W.S., Adams, P.J.H.M., Nachtegaal, G.H., Kentgens, A.P.M., Sommerdijk, N.A.J.M., and Nolte, R.J.M. (2001) *J. Polym. Sci. A Polym. Chem.*, **39**, 4255; Cornelissen, J.J.L.M., Graswinckel, W.S., Rowan, A.E., Sommerdijk, N.A.J.M., and Nolte, R.J.M. (2003) *J. Polym. Sci. A Polym. Chem.*, **41**, 1725; Cornelissen, J.J.L.M., Sommerdijk, N.A.J.M., and Nolte, R.J.M. (2002) *Macromol. Chem. Phys.*, **203**, 1625.
22 Razenberg, J.A.S.J., Vandermade, A.W., Smeets, J.W.H., and Nolte, R.J.M. (1985) *J. Mol. Catal.*, **31**, 271.
23 Samorì, P., Ecker, C., Goessl, I., de Witte, P.A.J., Cornelissen, J.J.L.M., Metselaar, G.A., Otten, M.B.J., Rowan, A.E., Nolte, R.J.M., and Rabe, J.P. (2002) *Macromolecules*, **35**, 5290.
24 de Witte, P.A.J., Castriciano, M., Cornelissen, J.J.L.M., Scolaro, L.M., Nolte, R.J.M., and Rowan, A.E. (2003) *Chem. Eur. J.*, **9**, 1775.
25 de Witte, P.A.J. (2004) Helical chromophoric nanowires. Ph.D. thesis. Radboud University Nijmegen, the Netherlands.
26 Gomar-Nadal, E., Mugica, L., Vidal-Gancedo, J., Casado, J., Navarrete, J.T.L., Veciana, J., Rovira, C., and Amabilino, D.B. (2007) *Macromolecules*, **40**, 7521.
27 Gomar-Nadal, E., Veciana, J., Rovira, C., and Amabilino, D.B. (2005) *Adv. Mater.*, **17**, 2095.
28 Wurthner, F. (2004) *Chem. Commun.*, 1564.
29 Hernando, J., de Witte, P.A.J., van Dijk, E.M.H.P., Korterik, J., Nolte, R.J.M., Rowan, A.E., Garcia-Parajo, M.F., and van Hulst, N.F. (2004) *Angew. Chem., Int. Ed.*, **43**, 4045.

30 de Witte, P.A.J., Hernando, J., Neuteboom, E.E., van Dijk, E.M.H.P., Meskers, S.C.J., Janssen, R.A.J., van Hulst, N.F., Nolte, R.J.M., Garcia-Parajo, M.F., and Rowan, A.E. (2006) *J. Phys. Chem. B*, **110**, 7803; Foster, S., Finlayson, C.E., Keivanidis, P.E., Huang, Y.-S., Hwang, I., Friend, R.H., Otten, M.B.J., Lu, L.-P., Schwartz, E., Nolte, R.J.M., and Rowan, A.E. (2009) *Macromolecules*, **42**, 2023.

31 Palermo, V., Otten, M.B.J., Liscio, A., Schwartz, E., de Witte, P.A.J., Castriciano, M.A., Wienk, M.M., Nolde, F., De Luca, G., Cornelissen, J.J.L.M., Janssen, R.A.J., Mullen, K., Rowan, A.E., Nolte, R.J.M., and Samorì, P. (2008) *J. Am. Chem. Soc.*, **130**, 14605.

32 Finlayson, C.E., Friend, R.H., Otten, M.B.J., Schwartz, E., Cornelissen, J.J.L.M., Nolte, R.L.M., Rowan, A.E., Samorì, P., Palermo, V., Liscio, A., Peneva, K., Mullen, K., Trapani, S., and Beljonne, D. (2008) *Adv. Funct. Mater.*, **18**, 3947.

33 Schwartz, E., Palermo, V., Finlayson, C.E., Huang, Y.-S., Otten, M.B.J., Liscio, A., Trapani, S., González-Valls, I., Brocorens, P., Cornelissen, J.J.L.M., Peneva, K., Müllen, K., Spano, F., Yartsev, A., Westenhoff, S., Friend, R.H., Beljonne, D., Nolte, R.J.M., Samorì, P., and Rowan, A.E. (2009) *Chem. Eur. J.*, **15**, 2536.

34 Shklovskii, B.I. and Efros, A.L. (1984) *Electronic Properties of Doped Semiconductors*, Springer, Berlin.

35 Dabirian, R., Palermo, V., Liscio, A., Schwartz, E., Otten, M.B.J., Finlayson, C.E., Treossi, E., Friend, R.H., Calestani, G., Mullen, K., Nolte, R.J.M., Rowan, A.E., and Samorì, P. (2009) *J. Am. Chem. Soc.*, **131**, 7055.

36 Tornoe, C.W., Christensen, C., and Meldal, M. (2002) *J. Org. Chem.*, **67**, 3057; Rostovtsev, V.V., Green, L.G., Fokin, V.V., and Sharpless, K.B. (2002) *Angew. Chem., Int. Ed.*, **41**, 2596.

37 Schwartz, E., Kitto, H.J., de Gelder, R., Nolte, R.J.M., Rowan, A.E., and Cornelissen, J.J.L.M. (2007) *J. Mater. Chem.*, **17**, 1876.

38 Kitto, H.J., Schwartz, E., Nijemeisland, M., Koepf, M., Cornelissen, J.J.L.M., Rowan, A.E., and Nolte, R.J.M. (2008) *J. Mater. Chem.*, **18**, 5615.

39 Lim, E., Tu, G., Schwartz, E., Cornelissen, J.J.L.M., Rowan, A.E., Nolte, R.J.M., and Huck, W.T.S. (2008) *Macromolecules*, **41**, 1945.

6
Functional Polyphenylenes for Supramolecular Ordering and Application in Organic Electronics
Martin Baumgarten and Klaus Müllen

6.1
Introduction

The benzene structure although it occurs as a quasi-omnipresent building block of organic chemistry in natural products and in drugs, it is also an important repeat unit of dyes and conjugated polymers. In the latter case, the key issue is the formation of extended π-systems as mobile electrons. This qualifies conjugated molecules not only to act as chromophores upon interaction with light but also to undergo electron transfer reactions. Is it logical that these molecules also serve as active components of electronic and optoelectronic devices [1], such as light emitting diodes (LEDs) [2], solar cells [3], and field-effect transistors (FETs) [1, 4]? The device function depends upon the solid state and thus upon the morphology of thin films, so it is of key importance to discriminate between functional properties being controlled by the molecular design and by the packing arrangements in the solid [5].

Benzene units can be arranged in a linear fashion to yield typical conjugated or conducting polymers [6], but they can also be extended, thus affording two-dimensional (2D) graphene-like molecules [7], or even into three dimensions yielding shape-persistent polyphenylene dendrimers [8]. The dimensionality of the benzene-based polyphenylenes is the determining factor for aggregation and self-assembly processes, for example, when charge carrier mobility is to be increased by establishing supramolecular order, one targets lamella-type packing with a side-by-side arrangement of linear chains, face-to-face-type stacking of disks, or even a three-dimensional order of polyphenylene dendrimers with interdigitation of dendrimer arms [9]. It should be noted that while the long-range supramolecular order can be a desirable structural feature, for example, for charge carrier transport, amorphous arrays with "noninteracting" conjugated molecules can be preferential, in order to avoid intermolecular exciton delocalization [10].

In organic photovoltaics, on the other hand, charge separation from a donor to an acceptor molecule is required with smooth charge transport to the electrodes, avoiding any recombination [3]. Here the major criterion is the transformation of light into current in which the molecular design plays a key role because the HOMO–LUMO

Functional Supramolecular Architectures. Edited by Paolo Samorì and Franco Cacialli
Copyright © 2011 WILEY-VCH Verlag GmbH & Co. KGaA, Weinheim
ISBN: 978-3-527-32611-2

bandgap determines the absorption characteristics and thus the light harvesting efficiency. The orbital levels of donors and acceptors must also be tuned for enabling an efficient light-induced charge transfer. Here again, however, the relative spatial arrangement of donor and acceptor molecules and the resulting morphologies of the films cannot be ignored because what is needed is a phase separation of donor and acceptor molecules to ensure separate percolation pathways for holes and electrons.

The orbital energies of 1D, 2D, or 3D polyphenylenes can be controlled not only by the topology but also by the size of the molecules. While size can be a desirable feature, in particular, for creating large areas of mobile π-electrons, the problems with solubility and meltability can severely hinder the desired film formation during the processing step. This is why substitution of the conjugated core, such as with alkyl chains, becomes a major issue that is also relevant for packing in the solid state. In addition, one must consider whether the alkyl chains compromise the extended π-conjugation.

While alkyl chains do not significantly change the relevant energy levels, substitution with electron-withdrawing or electron-donating substituents and additional incorporation of heteroatoms into the π-system can transform the polyphenylenes into electron donors or electron acceptors. Here again this is relevant not only for the formation of charges in the solid but also for the packing and thus for the kinetics of charge carrier transport.

When fabricating polymer-based devices, charge carrier mobilities in FETs or power efficiencies in solar cells depend not only upon the dimensionality of the polyphenylene but also upon the molecular weight, the amount of structural defects, and the impurities that can serve as traps for excitons or hamper the electron hopping processes [1]. Device behavior can thus vary significantly from batch to batch and this has stimulated the interest in monodisperse conjugated oligomers, both as active materials in their own right and as models for their polymeric counterparts [11]. These materials are monodisperse and defect free such that more reliable structure property relations can be obtained and single crystals can be grown.

This chapter intends to outline polyphenylenes as unique conjugated π-systems. Indeed, a good case can be made to emphasize the power of synthetic chemistry and exemplify the beauty by which clever manipulation of the molecular structure can allow one to control the electronic properties of the molecule. This ability at a molecular and supramolecular level by synthesis and processing will then pave the way to advanced electronics and sophisticated energy production by putting the molecules to work and utilizing their electronic function.

6.2
Conjugated Polymers

6.2.1
Linear and Ladderized Polyphenylenes

The formation of oligomeric and polymeric polyphenylene chains is strongly based on their well-defined accessibility. While unsubstituted poly-*para*-phenylenes **1**(PPP) usually contained a large number of defects, some well-defined oligomers became available and have been applied in OLEDs. The trick to overcome the synthetic difficulties and obtain reproducibility in the polymer synthesis of PPPs was based on two achievements: (i) introduction of solubilizing side chains and (ii) modern metal-catalyzed reactions such as Suzuki cross-coupling [12–15] and Yamamoto coupling [16, 17], which enabled a high-yield synthesis in order to obtain proper molecular weights.

The introduction of solubilizing side chains directly attached to the phenylene repeat units as in **2** (Scheme 6.1), however, dramatically increased the torsion (dihedral angle θ) between neighboring repeat units and thereby interrupted conjugation to such an extent that the bandgap increased and no fluorescence in the visible range occurred. Thus, a new synthesis was designed, where the neighboring phenyl units could be planarized while still carrying solubilizing side chains. In this vein, the fully ladderized poly(*para*-phenylene)s **3** (LPPPs) as demonstrated in Scheme 6.1 were developed [18, 19]. It turned out that this more planar alignment of neighboring benzene units led to a strong enhancement of the conjugation, which was indicated by bathochromically shifted optical absorptions, where the emissions often tend into the more green oriented spectral range [20, 21].

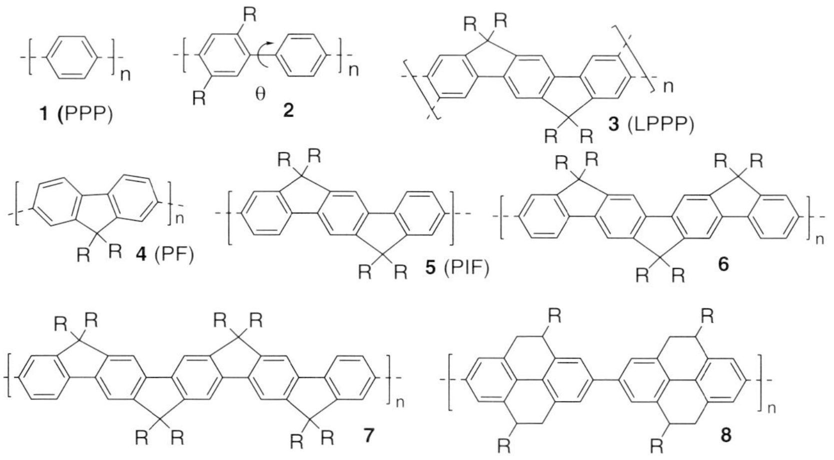

Scheme 6.1 Poly-*para*-phenylenes **1** and **2** (PPPs), ladderized poly-*para*-phenylenes **3** (LPPP), polyfluorenes **4** (PF), polyindenofluorene (**5**, PIF) ladderized polytetraphenylene **6**, ladderized polypentaphenylenes **7**, and polytetrahydropyrenes (**8**).

Since there was a strong need for stable blue emitters, partially ladderized PPPs (so-called step ladder polymers) [22, 23], where some phenylene units are planarized by additional linkages as in poly(2,7-fluorene)s **4**(PFs) [24–26], poly(indenofluorene)s **5** (PIFs) [26–28], poly(tetraphenylene)s **6** [29, 30], and poly(pentaphenylene)s **7** [31–33], raised further interest. The synthetic details of **3–7** are outlined in original manuscripts and earlier reviews, hence they are not described in more depth here [6, 10, 22, 34, 35]. Poly(indenofluorene)s **5**, which are intermediate in structure between poly(fluorene)s **4** (PF)s and fully ladderized poly(*para*-phenylene)s **3** (LPPPs), have been prepared and found to show PL maxima in solution around 430 nm, making them attractive candidates for use in blue OLEDs [30]. Unfortunately, they suffered from problems similar to that of **4** in that their electroluminescence (EL) emission rapidly turned green, which is now attributed to the formation of emissive ketone defects in **4** such as **9** [34, 35]. The reason for this could be attributed to small amounts of monoalkylated fluorenes (even below 1%), and heavy efforts were taken to avoid such impurities as new synthetic pathways of monomer preparation [2, 36]. One opportunity to overcome fluorenone defect formation could be complete arylene substitution (R = C_6H_4 alkyl) in the bridgehead position of partially ladderized PPPs. Two further examples were therefore considered, namely, the ladderized poly(tetraphenylene)s **6** and the poly(pentaphenylene)s **7** [31, 32]. In particular, the fully arylated poly(pentaphenylene)s finally displayed remarkably stable blue EL. Furthermore, a comparison of the novel fully aryl-substituted polymer with a partially aryl-substituted polymer revealed that the oxidative degradation-induced defect formation could be almost completely eliminated upon aryl substitution [31].

Another variation of the bridging of phenylene units was tried earlier by using dialkylated polytetrahydropyrenes **8**, which were achieved through Yamamoto coupling of the alkylated 2,7-dibromo-tetrahydropyrene monomers [37]. They however showed a bathochromically shifted fluorescence of their films compared to their solution counterparts indicating aggregation, even though they were used to make a blue emitting (λ_{max}em = 457 nm) device.

After the hydrocarbon polymers such as polyfluorenes **4** were synthesized from 2,7-dibromofluorene monomers via Yamamato polycondensation, their structural variation with heteroatoms at the bridgehead methylene position was introduced. Thus, electron-poor or electron-rich groups should allow an adjustment of their highest occupied molecular orbital (HOMO) and lowest unoccupied molecular orbital (LUMO). The corresponding poly(2,7-carbazole)s (**10**) were thus developed [38], where the carbazole unit is more electron rich than the fluorene moiety and therefore can serve as a superior donor. While the bromination of carbazole usually takes place in the 3,6-positions, the 2,7-dibromocarbazole monomers (**11**) were synthesized via a reductive Cadogan ring closure reaction [39] and decorated with swallow-tail branched alkyl chains (Scheme 6.2).

Polymerizations via Ni(COD)-mediated Yamamoto coupling yielded molecular weights of around 75 kg mol^{-1}. Based on 2,7-bridging, another analogue of poly-*para*-phenylene was born that was first considered for OLED applications. It turned out, however, that the fluorescence of polycarbazoles **10** is redshifted compared to those of polyfluorenes **4** and suffered from instability in OLED devices, presumably due to the high reactivity in 3,6-position. Thus, partially ladderized nitrogen-bridged poly(ladder-type tetraphenylene) **12** [29] was approached (Scheme 6.3) demonstrating a further increase in EL stability and luminance values compared to the corresponding hydrocarbon poly(tetraphenylene) **6** [32].

Since polycarbazoles **10** have a high-lying HOMO orbital, the question arose whether they may serve as suitable donor materials for photovoltaic devices in place

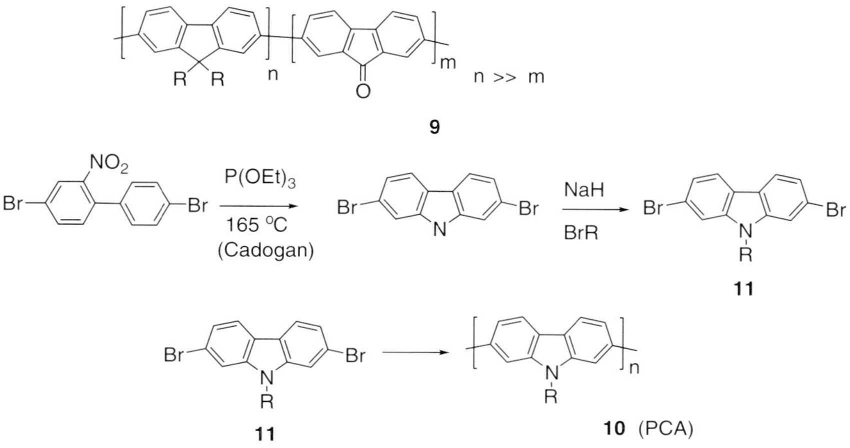

Scheme 6.2 Fluorenone defects in **9**, 2.7-polycarbazoles (**10**), and 2,7-dibromocarbazoles (**11**).

of the commonly used poly(3-hexylthiophene)s (P3HT). The bandgap of **10** is relatively large (~3 eV) such that a combination with an acceptor possessing stronger absorption in the visible range than fullerenes was tested for light harvesting. Perylenetetracarboxydiimide (**15**, PDI) seemed well suited (Scheme 6.4) and in first attempts, efficiencies of 0.6% under illumination with solar light were obtained [40]. Further comparisons with the partially ladderized N-containing polymers **12–14** as donors and PDI as the acceptor demonstrated that they were even better suited as the efficiency was raised up to 1.4% [41]. It could thus be confirmed that the results with

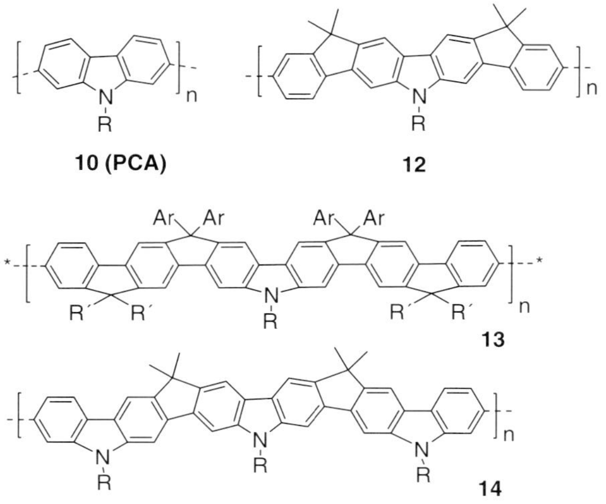

Scheme 6.3 Polycarbazoles (**10**), nitrogen-bridged poly(ladder-type tetraphenylene) (**12**), poly(ladder-type hexaphenylene) (**13**), and ladderized poly(triscarbazole)s (**14**).

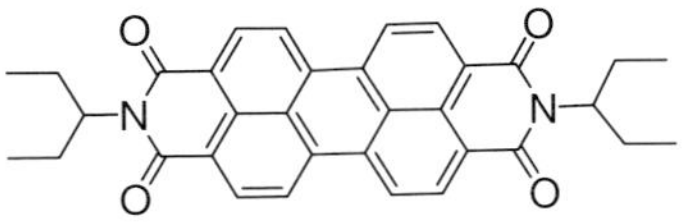

Scheme 6.4 PDI **15** and vinazene **16**.

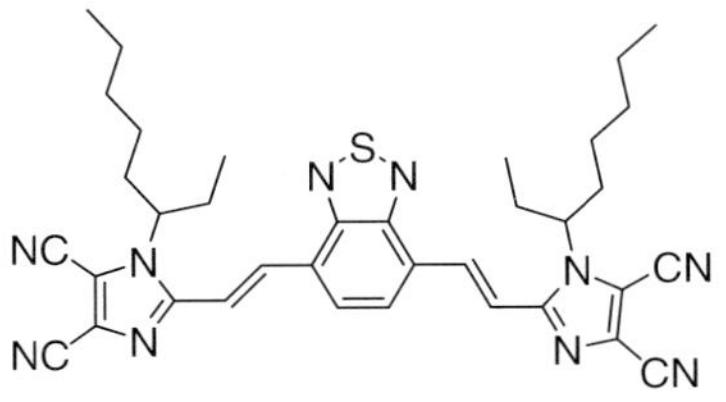

PDI were much better than those with the widely used acceptor (6,6)-phenyl-C_{60}-butyric acid methyl ester (PCBM) achieving only half of the demonstrated efficiencies (0.7%).

These results for the donor–acceptor pair polycarbazole/PDI were then compared with another small-molecule electron acceptor based on 2-vinyl-4,5-dicyanoimidazole (vinazene **16**) for use in solution-processed organic solar cells [42]. This material has a favorably located LUMO level of −3.6 eV and absorbs strongly in the visible spectrum up to 520 nm, attractive properties compared to PCBM. Vinazene **16** was blended with a poly(2,7-carbazole) donor chosen for its complementary absorption range and comparatively high-lying HOMO level of −5.6 eV and incorporated into bulk heterojunction devices. The best performing devices exhibited reasonable power conversion efficiencies of 0.75% and open-circuit voltages of more than 1.3 V, substantially higher than previously reported devices using small-molecule acceptors. Besides variation of the polyphenylene structures, the incorporation of new building blocks was also pursued.

6.2.2
Polycyclic Aromatic Hydrocarbon-Based Conjugated Polymers

While conjugated polymers made from polycyclic aromatic hydrocarbons (PAHs) attracted immense attention, it is interesting to note that PAHs, although holding great promise as chromophoric and electrophoric building blocks, have seldom been included in conjugated polymers.

Novel phenanthrene (**17**), anthracene (**18**), triphenylene (**19**), and even pyrene (**20**) containing polymers have recently been synthesized (Scheme 6.5). These not only gave rise to new photo- and electroluminescence properties but also highlighted the

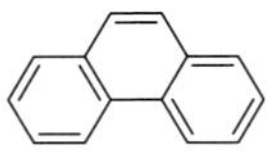

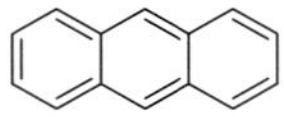

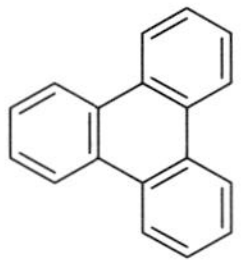

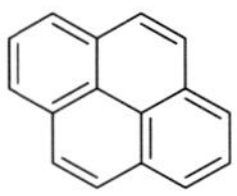

Scheme 6.5 Phenanthrene (**17**), anthracene (**18**), triphenylene (**19**), and pyrene (**20**).

importance of suitable functionalization and appropriate polymerization techniques for PAHs [43–49].

The poly(2,7-phenanthrylene)s **21** and poly(3,6-phenanthrylene)s **22** [43] can be considered as analogues of poly(phenylene)s and poly(phenylene vinylene)s (Scheme 6.6). The introduction of alkyl or aryl substituents in the 9,10-positions of the phenanthrene rendered these molecules soluble and processable. In the case of 2,7-linked polymers, dialkylsubstitution allowed strong unwanted aggregation indicated by a large bathochromic shift of the fluorescence of the film compared to their solution property, which could be avoided and suppressed by aryl substitution.

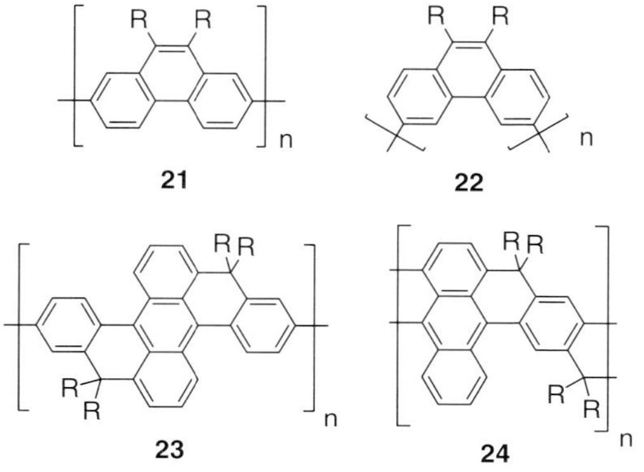

Scheme 6.6 Polyphenanthrylenes **21** and **22**, and step ladder and ladder poly(p-phenylene-alt-anthrylene)s **23** and **24**.

Step ladder and ladder poly(p-phenylene-alt-anthrylene)s **23** and **24** have been prepared to study their optoelectronic properties [45]. While phenylenes attached to anthracene induce a strong twist around the aryl–aryl bond, thereby hindering conjugation, the idea was to planarize them by introducing solubilizing groups at the bridging methylene carbons. This planarization led to a strong bathochromically shifted fluorescence with yellow emission for the step ladder polymer (λ_{max}em = 584 nm) and even red emission (λ_{max}em = 693 nm) for the ladderized polymer. Upon irradiation with visible light in air, however, both polymers underwent photooxygenation with colorless endoperoxides being formed. These peroxides, however, could reversibly be removed upon heat treatment.

As a novel building block for conjugated polymers, 6,11-dibromotriphenylenes **25** were developed (Scheme 6.7). They became available in just two synthetic steps in large amounts starting from the 2,7-dibromophenantrenedione and these triphenylenes were used for Suzuki–Miyaura copolymerizations with phenyldiboronic esters (**26**) and Yamamoto-type homopolymerizations (**27**) [46, 47].

Due to their narrow emission maxima in the blue spectral range between 418 and 430 nm in solution, they were also tested in OLED devices. Upon blending with electron transporting and hole injection material, an onset voltage of 4.6 V and luminance efficiencies of 0.73 cd A^{-1} were achieved [47]. These device properties were then compared with those of a completely new kind of functionalized

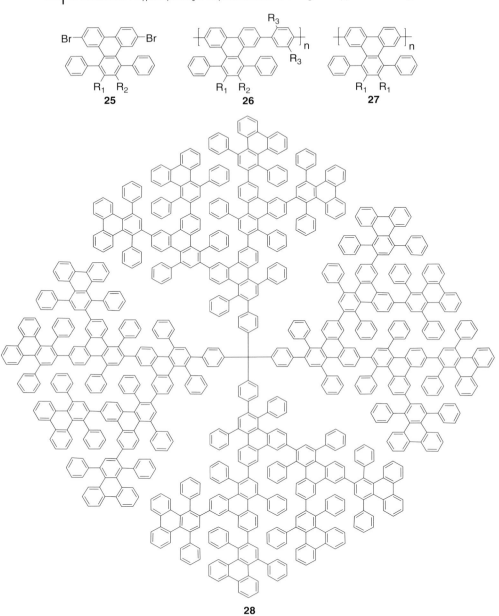

Scheme 6.7 Dibromotriphenylenes **25**, triphenylene polymers **26** and **27**, and polytriphenylene dendrimer **28**.

polyphenylene dendrimers, carrying triphenylene units in each branch, leading to so-called polytriphenylene dendrimers **28** [50]. Comparing their synthetic accessibility with those of the linear polymers, it should be noted that the monomeric precursors for both needed a three- to four-step synthesis, but dendrimers **28** could only be grown via a step-by-step Diels–Alder cycloaddition reaction being repeated for each generation [50]. In terms of structure, however, these dendrimers are more well defined, since they are monodisperse macromolecules of defined molecular weight, while the polymers still possess quite arbitrary number of repeat units with polydispersities ranging from 1.4–2. Devices based upon **28** emitted a sky blue electroluminescence with a maximum brightness of 300 cd m^{-2} at a bias voltage of 8 V and corresponding Commission Internationale de L'Eclairage (CIE) coordinates of (0.19, 0.18). The maximum of the electroluminescence spectrum peaked at 430 nm. For polymers **26** and **27**, the EL spectra exhibited two peaks at 428 and 456 nm and some broad tailing emission beyond 500 nm [47].

Thus, it still needs to be investigated whether these polymers, being accessible in a one-step synthesis from their monomeric precursor, or the monodisperse dendrimers, requiring a second generation at the minimum and demanding at least a three-step synthesis from the appropriate building block, are more powerful to be commercialized as blue emitters.

While the triphenylenes as shown above are relatively poor emitters, exhibiting increased quantum yield with an increasing number of polymeric repeat units or dendrimer generations (4–30%), pyrenes combine high photoluminescence efficiency with high charge carrier mobility [51]. Thus, regioselective chemical functionalizations of pyrene were developed, which allowed the preparation of poly(2,7-(4,5,9,10-phenylalkyl)pyrenylene)s (**29**) [48], poly(2,7-(4,5,9,10-tetraalkoxy)pyrenylene)s (**30**) [52], and poly(2-7-*tert*-butyl-1,3-pyrenylene)s (**31**) [49] as blue emitters (Scheme 6.8).

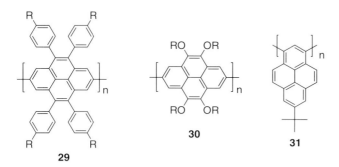

Scheme 6.8 Conjugated polymers **29–31** with pyrene in the main chain.

While the synthesis of the 4,5,9,10-tetraphenylated-2,7-dichloropyrene took five steps to build up the pyrene core [48], for **30** and **31** new synthetic routes were followed. The 2,7-dibromotetraalkoxypyrene for polymerization became available in two steps from the known pyrenetetraone upon bromination and alkoxylation.

For **31**, the *tert*-butylation of pyrene allowed the selective bromination at the active 1,3-positions [49].

Due to the unique substitution with bulky aryl groups at the 4,5,9,10-positions of pyrene in **29**, the polymer, although comprising large π-units, is readily soluble in common organic solvents. Polymer **29** showed a blue fluorescence in solution with an emission band maximum at $\lambda = 429$ nm, which fulfilled the requirements for a blue emitting polymer. Some additional redshifted emission bands in the solid film could be strongly reduced by blending with a nonconjugated polymer such as polystyrene.

For poly(tetraalkoxypyrenylene) **30**, it was very important to find out that the fluorescence maxima in solution and in the film were almost identical (peak shift <2 nm), such that the alkoxy groups seem to hinder aggregation. In the LED devices, however, a strong greenshift appeared and it could be proven by temperature-dependent emission measurements and upon UV irradiation that degradation occurred. Thus, the next target was polymer **31**, which due to the strong steric hindrance between the pyrenes in the chain (torsional angle about 70°) could be prevented from π-stacking. The emission maxima for the solid film occurred at 454 nm, while the blue turquoise electroluminescence peaked at 465 nm, but relatively high luminance values of 300 cd m^{-2} were measured at a voltage of 8 V. Of this series of pyrene polymers, the *tert*-butylated polymer **31** seemed to be the most stable and best blue emitter [49]. Surprisingly, its efficiency of 0.3 cd A^{-1} was still less than half of that found for the triphenylene polymer **27**.

So far we have outlined numerous approaches to blue emitters from linear conjugated polymers. While it was important to overcome the typical keto defects in alkylated polyfluorenes by arylation or insertion of nitrogen at the bridgehead position leading to polycarbazoles **9**, new PAHs were introduced as repeat units **21–31**. Up to this end, however, only one comparison between polydisperse conjugated polymers with sometimes undefined end groups carrying impurities from catalyst leftovers and a well-defined macromolecule prepared by catalyst free growth, namely, polytriphenylene dendrimer **28**, was made. It thus turned out to be important to compare or combine the advantages of both approaches, namely, the monodispersity and the unresolved question of end group contamination, by turning toward well-defined monodisperse conjugated polyphenylene macrocycles.

6.2.3
Polyphenylene Macrocycles

Fully conjugated macrocycles have gained tremendous attention since they can be considered as perfect polymer models without any endgroup defects. Shape-persistent macrocycles with porous cavities, capable of forming ordered columnar structures, have been synthesized both by choice of proper geometries for ring closure and by template-assisted reactions [53–56]. However, few of these macrocycles possessed an extended π-conjugation. A new approach toward cyclic conjugated oligomers was therefore developed by testing a concept of extended conjugation length using *para*-phenylene coupling units [57].

Often *meta*-phenylene units were applied for macrocycles since they facilitate the ring formation, but they interrupt the conjugation. A template-assisted synthesis of a monodisperse fully conjugated 2,7-carbazole-based macrocyclic dodecamer **32** has therefore been considered using triscarbazole precursors **33** and a tetraphenylporphyrin derivative **34** as a template (Scheme 6.9). The dichlorinated triscarbazoles **33** attached to core **34** were then cyclized via Yamamoto reaction to the corresponding dodecamer carbazole macrocycle **35**. This supermolecule **35**, where the porphyrin template is located in the center of the cavity of the carbazole macrocycle, showed a Förster energy transfer of 83% from the peripheral carbazole π-system to the central

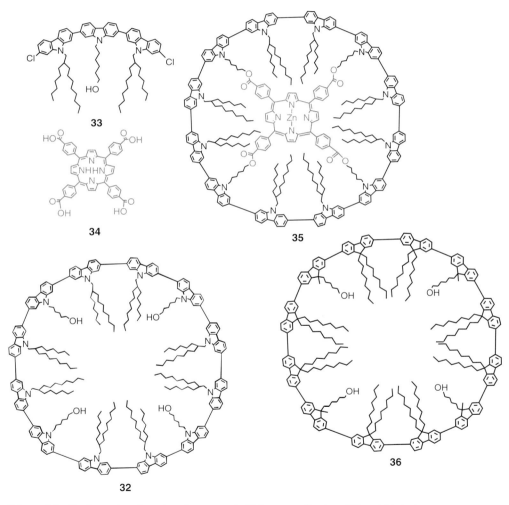

Scheme 6.9 The formation of a carbazole macrocycle **32** from esterification of **33** and **34** via ring-closed templated **35** and the fluorene macrocycle **36**.

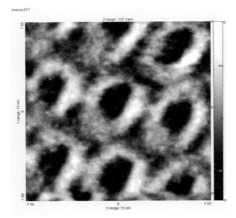 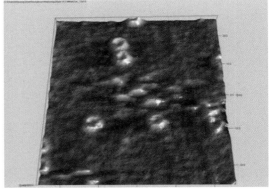

Figure 6.1 STM images of **32** as monolayer and isolated rings on HOPG surface.

porphyrin core. These elaborated materials thus give fascinating opportunities toward studying energy and charge transporting superstructures. For macrocycle **32**, 2D WAXS studies on extruded filaments demonstrated that the molecules self-assemble into hexagonal arrays of columns with a packing parameter of 4.7 nm between the columns [58]. Within the columns, the stacking distance was found to be 0.4 nm and a correlation between every fourth molecule along the column was found, that is, each macrocycle was rotated by ∼22.5° toward each other in agreement with the molecular structure where every third carbazole unit carried a hydroxyalkyl chain. In addition, monolayers of the macrocycles could be visualized (Figure 6.1) by scanning tunneling microscopy (STM) on highly oriented pyrolitic graphite (HOPG), which showed a well-ordered hexagonal packing of "face-on" structures [57].

Similar cyclic architectures constituted of 2,7-fluorene units have been synthesized (**36**) and fully characterized by NMR, optical absorption and fluorescence spectroscopy, and STM. Since oligo- and polyfluorenes represent efficient blue-emitting materials for OLEDs, such monodisperse macrocycles serve as attractive model compounds in order to gain a better understanding of linear architectures, for example, to study the impact of end groups on the optoelectronic properties of linear polyfluorenes. As for the carbazole macrocycle, the final ring closure via Yamamoto coupling, however, turned out to be quite inefficient, leading to very low yields of these macrocycles **32** and **36** (1–2%) [57, 58]. Therefore, it was difficult to prepare large amounts and a new synthetic pathway was developed.

Further synthetic extension combined with elaborate functional design via the former porphyrin template but with Sonogashira coupling of the endcapped diethynyltriscarbazoles macrocycle **37** was achieved (Scheme 6.10), where the sterically demanding solubilizing alkyl side chains pointed to the outside of the giant ring [59]. The template removal paved the way to a free cavity with a predefined size in carbazole macrocycle **38**, where combined with adequate processing techniques (physisorption from solution and gas-phase deposition) a hexabenzocoronene (HBC) molecule could be trapped (Scheme 6.11, Figure 6.2). Assuming that the guest

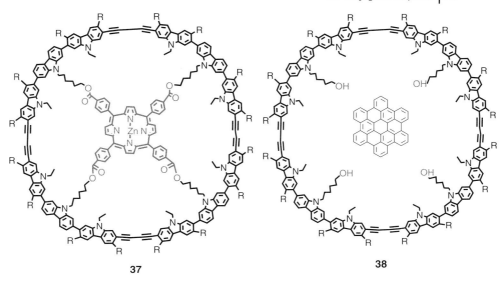

Scheme 6.10 Carbazoleethynyl macrocycles with porphyrin template **37** and with trapped HBC molecule **38**.

molecule would also be able to form columnar stacks, this new type of self-organization allowed columnar assemblies consisting of conjugated macrocycles with discotic pillars in their channels to form from the bulk. Such supramolecular structures have a high potential for electronic devices, such as light amplification using cascade energy transfer.

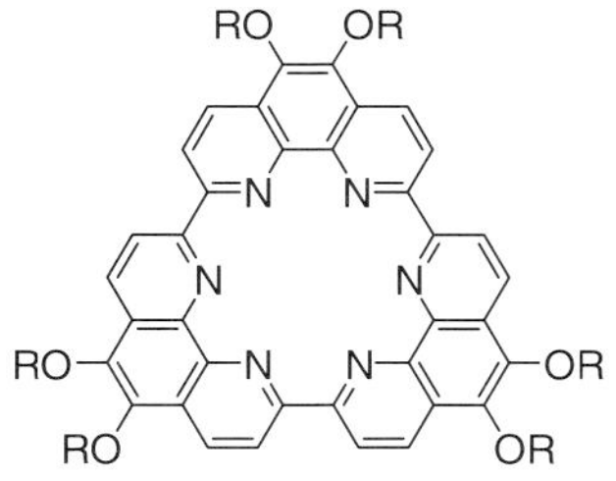

Scheme 6.11 Cyclotrisphenanthrolines **39a–39c**.

A quite different smaller cyclo-2,9-tris-1,10-phenanthroline hexaaza macrocycle **39** was prepared by cyclotrimerization of the 2,9-dichlorodialkoxyphenathroline precursor, which has further been vigorously studied not only for its phase-forming liquid crystallinity but also for its metal ion incorporation (Scheme 6.11) [60]. The alkoxy chain length was varied from octyl to hexadecyl in order to compare the discotic self-assembly behavior. The variation of the length of the alkyl chains led to severe changes of column formation with a π-stacking distance of 0.37 nm for **39a**. As anticipated, the intercolumnar distances increased from 3.32 over 3.75 up to 4.22 nm

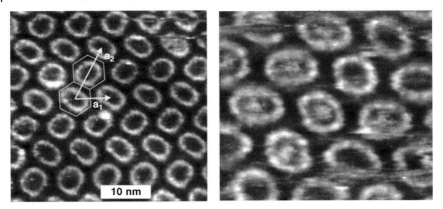

Figure 6.2 STMs of the empty macrocycle and inclusion of HBC **38**.

for **39a–39c** with increasing R showing overall hexagonal arrangement of the macrocycles. The interior is slightly larger than the one in standard porphyrin or phthalocyanine-based macrocycles, reaching opposite N–N distances of 0.55 nm. Thus, a number of rather large alkaline (Na^+, K^+) and transition metal ions (Ag^+, Pb^{2+}, Cd^{2+}, Zn^{2+}, and Cu^{2+}) were incorporated into the cycle and their binding evidenced by MALDI-TOF spectrometry. The replacement of sodium in the macrocycle upon addition of silver triflate was also proven by optical absorption, leading to a bathromic shift from 444 to 474 nm for the silver macrocycle, together with a strong decrease of intensity.

So far the macrocycles discussed above are appropriate models for the linear analogues, although they did not yet reach the conjugation length of the corresponding linear polymers. Due to their self-organization in stacked columns and their ability to uptake hosts or different metal ions to replace (such as sodium) in the center as shown for trisphenanthrolines **39**, they will be studied for host–guest recognition and for sensing applications.

6.2.4
Donors and Donor–Acceptor Approaches

6.2.4.1 Defined Oligomers of Heteroacenes

Usually fully planar conjugated molecules strongly tend to aggregate via strong π–π interactions and increase thereby the regularity in the thin-film channel layers. The gain in rigidity of the structure also allows the modification of the bandgap. Functional materials containing both electron-rich and ladderized systems can be used in OFETs or solar cell devices. In a simple two-step reaction involving triflic acid-induced ring closure reaction, two series of rigid, sulfur-bridged heteropentacenes **40** and **41** with different side groups were prepared and are shown in Scheme 6.12.

Depending on the nature of the alkyl chain substituents, the BBBT molecules arranged in the crystal in a typical "herringbone" structure like pentacene. The solution-processed OFETs gave initial hole mobilities of up to 0.01 $cm^2 V^{-1} s^{-1}$.

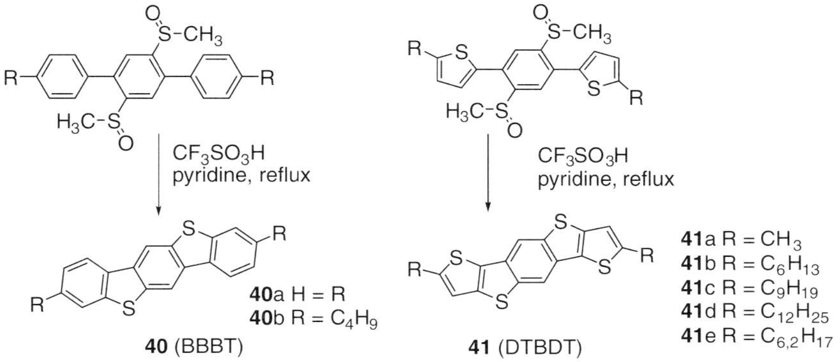

Scheme 6.12 Triflic acid-induced ring closure (CF$_3$SO$_3$H, P$_2$O$_5$, rt. pyridine, reflux) to full-ladder heteropentacene benzo[1,2-b:4,5-b']bis[b] benzo-thiophene (BBBT, **40**) and dithieno[2,3-d;2',3'-d'] benzo[1,2-b;4,5-b']dithiophene (DTBDT, **41**).

The other five-ring-fused pentacene analogue, with four symmetrically fused thiophene ring units (dithieno[2,3-d;2',3'-d0]benzo[1,2-b;4,5-b']dithiophene) (DTBDT), was achieved with even larger variety of alkyl substituents (**41a–41e**).

X-ray structural analysis revealed that the DTBDT core shows almost planar structures, like pentacene, while the hexyl chain on the thienyl α-position lies slightly outside the plane of the skeleton. The molecules form a layer-by-layer structure consisting of alternately stacked aliphatic layers and the DTBDT core, which is believed to enhance the charge carrier mobility (Figure 6.3).

A solution-processed OFET based on DTBDT with two linear hexyl chains (**41b**) was studied in more depth, paying particular attention to its film formation. It was found that much higher charge carrier mobilities of up to 1.7 cm^2 V^{-1} s^{-1} and on/off ratios of 10^7 were obtained from dip-coated rather than from spin-coated films (10^{-2} cm^2 V^{-1} s^{-1}). Dip coating means that the film was grown by immersing the transistor substrate in the polymer solution and slowly taking the substrate out at a specific rate. The X-ray diffraction (XRD) measurements of the dip-coated film showed much better ordering (Figure 6.3c) than the spin-casted films that exhibited a much larger number of reflections, indicating domains extending over several micrometers. The OFET performance was one of the most promising for small-molecule organic semiconductors processed from solution. Further options for structural variations involve the extension toward larger number of condensed rings leading to heteroheptacenes, where solubility problems rendered more bulky substituents necessary for solubilization and processing but hampered high ordering. Another approach currently under investigation is the replacement of some alkyl chains by alkylated aryl units, which was hypothesized to allow extension of conjugation and further fine-tuning of the bandgap.

6.2.4.2 Poly(benzodithiophene)s

The highly successful concept of using heteroacenes as semiconductors should be transferred to polymeric structures by incorporating benzodithiophene into a poly-

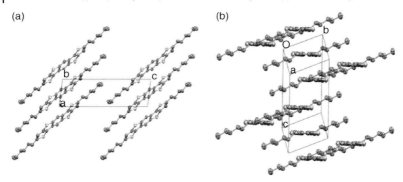

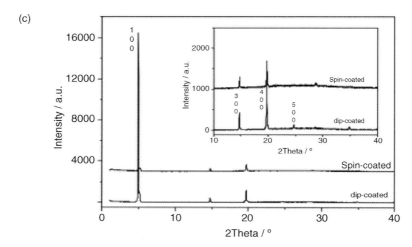

Figure 6.3 Crystal structure of **41**: (a) b-axis projection and (b) shifted cofacial packing in the layered structure. With an interplanar separation of 3.63 Å and side-by-side S–S contact of 3.73 Å, (c) XRD in reflection mode for the spin- and dip-coated films of **41b**. The reflections for the dip-coated film are assigned to the Miller indices.

thiophene chain (**42–44**, Scheme 6.13). These polybenzodithiophenes [61] allowed the fabrication of homogenous films by mass printing techniques, opening up the possibility to produce low-cost applications on a large scale. Extended π-systems as monomers tend to render barely soluble polymers, which cannot be easily processed

Scheme 6.13 Poly(benzodithiophene)s **42–44**.

otherwise into satisfactory devices. Therefore, a particular benzodithiophene was chosen to introduce curvature into the polymer backbone. The entropy gain of a curved chain increased the solubility, but the intramolecular interactions remained high enough for the necessary order in the film.

The starting point for the investigations was homopolymer **42** that is obtained by iron(III) chloride-induced oxidative polymerization. It turned out, however, that solubilizing alkyl chains at the benzo units are detrimental for efficient packing, leading to a large π–π distance of 0.43 nm as determined from X-ray diffraction analysis. The consequence is a relatively poor FET device from spin-casted films ($\mu_{sat} = 10^{-4}$ cm^2 V^{-1} s^{-1}, $I_{on}/I_{off} = 2.5 \times 10^2$).

To improve the packing behavior, additional thiophene units bearing alkyl chains were introduced, resulting in unsubstituted benzodithiophene **43**. Such a structure is obtained by a Stille polymerization using benzodithiophene bearing two trimethyltin groups and a dibrominated dialkyl-dithiophene as comonomer [61]. Good solubility was found (>20 mg ml^{-1}) together with much closer packing (0.37 nm). FETs prepared in the same way as for **42** gave mobilities of $\mu_{sat} = 0.13$ cm^2 V^{-1} s^{-1} and on–off ratios of $I_{on}/I_{off} = 1.8 \times 10^5$ with high reproducibility, which is very promising for further optimizations of the device structure.

In order to investigate the effect of the curved structure, the isomeric polymer **44** with a linear backbone was synthesized in the analogous Stille polymerization. It possessed an equal π-stacking distance, but gave much lower performance ($\mu_{sat} = 1.5 \times 10^{-2}$ cm^2 V^{-1} s^{-1}, $I_{on}/I_{off} = 9 \times 10^4$). In contrast to the curved **43**, the linear **44** precipitated upon cooling a hot solution. This fact obviously hindered the formation of homogenous films with a good contact to the dielectric.

The bent polymer **43** was also tested as substitute of P3HT in photovoltaic applications [62]. XRD measurements revealed that thermal treatment resulted in increased crystallinity of the bulk heterojunction network with the PCBM acceptor [70], and an overall conversion efficiency of 2.7% was found.

6.2.4.3 Poly(cyclopentadithiophene-benzothiadiazole)

The design of donor–acceptor copolymers is attractive for low bandgap polymers, where the long wavelength absorption is crucial for good efficiencies of heterojunction bulk solar cells. Keeping in mind that through interchain donor and acceptor interactions (π–π stacking), a certain degree of order can potentially be achieved by the polymer chains, they are also promising to study their charge carrier mobility. Therefore, the donor–acceptor cyclopenta-dithiophene benzothiadiazole (CDT-BTZ, **45**) copolymer was chosen (Scheme 6.14) and its OFET characteristics

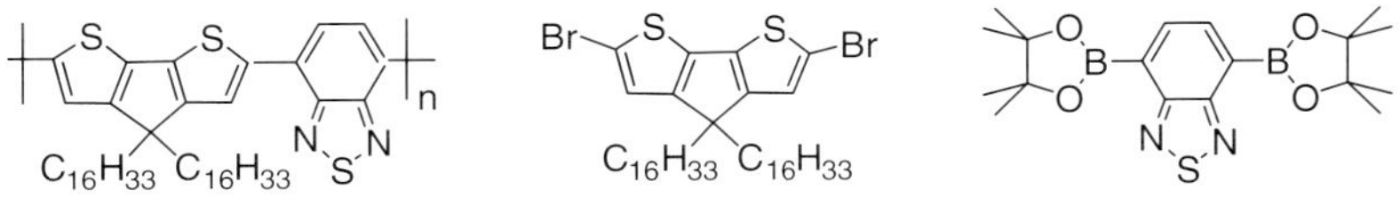

Scheme 6.14 CDT-BTZ copolymer **45** from dibromo-cyclopentadithiophene **46** and diboronic ester of benzothiadiazole **47**.

were studied [63]. Very similar copolymers with different side chains were reported by Konarka [64, 65] and applied with high efficiencies (5.5–6.5%) in organic photovoltaics.

The synthesis was performed by Suzuki–Miyaura copolymerization of the dibromo cyclopentadithiophene (**46**) with the diboronic ester of benzothiadiazole (**47**) to yield the corresponding copolymer **45**. C_{16} alkyl chains were introduced to the polymer backbone to guarantee solubility. When polymer **45** with number average molecular weights of around $10\,\text{kg mol}^{-1}$ was applied in bottom-gate bottom-contact OFETs by drop casting, hole mobilities of up to $0.17\,\text{cm}^2\,\text{V}^{-1}\,\text{s}^{-1}$ were reached. Surprisingly, X-ray diffraction of thin films of this polymer disclosed lack of high order that was not beneficial for charge transport. Therefore, this relatively high mobility was attributed to the interchain π-stacking distance of around 0.37 nm, evidenced by 2D WAXS of extruded CDT-BTZ (**45**) fibers.

Since charge carrier mobility typically scales with molecular weight, further optimizations of the synthetic protocol were tried that enabled us to raise the molecular weight of this polymer reaching a number average molecular weight of $M_n = 51\,\text{kg mol}^{-1}$. In this aspect, solution deposition via spin coating with top-contact FET geometry resulted in a further improved hole mobilitiy of up to $0.67\,\text{cm}^2\,\text{V}^{-1}\,\text{s}^{-1}$ [66]. Grazing incidence wide angle X-ray scattering (GIWAXS) on this spin-coated polymer layer revealed macroscopic order with lamellar-like packing of the polymer chains with similar π-stacking distance as the lower molecular weight case [67]. Thus, the additional presence of macroscopic arrangement in the elevated M_n films contributed to the increased transistor performance, underscoring the importance of both close intermolecular packing and macroscopic order.

Motivated by this excellent transistor performance, the polymer chains were directionally aligned with the aim to even further enhance macromolecular order and charge carrier mobility. This alignment was realized by dip coating the copolymers. Film morphology characterized by the majority of fibers consisting of polymer chains oriented with their backbone plane along the dip coating direction was observed by atomic force microscopy. Measuring the FET along this direction, an extraordinary charge carrier mobility of up to $1.4\,\text{cm}^2\,\text{V}^{-1}\,\text{s}^{-1}$ was obtained, one of the highest values for a polymer up to date. This further improvement in mobility is explained by the domination of charge transport along the oriented polymer chains that is faster than the intermolecular hopping mechanism via π–π stacking. Clearly, by controlling film morphology, excellent transistor performance was attained, rendering this donor–acceptor copolymer system particularly interesting for plastic electronics applications.

The conjugated polymer chains allowed the construction of intricate linear and cyclic conjugated structures possessing different kinds of functionalizations. Instead of entering other heteroatoms in heteroacenes or further modifying the donor–acceptor approach to tailor the bandgap, it is also possible to go for larger two-dimensional molecules. As such, graphene molecules (planar extended PAHs) possess an immense toolbox of functionalities for controlling and tweaking the desired electronic properties, and they form complex architectures on a supramolecular level.

6.3 Graphene Molecules and Their Alignment

Graphene molecules and graphene nanoribbons have recently garnered attention because they were recognized as promising building blocks for nanoelectronic and spintronic devices [68, 69]. Although even exfoliated single 2D graphene layers became available by mechanical techniques from highly ordered pyrolitic graphite (HOPG, bottom-down approach), the focus in this chapter will be on the bottom-up approach toward larger graphene entities [7, 70] via size extension of well-defined polycyclic hydrocarbons from suitable precursors and the control of their supramolecular ordering through π-stacking and local phase separation between flat rigid aromatic cores and flexible peripheral substitutents. These well-defined tailor-made graphene molecules offered a great opportunity to tune their physical properties upon variation of their chemical structures, leading to new nanoscopic objects, as will be shown below.

6.3.1 Extended Size Nanographene Molecules

Several years ago, a synthetic protocol was derived to prepare large and well-defined graphene-type molecules. The basic principle is demonstrated by the Scholl reaction of hexaphenylbenzene (**48**) to hexabenzocoronene (**49**) (Scheme 6.15), which was first reported in 1958 [71] and then further optimized [72]. Upon this optimization, $FeCl_3$ predissolved in nitromethane was used as both a Lewis acid and an oxidizing reagent mediating the intramolecular cyclodehydrogenation procedure (Scholl

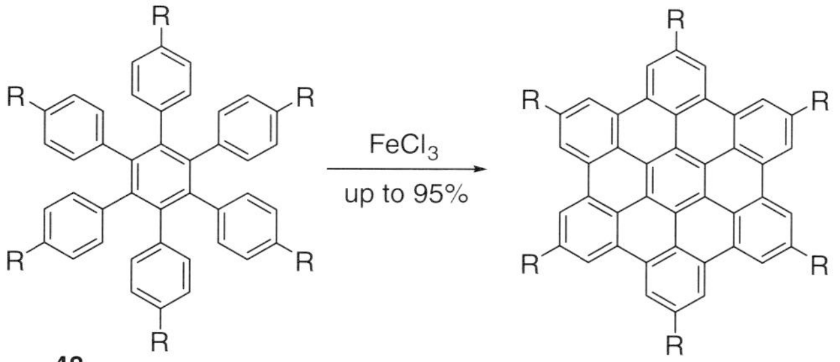

48 R = alkyl, alkoxy, thioether, ester amino, aryl, halogen, etc. **49**

Scheme 6.15 Cyclodehydrogenation of hexaphenylbenzene **48** to HBC **49**.

reaction) of well-suited twisted polyphenylene precursors. This procedure was successfully applied to prepare a whole zoo of different size and shape-defined graphene molecules [7, 70, 73–75] starting from HBC (**49**) to form a variety of substituted HBCs as presented in Scheme 6.16.

Using this procedure, many extended graphene molecules have been made available, and, upon carrying side chains or other functional groups, could undergo

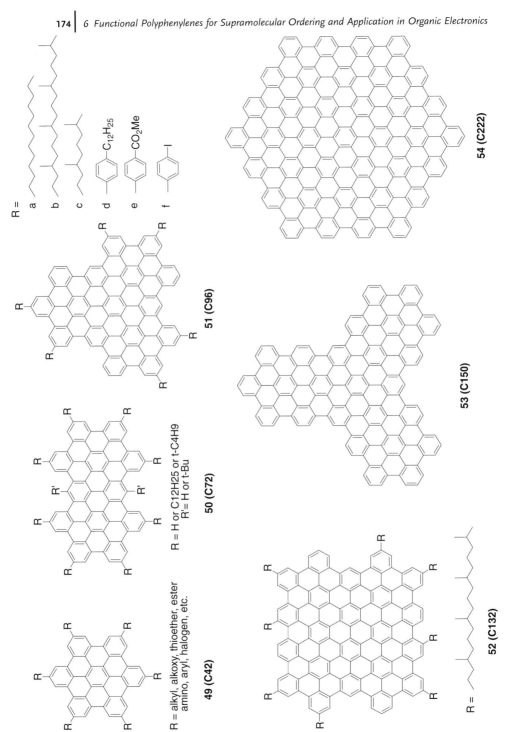

Scheme 6.16 Some extended planar hydrocarbons (**49–54**) C_{42}–C_{222}.

in-depth characterization and further solution handling (C_{42}–C_{132}, Scheme 6.17) [70, 74, 76]. They also served as a reference for comparing their properties with their nonsoluble counterparts that lack solubilizing groups. Although a complete cyclodehydrogenation of polyphenylene precursors demands careful control of the reaction conditions and their optimization, even very large molecules such as C_{150} and C_{222} could be successfully synthesized [77]. However, they were not soluble enough for standard molecular characterizations such as NMR spectroscopy and were identified by MALDI-TOF mass spectrometry and their surface alignment only.

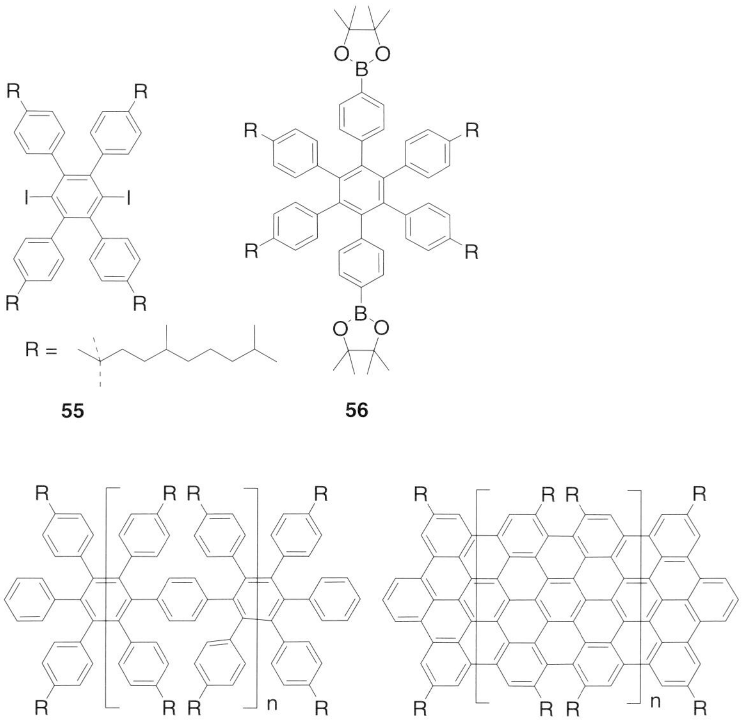

Scheme 6.17 Synthesis of nanoribbons by Suzuki–Miyaura coupling of 1,4-diidotetraphenylbenzene derivative **55** with diboronic acid esters **56** and **57** and cyclodehydrogenation yielding **58**.

At this point it should be mentioned that a crucial point of large polycyclic aromatic hydrocarbons is their processability that traditionally requires solubility or volatility; however, such large graphene molecules are neither soluble in common solvents nor sublimable without decomposition. Therefore, new processing techniques using the principles of mass spectroscopy were developed such as soft landing and pulsed laser deposition (PLD). In soft landing experiments, the nonvolatile and insoluble PAH molecules are transferred into the gas phase by a solvent-free sample preparation method for matrix-assisted laser desorption/ionization (MALDI) mass spectrometry [78]. The ions were accelerated and purified in the mass spectrometer by selection of their individual mass-to-charge ratios. Subsequently, the ions were decelerated to low velocities and "soft landed" on suitable substrates for thin film formation [79]. This procedure allows the separation of side products during synthesis that cannot be removed by conventional purification methods. Soft landing mass spectrometry, however, has a very low deposition rate and a more efficient approach turned out to be PLD. In the PLD, short laser pulses at low irradiation density are used to cause the phase transition of the molecules from the solid state to the gas phase. The molecules, having approximately supersonic speed after laser desorption, were deposited on a substrate surface in close vicinity to the thin film. By careful adjustment of the irradiation density, an intact "sublimation" of the large PAHs can be achieved that is not feasible by equilibrium processes like thermal evaporation [80]. In comparison to PAH soft landing where selected ions are deposited, PLD produces mainly neutral species in a significantly higher yield, thus predestining it for practical thin film applications such as organic electronics.

As shown above, the organic synthetic protocols leading to graphene-type molecules with different sizes have been well developed. With increasing molecular size, however, the synthetic method faced problems due to the limited solubility and the occurrence of side reactions. Thus, new approaches needed to be considered for larger well-defined distinct nanographene objects from the molecular level. While polymerizations of rigid planar subunits often suffer from low molecular weight and insolubility, we derived polymeric, dendritic, and hyperbranched polyphenylene precursors with solubilizing alkyl chains for polymerization that were then transformed into the target structures by cyclodehydrogenation-induced planarization.

6.3.2
Polymeric Extensions to Graphene Layers

Another approach toward growing nanographenes in 1D direction, which contribute to a better understanding of the chemical formation of graphenes, was recently developed. As illustrated in Scheme 6.18, the Suzuki–Miyaura coupling reaction of bis-boronic ester of hexaphenylbenzene **55** with 1,4-diiodo-2,3,5,6-tetraphenylbenzene **56** gave polymers **57** in 75% yield (Mn $\sim 14\,\text{kg mol}^{-1}$ with low polydispersity (Pd = 1.2) and with good solubility in commonly used solvents (Scheme 6.17) [81]. Subsequent intramolecular oxidative cyclodehydrogenation with $FeCl_3$ as an oxidative reagent provided graphene structure **58** as a black solid in 65% yield, which was still soluble in THF and dichloromethane. These soluble nanoribbons were fully characterized by UV-Vis absorption ($\lambda_{max} = 485$ nm) and MALDI-TOF mass measurement. These nano objects with lengths up to 12 nm could also be visualized by STM measurements.

The polycondensation of the appropriate monomers appeared to be the most promising route to continue such polymerization reactions for processable nanographenes. In that vein and as mentioned above, poly-zigzag HBCs (**59**, Scheme 6.18) were prepared via Yamamoto coupling of dibromo-triphenylene precursors (**60**) [47]

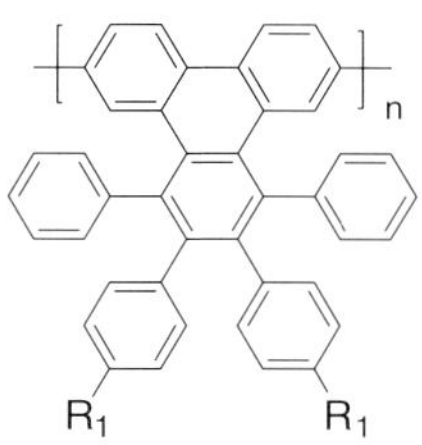

Scheme 6.18 High molecular weight graphene sheets **59**.

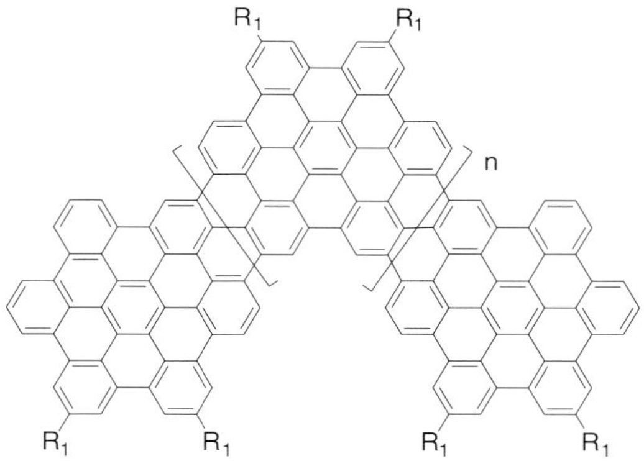

with molecular weights up to 100 kg mol^{-1}, which then could also be reacted via Scholl-type cyclodehydrogenation to fully planarized graphenes.

6.3.3
Variation of the Aromatic Core and Its Symmetry

So far we highlighted the possible molecular and polymeric approaches to giant graphene molecules, however there is also an intriguing influence on the electronic properties by subtle changes of the core periphery and its symmetry (Scheme 6.19). Upon using one phenanthrene unit in the precursor derivative, the introduction of partial zigzag periphery was enabled yielding C$_{44}$ (**61**) [82]. The introduction of one or two double bonds on the periphery of the Clar-type fully benzenoid hydrocarbon HBC as shown for C$_{44}$ (**61**) and C$_{46}$ (**62**) (Scheme 6.16) with partial zigzag structures had dramatic impact on their HOMO–LUMO bandgap. While for HBC the bandgap is 3.6 eV, it was lowered to 3.1 eV for C$_{44}$(**61**) and even to 2.6 eV for C$_{46}$ (**62**) [82].

The changed symmetry of the core also influences their bulk behavior. For instance, the novel dodecylphenyl-substituted *D*2 symmetric zigzag nanographenes (dibenzo[*hi,uV*]-phenanthro[3,4,5,6-*bcdef*]ovalene (**62**) opened the opportunity to manipulate the packing of these disks within the columnar superstructures [83]. For the first time, a modification of the columnar organization from helical to a staggered stacking has been achieved simply upon changing the substitution pattern. From 2D WAXS studies, it was concluded that **62a** and **62c** self-assemble into hexagonal columnar structures over the whole temperature range and are well aligned in the extrusion direction of the sample, characterized by distinct equatorial scattering intensities. A face-to-face π-stacking as close as 0.35 nm was measured. For **62a**, a helical packing with neighboring molecular rotation of 60° was found from meridional reflections related to an intracolumnar period of 1 nm, such that every fourth unit is back on top of the first. In sharp contrast, the columns formed by compound **62c** showed a rotation of 90° between the neighboring units (Figure 6.4),

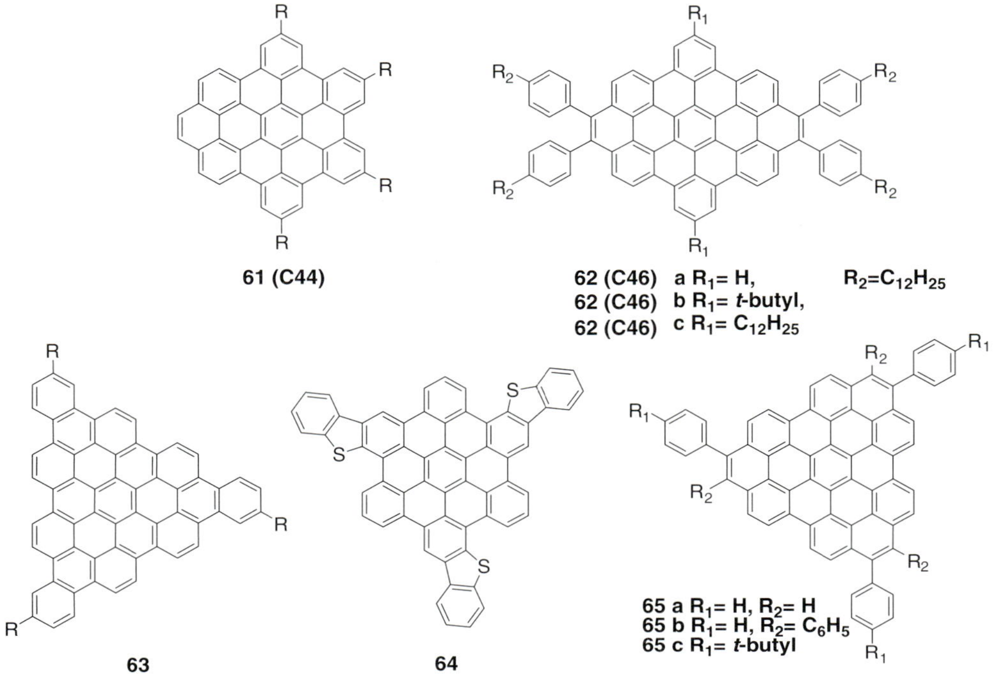

Scheme 6.19 Core and symmetry variation of some extended PAHs (**61–65**).

corresponding to the perpendicular stacking between every two neighboring molecules. Due to their bulkiness, *tert*-butyl substitutents hindered the intermolecular packing and no long-range ordering could be found for **62b**.

Some triangle-shaped benzenoid PAHs (**63** and **64**) with the same number of carbon atoms as fullerene C_{60} have been synthesized and studied aligned on highly ordered pyrolitic graphite (HOPG) surfaces by STM [84]. A monolayer of the

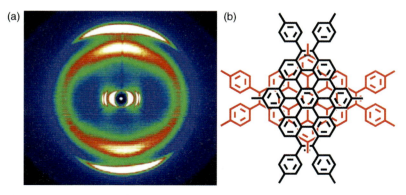

Figure 6.4 2D WAXS results (a) and sketch of stacking of **62c** (b).

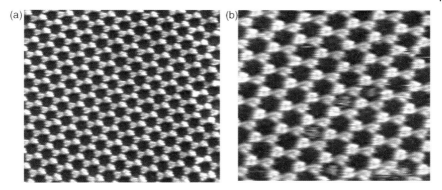

Figure 6.5 Honeycomb network as potential nanotemplate (a) and insertion of coronene (b).

thiophene-fused derivative **64** showed a new two-dimensional polycrystal network. The self-assembled molecules were all oriented in the same direction with the interstices filled by solvent molecules. For hexaphenyl-substituted C_{48} **65**, the so-called tris zigzag derivative, helical packing of columns was derived with every disk being rotated by 40°. At the liquid–HOPG interface, a honeycomb lattice was formed, demonstrating that the long alkyl chains dissolved on top, leaving holes between six molecules defining a circle that could be filled by appropriate targets (Figure 6.5). Typical example of using such honeycomb network as template is presented in Figure 6.5, the small coronene molecules can be entrapped in the cavities.

6.3.4
Influence of the Side Chains and Their Polarities

The different size and number of side chains on the hexabenzocoronene moiety allowed the control of its columnar packing (Scheme 6.20). While the disk-like molecules tended to aggregate strongly into extended ordered columns, as long as not too bulky side groups were present, they could be aligned on a surface in two very distinct ways, namely, "face-on" to a substrate or "edge-on," with the pending side groups attaching the surface [85, 86].

Such alignment control seemed important for the device applications since in OFETs, a major concern is a good contact between source and drain electrodes, best achieved with edge-on alignment to the surface, while for solar cell/photovoltaic applications, a face-on topography is favored, with good charge transport between a top and bottom electrode. Depending on the alkyl or alkoxy substituents grafted on the HBC core, the supramolecular organization could be controlled on the surface [86, 87]. While linear alkyl and alkoxy chains favored the edge-on alignment, the more bulky branched alkyl chains (swallow or dove tail) led to face-on alignment with homeotropic ordering sandwiched between two substrates.

Many further examples of HBCs with different substitution patterns have been examined as **67** and **68** [88–91]. For the C_3 symmetric arrangements with alternating

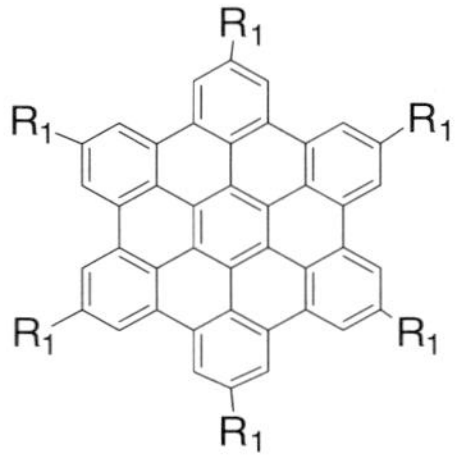

66a $R_1 = C_{12}H_{25}$
66b $R_1 = C_{6,2}$
66c $R_1 = C_{10,6}$
66d $R_1 = C_{14,10}$
66e $R_1 = C_6H_4C_{14,10}$
66f $R_1 = C_6H_4C_{12}H_{25}$

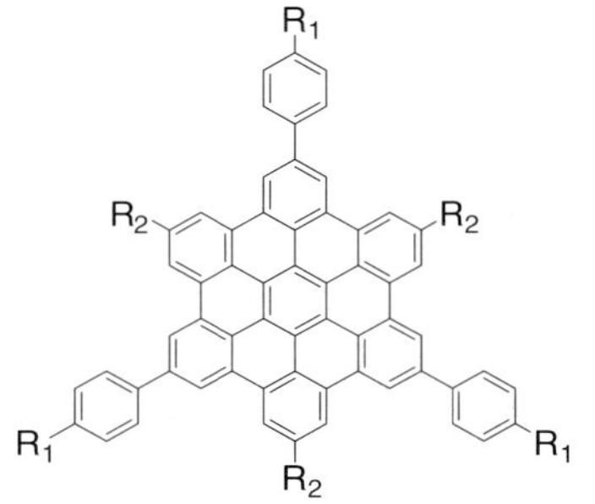

67a $R_1 = C_{8,2}$ $R_2 = C_6H_4\text{-COOMe}$
67b $R_1 = C_{8,2}$ $R_2 = C_6H_4\text{-CH}_2\text{COOMe}$
67c $R_1 = (OCH_2)_3OMe$ $R_2 = C_{8,2}$
67d $R_1 = (OCH_2)_3OMe$ $R_2 = C_{12}H_{25}$
67e $R_1 = (OCH_2)_3OMe$ $R_2 = C_6H_4\text{-}C_{12}H_{25}$

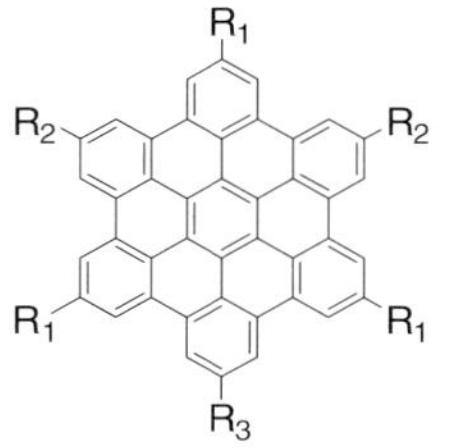

68a $R_1 = H$ $R_2 = R_3 = C_{12}H_{25}$
68b $R_1 = R_3 = H$ $R_2 = C_{12}H_{25}$
68c $R_1 = C_{8,2}$ $R_2 = R_3 = C_6H_4\text{-COOMe}$
68d $R_1 = C_{8,2}$ $R_2 = R_3 = C_6H_4\text{-CH}_2\text{COOMe}$
68e $R_1 = R_2 = C_{8,2}$ $R_3 = C\equiv C\text{-}C_6H_4CH_2SH$

Scheme 6.20 Various HBC derivatives **66–68**.

polar and apolar groups **67**, a stronger aggregation ability was found for **67a** than for **67b** carrying an additional methylene (CH$_2$) spacer group [89]. This could be explained by stronger dipole interactions when a polar group is directly attached to the phenylated core. From **67a**, therefore, submicrometer-sized fibers were also obtained, consisting of 50–100 bundles of stacked molecular wires (Figure 6.6).

The columnar arrangement of the disk molecules is mainly driven by π-stacking and local phase separation between a rigid core and flexible side chains, where the additional intermolecular dipole interactions for **67a** derived by electron-withdrawing ester groups play a further role in self-organization. Hence, the discotic molecules are aligned with their planes along the mechanical alignment direction. This spatial alignment under shearing has been so far only observed for high molecular weight main chain discotic polymers consisting of covalently linked triphenylenes, where the columns are aligned perpendicularly to the oriented polymer chains. Our case utilizes a more complex supramolecular approach based on noncovalent forces between monomers.

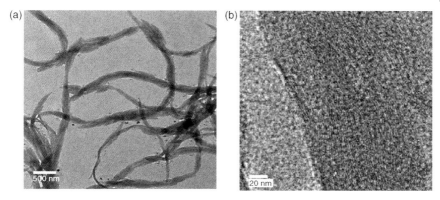

Figure 6.6 (a) Electron microscopy of **67a** fibers grown from THF:MeOH = 1:1 solution. (b) HRTEM image of a fiber displaying columnar structures of **67a**, electron diffraction pattern with reflections (inset) assigned to the π-stacking distance of 0.35 nm.

Additional strong influence was found by just varying the number of side chains comparing **66a** and **68a** and **68b** with six, three, or two alkyl chains, respectively [90]. A major influence on the isotropization temperature (Ti) was found, which was drastically reduced from 420 °C (**66a**) to 170 °C and 220 °C for **68a** and **68b**, respectively. Thus, the latter should be much favored for device fabrication under ambient annealing temperatures.

Furthermore, functional groups capable of forming hydrogen bonds can be added at the HBC periphery to enhance the self-assembly and degree of order within a single column. For instance, the attachment of amido (**69a** and **69b**) or ureido (**69c** and **69d**) groups to the HBC moieties (Figure 6.7) was thereby appealing due to their relatively strong and directional hydrogen bonds acting within the columns [92]. The hydrogen bonds effectively increased the aggregation tendency of these compounds in solution as a result of gel formation. The structural investigations indicated a strong influence of this noncovalent force on the thermotropic properties in the bulk. Interestingly, the typical columnar supramolecular arrangement of HBCs was either stabilized substantially (**69a–69d**) or suppressed by dominant hydrogen bonding interactions (**70**). For some of the compounds (**69a,69c**, and **69d**) [92], the supramolecular arrangement adopted in the liquid crystalline state was even retained after annealing, presumably owing to the reinforcement of the π-stacking interactions by the hydrogen bonds.

In order to achieve structurally perfect columnar ordering of discotics, especially on the surface with edge-on orientation that can provide ideal model for field-effect transistor device, the grafting of disk molecules on the substrate via covalent bonding seemed to be a promising approach. Along this line, we have employed a novel strategy to obtain HBC-based ultrathin films with both a high degree of ordering and a large π–π overlap between neighboring disks [93]. By anchoring the molecules rigidly to a metallic substrate via the formation of covalent S—Au bonds between thiol groups attached to the HBC core **68e** (see inset in Figure 6.8b) and the Au substrate, we not only yielded structurally perfect HBC columns but also a well-defined contact to a metal electrode (Figure 6.8). A variation of the molecular chain linking the HBC

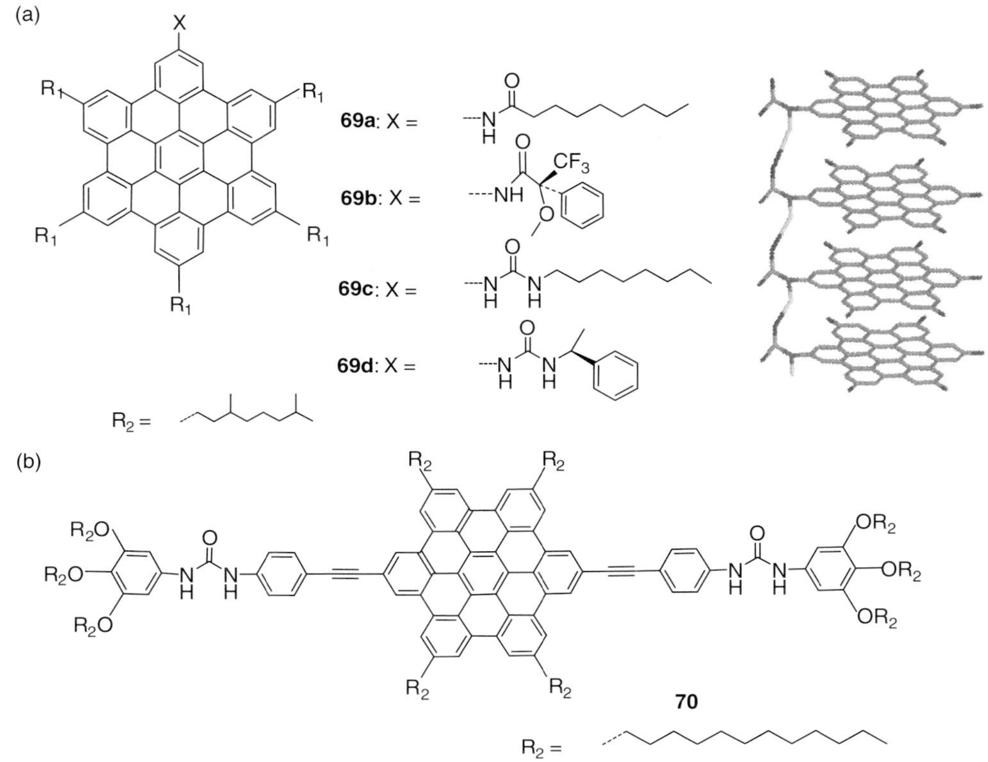

Figure 6.7 (a) Synthesized HBC derivatives bearing amido (**69a** and **69b**) and ureido groups (**69c,69d**, and **70**). (b) Schematic representation of assumed intracolumnar hydrogen bonding interactions and π-stacking of monosubstituted HBC moieties.

molecule and the thiolate anchor from aliphatic to aromatic type allowed to vary the electrical conductivity between the HBC core and the substrate and thus to study the relative importance of through bond transport within a single HBC thiolate and the lateral transport between adjacent HBC cores. From these experiments, electron mobilities of up to $\mu = 4.2\,\text{cm}^2\,\text{V}^{-1}\,\text{s}^{-1}$ have been derived [93].

As described above, the changes of the core and the periphery of the graphene molecules allow one to control the bandgap and their ordering and interfacing. The better alignment of HBC molecules with a single thiol anchor group to a gold surface even yielded extremely high mobilities, emphasizing that such PAHs are promising candidates for implementation in organic electronics. Besides the graphene disk-type molecules, an immense amount of effort went into extended dye molecules, which cover the whole range of the visible spectrum with their optical absorptions and the fluorescence even ranging into the NIR region. These so-called rylene dyes are usually good acceptors since they consist of condensed naphthalene units with tetracarboxydiimide acceptor units at the end.

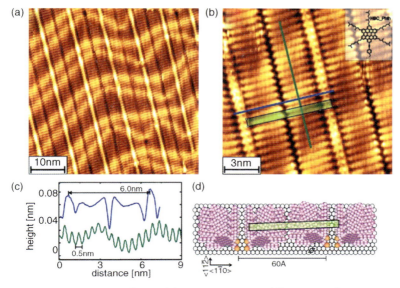

Figure 6.8 STM images of HBC-Phth **68e** on Au/mica at different magnifications (a) and (b) ($I = 150$ pA, $U = 0.1$V) with the molecular structure depicted in the inset. The corresponding line scans and a structural model are presented in (c) and (d).

6.4
Rylenes Dyes

Since the beginning of the twentieth century, perylene-based dyes belonged to the most important pigments and colorants, which have been further optimized for individual applications. It is their exceptional chemical, thermal, and photochemical stability together with high extinction coefficients in the visible range that led to manifold molecular variations and exploration in optoelectronic and photovoltaic devices.

6.4.1
Core-Extended Rylenes

Although a linear extension of the backbone up to pentarylene with *tert*-butyl solubilizing groups was demonstrated 20 years ago [94], it took quite a long development to transfer this knowledge to the synthesis of the more stable and variable tetracarboxydiimide derivatives **72** (Scheme 6.21) where the optical absorptions range into the NIR region for the penta- and hexarylene (tetracarboxydiimide) s **72d** and **72e** with $\lambda_{max} = 877$ and 950 nm [95], respectively, and with extremely high extinction coefficients ε (up to 300 000 $M^{-1} cm^{-1}$).

Although diisopropylphenyl substituents R had been often used, which partially hinder good packing, branched alkyl chains were introduced to increase solubility and lower the isotropization temperature of the liquid crystalline phase for better

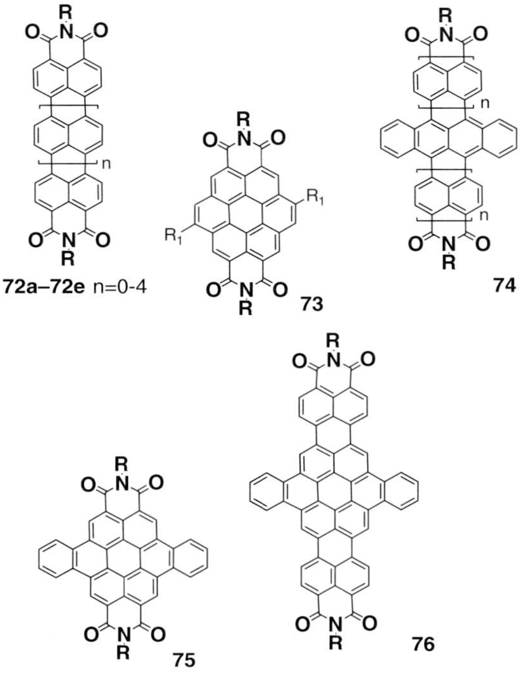

Scheme 6.21 Core extended rylenes 72–76.

processing. Thus, a comparison of perylene-, terrylene-, quaterrylene, and coronen-tetracarboxydiimide with N,N'-di(heptyloctyl) imides (**72a–72c** and **73**) was performed (Scheme 6.21) [96]. They all exhibited high thermal stabilities up to 450 °C and an identical columnar self-organization for the derivatives before their isotropization temperature. All four compounds showed columnar ordered structures with a periodicity of every fifth molecule along the columns (45° with respect to each other). However, differences in the packing alignment during film formation were observed. While the rylene diimides **72a–72d** led to an edge-on alignment of the molecules on the surface, the coronene diimide **73** with two extra linear alkyl chains R_1 favored face-on arrangement on the surface (flat) with homeotropic orientation.

Further core extensions by benzannelation have been pursued, with tetracene-like extension **74** [97] yielding a bathochromic shifted absorption ($\lambda_{max} = 1018$ nm) compared to the linear pentarylene ($\lambda_{max} = 880$ nm), while the dibenzocoronene **75** and the dinaphtho-quaterrylenbis(dicarboxymonoimide) **76** with pentacene-like cores [98] surprisingly provided a hypsochromic shift compared to the linear perylene and quaterrylene bis(dicarboximide)s **72a–72d**. However, their color stability in the yellow and red range was outstanding, while their ordering and self-assembly needs further investigation for use in FET or photovoltaic devices.

6.4.2
Rylenes for Ambipolar Field-Effect Transistors

For organic field-effect transistors, a number of electron-rich aromatic systems with good hole mobilities have been identified, among them pentacene being the benchmark organic semiconductor for OFETs. However, only a limited number of molecules for n-type conduction have been found, some of them containing perylenbis(dicarboxyimide)s, which formed liquid crystalline phases and reached electron mobilities up to $2.1\,cm^2\,V^{-1}\,s^{-1}$ [99, 100].

Ambipolar organic field-effect transistors are of special interest owing to their application in complementary circuits or light-emitting field-effect transistors. In principle, such materials may be considered to have been constructed from mixtures of donor and acceptor compounds or as bilayers, but then they would be difficult to be addressed. The most straightforward and promising approach is the search for molecules that fulfill electron as well as hole transport in a single layer that can be cast from solution. Although PDI is known to be a good acceptor and thus suited for electron mobilities, the extended terrylene- and quaterrylene-tetracarboxidiimides (TDI, QDI, Scheme 6.22) possess similar acceptor properties, however, the HOMO levels are increased that make them promising candidates for this approach. The swallow tail derivatives of TDI and QDI were thus tested in OFET devices after drop casting their solution on bottom-contact bottom-gate transistors with a hexamethyl-silazane-treated SiO_2 surface. Decent hole and electron mobilities with clear ambipolar behavior could only be found for the quaterrylene case since the output curves biased both in the positive and negative regimes showed an increase in source–drain current I_{SD} at low gate voltages [101]. The electron mobilities reached $1.5\,10^{-3}\,cm^2\,V^{-1}\,s^{-1}$, while the hole mobilities were slightly lower ($10^{-3}\,cm^2\,V^{-1}\,s^{-1}$).

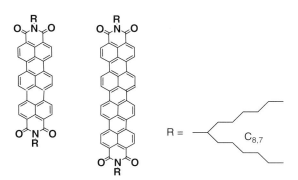

Scheme 6.22 TDI and QDI with branched heptyloctyl chains ($C_{8,7}$) for FET studies.

So far only one drawback was found that upon further ordering after annealing, the hole mobilities were lost, possibly due to columnar reorganization that hindered intercolumnar charge transfer. Thus, ongoing research is directed toward the application of less spatially demanding alkyl chains and extended rylenes as terrylene bearing additional cyanide acceptor groups at the bay positions.

6.4.3
Rylenes for Solar Cell Applications

Due to their acceptor strength and electron mobility values, the rylenes, especially PDI derivatives, have been recognized also as promising materials for photovoltaic applications in not only the typical heterojunction cells but also for dye-sensitized solar cells (DSSCs). As mentioned for the polycarbazoles, the use of PDI **15** with branched alkyl chains as the acceptor led to even better results than using PCBM as the acceptor. A similar finding was made using liquid crystalline hexabenzocoronene (**66e**) as the donor and PDI **15** as the acceptor, which formed thin films with segregated donor and acceptor moieties. Upon blending these two liquid crystalline materials in a 40:60 ratio into a two-phase photovoltaic material, high external quantum efficiencies (EQE) of 34% were reached at a wavelength of 490 nm [102–104]. The high efficiencies were concluded to originate from an efficient charge separation between the hexabenzocoronene and the perylene together with an efficient charge transport through vertically segregated stacks of π-systems. Follow-up studies of photovoltaic devices built from PDI and HBCs with different lengths of branched alkylated chains $C_{6,2}$, $C_{10,6}$, and $C_{14,10}$ (**66b–66d**) were performed. These studies demonstrated that shorter side chains such as ethyl-hexyl (**66b**) lead to higher crystallinity and shorter intercolumnar distances of the liquid crystalline stacks and are promising for high efficiencies, while longer branched alkyl chains revealed much lower performances [105].

Besides the above-outlined increase of the rylene cores leading to absorptions in the NIR region, it was quite recently demonstrated that the variation of donor–acceptor substituents at the perylene core enabled to produce dyes with all possible colors of the visible range from yellow to black and to fine-tune them by control of the HOMO and LUMO properties [106]. As basic structure, therefore, have been used the perylenedicarboxymonoimides **79** that were further substituted in the bay-(1,6,7,12) and peri-positions (9,10) (Scheme 6.23). The advantage of the small core over the synthetically demanding size extension already discussed is that many further derivatives could be prepared in a two-step synthesis covering all color needs with high extinction coefficients.

The use of strong donors in the peri-position together with the dicarboxyimide acceptor resulted in new push–pull systems and the optical measurements indicated a strong intramolecular charge transfer with a high extinction coefficient, far redshifted beyond the typical perylene absorptions. However, if the donor is strongly twisted as in the phenothiazine case **81**, no orbital partitioning between HOMO and LUMO occurs, reducing the charge transfer to nearly zero and leaving the individual chromophore with an absorption around 515 nm [106].

Many of these dyes have been tested as sensitizers in dye-sensitized solar cells, where the dyes are used to cover the metal oxide film (TiO_2 or ZnO). Taking advantage of the gained knowledge on the optimum sunlight absorption, a highly efficient novel perylene sensitizer **79** was engineered that showed an incident monochromatic photon-to-current conversion efficiency of 87% as well as a power conversion efficiency of 7.2% under standard AM 1.5 solar conditions [107].

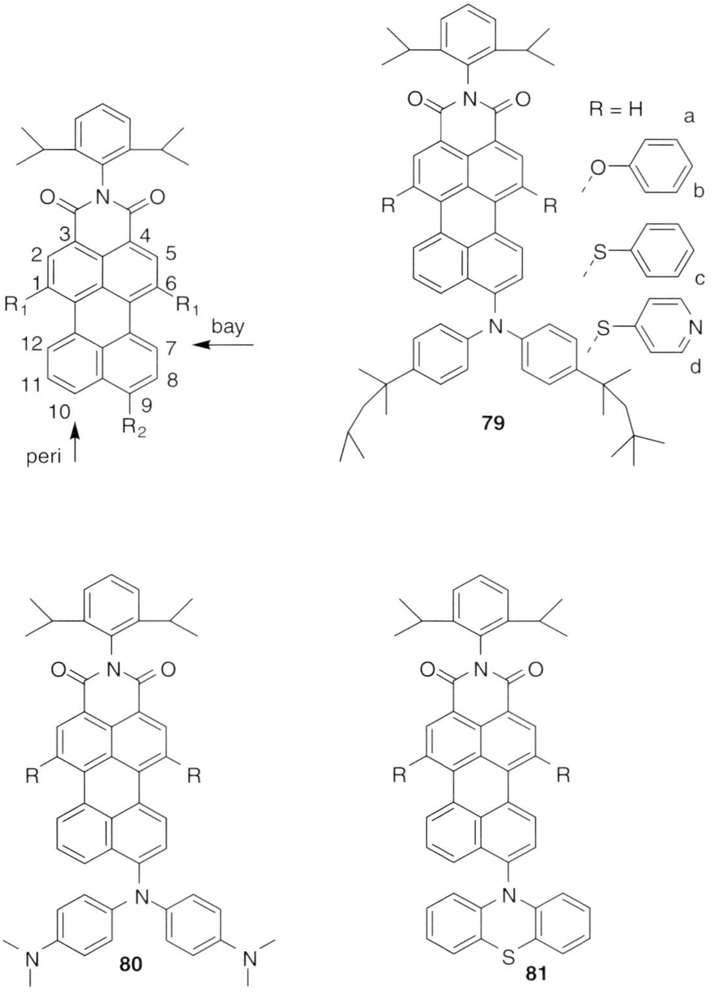

Scheme 6.23 Perylene derivatives for DSSCs.

While rylene dyes have proven to be highly valuable for many applications, they could also be introduced into more rigid scaffolds and even into the core of polyphenylene dendrimers, which can now be followed as the extension of dimension.

6.5
Dendritic Polyphenylenes: The Three-Dimensional Case

It has been shown above that benzene units can serve as conjugated repeat units of both chain and disk-type conjugated systems leading to polyphenylenes and polycyclic aromatic hydrocarbons or graphenes. Phenyl substitution at the benzene core can alternatively lead to propeller-shape repeat units such as a pentaphenyl or hexaphenylbenzene. We have indeed introduced a unique type of dendrimer that is made from twisted, tightly packed benzene rings where the branching points consist of multiple phenyl-substituted benzene ring (Scheme 6.24) [8, 108, 109].

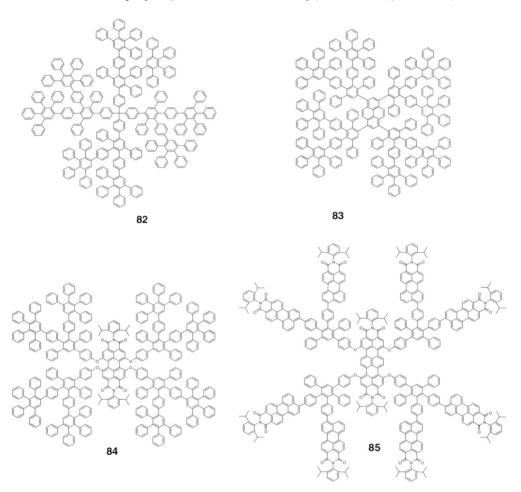

Scheme 6.24 Polyphenylene dendrimer **82**, and some dye-incorporated dendrimers **83–85**.

The discussion of the synthesis of these polyphenylene dendrimers is beyond the scope of this text, but some typical cases are shown in Scheme 6.24 [8, 110–112]. Thus, it is important to mention a few points: (i) The dye version of dendrimers grown by repetitive Diels–Alder reactions ensured a monodisperse character and high structural perfection. Indeed, even a dendrimer with a molecular weight above 500 000 Da was made as a monodisperse three-dimensional polymer [113, 114] and was convincingly proven by sophisticated mass spectrometric techniques. Remarkably, even such high molecular weight dendrimers are still soluble in common organic solvents. What is more important, however, is that due to the structural design, the conformational mobility is largely restricted and there is no significant back bending of the dendron arms. This is a severe difference compared to other dendrimers that are composed of flexible dendrons [115], where the lack of shape persistence does not allow a perfect site definition of functional groups within or outside the dendritic environment.

While the dendrimers are well soluble in standard organic solvents such as tetrahydrofurane and chloroform, the addition of hydrophobic solvents such as hexane can provoke solvophobic effects as a result of which discrete aggregates are formed that can be seen not only by mass spectrometry but also by atomic force microscopy [9, 114]. From a synthetic point of view, upon varying the choice of the core and of the branching substituent, the overall shape of the dendrimer and the intensity of packing of benzene rings can be modified. More important, the dendrimer synthesis can be further modified to incorporate functional units such as fluorophores into the core, into the scaffold, or at the rim of the dendrimer [110, 116]. This has significant consequences because the structure of the dendrimers containing twisted benzene rings does not establish a pronounced chromophoric character. Cooperation of different dyes at well-defined locations leads to identical or different dyes with controllable distance and orientation. It has been particularly successful to choose fluorophores from the rylene family mentioned above. With these dyes, it is possible to study vectorial and degenerate energy transfer as well as electron transfer processes in a geometrically defined nanoenvironment. These studies have included investigations at the single molecular level [117, 118]. It is clear that such energy and electron transfer investigations provide important model cases or multichromophoric supramolecular assemblies. The structural versatility of functionalized polyphenylene dendrimers is further cooperated by the synthesis of water-soluble analogues that can then serve as bioassays [119, 120] or for core-shell systems with polyelectrolyte shells that hold potentials in the study of gene transfection [121].

6.6 Conclusion and Outlook

The above text has convincingly demonstrated the power of synthesis-driven approaches toward electronic properties and device performance. Key issues are the careful control of "structure" at the atomic, molecular, and supramolecular levels. While typical molecular properties such as topology and size of the π-system and the presence of heteroatoms or substituents are known to enable control over, for example, colorant and semiconductor behavior, the synthetic approach also allows one to encode those structural features that are relevant since one can allow or disallow a formation of lamellae from conjugated chains or favor a "face-on" over an "edge-on" arrangement of discotic molecules. While emphasizing the power of such text it should not be overlooked that we are still far away from devices by design. Thus, the discontinuity in a columnar stack serves as a scattering side for a charge carrier, although this defect hardly shows in methods of structural elucidation. Further, when putting a semiconductor molecule to work in a FET and focusing on supramolecular order and bulk, one should nevertheless be aware of the fact that what really counts are the first few layers close to the insulator surface. The situation becomes even more complex for a bulk heterojunction solar cell in view of the complexity of the ongoing processes such as exciton diffusion and dissociation. The somewhat pessimistic view, however, should not discourage the researchers still foreign to the field in spite of the complexity of the problem; there are still tremendous and rewarding tasks for synthesis processing and device fabrication. Key issues need to be addressed from a fundamental point of view, although of course this area of research holds enormous market expectations. It is clear that the synthetic chemist, although in most cases performing synthetic reactions in solution, cannot restrict his attention to the behavior of a solubilized compound but has to focus on the solid state and on interphases. The closely related issue is to bring miniaturization to an extreme and look at the electronic behavior of the single molecules as a kind of separate individual of functional entities. Indeed, studying electron or energy transfer at the single molecular level is of great value, since one avoids statistics of ensemble behavior. It is clear, however, that nanoscientific attempts at characterizing single molecules do not solve the problems of organic electronics as far as device fabrication is concerned. Nevertheless, the field is one of beauty and elegance that requires an enormous range of expertise and competence. It should not be forgotten, however, that all materials research starts with proper, well-defined material synthesis.

Acknowledgments

Some coworkers helped us in finalizing and proofreading this chapter. We thank Don Cho, Tianshi Qin, Joachim Räder, Ralph Rieger, Xinliang Feng, and Nok Tsao. European project funding is gratefully acknowledged as ONE-P, SUPRAMATES, SONS, SENSORS, and SUPERIOR.

References

1 Shirota, Y. and Kageyama, H. (2007) *Chem. Rev.*, **107**, 953.
2 Grimsdale, A.C., Chan, K.L., Martin, R.E., Jokisz, P.G., and Holmes, A.B. (2009) *Chem. Rev.*, **109**, 897.
3 Gunes, S., Neugebauer, H., and Sariciftci, N.S. (2007) *Chem. Rev.*, **107**, 1324.
4 Braga, D. and Horrowitz, G. (2009) *Adv. Mater.*, **21**, 1.
5 Schmaltz, B., Weil, T., and Müllen, K. (2009) *Adv. Mater.*, **21**, 1067.
6 Grimsdale, A.C. and Müllen, K. (2007) *Macromolecules*, **4**, 2225.
7 Wu, J.S., Pisula, W., and Müllen, K. (2007) *Chem. Rev.*, **107**, 718.
8 Bauer, R.E., Grimsdale, A.C., and Müllen, K. (2005) *Top. Curr. Chem.*, **245**, 253.
9 Clark, C.G., Jr., Wenzel, R.J., Andreitchenko, E.V., Steffen, W., Zenobi, R., and Müllen, K. (2007) *New J. Chem.*, **31**, 1300.
10 Grimsdale, A.C. and Müllen, K. (2005) *Angew. Chem., Int. Ed.*, **44**, 5592.

11 Müllen, K. and Wegner, G. (1998) *The Oligomer Approach*, Wiley-VCH Verlag GmbH.
12 Miyaura, N., Ishiyama, T., Sasaki, H., Ishikawa, M., Satoh, M., and Suzuki, A. (1989) *J. Am. Chem. Soc.*, **111**, 314.
13 Ishiyama, T., Nishijima, K., Miyaura, N., and Suzuki, A. (1993) *J. Am. Chem. Soc.*, **115**, 7219.
14 Ishiyama, T., Abe, S., Miyaura, N., and Suzuki, A. (1992) *Chem. Lett.*, 691.
15 Ohe, T., Miyaura, N., and Suzuki, A. (1993) *J. Org. Chem.*, **58**, 2201.
16 Yamamoto, T., Morita, A., Miyazaki, Y., Maruyama, T., Wakayama, H., Zhou, Z., Nakamura, Y., Kanbara, T., Sasaki, S., and Kubota, K. (1992) *Macromolecules*, **25**, 1214.
17 Yamamoto, T. and Yamamoto, A. (1977) *Chem. Lett.*, 353.
18 Scherf, U. and Müllen, K. (1991) *Makromol. Chem. Rapid Commun.*, **12**, 489.
19 Scherf, U. and Müllen, K. (1992) *Macromolecules*, **25**, 3546.
20 Lemmer, U., Heun, S., Mahrt, R.F., Scherf, U., Hopmeier, M., Siegner, U., Gobel, E.O., Müllen, K., and Bassler, H. (1995) *Chem. Phys. Lett.*, **240**, 373.
21 Mahrt, R.F., Panck, T., Lemmer, U., Siegner, U., Hopmeier, M., Hennig, R., Bassler, H., Gobel, E.O., Bolivar, P.H., Wegmann, G., Kurz, H., Scherf, U., and Müllen, K. (1996) *Phys. Rev. B Condens. Matter*, **54**, 1759.
22 Scherf, U. and Müllen, K. (1997) in *Photonic and Optoelectronic Polymers* (eds S.A. Jenekhe and K.J. Wynne), vol. 672, American Chemical Society, p. 358.
23 Setayesh, S., Marsitzky, D., Scherf, U., and Müllen, K. (2000) *Comptes Rendus De L Academie Des Sciences Serie Iv Physique Astrophysique*, **1**, 471.
24 Fukuda, M., Sawada, K., and Yoshino, K. (1993) *J. Polym. Sci. A Polym. Chem.*, **31**, 2465.
25 Funaki, H., Aramaki, K., and Nishihara, H. (1995) *Synth. Met.*, **74**, 59.
26 Keivanidis, P.E., Jacob, J., Oldridge, L., Sonar, P., Carbonnier, B., Baluschev, S., Grimsdale, A.C., Müllen, K., and Wegner, G. (2005) *ChemPhysChem*, **6**, 1650.
27 Rumi, M., Zerbi, G., Scherf, U., and Reisch, H. (1997) *Chem. Phys. Lett.*, **273**, 429.
28 Silva, C., Russell, D.M., Stevens, M.A., Mackenzie, J.D., Setayesh, S., Müllen, K., and Friend, R.H. (2000) *Chem. Phys. Lett.*, **319**, 494.
29 Mishra, A.K., Graf, M., Grasse, F., Jacob, J., List, E.J.W., and Müllen, K. (2006) *Chem. Mater.*, **18**, 2879.
30 Jacob, J., Zhang, J.Y., Grimsdale, A.C., Müllen, K., Gaal, M., and List, E.J.W. (2003) *Macromolecules*, **36**, 8240.
31 Jacob, J., Sax, S., Gaal, M., List, E.J.W., Grimsdale, A.C., and Müllen, K. (2005) *Macromolecules*, **38**, 9933.
32 Jacob, J., Sax, S., Piok, T., List, E.J.W., Grimsdale, A.C., and Müllen, K. (2004) *J. Am. Chem. Soc.*, **126**, 6987.
33 Laquai, F., Mishra, A.K., Ribas, M.R., Petrozza, A., Jacob, J., Akcelrud, L., Müllen, K., Friend, R.H., and Wegner, G. (2007) *Adv. Funct. Mater.*, **17**, 3231.
34 Grimsdale, A.C. and Müllen, K. (2006) *Adv. Polym. Sci.*, **199**, 1.
35 Grimsdale, A.C. and Müllen, K. (2008) *Polyfluorenes*, **212**, 1.
36 Cho, S.Y., Grimsdale, A.C., Jones, D.J., Watkins, S.E., and Holmes, A.B. (2007) *J. Am. Chem. Soc.*, **129**, 11910.
37 Kreyenschmidt, M., Uckert, F., and Müllen, K. (1995) *Macromolecules*, **28**, 4577.
38 Dierschke, F., Grimsdale, A.C., and Müllen, K. (2003) *Synthesis (Stuttg)*, 2470.
39 Cadogan, J.I.G., Cameronw, M., Mackie, R.K., and Searle, R.J.G. (1965) *J. Chem. Soc.*, 4831.
40 Li, J., Dierschke, F., Wu, J., Grimsdale, A.C., and Müllen, K. (2006) *J. Mater. Chem.*, **16**, 96.
41 Pisula, W., Mishra, A.K., Li, J., Baumgarten, M., and Müllen, K. (2008) in *Org. Photovoltaics* (eds. C. Brabec, V. Dyakanov, and U. Scherf), Wiley-VCH, p. 93.
42 Zi En, O., Teck Lip, T., Shin, R.Y.C., Zhi Kuan, C., Kietzke, T., Sellinger, A., Baumgarten, M., Müllen, K., and de Mello, J.C. (2008) *J. Mater. Chem.*, 4619.
43 Yang, C.D., Scheiber, H., List, E.J.W., Jacob, J., and Müllen, K. (2006) *Macromolecules*, **39**, 5213.
44 Yang, C.D., Jacob, J., and Müllen, K. (2006) *Macromol. Chem. Phys.*, **207**, 1107.

45 Yang, C., Jacob, J., and Müllen, K. (2006) *Macromolecules*, **39**, 5696.
46 Saleh, M., Baumgarten, M., Mavrinskiy, A., Schafer, T., and Müllen, K. (2010) *Macromolecules*, **43**, 137.
47 Saleh, M., Park, Y.-S., Baumgarten, M., Kim, J.-J., and Müllen, K. (2009) *Macromol. Rapid Commun.*, **30**, 1279.
48 Kawano, S.I., Yang, C., Ribas, M., Baluschev, S., Baumgarten, M., and Müllen, K. (2008) *Macromolecules*, **41**, 7933.
49 Figueira-Duarte, T.M., Del Rosso, P.G., Trattnig, R., Sax, S., List, E.J.W., and Müllen, K. (2010) *Adv. Mater.*, **22**, 990.
50 Qin, T., Zhou, G., Scheiber, H., Bauer, R.E., Baumgarten, M., Anson, C.E., List, E.J.W., and Müllen, K. (2008) *Angew. Chem., Int. Ed.*, **47**, 8292.
51 Winnik, F.M. (1993) *Chem. Rev.*, **93**, 587.
52 Kawano, S.I., Baumgarten, M., Müllen, K., Schäfer, T., Murer, P., and Saleh, M. (2010) WO 2010/006852A1.
53 Moore, J.S. (1996) *Curr. Opin. Solid State Mater. Sci.*, **1**, 777.
54 Zhang, W. and Moore, J.S. (1996) *J. Am. Chem. Soc.*, **126**, 12796.
55 Mena-Osteritz, E. and Bauerle, P. (2001) *Adv. Mater.*, **13**, 243.
56 Hoger, S. (2004) *Chem. Eur. J.*, **10**, 1320.
57 Jung, S.H., Pisula, W., Rouhanipour, A., Rader, H.J., Jacob, J., and Müllen, K. (2006) *Angew. Chem., Int. Ed.*, **45**, 4685.
58 Simon, S.C., Schmaltz, B., Rouhanipour, A., Raeder, H.J., and Müllen, K. (2009) *Adv. Mater.*, **21**, 83.
59 Schmaltz, B., Rouhanipour, A., Raeder, H.J., Pisula, W., and Müllen, K. (2009) *Angew. Chem., Int. Ed.*, **48**, 720.
60 Schwab, M.G., Takase, M., Mavrinskiy, A., Pisula, W., Wu, D.Q., Feng, X., Mali, K., deFeyter, S., and Müllen, K. (2010) to be published.
61 Rieger, R., Beckmann, D., Pisula, W., Steffen, W., Kastler, M., and Müllen, K. (2010) *Adv. Mater.*, **22**, 83.
62 Liu, M., Rieger, R., Li, C., Menges, H., Kastler, M., Baumgarten, M., and Müllen, K. (2010) *ChemSusChem*, **3**, 106.
63 Zhang, M., Tsao, H.N., Pisula, W., Yang, C.D., Mishra, A.K., and Müllen, K. (2007) *J. Am. Chem. Soc.*, **129**, 3472.
64 Muhlbacher, D., Scharber, M., Morana, M., Zhu, Z.G., Waller, D., Gaudiana, R., and Brabec, C. (2006) *Adv. Mater.*, **18**, 2931.
65 Soci, C., Hwang, I.W., Moses, D., Zhu, Z., Waller, D., Gaudiana, R., Brabec, C.J., and Heeger, A.J. (2007) *Adv. Funct. Mater.*, **17**, 632.
66 Tsao, H.N., Rader, H.J., Pisula, W., Rouhanipour, A., and Müllen, K. (2008) *Physica Status Solidi A Appl. Mater. Sci.*, **205**, 421.
67 Tsao, H.N., Cho, D., Andreasen, J.W., Rouhanipour, A., Breiby, D.W., Pisula, W., and Müllen, K. (2009) *Adv. Mater.*, **21**, 209.
68 Geim, A.K. and Novoselov, K.S. (2007) *Nat. Mater.*, **6**, 183.
69 Zhang, Y.B., Tang, T.T., Girit, C., Hao, Z., Martin, M.C., Zettl, A., Crommie, M.F., Shen, Y.R., and Wang, F. (2009) *Nature*, **459**, 820.
70 Watson, M., Fechtenkoetter, A., and Müllen, K. (2001) *Chem. Rev.*, **101**, 1267.
71 Halleux, A., Martin, R.H., and King, G.S.D. (1958) *Helv. Chim. Acta*, **41**, 1177.
72 Herwig, P., Kayser, C.W., Müllen, K., and Spiess, H.W. (1996) *Adv. Mater.*, **8**, 510.
73 Berresheim, A.J., Muller, M., and Müllen, K. (1999) *Chem. Rev.*, **99**, 1747.
74 Simpson, C.D., Wu, J., Watson, M.D., and Müllen, K. (2004) *J. Mater. Chem.*, **14**, 494.
75 Wu, J.S., Grimsdale, A.C., and Müllen, K. (2005) *J. Mater. Chem.*, **15**, 41.
76 Zhi, L.J. and Müllen, K. (2008) *J. Mater. Chem.*, **18**, 1472.
77 Simpson, C.D., Brand, J.D., Berresheim, A.J., Przybilla, L., Rader, H.J., and Müllen, K. (2002) *Chem. Eur. J.*, **8**, 1424.
78 Rader, H.J., Spickermann, J., Kreyenschmidt, M., and Müllen, K. (1996) *Macromol. Chem. Phys.*, **197**, 3285.
79 Raeder, H.J., Rouhanipour, A., Talarico, A.M., Palermo, V., Samorì, P., and Müllen, K. (2006) *Nat. Mater.*, **5**, 276.
80 Rouhanipour, A., Roy, M., Feng, X., Raeder, H.J., and Müllen, K. (2009) *Angew. Chem., Int. Ed.*, **48**, 4602.
81 Yang, X., Dou, X., Rouhanipour, A., Zhi, L., Raeder, H.J., and Müllen, K. (2008) *J. Am. Chem. Soc.*, **130**, 4216.

82 Kastler, M., Schmidt, J., Pisula, W., Sebastiani, D., and Müllen, K. (2006) *J. Am. Chem. Soc.*, **128**, 9526.

83 Feng, X., Pisula, W., and Müllen, K. (2007) *J. Am. Chem. Soc.*, **129**, 14116.

84 Feng, X.L., Wu, J.S., Ai, M., Pisula, W., Zhi, L.J., Rabe, J.P., and Müllen, K. (2007) *Angew. Chem., Int. Ed.*, **46**, 3033.

85 Kastler, M., Pisula, W., Wasserfallen, D., Pakula, T., and Müllen, K. (2005) *J. Am. Chem. Soc.*, **127**, 4286.

86 Pisula, W., Tomovic, Z., Simpson, C., Kastler, M., Pakula, T., and Müllen, K. (2005) *Chem. Mater.*, **17**, 4296.

87 Pisula, W., Tomovic, Z., El Hamaoui, B., Watson, M.D., Pakula, T., and Müllen, K. (2005) *Adv. Funct. Mater.*, **15**, 893.

88 Feng, X., Pisula, W., Kudernac, T., Wu, D., Zhi, L., De Feyter, S., and Müllen, K. (2009) *J. Am. Chem. Soc.*, **131**, 4439.

89 Feng, X., Pisula, W., Zhi, L., Takase, M., and Müllen, K. (2008) *Angew. Chem., Int. Ed.*, **47**, 1703.

90 Feng, X.L., Pisula, W., Ai, M., Groper, S., Rabe, J.P., and Müllen, K. (2008) *Chem. Mater.*, **20**, 1191.

91 Feng, X., Pisula, W., Takase, M., Dou, X., Enkelmann, V., Wagner, M., Ding, N., and Müllen, K. (2008) *Chem. Mater.*, **20**, 2872.

92 Dou, X., Pisula, W., Wu, J., Bodwell, G.J., and Müllen, K. (2008) *Chem. Eur. J.*, **14**, 240.

93 Kaefer, D., Bashir, A., Dou, X., Witte, G., Müllen, K., and Woell, C. (2010) *Adv. Mater.*

94 Bohnen, A., Koch, K.H., Luttke, W., and Müllen, K. (1990) *Angew. Chem., Int. Ed.*, **29**, 525.

95 Pschirer, N.G., Kohl, C., Nolde, F., Qu, J., and Müllen, K. (2006) *Angew. Chem., Int. Ed.*, **45**, 1401.

96 Nolde, F., Pisula, W., Muller, S., Kohl, C., and Müllen, K. (2006) *Chem. Mater.*, **18**, 3715.

97 Avlasevich, Y. and Müllen, K. (2006) *Chem. Commun.*, 4440.

98 Avlasevich, Y., Mueller, S., Erk, P., and Müllen, K. (2007) *Chem. Eur. J.*, **13**, 6555.

99 Tatemichi, S., Ichikawa, M., Koyama, T., and Taniguchi, Y. (2006) *Appl. Phys. Lett.*, **89**, 112108.

100 Tatemichi, S., Ichikawa, M., Kato, S., Koyama, T., and Taniguchi, Y. (2008) *Phys. Status Solidi Rapid Res. Lett.*, **2**, 47.

101 Tsao, H.N., Pisula, W., Liu, Z., Osikowicz, W., Salaneck, W.R., and Müllen, K. (2008) *Adv. Mater.*, **20**, 2715.

102 Schmidt-Mende, L., Fechtenkoetter, A., Müllen, K., Moons, E., Friend, R.H., and MacKenzie, J.D. (2001) *Science*, **293**, 1119.

103 Schmidt-Mende, L., Fechtenkotter, A., Müllen, K., Friend, R.H., and MacKenzie, J.D. (2002) *Physica E*, **14**, 263.

104 Schmidt-Mende, L., Watson, M., Müllen, K., and Friend, R.H. (2003) *Mol. Cryst. Liq. Cryst.*, **396**, 73.

105 Li, J., Kastler, M., Pisula, W., Robertson, J.W.F., Wasserfallen, D., Grimsdale, A.C., Wu, J., and Müllen, K. (2007) *Adv. Funct. Mater.*, **17**, 2528.

106 Li, C., Schoneboom, J., Liu, Z.H., Pschirer, N.G., Erk, P., Herrmann, A., and Müllen, K. (2009) *Chem. Eur. J.*, **15**, 878.

107 Li, C., Liu, Z.H., Schoneboom, J., Eickemeyer, F., Pschirer, N.G., Erk, P., Herrmann, A., and Müllen, K. (2009) *J. Mater. Chem.*, **19**, 5405.

108 Morgenroth, F. and Müllen, K. (1997) *Tethedron*, **53**, 15349.

109 Wiesler, U.M., Weil, T., and Müllen, K. (2001) in *Dendrimers III: Design, Dimension, Function* (ed. F. Vögtle), vol. 212, Springer, p. 1.

110 Weil, T., Wiesler, U.M., Herrmann, A., Bauer, R., Hofkens, J., De Schryver, F.C., and Müllen, K. (2001) *J. Am. Chem. Soc.*, **123**, 8101.

111 Weil, T., Reuther, E., Beer, C., and Müllen, K. (2004) *Chem. Eur. J.*, **10**, 1398.

112 Bernhardt, S., Kastler, M., Enkelmann, V., Baumgarten, M., and Müllen, K. (2006) *Chem. Eur. J.*, **12**, 6117.

113 Andreitchenko, E.V., Clark, C.G., Jr., Bauer, R.E., Lieser, G., and Müllen, K. (2005) *Angew. Chem., Int. Ed.*, **44**, 6348.

114 Clark, C.G., Jr., Wenzel, R.J., Andreitchenko, E.V., Steffen, W., Zenobi, R., and Müllen, K. (2007) *J. Am. Chem. Soc.*, **129**, 3292.

115 Mourey, T.H., Turner, S.R., Rubinstein, M., Frechet, J.M.J., Hawker, C.J.,

and Wooley, K.L. (1992) *Macromolecules*, **25**, 2401.
116 Oesterling, I. and Müllen, K. (2007) *J. Am. Chem. Soc.*, **129**, 4595.
117 Cotlet, M., Vosch, T., Masuo, S., Sauer, M., Müllen, K., Hofkens, J., and De Schryver, F. (2004) *Proc. SPIE*, **5322**, 20.
118 Cotlet, M., Vosch, T., Habuchi, S., Weil, T., Müllen, K., Hofkens, J., and De Schryver, F. (2005) *J. Am. Chem. Soc.*, **127**, 9760.
119 Yin, M.Z., Shen, J., Gropeanu, R., Pflugfelder, G.O., Weil, T., and Müllen, K. (2008) *Small*, **4**, 894.
120 Yin, M., Kuhlmann, C.R.W., Sorokina, K., Li, C., Mihov, G., Pietrowski, E., Koynov, K., Klapper, M., Luhmann, H.J., Müllen, K., and Weil, T. (2008) *Biomacromolecules*, **9**, 1381.
121 Yin, M.Z., Ding, K., Gropeanu, R.A., Shen, J., Berger, R., Weil, T., and Müllen, K. (2008) *Biomacromolecules*, **9**, 3231.

7
Molecular Tectonics: Design of Hybrid Networks and Crystals Based on Charge-Assisted Hydrogen Bonds

Sylvie Ferlay and Mir Wais Hosseini

7.1
Introduction

Molecular crystals are periodic three-dimensional arrangement of molecules in the solid state. This category of materials is defined by the chemical nature of the molecular components constituting the crystal and interactions between them. The crystallization process leading to the organization of molecules in ordered periodic architectures in the solid state is a typical example of self-assembly [1]. The latter, governed by the rules of supramolecular chemistry [2], leads to spontaneous generation of complex molecular architectures from molecular units through reversible intermolecular interactions. By analogy, whereas molecular chemistry deals with the design and the synthesis of individual units constituting the crystal, their self-assembly [3], through diverse intermolecular interactions, results from concepts developed in supramolecular chemistry. The area of solid-state synthesis dealing specifically with the design and formation of molecular crystals is called *crystal engineering* [4]. Within the vast area of crystal engineering, the *molecular tectonics* [5] approach is based on the construction and analysis of molecular crystals, seen as periodic architectures called molecular networks [6], individual building blocks named tectons, and their mutual interactions through specific recognition patterns called supramolecular synthons [7]. Thus, for this approach, the main goal is the design and formation of molecular networks [8] and their packing in the crystalline phase using structurally defined and energetically programmed molecular tectons [9]. A *tecton* is defined as an active molecular construction unit or building block bearing recognition sites [10]. Thus, molecular networks may be seen as molecular assemblies formed between complementary molecular tectons capable of mutual interactions through one or several molecular recognition events, which define structural nodes of the architecture. The dimensionality of molecular network (1D, 2D, or 3D) depends on the number of translations operating at the level of the assembling nodes. Molecular networks may be formed in any type of condensed media such as solution, gel, or solid state. Within the last category, owing to the presence of short- and long-range order, the crystalline phase has been extensively

Functional Supramolecular Architectures. Edited by Paolo Samorì and Franco Cacialli
Copyright © 2011 WILEY-VCH Verlag GmbH & Co. KGaA, Weinheim
ISBN: 978-3-527-32611-2

used because of the possibility of accurate structural studies using X-ray diffraction methods.

Among many factors governing the formation of crystals, recognition patterns or supramolecular synthons, which appear during the crystallization process, are of prime importance allowing, to a certain extent, to predict the final crystal structure [11]. This aspect has been theoretically investigated using different approaches [12]. However, it is worth noting that owing to our limited knowledge of all subtle intermolecular interactions governing the formation of the crystalline phase, the complete understanding of the formation of molecular networks and their packing leading to the crystal remains so far unreachable [13]. Nonetheless, our current level of knowledge allows to control some of the intermolecular interactions with precision, that is, the appearance of recognition patterns with a good degree of reliability, by properly designing molecular tectons. Thus, one may predict the formation of molecular networks and in some cases their packing in the crystalline phase.

The three main features governing the design of molecular networks are (i) the design of recognition patterns linking a self-complementary or two or several complementary tectons, (ii) the geometrical aspects dealing with the localization of the interaction sites, and (iii) the nature of intermolecular forces allowing the interconnection of tectons. The latter, by principle, must be reversible in order to allow self-repairing processes to take place during the construction events. Different reversible intermolecular interactions such as H-bonding [14] and coordination bonding [15] have been widely explored over the last two decades. Other forces such as π-stacking interactions [16] or inclusion processes based on van der Waals interactions [17] have also been used for the construction of inclusion molecular networks in the solid state.

For the design of the structural nodes, a possible reason for the extensive use of H-bond interaction is its rather directional nature [18]. In terms of strength, H-bond ranges from weak (about a fraction of $kcal\,mol^{-1}$) to moderate interactions (about $5-10\,kcal\,mol^{-1}$) [19]. Consequently, in order to increase the robustness of the architecture, one needs to enhance the interaction energy between the tectons. This may be achieved by combining H-bonding with less directional but more energetic electrostatic charge–charge interactions. This type of recognition pattern is called *charge-assisted H-bonds* (CAHBs) [20]. It results from the interaction between charged hydrogen bond donors (D) and acceptors (A) leading to the ($^+$D–H $\cdots$ A$^-$) supramolecular synthon.

The majority of molecular networks reported to date are based on nonionic H-bonds [14]. However, several cases of charge-assisted hydrogen bonding have been also described in the literature. Some of these cases will be presented in the following section.

Although at the initial steps of this approach, investigators were interested in the understanding and control of structural features governing the formation of molecular networks in the crystalline phase, that is, mainly the connectivity patterns, since a decade, considerable effort has been made in obtaining functional molecular materials with specific properties in the area of, for example, optics [21], magnetism [22], or porosity [23].

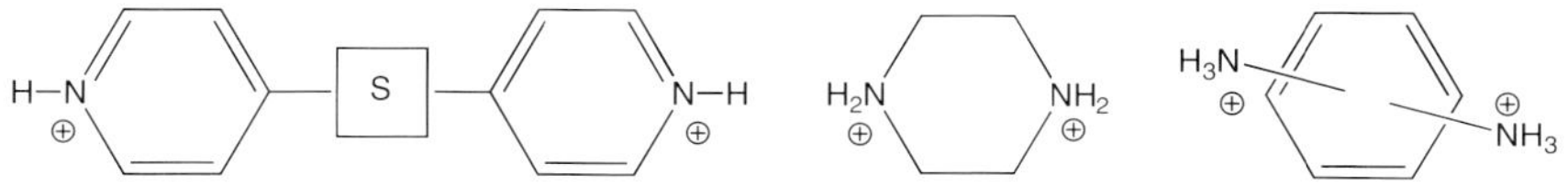

Figure 7.1 Selected organic dications usually used in the formation of CAHB networks.

7.2
Examples of Robust Charge-Assisted H-Bonded (CAHB) Networks

In recent years, only few examples of CAHB networks have been reported in the literature. We will give here a short (nonexhaustive) overview of described cases. It is interesting to note that most of the networks are based on $^+$N-H $\cdots$ O$^-$, $^+$N-H $\cdots$ N$^-$, or $^+$N-H $\cdots$ X$^-$ type of recognition patterns. Pyridinium, ammonium, amidinium, and guanidinium derivatives, all bearing a protonated N-atom, are among the most used hydrogen bond donors. Some of the used dicationic pyridinium and ammonium units are depicted in Figure 7.1.

7.2.1
Organic Networks

A new type of charge-assisted purely organic network was proposed by Ward using a combination of the guanidinium group as the cationic tecton and monosulfonate and disulfonate derivatives as anionic partners. This combination usually forms a (6,3) bidimensional network (a sheet) in which each ion participates in six $^+$N-H $\cdots$ O$^-$ hydrogen bonds. The organic substituent (R group) of the sulfonate moiety acts as a pillar supporting the inclusion cavities between the sheets. By changing the nature of the R group, a rich family of architectures has been obtained (Figure 7.2) [24]. Furthermore, the same group has elegantly demonstrated the possibility of modulating the pore size through the nature and length of the pillar and thus creating porous materials able of including specific guest molecules [25].

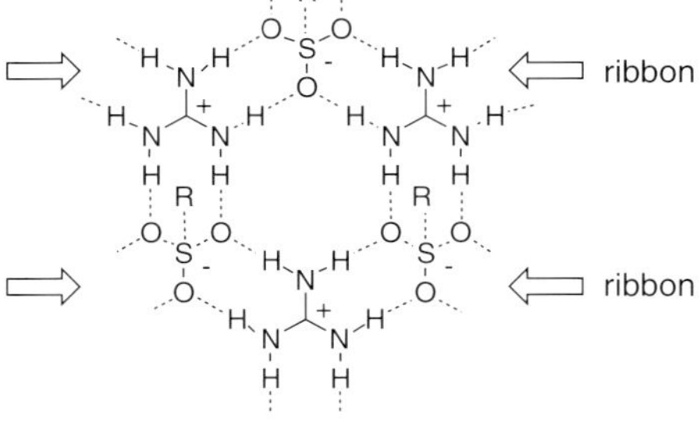

Figure 7.2 Quasi-hexagonal (6,3) 2D network formed upon combining the guanidinium cation with sulfonate anions. Reproduced with permission from Ref. [24a]. Copyright 2009 American Chemical Society.

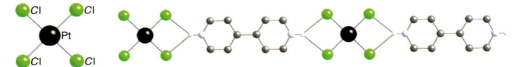

Figure 7.3 Tetrachloroplatinate (left) and a 1D CAHB network (right) formed in the presence of 4,4′-bipyridine.

7.2.2
Inorganic Networks

A new class of ionic metalloorganic architectures have been developed recently by G. Orpen and coworkers using perhalogenometallate complexes such as (i) square planar or tetrahedral $[M^{II}X_4]^{2-}$ (X = Cl, M = Pd, Pt, Co, Zn, Hg, Mn, Cd, and Pb; X = Br, M = Pd, Co, Zn, and Mn); (ii) octahedral $[M^{IV}Cl_6]^{2-}$ (M = Os and Pt); (iii) trigonal bipyramidal $[Fe^{III}Cl_5]^{2-}$; (iv) planar $[Cu^{II}{}_2Cl_6]^{2-}$, and (v) square pyramidal $[Sb^{III}Cl_5]^{2-}$ anionic complexes. The combination of these anions with pyridinium or bipyridine cations leads to a variety of architectures [26]. The $^+$N–H $\cdots$ X$^-$ recognition pattern is responsible for the formation of networks. The connectivity of the network is governed by the shape and orientation of the H-bonds taking place between anions and cations. An example of 1D H-bonded network is shown in Figure 7.3.

In order to extend the variety of networks, the nature of the pyridinium derivative was varied [27]. A similar approach has also been followed by Brammer and coworkers [28].

The strategy was further extended to other hydrogen bond acceptors such as metaldithiolate, metal 1,2-dithiomaleonitrile, tetracyanometallate, and thiocyanatometallate complexes [29].

Other examples involving hybrid organic/inorganic CAHB networks (with both organic and metallic tectons) were recently reported by Sutter and coworkers [30]. For example, using the $^+$N–H $\cdots$ O$^-$ -type recognition pattern, combinations of zirconiumtetraoxalate $[Zr^{IV}(C_2O_4)_4]^{4-}$ with different dicationic units afford different extended networks. The same type of approach was also reported by Sevov and coworkers [31]. They used metal complexes bearing a diamino and oxalate chelates (Figure 7.4) as H-bond donors and sulfonate derivatives as H-bond acceptors.

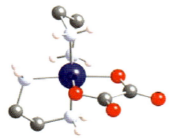

Figure 7.4 Building block [Co(en)$_2$ox] used for the formation of CAHB networks.

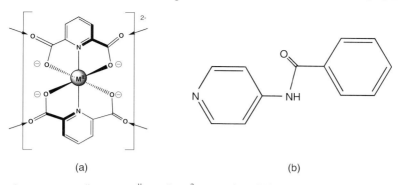

Figure 7.5 Metallatecton [M^{II}(2,6pdca)$_2$]$^{2-}$ (a) and pyridyl benzamide-based ligand (b) used for the construction of hybrid organic/inorganic H-bonded networks.

Hybrid networks, based on organic tectons offering structural flexibility and inorganic tectons imposing rigidity and bringing physical properties, may lead to functional crystals. This transition from structural to functional aspects remains so far almost unachieved and only very few examples have been reported till now. These cases will be presented hereafter and in the section dealing with our own investigations using bisamidinium-type tectons.

The first examples were given by Palmore and coworkers, as well as by McDonald et al.. They combined the imidazolium cation with metallopyridinecarboxylate complexes (Figure 7.5a) [32]. Stang and coworkers used pyridyl benzamide (Figure 7.5b) as a self-complementary unit acting as a ligand toward metal cations and as both H-bond donor and acceptor [33]. In both cases, magnetic networks as well as layered crystalline materials with possible applications in optics have been obtained.

Finally, the last example by Dalrymple and Shimizu [34] describes the formation of highly porous materials. This group used the formation of $^+$N–H$\cdots$O$^-$-type recognition pattern generated upon combining organosulfonates and metalloamino-based tectons.

We have described here several examples reported in the literature dealing with combinations of H-bond donors and acceptors leading to the formation of a variety of architectures based on charge-assisted H-bonded networks. In the following section, we shall report on our own extensive studies dealing with amidinium-based tectons.

7.3
Charge-Assisted H-Bonded Networks Based on Amidinium Tectons

7.3.1
Amidinium-Based Tectons

Bisamidinium dications (Figure 7.6) are interesting candidates as symmetric building blocks for the design of CAHB molecular networks. Indeed, due to the presence

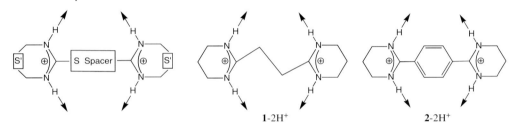

Figure 7.6 Schematic representation of the bisamidinium cation used for the formation of CAHB networks.

of four acidic N–H protons, oriented in a divergent fashion, this type of cationic unit may act as a tetra-H-bond donor. The distance between two nitrogen atoms on the same face of the building block can be adjusted by the nature of the spacer S. For example, although for the short ethyl spacer (**1**-2H$^+$), a distance of about 5.2 Å between the two N-atoms located on the same face may be imposed [35], this distance may be increased to 7.0 Å by using the phenyl spacer (**2**-2H$^+$) [36]. Furthermore, distance between the two N-atoms of the two amidinium groups may be finely tuned by varying the size of the cyclic amidinium moieties through the nature and length of the spacer S' (5-, 6-, and 7-member rings). By imposing all these metrics, one may control the mode of recognition of different anions by the dicationic building block. For example, one may impose a bisdihapto mode (two dihapto modes of H-bonding, one on each side) or a dihapto and bismonohapto or even a tetra-monohapto mode of H-bonding.

7.3.2
Organic Networks Based on Amidinium Tectons

It has been shown that symmetric bisamidinium cations lead to the formation of purely organic CAHB networks with various and controlled dimensionality when combined with carboxylates as well as sulfonate and phosphonate derivatives [37].

Indeed, tecton **1**-2H$^+$ (Figure 7.6) for which the spacer S is an ethyl group, owing to the distance separating the two amidinium moieties, is particularly well suited for the recognition of carboxylate anion through a dihapto mode of H-bond of the $^+$N–H···O$^-$ type. Using this type of recognition pattern, a series of molecular networks with tunable dimensionality have been generated [36, 38]. Two examples, describing the formation of a 1D network with benzene 1,4-dicarboxylate and of a 2D network with fumarate, are given in Figure 7.7a and b, respectively.

Other examples of nonmetallic networks generated upon combining bisamidinium cations with phosphate derivatives have been also reported [39].

Another example of purely organic networks containing bisamidinium-type tectons is based on $^+$N–H···N$^-$ recognition pattern. This was achieved with an organic tectons offering terminal cyano groups (2,2'-(1,4-phenylene)pyrimidine^{2+}) such as polycyano anions (Figure 7.8) [40]. The anion, possessing a C$_3$ symmetry, due to metric constraints, acts as a tetra-H-bond acceptor and the 1D network is formed

Figure 7.7 1D and 2D H-bonded networks formed with benzene 1,4-dicarboxylate (a) and fumarate (b) anions.

through only four H-bonds per hexacyano unit. In other words, of the six cyano groups, two are not involved in the formation of the 1D network.

An additional example of purely organic functional 1D network based on a bisamidinium cation combined with a benzenecarboxylate derivative bearing the nitronyl nitroxide spin careers was reported [41]. The 1D network obtained displays weak ferromagnetic interaction. Finally, by combining bisamidinium cations with benzene carboxylate derivatives bearing mesogenic groups, liquid crystalline-type materials have been obtained [42].

7.3.3
Inorganic Networks Based on Amidinium Tectons

Few examples of hybrid networks generated upon combining bisamidinium-based tectons with organometallic anionic tectons, also called metallatectons, are given in the following section.

7.3.3.1 Metallatectons Bearing a Carboxylate at Their Periphery

Exploiting the propensity of bisamidinium-based tecton **1**-2H$^+$ to recognize carboxylate anions through a dihapto mode of H-bonding of the $^+$N–H$\cdots$O$^-$ type, a series of 1D networks based on ferrocenyl derivatives bearing two carboxylate moieties at their periphery were obtained (Figure 7.9) [43].

Figure 7.8 1D CAHB network formed between **1**-2H$^+$ and hexacyanocyclopropanediide^{2-}.

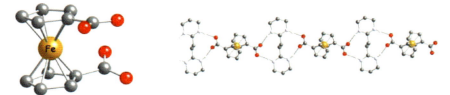

Figure 7.9 Ferrocenedicarboxylate anion (left) and 1D CAHB network (right) formed using the tecton **1**-2H$^+$.

A further extension consisting in combining charge-assisted hydrogen bonding with coordination bond formation was also achieved. Thus, 2,2′-bipyridine or phenanthroline derivatives bearing two carboxylate groups have been used as both coordinating ligand toward transition metal cations and as H-bond acceptor units. In the presence of silver cation, these ligands form 2/1 (L/M) complexes bearing four carboxylate moieties oriented in a divergent fashion. These species behave as metallatectons and in the presence of dicationic organic tectons such as **1**-2H$^+$, they lead to the formation of cationic 2D networks. The formation of the latter results from the recognition of the dicarboxylate moieties by the bisamidinium tecton through a dihapto mode of H-bonding. This is an elegant example of a three-component system (Ag, L, and bisamidinium) combining two different types of recognition events and thus two different connectivities (Figure 7.10) [44].

7.3.3.2 Oxalatometallate as H-Bond Acceptor

As already demonstrated [30], oxalatometallate complexes may be used as hydrogen bond acceptors. The combination of octahedral bisoxalatometallate [Cr(ox)$_2$bipy]$^{2-}$ [45] with **2**-2H$^+$ (Figure 7.6) [46] leads to a 1D network (Figure 7.11). The formation of the network results from the recognition of the terminal oxygen

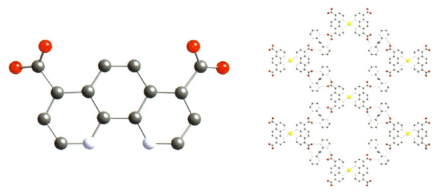

Figure 7.10 4,9 phenanthrolinedicarboxylate L (left) and 2D CAHB network (right) formed between **1**-2H$^+$, L, and silver cation.

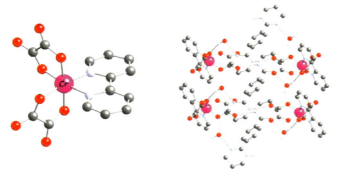

Figure 7.11 [Cr(ox)$_2$bipy]$^-$ complex (left) and 1D CAHB network (right) formed in the presence of **1**-2H$^+$.

atoms of the metal complex by the amidinium moiety through $^+$N–H$\cdots$O$^-$-type interactions.

7.3.3.3 Thiocyanatometallate as H-Bond Acceptor

For the formation of H-bonded networks, combinations of cationic tectons with other metal complexes bearing negatively charged terminal N-atoms at their periphery have been also investigated. In particular, thiocyanatometallate [M(SCN)$_4$]$^{2-}$ (M = Pd or Hg) and isothiocyanatometallates [Cu(NCS)$_4$]$^{2-}$ have been used. When combined with **2**-2H$^+$ dication, different molecular networks based on $^+$N–H$\cdots$N$^-$ or $^+$N–H$\cdots$S$^-$ type of interaction patterns have been obtained [47]. The dimensionality of the networks depends on the orientation of the lone pairs of the anionic moieties. Thus, 1D architectures with [M(SCN)$_4$]$^{2-}$ (M = Pd or Hg) and 2D networks with [Cu(NCS)$_4$]$^{2-}$ have been generated (Figure 7.12).

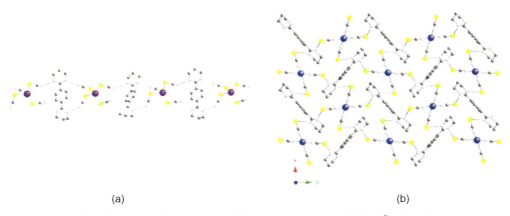

Figure 7.12 (a) 1D CAHB network formed when **2**-2H$^+$ was combined with [M(SCN)$_4$]$^{2-}$ (M = Pd or Hg). (b) 2D CAHB network formed between **2**-2H$^+$ and [Cu(NCS)$_4$]$^{2-}$.

Using $^+$N–H···X$^-$-type interactions, the formation of H-bonded networks was demonstrated by combining other metallatectons such as halogenometallates and bisamidinium-type tectons. These architectures are not discussed here.

The use of cyanometallate anions in conjunction with a variety of dicationic bisamidinium-type tectons has generated a massive collection of networks and properties. The following sections will extensively discuss the obtained results.

7.3.3.4 Polycyanometallate as H-Bond Acceptor

Due to their tunability and stability, cyanometallate anions are interesting H-bond acceptor tectons. Indeed, this class of compounds offers a large variety of charge (-1, -2, -3, -4) and geometry (linear, square planar, octahedral) around the metal center. Among the stable metal cyanide complexes, those shown in Figure 7.13 have been used as building blocks for the formation of hybrid molecular networks in the presence of organic dicationic tectons. It is worth noting that with respect to the localization of the protons, owing to the difference in pKa values between cyanometallates $[M^I(CN)_2]^-$ (M = Ag and Au), $[M^{II}(CN)_4]^{2-}$ (M = Ni, Pd, and Pt), $[Fe^{II}(CN)_5NO]^{2-}$, $[M^{III}(CN)_6]^{3-}$ (M = Fe, Co, and Cr), and $[M^{II}(CN)_6]^{4-}$ (M = Fe and Ru) anions (Figure 7.13) and unprotonated bisamidines, the acidic protons should be localized on the nitrogen atoms of bisamidinium cations, leading to a reliable $^+$N–H···N$^-$ recognition patterns.

The N–N distances in cyanometallates are in the range of 3.9–4.7 Å for $[M^{II}(CN)_4]^{2-}$, $[Fe^{II}(CN)_5NO]^{2-}$, $[M^{III}(CN)_6]^{3-}$, and $[M^{II}(CN)_6]^{4-}$, which impose a rather long distance between the two amidinium moieties if a dihapto mode of H-bonding is desired. This may be achieved by using the phenyl group as a spacer (compound 2-2H$^+$, Figure 7.14). For $[M^I(CN)_2]^-$, the N–N distances is in the range of 6.2–6.6 Å. However, for this type of anion, the formation of extended networks, instead of a dihapto mode of H-bonding, requires monohapto modes of connectivity.

In the following, the formation of molecular networks will be discussed based on (i) the stoichiometry between organic and inorganic tectons; (ii) the metrics and the

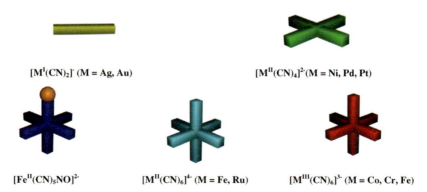

Figure 7.13 Schematic representation of the polycyanometallate acting as H-bond acceptors.

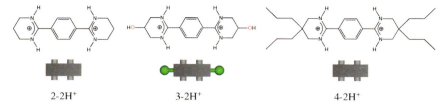

Figure 7.14 Bisamidinium cations **2**-2H$^+$, **3**-2H$^+$, and **4**-2H$^+$ used for the formation of CAHB networks when combined with cyanometallates.

geometry around the metal center and the geometry of the organic tecton, and (iii) the influence of the nature of cations as well as of solvent molecules on the packing of networks and the formation of crystalline materials.

7.4
Charge-Assisted H-Bonded Networks Based on Amidinium and Polycyanometallate Tectons

Here, we shall focus on the three most illustrative bisamidinium cations **X**-2H$^+$ (X = 2–4) (Figure 7.14). Results obtained with many other bisamidinium-based tectons will not be presented. Owing to the presence of the phenyl group as spacer connecting the two cyclic amidinium moieties, these tectons **X**-2H$^+$ (X = 2–4) are not only stable toward hydrolysis but, because of the imposed distance between two acidic H-atoms on each face of the tecton, they also interact with tetra- and hexa-cyanometallate anions through a dihapto mode of H-bonding.

All networks presented below are air stable and have been studied in the crystalline state using X-ray diffraction on single crystals.

7.4.1
Octahedral Cyanometallate Anions

7.4.1.1 [M^{III}(CN)$_6$]$^{3-}$ and 2-2H$^+$ or 3-2H$^+$ (M = Fe, Co, and Cr)

For charge neutrality reasons, the combination of dicationic tectons **2**-2H and **3**-2H with [M^{III}(CN)$_6$]$^{3-}$ (M = Fe, Co, and Cr) should lead to crystals of general formula (**X**-2H$^+$)$_3$([M^{III}(CN)$_6$]$^{3-}$)$_2$ (X = **2** or **3**). Because of the dihapto or chelate mode of H-bonding between **X**-2H$^+$ and the octahedral [M^{III}(CN)$_6$]$^{3-}$ anion, a supramolecular chirality taking place within the second coordination sphere around the metal is expected (Figure 7.15). Indeed, by analogy with Δ and Λ type of chirality defined for octahedral complexes surrounded by two or three bidentate chelating ligands, within the second coordination sphere, the formation of a chelate through a dihapto mode of H-bonding between the dicationic tecton **X**-2H$^+$ and [M^{III}(CN)$_6$]$^{3-}$ generates the chirality of the Δ' and Λ' type. This type of chirality, although based on H-bonding chelates, may be seen as analogous to the trisoxalatometallate-extended architectures [48].

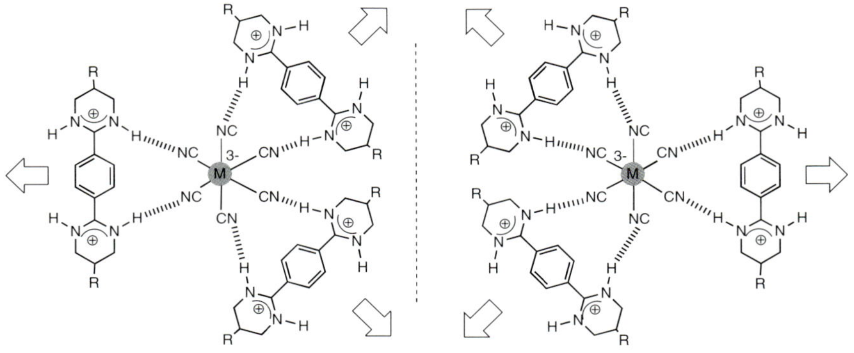

Figure 7.15 Schematic representation of supramolecular chirality of the Δ' and Λ' type obtained upon binding within the second coordination sphere of octahedral $[M^{III}(CN)_6]^{3-}$ anion by dicationic **X**-2H$^+$ during the formation of $(\mathbf{X}\text{-}2\mathrm{H}^+)_3([\mathrm{M}^{III}(\mathrm{CN})_6]^{3-})_2$ networks (**X** = 2 or 3).

The 2D honeycomb-type neutral 2D charge-assisted H-bonded networks formed upon combining the cationic organic tecton with $[M^{III}(CN)_6]^{3-}$ (M = Co, Cr, and Fe) present deformed hexagonal cavities [49]. The crystal is composed of dications **X**-2H$^+$, two $[M^{III}(CN)_6]^{3-}$ dianions, and water molecules (Figure 7.16). The 2D networks are formed by interconnection of dicationic and dianionic units through dihapto mode of H-bonding (N $\cdots$ N distance varying between 2.8 and 3.0 Å). The connectivity mode and the stoichiometry (3/2) between cations and anions lead to the generation of deformed hexagonal channels filled with seven water molecules in the case of $(\mathbf{2}\text{-}2\mathrm{H}^+)_3([\mathrm{M}^{III}(\mathrm{CN})_6]^{3-})_2$ (M = Fe and Co), six water molecules in the case of $(\mathbf{2}\text{-}2\mathrm{H}^+)_3([\mathrm{Cr}^{III}(\mathrm{CN})_6]^{3-})_2$, and eight water molecules in the case of $(\mathbf{3}\text{-}2\mathrm{H}^+)_3([\mathrm{M}^{III}(\mathrm{CN})_6])_2$ (M = Fe, Cr, and Co) (Figure 7.17). The water molecules form H-bonded networks parallel to the axis of the channels. Due to the presence of

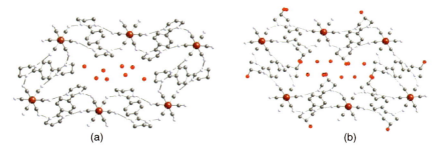

Figure 7.16 Portions of the X-ray structure of $(\mathbf{X}\text{-}2\mathrm{H}^+)_3([\mathrm{M}^{III}(\mathrm{CN})_6]^{3-})_2$ showing the alternate positioning of Δ' and Λ' supramolecular enantiomers and the inclusion of water molecules and a view of the H-bonded water polymer: (a) with (**2**-2H$^+$) and (b) with (**3**-2H$^+$). H-atoms are omitted for clarity.

7.4 Charge-Assisted H-Bonded Networks Based on Amidinium and Polycyanometallate Tectons

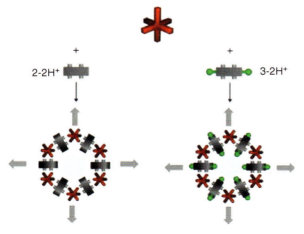

Figure 7.17 Schematic representation of the 2D H-bonded network formed upon self-assembly of the dicationic **X**-2H$^+$ (X = **2** or **3**) with [M^{III}(CN)$_6$]$^{3-}$ (M = Fe, Cr, or Co).

additional H-bond donors in the case of (**3**-2H$^+$)$_3$([M^{III}(CN)$_6$]), a strong H-bonded network is formed between hydroxyl groups and the water molecules.

7.4.1.2 [M^{III}(CN)$_6$]$^{3-}$ and 4-2H$^+$ (M = Fe, Co, and Cr)

When **4**-2H$^+$ is combined with [M^{III}(CN)$_6$]$^{3-}$ (M = Fe, Co, and Cr), the resulting network has the same connectivity as those observed for (**X**-2H$^+$)$_3$([M^{III}(CN)$_6$]$^{3-}$)$_2$ (X = **2** or **3**) (isostructural crystals) [50]. Interestingly, in this case, the hexagons are not deformed and no solvent molecules are present in the crystal. The water molecules present in previously discussed crystals based on tectons (**2**-2H$^+$) and (**3**-2H$^+$) are replaced by alkyl chains of the organic tecton (**4**-2H$^+$) (Figure 7.18).

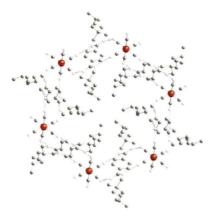

Figure 7.18 A portion of the X-ray structure of (**4**-2H$^+$)$_3$([M^{III}(CN)$_6$]$^{3-}$)$_2$ (M = Fe, Co, or Cr).

7.4.1.3 $[M^{II}(CN)_6]^{4-}$ with 2-2H$^+$ or 4-2H$^+$ (M = Fe and Ru)

Dealing with octahedral $[M^{II}(CN)_6]^{4-}$ anionic complexes, again based on charge neutrality, for combinations of **2**-2H$^+$ or **4**-2H$^+$ with Li$_4$[M^{II}(CN)$_6$] and (NH$_4$)$_4$[M^{II}(CN)$_6$] salts, crystals of formula (**X**-2H$^+$)$_2$([M^{II}(CN)$_6$]$^{4-}$) (X = **2** or **4**) are obtained [51]. The recognition pattern between anion and cation is rather different when compared to the previous cases discussed above. The crystal is composed of two **2**-2H$^+$ (or **4**-2H$^+$) dications, one [M^{II}(CN)$_6$]$^{4-}$ (M = Fe or Ru) anion, and eight H$_2$O molecules. Each [M^{II}(CN)$_6$]$^{4-}$ (M = Fe or Ru) complex is surrounded by four bisamidinium **2**-2H$^+$. For both cases, a neutral 2D network is obtained. The latter results from the interconnection of 1D networks, formed by H-bonds in a dihapto mode between two organic cations and four cyanide groups located at the square base of the octahedron, by a transversal organic tectons interacting through a bismonohapto mode of H-bonding with the remaining two cyanides occupying the apical positions (Figures 7.19 and 7.20). The remaining two H-bond donor N–H sites of the transversal organic tectons are hydrogen bonded to water molecules. For both cases, the water molecules form branched hexagons composed of eight solvent molecules, which are located between 2D networks.

7.4.1.4 $[M^{II}(CN)_6]^{4-}$ with 3-2H$^+$ (M = Fe and Ru)

Surprisingly, the connectivity mode for the combination of [M^{II}(CN)$_6$]$^{4-}$ (M = Fe or Ru) with **3**-2H$^+$ is different from the one of (**2**-2H$^+$)$_2$([M^{II}(CN)$_6$]$^{4-}$). In this case, as expected, although the formula is (**3**-2H$^+$)$_2$([M^{II}(CN)$_6$]) (M = Fe or Ru), the recognition pattern leads to a 1D architecture. Each metallic anion is surrounded, in its square base, by two organic tectons with a dihapto mode of recognition on one side and a monohapto mode on the other face with the apical cyanide of an adjacent [M^{II}(CN)$_6$]$^{4-}$ anion (Figure 7.21). This connectivity mode leads to a 1D system for

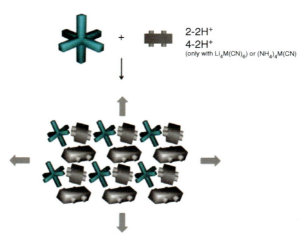

Figure 7.19 Schematic representation of a 2D network formed between octahedral [M^{II}(CN)$_6$]$^{4-}$ (M = Fe or Ru) anionic complexes and dicationic tectons (**X**-2H$^+$, X = **2** or **4**).

7.4 Charge-Assisted H-Bonded Networks Based on Amidinium and Polycyanometallate Tectons

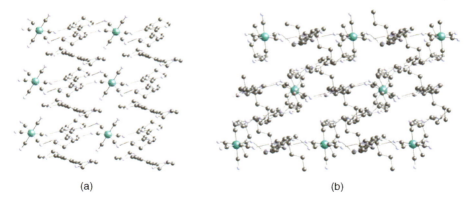

Figure 7.20 Portions of the X-ray structure of 2D H-bonded network (**X**-2H$^+$)$_2$([M^{II}(CN)$_6$]$^{4-}$) (X = **2** or **4**). The 2D network is formed through the interconnection of anionic 1D networks by **X**-2H$^+$ through a monohapto mode of H-bonding with the remaining two CN groups in axial positions of [M^{II}(CN)$_6$]$^{4-}$: (a) with (**2**-2H$^+$) and (b) with (**4**-2H$^+$). Water molecules are omitted for clarity.

which two cations are parallel with a distance of 4.8 Å between their centroids. Each cation is H-bonded only to three cyanide groups. These 1D systems are also surrounded by water molecules (six per formula unit) with a strong hydrogen bond network between hydroxyl groups and water molecules.

7.4.1.5 [M^{II}(CN)$_6$]$^{4-}$ with 4-2H$^+$ (M = Fe and Ru)

Interestingly, when **4**-2H$^+$ is combined with other X$_4$[Fe^{II}(CN)$_6$] salts (X = K, Na, Rb, and Cs), crystals of general formula X$_2$(**4**-2H$^+$)$_3$([Fe^{II}(CN)$_6$]$^{4-}$)$_2$ (3/2 stoichiometry) are obtained instead of the expected 2/1 stoichiometry ((**4**-2H$^+$)$_2$([Fe^{II}(CN)$_6$]$^{4-}$)) [52]. These crystals are isostructural to (**4**-2H$^+$)$_3$([M^{III}(CN)$_6$]$^{3-}$)$_2$ (see Section 7.4.1.2) with the same recognition pattern between cations and anions. The alkali cations occupy a site located between the [Fe^{II}(CN)$_6$]$^{4-}$ anions, forming a 1D array parallel to the *c*-axis, with short C≡N–X distances (Figure 7.22, right). The increase in the *c* parameter from 6.95 Å for Na$^+$ to 8.37 Å for Cs$^+$ is accompanied by a less pronounced decrease

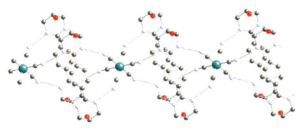

Figure 7.21 A portion of the X-ray structure of 1D H-bonded network (**3**-2H$^+$)$_2$([M^{II}(CN)$_6$]$^{4-}$) (M = Fe or Ru).

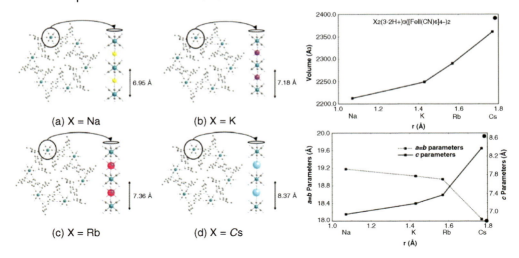

Figure 7.22 *Left*: A portion of X-ray structures obtained for $X_2(4\text{-}2H^+)_3([Fe^{II}(CN)_6]^{4-})$, viewed in the (001) plane and along the *c*-axis (X = Na (a), K (b), Rb (c), and Cs (d)). *Right*: Plots of the variation of the cell parameters (i) and cell volume (ii) for $X_2(4\text{-}2H^+)_3([Fe^{II}(CN)_6]^{4-})_2$ (X = Na, K, Rb, and Cs) (v). The same parameters for $(4\text{-}2H^+)_3([Fe^{III}(CN)_6]^{3-})_2$ (λ) are also given for the sake of comparison.

of *a*, *b* parameters, as shown in Figure 7.22, left. The volume of the cell is enhanced by increasing the radii of the inserted alkali cation. It is worth noting that for $Cs_2(4\text{-}2H^+)_3([Fe^{II}(CN)_6]^{4-})_2$, the obtained metrics are the closest ones to those observed for $(4\text{-}2H^+)_3([Fe^{III}(CN)_6]^{3-})_2$.

7.4.2
Pentadentate Cyanometallate Anions, $[Fe^{II}(CN)_5NO]^{2-}$ and 2-2H$^+$

By analogy with $[M^{III}(CN)_6]^{3-}$ anions (Section 7.4.1.1), octahedral nitroprussiate $[Fe^{II}(CN)_5NO]^{2-}$ is an interesting anionic tecton allowing to study several issues related to supramolecular chirality, that is, chirality taking place within the second coordination sphere around the metal center. Due to the presence of five CN$^-$ groups in $[Fe^{II}(CN)_5NO]^{2-}$, which may be engaged in H-bond patterns, the formation of achiral or chiral architectures may be envisaged. Indeed, if the dihapto mode of H-bonding of two bisamidinium moieties occurs in the basal plane of $[Fe^{II}(CN)_5NO]^{2-}$ (Figure 7.23, top), no chirality may be expected (presence of a plane of symmetry). However, for the other possibility, a chiral architecture may be formed (Figure 7.23, bottom).

The combination of $[Fe^{II}(CN)_5NO]^{2-}$ with 2-2H$^+$ leads to a neutral 1D network of formula $(2\text{-}2H^+)([Fe^{II}(CN)_5NO]^{2-})$ resulting from interconnection of the anionic units by the tecton 2-2H$^+$ through a dihapto mode of H-bonding [53]. The chelate mode of H-bonding takes place on both sides of the tecton 2-2H$^+$ with two CN$^-$ units

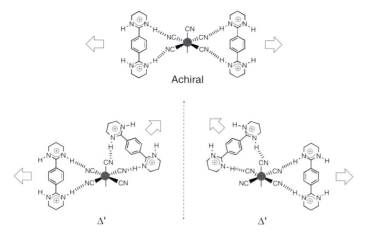

Figure 7.23 Schematic representation of the octahedral complex [Fe^{II}(CN)$_5$NO]$^{2-}$ and its double chelation by the dicationic tectons **2**-2H$^+$ through a dihapto mode of H-bonding either with two adjacent cyanides within the square base or with one CN$^-$ anion located within the square base and the other occupying an apical position leading thus to supramolecular chirality of the Δ' and Λ' type.

located at the square base of the octahedron leading thus to an achiral unit. Within each neutral 1D networks, the NO units are disposed in an alternate fashion, with an unusual short d_{O-O} distance of about 2.71 Å between NO groups facing each other in consecutive networks. The packing of the 1D networks in space, leading to the formation of the solid, generates channels, which are filled with water molecules forming polymeric chains (Figure 7.24).

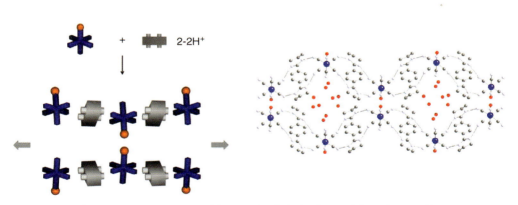

Figure 7.24 *Left*: Schematic representation of the 1D H-bonded achiral network formed upon self-assembly of the dicationic **2**-2H$^+$ and [Fe^{II}(CN)$_5$NO]$^{2-}$. *Right*: A portion of the X-ray structure of (**2**-2H$^+$)([Fe^{II}(CN)$_5$NO]$^{2-}$).

7.4.3
Square Planar Cyanometallate Anions

7.4.3.1 $[M^{II}(CN)_4]^{2-}$ and 2-2H$^+$ or 4-2H$^+$ (M = Ni, Pd, and Pt)

Owing to the metric (distance between two acidic hydrogen atoms located on the same face of the tecton) and charge (dication) of the **X**-2H$^+$ (**X** = **2** or **4**) for the combination of these two tectons with $[M^{II}(CN)_4]^{2-}$ (M = Ni, Pd, or Pt) a 1/1 stoichiometry and a dihapto mode of recognition may be expected. In other terms, for such a network, each dication **X**-2H$^+$ should be surrounded by two dianions and, conversely, each dianion $[M^{II}(CN)_4]^{2-}$ should be in interactions with two dications through a dihapto mode of H-bonding.

The predicted networks were indeed observed for combinations of **2**-2H$^+$ or **4**-2H$^+$, $[M^{II}(CN)_4]^{2-}$ with M = Pd, Pt, and Ni (crystals of general formula (**X**-2H$^+$) ($[M^{II}(CN)_4]^{2-}$) (X = **2** or **4**, M = Ni, Pd, and Pt)) [50, 51]. In all cases, X-ray diffraction on single crystals (Figure 7.25) revealed that all the crystalline materials are isostructural. The 1D networks, of general formula (**X**-2H$^+$)($[M^{II}(CN)_4]^{2-}$), resulting from a dihapto mode of H-bonding between **X**-2H$^+$ and the square anions, are packed in a parallel fashion in the crystal. The absence of solvent molecules in the lattice indicates that the cohesion of crystals is ensured by the two complementary components.

Interestingly, as expected by the design of cation **4**-2H$^+$, the presence of propyl chains controls the spacing between the 1D networks. Indeed, whereas for the 1D networks obtained with $[M(CN)_4]^{2-}$ anions and **2**-2H$^+$, the distance between two consecutive networks in the yOz plane is the range of 6.40 and 6.44 Å, in the case of **4**-2H$^+$ the spacing is increased and it ranges from 8.41 Å for Ni to 8.28 and 8.26 Å for Pd and Pt, respectively.

This example clearly illustrates the control of the packing of networks through the design of the organic tecton.

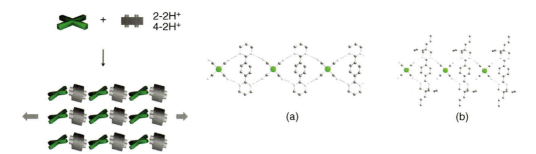

Figure 7.25 Left: Schematic representation of the 1D H-bonded network formed upon self-assembly of the dicationic **X**-2H$^+$ (X = **2** or **4**) and $[M^{II}(CN)_6]^{2-}$ (M = Ni, Pd, or Pt) ensured by strong H-bonds through a dihapto mode of interaction. Right: Portions of the X-ray structure of (a) (**2**-2H$^+$)($[M^{II}(CN)_4]^{2-}$) and (b) (**4**-2H$^+$)($[M^{II}(CN)_4]^{2-}$) (M = Ni, Pd, and Pt).

7.4 Charge-Assisted H-Bonded Networks Based on Amidinium and Polycyanometallate Tectons

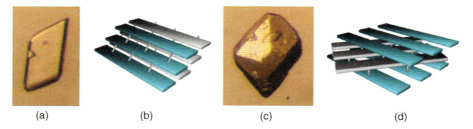

Figure 7.26 Pictures of (a) polymorph α, obtained for the combination of tecton **3**-2H$^+$ with [M^{II}(CN)$_4$]$^{2-}$ (M = Ni, Pd, and Pt), and (c) polymorph β, obtained only for (**3**-2H$^+$) ([Pt^{II}(CN)$_4$]$^{2-}$). (b) and (d) represent schematically the arrangement of the networks in the α- and β-phases, respectively.

7.4.3.2 [M^{II}(CN)$_4$]$^{2-}$ and 3-2H$^+$ (M = Ni, Pd, and Pt)

Since the organic tecton **3**-2H$^+$ offers additional H-bond donor sites (overall six acidic hydrogen atoms), one may expect different types of connectivity with [M^{II}(CN)$_4$]$^{2-}$ anions [38f]. Indeed, the combination of **3**-2H$^+$ with [M^{II}(CN)$_4$]$^{2-}$ leads to polymorphs, having the same formula (**3**-2H$^+$)([M^{II}(CN)$_4$]$^{2-}$) (M = Ni, Pd, and Pt) [54].

Whereas in the case of [Ni(CN)$_4$]$^{2-}$ and [Pd(CN)$_4$]$^{2-}$, plate shape crystals (α-phase) were obtained (Figure 7.26), for [Pt(CN)$_4$]$^{2-}$, a mixture of plate (α-phase) and prismatic shape (β-phase) (Figure 7.26) crystals were observed. For the α-phase, the crystals are composed of **3**-2H$^+$ dications and [M(CN)$_4$]$^{2-}$ (M = Ni^{2+}, Pd^{2+}, and Pt^{2+}) dianions, forming as previously described for **2**-2H$^+$ and **4**-2H$^+$ cations, neutral 1D network through a dihapto mode of H-bonding (d_{N-N} in the range of 2.83 and 2.90 Å) without any solvents molecules (Figure 7.27).

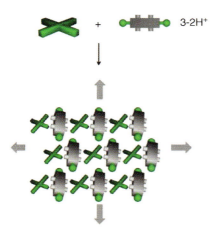
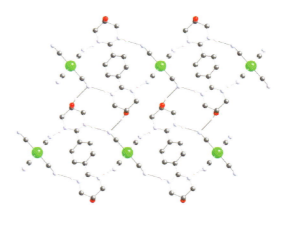

Figure 7.27 *Left*: Schematic representation of the 2D H-bonded network formed upon self-assembly of the dicationic **3**-2H$^+$ and [M^{II}(CN)$_6$]$^{2-}$ (M = Ni, Pd, or Pt) ensured by interconnection between the 1D arrays through OH···O interactions. *Right*: A portion of the X-ray structure of (**3**-2H$^+$)([M^{II}(CN)$_4$]$^{2-}$) (M = Ni, Pd, and Pt).

Owing to the presence of hydroxyl groups within the backbone of the cations, the OH groups behave as secondary H-bond donor sites and the overall structure is a 2D network. The two OH groups in axial position interconnect the 1D networks, disposed in parallel fashion, through H-bonds of the OH–N≡C type (d_{N-O} in the range of 2.95, and 2.98 Å). Among the four CN groups connected to the metal center, only two are involved in H-bond with the OH groups (Figure 7.27). In the case of [Pt(CN)$_4$]$^{2-}$, in addition to the α-phase, another morphology (β-phase) was observed (Figure 7.26). The difference with the α-phase relies on the connection of the formed 1D arrays through H-bonds. Here, the second type of H-bond is of OH–O type and takes place between OH moieties belonging to consecutive planes ($d_{OH-O} = 2.88$ Å), leading to a tilt angle of about 60° between two consecutive 1D arrays (Figure 7.26). Because of this particular orientation of consecutive layers, the overall architecture may be considered a 3D hydrogen-bonded network.

7.4.4
Linear Cyanometallate Anions [M^I(CN)$_2$]$^-$ (M = Ag and Au)

Due to the linear geometry of the [M^I(CN)$_2$]$^-$ (M = Ag or Au) anions and the rather long distance between the two N-atoms of the cyano groups located at the extremities of the anion, the combination of such anions with bisamidinium cation should, in principle, lead to a tetrakis-monohapto mode of H-bonding between the anion and the cation. The combination of **2**-2H$^+$ with [M^I(CN)$_2$]$^-$ (M = Ag and Au) leads to crystals exclusively composed of **2**-2H$^+$ dication and [M^I(CN)$_2$]$^-$ (M = Ag and Au) anion with a (**2**-2H$^+$)([M^I(CN)$_2$]$^-_2$) stoichiometry [55]. Each face of the tecton **2**-2H$^+$ is connected to two [M^I(CN)$_2$]$^-$ anions imposing thus a Au–Au and Ag–Ag distances of 3.33 and 3.37 Å, respectively. The connectivity pattern, resulting from strong H-bonds between donor and acceptor sites as well as d^{10}–d^{10} interactions is given in Figure 7.28. The same type of pattern is observed for **1**-2H$^+$ and **4'**-2H$^+$ cation bearing hexyl chains in its periphery (Figure 7.28).

Other polymorphs with different connectivity between the tectons are also observed. They will not be discussed here.

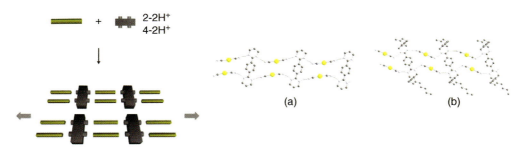

Figure 7.28 Left: Schematic representation of the 1D H-bonded network formed upon self-assembly of the dicationic **X**-2H$^+$ (X = **2** or **4'**) and [M^I(CN)$_2$]$^-$ (M = Ag or Au). Right: Portions of the X-ray structure of (a) (**2**-2H$^+$) ([M^I(CN)$_2$]$^-$)$_2$ (M = Ag or Au) and (b) (**4'**-2H$^+$) ([M^I(CN)]$^-$)$_2$ (M = Ag or Au).

7.5
Properties of Charge-Assisted H-Bonded Networks Based on Amidinium Tectons

Exploiting specific properties resulting from the interconnection of tectons into networks and the packing of these periodic architectures into crystals is a matter of current interest [56]. In the following section, the solid-state luminescent properties of a series of H-bonded architectures based on dicationic tectons and dicyanometallate anions are described. Furthermore, a series of 2D assemblies offering channels occupied by solvent molecules will be presented and their solvation/desolvation processes demonstrating their porous nature will be described.

7.5.1
Luminescence

Crystals containing $[M^I(CN)_2]^-$-type anion were shown to exhibit emission. For example, luminescent property of $KAg^I(CN)_2 \cdot H_2O$ has been studied in the solid state [57]. More recently, examples of luminescent extended H-bonded architectures have been described [58].

Interestingly, colorless crystals obtained by combining **2**-2H$^+$ with $[M^I(CN)_2]^-$ (M = Ag or Au) were found to be strongly luminescent at room temperature (Figure 7.29). Their excitation at 370 nm generates a blue emission ($\lambda_{max} = 430$ nm) [55]. The luminescence is attributed to aurophilic or argentophilic interactions resulting from rather short M–M distances. The latter parameter can be modulated within the backbone of the dicationic organic tecton through the variation of the length of the spacer connecting the two amidinium moieties.

7.5.2
Porosity

Porosity is one of the interesting features associated with molecular crystals. Although the majority of porous crystals derives from coordination networks or MOFs, few examples of crystals based on robust H-bonded networks have been reported [10, 34, 59].

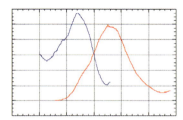

Figure 7.29 An image under fluorescent microscope of a luminescent crystal of (**2**-2H$^+$) ([Ag^I(CN)$_2$]$_2$ anion (left) and its excitation and emission spectra at 298 K (right) in the solid state.

Some of the hybrid 2D networks presented above display channels and cavities filled with solvent molecules. In particular, the channels formed by H-bonded networks of the type $(\mathbf{X}\text{-}2\text{H}^+)_3([\text{M}^{\text{III}}(\text{CN})_6]^{3-})_2$ ($X = \mathbf{2}$ or $\mathbf{3}$) are filled with water molecules. The dehydration of these systems was induced thermally upon heating the crystal to 180 °C and the resulting dehydrated samples were studied by XRD on single crystal. Several dehydration–rehydration cycles were performed and in all cases the initial number of water molecules (7 for $\mathbf{2}\text{-}2\text{H}^+$ and 8 for $\mathbf{3}\text{-}2\text{H}^+$) was observed by TGA analysis. In all cases, the dehydration–rehydration event takes place via single crystal-to-single crystal (SCSCT) transformation (Figure 7.30). This is a rare example of SCSCT based on H-bond networks [49b].

All crystals are thermally robust and their decomposition temperature is in the range of 240–300 °C depending on the metal cation used. As expected through the

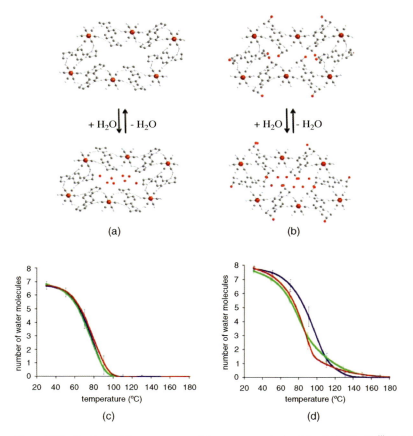

Figure 7.30 Portions of X-ray structure of the hydrated and dehydrated forms obtained through single crystal to single crystal transformation of isostructural crystals of $(\mathbf{X}\text{-}2\text{H}^+)_3([\text{M}^{\text{III}}(\text{CN})_6]^{3-})_2$ (M = Fe, Co, and Cr): (a) $X = \mathbf{2}$ and (b) $X = \mathbf{3}$. At the bottom, TGA curves obtained for $(\mathbf{2}\text{-}2\text{H}^+)_3([\text{M}^{\text{III}}(\text{CN})_6]^{3-})_2$ (c) and $(\mathbf{3}\text{-}2\text{H}^+)_3([\text{M}^{\text{III}}(\text{CN})_6]^{3-})_2$ (d) (M = Fe in red, Co in blue, and Cr in green) showing the release of 7 and 8 H$_2$O molecules for $\mathbf{2}\text{-}2\text{H}^+$ and $\mathbf{3}\text{-}2\text{H}^+$, respectively.

design of positively charged tectons **2-2H$^+$** and **3-2H$^+$**, the introduction of two OH groups as additional H-bond donor/acceptor sites in **3-2H$^+$** interferes with the release of water molecules with a ΔT of about 40 °C between **2-2H$^+$** lacking the OH moieties and **3-2H$^+$**. This is related to the possibility of controlling the hydrophilic nature of the channels within the network through the decoration of the tecton's backbone with appropriate auxiliary groups (here hydroxyl groups) and thus the interactions between water molecules and the interior of the channels.

This has been demonstrated using a ternary system (solid solution, see Section 7.6.1.2) composed of the same anion ($[Cr^{III}(CN)_6]^{3-}$) and two different cations (**2-2H$^+$**) and (**3-2H$^+$**). The composition of the solid solution of formula (**2-2H$^+$**)$_{3y}$(**3-2H$^+$**)$_{3(1-y)}$($[Cr^{III}(CN)_6]^{3-}$)$_2$ was varied by changing y between 0 and 1 [60]. In Figure 7.31 are shown the thermal behavior when varying the ratio of the two cations. This study clearly demonstrates the possibility of finely tuning the temperature of water release through the nature of tectons.

The gas adsorption properties of the (**X-2H$^+$**)$_3$([M^{III}(CN)$_6$]$^{3-}$)$_2$ (M = Fe, Co, and Cr) compounds were also studied, unfortunately, no adsorption of N$_2$ or H$_2$ was detected.

7.5.3
Redox Properties in the Solid State

Besides the exchange of guest molecules in porous frameworks, one may perform postsynthetic transformation (modification of the preformed crystalline material) using chemical or redox processes. This approach requires geometrically suitable and robust but flexible crystals. Very few examples of such transformations have been reported [61].

Using tecton **4**, isostructural crystals of (**4-2H$^+$**)$_3$($[Fe^{III}(CN)_6]^{3-}$)$_2$ and X$_2$(**4-2H$^+$**)$_3$($[Fe^{II}(CN)_6]^{4-}$)$_2$ (X = K, Na, Rb, and Cs) corresponding to two different oxidation states Fe(II) and Fe(III) have been formed (see Section 7.4.1.5).

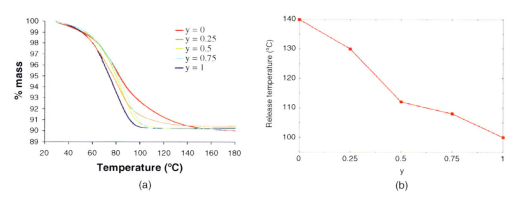

Figure 7.31 (a) TGA curves of molecular alloys (**2-2H$^+$**)$_{3y}$(**3-2H$^+$**)$_{3(1-y)}$(Cr^{III}(CN)$_6$]$^{3-}$)$_2$ showing the fine-tuning of water release temperature. (b) Plot of the temperature of water release for (**2-2H$^+$**)$_{3y}$(**3-2H$^+$**)$_{3(1-y)}$(Cr^{III}(CN)$_6$])$_2$ versus y.

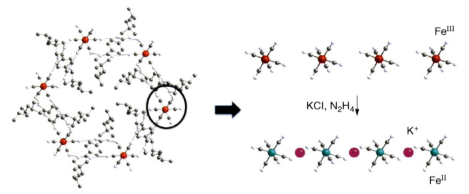

Figure 7.32 Portions of X-ray structure of **(4-2H$^+$)$_3$([Fe^{III}(CN)$_6$]$^{3-}$)$_2$** when treated with N$_2$H$_4$ and KCl, leading to K$_2$(4-2H$^+$)$_3$([Fe^{II}(CN)$_6$]$^{4-}$)$_2$.

7.5.3.1 Reduction in the Solid State

Taking advantage of isostructurality and close metrics of the formed compounds, a postsynthetic transformation of crystals of **(4-2H$^+$)$_3$([Fe^{III}(CN)$_6$]$^{3-}$)$_2$** was demonstrated using the reduction of Fe(III) into Fe(II). Crystals of **(4-2H$^+$)$_3$([Fe^{III}(CN)$_6$]$^{3-}$)$_2$** were reduced by a mixture of hydrazine and K$^+$ cation without the loss of crystallinity. Indeed, the reduction process produced a polycrystalline powder of K$_2$(4-2H$^+$)$_3$([Fe^{II}(CN)$_6$]$^{4-}$)$_2$ [62]. The reduction was monitored by XRPD as well as by IR. The reduction process corresponds to the injection of one electron per metallic center and the incorporation, between consecutive 2D formed H-bond networks, of one K$^+$ cation for each electron exchanged. The process reported here is a rare example of postcrystallization transformation taking place on molecular crystals composed of robust H-bonded molecular networks (Figure 7.32).

7.5.3.2 Formation of Mixed Valence Compounds

A crystalline mixed valence phase starting from a mixture of Cs$_4$Fe^{II}(CN)$_6$ and Cs$_3$Fe^{III}(CN)$_6$ with **4-2H$^+$** was also obtained. The crystals obtained were analyzed by XRD that revealed the formation of a mixed valence Fe(II)/Fe(III) solid solution with 83/17 ratio ((Cs$_2$(4-2H$^+$)$_3$[Fe^{II}(CN)$_6$]$_2^{4-}$)$_{0.83}$((4-2H$^+$)$_3$-[Fe^{III}(CN)$_6$]$_2^{3-}$)$_{0.17}$) (see Figure 7.33) [52]. Within the solid solution, the Fe(II) and Fe(III) cations are randomly distributed within the crystal. The presence of both oxidation states was further confirmed by IR spectroscopy.

7.6
Design of Crystals Based on CAHB Networks

In Section 7.4.1, a series of crystalline materials based on combinations of bisamidinium cations and polycyanometallate anions were described. A large number of obtained crystals were found to be isostructural. Taking advantage of the isostructur-

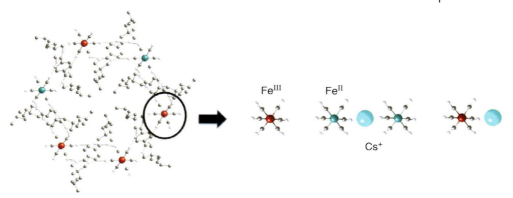

Figure 7.33 Portions of X-ray structure of $((Cs_2(4\text{-}2H^+)_3[Fe^{II}(CN)_6]^{4-}{}_2)_{0.83}((4\text{-}2H^+)_3\text{-}[Fe^{III}(CN)_6]_2{}^{3-})_{0.17})$ with a random occupation of Cs^+ cation as well as distribution of Fe(II) and Fe(III) within the unit cell.

ality of some of the generated architectures, crystals with new shape and composition have been prepared. In particular, solid solutions as well as crystals of crystals, formed by epitaxial growth of crystals on seed crystals, have been obtained.

7.6.1
Solid Solutions

A crystalline solid solution is a mixture of at least two crystalline solids that coexist as a new crystalline solid. When the components of the solid solution are molecular entities, these systems are called *molecular alloys* [63]. The majority of reported molecular solid solutions are inorganic in nature. However, few examples of organic [64] or hybrid organic/inorganic materials have been reported [63a,b].

7.6.1.1 Solid Solutions Using $(2\text{-}2H^+)_2([M^{II}(CN)_6])(M = Fe$ and $Ru)$

The first reported solid solution was based on $(2\text{-}2H^+)_2([M^{II}(CN)_6]^{4-})$ (M = Fe or Ru) [65]. As shown in Figure 7.34, the combination of $2\text{-}2H^+$ with an equimolar mixture of $[Fe^{II}(CN)_6]^{4-}$ and $[Ru^{II}(CN)_6]^{4-}$ leads to an homogenous phase of formula $((2\text{-}2H^+)_2[Ru(CN)_6]^{4-}{}_{0.5}[Fe(CN)_6]^{4-}{}_{0.5})$. The latter crystalline phase displays an intermediate color between the pure $((2\text{-}2H^+)_2[Ru(CN)_6]^{4-})$ and the pure $((2\text{-}2H^+)_2[Fe(CN)_6]^{4-})$. As expected, crystallographic parameters following the Vegard's law [66] are also intermediate.

7.6.1.2 Solid Solutions Using $(X\text{-}2H^+)_3([M^{III}(CN)_6])_2$ (X = 2 or 3, M = Fe, Cr, or Co)

The example described above was based on a ternary system (one dication and two anions). More sophisticated and tunable systems have been generated using 2D porous systems using two different organic tectons and three different anions (formula $(X\text{-}2H^+)_3([M^{III}(CN)_6]^{3-})_2$ (X = 2 or 3, M = Fe, Cr, or Co). Based on crystallographic parameters given in Table 7.1, systems based on up to five components have been generated [60]. As an example shown in Figure 7.35, while keeping

Figure 7.34 Pictures of crystals of $(2\text{-}2H^+)_2(Ru^{II}(CN)_6]^{4-})$ (a), the solid solution $((2\text{-}2H^+)_2[Ru(CN)_6]^{4-}{}_{0.5}[Fe(CN)_6]^{4-}{}_{0.5})$ (b), and $(2\text{-}2H^+)_2(Fe^{II}(CN)_6]^{4-})$ (c).

the same cationic tecton, an intermediate case has been obtained by varying the anionic part of the system. Thus, a solid solution of the following formula $(3\text{-}2H^+)_3([Fe(CN)_6]^{3-}{}_{2/3}[Co(CN)_6]^{3-}{}_{2/3}[Cr(CN)_6]^{3-}{}_{2/3})$ was prepared.

7.6.2
Epitaxial Growth of Isostructural Crystals

7.6.2.1 Principles and Examples

Novel approaches for the fabrication of patterned and bulk crystals [67] have been developed recently. These investigations are motivated by the use of these crystalline materials in optical applications such as electro-optics or as photorefractive materials [68]. Only a limited number of monocrystalline multilayers based on combination of coordination and hydrogen bonding have been reported so far [33, 34]. Recently, few examples based on pure coordination networks have been published showing the increasing interest in this type of approach [69].

As stated above, using the isostructurality between a series of crystals based on the combination of amidinium-based tectons and cyanometallates (see Section 7.4.1), crystals of crystals also called crystalline molecular alloys have been generated by 3D epitaxial growth.

The first example of crystals of crystals reported was based on (2-2H$^+$)$_3$([M^{III}(CN)$_6$]$^{3-}$$_2$) (M = Fe and Co) system [63b]. However, due to the fragility of the crystals, the more robust system based on (2-2H$^+$)$_2$([M^{II}(CN)$_6$]$^{4-}$) (M = Fe and Ru) was further investigated. A greater difference in color between [Fe^{II}(CN)$_6$]$^{4-}$) (orange) and [Ru^{II}(CN)$_6$]$^{4-}$) (colorless) complexes constitutes another advantage allowing a better visualization of the growth process. Using this system, a monocrystalline layer of (2-2H$^+$)$_2$([Fe^{II}(CN)$_6$]$^{4-}$)] (orange) on a single crystal of (2-2H$^+$)$_2$[Ru^{II}(CN)$_6$]$^{4-}$) (colorless) was grown through epitaxial monocrystal growth (see Figure 7.36) [65]. The growth of successive generations was pursued up to generation III (Figure 7.36). The growth process could be performed independent of the starting seed crystal.

Other interesting examples are related to solid solutions (see Section 7.6.1). We have recently shown the possibility to growing crystals of crystals using a single crystal based

Table 7.1 Short crystallographic parameters for $(X-2H^+)_3\{[M^{III}(CN)_6]^{3-}\}_2$ (M = Fe, Co, and Cr) (X = **2** or **3**) recorded at 173 K.

Formula	$(X-2H^+)_3\{[Cr(CN)_6]^{3-}\}_2$		$(X-2H^+)_3\{[Fe(CN)_6]^{3-}\}_2$		$(X-2H^+)_3\{[Co(CN)_6]^{3-}\}_2$	
	X = 2	X = 3	X = 2	X = 3	X = 2	X = 3
Mwt	1275.37	1389.39	1283.07	1397.09	1289.23	1403.25
Crystal system	Monoclinic	Monoclinic	Triclinic	Monoclinic	Triclinic	Monoclinic
Space group	$P2(1)/n$	$P2(1)/n$	P-1	$P2(1)/n$	P-1	$P2(1)/n$
a (Å)	7.1142(4)	7.1190(2)	7.0949(3)	7.0978(5)	7.0982(4)	7.1180(10)
b (Å)	21.4319(13)	22.3250(6)	12.3765(5)	22.2254(16)	12.3509(7)	22.252(4)
c (Å)	20.9403(13)	21.0110(6)	17.9534(7)	20.6357(12)	17.8586(10)	20.508(4)
α (deg)	90	90	83.580(2)	90	83.535(2)	90
β (deg)	91.702(2)	92.8190(17)	87.6101(10)	92.363(3)	87.4280(10)	92.239(6)
γ (deg)	90	90	83.9340(10)	90	83.967(2)	90
V (Å³)	3191.4(3)	3335.27(16)	1557.12(11)	3252.5(4)	1546.21(15)	3245.8(10)

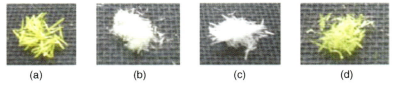

Figure 7.35 Pictures of pure crystals of $(3\text{-}2H^+)_3([Fe(CN)_6]^{3-}{}_2)$ (a), $(3\text{-}2H^+)_3([Cr(CN)_6]^{3-}{}_2)$ (b), $(3\text{-}2H^+)_3([Co(CN)_6]^{3-}{}_2)$ (c), and the ternary crystalline alloy $(3\text{-}2H^+)_3([Fe(CN)_6]^{3-}{}_{2/3}[Co(CN)_6]^{3-}{}_{2/3}[Cr(CN)_6]^{3-}{}_{2/3})$ (d).

on a solid solution [60]. With systems based on $(3\text{-}2H^+)_3([M^{III}(CN)_6]^{3-})_2$ (M = Fe, Cr, or Co) combinations, the growth of a solid solution crystal of $(2\text{-}2H^+)_3([Cr^{III}(CN)_6]^{3-})_2$ on $(3\text{-}2H^+)_3([Cr^{III}(CN)_6]^{3-}[Fe^{III}(CN)_6]^{3-})$ or on $(3\text{-}2H^+)_3([Co^{III}(CN)_6]^{3-}[Fe^{III}(CN)_6]^{3-})$ was performed, as shown in Figure 7.37.

7.6.2.2 Study of the Interface

The growth mechanism of composite crystals was studied by investigating the nature of the interface on robust $(2\text{-}2H^+)_2([M^{II}(CN)_6]^{4-})$ (M = Fe and Ru) systems. In order to determine the nature of the metal distribution perpendicular to the interface, the

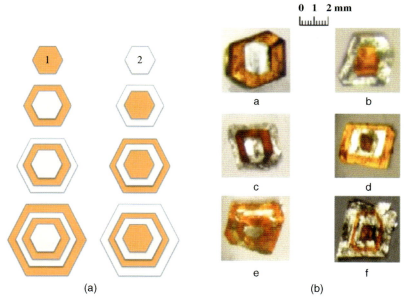

Figure 7.36 (a) Schematic representation of slices of composite crystals and their consecutives generations. The procedure starts with the pure crystals of $(2\text{-}2H^+)_2([Ru^{II}(CN)_6]^{4-})$ or $(2\text{-}2H^+)_2([Fe^{II}(CN)_6]^{4-})$ and the following generations are obtained, in an iterative fashion, by immersing the crystals of precedent generation into a solution containing tecton $2\text{-}2H^+$ and the other cyanometallate anion. (b) Pictures of slices of the different composite crystals.

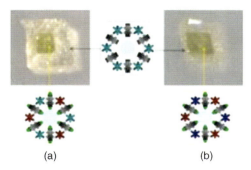

Figure 7.37 Slices of composite crystals obtained by epitaxial growth of $(2\text{-}2H^+)_3([Cr^{III}(CN)_6]^{3-})_2$ on $(3\text{-}2H^+)_3([Cr^{III}(CN)_6]^{3-}[Fe^{III}(CN)_6]^{3-})$ (a) or on $(3\text{-}2H^+)_3([Co^{III}(CN)_6]^{3-}[Fe^{III}(CN)_6]^{3-})$ (b). Reproduced with permission from Ref. [60]. Copyright 2009 Royal Chemical Society.

latter was studied by both TEM and FIB/STEM microscopies using energy dispersive spectroscopy (EDS) (see Figure 7.38). The study revealed a diffuse interface between the two isostructural phases extending over about 0.7 mm [70], implying probably a partial redissolution of the surface during the growth process of the pure phase on the preformed crystal.

In order to elucidate this mechanism, using AFM, some additional characterizations were performed on the same system.

7.6.2.3 *In Situ* Study of the Growth

Real-time *in situ* atomic force microscopy (AFM) was performed on different crystals of $(2\text{-}2H^+)_2([Fe^{II}(CN)_6]^{4-})$ and $(2\text{-}2H^+)_2([Ru^{II}(CN)_6]^{4-})$, and the perfectly epitaxial growth could also be demonstrated at the atomic level (see Figure 7.39). The crystal topography and roughness under various conditions revealed that the interface observed during epitaxial crystals growth depends on the nature of the surface (roughness) and on the way the crystal are dried before immersion in the media [71]. The roughness of the growth interface can be regulated using specific growth

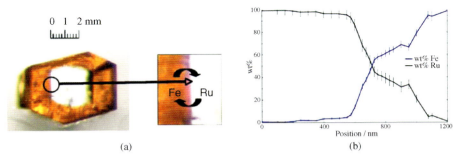

Figure 7.38 (a) Heterocrystal formed through epitaxial growth of a shroud of $(2\text{-}2H^+)_2([Fe^{II}(CN)_6]^{4-})$ (orange) on a core seed crystal of $(2\text{-}2H^+)_2([Ru^{II}(CN)_6]^{4-})$ (colorless). (b) EDS profiles of the Fe and Ru content across the heterocrystal interface, expressed as count rate for Fe–K and Ru–L signals and weight percent.

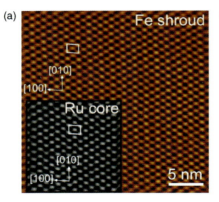 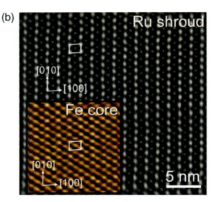

Figure 7.39 (a) (2-2H$^+$)$_2$([Fe^{II}(CN)$_6$]$^{4-}$) over layer grown on the (001) face of a (2-2H$^+$)$_2$([Ru^{II}(CN)$_6$]$^{4-}$) crystal. (b) (2-2H$^+$)$_2$([Ru^{II}(CN)$_6$]$^{4-}$) over layer grown on the (001) face of a (2-2H$^+$)$_2$([Fe^{II}(CN)$_6$]$^{4-}$) crystal. Both cases display an extraordinary degree of epitaxial alignment between the core and the shroud, the sides of the rectangular crystals. Reproduced with permission from Ref. [71]. Copyright 2009 American Chemical Society.

protocols (seeding time, drying method) that minimize the intermixing of the two compounds.

7.7
Conclusion

Using concepts developed in the field of molecular recognition of anions in solution, cationic molecular tectons capable of forming CAHB molecular networks in the crystalline phase by iteration of the recognition pattern were designed. Among the many examples of H-bonded networks reported in the literature, the use of bisamidinium derivatives as dicationic tectons and polycyanometallates of different charge and geometry as anionic tectons appeared as the most effective system. Indeed, the design of robust and reliable mode of recognition between the amidinium acidic hydrogen atoms as donors and cyanide groups bound to metal centers as acceptors not only allowed to generate a large variety of networks with different dimensionality but also afforded functional crystals displaying either optical properties or porosity. Using the isostructurality between several crystals, new materials such as solid solutions combining up to five different components have been fabricated. Furthermore, crystalline materials composed of metal cyanide complexes with two different oxidation states, mainly Fe(II) and Fe(III) (mixed valence systems), have been obtained. Using a reductive process, the possibility of postcrystallization modifications was demonstrated by transforming Fe(III)-based crystals into Fe(II)-based crystalline materials. Finally, new classes of H-bonded crystals based on 3D epitaxial growth of crystalline layers on seed crystals (crystals of crystals) afforded hierarchical molecular alloys with tunable parameters. A step toward bridging the gap between structural aspects of crystal engineering and the design and control of physical properties (functional crystalline materials) was taken that should lead to new solid state materials displaying predicted and tailored properties.

Acknowledgments

Many thanks to Dr. P. Dechambenoit, Dr. C. Paraschiv, Dr. G. Marinescu, Dr. O. Félix, Dr. A. De Cian, Professor J.-M. Planeix, and N. Kyritsakas who performed part of the work. Professor M.D. Ward (NYU) is also kindly acknowledged for AFM measurements and Professor E. Brès (Université de Lille) for electronic microscopy. We also acknowledge financial support from the Université de Strasbourg, Institut Universitaire de France, The Frontier Research in Chemistry (FRC), Strasbourg, the CNRS, the Ministry of Research and Education, and Marie Curie Est Actions FUMASSEC Network (Contract No. MEST-CT-2005-020992).

References

1 (a) Whitesides, G.M., Mathias, J.P., and Seto, T. (1991) Molecular self-assembly and nanochemistry: a chemical strategy for the synthesis of nanostructures. *Science*, **254**, 1312–1319; (b) Lindsey, J.S. (1991) Self-assembly in synthetic routes to molecular devices. Biological principles and chemical perspectives: a review. *New J. Chem.*, **15**, 153–180.

2 Lehn, J.-M. (1995) *Supramolecular Chemistry. Concepts and Perspectives*, Wiley-VCH Verlag GmbH, Weinheim.

3 (a) Schmidt, G.M. (1971) Photodimerization in the solid state. *Pure Appl. Chem.*, **27**, 647–678; (b) Dunitz, J.D. (1991) Phase transitions in molecular crystals from a chemical viewpoint. *Pure Appl. Chem.*, **63**, 177–185; (c) Dunitz, J.D. (1996) *Perspectives in Supramolecular Chemistry*, vol. 2 (ed. G.R. Desiraju), John Wiley & Sons, Inc., New York.

4 (a) Aakeröy, C.B. (1997) Crystal engineering: strategies and architectures. *Acta Crystallogr.*, **B53**, 569–586; (b) Braga, D. and Grepioni, F. (2000) Intermolecular interactions in nonorganic crystal engineering. *Acc. Chem. Res.*, **33**, 601–608; (c) Moulton, B. and Zaworotko, M.J. (2001) From molecules to crystal engineering: supramolecular isomerism and polymorphism in network solids. *Chem. Rev.*, **101**, 1629–1658; (d) Braga, D., Desiraju, G.R., Miller, J.S., Orpen, A.G., and Price, S.L. (2002) Innovation in crystal engineering. *CrystEngComm*, **4**, 500–509; (e) Sommerdijk, N.A.J.M. (2003) Crystal design and crystal engineering. *Angew. Chem., Int. Ed.*, **42**, 3572–3574; (f) Braga, D., Brammer, L., and Champness, N.R. (2005) New trends in crystal engineering. *CrystEngComm*, **7**, 1–19; (g) Brammer, L. (2004) Developments in inorganic crystal engineering. *Chem. Soc. Rev.*, **33**, 476–489.

5 (a) Simard, M., Su, D., and Wuest, J.D. (1991) Use of hydrogen bonds to control molecular aggregation. Self-assembly of three-dimensional networks with large chambers. *J. Am. Chem. Soc.*, **113**, 4696–4698; (b) Mann, S. (1993) Molecular tectonics in biomineralization and biomimetic materials chemistry. *Nature*, **365**, 499–505.

6 Hosseini, M.W. (2004) Reflexion on molecular tectonics. *CrystEngComm*, **6**, 318–322.

7 Desiraju, G.D. (1989) *Crystal Engineering: The Design of Organic Solids*, Elsevier, New York.

8 Hosseini, M.W. (2005) Molecular tectonics: from simple tectons to complex molecular networks. *Acc. Chem. Res.*, **38**, 313–323.

9 Hosseini, M.W. (2005) Self-assembly and generation of complexity. *Chem. Commun.*, 5825–5829.

10 Brunet, P., Simard, M., and Wuest, J.D. (1997) Molecular tectonics. Porous hydrogen-bonded networks with unprecedented structural integrity. *J. Am. Chem. Soc.*, **119**, 2737–2738.

11 Dunitz, J.D. (2003) Are crystal structures predictable? *Chem. Commun.*, 545–548.

12 (a) Price, S.L. (2004) Quantifying intermolecular interactions and their use in computational crystal structure prediction. *CrystEngComm*, **6**, 344–354; (b) Dunitz, J.D. and Gavezzoti, A. (2005) Molecular recognition in organic crystals: directed intermolecular bonds or nonlocalized bonding. *Angew. Chem., Int. Ed.*, **44**, 1766–1787.

13 Gavezzotti, A. (1994) Are crystal structures predictable? *Acc. Chem. Res.*, **27**, 309–314.

14 (a) Ermer, O. (1988) Five-fold diamond structure of adamantane-1,3,5,7-tetracarboxylic acid. *J. Am. Chem. Soc.*, **110**, 3747–3754; (b) Aakeröy, C.B. and Seddon, K.R. (1993) The hydrogen bond and crystal engineering. *Chem. Soc. Rev.*, **22**, 397–407; (c) Subramanian, S. and Zaworotko, M.J. (1994) Exploitation of the hydrogen bond: recent developments in the context of crystal engineering. *Coord. Chem. Rev.*, **137**, 357–401; (d) Lawrence, D.S., Jiang, T., and Levett, M. (1995) Self-assembling supramolecular complexes. *Chem. Rev.*, **95**, 2229–2260; (e) Stoddart, J.F. and Philip, D. (1996) Self-assembly in natural and unnatural systems. *Angew. Chem., Int. Ed. Engl.*, **35**, 1154–1196; (f) Fredericks, J.R. and Hamilton, A.D. (1996) in *Comprehensive Supramolecular Chemistry*, vol. 9 (series eds J.L. Atwood, J.E. Davis, D.D. Macnico, and F. Vögtle, eds J.P. Sauvage and M.W. Hosseini), Elsevier, Oxford, pp. 565–594.

15 (a) Batten, S.R. and Robson, R. (1998) Interpenetrating nets: ordered periodic entanglement. *Angew. Chem., Int. Ed.*, **37**, 1460–1494; (b) Blake, A.J., Champness, N.R., Hubberstey, P., Li, W.-S., Withersby, M.A., and Schröder, M. (1999) Inorganic crystal engineering using self-assembly of tailored building-blocks. *Coord. Chem. Rev.*, **193**, 117–138; (c) Hosseini, M.W. (1999) *NATO ASI Series, Series C*, vol. 538 (eds D. Braga, F. Grepiono, and G. Orpen), Kluwer, Dordrecht, The Netherlands, pp. 181–208;(d) Eddaoui, M., Moler, D.B., Li, H., Chen, B., Reineke, T.M., O'Keeffe, M., and Yaghi, O.M. (2001) Modular chemistry: secondary building units as a basis for the design of highly porous and robust metal–organic carboxylate frameworks. *Acc. Chem. Res.*, **34**, 319–330; (e) Swiergers, G.F. and Malefetse, T.J. (2000) New self-assembled structural motifs in coordination chemistry. *Chem. Rev.*, **100**, 3483–3538; (f) Janiak, C. (2003) Engineering coordination polymers towards applications. *Dalton Trans.*, 2781–2804; (g) Ferey, G. (2008) Hybrid porous solids: past, present, future. *Chem. Soc. Rev.*, **37**, 191–214.

16 Hill, D.J., Moi, M.J., Prince, R.B., Hughes, T.S., and Moore, J.S. (2001) A field guide to foldamers. *Chem. Rev.*, **101**, 3893–4011.

17 Hosseini, M.W. and De Cian, A. (1998) Crystal engineering: molecular networks based on inclusion phenomena. *Chem. Commun.*, 727–734.

18 (a) Jeffrey, G.A. (1997) *An Introduction to Hydrogen Bonding*, Oxford University Press, Oxford;(b) Taylor, R. and Kennard, O. (1984) Hydrogen-bond geometry in organic crystals. *Acc. Chem. Res.*, **17**, 320–326; (c) Etter, M.C. (1990) Encoding and decoding hydrogen-bond patterns of organic compounds. *Acc. Chem. Res.*, **23**, 120–126.

19 Steiner, T. (2002) The hydrogen bond in the solid state. *Angew. Chem., Int. Ed.*, **41**, 48–76.

20 Ward, M.D. (2005) Design of crystalline molecular networks with charge-assisted hydrogen bonds. *Chem. Commun.*, 5838–5842.

21 (a) Sliwa, M., Letard, S., Malfant, I., Nierlich, M., Lacroix, P.G., Asahi, T., Masuhara, H., Yu, P., and Nakatani, K. (2005) Design, synthesis, structural and nonlinear optical properties of photochromic crystals: toward reversible molecular switches. *Chem. Mater.*, **17**, 4727–4735; (b) Datta, A. and Pati, S.K. (2006) Dipolar interactions and hydrogen bonding in supramolecular aggregates:

understanding cooperative phenomena for 1st hyperpolarizability. *Chem. Soc. Rev.*, **35**, 1305–1323; (c) Cariati, E., Macchi, R., Robert, D., Ugo, R., Galli, S., Casati, N., Macchi, P., Sironi, A., Bogani, L., Caneschi, A., and Gatteschi, D. (2007) Polyfunctional inorganic–organic hybrid materials: an unusual kind of NLO active layered mixed metal oxalates with tunable magnetic properties and very large second harmonic generation. *J. Am. Chem. Soc.*, **129**, 9410–9420.

22 (a) Miller, J.S. (2000) Organometallic- and organic-based magnets: new chemistry and new materials for the new millennium. *Inorg. Chem.*, **39**, 4392–4408; (b) Andruh, M. (2007) Oligonuclear complexes as tectons in crystal engineering: structural diversity and magnetic properties. *Chem. Commun.*, 2565–2567; (c) Maspoch, D., Ruiz-Molina, D., and Veciana, J. (2007) Old materials with new tricks: multifunctional open-framework materials. *Chem. Soc. Rev.*, **36**, 770–818.

23 (a) Ockwig, N.W., Delgado-Friedrichs, O., O'Keefe, M., and Yaghi, O.M. (2005) Reticular chemistry: occurrence and taxonomy of nets and grammar for the design of frameworks. *Acc. Chem. Res.*, **38**, 176–182; (b) Rosseinsky, M.J. (2004) Recent developments in metal–organic framework chemistry: design, discovery, permanent porosity and flexibility. *Microporous. Mesoporous. Mater.*, **73**, 15–30; (c) Kitagawa, S., Kitaura, R., and Noro, S. (2004) Functional porous coordination polymers. *Angew. Chem., Int. Ed.*, **43**, 2334–2375; (d) Férey, G. (2005) Crystallized frameworks with giant pores: Are there limits to the possible? *Acc. Chem. Res.*, **38**, 217–225.

24 (a) Russell, A., Etter, M.C., and Ward, M.D. (1994) Layered materials by molecular design: structural enforcement by hydrogen bonding in guanidinium alkane- and arenesulfonates. *J. Am. Chem. Soc.*, **116**, 1941–1952; (b) Russell, V.A., Evans, C.C., Li, W., and Ward, M.D. (1997) Nanoporous molecular sandwiches: pillared two-dimensional hydrogen-bonded networks with adjustable porosity. *Science*, **276**, 575–579; (c) Holman, K.T., Pivovar, A.M., Swift, J.A., and Ward, M.D. (2001) Metric engineering of soft molecular host frameworks. *Acc. Chem. Res.*, **34**, 107–118; (d) Holman, K.T., Pivovar, A.M., and Ward, M.D. (2001) Engineering crystal symmetry and polar order in molecular host frameworks. *Science*, **294** 1907–1911; (e) Custelcean, R. and Ward, M.D. (2005) Chiral discrimination in low-density hydrogen-bonded frameworks. *Cryst. Growth Des.*, **5**, 2277–2287; (f) Liu, Y. and Ward, M.D. (2009) Molecular capsules in modular frameworks. *Cryst. Growth Des.*, **9**, 3859–3861.

25 (a) Martin, S.M., Yonezawa, J., Horner, M.J., Macosko, C.W., and Ward, M.D. (2004) Structure and rheology of hydrogen bond reinforced liquid crystals. *Chem. Mater.*, **16**, 3045–3055; (b) Horner, M.J., Holman, K.T., and Ward, M.D. (2007) Architectural diversity and elastic networks in hydrogen-bonded host frameworks: from molecular jaws to cylinders. *J. Am. Chem. Soc.*, **129**, 14640–14660; (c) Comotti, A., Bracco, S., Sozzani, P., Hawxwell, S.M., Hu, C., and Ward, M.D. (2009) Guest molecules confined in amphipathic crystals as revealed by X-ray diffraction and MAS NMR. *Cryst. Growth Des.*, **9**, 2999–3002; (d) Soegiarto, A.C. and Ward, M.D. (2009) Cooperative polar ordering of acentric guest molecules in topologically controlled host frameworks. *Cryst. Growth Des.*, **9**, 3803–3815.

26 (a) Lewis, G.R. and Orpen, A.G. (1998) A metal-containing synthon for crystal engineering: synthesis of the hydrogen bond ribbon polymer [4,4-H$_2$bipy][MCl$_4$] (M = Pd, Pt). *Chem. Commun.*, 1873–1874; (b) Lewis, A.L., Orpen, A.G., Rotter, S., Starbuck, J., Wang, X.M., Rodriguez-Martin, Y., and Ruiz-Perez, C. (2000) Organic–inorganic hybrid solids: control of perhalometallate solid state structures. *J. Chem. Soc., Dalton Trans.*, 3897–3905; (c) Dolling, B., Gillon, A.L., Orpen, A.G., Starbuck, J., and Wang, X.-M. (2001) Homologous families of chloride-rich 4,4-bipyridinium salt structures. *Chem. Commun.*, 567–568; (d) Angeloni, A., Crawford, P.C., Orpen,

A.G., Podesta, T.J., and Shore, B.J. (2004) Does hydrogen bonding matter in crystal engineering? Crystal structures of salts of isomeric ions. *Chem. Eur. J.*, **10**, 3783–3791.

27 (a) Podesta, T.J. and Orpen, A.G. (2005) Tris(pyridinium)triazine in crystal synthesis of 3-fold symmetric structures. *Cryst. Growth Des.*, **5**, 681–693; (b) Adams, C.J., Crawford, P.C., Orpen, A.G., Podesta, T.J., and Salt, B. (2005) Thermal solid state synthesis of coordination complexes from hydrogen bonded precursors. *Chem. Commun.*, 2457–2458; (c) Adams, C.J., Angeloni, A., Orpen, A.G., Podesta, T.J., and Shore, B. (2006) Crystal synthesis of organic–inorganic hybrid salts based on tetrachloroplatinate and -palladate salts of organic cations: formation of linear, two-, and three-dimensional NH···Cl hydrogen bond networks. *Cryst. Growth Des.*, **6**, 411–422; (d) Adams, C.J., Crawford, P.C., Orpen, A.G., and Podesta, T.J. (2006) Cation and anion diversity in [M(dithiooxalate)$_2$]$^{2-}$ salts: structure robustness in crystal synthesis. *Dalton Trans.*, 4078–4092.

28 (a) Rivas, J.C.M. and Brammer, L. (1998) Self-assembly of 1-D chains of different topologies using the hydrogen-bonded inorganic supramolecular synthons N−H···Cl$_2$M or N−H···Cl$_3$M. *Inorg. Chem.*, **37**, 4756–4757; (b) Zordan, F., Mınguez Espallargas, G., and Brammer, L. (2006) Unexpected structural homologies involving hydrogen-bonded and halogen-bonded networks in halopyridinium halometallate salts. *CrystEngComm*, **8**, 425–431.

29 (a) Podesta, T.J. and Orpen, A.G. (2002) Use of the Ni(dithiooxalate)$_2^{2-}$ unit as a molecular tecton in crystal engineering. *CrystEngComm*, **4**, 336–342; (b) Crowford, P.C., Gillon, A.L., Green, J., Orpen, A.G., Podesta, T.J., and Pritchard, S.V. (2004) Synthetic crystallography: synthon mimicry and tecton elaboration in metallate anion salts. *CrystEngComm*, **6**, 419–428.

30 Imaz, I., Thillet, A., and Sutter, J.-P. (2007) Charge-assisted hydrogen-bonded assemblage of an anionic $\{M(C_2O_4)_4\}^{4-}$ building unit and organic cations: a versatile approach to hybrid supramolecular architectures. *Cryst. Growth Des.*, **7**, 1753–1761.

31 (a) Wang, X.-Y. and Sevov, S.C. (2007) Hydrogen-bonded host frameworks of cationic metal complexes and anionic disulfonate linkers: effects of the guest molecules and the charge of the metal complex. *Chem. Mater.*, **19**, 4906–4912; (b) Wang, X.-Y. and Sevov, S.C. (2008) Synthesis, structures, and magnetic properties of metal-coordination polymers with benzenepentacarboxylate linkers. *Inorg. Chem.*, **47**, 1037–1043; (c) Wang, X.-Y. and Sevov, S.C. (2008) A series of guest-defined metal-complex/disulfonate frameworks of hydrogen-bonded [Co(en)$_2$(ox)]$^+$ and 2,6-naphtalenedisulfonate. *Cryst. Growth Des.*, **8**, 1265–1270; (d) Forbes, T.Z. and Sevov, S.C. (2009) Metal–organic frameworks with direct transition metal–sulfonate interactions and charge-assisted hydrogen bonds. *Inorg. Chem.*, **48**, 6873–6878.

32 (a) MacDonald, J.C., Dorrestein, P.C., Pilley, M.M., Foote, M.M., Lundburg, J.L., Henning, R.W., Schutlz, A.J., and Manson, J.L. (2000) Design of layered crystalline materials using coordination chemistry and hydrogen bonds. *J. Am. Chem. Soc.*, **122**, 11692–11702; (b) Luo, T-J.M., MacDonald, J.C., and Palmore, G.T.R. (2004) Fabrication of complex crystals using kinetic control, chemical additives, and epitaxial growth. *Chem. Mater.*, **16**, 4916–4927.

33 Noveron, J.C., Lah, M.S., Del Sesto, R.E., Arif, A.M., Miller, J.S., and Stang, P.J. (2002) Engineering the structure and magnetic properties of crystalline solids via the metal-directed self-assembly of a versatile molecular building unit. *J. Am. Chem. Soc.*, **124**, 6613–6625.

34 Dalrymple, S.A. and Shimizu, G.K.H. (2007) Crystal engineering of a permanently porous network sustained exclusively by charge-assisted hydrogen bonds. *J. Am. Chem. Soc.*, **129**, 12114–12116.

35 Brand, G., Hosseini, M.W., Ruppert, R., De Cian, A., Fischer, J., and Kyritsakas, N. (1995) A molecular approach to organic solids: molecular recognition of arene

sulfonate by a bis-cyclic amidinium in the solid state. *New J. Chem.*, **19**, 9–13.

36 Félix, O., Hosseini, M.W., De Cian, A., and Fischer, J. (1998) Molecular tectonics VIII: formation of 1D and 3D networks based on the simultaneous use of hydrogen bonding and electrostatic interactions. *New J. Chem.*, **22**, 1389–1393.

37 Hosseini, M.W. (2003) Molecular tectonics: from molecular recognition of anions to molecular networks. *Coord. Chem. Rev.*, **240**, 157–166.

38 (a) Hosseini, M.W., Ruppert, R., Schaeffer, P., De Cian, A., Kyritsakas, N., and Fischer, J. (1994) A molecular approach to the solid state synthesis: prediction and synthesis of self-assembled infinite rods. *Chem. Commun.*, 2135–2136; (b) Brand, G., Hosseini, M.W., Ruppert, R., De Cian, A., Fischer, J., and Kyritsakas, N. (1995) A molecular approach to organic solids: molecular recognition of arene sulfonate by a bis-cyclic amidinium in the solid state. *New J. Chem.*, **19**, 9–13; (c) Félix, O., Hosseini, M.W., De Cian, A., and Fischer, J. (1997) Molecular tectonics IV: molecular networks based on hydrogen bonding and electrostatic interactions. *Tetrahedron Lett.*, **38**, 1755–1758; (d) Félix, O., Hosseini, M.W., De Cian, A., and Fischer, J. (1997) Molecular tectonics V: molecular recognition in the formation of molecular networks based on hydrogen bonding and electrostatic interaction. *Tetrahedron Lett.*, **38**, 1933–1936; (e) Félix, O., Hosseini, M.W., De Cian, A., and Fischer, J. (1997) Molecular tectonics III: the simultaneous use of H-bonding and charge–charge interactions for the self-assembly of fumaric acid and cyclic bisamidinium into one- and two-dimensional molecular networks. *Angew. Chem., Int. Ed. Engl.*, **36**, 102–104; (f) Félix, O., Hosseini, M.W., De Cian, A., and Fischer, J. (2000) Crystal engineering of 2-D hydrogen bonded molecular networks based on the self-assembly of anionic and cationic modules. *Chem. Commun.*, 281–282; (g) Félix, O., Hosseini, M.W., and De Cian, A. (2001) Design of 2-D hydrogen bonded molecular networks using pyromellitate dianion and cyclic bisamidinium dication as complementary tectons. *Solid State Sci.*, **3**, 789–793.

39 (a) Hosseini, M.W., Brand, G., Schaeffer, P., Ruppert, R., De Cian, A., and Fischer, J. (1996) A molecular approach to solid structures II: prediction and synthesis of sheets through assembly of complementary molecular units. *Tetrahedron Lett.*, **37**, 1405–1408; (b) Ball, V., Planeix, J.-M., Félix, O., Hemmerle, J., Schaaf, P., Hosseini, M.W., and Voegel, J.-C. (2002) Molecular tectonics: abiotic control of hydroxyapatite crystals morphology. *Cryst. Growth Des.*, **2**, 489–492; (c) Planeix, J.-M., Jaunky, W., Duhoo, T., Czernuszka, J., Hosseini, M.W., and Brès, E.F. (2003) A molecular tectonics crystal engineering approach for building organic inorganic composites. Potential application to the growth control of hydroxyapatite crystals. *J. Mater. Chem.*, **13**, 2521–2524.

40 Shacklady, D.M., Lee, S.-O., Ferlay, S., Hosseini, M.W., and Ward, M.D. (2005) Translational design and bimodal assembly of hydrogen-bonded networks. *Cryst. Growth Des.*, **5**, 995–1003.

41 Félix, O., Hosseini, M.W., De Cian, A., Fischer, J., Catala, L., and Turek, P. (1999) Synthesis, structural and EPR analysis of nitronyl-nitroxide labelled isophthalic acid. *Tetrahedron Lett.*, **41**, 2943–2946.

42 Hosseini, M.W., Tsiourvas, D., Planeix, J.-M., Sideratou, Z., Thomas, N., and Paleos, C.M. (2004) Design of liquid crystals based on a combination of hydrogen bonding and ionic interactions. *Collec. Czech. Chem. Commun.*, **69**, 1161–1168.

43 Braga, D., Maini, L., Grepioni, F., De Cian, A., Felix, O., Fischer, J., and Hosseini, M.W. (2000) Charge-assisted N-H–O and O-H–O hydrogen bonds control the supramolecular aggregation of ferrocene dicarboxylic acid and bis-amidines. *New J. Chem.*, **24**, 547–553.

44 Carpanese, C., Ferlay, S., Kyritsakas, N., Henry, M., and Hosseini, M.W. (2009) Molecular tectonics: design of 2-D networks by simultaneous use of charge-assisted hydrogen and coordination bonds. *Chem. Commun.*, 2514–2516.

45 Paraschiv, C., Ferlay, S., Hosseini, M.W., Kyritsakas, N., Planeix, J.-M., and Andruh, M. (2007) Hybrid organic/inorganic H-bonded supramolecular networks including [Cr(ox)$_2$L]$^-$ (L = 1,10-phenanthroline and 2,2$^-$-bipyridine) tectons. *Rev. Roum. Chim.*, **52**, 101–104.

46 Félix, O., Hosseini, M.W., De Cian, A., and Fischer, J. (1997) A molecular approach to organic solids: synthesis of phenyl di- and tri-carbo xamidines. *New J. Chem.*, **21**, 285–288.

47 Marinescu, G., Ferlay, S., Kyritsakas, N., and Hosseini, M.W. (2008) Molecular tectonics: design and generation of charge assisted H bonded hybrid molecular networks based on amidinium cations and thio- or isothio-cyanatometallates. *Dalton Trans.*, 615–619.

48 (a) Decurtins, S., Schmalle, H.W., Oswald, H.R., Linden, A., Enseling, J., and Gütlich, P. (1994) A polymeric two-dimensional mixed-metal network. Crystal structure and magnetic properties of {[P(Ph)$_4$][MnCr(ox)$_3$]}. *Inorg. Chim. Acta*, **216**, 65–73; (b) Decurtins, S., Schmalle, H.W., and Pellaux, R. (1998) Polymeric two- and three-dimensional transition-metal complexes comprising supramolecular host–guest systems. *New J. Chem.*, 117–121.

49 (a) Ferlay, S., Félix, O., Hosseini, M.W., Planeix, J.-M., and Kyritsakas, N. (2002) Second sphere supramolecular chirality: racemic hybrid H-bonded 2-D molecular networks. *Chem. Commun.*, 702–703; (b) Dechambenoit, P., Ferlay, S., Hosseini, M.W., and Kyritsakas, N. (2008) Molecular tectonics: control of reversible water release in porous charge-assisted H-bonded networks. *J. Am. Chem. Soc.*, **130**, 17106–17113.

50 Dechambenoit, P., Ferlay, S., Hosseini, M.W., Planeix, J.-M., and Kyritsakas, N. (2006) Molecular tectonics: control of packing of hybrid 1-D and 2-D H-bonded molecular networks formed between bisamidinium dication and cyanometallate anions. *New J. Chem.*, **30**, 1403–1410.

51 Ferlay, S., Bulach, V., Félix, O., Hosseini, M.W., Planeix, J.-M., and Kyritsakas, N. (2002) Molecular tectonics and supramolecular chirality: rational design of hybrid 1-D and 2-D H-bonded molecular networks based on bis-amidinium dication and metal cyanide anions. *CrystEngComm*, 447–453.

52 Dechambenoit, P., Ferlay, S., Hosseini, M.W., and Kyritsakas, N. (2010) Molecular tectonics: crystal engineering of mixed valence Fe(II)/Fe(III) solid solutions. *Chem. Commun.*, **46** (6), 868–870.

53 Ferlay, S., Holakovsky, R., Hosseini, M.W., Planeix, J.-M., and Kyritsakas, N. (2003) Charge assisted chiral hybrid H-bonded molecular networks. *Chem. Commun.*, 1224–1225.

54 Dechambenoit, P., Ferlay, S., Hosseini, M.W., and Kyritsakas, N. (2007) Molecular tectonics: polymorphism and enhancement of networks dimensionality by combination of primary and secondary H-bond sites. *Chem. Commun.*, 4626–4627.

55 Paraschiv, C., Ferlay, S., Hosseini, M.W., Bulach, V., and Planeix, J.-M. (2004) Molecular tectonics: design of luminescent H-bonded molecular networks. *Chem. Commun.*, 2270–2271.

56 (a) Braga, D., Grepioni, F., and Orpen, A.G. (eds) (1999) *NATO ASI Series*, vol. 538, Kluwer, Dordrecht, The Netherlands; (b) Seddon, R.K. and Zaworotko, M. (1999) in *NATO ASI Series C*, vol. 539 (eds D. Braga, F. Grepioni, and G. Orpen), Kluwer, Dordrecht, The Netherlands.

57 (a) Nagasundaram, N., Roper, G., Biscoe, J., Chai, J.W., Patterson, H.H., Blomg, N., and Ludi, A. (1986) Single-crystal luminescence study of the layered compound potassium dicyanoaurate. *Inorg. Chem.*, **25**, 2947–2951; (b) Rawashdeh-Omary, M.A., Omary, M.A., and Patterson, H.H. (2000) Oligomerization of Au(CN)$_2^-$ and Ag(CN)$_2^-$ ions in solution via ground-state aurophilic and argentophilic bonding. *J. Am. Chem. Soc.*, **122**, 10371–10380; (c) Stender, M., Olmstead, M.M., Balch, A.L., Riosb, D., and Attar, S. (2003) Cation and hydrogen bonding effects on the self-association and luminescence of the dicyanoaurate ion, [Ag(CN)$_2$]$^-$ *Dalton Trans.*, 4282–4287.

58 (a) Guo, J., Wong, W.-K., and Wong, W.-Y. (2005) Synthesis and structural characterization of some arylamidinium diphenylphosphinates: formation of one-, two-, and three-dimensional networks by charge-assisted hydrogen bonds. *Polyhedron*, **24**, 927–939; (b) Yoshida, Y., Fujii, J., and Saito, G. (2006) Dicyanoaurate(I) salts with 1-alkyl-3-methylimidazolium: luminescent properties and room-temperature liquid forming. *J. Mater. Chem.*, **16**, 724–727.

59 (a) Yaghi, O.M., Li, H., and Groy, T.L. (1996) Construction of porous solids from hydrogen-bonded metal complexes of 1,3,5-benzenetricarboxylic acid. *J. Am. Chem. Soc.*, **118**, 9096–9101; (b) Aakeröy, C.B., Beatty, A.M., and Leinen, D.S. (1999) A versatile route to porous solids: organic–inorganic hybrid materials assembled through hydrogen bonds. *Angew. Chem., Int. Ed.*, **38**, 1815–1819; (c) Maspoch, D., Domingo, N., Ruiz-Molina, D., Wurst, K., Tejada, J., Rovira, C., and Veciana, J. (2004) A robust nanocontainer based on a pure organic free radical. *J. Am. Chem. Soc.*, **126**, 730–731; (d) Uemura, K., Kitagawa, S., Fukui, K., and Saito, K. (2004) A contrivance for a dynamic porous framework: cooperative guest adsorption based on square grids connected by amide–amide hydrogen bonds. *J. Am. Chem. Soc.*, **126**, 3817–3828; (e) Demers, E., Maris, T., and Wuest, J.D. (2005). Molecular tectonics. Porous hydrogen-bonded networks built from derivatives of 2,2′,7,7′-tetraphenyl-9,9′-spirobi[9H-fluorene]. *Cryst. Growth Des.*, **5**, 1227–1235; (f) Trolliet, C., Poulet, G., Tuel, A., Wuest, J.D., and Sautet, P. (2007) A theoretical study of cohesion, structural deformation, inclusion, and dynamics in porous hydrogen-bonded molecular networks. *J. Am. Chem. Soc.*, **129**, 3621–3626; (g) Stephenson, M.D. and Hardie, M.G. (2007) Network structures with 2,2′-bipyridine-3,3′-diol: a discrete Co (III) complex that forms a porous 3-D hydrogen bonded network, and Cu(II) coordination chains. *CrystEngComm*, **9**, 496–502.

60 Dechambenoit, P., Ferlay, S., Hosseini, M.W., and Kyritsakas, N. (2009) Playing with isostructurality: from tectons to molecular alloys and composite crystals. *Chem. Commun.*, 1559–1561.

61 (a) Choi, H.J. and Suh, M.P. (2004) Dynamic and redox active pillared bilayer open framework: single-crystal-to-single-crystal transformations upon guest removal, guest exchange, and framework oxidation. *J. Am. Chem. Soc.*, **126**, 15844–15851; (b) Ritchie, C., Streb, C., Thiel, J., Mitchell, S.G., Miras, H.N., Long, D.-L., Boyd, T., Peacock, R.D., McGlone, T., and Cronin, L. (2008) Reversible redox reactions in an extended polyoxometalate framework solid. *Angew. Chem., Int. Ed.*, **47**, 6881–6884.

62 Dechambenoit, P., Ferlay, S., Hosseini, M.W., and Kyritsakas, N. (2009) In situ reduction of Fe(III) into Fe(II): an example of post-crystallisation transformation. *Chem. Commun.*, 6798–6800.

63 (a) Braga, D., Cojazzi, G., Paolucci, D., and Grepioni, F. (2001) A remarkable water-soluble (molecular) alloy with two tunable solid-to-solid phase transitions. *Chem. Commun.*, 803–804; (b) Ferlay, S. and Hosseini, M.W. (2004) Crystalline molecular alloys. *Chem. Commun.*, 787–786; (c) Sada, K., Inoue, K., Tanaka, T., Epergyes, A., Tanaka, A., Tohnai, N., Matsumoto, A., and Miyata, M. (2005) Multicomponent organic alloys based on organic layered crystals. *Angew. Chem., Int. Ed.*, **44**, 7059–7052.

64 (a) Maroncelli, M., Strauss, H.L., and Snyder, R.G. (1985) Structure of the n-alkane binary solid n-$C_{19}H_{40}$/n-$C_{21}H_{44}$ by infrared spectroscopy and calorimetry. *J. Phys. Chem.*, **89**, 5260–5267; (b) Dorset, D.L. (1990) Direct structure analysis of a paraffin solid solution. *Proc. Natl. Acad. Sci. USA*, **87**, 8541–8544; (c) Sirota, E.B., King, H.E., Jr., Shao, H.H., and Singer, D.M. (1995) Rotator phases in mixtures of n-alkanes. *J. Phys. Chem.*, **99**, 798–808; (d) Mondieig, D., Espeau, P., Robles, L., Haget, Y., Oonk, H.A.J., and Cuevas-Diarte, M.A. (1997) Mixed crystals of n-alkane pairs. A global view of the thermodynamic melting properties. *J. Chem. Soc., Faraday Trans.*, **93**, 3343–3346.

65 Dechambenoit, P., Ferlay, S., and Hosseini, M.W. (2005) From tectons to composite crystals. *Cryst. Growth Des.*, **5**, 2310–2312.

66 Vegard, L. (1921) Gitterkonstanten von mischkristallen. *Z. Phys.*, **5**, 17–26.

67 (a) Bonafe, S.J. and Ward, M.D. (1995) Selective nucleation and growth of an organic polymorph by ledge-directed epitaxy on a molecular crystal substrate. *J. Am. Chem. Soc.*, **117**, 7853–7861; (b) Mitchell, C.A., Yu, L., and Ward, M.D. (2001) Selective nucleation and discovery of organic polymorphs through epitaxy with single crystal substrates. *J. Am. Chem. Soc.*, **123**, 10830–10839.

68 (a) Swalen, J.D. (1991). Materials assembly and formation using engineered polypeptides. *Annu. Rev. Mater. Res.*, **34**, 373–408; (b) Forrest, S.R. (1997) Ultrathin organic films grown by organic molecular beam deposition and related techniques. *Chem. Rev.*, **97**, 1793–1896; (c) Hooks, D., Fritz, T., and Ward, M.D. (2001) Epitaxy and molecular organization on solid substrates. *Adv. Mater.*, **13**, 227–241.

69 (a) Furukawa, S., Hirai, K., Nakagawa, K., Takashima, Y., Matsuda, R., Tsuruoka, T., Kondo, M., Haruki, R., Tanaka, D., Sakamoto, H., Shimomura, S., Sakata, O., and Kitagawa, S. (2008) Heterogeneously hybridized porous coordination polymer crystals: fabrication of heterometallic core–shell single crystals with an in-plane rotational epitaxial relationship. *Angew. Chem., Int. Ed.*, **47**, 1766–1770; (b) Catala, L., Brinzei, D., Prado, Y., Gloter, A., Stéphan, O., Rogez, G., and Mallah, T. (2009) Core-multishell magnetic coordination nanoparticles: toward multifunctionality on the nanoscale. *Angew. Chem., Int. Ed.*, **48**, 183–187; (c) Furukawa, S., Hirai, K., Takashima, Y., Nakagawa, K., Kondo, M., Tsuruoka, T., Sakata, O., and Kitagawa, S. (2009) A block PCP crystal: anisotropic hybridization of porous coordination polymers by face-selective epitaxial growth. *Chem. Commun.*, 5097–5099; (d) Koh, K., Wong-Foy, A.G., and Matzger, A.J. (2009) MOF@MOF: microporous core–shell architectures. *Chem. Commun.*, 6162–6164.

70 Brès, E.F., Ferlay, S., Dechambenoit, P., Leroux, H., Hosseini, M.W., and Reyntjens, S. (2007) Playing with isostructurality: from tectons to molecular alloys and composite crystals. *J. Mater. Chem.*, **17**, 1559–1561.

71 Olmsted, B.K., Ferlay, S., Dechambenoit, P., Hosseini, M.W., and Ward, M.D. (2009) Microscopic topography of heterocrystal interfaces. *Cryst. Growth Des.*, **6**, 2841–2847.

8
Synthesis and Design of π-Conjugated Organic Architectures Doped with Heteroatoms

Simon Kervyn, Claudia Aurisicchio, and Davide Bonifazi

8.1
Introduction

"*Many natural phenomena exhibit a dependence of a periodic character. Thus the phenomena of day and night and the seasons of the year, and vibrations of all kinds, exhibit variations of a periodic character in dependence on time and space*" (Dimitri Ivanovich Mendeleev, 1889). This is as much true for materials' properties (Figure 8.1).

Among all elements, carbon is certainly a singular element because of its rather unique ability to exceptionally form bonding interaction with itself and any other atom of the periodic table, thus giving rise to molecules capable of generating materials or even life. The abundance and availability of carbon as a natural element together with the discovery of fullerenes in 1985 by Sir Harold W. Kroto, Robert F. Curl, and the late Richard E. Smalley (Chemistry Nobel Prize winners in 1996) [1] and that of other carbon nanostructures such as nanohorns, graphenes, and other less-common carbon nanoobjects (nanoonions, nanobuds, peapods, nanocups, and nanotori) brought carbon-based materials a lot of attention from the scientific community provoking an unabated interest as testified by the unprecedented scientific production during the last 15 years [2]. This amount of work stems from the potential applications that these carbon nanostructures have in different fields ranging from biological and biomedical applications [3] to materials science and molecular electronics [4], those further spurred on by the rapid technological progresses in their fabrication and manipulation. Carbon atoms show four valence states, three sp^2-orbitals form a σ state with three neighboring carbon atoms (or a hydrogen), and one p-orbital develops into delocalized π and $π^*$ states that form the highest occupied molecular orbitals, HOMOs (or valence band), and the lowest unoccupied molecular orbitals, LUMOs (or conduction band), respectively. As Primo Levi said in his book *Periodic Table*, "... *conquering matter is to understand it, and understanding matter is necessary to understanding the universe and ourselves*" changing little variables, one of the main strategies to tune the properties of carbon-based

Functional Supramolecular Architectures. Edited by Paolo Samorì and Franco Cacialli
Copyright © 2011 WILEY-VCH Verlag GmbH & Co. KGaA, Weinheim
ISBN: 978-3-527-32611-2

```
                    Ti = 50      Zr =  90     ?  = 180
                    V  = 51      Nb =  94     Ta = 182
                    Cr = 52      Mo =  96     W  = 186
                    Mn = 55      Rh = 104,4   Pt = 197,4
                    Fe = 56      Ru = 104,4   Ir = 198
                Ni = Co = 59     Pd = 106,6   Os = 199
H = 1               Cu = 63,4    Ag = 108     Hg = 200
    Be =  9,4  Mg = 24   Zn = 65,2   Cd = 112
    B  = 11    Al = 27,4  ? = 68     Ur = 116     Au = 197?
    C  = 12    Si = 28    ? = 70     Sn = 118
    N  = 14    P  = 31   As = 75     Sb = 122     Bi = 210?
    O  = 16    S  = 32   Se = 79,4   Te = 128?
    F  = 19    Cl = 35,5 Br = 80     J  = 127
Li = 7 Na = 23 K  = 39   Rb = 85,4   Cs = 133     Tl = 204
               Ca = 40   Sr = 87,6   Ba = 137     Pb = 207
               ?  = 45   Ce = 92
               ?Er = 56  La = 94
               ?Yt = 60  Di = 95
               ?In = 75,6 Th = 118?
```

Figure 8.1 First classification of the Mendeleev's Periodic Table (1869).

materials is based on the insertion of electron-deficient or electron-rich heteroatoms into the carbon network that strongly affects the frontier orbitals, introducing vacancies (or holes) or nonbonding states (electrons) thus dramatically altering their electronic properties. While the insertion of group III elements, such as boron or aluminum atoms, causes the appearance of electronic vacancies, the insertion of group VI elements electronically enriches the molecule inserting nonbonding states. This chapter gives a tutorial account of the latest discoveries reported by chemists, *"the transformer of the matter,"* on the design and synthesis of functional molecular modules containing B, S, Se, and Te atoms showing fascinating electronic and supramolecular properties suitable for materials science applications, such as sensors, photonic materials, host–guest complexes, self-assembled two-dimensional (2D) and three-dimensional (3D) architectures in the bulk or at interfaces.

8.2
Boron

Structures containing boron atoms have been known for decades [5]. However, their use in electronic materials is relatively new. When a boron atom adopts a position in the molecule formerly occupied by a carbon atom, it is electronically equivalent to introducing a trivalent carbocationic species, bearing an empty p-orbital perpendicular to the plane of the molecule. Because of their electronic deficiency, boron atoms can operate as electrophilic centers accommodating nucleophilic anionic species or accepting charges from electron donating bonding units, ultimately giving rises to intramolecular charge transfer phenomenon and serving as hole carriers. Taking advantage of these properties, boron-containing organic architectures turned out to be promising candidates as (i) molecular sensors for anion detection (mainly fluorides and cyanides), (ii) organic materials for nonlinear optics (NLO) applications, and (iii) hole transporters and/or emitters for organic light-emitting diodes (OLEDs) [6]. In recent years, the field of boron-containing organic structures has become too vast to be covered in a chapter and thus we restrict our discussion to organic molecules in which the boron atom is connected to at least an aryl unit or inserted into an aromatic structure. In this section, the large number of boron dipyrromethenyl (BODIPY) derivatives would not be treated and the interested reader may refer to the specialized literature [7].

8.2.1
Boron Connected to Aryl Unit

8.2.1.1 Boron Derivatives Used as Chemical Sensors

Owing to the presence of an empty p-orbital, boron atoms are usually protected by hindered bulky substituents, which provide kinetic stability. Usual aryl substituents for boron-containing derivatives are mesityl, durene, and 2,4,6-tri-i-propylphenyl moieties bearing *ortho* groups surrounding the boron-centered electronic vacancy allowing only small anions (e.g., fluoride or cyanide ions) coordinating with the boron atoms. This selectivity has been exploited to build two types of luminescent chemical sensors: "turnoff" or "turnon" systems. For instance, in turnoff systems anion binding hampers the formation of any intramolecular charge transfer states quenching their emissive properties, whereas anion complexation in turnon sensors exalts their emissive properties (mainly those deriving from $\pi \rightarrow \pi^*$ transitions). In this respect, boron–aryl conjugates **1–3** displayed interesting properties.

The synthesis pathways are shown in Schemes 8.1–8.3. While the boron-phenyl derivatives bearing three identical appends are obtained by reaction of the corresponding ArLi on BF_3 (Scheme 8.3), boron compounds bearing two identical aryl groups (such as mesityl groups like in molecules **1** and **3**) are synthesized by nucleophilic substitution reaction of the ArLi derivative with Mes_2BF (Schemes 8.1 and 8.2). Molecule **1** has been used as a turnoff sensor to detect fluoride anions (sensitivity down to 0.02 mM) [8]. This compound, synthesized from precursor **2** by metal–halogen exchange reaction followed by a substitution on Mes_2BF, has also been used as electron transport material and blue light emitter in OLEDs due to its

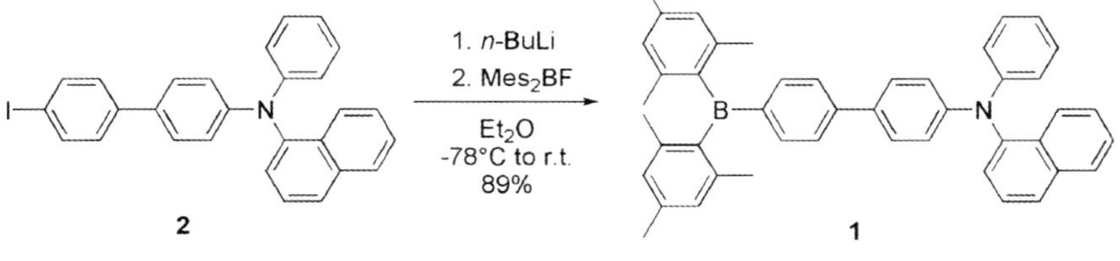

Scheme 8.1 Synthesis of fluoride anion molecular sensor **1**.

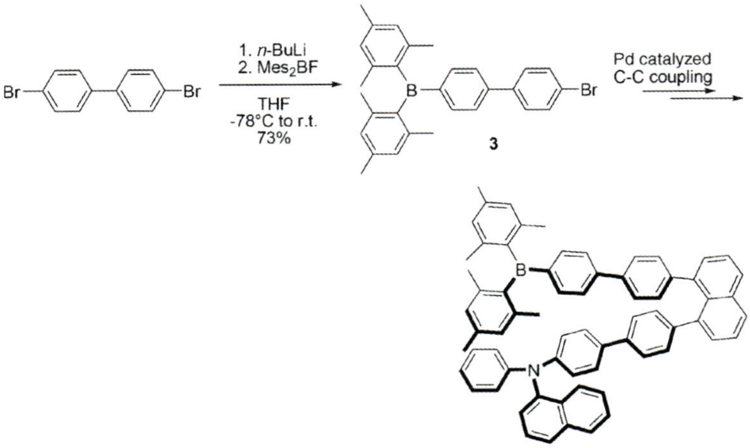

Scheme 8.2 Synthesis of compound **4**.

unique photophysical properties ($\Phi = 0.95$, $\lambda_{em} = 513$ nm, $\lambda_{ex} = 377$ nm in CH$_3$CN; $\Phi = 0.31$, $\lambda_{em} = 451$ nm, $\lambda_{ex} = 390$ nm in solid state) [9].

Molecule **3** (Scheme 8.2), synthesized with a similar procedure as followed for preparing **1**, undergoes Pd-catalyzed carbon–carbon cross coupling reaction with 1,8 diiodonaphthalene yielding molecule **4** ($\Phi_P = 0.98$, $\lambda_{em} = 414$ nm, $\lambda_{ex} = 340$ nm in CH$_2$Cl$_2$). In this system, the amine function interacts with the boron-centered electronic vacancy, allowing charge separation in ground state. Upon addition of fluoride ions, the through-space ($\sim$10 Å) charge transfer interaction is broken along with its fluorescence properties, thus activating only aminoaryl-centered $\pi \rightarrow \pi^*$ transitions ($\lambda_{em} = 443$ nm, $\lambda_{ex} = 340$ nm in CH$_2$Cl$_2$, Figure 8.2),

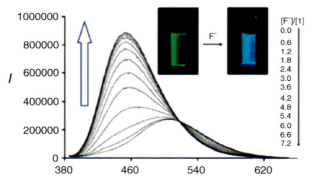

Figure 8.2 Fluorescence titration spectra of molecule **4** upon addition of tetrabutylammoniumfluoride (TBAF). Adapted with permission from Wiley-VCH Verlag GmbH & Co. KGaA. Copyright 2006.

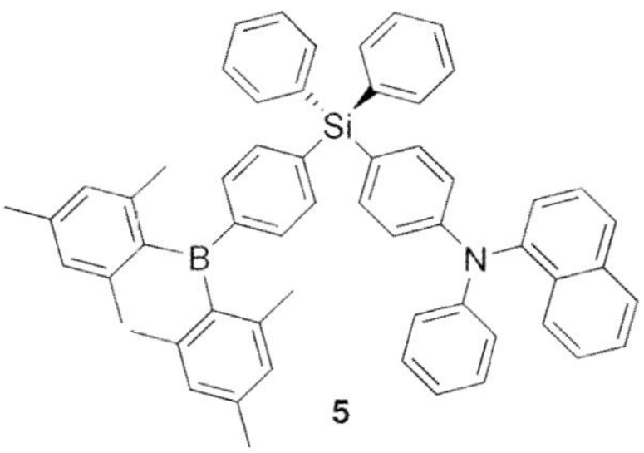

Figure 8.3 Fluoride anion molecular sensor **5**.

displaying a "turnon" sensing activity of compound **4** (fluoride detection limit: 0.020 mM) [8].

The distance between the donor and the acceptor moieties in these nonconjugated systems can be tuned using a flexible tetrahedral silane linker [10]. The great distance (~10 Å) between the electron donating and accepting units and the rotational freedom around the carbon–silicon bond diminish the through-space charge transfer process, and thus the enhancement of the emissive properties of the $\pi \rightarrow \pi^*$ excited states (Figure 8.3). The binding of the fluoride is also stronger in molecule **5** (Figure 8.4) compared to molecule **4** due to the reduced sterical hindrance between the donor and the acceptor moieties.

When more than one donating unit are conjugated with the boron center, the *para* substitution leads to both a stronger charge transfer interaction and a weaker electron acceptor ability compared to its *meta* regioisomer [11]. This is the case of molecule **6** (Scheme 8.3) that has been used as emitter ($\Phi = 0.67$, $\lambda_{em} = 476$ nm, $\lambda_{abs} = 385$ nm, in DMSO) and hole injection material in OLED [12]. Boron derivative **6**, substituted by identical groups, is obtained from the nucleophilic addition of lithiated derivative **7** to BF$_3$.

The wavelength of emission of the $\pi \rightarrow \pi^*$ transition can be redshifted upon increasing the distance between the donor and the acceptor moieties. The drawback

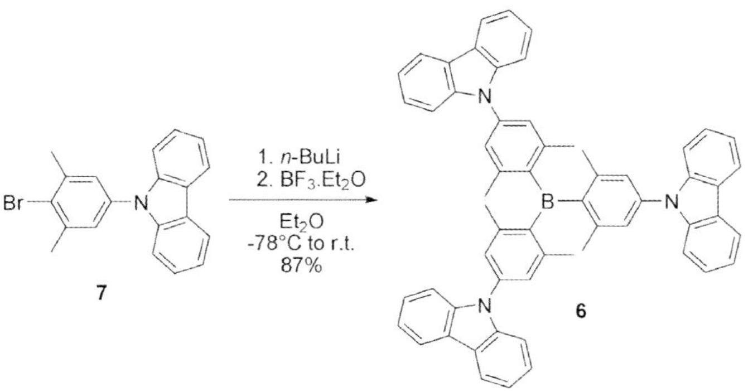

Scheme 8.3 Synthesis of a C$_3$ symmetry derivative **6**.

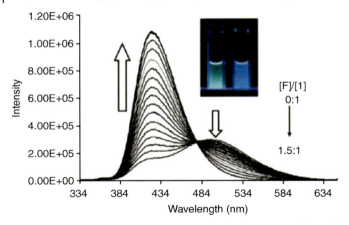

Figure 8.4 Fluorescence titration spectra of molecule **5** upon addition of TBAF. Adapted with permission from Wiley-VCH Verlag GmbH & Co. KGaA. Copyright 2009.

of this strategy is the decreasing efficiency of the through-bond charge transfer process with the distance [13]. Boron atoms in molecule **8** (Figure 8.5) are conjugated with the π electrons of the aryls groups. As a consequence, the $\pi \rightarrow \pi^*$ electronic transition ($\lambda_{em} = 450$ nm, $\lambda_{abs} = 432$ nm, in acetone) is quenched upon coordination of the first equivalent of fluoride to the boron atom center. The negative-charged tetravalent boron atom acts as a weak donating moiety, thus leading to a weak through-bond charge transfer process, despite the long distance (~20 Å). The addition of the second equivalent of fluoride ions quenches the fluorescence and activates the $\pi \rightarrow \pi^*$ transition ($\lambda_{em} = 450$ nm, $\lambda_{abs} = 432$ nm in acetone). Interestingly, an analogue of molecule **4**, bearing two boron atoms, was also used as a turnoff sensor in contrast to compound **8** (sensitivity down to 0.001 mM).

The use of mesityl and duryl protecting groups, substituted by the weak donating methyl groups, reduces the Lewis acidity of boron. One way to overcome this drawback is to substitute the anthryl moiety with several boron atoms (molecule **9**, Scheme 8.4) [14]. Despite the fact that the interplanar angle between the boron plane and the anthracene plane is 53°, a weak electronic conjugation is effective as an important bathochromic shift was observed in the UV-Vis spectra of compound **9**

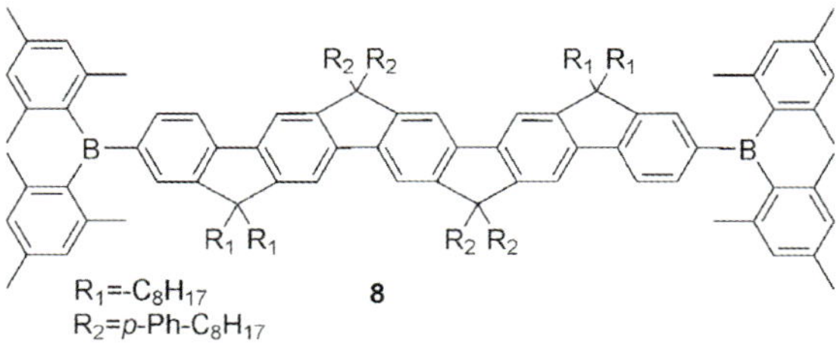

Figure 8.5 Fluoride anion molecular sensor **8**.

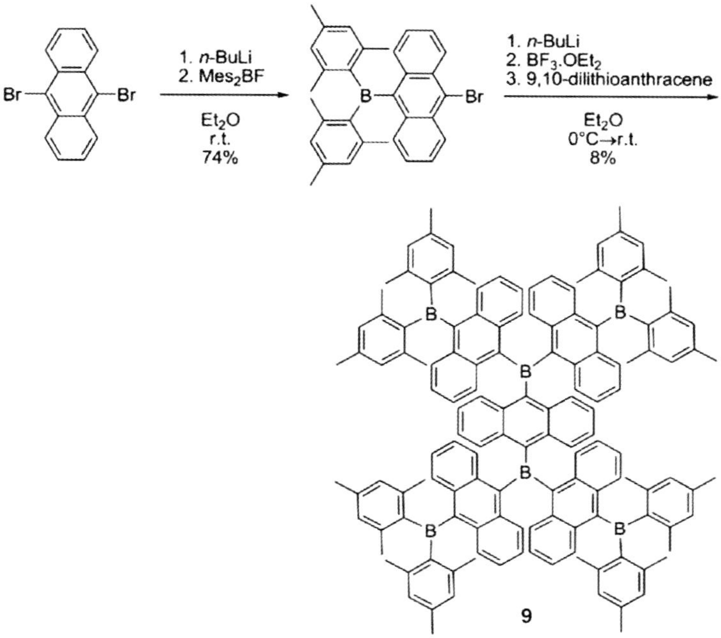

Scheme 8.4 Synthesis of the extended boron aryl conjugate **9**.

($\lambda_{abs} = 535$ nm, $\Delta\lambda = 65$ nm, in THF) compared to the trianthrylboron derivative ($\lambda_{abs} = 470$ nm, in THF).

Another strategy to enhance the Lewis acidity of boron is to connect more inductive withdrawing aryl units such as bipyridyl units, which can be linked to metals to further tune the electronic properties, such as molecule **10** (Figure 8.6), a very sensitive sensor (the color change first from red to orange after the addition of the first equivalent of fluorine, then to yellow upon the addition of the second equivalent, 0.01 mM in CH_2Cl_2) [15]. The coordination of the metal

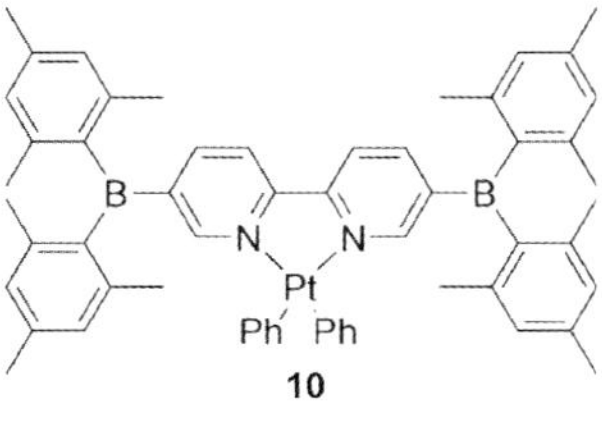

Figure 8.6 Chemical structure of Pt-coordination complex **10** [15] bearing two boryl functions.

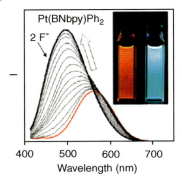

Figure 8.7 Fluorescence titration spectra of molecule **10** upon addition of TBAF. Adapted with permission from the American Chemical Society. Copyright 2007.

planarizes the molecule, enhancing the π-conjugation. The π–π* transition is even more enhanced during phosphorescence titration experiment (Figure 8.7) of molecule **10** in air.

Finally, fluoride anions can be coordinated by two boron atoms (B–F distance: 1.60 Å, according to X-ray measurement) to improve its binding. In this respect, molecule **11** (Scheme 8.5) was found to be very sensitive to fluoride anions (5×10^{-6} mM), but it showed only turnoff behavior for the fluorescence signal (Figure 8.8) [16].

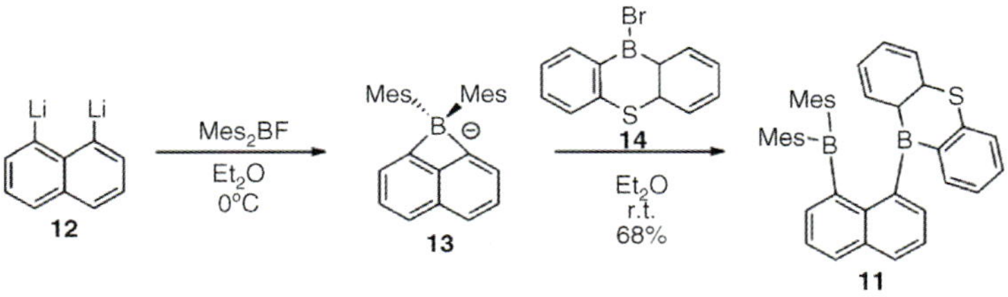

Scheme 8.5 Synthesis of bisboron derivative **11**.

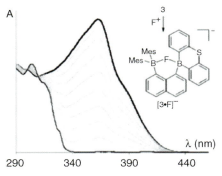

Figure 8.8 Fluorescence titration spectra of molecule **11** upon addition of TBAF. Adapted with permission from Wiley-VCH Verlag GmbH & Co. KGaA. Copyright 2009.

The synthesis of molecule **11** is displayed in Scheme 8.5. Compound **12**, obtained after lithium–halogen exchange from the corresponding dibromide derivative, was added to Mes$_2$BF to give compound **13**. Subsequently, borane **13** undergoes a ring-opening reaction when mixed with compound **14** to afford desired product **11** in good yield (68%). 10-bromo-9-thia-10-boraanthracene derivative **14** (Scheme 8.6) was prepared by reaction of disilyl compound **15** with BBr$_3$ [17].

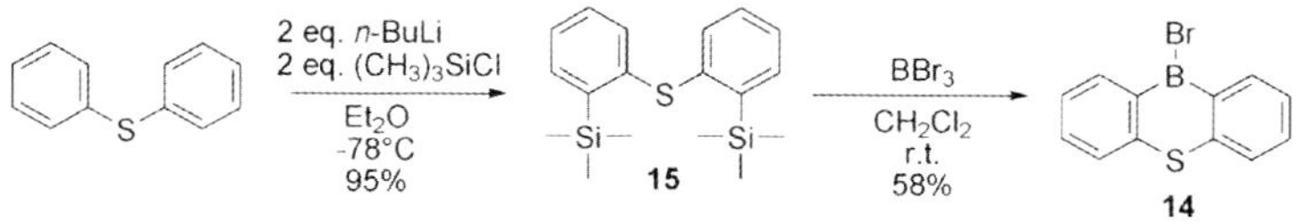

Scheme 8.6 Synthesis of bromo-thia-boraanthracene derivative **14**.

Beside fluoride detection, a few examples of cyanide sensors have also been developed. For example, in water, molecule **16** (Figure 8.9) is very sensitive to cyanide ions ($\sim 10^8$ M^{-1}, $\lambda_{em} = 460$ nm, $\lambda_{ex} = 300$ nm) and nearly inactive toward fluoride anions (4% of fluorescence quenching upon addition of 15 eq.) because of the high hydration enthalpy. The presence of the sulfonium functional group increases the electrophilicity of the boron atom center by the attractive inductive effect and further stabilizes the cyanide complex by bonding and back-bonding interactions with the CN triple bond [18]. An enhancement of the electrophilicity of boron has also been shown by ammonium-bearing molecular sensor **17**, which binds both fluoride and cyanide ions in organic solvents but shows an enhanced selectivity toward cyanides (3.9×10^8 M^{-1}) in aqueous solutions. Interestingly, the *ortho* regioisomer of molecule **17** revealed to be selective only toward the recognition of fluoride ions as the great steric hindrance around the boron atom center prevents from binding the larger ions, such as the cyanide ones [19].

Boronic acid derivatives have been used to detect sugars, using the alcohol functionality to reversibly form boronic esters, thus leading to a quenching of the emissive properties. In particular, in the example shown in Figure 8.9, the distance between the two boron centers makes molecule **18** very selective toward sialic acid [20].

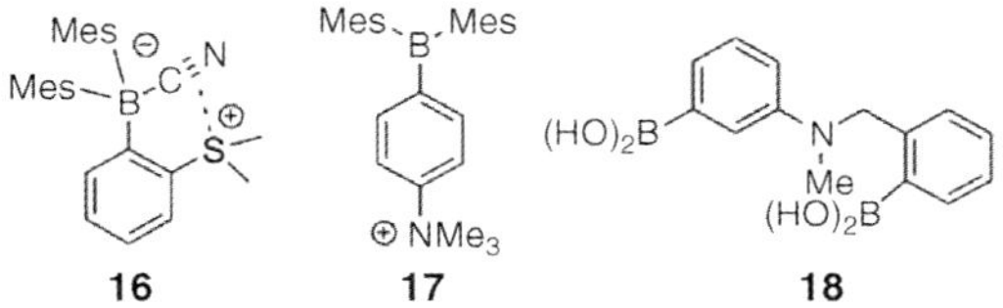

Figure 8.9 Examples of cyanide (**16**, **17**) and sugar (**18**) sensors.

8.2.1.2 Covalent Organic Frameworks

This reversible boronic ester formation has been exploited to build SCOFs (self-assembled covalent organic frameworks), porous 2D and 3D crystalline materials, the structure of which depends on the geometrical properties of the organic angular unit [21]. The crystalline structures have high pore volume (0.7907 cm^{3}g^{-1}), thus utilizable as gas storages or as hosts for luminescent molecules [22]. For example, bidimensional networks constituted of hexameric substructures are shown in Scheme 8.7. In particular, each hexameric unit piles through π–π stacking interactions (intersheets distance: 3.4 Å) to give an anisotropic conductive and fluorescent (λ_{em} = 484 nm, λ_{ex} = 414 nm, in solid state) porous material [23, 24]. In order to have a regular network, the condensation reaction was carried out under reversible conditions (as suspension in apolar solvent at 120°) to allow a self-healing process and thus the condensation of molecule **19** into the regular porous network. Notably, when the crystalline structure is excited by polarized light, the emission light is depolarized due to the exciton migration within the lifetime of the molecular excited state [23]. The same strategy was also used to construct fluorescent self-assembling polymers [25].

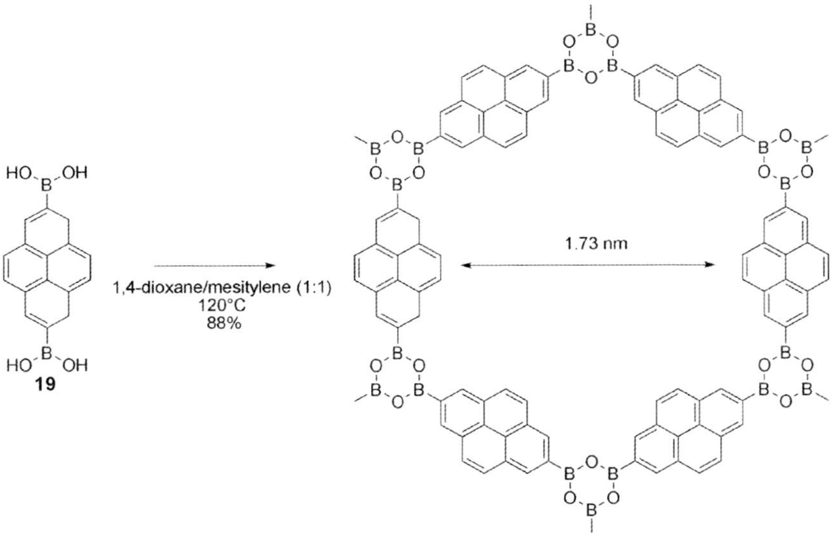

Scheme 8.7 Formation of a pyrenyl COF material.

In another example, SCOFs formed under UHV conditions on Ag(111) surfaces were reported by Zwaneveld et al. exploiting condensation reactions between boronic acid derivatives [26]. Condensation reactions involving 1,4-benzenediboronic acid led to six-member B$_3$O$_3$ rings (Figure 8.10) arranged into ordered porous hexagonal arrays with cavity diameters of 15 ± 1 Å (sublimation of benzenediboronic at 300–500 K, surface coverage ranging from <1% to near-complete monolayer). The stability of the assembly was tested by thermal

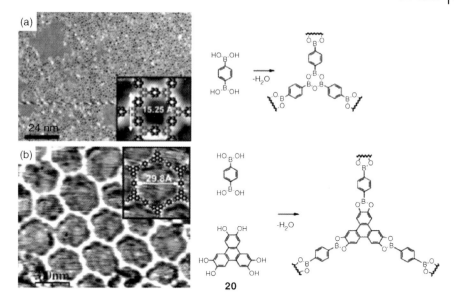

Figure 8.10 (a) STM image of a covalent network formed upon deposition of 1,4-benzenediboronic acid on Ag(111) at near-complete monolayer coverage; (b) STM image of the SCOF formed upon codeposition of 1,4-benzenediboronic acid and hexahydroxytriphenylene **20** on Ag(111). The insets in both images show the theoretical pore diameter obtained by DFT. Adapted with permission from the American Chemical Society. Copyright 2008.

annealing at 750 K. No degradation was observed after 5 min; however, following a prolonged period of annealing (12 h), a significant degradation occurred and only small islands of the intact assembly were observed. Hexagonal networks with larger pore sizes (29 Å) were also obtained upon codeposition of benzenediboronic acid and 2,3,6,7,10,11-hexahydroxytriphenylene **20**, and once again annealing at 750 K did not degrade the network, thus confirming the covalent nature of the assembly (Figure 8.10).

8.2.1.3 Boron Connected to Electron-Rich Aryls Groups

Boron moieties have also been introduced in polymers. In the example reported in Scheme 8.8, the reaction of fluorenobisstannate **21** with fluorenobisbromine **22** affords polymer **23**, in which a linking boryl group has been introduced in the polymer backbone. The electronic properties of compound **23** could be tuned by substituting the boron ligand (from **23** to **24–26**). The boron atoms in polymers **24** and **25** are functionalized by aryl groups, of which the mesityl functions revealed to be not bulky enough to protect structure **24** from any degradation upon exposure to air, suggesting that *i*-propyl *ortho* substituents are necessary for its protection. Fluorescence titrations of polymer **25** show a similar detection behavior to the bisboron molecular ladder **8**. Namely, after the addition of the

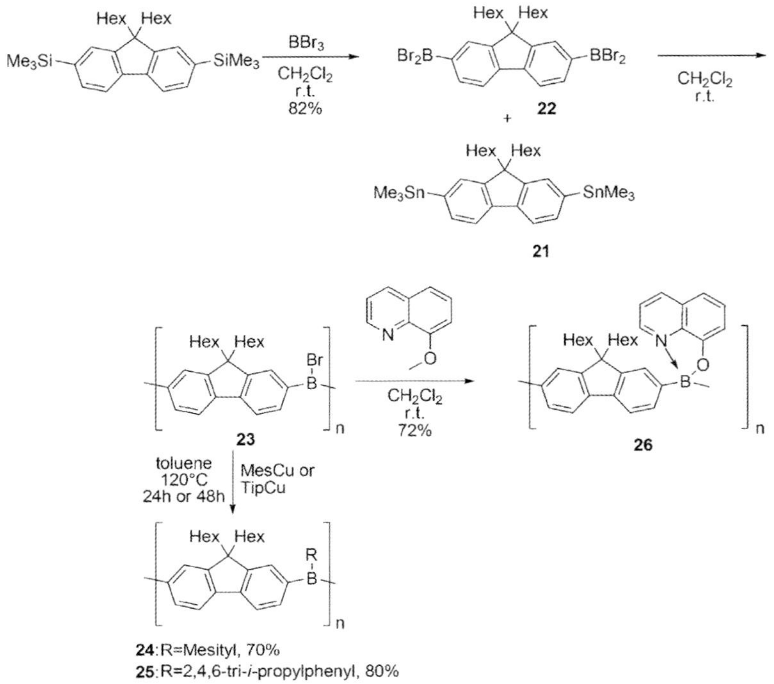

Scheme 8.8 Synthesis of fluorenyl-based polymers **24**, **25**, and **26** linked by doping boron atoms.

first equivalent of fluoride anions, a charge transfer process occurs between the tetracoordinated boron and the electron-deficient center, showing a turnon behavior of the fluorescence signal. Compared to **25** ($\Phi = 0.81$, $\lambda_{em} = 399$, 423, 447 nm, $\lambda_{ex} = 371$ nm, in CH_2Cl_2), polymer **26** revealed to be weakly fluorescent ($\Phi = 0.21$, $\lambda_{em} = 504$, 338 nm, $\lambda_{ex} = 313$ nm, in CH_2Cl_2) and its UV absorption spectra profile typical of that of a quinoline molecule. From this data, one can conclude that the complexation of the boron by the lone pair of the nitrogen disrupts π-conjugation along the polymer chain, thus leading to a quenching of the luminescence properties [27].

Another strategy of doping polymeric structures is to laterally append a substituted boryl moiety [28]. The boryl substituents were found to influence the electronic properties of the material if equipped at the *para* or *ortho* positions with respect to the polymer chain [28, 29]. The boryl substituents usually have been connected to electron-rich substituents such as thiophene (see molecule **27**) and ferrocenyl derivatives. For example, molecule **27** (Scheme 8.9) has been prepared by *ortho* lithiation of precursor **28** followed by a nucleophilic addition to Mes_2BF. Although the compound has shown interesting NLO properties ($\beta = 37 \times 10^{-30}$ esu, $\mu\beta = 148 \times 10^{-8}$, in acetone), it easily degrades within hours upon exposure to light [30], thus showing a limited applicability.

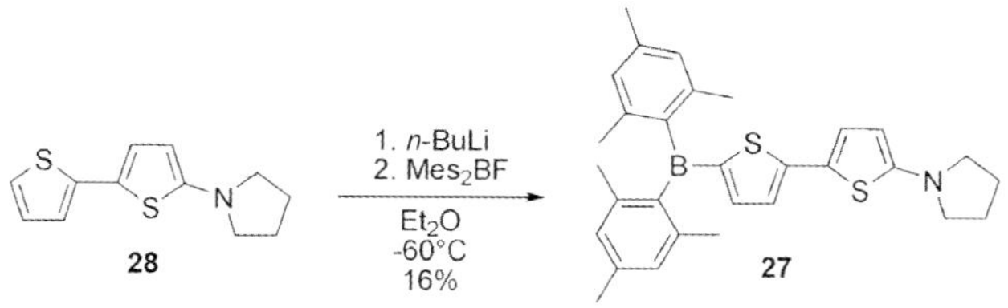

Scheme 8.9 Synthesis of a bisthiophene moiety connected to a boryl functional center, molecule **27**.

Substituents other than mesityl, duryl, 2,4,6-tri-i-propylphenyl, and anthryl functions can be used, but the resulting polymers are often instable in air. Polymer **29** (Scheme 8.10) was synthesized from bis-stannate derivative **30** and $C_6F_5BBr_2$. The attachment of electron-withdrawing fluorobenzene substituents led to a bathochromic shift of the fluorescence spectra of polymer **29** ($\Phi = 0.15$, $\lambda_{em} = 529$ nm, $\lambda_{ex} = 413$ nm, in CH_2Cl_2) compared to the same structure equipped with 4-i-propylphenyl appends ($\Phi = 0.21$, $\lambda_{em} = 491$ nm, $\lambda_{ex} = 391$ nm, in CH_2Cl_2). Electron-rich ferrocenyl moieties have also been connected to the starting thiophene derivative **30** to afford polymer **32**, using reaction conditions similar to those used for preparing **29**. Both polymers **29** and **32** have been employed for sensing aromatic amines such as pyridines and picolines. UV and fluorescence (Figure 8.11a and b, respectively) titration experiments of samples containing polymer **29** or **32** were used to test the binding capabilities of the polymeric architectures. In particular, the changes in the absorbance and emission spectra profiles for molecule **29** showed a strong

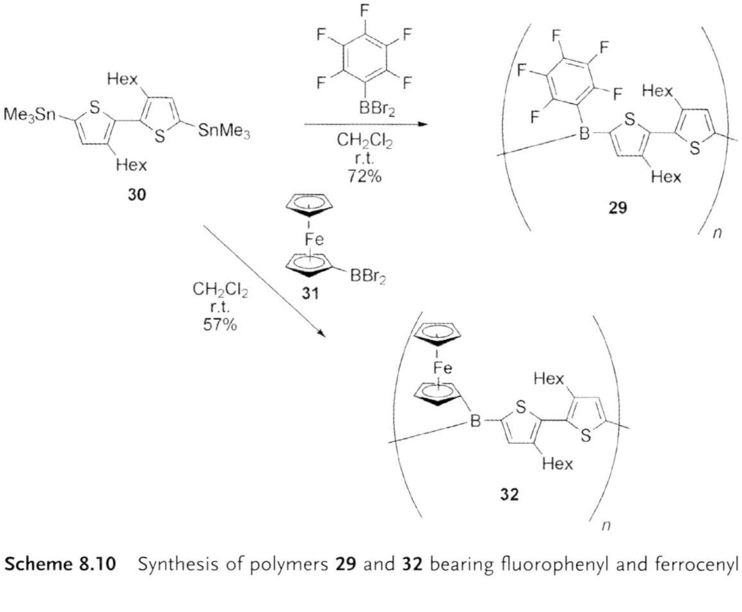

Scheme 8.10 Synthesis of polymers **29** and **32** bearing fluorophenyl and ferrocenyl moities.

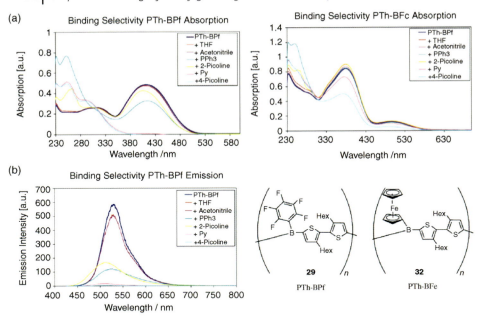

Figure 8.11 (a) Absorbance (for **29** and **32**) and (b) fluorescence (**29**) titration spectra upon addition of various aromatic amines. Adapted with permission from American Chemical Society Copyright 2005.

association with both amines (pyridine and 4-picoline), whereas the response of **32** revealed to be lower, probably because of the reduced electrophilic character of the boron atom center [31].

Boron-doped ferrocenyl polymers have also been synthesized. As a model compound, diferrocenyl borane derivative **34** has been synthesized (Scheme 8.11) to probe the electronic properties [32]. Boron ferrocenyl derivative **31**, prepared in excellent yield (89%) using a well-known procedure [33], was dimerized via homocoupling reaction using HSiEt$_3$ to give bisferrocenyl derivative **33**. Subsequent nucleophilic addition of a mesityl cuprate led to molecule **34**, in which the mesityl group surrounds the boron atom center. The two ferrocenyl entities were found to electronically communicate via the empty p-orbital of the boron atoms, as shown by the presence of two reversible ferrocenyl-centered redox waves measured by cyclovoltammetry ($E^{\circ\prime} = 45$ mV and 467 mV, at r.t. in CH$_2$Cl$_2$ with NBu$_4$[B(C$_6$F$_5$)$_3$] as supporting electrolyte using decamethylferrocene as internal standard) [33]. Cyclovoltammetry measurement of polymer **35** (Scheme 8.12), synthesized by a similar pathway as that used for model ferrocenyl dimer **34**, exhibits a high amount of splitting between the ferrocenyl-centered redox waves ($\Delta E^{\circ\prime} = 705$ mV) as a consequence of the extended conjugation.

With the aim of studying a less hindered polymer as an electron-deficient analogue of all-carbon structure, bisferrocenyl molecule **37** (Scheme 8.13) was

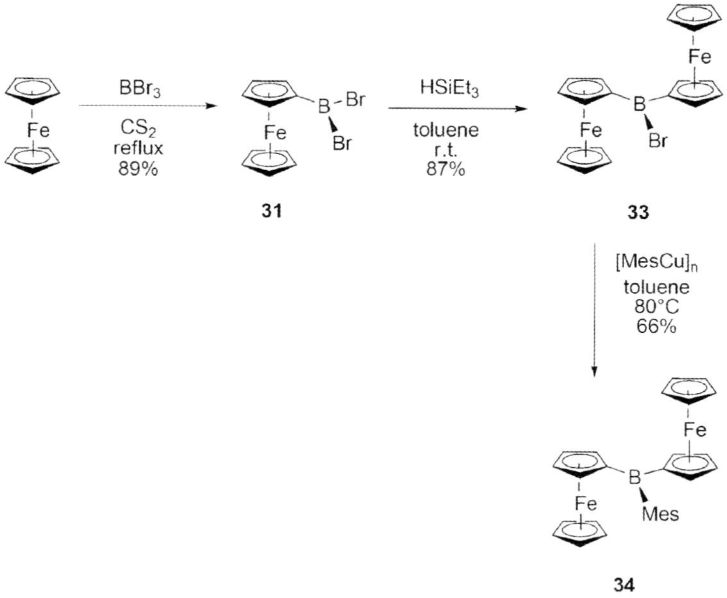

Scheme 8.11 Synthesis of a protected boron bisferrocenyl derivative **34**.

synthesized. The synthesis is displayed in Scheme 8.13. Boronic ester **38** could be reduced to give molecule **39** that dimerized to bisferrocenyl derivative **37** upon addition of Me₃SiCl. Unfortunately, bisferrocenyl molecule **37** revealed to be less stable compared to molecule **34** and thus limited investigations have been conducted further [34].

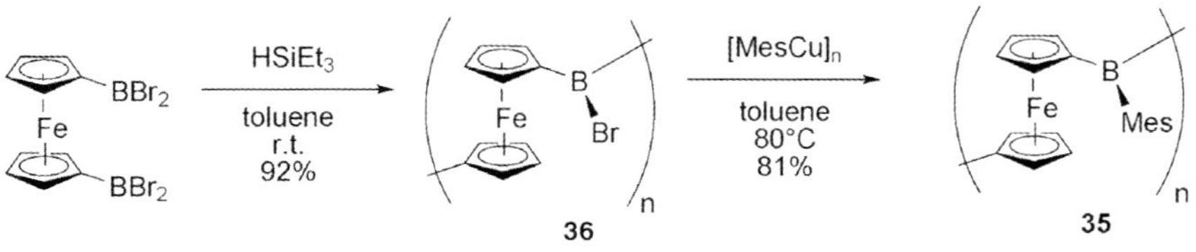

Scheme 8.12 Synthesis of boron-doped ferrocenyl polymer **35**.

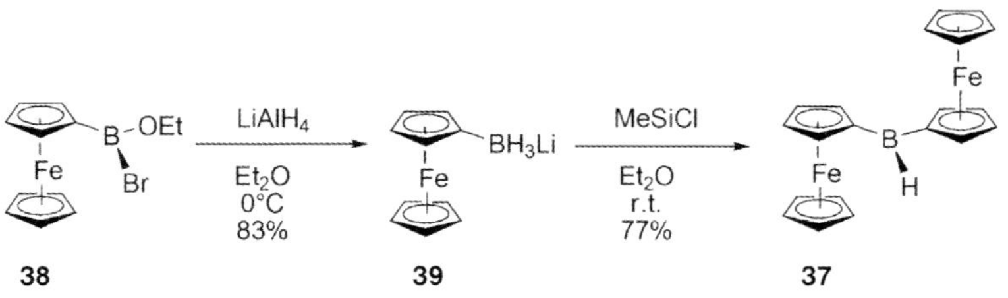

Scheme 8.13 Synthesis of bisferrocenyl borane **37**.

All the molecular structures described so far in this chapter have been mainly used as molecular sensors to detect fluoride anions, but no example of molecular sensors resistant to detection of HF has been reported. In this respect, molecule **31** (Scheme 8.14), obtained by electrophilic substitution described in Scheme 8.11, was transformed into protected boronic ester **40**, which was thought to be resistant in the presence of HF. However, it was observed that molecule **40** degraded upon addition of HF acid despite the extra stabilization of the boron center by the nitrogen lone pair.

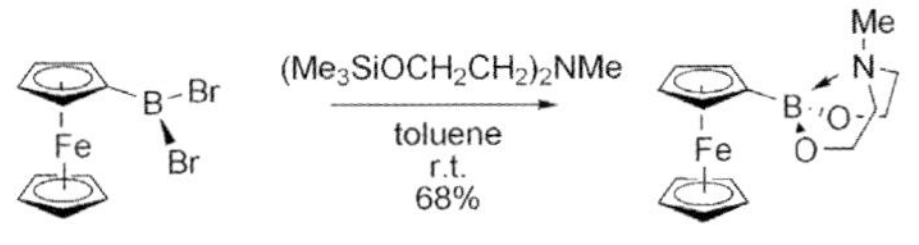

Scheme 8.14 Synthesis of protected boron ferrocenyl derivative **40**.

In order to have stable HF sensors, amino acid **41** (Scheme 8.15) was synthesized through *ortho*-lithiation of the starting ferrocene derivative followed by a nucleophilic substitution reaction with $B(OC_4H_9)_3$ followed by hydrolysis during the workup [35]. Addition of three equivalents of HF gives molecule **42**, which was found to dimerize in the solid state (Scheme 8.15) through the formation of intermolecular hydrogen-bonding interactions (hydrogen–fluoride contact distance amounting to 2.204 Å) [36]. The HF selectivity of **42** was investigated through electrochemical

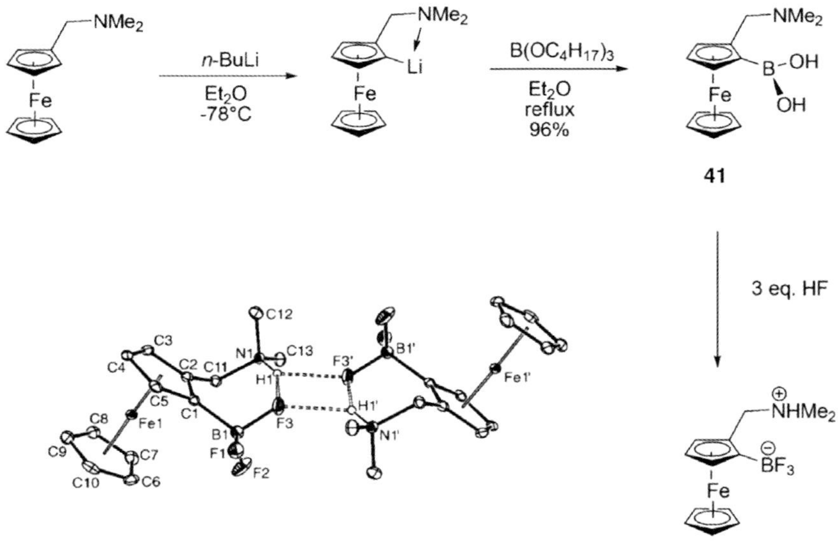

Scheme 8.15 Synthesis and X-ray crystal structure of the HF molecular sensor **42**. Adapted with permission from Wiley-VCH Verlag GmbH & Co. KGaA. Copyright 2005.

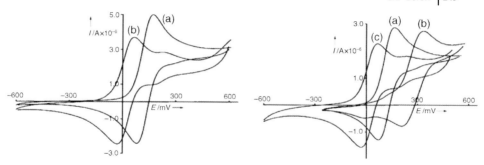

Figure 8.12 *Left*: Cyclic voltammograms of **41** (a) and **42** (b). *Right*: Cyclic voltammograms of **41** (a), **41** after addition of HCl (b), and **41** after subsequent addition of nBu₄NF (c). Adapted with permission from Wiley-VCH Verlag GmbH & Co. KGaA. Copyright 2005.

analysis, resulting in opposite responses for HF compared to other acids, such as HCl. Thus, the voltammograms of **41** and **42** present a cathodic shift of −80 mV after HF addition ($E_{1/2}$ = + 37.5 mV (85 mV) for **41** and −43 mV (75 mV) for **42**; Figure 8.12 left side). This behavior is consistent with a three-coordinate electron-withdrawing boronic acid group into a four-coordinate electron-donating boronate. The reversed situation is found when three equivalents of HCl are added to molecule **41** (Figure 8.12 right side). In this case, a significant anodic shift is detected (+149.5 mV) suggesting a possible protonation of the dimethylamino group without any coordination of the boron center. In addition, the reaction with an excess of fluoride, accomplished by nBu₄NF (Figure 8.12), leads to a voltammetric curve resembling that observed for the authentic sample of **42**, proving then the intrinsic HF selectivity of the boronic acid derivative.

8.2.1.4 Vinyl Borane

Vinyl borane derivatives have been synthesized by hydroboration reaction of the corresponding ethynyl derivatives. Compound **43** (Scheme 8.16) adds to the triple bond of pyrene derivative **44** to form *trans* product in low yield **45** (25%) due to the presence of sterically hindered 2,4,6-tri-*i*-propyl moieties. The bispyrene compound **45** is fluorescent both in solution (λ_{em} = 490 nm, λ_{ex} = 437 nm, in CH_2Cl_2) and in solid state (λ_{em} = 530 nm, λ_{ex} = 437 nm) [37].

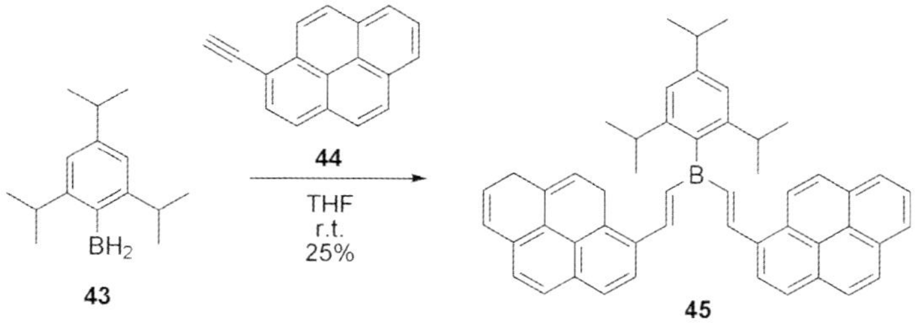

Scheme 8.16 Synthesis of bisvinyl bispyrenyl borane **45**.

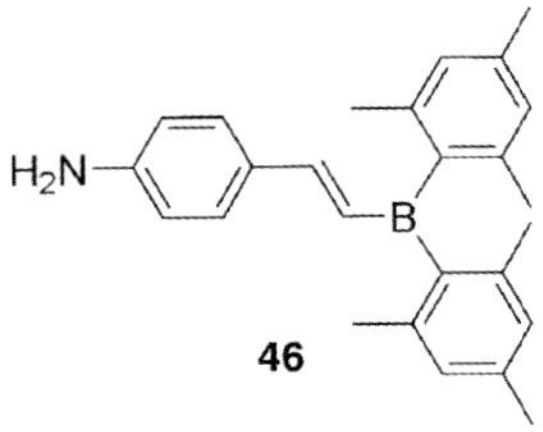

Figure 8.13 Vinyl borane derivative **46**.

Other groups have used the empty p-orbital of the boron atom to design efficient donor acceptor systems with nonlinear optical properties. In a similar synthesis, molecule **46** (Figure 8.13), stable in air, was synthesized from the corresponding acetylenic derivative [38]. Nonlinear properties, measured by the second-order molecular hyperpolarizabilities (β), revealed to be large in solution (18×10^{-30} esu). As the molecule and its derivatives do not crystallize in centrosymmetric spatial groups, they revealed to be promising candidates for engineering novel materials for NLO applications.

Polymer **47** (Figure 8.14), displaying a π-conjugated backbone doped with vinyl-boron atoms and a transition metal atom, has also been prepared showing an intense electronic $d_\pi – p_\pi^*$ transition ($\lambda_{abs} = 514$ nm, redshifted by 141 nm with respect to the monomer) and strong emission ($\lambda_{em} = 435$ nm with $\lambda_{ex} = 350$ nm and $\lambda_{em} = 592$ nm with $\lambda_{ex} = 500$, in CHCl$_3$) [39]. For the nondoped polymer **48** (Figure 8.15), the emission spectrum ($\lambda_{em} = 441$ nm, $\lambda_{ex} = 350$ nm in CHCl$_3$) shows a less pronounced bathochromic shift ($\Delta\lambda = 38$ nm in CHCl$_3$) due to the extended conjugation compared to the monomer [40].

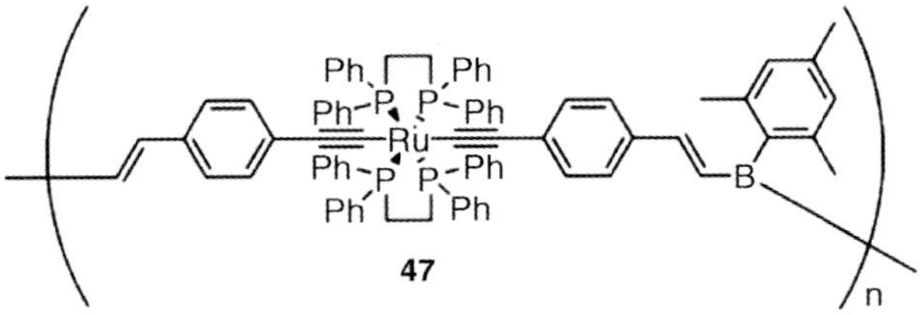

Figure 8.14 Vinyl borane polymer **47** doped with a transition metal.

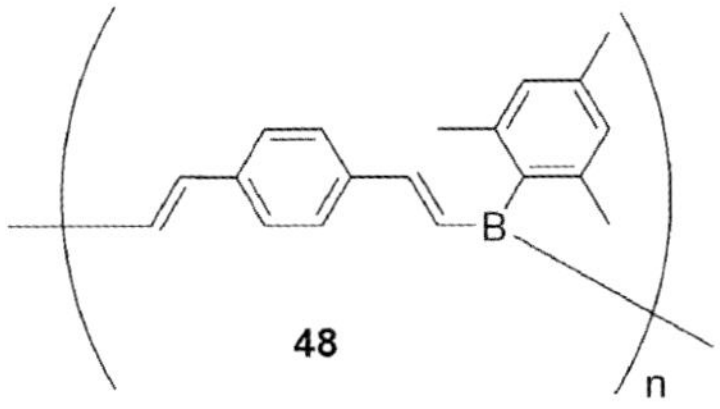

Figure 8.15 Vinyl borane polymer **48**.

8.2.2
Boron Inserted into Aryl Unit

8.2.2.1 Borabenzene

Borabenzene has been synthesized and used as a complex for transition metal decades ago [41]. The stannohydration of 1,6-diethynyl **49** (Scheme 8.17) gave rise to cyclic compound **50**. The latter is transformed into cyclic borane **51** by stannate–boron exchange reaction. Deprotonation reaction of **51** with *t*–BuLi leads to borabenzene derivative **52**, displaying a strong aromatic character [42].

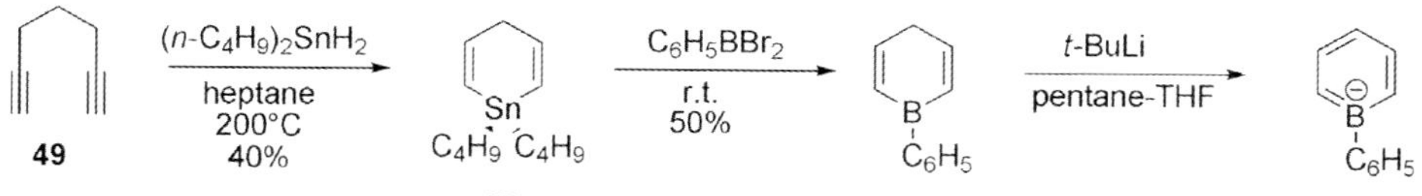

Scheme 8.17 Synthesis of borabenzene **52**.

Recently, the interest for molecule **52** and its derivatives has been renewed by its use in a chiral Lewis acid complex **53** (Scheme 8.18) [43]. This was obtained using a similar synthesis strategy to that reported in Scheme 8.17, followed by complexation of Cr(0) metal atom to intermediate **54**.

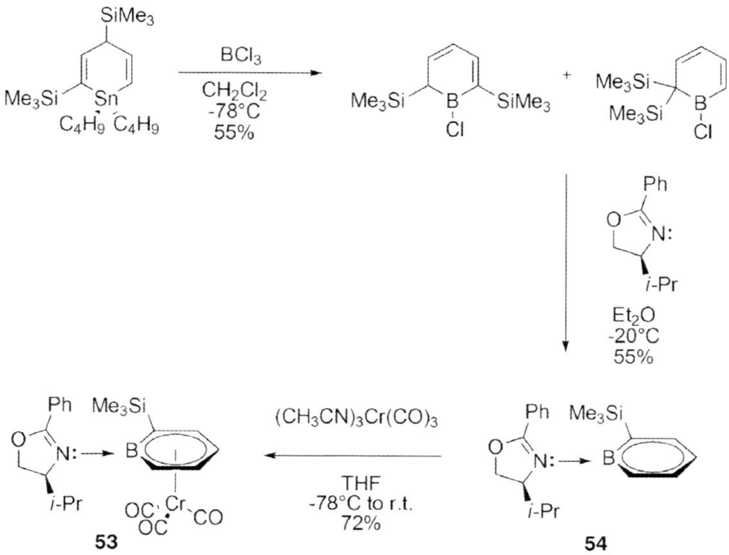

Scheme 8.18 Synthesis of chiral borabenzene complex **53**.

Borabenzene can also be connected to a phenyl ring through a double bond, forming a borastilbene. The latter is being used as a ligand for the catalyst in ethylene polymerization [44]. The synthesis of borastilbene **55** (Scheme 8.19) can be achieved by transmetalation reaction of the styryl functional group connected to a zirconium complex (or transmetalation of the stannate derivative, although this synthesis route seems to give rise to a complicated mixture of products) [45]. After reaction with metallic Na, compound **55** gave fluorescent boratastilbene **56**. The UV-Vis absorption analysis ($\lambda_{abs} = 357$ nm, in THF) revealed the presence of a shoulder peak at 400 nm that appears upon dissolution in THF and diminishes with increasing solution concentration. Upon addition of two equivalents of dibenzo-18-crown-6, the absorption peaks revealed to be redshifted ($\lambda_{abs} = 407$ nm; $\Phi = 0.68$, $\lambda_{em} = 518$ nm, in THF) suggesting the presence of molecular aggregates.

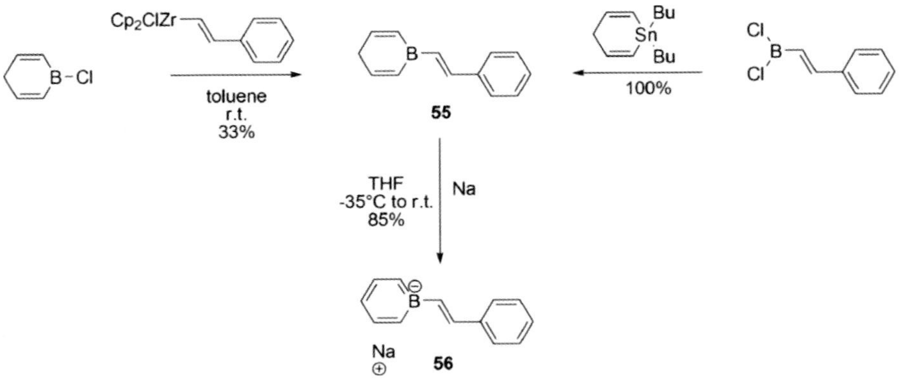

Scheme 8.19 Synthesis of boratastilbene **56**.

The absorption and emission spectra of the related bisborastilbene **57** ($\Phi = 0.16$, $\lambda_{em} = 604$ nm and $\lambda_{ex} = 396$ nm, in THF) upon addition of dibenzo-18-crown-6 also showed a reduction in the HOMO–LUMO gap, as observed for the specie **58** ($\Phi = 0.44$, $\lambda_{em} = 534$ nm and $\lambda_{ex} = 447$ nm, in THF), Scheme 8.20 [46].

8.2.2.2 Boranaphathalene and Higher Acenes
Different isomers of boranaphthalene were synthesized. The first approach employs the well-explored stannate–boron exchange to synthesize 1-boranaphthalene **59** (Scheme 8.21), also used as a ligand for transition–metal coordination complexes [47].

The second synthetic approach implies the ring closure of molecule **60** (Scheme 8.22) with BCl$_3$ to give a mixture of isomers. The crude product was directly subjected to the next reactions to give 2-boranaphthalene **61** in good yields (85%) [48].

In an analogy to the boranaphthalene derivatives, a few synthetic pathways have also been developed to form 9-boraanthracene derivatives, the electron-deficient analogues of anthracene. They can be formed by metal–boron exchange from the corresponding stannate derivative [49]. Stannate **62** (Scheme 8.23) reacts with BCl$_3$ to

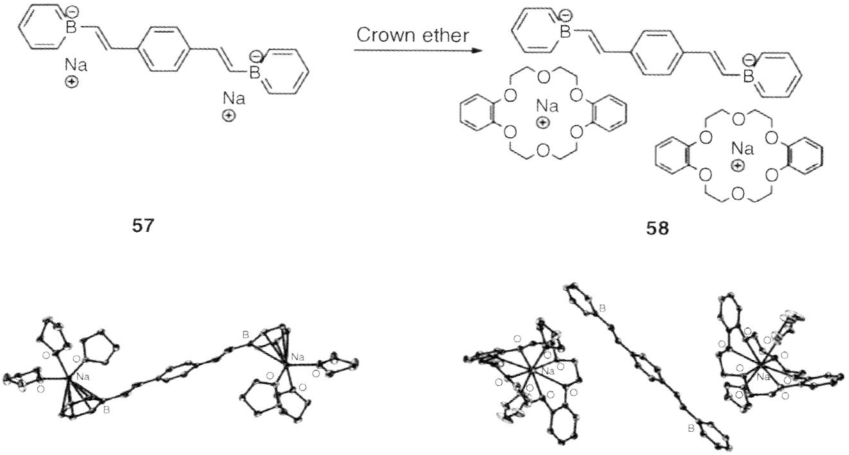

Scheme 8.20 Schematic representation and X-Ray analysis of the influence of the counterion on the properties of bisborastilbene **57**. Adapted with permission from the American Chemical Society. Copyright 2000.

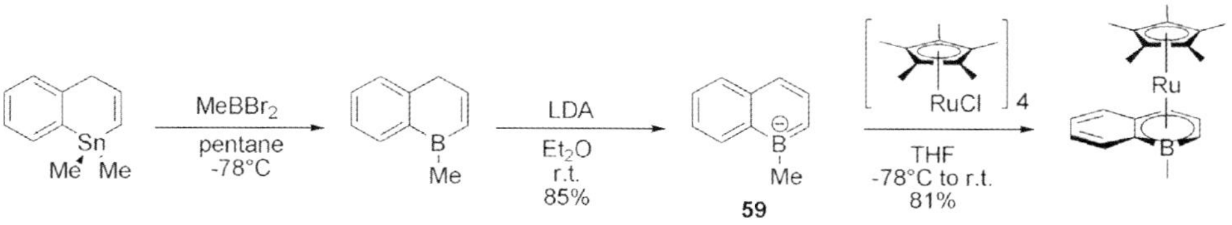

Scheme 8.21 Synthesis of 1-boranaphthalene **59**.

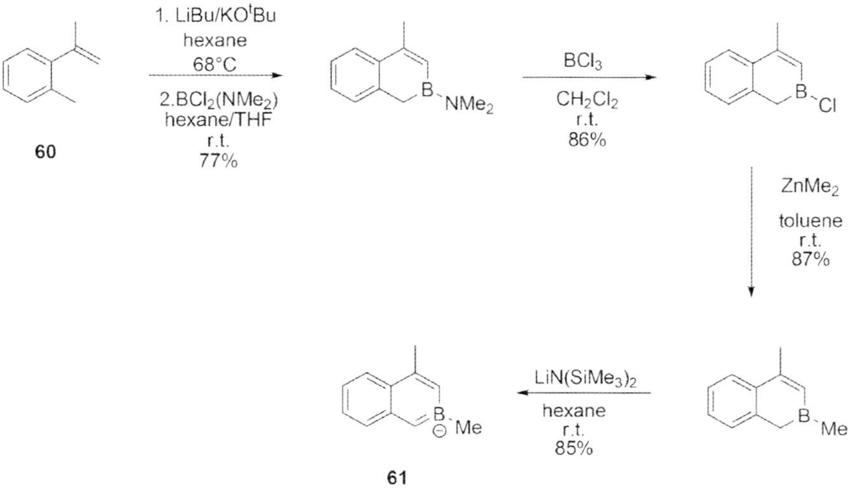

Scheme 8.22 Synthesis of 2-boranaphthalene **61**.

give boron derivative **63**, which can be substituted by 1,3-dimesityl-4,5-dihydroimidazol-2-ylidene **64** to yield compound **65**. The latter reacts with a base to form aromatic fluorescent compound **66** ($\lambda_{em} = 597$ nm, $\lambda_{ex} = 485$ nm, in CH_2Cl_2). Using the same synthetic pathway, molecule **67** was obtained. Weak solvatochromism was also observed (12 nm hypsochromic shift in CH_3CN). Notably, both compounds **66** and **67** react quickly with O_2 displaying a great potential as molecular oxygen sensors [49].

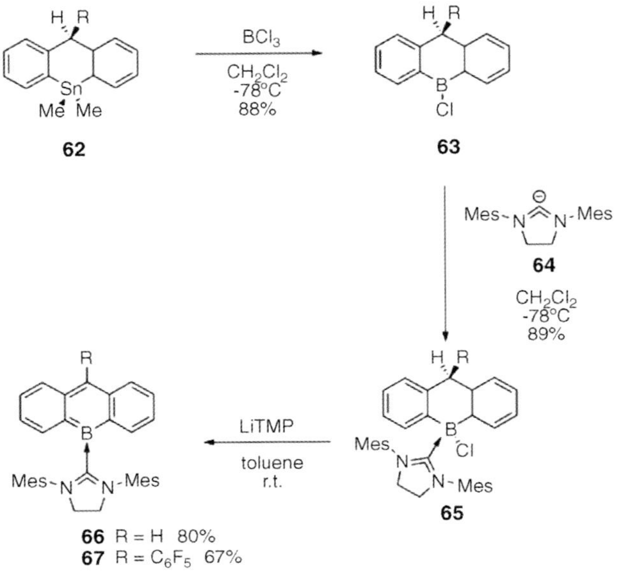

Scheme 8.23 Synthesis of boraanthracenes **66** and **67**.

9,10-Diboraanthracenes have been synthesized upon dimerization of compound **68** (Scheme 8.24) in the presence of BBr_3. Compound **69**, highly sensitive to moisture, is transformed into the molecule **70** in excellent yield (93%) upon addition of Me_3SiNMe_2 in toluene [50]. Molecule **69** was further used to prepare fluorescent polymer **71** (Scheme 8.24, $\Phi = 0.09$, $\lambda_{em} = 518$ nm, $\lambda_{ex} = 410$ nm, UV $\lambda_{abs} = 410$, 349, 290 nm, in toluene). Reaction of **69** with $HSiEt_3$ led to diboraanthracene hydride polymer **72** (held by agostic bonding interaction between the hydride functions and the boron atom centers), which was further transformed into polymer **72**, via an hydroboration reaction with 1,4-diethynylphenyl **73** [51].

Following a similar synthetic strategy to that used for preparing the diboraanthracene derivatives, a highly fluorescent ($\lambda_{em} = 464$, 435, and 410 nm, $\lambda_{ex} = 330$ nm, UV $\lambda_{abs} = 407$, 385, 366, 328, 302, and 268 nm, in cyclohexane) diborapentacene **74**

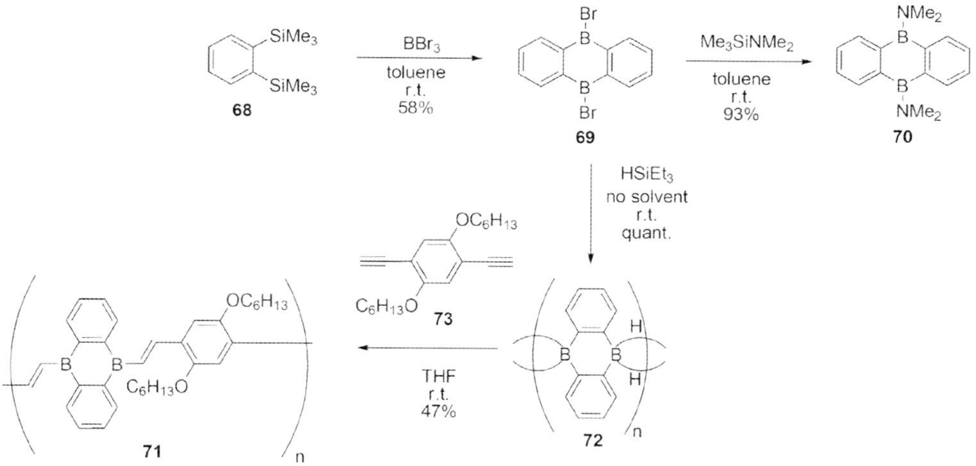

Scheme 8.24 Synthesis of 9,10-diboraanthracene **71**.

(Scheme 8.25) has also been synthesized [52]. Interestingly, no π–π stacking interactions were observed in the crystal organization of **74** probably because of the presence of the mesityl moieties (typical π-stacks separated 3.50 Å were observed for the dimethyl analogue).

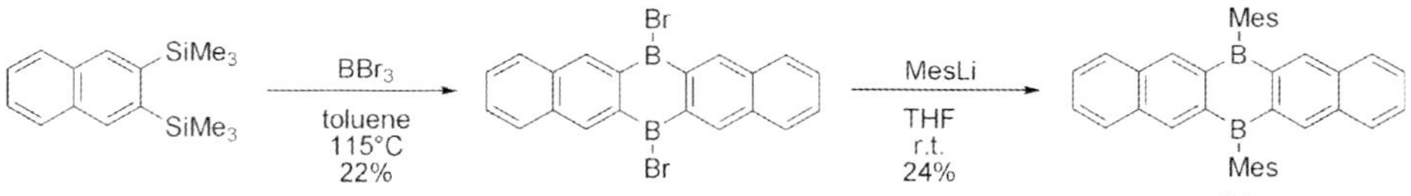

Scheme 8.25 Synthesis of a diboropentacene **74**.

8.2.2.3 Borepin

Larger aromatic cycles such as the borepin **75** (a neutral heterocycle isoelectronic to the tropylium cation) have also been synthesized (Scheme 8.26). Deprotonation of stannate derivative **76** followed by the reaction with MeLi gives stannepin **77**. Reaction of molecule **77** with MeBBr$_2$ affords borepin **78** in good yield while, under the same conditions, MesBBr$_2$ does not afford any product probably due to the steric hindrance of the mesityl group. In this case, an alternative synthetic pathway involving a stannate–boron exchange reaction with BCl$_3$ (giving chloroborepin **79**) followed by a nucleophilic addition of MesLi has been employed to prepare mesityl-bearing protected borepin **75** [53].

Using a similar strategy, fluorescent ($\Phi = 0.70$, $\lambda_{em} = 400$ nm and $\lambda_{ex} = 260$ nm, in CH$_2$Cl$_2$) borepin **80** was also obtained in good yield (58%). After a Li–halogen exchange reaction performed with n-BuLi, compound **81** (Scheme 8.27) was reacted with Me$_2$SnCl$_2$ to afford derivative **82**, which gave arylborepin **80** after reaction with

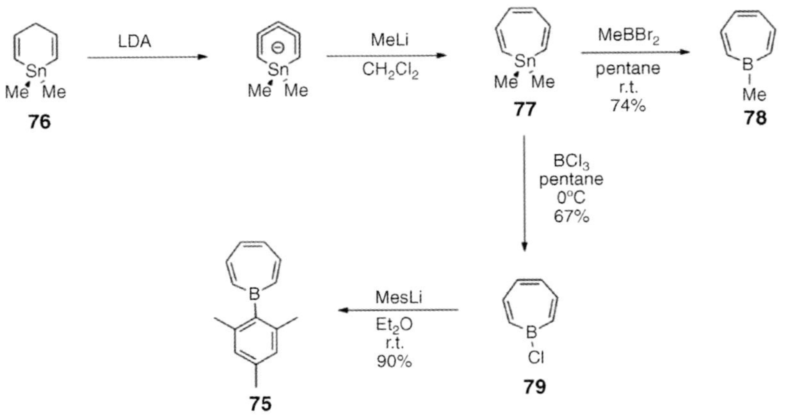

Scheme 8.26 Synthesis of methyl- (**78**) and mesityl-borepin (**75**) derivatives.

BCl$_3$ and MesLi using the same synthesis pathway as that outlined in Scheme 8.26 [54]. The conjugation of the empty p-orbital of the boron with the π-orbital reduces the Lewis acidity of compound **80** as demonstrated by competition binding NMR experiments performed with dimethylamino pyridine.

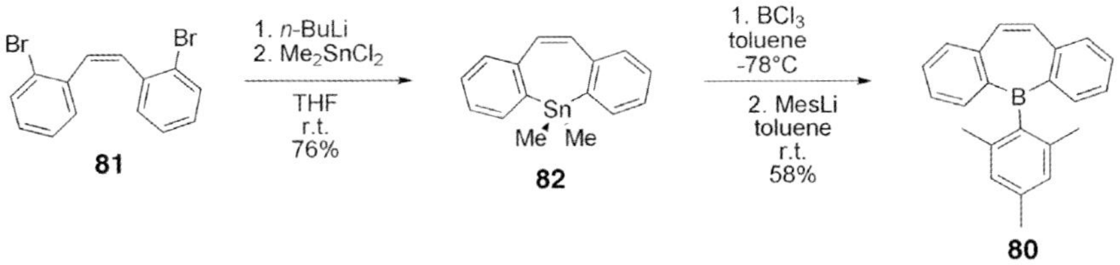

Scheme 8.27 Synthesis of aryl borepin **80**.

Finally, borepin cycles can be obtained from borole **83** (Scheme 8.28) upon thermal cycloaddition reaction with bisphenylacetylene in toluene [55]. The Diels–Alder cycloaddition of **83** with triple bond affords boron-bridged molecule **84**, which after prolonged heating, undergoes a 1,3-suprafacial-sigmatropic rearrangement followed by a 1,6-disrotatory electrocyclic ring opening to afford borepin **85** in high yield (84%) [56].

The Lewis acidity of boroles, heterocyclic analogues of the antiaromatic cyclopentadiene cation, is enhanced compared to the borepin. Their crystal structures reveal a planar ring with alternated bond length, whether this is due to either a conjugation of the empty p-orbital with the π electron or a packing effect is subject to debate [57, 58]. With the aim to investigate the Lewis acidity capabilities of highly electron-poor boron centers, perfluoroaryl borole **86** (Scheme 8.29) was also synthesized [58]. Reaction of

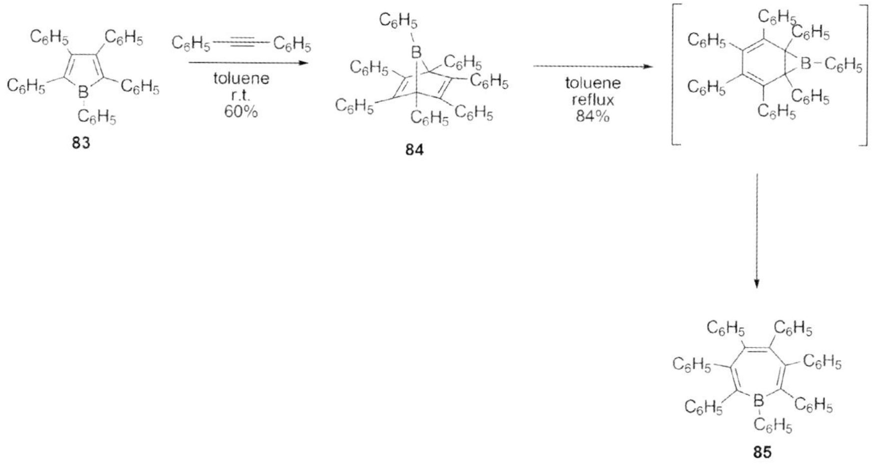

Scheme 8.28 Synthesis of borepin **85** from borole **83**.

stannate **87** in neat BBr$_3$ led to molecule **89** that, after treatment with Zn(C$_6$F$_5$)$_2$, gave rise to desired borole **86** in high yield (80%). When one equivalent of CH$_3$CN is added to the solution of borole **84** and B(C$_6$H$_5$)$_3$, it binds exclusively to borole (as determined by ^{1}H-NMR spectroscopy), providing good evidence of its dramatically enhanced Lewis acidity. Arylborole reacts with potassium in THF to form stable aromatic dianions [59].

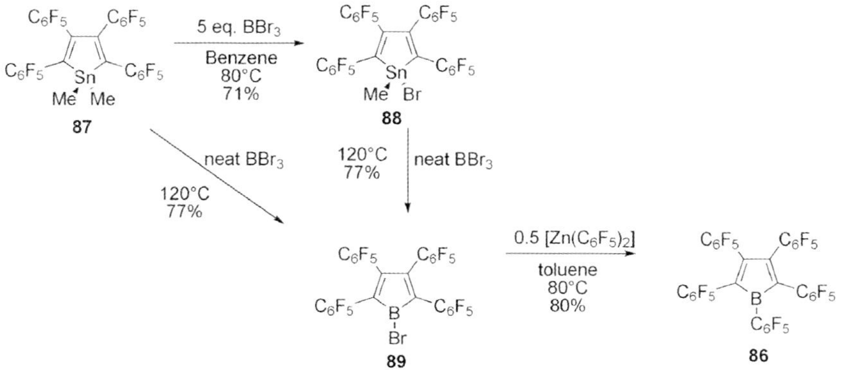

Scheme 8.29 Synthesis of perfluorinated borole **86**.

8.3
Sulfur, Selenium, and Tellurium

Till a few decades ago, among the so-called *chalcogen* atoms only sulfur was intensively utilized in organic synthesis as a versatile reactive element. Several years later, selenium and, subsequently, tellurium chemistry has rapidly developed due to the growing interest in the chemical reactivity of chalcogen derivatives. In contrast to the electron-deficient boron atom, the elements of the group VI are strong electronic donors if connected to or inserted into π-conjugated structures. This leads to unusual electronic properties if, for example, they are combined with acceptor moieties, enhancing the charge transfer properties of the molecule. Belonging to the same group as that of oxygen-, sulfur-, selenium-, and tellurium-based structures should exhibit many similar characteristics. However, significant differences exist among them. These are mainly due to the increase in the number of electrons in the outer shells and hence of the atom size, rendering the heteroatoms more polarizable. The high polarizability and the manifested electron-rich characters are responsible for the whole pattern of different electronic and structural properties among the organochalcogen compounds.

In view of supramolecular abilities of sulfur-bearing organic architectures, one of the most explored research fields is provided by the chemistry of thiophene and the series of its derivatives. In the next section, the synthesis and the properties of several thiophene-based molecules have been reviewed.

Adds two or three spatulae of sulfur
into the improvised labware,
sets fire and encases the can and the pristine sphere:
sulfur, crowns of six bonded atoms,
starts melting.
(A. Terron "à mon seul désir," El Tall, Palma de Mallorca, 2007)

Different from sulfur, the potentialities of selenium and tellurium as doping elements for supramolecular architectures have been unknown for many years due to the difficulties in finding high-yielding synthesis methods. For this reason, the employment of these two elements in organic supramolecular chemistry represents a challenge not only from the synthetic point of view but also for the discovery of new and fascinating properties. A brief overview is given in this section on the latest discoveries on selenium- and tellurium-doped supramolecular architectures, starting from macrocyclic structures and concluding with aromatic annulated frameworks.

8.3.1
Sulfur

8.3.1.1 **Oligothiophene-Based Structures**
The electron-rich character of thiophene rings makes their derivatives suitable structures for a large variety of applications in electronic devices as, for example, conducting materials [60]. Among all, poly(thiophene)s and oligo(thiophene)s have been the most widely studied because of their conducting properties as highly ordered polymers for organic electronic devices such as light-emitting diodes [61], field-effect transistors [62], and organic solar cells [63], to name a few. The versatile

synthesis of these classes of compounds has been intensively reviewed over the past decades [64], and it is not the purpose of this chapter to review them. In this section, we have focused our attention on the latest designing and synthetic developments along with the supramolecular properties of such molecules, starting from oligothiophene derivatives in which thiophene rings are connected through one or two Csp^2–Csp^2 bonds and the fundamental thiophene units are fused into rigid structures.

Oligothiophenes Connected through Csp^2–Csp^2 Bonds Among oligothiophenes, particular attention has been given to their dendritic derivatives. Advincula and coworkers described one of the first examples of the synthesis of a novel thiophene-based architecture, dendrimer **90**, composed only of thiophene units (Figure 8.16) [65]. More recently, the same group explored the properties of molecule **90** and analogues, describing their exceptional photophysical properties and its supramolecular assembly [66]. In fact, the unsymmetrical dendrimer **90** was found to show an intense broadband in the absorption spectrum, ranging from 495 nm up to the UV region (THF). The increase of the degree of generation led to a bathochromic shift of the low-energy absorption bands, showing the potential to serve as light-harvesting materials.

Together with the photophysical properties, the investigation on the self-assembly of **90** on surfaces at the solid–liquid interface evidenced the presence of noncovalent interactions controlling the formation of nanostructures of **90**. On mica surface, AFM (atomic force microscopy) measurements displayed bright spots recognized as globular aggregates (Figure 8.17a). This behavior can be explained by considering

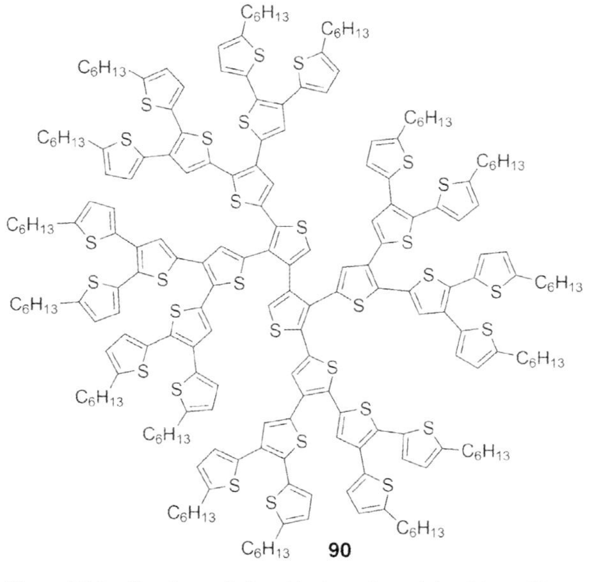

Figure 8.16 Structure of oligothiophene-based dendrimer **90**.

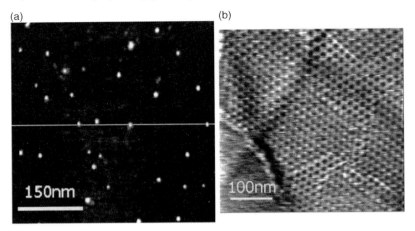

Figure 8.17 (a) AFM topographic image on mica substrate, evidencing globular aggregates of **90**; (b) STM image of the self-assembled hexagonal structures of **90** on HOPG ($I = 1nA$, $U = 0.1V$). Adapted with permission from the American Chemical Society. Copyright 2004.

the hydrophobic intermolecular interactions between the molecules on the hydrophilic substrate of mica. In addition, the assembly on HOPG (highly ordered pyrolytic graphite) was also studied by STM (scanning tunneling microscope). Figure 8.17b shows the STM image of dendrimer **90** organized in hexagonal-type motifs, where the easily discernible bright spots are assigned to the thiophene units.

Conjugated dendrimers present several advantages over the linear analogues in terms of applications in optoelectronics. In fact, the particular shape enhances the solubility minimizing π–π stacking and therefore improves the preparation of nanocomposite materials. The Advincula group further exploited a convergent synthetic strategy to prepare molecule **90**, starting from the synthesis of the dendritic substructures, called dendrons, and coupling them in the second step to form the final dendrimers (Schemes 8.30 and 8.31). In the first step, thiophene was alkylated in position 2 using *n*-BuLi and *n*-bromohexane, giving compound **91** that was subsequently brominated with NBS/DMF leading to 2-bromo-5-hexylthiophene **92**. The reaction of **92** with 2,3-dibromothiophene under Kumada reaction conditions, gave coupling product **93** in excellent yield (90%). This was allowed, then, to react with *n*-BuLi and Bu$_2$SnCl, giving intermediate **94**. Without purification, the latter was subsequently coupled with 2,3-dibromothiophene following a Stille-type reaction protocol and reacted with *n*-BuLi and Bu$_3$SnCl to give the first dendritic substructure **95**. At last, direct Stille-type coupling reaction of dendron **95** with central core **96** (synthesized from the homocoupling of two molecules of 2,3-dibromothiophene) afforded desired compound **90**, with a 58% yield (Scheme 8.31).

In parallel, Bäuerle and his group described the liquid–solid interface assembly of a cyclo[12]thiophene derivative **97** on HOPG investigated by STM technique

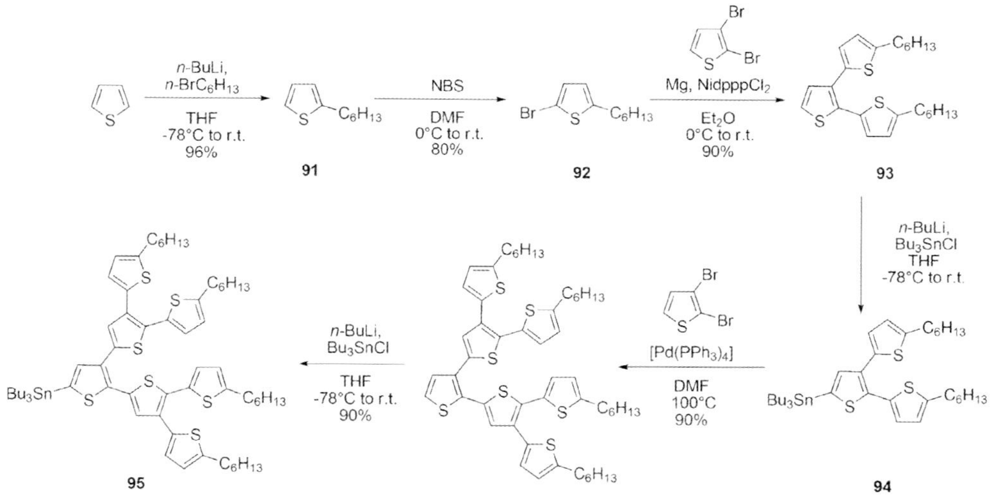

Scheme 8.30 Synthesis of dendritic substructure **95**.

(Figure 8.18) [67]. With respect to the previous case, α-conjugated macrocyclic oligothiophene can selectively nest guest molecules within their cavities. Self-assembly of macrocycle **97** on HOPG yielded hexagonal-type patterns. Moreover, the molecules cover a large surface area ($\geq 1 \times 1\ \mu m^2$) leading to the formation of a highly ordered unique domain. This approach could lead to an infinite number of

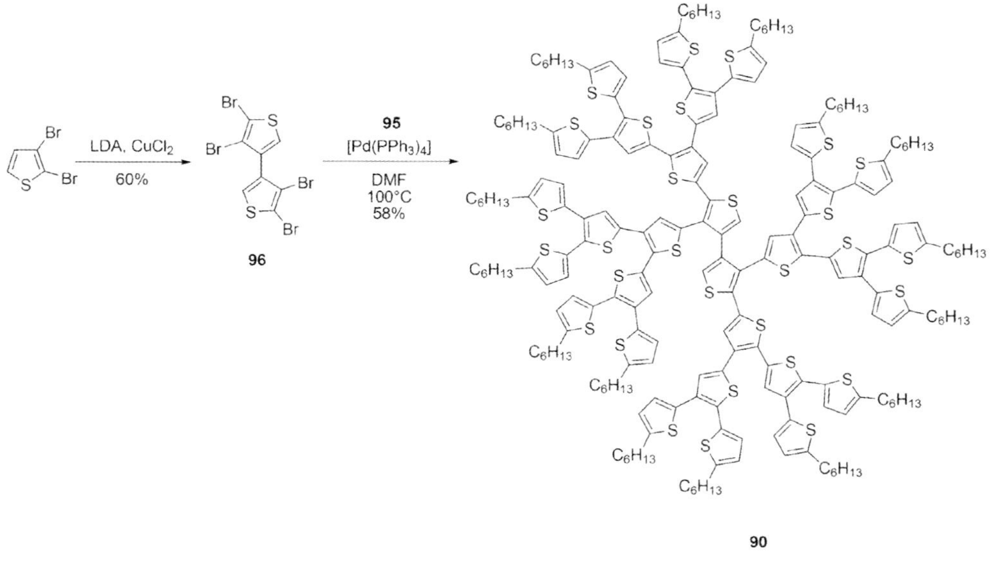

Scheme 8.31 Synthesis route leading to thiophene-based dendrimer **90**.

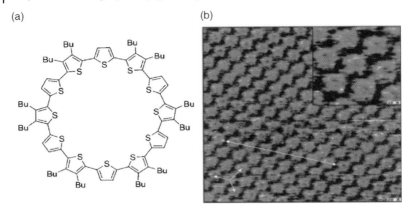

Figure 8.18 (a) Cyclo[12]thiophene compound **97**; (b) STM image of 2D hexagonal pattern of molecules of **97** adsorbed on HOPG ($I = 24$ pA, $U = -430$ mV). Adapted with permission from Wiley-VCH Verlag GmbH & Co. KGaA. Copyright 2009.

applications in molecular electronics, such as thin-film electronic devices, as it can give valuable information about intermolecular and molecule–substrate interactions [68]. In addition, the possibility of preparing other cyclo[n]thiophene compounds with $n \neq 12$, thus showing tunable cavity sizes, could let us understand how to engineer programmed molecular modules generating predictable structural assemblies with necessary electronic properties for future applications.

In view of the envisaged synthetic strategy, a combination of oligomerization steps and cyclization reactions has been employed to prepare compound **97** as displayed in Scheme 8.32. Iodination of unsubstituted tri-thiophene derivative with Hg(OAc)$_2$ and I$_2$ led to bis-iodo intermediate **98** in 85% yield. The two iodinated positions were then functionalized with TMSA (trimethylsilylacetylene) under the standard conditions of the Sonogashira cross-coupling reaction, leading to product **99**. After the quantitative cleavage of the silylated protecting group, the free acetylene intermediate underwent the macrocyclization reaction. For oxidative coupling, the authors carried out several different procedures, ranging from Glaser-type reaction to Eglington- or Hay-type coupling. None of them was found to be effective on this kind of substrate, as they obtained mixtures of cyclic and linear polythiophene aggregates.

Thus, macrocycle **101** was obtained in a low yield (2%) through a modified Eglington–Glaser procedure using a mixture of CuCl and CuCl$_2$ in pyridine at room temperature. The low yield obtained in this transformation is due to the fact that large cycles are formed in only one step involving more than one reaction in different sites of the substrate. Ultimately, molecule **101** underwent nucleophilic attacks by sodium sulfide achieving cyclo[12]thiophene product **97** with a 23% yield.

Fused Oligothiophenes The main limitation of the previously described mono-linked oligothiophenes lies in the structural deviation from planarity, which dra-

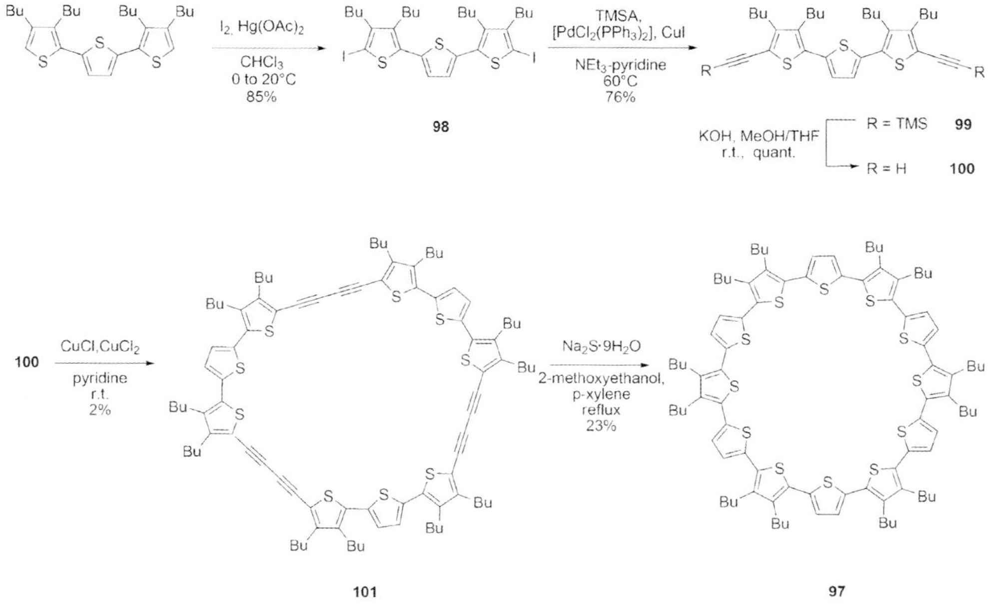

Scheme 8.32 Synthesis pathway leading to cyclo[12]thiophene **97**.

matically reduces the extension of the π-conjugation. Conversion of these structures into more rigid derivatives would make these oligothiophene derivatives more suitable as semiconductors, as the intrinsic conductivity increases with the enhancement of the π–π conjugation. Innovative structural engineering in this context would be the preparation of fused oligothiophene derivatives, in which thiopenyl units are doubly linked to neighboring rings. This would enhance the high structural planarity, producing extended flat pathways for π-conjugation, lowering the molecular HOMO–LUMO gap, and thus giving rise to unique conducting properties. In this respect, Takimiya and coworkers described the synthesis and properties of thiophene–benzene annulated acenyl-like compound, **102** (Scheme 8.33) [69]. The employed synthetic strategy involves two intramolecular cycloadditions: an electrophilic cyclization forming the intermediate with the three fused central rings, followed by a Bergman-type cycloaromatization [70], which ultimately leads to the final molecule with five fused aromatic rings. According to this strategy, the readily available 1,4-dibromo-2,5-bis(trimethylsilylethynyl)benzene reacted with *t*-BuLi and Me$_2$S$_2$ to give compound **103** in 50% yield. Subsequently, intermediate **103** underwent the first cycloaddition reaction following the Larock protocol [71] to fuse thiophene moieties to a central aromatic ring as in the benzo-dithiophene derivative **104**. The TMS (trimethylsilyl) protecting group was successfully (98% yield) substituted with iodine leading to tetraiodo conjugate **105**, which was successively coupled to TMSA under Pd(II)-catalyzed cross-coupling reaction conditions. Employing *in situ* prepared sodium telluride, a second cyclization reaction has been induced leading to the conjugate **102**, which was subsequently extended

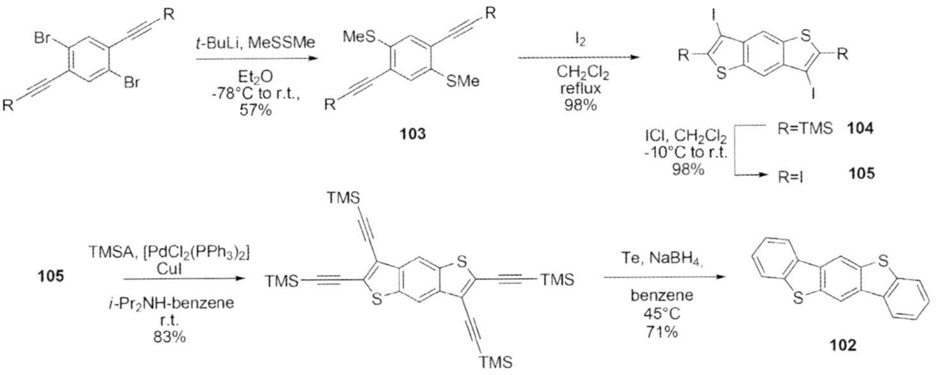

Scheme 8.33 Synthesis pathway leading to benzo[1,2-b:4,5-b']bis[b]benzothiophene **102**.

through the formation of two peripheral aromatic rings obtaining the desired molecule in very good yield, 71%, and inducing the removal of trimethylsilyl group at the same time.

The fully fused aromatic structure **102** induced, according to the authors, a slight but significant difference in the UV-Vis absorption spectrum compared to the estimated value for parent compound **104**. In particular, the whole envelope of absorption bands of the five fused aromatic rings derivative is shifted to longer wavelengths (**102**: $\lambda_{abs} = 369$ nm; **104**: $\lambda_{abs} = 335$ nm, in CH_2Cl_2). This behavior arises from an evident π-extension in the electronic structure and hence a decrease in the molecular HOMO–LUMO gap. Although the extension of planarity in polyaromatic compounds is thought to increase the π-conjugation character of molecules, exceptions are also likely. To this end, Wu and Baldridge reported the synthesis and UV-Vis characterization of a twisted trithiophene derivative, molecule **106**, with three methyl-substituted thiophene rings connected to a central benzene core (Scheme 8.34) [72]. The presence of neighboring methyl substituents does not allow the molecule to adopt a planar conformation, and a strong distortion from the planarity forcing the methyl moieties in a pseudo-anti position has been observed upon the crystal structure analysis (Figure 8.19). Surprisingly, the UV absorption spectrum of molecule **106** in CH_2Cl_2 shows UV-centered electronic transition with an absorption trend very similar to that of unsubstituted precursor **107** (**106**: $\lambda_{abs} = 264$ nm, log $\varepsilon = 4.91$; **107**: $\lambda_{max} = 248$ nm, log $\varepsilon = 4.88$), supporting the

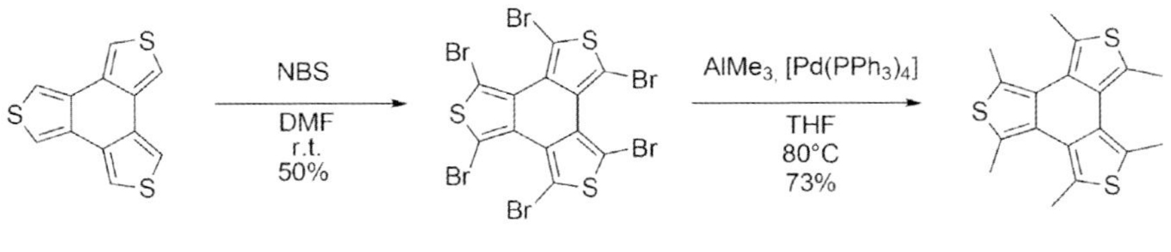

Scheme 8.34 Preparation of hexamethylbenzo[1,2-c:3,4-c':5,6-c'']trithiophene **106**.

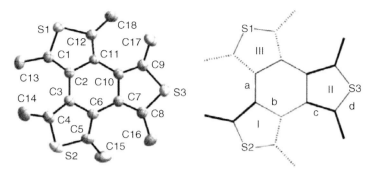

Figure 8.19 Crystal structure showing the twisted angles adopted by the fused thiophene rings in molecule **106**. Adapted with permission from The Royal Society of Chemistry. Copyright 2009.

hypothesis that the poor conjugation between the thiophene rings and the weak electron-donating effect of the methyl groups are responsible for this behavior.

The three-step synthetic procedure for preparing trithiophene molecule **106** is outlined in Scheme 8.34. Starting from benzo[1,2-c:3,4-c′:5,6-c″]trithiophene **107**, bromination reaction using an excess of NBS in DMF led to hexabrominated intermediate **108**, which was subsequently methylated with AlMe$_3$ exploiting Pd (0)-catalyzed coupling reaction giving **106** as desired product in 73% yield.

In view of the new fused oligothiophene derivatives, much attention is now being given to polythiacirculene as a new class of compounds. They consist of annelated thiophenic rings with the formula $(C_2S)_n$ for any number of rings (n), and for this reason they are classified as a novel form of carbon disulfide. A great contribution to this field has come form Dopper and Wynberg in the early 1970s [73]. They synthesized the first examples of S-containing heterocirculenes, reporting their first photophysical properties. Recently, the development of these molecules has increased due to their great potential applications in the entire field of electronics [74]. In this respect, the Nenajdenko group was the first to describe the synthesis of octathio[8]circulene **109** (Scheme 8.35) [75], renamed as "sulflower" referring to its similarity in shape to a flower. The group developed a short and effective strategy to get their compound, starting from tetrathiophene **110**, which had already been synthesized by the same group. Treatment of precursor **110** with 16 equivalents of LDA and elemental S at room temperature, followed by acidification with an aqueous

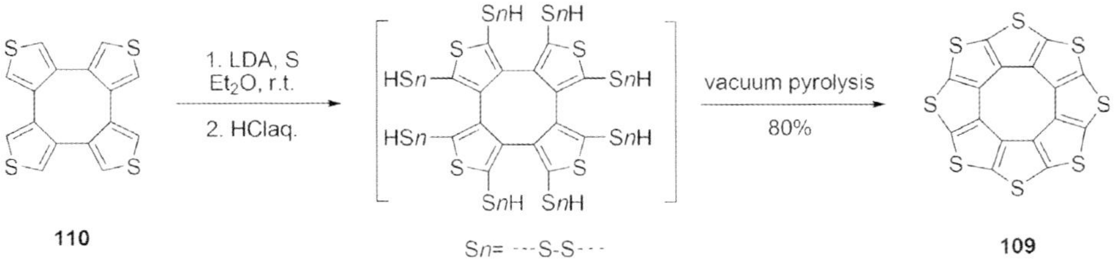

Scheme 8.35 Synthesis of "sulflower" **109**.

solution of HCl, led to the precipitation of a polythiol intermediate. The precipitate was subjected to vacuum pyrolysis, giving the desired octathio[8]circulene **109** with an overall yield of 80%.

The excellent stability, together with the great charge carrier mobility shown by these molecules, prompted the same group to investigate, a few years later, the properties of thin films of heterocirculene **109**, as conducting materials [76]. In solid state, these systems are packed in π-stacked columns showing a strong charge transport within the column but little or no electronic communication between the columns. With the introduction of heavier atoms, such S, in the circulene core, greater interaction between columns is provided by close S⋯S contacts (3.25 Å) improving the 3D charge transport. Considering these interactions, they fabricated one of the first models of thin-film transistor of **109** deposited on Si/SiO$_2$ substrate prepatterned with Au circular electrodes, studying the charge mobility as a function of the morphology of these systems on the surface. The performance of the resulting organic field effect transistor (OFET) involving the "sulflower" as organic semiconductor, was measured in a N$_2$-saturated atmosphere. The **109**-based thin film showed a high hole field effect mobility and on/off current ratio of about $9 \times 10^{-3}\,cm^2\,V^{-1}\,s^{-1}$ and 10^6, respectively. Moreover, a higher threshold voltage, from -40 to $-46\,V$, was observed for this device. These phenomena could be explained considering the very low HOMO of **109** ($-5.7\,eV$) and the resulting hole–injection barrier at the Au electrode. In parallel, the analysis by AFM measurements under vacuum conditions, at different times, has been performed to study the morphology of the system (Figure 8.20). In the films, the authors observed one-dimensional deposition of "sulflower" on the substrate, which should suggest good ability in charge mobility (Figure 8.20a). Increasing the surface coverage, the thin films display needle-like arrangements, keeping up growing in one dimension (Figure 8.20b) until they reach the point in which the surface is fully covered and the linear disposition is no longer accessible. Therefore, a vertical growth, perpendicular to the substrate, is observed. This, unfortunately, prevents the formation of long conducting films, causing serious limitations to the charge mobility of **109** (Figure 8.20c). Mixed selena-thiacirculene derivatives have been also synthesized as reported in Section 8.4.

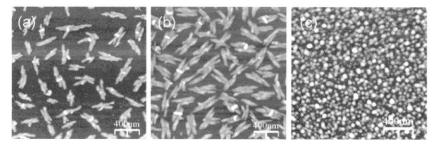

Figure 8.20 AFM images showing the growth of thin films of **109** on Si/SiO$_2$ substrate at different times (a) 1 min, (b) 2 min, and (c) 10 min. Adapted with permission from The Royal Society of Chemistry. Copyright 2008.

8.3.2
Selenium and Tellurium

8.3.2.1 Selenium-Doped Macrocyclic Structures

Over the past decades, a great attention has been given to the chemistry of macrocyclic architectures for their receptor properties [77]. Among these, carbon-bridged calixarene [78] and cyclophane [79] derivatives are well known in the literature. Despite this fact, only few examples of heterocalixarene [80] and heterocyclophane [81] compounds have been reported, presenting S and Se atom centers inserted in the framework. One of the latest example of synthesis of heterocalixarenes has been presented by the groups of Smet and Dehaen [82], which reported the synthesis of several new homoselenacalix[n]arenes (n = 3–7). Although the properties of these structures have not yet been reported, these molecules could potentially act as molecular receptors to electronically deficient molecular guest due to the donating character of the chalcogen atoms together with the π-interactions provided by the aromatic units. Reaction of 1,3-bis(bromomethyl)-5-t-butyl-2-methoxybenzene **111** with one equivalent of a freshly prepared solution of NaSeH in THF yielded a complex mixture of five different selenacalixarene derivatives, **112a–e**, in a 37 : 20 : 14 : 8 : 7 ratio, respectively (Scheme 8.36, path (a)). Synthesis of small-ring selenacalixarene compounds, such as macrocycle **112b**, could be selectively achieved via a [2 + 2] macrocyclization mechanism. In this case, the reaction of compound **111** in the presence of bis-selenocyanate compound **113** reduced *in situ* with NaBH$_4$, under high dilution conditions, afforded selenacalixarene **112b** in 67% yield mixed with **112d** (18% yield) and **112e** (9%) (Scheme 8.36, path (b)).

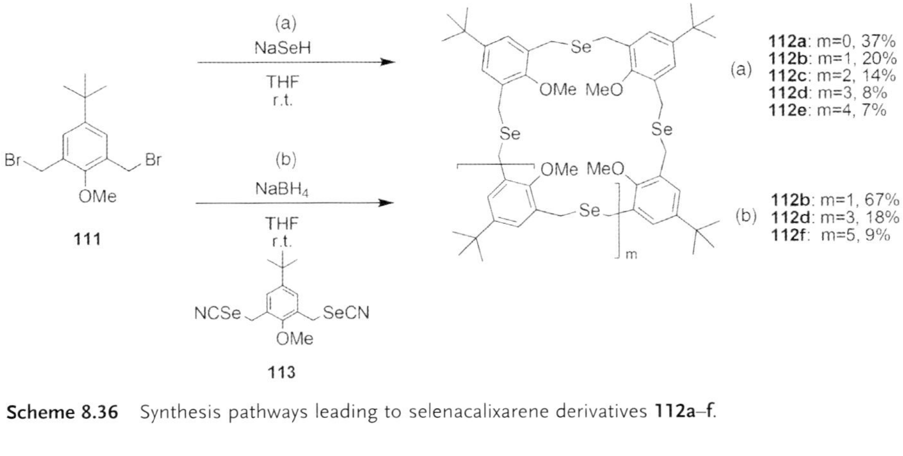

Scheme 8.36 Synthesis pathways leading to selenacalixarene derivatives **112a–f**.

Another interesting contribution to the chemistry of Se-containing macrocyclic molecules has been made by Gleiter and coworkers [83]. In particular, they reported the synthesis of macrocyclic cyclophanes **114a–d**, consisting of two substituted 1,4-diethynylphenyl moieties, linked to two selenium-bearing alkylic chains. The synthesis pathway is outlined in Scheme 8.37. Selenacyclophanes **114a–d** were obtained from the reaction between 1,4-diethynyl-2,5-di(*n*-propyl)benzene **115**, after deprotonation with LiHMDS, and diselenocyanato alkylic compounds **116a–d**.

115

116a: n=3
116b: n=4
116c: n=5
116d: n=6

114a: n=3, 4%
114b: n=4, 3%
114c: n=5, 9%
114d: n=6, 3%

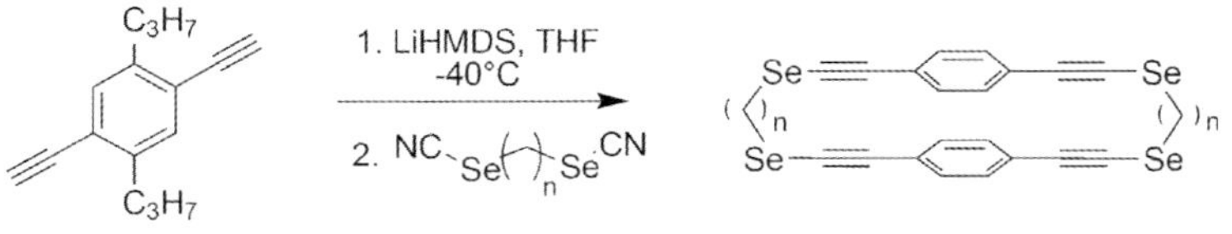

Scheme 8.37 Synthesis of selenacyclophane compounds **114a–d**.

As in the case of molecules **112a–e**, no investigation on electronic properties has been reported so far, apart from the structural studies via X-ray analysis. These molecules showed the presence of different molecular packing modes, depending on the length of the spacing alkylic chain between the selenium atoms. Moreover, close Se · · · Se interactions giving rise to unusual 3D arrays in the crystal are observed. Figure 8.21 shows the crystal packing of molecule **114c** ($n = 5$). The macrocycle creates in the solid state the columnar structures formed by the π–π interactions between the aromatic rings and by Se · · · Se vertical contacts. In addition, intercolumnar Se · · · Se contacts, displaying distances of about 4.015 Å, have also been observed.

The Se · · · Se interactions at the solid state were already observed by the same authors few years earlier in a 3D crystal organization [84]. In this case, they took into account the self-organization of Se-doped cyclic alkynes, and they observed, through X-ray analysis, a similar behavior to that shown by **114c**, with the cyclic alkynes arranged in columnar stacks closely packed via Se · · · Se short inter- and intracolumn contacts. Concerning the preparation (Scheme 8.38) [85], among the large number of cyclic alkynes they synthesized, the best result obtained in terms of yield of the final molecule was achieved with symmetrical compound **117** (Scheme 8.38), following the straightforward protocol similar to that illustrated for compounds **114**.

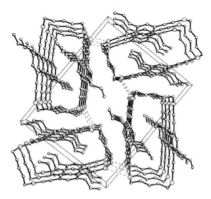

Figure 8.21 Crystal packing of macrocycle **114c** at the solid state. Adapted with permission from the American Chemical Society. Copyright 2008.

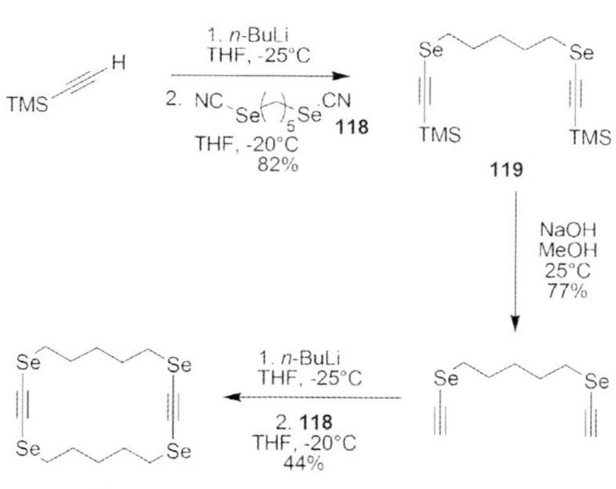

Scheme 8.38 Synthesis of tetraselenadiyne macrocycle **117**.

Treatment of TMSA with *n*-BuLi and subsequent coupling with diselenocyanatopentane derivative **118** afforded intermediate **119** in very good yield (82%). Cleavage of the TMS protecting groups under basic conditions followed by a second nucleophilic addition of lithium acetylides to a second equivalent of diselenocyanatopentane under dilution conditions yielded desired macrocycle **117** (44%).

8.3.2.2 Selenium- and Tellurium-Annulated Aromatic Structures

Besides the electronic properties, chalcogen-doped π-conjugated heteroarene structures are of special interest in supramolecular chemistry as they can display a variety of intermolecular recognition interactions such as heteroatom–heteroatom interactions, which can template and direct the formation of self-assembled structures displaying important optoelectronic properties applicable to various electronic devices, such as OLEDs, solar cells, and NLO materials. One of the main interesting approaches of replacing an aromatic carbon atom with Se and/or Te atoms was reported several years ago by the group of Suzuki and Ohta [86]. Their strategy dealt with the preparation of a reactive homogeneous mixture, formed by the addition of Na_2Te and CuI in 1:2 ratio to NMP (*N*-methylpyrrolidone). This black dense solution, acting as a mild telluration agent, reacted at high temperature (150 °C) with diiodo precursor **120** (Scheme 8.39) leading to tellura-derivatized intermediate **121** (from 31 to 52% yields). Finally, molecule **121** underwent a mono reduction in the presence of $H_2NNH_2 \cdot H_2O$ in a solution of EtOH and THF, isolating the desired dibenzotelluraphene product **122** with a satisfactory yield (55%). This strategy was also employed

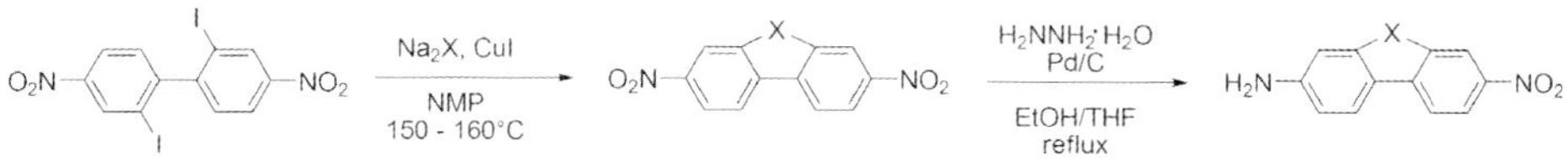

Scheme 8.39 Synthesis of dibenzotelluraphene **122** and dibenzoselenaphene **123** derivatives.

by the same group to synthesize dibenzoselenaphene derivative **123**. In order to explore the effect of the heavier atoms on the NLO properties, measurements of second-order and third-order NLO properties have been performed. In particular, high values for the second-order molecular hyperpolarizability (1.4×10^{-31} and 1.0×10^{-31} esu for **121** and **122**, respectively) were measured for both compounds (the reference C-based atom displayed a value for γ equal to 5.1×10^{-32} esu), thus confirming the positive effect of the heavy atom on the NLO properties.

Following the same trend, Wang and coworkers recently investigated the effect of elemental Se on perylene-type structures [87]. They set up a one-pot synthetic methodology based on the incorporation of Se into the nitro-substituted perylene scaffolds (Scheme 8.40). Hence, starting form 1-nitroperylene **124**, obtained by a nitration reaction of an unsubsustituted perylene molecule, perylo[1,12-*b,c,d*]selenaphene **125** was obtained in good yield (65%) in NMP at 180 °C. The same procedure was applied to synthesize the parent sulfur derivative, perylo[1,12-*b,c,d*]thiaphene **126**.

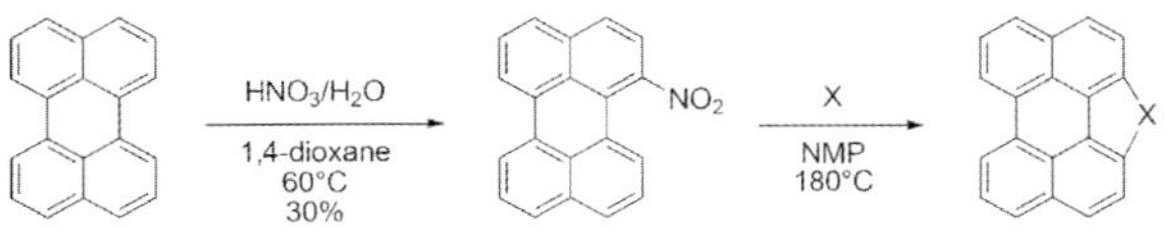

Scheme 8.40 Two-step synthesis of selena- and thiaperylene compounds **125** and **126**.

In addition to the easy synthesis reported in this work, UV-Vis measurements of **125** and similar compounds such as **126**, and perylene used as reference compound, were carried out. Specifically, a weak bathochromic shift was measured going from perylo[1,12-*b,c,d*]thiaphene to its Se-doped derivative **125** (**126**: $\lambda_{abs} = 412.0$ nm; **125**: $\lambda_{abs} = 416.5$ nm, in CHCl$_3$), in agreement with the higher electron-donor character exhibited by the selenium atom. On the contrary, these absorption values were found to be slightly blueshifted compared to the value measured for the unsubstituted perylene ($\lambda_{abs} = 438.5$ nm, in CHCl$_3$). This behavior can be explained considering the aromatic character of the latter compared to its parent compound. A similar synthetic procedure was exploited by the Takimiya group to build novel promising optoelectronic materials such as benzo[*b*]thiaphene and benzo[*b*]selenaphene derivatives [88]. For instance, addition of selenium powder to a solution of a bromo-substituted trimethylsilylethynylphenyl derivative in NMP led to the halogen–chalcogen exchange reaction followed by the nucleophilic attack of Se on the triple bond and the *in situ* cleavage of the TMS protecting group that afforded the desired benzo[1,2-*b*:3,4-*b'*-5,6-*b''*]triselenaphene **127**, as depicted in Scheme 8.41.

A different synthetic approach for introducing heavier chalcogen atoms into aromatic backbones was reported by the group of Bendikov. In two distinct works, the group employed a new one-pot synthetic methodology to prepare 3,4-dimethox-

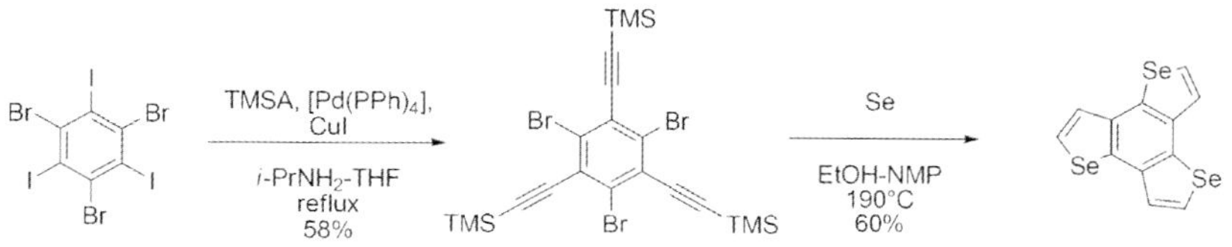

Scheme 8.41 Synthesis of triselena-annulated compound **127**.

yselenaphene **128** [89] and 3,4-dimethoxytelluraphene **129** [90]. The synthesis of these molecules is outlined in Scheme 8.42. The authors envisaged the use of 2,3-dimethoxy-1,3-butadiene **130** as precursor material. Thus, butadiene derivative **130** reacted separately, in the presence of NaOAc in hexane, with SeCl$_2$ at −78 °C and with freshly prepared TeCl$_2$ at room temperature affording desired products **128** and **129**, in 42 and 5% yield, respectively. As in the case of selenacyclophane architectures (see Section 8.2.1.1), X-ray crystal analysis showed that the five-member rings intermolecularly interact with each other via Se···Se or Te···Te contacts (heteroatom distances are 3.80–4.04 Å and 3.78 Å for **129** and **128**, respectively), showing the shorter distances for the Te···Te. Despite the large amount of publications based on thiophene compounds, involving oligo and polythiophenes as described in Section 8.2.1.1, the electronic and supramolecular arrangements of selena and tellura parent monomers along with their polymeric derivatives have not yet been fully explored, and it is still a topic under investigation.

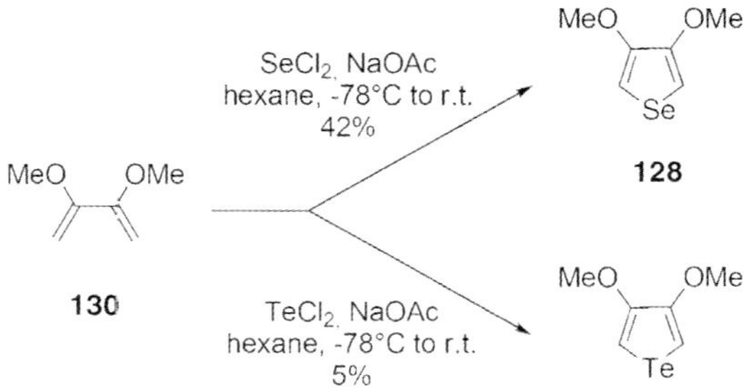

Scheme 8.42 Synthesis of 3,4-dimethoxyselenaphene **128** and 3,4-dimethoxytelluraphene **129**.

8.4
Miscellaneous

In this section, examples of miscellaneous molecules containing multinuclear heteroatoms in the same fundamental unit are described. The first example, which merits particular attention, is of 2,2′-bis(1,3-dithiolylidene), commonly called tetrathiofulvalene (TTF). Although the discovery of the first TTF-based organic conductor dates back to 1972, the chemistry of this heteroaromatic compound is in continuous expansion and many TTF derivatives have been synthesized in the last few years [91]. The unusual great stability together with electron-donating capabilities have made such sulfur-doped scaffolds exceptionally functional molecules for many applications, mainly for organic conductors in donor–acceptor π-conjugated systems and the interested reader may refer to the specialized literature [92]. Despite the numerous reports, only a few examples are reported in which the redox abilities of such compounds are responsible of molecular motions in supramolecular architectures. One of the latest research in this area is that of Azov and coworkers [93] who described the synthesis of redox-switching TTF-based molecular tweezers (**131a–c**, Scheme 8.43). In order to increase the solubility properties, the TTF boards were peripherally functionalized with bulky 3,5-di-*t*-butylbenzylthio groups. Thus, as reported in Scheme 8.43, starting material **132**, obtained from bis(tetraethylammonium)-bis(1,3-dithiole-2-thione-4,5-dithiol) zinc complex **133** and 3,5-di-*t*-butylbenzylbromide in 90% yield, was treated with oxo-derivative **134** and P(OEt)$_3$, leading to molecule **135**, key intermediate in the synthesis of **131a–c**. Subsequently, molecule **135** was reacted with the differently substituted aromatic methylene bromides **136a–c**, yielding final molecules **131a–c**. Two types of molecular tweezers, **131a** and **131c**, consisting of benzene and naphthalene spacing cores, were synthesized so that the two TTFs were differently spaced in the molecule. ^{1}H-NMR-based titration experiments using TCNQ (tetracyanoquinodimethane), TNF (2,4,7-trinitro-9-fluorenylidenemalonitrile), and TCNB (tetracyanobenzene) as guest molecules in CHCl$_3$ showed fast host–guest exchange equilibrium. For TNF association, constants of 16 and 22 M^{-1} for **131a** and **131c**, respectively, were measured, whereas none or weak association (**131c**: $K_a = 6M^{-1}$) was found with TCNB or TCNQ, suggesting weak interactions. However, it is thought that over a redox stimulus the conformation of the molecule, which should be in a "closed-type" shape in the neutral and monooxidized state, undergoes an opening process upon double electrochemical oxidation reaction as a consequence of the coulombic repulsions between the positive-charged TTF moieties (see Figure 8.22).

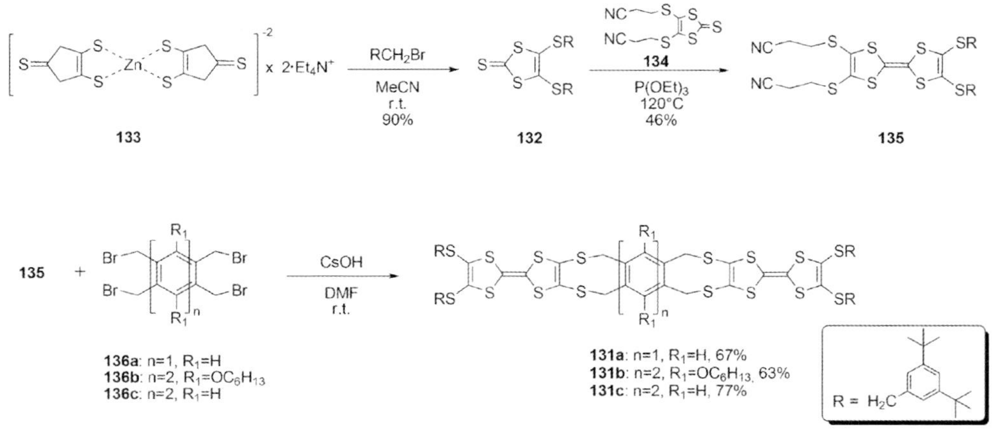

Scheme 8.43 Synthesis pathway to molecular tweezers **131a–c**.

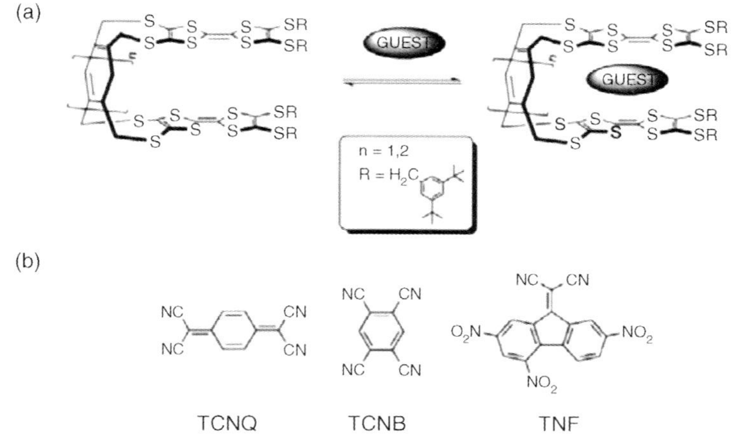

Figure 8.22 (a) General scheme for the host–guest complexation equilibrium involving TTF-based molecular tweezers and (b) electronically deficient guest molecules.

With respect to molecule **109**, the same group reported the synthesis of a hybrid version of the flower-type structure called "selenasulflower" (molecule **137** in Figure 8.23a) [75, 94]. The same synthetic methodology as that reported for molecule **109** (Scheme 8.35) was employed for the synthesis of tetrathiatetraselenacirculene **137**, using Se powder at the place of elemental S (yield 50%). Codeposition of 1,3,5-tricarboxylic acid (TMA) and circulene **137** on HOPG surface featured the formation of hexagonal-like pores (formed by the intermolecular H-bonding interactions between the octanoic acid and the TMA) filled by molecules of selenacirculene (Figure 8.23b) at the solid–liquid interface (solvent: octanoic acid), as imaged by STM. Figure 8.23c also shows a high-resolution STM image of TMA-**137** interaction on HOPG.

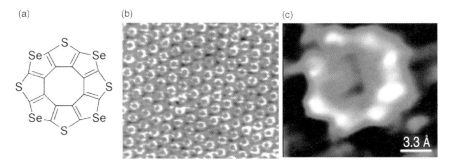

Figure 8.23 (a) Structure of tetrathiatetraselenacirculene **137**; (b) STM image of a **137**-TMA host–guest interaction on HOPG ($I = 0.3$ nA, $U = 0.3$ V); (c) high-resolution STM image of **137**-TMA host-guest interaction on HOPG ($I = 0.3$ nA, $U = 0.6$ V). Adapted with permission from The Royal Society of Chemistry. Copyright 2009.

Another interesting example of fused aromatic rings containing sulfur and selenium atoms was presented by the Yamaguchi group, which reported the synthesis [95] and the electronic properties [96] of S- and Se-doped acene-like structures. The path leading to the synthesis of derivative **138** is shown in Scheme 8.44. Following an elegant one-pot reaction protocol, the reaction of *t*-BuLi with bis(2-bromobenzothiophen-1-yl)diacetylene precursor **139**, followed by the sequential addition of Se and aq. NaOH and [$K_3Fe(CN)_6$], gives rise to product **138** in good yield. This remarkable transformation involves sequential Br/Li and then Li/Se exchange leading to the diselenolate intermediate **140**. Each of the two selenolate functions adds regioselectively to the C–C triple bonds attached to thiophene to produce, by cyclization, a fused selenophene ring bearing a carbanion at position 3 of the newly-formed ring. This then reacts with the excess of selenium present in the medium to generate molecule **141**. Potassium ferricyanate achieves the oxidative step leading to diselenide **138**. This adduct undergoes a Li/Na exchange and is finally oxidized by [$K_3Fe(CN)_6$], leading to the final molecule **138**. The UV-Vis characterization of this compound revealed an absorption spectrum featuring a well-resolved vibronic structure that covers a large wavelength range of absorption ($\Delta\lambda_{abs} = 130$ nm) due to the high rigidity of the molecule.

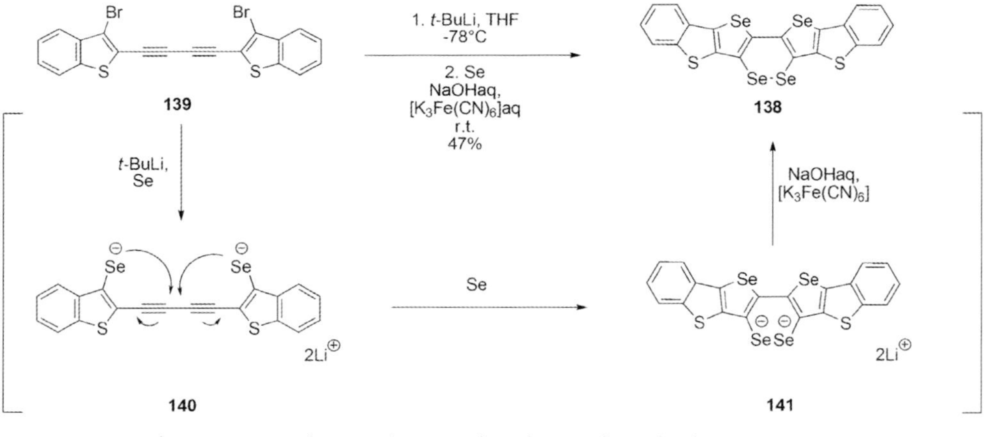

Scheme 8.44 Synthesis and supposed mechanism for molecule **138**.

An appealing aspect is provided by the introduction of a chalcogen atom into electron-deficient architectures, such as boron-doped aromatic structures [97]. Here, we present two examples of mixed molecules characterized by the presence of both boron and chalcogen atoms. Kawashima and his group recently reported the synthesis and photophysical characterization of dibenzochalcogenaborins, analogues of anthracenyl derivatives bearing a boron and a chalcogen atom in positions 9 and 10, respectively, [98]. The synthesis leading to dibenzochalcogenaborins **142a–c** is presented in Scheme 8.45. The classical reaction of *t*-BuLi with ether **143a**, thioether **143b**, and selenoether **143c**, subsequently followed by an intramolecular cyclization reaction in the presence of dimethyl mesitylboronate, used as boron protecting group, afforded the three desired dibenzochalcogenaborins, **142a–c** (with a yield of 39, 23, and 24%, respectively).

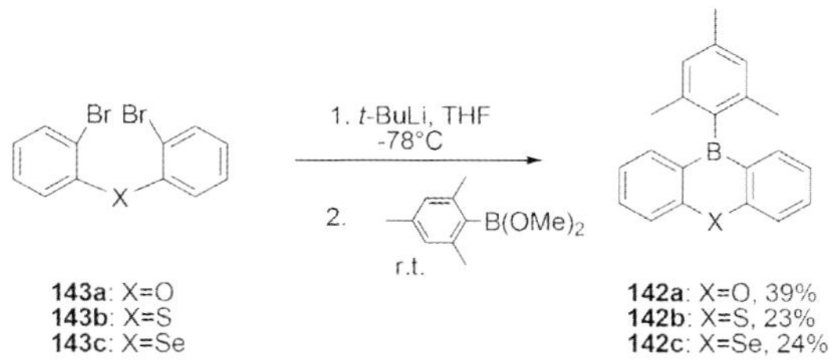

Scheme 8.45 One-pot reaction to produce chalcogenaborins **142a–c**.

The UV-Vis and fluorescence spectra of the dibenzochalcogenaborins are shown in Figure 8.24a and b. The absorption spectra revealed the presence of two distinct electronic transition bands for each compound. The longer wavelength bands are assigned to an intramolecular charge transfer state, which correspond to the ground-state charge separation occurring between the chalcogen atom and the boron atom. Passing from **142a** to **142b** to **142c**, a redshift behavior is observed evidencing a "heavy atom" effect, which decreases the molecular HOMO–LUMO gap. The high-energy absorption bands, which correspond to electronic transition centered on the aromatic ring, also present a bathochromic shift (**142a**: $\lambda_{abs} = 447$, 320 nm; **142b**: $\lambda_{abs} = 382$, 326 nm; **142c**: $\lambda_{abs} = 392$, 331 nm; in cyclohexane). The same trend is observed in the fluorescence spectra (Figure 8.11b, **142a**: $\lambda_{em} = 377$ nm; **142b**: $\lambda_{em} = 412$ nm; **142c**: $\lambda_{em} = 427$ nm; in cyclohexane), again reflecting the chalcogen atom effect.

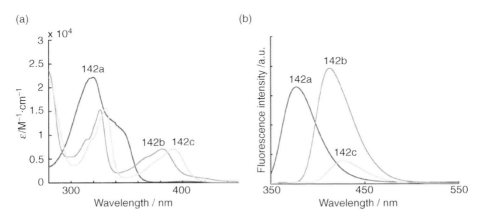

Figure 8.24 (a) UV-Vis absorption and (b) fluorescence spectra of chalcogenaborins **142a-c** in cyclohexane at 298 K. Adapted with permission from the Wiley-VCH Verlag GmbH & Co. KGaA. Copyright 2009.

Another contribution describes an interesting chemistry of diborin derivatives. Among them, the interesting dianion of 1,2-dihydro-1,2-diborin has to be noticed [99]. The synthesis of fused 1,2-dihydro-1,2-diborin starts from the dibromobithiophene **144** (Scheme 8.46) that reacts with the dibromoborane derivative to form fluorescent ($\Phi = 0.16$, $\lambda_{em} = 465$ nm and $\lambda_{abs} = 386$ nm, in THF) polycyclic derivative **145**. After reduction with K, the absorption spectra show a large bathochromic shift dependent on the counter cation ($\lambda_{abs} = 561$ nm with Na and $\lambda_{abs} = 490$ nm with Li, in THF) as observed in the case of the borastilbene **55**. The crystal structure and the calculated NICS parameters (-7.9 and -3.2 ppm for diborin and thiophene rings, respectively) suggest a peripheral ring current rather than a classical benzene-type aromaticity [100].

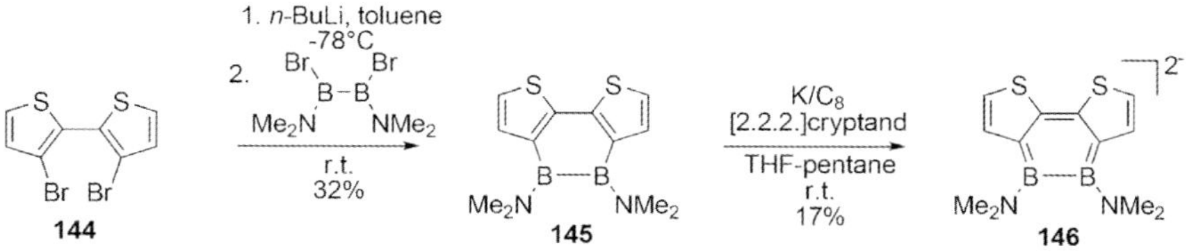

Scheme 8.46 Synthesis of the polycyclic S-doped diboron derivative **146**.

8.5
Conclusions

The effect of introducing electron-deficient, B, and electron-rich elements, S, Se, and Te, into carbon-based architectures has been reviewed in this chapter. Representative synthetic methodologies have been described, highlighting the relationship between structural design and electronic and functional properties. Specifically, the insertion of boron is manifested by marked changes in the electron-accepting properties of the molecule, evidenced by a pronounced tuning of the photophysical behavior as it can accommodate anionic species or accepting charges from electron-donating bonding units, ultimately giving rise to intramolecular charge transfer phenomenon, often leading to highly fluorescent materials. Taking advantage of these properties, boron-containing organic architectures turned out to be promising candidates as molecular sensors, organic materials for NLO and OLED applications. The replacement of carbon atom by S, Se, or Te not only affects the HOMO energy level rendering the molecule electronically rich but also can lead to new supramolecular architectures exploiting *chalcogen–chalcogen* interactions. In this respect, and different from S, the potentialities of Se and Te as doping elements for supramolecular architectures have been unknown for many years due to the difficulties in finding high-yielding synthesis methods. For this reason, the employment of these two elements in organic supramolecular chemistry represents a challenge not only from the synthesis point of view but also for the discovery of new and fascinating properties. Although no final system for any applications has been created, yet these examples illustrate how carrying out innovative and imaginative research only by replacing C-atoms with heteroatom centers has increased in its complexity creeping ever closer to the ultimate goal of producing real-world technologies, showing its limits only in the scientist's imagination.

References

1 Kroto, H.W., Heath, J.R., O'Brien, S., Curl, R.F., and Smalley, R.E. (1985) *Nature*, **318**, 162.
2 Bonifazi, D., Enger, O., and Diederich, F. (2007) *Chem. Soc. Rev.*, **36**, 390; Singh, P., Campidelli, S., Giordani, S., Bonifazi, D., Bianco, A., and Prato, M. (2009) *Chem. Soc. Rev.*, **38**, 2214.
3 Kostarelos, K., Bianco, A., and Prato, M. (2009) *Nat. Nanotechnol.*, **4**, 627; Prato, M., Kostarelos, K., and Bianco, A. (2008) *Acc. Chem. Res.*, **41**, 60; Tasis, D., Tagmatarchis, N., and Bianco, A. (2006) *Chem. Rev.*, **106**, 1105.
4 Diederich, F. and Gomez-Lopez, M. (1999) *Chem. Soc. Rev.*, **28**, 263; Guldi, D.M., Illescas, B.M., and Atienza, C.M. (2009) *Chem. Soc. Rev.*, **38**, 1587; Guldi, D.M., Rahman, G.M.A., and Zerbetto, F. (2005) *Acc. Chem. Res.*, **38**, 871; Thilgen, C., and Diederich, F. (2006) *Chem. Rev.*, **106**, 5049.
5 Grimes, R.N. (1992) *Chem. Rev.*, **92**, 251; Jemmis, E.D. and Jayasree, E.G. (2003) *Acc. Chem. Res.*, **36**, 816; Kaim, W. and Hosmane, N.S. (2010) *J. Chem. Sci.*, **122**, 7; Miyaura, N. and Suzuki, A. (1995) *Chem. Rev.*, **95**, 2457.
6 Entwistle, C.D. and Marder, T.B. (2004) *Chem. Mater.*, **16**, 4574; Wang, S.N. (2001) *Coord. Chem. Rev.*, **215**, 79.
7 Loudet, A. and Burgess, K. (2007) *Chem. Rev.*, **107**, 4891; Gilles, U., Raymond, Z., and Anthony, H. (2008) *Angew. Chem. Int. Ed. Eng.*, **47**, 1184.
8 Liu, X.Y., Bai, D.R., and Wang, S.N. (2006) *Angew. Chem. Int. Ed. Eng.*, **45**, 5475.
9 Jia, W.L., Feng, X.D., Bai, D.R., Lu, Z.H., Wang, S., and Vamvounis, G. (2005) *Chem. Mater.*, **17**, 164.
10 Dong-Ren, B., Xiang-Yang, L., and Suning, W. (2007) *Chem. Eur. J.*, **13**, 5713.
11 Pron, A., Zhou, G., Norouzi-Arasi, H., Baumgarten, M., and Mullen, K. (2009) *Org. Lett.*, **11**, 3550.
12 Stahl, R., Lambert, C., Kaiser, C., Wortmann, R., and Jakober, R. (2006) *Chem. Eur. J.*, **12**, 2358.
13 Zhou, G., Baumgarten, M., and Mullen, K. (2008) *J. Am. Chem. Soc.*, **130**, 12477.
14 Yamaguchi, S., Akiyama, S., and Tamao, K. (2000) *J. Am. Chem. Soc.*, **122**, 6335.
15 Sun, Y., Ross, N., Zhao, S.-B., Huszarik, K., Jia, W.-L., Wang, R.-Y., Macartney, D., and Wang, S. (2007) *J. Am. Chem. Soc.*, **129**, 7510.
16 Sole, S. and Gabbai, F.P. (2004) *Chem. Commun.*, 1284; Hudnall, T.W., Chiu, C.-W., and GabbaÎ, F.O.P. (2009) *Acc. Chem. Res.*, **42**, 388; Hudson, Z.M. and Wong, S. (2009) *Acc. Chem. Res.*, **42**, 1584.
17 Hoefelmeyer, J.D. and Gabbai, F.P. (2002) *Organometallics*, **21**, 982.
18 Kim, Y., Zhao, H., and Gabbai, F.P. (2009) *Angew. Chem. Int. Ed. Eng.*, **48**, 4957.
19 Hudnall, T.W. and Gabbai, F.P. (2007) *J. Am. Chem. Soc.*, **129**, 11978.
20 Levonis, S.M., Kiefel, M.J., and Houston, T.A. (2009) *Chem. Commun.*, 2278.
21 El-Kaderi, H.M., Hunt, J.R., Mendoza-Cortes, J.L., Cote, A.P., Taylor, R.E., O'Keeffe, M., and Yaghi, O.M. (2007) *Science*, **316**, 268.
22 Mastalerz, M. (2008) *Angew. Chem. Int. Ed. Eng.*, **47**, 445.
23 Wan, S., Guo, J., Kim, J., Ihee, H., and Jiang, D.L. (2009) *Angew. Chem. Int. Ed. Eng.*, **48**, 5439.
24 Mastalerz, M. (2008) *Angew. Chem. Int. Ed. Eng.*, **47**, 445; Wan, S., Guo, J., Kim, J., Ihee, H., and Jiang, D.L. (2008) *Angew. Chem. Int. Ed. Eng.*, **47**, 8826.
25 Niu, W., Smith, M.D., and Lavigne, J.J. (2006) *J. Am. Chem. Soc.*, **128**, 16466.
26 Zwaneveld, N.A.A., Pawlak, R., Abel, M., Catalin, D., Gigmes, D., Bertin, D., and

Porte, L. (2008) *J. Am. Chem. Soc.*, **130**, 6678.
27 Li, H. and Jakle, F. (2009) *Angew. Chem. Int. Ed. Eng.*, **48**, 2313.
28 Elbing, M. and Bazan, G.C. (2008) *Angew. Chem. Int. Ed. Eng.*, **47**, 834.
29 Reitzenstein, D.R. and Lambert, C. (2009) *Macromolecules*, **42**, 773.
30 Branger, C., Lequan, M., Lequan, R.M., Barzoukas, M., and Fort, A. (1996) *J. Mater. Chem.*, **6**, 555.
31 Sundararaman, A., Victor, M., Varughese, R., and Jakle, F. (2005) *J. Am. Chem. Soc.*, **127**, 13748.
32 Renk, T., Ruf, W., and Siebert, W. (1976) *J. Organomet. Chem.*, **120**, 1.
33 Julia, B.H., Matthias, S., Yang, Q., Anand, S., Frieder, J., Tonia, K., Michael, B., Hans-Wolfram, L., Max, C.H., and Matthias, W. (2006) *Angew. Chem. Int. Ed. Eng.*, **45**, 920.
34 Scheibitz, M., Bats, J.W., Bolte, M., Lerner, H.-W., and Wagner, M. (2004) *Organometallics*, **23**, 940.
35 Marr, G., Moore, R.E., and Rockett, B.W. (1968) *J. Chem. Soc. C*, 24.
36 Christopher, B., Simon, A., Ian, A.F., Cameron, J., and Li-Ling, O. (2005) *Angew. Chem. Int. Ed. Eng.*, **44**, 3606.
37 Nagata, Y. and Chujo, Y. (2009) *J. Organomet. Chem.*, **694**, 1723.
38 Yuan, Z., Taylor, N.J., Marder, T.B., Williams, I.D., Kurtz, S.K., and Cheng, L.T. (1990) *Chem. Commun.*, 1489.
39 Matsumi, N., Chujo, Y., Lavastre, O., and Dixneuf, P.H. (2001) *Organometallics*, **20**, 2425.
40 Matsumi, N., Naka, K., and Chujo, Y. (1998) *J. Am. Chem. Soc.*, **120**, 5112.
41 Ashe, A.J., Meyers, E., Shu, P., Von Lehmann, T., and Bastide, J. (1975) *J. Am. Chem. Soc.*, **97**, 6865; Ashe, A.J. and Shu, P. (1971) *J. Am. Chem. Soc.*, **93**, 1804; Bazan, G.C., Rodriguez, G., Ashe, A.J., Al-Ahmad, S., and Muller, C. (1996) *J. Am. Chem. Soc.*, **118**, 2291.
42 Ashe, A.J. and Shu, P. (1971) *J. Am. Chem. Soc.*, **93**, 1804.
43 Tweddell, J., Hoic, D.A., and Fu, G.C. (1997) *J. Org. Chem.*, **62**, 8286.
44 Lee, B.Y. and Bazan, G.C. (2002) *J. Organomet. Chem.*, **642**, 275.
45 Lee, B.Y., Wang, S., Putzer, M., Bartholomew, G.P., Bu, X., and Bazan, G.C. (2000) *J. Am. Chem. Soc.*, **122**, 3969.
46 Lee, B.Y. and Bazan, G.C. (2000) *J. Am. Chem. Soc.*, **122**, 8577.
47 Ashe, A.J., Fang, X., and Kampf, J.W. (1999) *Organometallics*, **18**, 466.
48 Herberich, G.E., Cura, E., and Ni, J. (1999) *Inorg. Chem. Commun.*, **2**, 503.
49 Wood, T.K., Piers, W.E., Keay, B.A., and Parvez, M. (2009) *Angew. Chem. Int. Ed. Eng.*, **48**, 4009.
50 Bieller, S., Zhang, F., Bolte, M., Bats, J.W., Lerner, H.-W., and Wagner, M. (2004) *Organometallics*, **23**, 2107.
51 Lorbach, A., Bolte, M., Li, H.Y., Lerner, H.W., Holthausen, M.C., Jakle, F., and Wagner, M. (2009) *Angew. Chem. Int. Ed. Eng.*, **48**, 4584.
52 Chen, J., Kampf, J.W., and Ashe, A.J. (2008) *Organometallics*, **27**, 3639.
53 Ashe, A.J., Klein, W., and Rousseau, R. (1993) *Organometallics*, **12**, 3225.
54 Lauren, G.M., Warren, E.P., and Masood, P. (2009) *Angew. Chem. Int. Ed. Eng.*, **48**, 6108.
55 Eisch, J.J. and Galle, J.E. (1975) *J. Am. Chem. Soc.*, **97**, 4436.
56 Eisch, J.J., Galle, J.E., Shafii, B., and Rheingold, A.L. (1990) *Organometallics*, **9**, 2342.
57 Holger, B., Israel, F., Gernot, F., and Thomas, K. (2008) *Angew. Chem. Int. Ed. Eng.*, **47**, 1951.
58 Fan, C., Piers, W.E., and Parvez, M. (2009) *Angew. Chem. Int. Ed. Eng.*, **48**, 2955.
59 So, C.-W., Watanabe, D., Wakamiya, A., and Yamaguchi, S. (2008) *Organometallics*, **27**, 3496.
60 McCullough, R.D. (1998) *Adv. Mater.*, **10**, 93.
61 Andersson, M.R., Thomas, O., Mammo, W., Svensson, M., Theander, M., and Inganas, O. (1999) *J. Mater. Chem.*, **9**, 1933.
62 Sakai, N., Prasad, G.K., Ebina, Y., Takada, K., and Sasaki, T. (2006) *Chem. Mater.*, **18**, 3596.
63 Hou, J.H., Tan, Z.A., Yan, Y., He, Y.J., Yang, C.H., and Li, Y.F. (2006) *J. Am. Chem. Soc.*, **128**, 4911.

64 Osaka, I. and McCullough, R.D. (2008) *Acc. Chem. Res.*, **41**, 1202; Perepichka, I.F., Perepichka, D.F., Meng, H., and Wudl, F. (2005) *Adv. Mater.*, **17**, 2281.

65 Xia, C.J., Fan, X.W., Locklin, J., and Advincula, R.C. (2002) *Org. Lett.*, **4**, 2067.

66 Xia, C.J., Fan, X.W., Locklin, J., Advincula, R.C., Gies, A., and Nonidez, W. (2004) *J. Am. Chem. Soc.*, **126**, 8735.

67 Kromer, J., Rios-Carreras, I., Fuhrmann, G., Musch, C., Wunderlin, M., Debaerdemaeker, T., Mena-Osteritz, E., and Bäuerle, P. (2000) *Angew. Chem. Int. Ed. Eng.*, **39**, 3481.

68 Azumi, R., Gotz, G., Debaerdemaeker, T., and Bäuerle, P. (2000) *Chem. Eur. J.*, **6**, 735.

69 Ebata, H., Miyazaki, E., Yamamoto, T., and Takimiya, K. (2007) *Org. Lett.*, **9**, 4499.

70 Grissom, J.W., Gunawardena, G.U., Klingberg, D., and Huang, D.H. (1996) *Tetrahedron*, **52**, 6453.

71 Yue, D.W. and Larock, R.C. (2002) *J. Org. Chem.*, **67**, 1905.

72 Wu, Y.T., Tai, C.C., Lin, W.C., and Baldridge, K.K. (2009) *Org. Biomol. Chem.*, **7**, 2748.

73 Dopper, J.H. and Wynberg, H. (1972) *Tetrahedron Lett.*, **13**, 763; Dopper, J.H. and Wynberg, H. (1975) *J. Org. Chem.*, **40**, 1957.

74 Chernichenko, K.Y., Balenkova, E.S., and Nenajdenko, V.G. (2008) *Mendeleev Commun.*, **18**, 171; Fujimoto, T., Suizu, R., Yoshikawa, H., and Awaga, K. (2008) *Chem. Eur. J.*, **14**, 6053; Miyasaka, M., Rajca, A., Pink, M., and Rajca, S. (2005) *J. Am. Chem. Soc.*, **127**, 13806.

75 Chernichenko, K.Y., Sumerin, V.V., Shpanchenko, R.V., Balenkova, E.S., and Nenajdenko, V.G. (2006) *Angew. Chem. Int. Ed. Eng.*, **45**, 7367.

76 Dadvand, A., Cicoira, F., Chernichenko, K.Y., Balenkova, E.S., Osuna, R.M., Rosei, F., Nenajdenko, V.G., and Perepichka, D.F. (2008) *Chem. Comm.*, 5354.

77 An, H.Y., Bradshaw, J.S., and Izatt, R.M. (1992) *Chem. Rev.*, **92**, 543; Schramm, M.P., Hooley, R.J., and Rebek, J. (2007) *J. Am. Chem. Soc.*, **129**, 9773; Soncini, P., Bonsignore, S., Dalcanale, E., and Ugozzoli, F. (1992) *J. Org. Chem.*, **57**, 4608.

78 de Namor, A.F.D., Cleverley, R.M., and Zapata-Ormachea, M.L. (1998) *Chem. Rev.*, **98**, 2495; Ibach, S., Prautzsch, V., Vogtle, F., Chartroux, C., and Gloe, K. (1999) *Acc. Chem. Res.*, **32**, 729; Ikeda, A. and Shinkai, S. (1997) *Chem. Rev.*, **97**, 1713.

79 Hopf, H. (2008) *Angew. Chem. Int. Ed. Eng.*, **47**, 9808.

80 Morohashi, N., Narumi, F., Iki, N., Hattori, T., and Miyano, S. (2006) *Chem. Rev.*, **106**, 5291.

81 Mascal, M., Kerdelhué, J.-L., Blake, A.J., and Cooke, P.A. (1999) *Angew. Chem. Int. Ed. Eng.*, **38**, 1968.

82 Thomas, J., Maes, W., Robeyns, K., Ovaere, M., Van Meervelt, L., Smet, M., and Dehaen, W. (2009) *Org. Lett.*, **11**, 3040.

83 Werz, D.B., Fischer, F.R., Kornmayer, S.C., Rominger, F., and Gleiter, R. (2008) *J. Org. Chem.*, **73**, 8021.

84 Werz, D.B., Staeb, T.H., Benisch, C., Rausch, B.J., Rominger, F., and Gleiter, R. (2002) *Org. Lett.*, **4**, 339.

85 Werz, D.B., Gleiter, R., and Rominger, F. (2002) *J. Org. Chem.*, **67**, 4290.

86 Suzuki, H., Nakamura, T., Sakaguchi, T., and Ohta, K. (1995) *J. Org. Chem.*, **60**, 5274.

87 Jiang, W., Qian, H.L., Li, Y., and Wang, Z. (2008) *J. Org. Chem.*, **73**, 7369.

88 Kashiki, T., Shinamura, S., Kohara, M., Miyazaki, E., Takimiya, K., Ikeda, M., and Kuwabara, H. (2009) *Org. Lett.*, **11**, 2473.

89 Patra, A., Wijsboom, Y.H., Zade, S.S., Li, M., Sheynin, Y., Leitus, G., and Bendikov, M. (2008) *J. Am. Chem. Soc.*, **130**, 6734.

90 Patra, A., Wijsboom, Y.H., Leitus, G., and Bendikov, M. (2009) *Org. Lett.*, **11**, 1487.

91 Fabre, J.M. (2004) *Chem. Rev.*, **104**, 5133.

92 Bendikov, M., Wudl, F., and Perepichka, D.F. (2004) *Chem. Rev.*, **104**, 4891.

93 Skibinski, R.G.M., Lork, E., and Azov, V.A. (2009) *Tetrahedron*, **65**, 10834.

94 Ivasenko, O., MacLeod, J.M., Chernichenko, K.Y., Balenkova, E.S., Shpanchenko, R.V., Nenajdenko, V.G., Rosei, F., and Perepichka, D.F. (2009) *Chem. Commun.*, 1192.

95 Okamoto, T., Kudoh, K., Wakamiya, A., and Yamaguchi, S. (2005) *Org. Lett.*, **7**, 5301.
96 Osuna, R.M., Ortiz, R.P., Okamoto, T., Suzuki, Y., Yamaguchi, S., Hernandez, V., and Navarrete, J.T.L. (2007) *J. Phys. Chem. B*, **111**, 7488.
97 Siebert, W. (2009) *J. Organomet. Chem.*, **694**, 1718.
98 Kobayashi, J., Kato, K., Agou, T., and Kawashima, T. (2009) *Chem. Asian J.*, **4**, 42.
99 Herberich, G.E., Hessner, B., and Hostalek, M. (1986) *Angew. Chem. Int. Ed. Eng.*, **25**, 642.
100 Wakamiya, A., Mori, K., Araki, T., and Yamaguchi, S. (2009) *J. Am. Chem. Soc.*, **131**, 10850.

Part Three
Nanopatterning and Processing

9
Functionalization and Assembling of Inorganic Nanocontainers for Optical and Biomedical Applications
André Devaux, Fabio Cucinotta, Seda Kehr, and Luisa De Cola

9.1
Introduction

Nanotechnology enables the fabrication of functional nanometer-scale devices and structures with unique properties in order to observe, characterize, and control phenomena at the nanometer scale. Functional nano- and microparticles are of increasing interest in a variety of scientific fields such as cell biology, biotechnology, diagnostics, nanoanalytics, and pharmaceutics because of their optical, electronic, and magnetic properties [1]. Many scientific strategies have focused on the development of smart functional nanomaterials in order to mimic nanoscale biological entities. Design of multifunctional controlled systems at the same scale as nature (e.g., mitochondria, DNA, cells) will provide a very efficient approach to the production of chemicals, energy, and materials and allow control over the vesicular trafficking in a cell with controlled release of encapsulated contents upon an external trigger.

Among various multifunctional materials, nanocontainers are promising candidates as controlled nanodevices and biomedical systems because of their ability to enclose molecules; they can be used as chambers for controlled chemical reactions and as transporters for drugs or genes. Several types of nanocontainers have been described in the literature such as polymer micelles [2], vesicles [3], liposomes [4], metal–organic frameworks (MOFs) [5], carbon nanotubes [6], and mesoporous silica nanoparticles [7]. These nanocontainers can protect proteins, enzymes, peptides, hormones, drugs, metabolites, or molecules against chemical and biological degradation. In addition, the nanocontainers can allow targeting of a specific compartment, such as in a cell where the encapsulated molecules can be released in a controlled manner [8]. Nanocontainers such as polymer micelles, vesicles, and liposomes have many interesting applications and have been extensively used for drug delivery [9]. MOFs have increasingly gained in popularity as gas sorption media due to their high porosity and thermal stability [10]. Carbon nanotubes have diverse application possibilities such as electronics, hydrogen storage devices, agents for drug delivery, light-emitting devices, and chemical/biological sensors [11]. Mesoporous silica nanoparticles have a wide variety of applications, including adsorption, catalysis, sensing, separation, and drug and gene delivery [12].

Functional Supramolecular Architectures. Edited by Paolo Samorì and Franco Cacialli
Copyright © 2011 WILEY-VCH Verlag GmbH & Co. KGaA, Weinheim
ISBN: 978-3-527-32611-2

Traditional nanocarriers such as liposomes and polymer micelles are highly biocompatible materials, but such materials have limited mechanical stabilities. Although carbon nanotubes have many interesting properties, including their one-dimensional hollow structure, high tensile strength paired with flexibility, large surface area, and chemical stability [13], their toxicity restricts their biomedical applications. MOFs that combine magnetic and anisotropic properties with high emission quantum yields under ambient conditions [14] have low hydrolytic stability. Decomposition of the MOF framework occurs rapidly if the gas or liquid phase contains a small percentage of H_2O [15]. Therefore, the use of MOFs in catalytic oxygenation reactions where water constitutes a major reaction product is limited. In addition, the introduction of functional groups directly into the MOF framework is difficult since the majority of MOFs are formed via solvothermal reactions that typically employ high temperatures and elevated pressures [16]. Unlike these nanocontainers, mesoporous silica particles (Si-MPs) provide a rigid framework with a porous reservoir that can encapsulate a large amount of guest molecules [17]. They show unique properties such as high and homogeneous pore volume, biocompatibility, high thermal stability, and large surface area that can be easily functionalized using a wealth of previously developed silane chemistries [18]. Moreover, it is possible to decorate the external surface of these nanocontainers with functional groups that are able to change their properties via external physical or chemical stimuli such as pH [19], temperature [20], redox potential [21], light [22], or small molecules [23] leading to the "opening" or "closing" of the pores.

One class of well-known silica-based nanocontainers are classical zeolites [24], which can be found as natural minerals or can be artificially synthesized [25]. There are about 40 natural zeolites known since the discovery of *stilbite* in 1756 by the Swedish mineralogist A.F. Cronstedt. He coined the name "zeolite" from the Greek words "boiling stone" upon observation that the mineral boils (looses water) while heated. Although zeolite molecular sieves are restricted to a pore size of around 15 Å, synthetic zeolites have two important advantages: easy control of their size and morphology along with their highly selective functionalization property. Also, it is worth noting that among the porous materials, which are mostly amorphous, zeolites are crystalline with well-defined cavities and that perhaps their crystallinity could have an important role in zeolite biointeractions.

In this chapter, we will discuss the basic properties and applications of inorganic silica-based nanocontainers with a focus on zeolite L nanocrystals, even though we are very much aware of the use of many other porous structures that can play a similar role in what will be discussed here [26]. Also, due to space restrictions, we will mostly describe examples from our own work, putting it in perspective as well as comparing it with related systems.

Zeolite L is a synthetic zeolite with hexagonal symmetry. Zeolite L crystals are built from *cancrinite* cages linked in plane to form double six-rings, which are stacked into chains. Crystals can be synthesized in different sizes and aspect ratios (length/diameter) (see Sections 9.2 and 9.3). Due to their porous structure, zeolites are able to host a variety of guest molecules. Their geometrical properties allow the study of energy transfer processes within dye–zeolite L composites (Section 9.4). Chemical functionalization of zeolite L crystals is carried out with silane chemistry, and either the whole external surface or just the channel entrances of the crystals can be functionalized selectively (Section 9.3). Furthermore, zeolite L crystals can self-assemble in solution and on surfaces (Section 9.5) and they have many potential optical and biomedical applications (Section 9.6).

9.2
Zeolite L as Inorganic Nanocontainers

Supramolecular chemistry is a very versatile and powerful concept for the design and realization of multifunctional materials. This field of chemistry, which focuses on chemical systems made up of discrete numbers of assembled molecular subunits or components, has led to the development of materials with functionalities beyond the reach of single component [27]. Among the techniques of this field, host–guest chemistry offers a beautiful and convenient way to arrange molecules, metal complexes and clusters, quantum dots, or even small biomolecules in porous host materials. These host–guest composites exhibit a high internal ordering, which can lead to fascinating new functionalities such as anisotropic optical properties [28], efficient energy or electron transport [29, 30], or even controlled release of guest molecules [31]. Ordered host–guest materials can, in turn, be assembled on a larger scale to produce systems that exhibit organizational patterns from the molecular to the macroscopic scale. In this section, we will develop a general model for porous host materials and discuss how each of its components can be employed to achieve further functionalities. The concepts developed here will be illustrated by zeolite L, a microporous aluminosilicate featuring a one-dimensional channel system. It should, however, be stressed that the concepts and ideas developed here can be transferred to any other porous materials.

9.2.1
General Concept for Nanocontainers

A wide variety of porous materials are known, including classical zeolites (microporous aluminosilicates) [28], other zeolite-like materials such as microporous aluminum– or galium–aluminum phosphates (ALPOs and GALPOs) [32], metal–organic frameworks [33], mesoporous silica or organosilicates [34, 35], porous crystals composed of complex salts [36], and even single-walled carbon nanotubes (SW-CNT). Many of these micro- or mesoporous materials have already been used for host–guest chemistry. The short list given here should by no means be taken as being exhaustive. Zeolite-based materials are more advanced with respect to the organizational level and to the realization of practical applications than materials based on mesoporous hosts [37, 38]. The latter are, however, catching up [35, 39]. Despite the obvious differences between microporous and mesoporous host–guest materials, many similarities and analogies exist between them [38].

All these systems have widely different morphologies and intrinsic properties. In a most general approximation, however, these porous materials can be divided into three main units: the internal cavities, the pore openings, and the external surface. This is illustrated in Figure 9.1a for a material with a one-dimensional channel system. We will first derive the importance of each unit and its uses in a very broad way and then follow up with a description of the practical example material, zeolite L. The strategies for making use of these diverse units will be discussed in Sections 9.3 and 9.4.

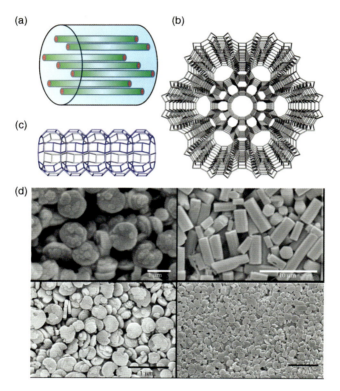

Figure 9.1 Porous host materials. (a) General model for a porous host, which can be divided into three main functional units: the internal channel system with its interior walls (in green), the pore openings that allow the control of communication between the interior and exterior (marked in red), and the external surface of the host that can be further functionalized (light blue). Many host materials feature cavities or channels that are connected in two or three dimensions. (b) Top view of the structure of zeolite L, illustrating its hexagonal framework. It shows a channel surrounded by six neighboring channels. The center-to-center distance between two channels is 1.84 nm. (c) Side view of a channel that consists of 0.75 nm long unit cells with a van der Waals opening of 0.71 nm at the smallest and 1.26 nm at the widest place. (d) SEM image of zeolite L crystals with different sizes and morphologies.

9.2.1.1 The Internal Cavity System

The internal channel or cavity system of a porous material can host different kinds of guest species. Depending on the cavity size and the dimensionality of the channel system, geometrical constraints can be imposed by the hosts that, in turn, lead to a preferred spatial arrangement of the guests. The large variety in pore structure and morphologies provided by different types of molecular sieves offers many possibilities for the design of host–guest materials with specific properties [26a]. The dimensionality of the internal cavity system (i.e., how the free spaces are interconnected) has to be considered when choosing a porous material for specific applications. One-dimensional channel systems are the simplest case and are very desirable when energy or electron transport is of interest. The low dimensionality favors such transfer mechanisms and allows very efficient and fast transfers [40]. Channel systems connected in two or three dimensions make it easier for guests to travel around in the hosts, since there are less risks of channel blocking through aggregate formation. On the other hand, the anisotropy of the system will decrease with increasing dimensionality as well as the control of release of guests from the channels.

The properties of molecules, complexes, and clusters in cavities and channels – outside the field of catalysis – have been investigated by several authors [34, 41, 42]. Inserted species usually exhibit increased chemical or thermal stability when embedded into an inorganic host matrix. The host material can protect the guests by providing a nonreactive environment, by hindering contact with external reactive species, or by simply preventing molecular isomerizations needed for decomposition processes [43].

Another important issue is the chemical properties of the internal cavity walls, as this can play a role in guest properties or even control host–guest interactions. Mesoporous materials are especially interesting, since their relatively large cavities are well suited for chemically modifying their internal wall structure by appropriate chemistries. A good example of this concept is the tuning of the polarity in the mesoporous silica by grafting hydrophobic molecules to their inner surfaces [44]. Other chemical properties that can be influenced include, but are not limited to, inner pH value, ionic composition and strength, as well as coadsorbed solvent or gas molecules [28, 45].

In summary, one can adapt the intrinsic properties of the cavities of a chosen host material to fit the requirements of the guest species or the intended applications by applying appropriate chemistry. The host material is not only responsible for providing a scaffold to organize guest species but also for protecting them from the environment. Furthermore, it has been shown that the walls of the host can contain active "tiles" – such as organic chromophores – in order to tune the guest behavior [35].

9.2.1.2 The Pore Openings

Due to their location between the inside and outside of the host material, the pore openings play an essential role in applications of host–guest materials. They are responsible for the entry or release of guest species and control over their structure is required for many applications, such as drug delivery [46].

An interesting concept that was first developed for zeolite L is the so-called stopper molecule [47]. These specific closure molecules, which can only partly enter the channels, consist typically of a head and a tail. Due to size restrictions, only the tail can enter the channel. Depending on their nature, these stopcocks can be bound by physisorption, electrostatical interaction, or covalent bonding. Since these molecules are located at the interface between the interior of a zeolite L crystal and the surroundings, they can be considered as mediators for communication between dye molecules inside the nanochannels and objects outside the crystals. Stopper molecules can also be used to prevent penetration of small molecules, such as oxygen and water, into the channels or to hinder encapsulated dye molecules from leaving the channels. This type of pore opening functionalization is obviously less suited for mesoporous materials, where a complete sealing of the pores by a single molecular entity is more complicated.

However, functional groups attached to the entrances of the host material can be imagined to fulfill other roles, such as promoting or hindering the diffusion of selected molecular species [48], injecting or collecting energy or electrons [28, 49], serving as molecular valves that will open or close under selected conditions [50], or even chemically modifying the guests as they leave the host system. It is important to note that any chemistry to be employed for the modification of pore openings has to be very specific and spatially well controlled. Otherwise, one ends up with functional groups grafted over the whole host material.

9.2.1.3 The External Surface

The external surface of the host is another useful area for adding further functionalities. This surface plays an important role in how the host–guest material will interact with its environment. Therefore, most of these modifications aim at improving the dispersability or biocompatibility of the material [51, 52]. Functionalization of the external surface can also be envisaged to enable targeting of biomolecules, bacteria, or cells [53], to add chemical, luminescent, magnetic, or radioactive labels [54], or even to promote self-assembly of host–guest objects into well-defined macroscopic structures on various supports [55–58]. Making such use of a host's external surface extends not only the range of potential applications but also allows the realization of molecular patterns that reach from the nano to the micro or even to the macro scale. The organization of quantum-sized particles, nanotubes, and microporous materials has been studied on different surfaces and used in science, technology, diagnostics, and medicine [42, 59–67]. Size, shape, and surface composition of the objects, as well as the properties of the surface on which they should be organized, play a decisive role and in some cases determine not only the quality of the self-assembly but also its macroscopic properties. Self-assembly strategies make hierarchical ordering of materials attractive by avoiding expensive techniques such as photolithography.

9.2.2
A Practical Sample Material: Zeolite L

The concepts discussed above will be illustrated by zeolite L, as it has proven to be a versatile host material for the organization of a large variety of guest molecules [28, 68]. The reasonings and methods presented here and in the following sections are, however, also valid for any other type of host material. Zeolite L crystals can be synthesized with different morphologies and sizes varying from about 30 nm up to 10 000 nm, meaning that a volume range of about 7 orders of magnitude can be covered [69, 70]. Figure 9.1b–d) depicts the structure and morphology of zeolite L. The primary building unit of the framework consists of TO_4 tetrahedrons where T represents either Al or Si. The secondary building unit, the cancrinite cage, is made up of 18 corner-sharing tetrahedrons. These cages are stacked into columns, which are then connected by means of oxygen bridges in the a–b plane to form a one-dimensional channel system running parallel to the crystal c-axis. The channel system exhibits a hexagonal symmetry. The molar composition of zeolite L is $(M^+)_9[(AlO_2)_9(SiO_2)_{27}] \times nH_2O$, where M^+ are monovalent cations, which compensate for the negative charge resulting from the aluminum atoms. The variable n is equal to 21 in fully hydrated materials, and equal to 16 for crystals equilibrated at about 22% relative humidity [24b,26b,71].

Zeolite L can be considered as consisting of a bunch of strictly parallel channels as shown in Figure 9.1b [68, 72]. The channels have a smallest free diameter of about 0.71 nm, while the largest diameter in the crystal is 1.26 nm. The center-to-center distance between two channels is 1.84 nm. Each zeolite L crystal consists of a large number of channels (n_{ch}) that can be estimated by Equation 9.1:

$$n_{ch} = 0.267(d_Z)^2 \tag{9.1}$$

where d_Z represents the diameter of the crystal in nanometer. For example, a crystal with a diameter of 600 nm features nearly 100 000 strictly parallel channels. The ratio of void space available in the channels with respect to the total crystal volume is about 0.26. An important consequence of the void space ratio is that zeolite L allows, through geometrical constraints, the realization of extremely high concentrations of well-oriented molecules that behave essentially as monomers. A 30 nm × 30 nm crystal can take up to nearly 5000 dye molecules that occupy 2 unit cells, while a 60 nm × 60 nm crystal can host nearly 40 000 of the same guest type. The dye concentration of a dye–zeolite material $c(p)$ can be expressed as a function of the loading p as follows:

$$c(p) = 0.752 \frac{p}{n_s} \left(\frac{mol}{l}\right) \tag{9.2}$$

where n_s indicates the number of unit cells that form a site and are occupied by one guest. The value of n_s is usually an integer, but this must not necessarily be so. The loading, or occupation probability, p is defined as follows:

$$p = \frac{\text{number of occupied sites}}{\text{total amount of sites}} \tag{9.3}$$

Zeolite L is a versatile host material allowing the design and preparation of a variety of highly organized host–guest systems. The "cornerstones" in the development of these composite materials were the finding that dye-loaded crystals with different well-defined domains can be prepared [73], the invention of the stopcock principle [74], the discovery of quasi-1D energy transfer [29], the preparation of unidirectional energy transfer material [40], and finding ways to create hybrid materials fully transparent in the visible range [52]. The latter is important for spectroscopic investigations, since zeolite crystals exhibit considerable light scattering due to their size and refractive index between 1.4 and 1.5.

9.3
Functionalization of Zeolites: Host–Guest Chemistry and Surface and Channel Functionalizations

9.3.1
Inorganic–Organic Host–Guest Systems

In inorganic–organic host–guest systems, the useful properties from two distinct components can be combined to form one material. As a typical example, fluorescent organic chromophores can be employed for their particular photophysical features that would be suitable to design an artificial antenna system. Embedding these dyes into inorganic host structures leads to several advantages concerning these properties, as will be discussed in this section.

Zeolite L stabilizes the incorporated dyes, thereby yielding composite materials with improved thermal, mechanical, and photochemical resistance, as well as offering an optical transparency similar to quartz [34a]. For example, dyes aggregate in solution above certain concentrations and, in consequence, new nonradiative pathways can be created. Inserting the chromophores into an appropriate host material not only prevents them from building dimers but also leads to a strict orientation of the dyes, thus enhancing electronic excitation energy transfer.

Moreover, inclusion of organic molecules into inorganic structures can prevent photoinduced degradation. This process rapidly destroys fluorescent dyes, since, the dyes become more reactive when promoted to an excited state and can undergo irreversible redox processes leading to decomposition [75]. Furthermore, when the host is able to impose a strict spatial restriction to the guest molecules, internal rotational relaxation modes are hindered, which is expected to decrease the nonradiative deactivation of the excited states [76].

The host material does not only improve the properties of the guest but also shows another useful feature in that the dimensions can be opportunely tuned and the outer surface can be chemically functionalized in order to interface the system to a specific environment. The solubility of the guest molecules will not hinder or restrict their use in specific conditions since the host framework will define the overall solubility.

All these aspects show that inserting organic chromophores into an inorganic host leads to useful new materials with interesting characteristics. Molecular sieves are widely studied as inorganic host materials [34c, 77–80]; among them, zeolites and metal–organic frameworks undoubtedly represent the most promising choices in terms of possible applications.

9.3.2
Dye Loading of Zeolites

Because of their well-defined internal structure with uniform cages, cavities, or channels, zeolites are extremely suitable host materials [81–83]. With respect to organic frameworks, they have the advantage of high mechanical and thermal stability. Among the existing channel- and cage-type materials, zeolites are the most appealing porous materials.

The crystalline and regular structures of zeolites represent a perfect host environment for several kinds of guests [24b]. The channel system can be one dimensional, as in zeolite L, with channels extending along a crystal, or two–three dimensional, with intersecting channels and cages, as is found in zeolite X and Y. Many different zeolite framework structures with various geometrical properties are known and can be selected for specific applications [26a].

One-dimensional cavity systems are more desirable for achieving optimal control of the spatial organization, insertion, and release of guest molecules. A wide spectrum of zeolite guest molecules has been studied. Once incorporated into these hosts, molecules or clusters undergo changes in properties and reactivity that can lead to very interesting applications such as gas separation [84], selective catalysis [85], removal of pollutants and ion exchangers [86], data storage [87], quantum electronics and nonlinear optics [88], chemoselective devices [89], nano reaction chambers [90], or energy conversion systems [91].

In addition, zeolite can impose geometrical constraints on guest molecules that are related to its pore structure, resulting in extremely anisotropic materials. Chromophore-loaded zeolites have been investigated for different purposes such as interfacial electron transfer [92, 93], microlasers [94], second harmonic generation (SHG) [95, 96], frequency doubling, and optical bistabilities giving rise to spectral hole burning [97].

9.3.2.1 Inserting Molecules

The four main ways to incorporate organic molecules in the zeolite cavities are as follows: adsorption from the liquid phase [98–100] or the gas phase [101, 102], ion exchange [103], crystallization inclusion [104], and an *in situ* synthesis in the zeolite cages (also known as ship-in-a-bottle synthesis) [105].

The last two methods are suitable for the insertion of guest molecules that exceed the free pore diameter of the zeolite. Adsorption from liquid or gas phase and ion exchange are reversible methods, which means that the molecules can not only enter the zeolite but can also leave it in the same way. Some typical dye molecules that were inserted into zeolite L are shown in Figure 9.2.

- **Neutral Molecules:** Neutral molecules can be inserted into zeolite L by sorption. This can be carried out from a solution or the gas phase [34a]. Insertion from the gas phase is carried out in two steps. First, zeolites must be dried under vacuum in order to empty adsorption sites that under ambient conditions are occupied by water molecules. Second, neutral dye molecules can be inserted from the gas

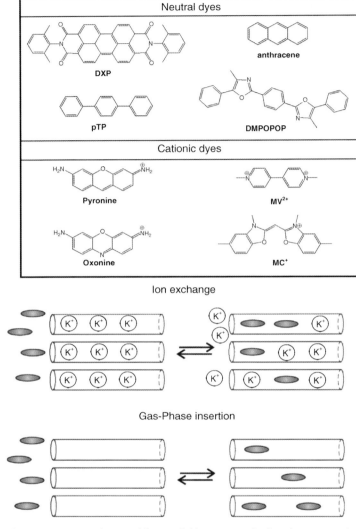

Figure 9.2 *Upper:* Selection of dyes (and abbreviations) that have been inserted in zeolite L. *Lower:* Schematic representation of the ion exchange and gas-phase loading equilibriums.

phase, as long as their size is smaller than the free pore diameter. Because the dye molecules are physisorbed, they can be relatively easily extracted or displaced by other molecules such as water.

- **Cationic Molecules:** Cationic guest molecules can be inserted into zeolite L by exchanging the charge compensating metal ions that are present in the zeolite channels (generally K^+ or Na^+). Usually ion exchange is carried out in aqueous solution, but can, in some cases, also be performed in nonaqueous environments [106–109]. The reaction can be divided into two steps. First, the dyes are adsorbed on the external zeolite surface and, second, the dyes enter the zeolite. The overall ion exchange equilibrium can be expressed by the following equation:

$$yD^+ (aq) + M^+ (Z)_x \leftrightarrow M^+ (Z)_{x-y} D^+ (Z)_y + yM^+ (aq) \quad (9.4)$$

where D^+ and M^+ represent, respectively, the dye molecule and the monovalent metal ion, while (Z) and (aq) denote whether the species are located in the zeolite or in aqueous solution, respectively.

This equilibrium can be influenced in favor of the dye-loaded zeolite, $M^+ (Z)_{x-y} D^+ (Z)_y$, by using an ionophore cryptand to trap the exchanged metal ions, yM^+ (aq) [109–111]. Because of space restrictions in the zeolite channels, dye adsorption results in a decrease of the overall entropy of the system. Therefore, it is very difficult to reach complete loading (i.e., an occupation probability of 1).

In general, one can state that the dyes should have higher affinity for the interior of the zeolite L than for the solvent environment for efficient insertion. If, for example, the dye is extremely soluble in a specific solvent, it will remain in solution rather than enter the zeolite channels. In such cases, it is advisable to search for a solvent in which the dye is less soluble for the insertion reaction.

For the dyes listed in the left part of Figure 9.2, the spectroscopic properties in the zeolite channels are comparable to those in diluted solution at low loadings. Small spectral shifts are observed, but no structural changes arise upon insertion. The latter observation is a sign that no dimers are formed in the zeolite L channels. However, aggregate formation can be observed when using zeolite types that have larger channels or cages, when smaller molecules are inserted, or when a very high concentration in the pores is reached [99, 112, 113].

9.3.2.2 Doping of Zeolites with Rare Earth Ions

The majority of the work on inorganic–organic host–guest systems has dealt with the optical activation of zeolites with organic dyes, however, the range of guests can be extended from uncoordinated rare earth ions to inorganic and metal–organic complexes. Generally, the efficiency of rare earth-based emitters becomes problematic with increasing wavelengths, as the emissive states are increasingly liable to deactivation via vibrational states. This is particularly pronounced for environments that exhibit high-energy phonons that are brought about, for example, by −CH and −OH groups, or even by coordinated water. By doping zeolites X and Y with lanthanides, Kynast and coworkers reported enhanced luminescence for ultraviolet (UV) [114], near-infrared (NIR) [115], and even vacuum ultraviolet (VUV) [116] excitations. Moreover, as opposed to typical, larger pore zeolites (e.g., up to about 1200 pm in zeolites X and Y), sodalites with smaller cage diameters (approximately 920 pm, as calculated by Kondo [117]) can completely prevent the admission of water and fulfill the coordination sphere of the rare earth ion by the oxygen atoms present in the neighboring sites, thus giving rise to a significant emission increase, as reported for Eu^{3+} and Nd^{3+} [118].

9.3.3
Functionalization of the External Surface

This section focuses on ways to reach a second stage of organization by functionalizing either the whole external surface or just the channel entrances of the cylindrical crystals with covalently bound dye molecules.

9.3.3.1 Reactions with the Whole External Surface of Zeolite L

Chemical functionalization of the zeolite L crystals is carried out with silane chemistry that has proven to be a versatile technique. The most common procedure exploits the reactivity of the hydroxyl groups that are present on the outer zeolite surface with the siloxyl moieties of a wide variety of grafting molecules. Various zeolite materials have been chemically modified with different functionalities such as amino, epoxide, and aldehyde, with polymers or DNA bases [66].

Introducing new chemical groups to the external surfaces provides more possibilities for further applications. One of the important milestones to be achieved, in order to make these materials industrially viable, is dispersion stability in commonly used solvents. For example, the use of alkoxysilane derivatives with a hydrophobic part has been explored in order to obtain transparent suspensions in nonpolar solvents. Alternatively, derivatives bearing polyglycole, PEG, or a positive charge lead to solubilization in water [119].

Covalently binding fluorescent dye molecules to the external zeolite L surface can be performed using two different approaches:

- The desired dye molecule is first reacted with a linking reagent; the whole reactive unit consisting of dye molecule plus linker can then be bound to the external zeolite L surface.
- Alternatively, the external surface of zeolite L can be first decorated with a linking reagent; afterward, a fluorescent dye molecule can bind to this linking reagent, yielding zeolite L crystals functionalized with the desired dye molecule.

A similar procedure was followed to covalently bind a rhodamine dye to mesoporous MCM-41 and a silylated coumarin to zeolite L [120].

9.3.3.2 Selective Functionalization of the Channel Entrances

All the aforementioned modifications are performed in a nonspecific way regarding the topology of the crystal's outer surface. The chemical and physical properties of the basal (bearing the channel entrances) and coat surfaces of a zeolite L crystal are quite different. This difference in reactivity can be used for selective functionalization of only the channel entrances and is particularly important for further use of the modified material.

The modification can be performed using so-called stopcock molecules [74b, 106, 121]. These molecules bear a reactive tail that can penetrate the channel entrances and a head that is too large to enter the channel. The stopcock principle and its use are depicted in Figure 9.3.

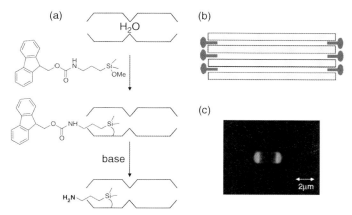

Figure 9.3 (a) Schematic procedure for the selective amino functionalization of the zeolite channel entrances. (b) The stopcock concept: a molecule that can partially penetrate the channels. Luminescent stopcocks can participate in FRET processes, driving the excitation energy into or out of the zeolites. (c) Fluorescence microscopy of a zeolite crystal whose channel entrances were functionalized with emitting dyes.

The resulting modified zeolites possess free amino groups at the pore openings and offer the possibility to use this group as a precursor for attaching any amino-reactive molecule, as well as fluorescent small dyes and metal complexes. Such molecules can act as donors or acceptors in FRET processes [48, 122]. Fluorescent tails are protected by the zeolite channel walls, increasing the photostability of the material. The advantage of fluorescent heads, however, is that the chromophore is closer to the outer environment, which makes it easier to transfer the electronic excitation energy from the crystals to an external device. The interaction between the stopcock molecules and the zeolite channels can be of electrostatic or covalent nature.

9.3.4
More Complex Functionalizations

By combining both aforementioned approaches, a bifunctional material can be realized. The coat and the base are modified with two different fluorescent molecules, leading to what is known as an orthogonally functionalized system for potential imaging applications, as shown in Figure 9.4 [123].

It is also possible to avoid leakage of dye molecules from the pores of zeolites by performing a complete coating of the crystal outer surface by either building up silica shells or through polyelectrolyte layer-by-layer (LBL) techniques. These techniques result in fluorescent anisotropic cored–isotropic shelled microcontainers, as recently demonstrated by De Cola and coworkers [124], which is depicted in Figure 9.5.

9 Functionalization and Assembling of Inorganic Nanocontainers

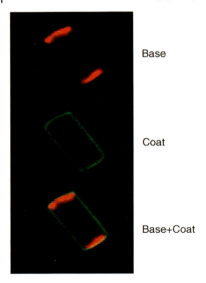

Figure 9.4 Fluorescence microscopy pictures of a zeolite L crystal functionalized with two different fluorescent dyes in a spatially resolved manner. The selective derivatization of the channel entrances and of the coat leads to an interesting bifunctional material with potential uses for imaging technologies [123].

9.3.5
Metal–Organic Frameworks as Host Materials

Among the aforementioned porous materials, metal–organic frameworks are currently of great interest and importance [125] due to their novel coordination structures, diverse topologies, and potential applications in gas storage [126], separation [127], catalysis [128], drug delivery [129], molecular recognition [130], luminescence [131], magnetism [132], and conductivity [133]. Basically, MOFs result from the tight association and three-dimensional covalent connection of inorganic clusters

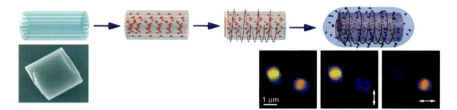

Figure 9.5 Schematic representation of the silica coating procedure followed for the preparation of multifluorescent dye-loaded zeolite L crystals with luminescent anisotropic cores and isotropic shells. On the right bottom part of the figure, the fluorescence microscopy images show the polarized emission from the inserted dye (DXP) and the isotropic luminescence of the dye on the outer shell [124].

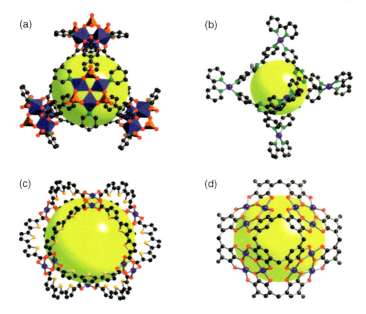

Figure 9.6 Examples of MOFs. (a) The tetrahedral structure of IRMOP-51 (Fe blue, S orange, O red, C black, and N green). (b) The tetrahedral cage [(bpy)$_6$Pd$_6$(tpt)$_4$]$^{12+}$ (bpy = 2,2′-bipyridine, tpt = 2,4,6-tris(4-pyridyl)-triazine) (Pd blue, C black, and N green). (c) The octahedral structure of MOP-28 (Cu blue, C black, O red, and S orange). (d) A rhombic dodecahedron composed by eight trivalent planar Cu$_3$SO$_3$ units and six tetravalent pyrogallo[4]arene units (Cu blue, C black, and O red). The yellow spheres delineate the empty space at the center of the cage [125d].

and organic moieties to build up open frameworks. The basic building block is a symmetric organic molecule, which binds metal ions to form a regular layered structure (see Figure 9.6).

Such materials display cavities or channels in which other metal ions, polyanions, drugs (such as ibuprofen [129a]), and guest aromatic molecules can be hosted [134]. The properties of MOFs often resemble those of classical zeolites. The pioneering work of Yaghi *et al.* in this emerging field has shown that MOFs can adsorb gases and allow shape-selective catalysis [95, 135]. The existence of coordinatively unsaturated metal sites provides an intrinsic chelating property with electron-rich functional groups. This feature offers a powerful way to selectively functionalize unsaturated sites in MOFs [134b]. Moreover, surface amine grafting provides a general way for selective functionalization of their surfaces.

9.4
Photoinduced Processes in Zeolites

In the previous two sections, we presented a general concept and strategies for the use of porous host materials, as well as their functionalization. The present section mainly discusses some of the important photoinduced processes that can occur in such host–guest materials. The given practical examples will, once again, be mostly focused on zeolite L as a host material. However, these processes can also be observed or realized with any other type of porous host material.

9.4.1
Förster Resonance Energy Transfer

Due to their geometrical properties, dye–zeolite L composites are well suited for the study of energy transfer processes. We will give here only a short overview of the theoretical background of Förster resonance energy transfer (FRET), followed by an experimental case study. A more thorough mathematical treatment of this process was recently published [28]. The FRET process is a powerful tool employed to determine inter- or intramolecular distances over the range of 1.5–10 nm. Its main application areas are protein chemistry or as a distance-dependent sensor in biochemistry. Another field of application is the preparation of light-harvesting materials for solar energy conversion [37].

The understanding of the FRET process is very advanced and goes back to the pioneering work of Theodor Förster [136]. A chromophore can be considered to consist of a positively charged backbone with some delocalized electrons. When a photon is absorbed by the molecule, its energy is transformed into kinetic energy of one of these delocalized electrons. The oscillating electromagnetic field caused by the rapid movement of the electrons can interact with a neighboring acceptor molecule A, as long as it bears states being in resonance with the excited state of the donor D^*. Figure 9.7a illustrates these processes in an artificial antenna material. The material is prepared by inserting two types of molecules into the channels of a porous material. The donors (represented as green ovals) are selectively excited by photon absorption. In a first step, their electronic excitation energy migrates to an unexcited neighbor. After a series of such steps, the electronic excitation reaches a luminescent trap (acceptor molecule) and is then released as fluorescence.

This radiationless electronic excitation energy transfer is solely due to the very weak near-field interaction between excited configurations of the initial state $(D^* \cdots A_i)$ and of the final state $(D \cdots A_i^*)$. The Förster mechanism involves no orbital overlap between the donor and acceptor molecules, meaning that no electron transfer occurs. The donor D and the acceptor A can be of the same type or of different species. The rate constant k_{EnT} for the transfer from one electronic configuration to the other can be expressed as follows:

$$k_{\text{EnT}} = \text{TF} \frac{\varkappa_{D^*A}^2}{n^4 R_{DA}^6} \frac{\phi_{D^*}}{\tau_{D^*}} J_{D^*A} \tag{9.5}$$

where $\varkappa_{D^*A}^2$ describes orientation between the electronic transition dipole moments of the donor and the acceptor and a general orientation situation is sketched in Figure 9.7b. The efficiency of the FRET process depends strongly on the relative orientation of both transition dipole moments and $\varkappa_{D^*A}^2$ can take values between 0 and 4. The lowest value is obtained when they are perpendicular to each other, while

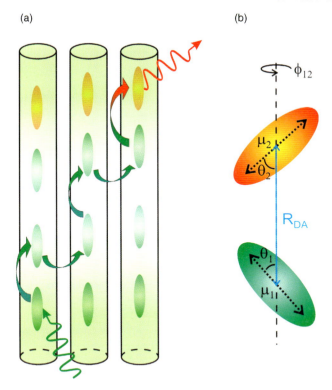

Figure 9.7 Förster resonance energy transfer (FRET). (a) Scheme of an artificial photonic antenna, where chromophores are embedded in the host channels. The donor dyes (green ovals) are selectively excited by light absorption and transport their excitation energy via FRET first among themselves before transferring it to the acceptors (orange ovals) shown at the ends of the channels. (b) Definition of the geometric factors involved in the computation of the orientation factor $\varkappa^2_{D^*A}$ and the interchromophore distance R_{DA}. The angles describing the relative orientation of the electronic transition dipoles of two molecules μ_1 and μ_2, are indicated as θ_1 and θ_2. The subscripts 1 and 2 not only refer to the corresponding electrons but are also used to identify the two species.

the maximum is reached for a collinear arrangement. Another important factor is the intermolecular distance between the donor and acceptor, as is indicated by the factor $1/R_{DA}^6$. The inverse sixth power dependence on intermolecular separation makes FRET useful over distances in the range of 1.5–10 nm. The terms Φ_{D^*} and τ_{D^*} specify the emission quantum yield and excited state lifetime of the donor in absence of an acceptor. J_{D^*A} represents the spectral overlap integral between the acceptor absorption and the donor emission bands, taking into account the resonance condition. The environment of the donor–acceptor pair comes into play by the refractive index n. The Theodor Förster constant (TF) is an empirical value and is given by

$$\text{TF} = \frac{9000 \ln(10)}{128\pi^5 N_L} \quad \text{or} \quad \text{TF} = 8.785 \cdot 10^{-25} \text{ mol} \quad (9.6)$$

where N_L is Avogadro's number. An important experimental parameter that helps to distinguish between FRET and trivial energy transfer processes – in which a photon is emitted by the donor and subsequently absorbed by the acceptor – is the donor lifetime. In the presence of an acceptor, the donor lifetime will be shorter, since the FRET process is a new deactivation pathway.

A nice case study for the FRET process is the Py^+, Ox^+–zeolite L system [137]. The strongly fluorescent dyes Py^+ and Ox^+ enter the zeolite channels at about the same speed, providing a homogeneous distribution throughout the crystal. The two dyes form an excellent donor–acceptor pair, due to the large overlap between the Py^+ fluorescence and the Ox^+ absorption spectra. The energy transfer process can be observed on the single-crystal scale, by following the diffusion of the random dye mixture as a function of time [138]. The fluorescence microscopy images of a Py^+, Ox^+–zeolite L sample given in Figure 9.8 illustrate the process very nicely. The images where taken after loading the crystals for 20 min (1), 60 min (2), 470 min (3a, 3b), and 162 h (4), respectively. All samples were excited at 485 nm, where Py^+ has a strong absorbance and Ox^+ has negligible absorption. The only exception is 3b where Ox^+ was specifically excited. The samples with short equilibrium times exhibit an orange to yellow luminescence, which is due to a mixture of green Py^+ and red Ox^+ fluorescence. This means that energy transfer is significant in this case. After a diffusion time of 162 h, the dyes are so far apart from each other that upon selective excitation of Py^+, only its green emission can be observed. This type of experiment also allows the estimation of the diffusion constants for the molecular species in the porous materials by monitoring the reduction of FRET efficiency over time. The time required for the two dyes to sufficiently diffuse far apart for FRET to become negligible is quite long (at least 162 h). This diffusion speed depends strongly on the guest type and shape and can be much faster. Good examples for fast diffusion in zeolite L are earth alkali ions or the cationic dye methylviologen, where in both cases full loading and homogeneous distribution can be reached in less than 3 h [139].

The FRET process can be extended from the inside of a crystal to its environment (or vice versa) by means of stopcock molecules. Since these chromophores are located at the zeolite basal surfaces, they can promote communication between dyes in the material and the outside world. A nice example of such a system, which was recently reported, consists of a dye-loaded, stopcock-modified zeolite L crystal that has been embedded into an electroluminescent polymer fiber [138b]. Electronic excitation energy from the polymer is first transferred to the stopcock molecules and then to the chromophores located deeper in the channels. Other examples will be discussed in the frame of possible applications (see Section 9.6.1)

9.4.2
Electron Transfer

Another process in which guests organized in porous host materials can be advantageous is electron and charge transfer [140]. In contrast to the FRET process

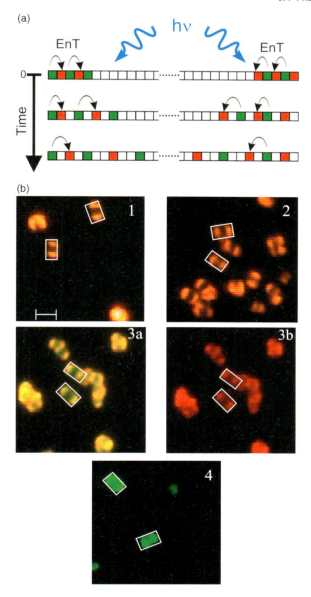

Figure 9.8 FRET case study on a mixed Py^+, Ox^+–zeolite L composite. (a) Schematic representation of the situation. (b) Fluorescence microscopy pictures, visualizing the diffusion of Ox^+ and Py^+ in zeolite L, taken after 20 min (1), 60 min (2), 470 min (3a, 3b), and 162 h (4), respectively. Py^+ was specifically excited, with the exception of 3b where Ox^+ was specifically excited. Two crystal images are framed. The scale given in 1 corresponds to a length of 1.5 μm. With increasing equilibration time, the distance between donor and acceptor increases and FRET becomes less efficient until only the green donor emission can be observed [137, 138].

outlined above, electron transfer or exchange requires the involved molecules to be close together. Due to their specific characteristics outlined in the previous sections, porous host materials such as zeolites or (organo) silicates have gathered much interest in the electrochemical community over the past 20 years, and interested readers are referred to a recent review by Walcarius and to the works of Mallouk and coworkers for detailed discussion on electron transfer reactions in zeolites [141]. The main application of inorganic nanocontainers in electron transfer lay in the preparation of so-called zeolite-modified electrodes (ZMEs) for analytical purposes [141a].

The main interest in using microporous materials, such as zeolites, for electrode modification lies in their combined size selectivity and ion exchange capabilities [141]. This characteristic is of interest for selective recognition or discrimination at the electrode surface as well as for the immobilization of redox guests. The immobilization can take place either through ion exchange or through "ship-in-a-bottle" synthesis. On the other hand, mesoporous materials offer several interesting characteristics: their larger cavities can behave as nanoreactors [142] and their highly ordered and larger pore structure allows a better accessibility of binding sites, leading to faster mass transport processes [143]. Mesoporous materials can also immobilize much larger organofunctional groups, even going into the realm of nanobioencapsulation that could result in the development of electrochemical biosensors. Both types of porous host materials can also be envisaged for the realization of intraparticle electron transfer chains.

Despite their many desirable properties, zeolites and mesoporous silicates or organosilicates are electric insulators. Their use in electrochemical systems requires specific contacting schemes, such as the use of liquid mercury or adaptive electrodes, for example eutectic gallium indium (EGaIn) electrodes. Alternatively, the silicate can be confined within or to the surface of an electrode [144]. Several strategies have been developed that can be divided into four main families: First, self-assembling the host particles as mono- or multilayers on electrodes [56]. These layers are usually either coated with a thin polymer film or copolymerized to ensure mechanical stability. Second, dispersing the nanocontainers into a composite conducting matrix, such as carbon paste [145]. Thick-film screen-printed ZMEs also belong to this class. Third, immobilizing a carbon particle mixture on conducting surfaces by applying mechanical pressure or by preparing a nanocontainer pellet that is then inserted between two electrodes [146]. Fourth, growing membranes of the desired host material on the conducting surface [147].

In summary, host–guest materials based on porous nanocontainers are a promising concept for the development of electrochemical or photoelectron-active systems. The electronically insulating characteristics of some typical host materials can present a serious drawback for some applications, but other materials such as MOFs might help to overcome this limitation. It is expected that systems capable of integrating molecular recognition, catalysis, and signal transduction can be realized by organizing diverse guest species in the selected nanocontainers, leading to novel electrochemical sensors or devices.

9.4.3
Aggregates Formation in Nanochannels

The geometrical constraints imposed by a host material lead to a highly ordered and well-defined arrangement of the guests in the channels. Inserted organic molecules are usually too large to overtake or stack with molecules already present in the channels, making it possible to create systems with two or more defined domains containing only one type of guest. Figure 9.9 shows a schematic representation of a one-dimensional channel material with different types of molecular arrangements.

If the guests are much smaller with respect to the cavities, such as naphthalene in zeolite L, they can stack and form dimers [102]. Molecules with appropriate size and shape can come close enough and thus significant coupling of their electronic transition dipole moments can occur. One example that has been recently reported in detail is obtained by inserting the cationic dye Py^+ into zeolite L [148a]. At higher concentrations, the spatially resolved luminescence spectra recorded at the channel entrances show a new band at lower energy. This band, along with its lifetime, is identical to that observed for Py^+ aggregates. Due to the insertion method, dye molecules will accumulate in the region of the channel entrances before they can diffuse further in. This leads to a kind of "traffic jam" where molecules are packed closely enough to show significant exciton splitting. As an illustration of this effect, we show in Figure 9.10 a series of fluorescence microscopy images of zeolite L crystals with increasing loading concentrations of green emitting perylene dyes (DXP) [148b]. In the first sample (a), the dye concentration is low and thus they are far apart such that no exciton splitting appears. The green emission of the monomeric species can be observed. When the concentration is increased (b), the dyes become more closely packed and the coupling between them is significant. The fluorescence microscopy image in Figure 9.10b clearly shows the red light stemming from

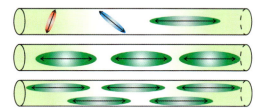

Figure 9.9 Schematic representation of a one-dimensional channel material loaded with different types of guest molecules. The orientation of the electronic transition dipole moment is indicated by a double arrow. *Top channel*: The red molecule is small enough to fit into a single unit cell and can arrange itself nearly perpendicular to the channel axis. The blue molecule is slightly larger and thus is oriented at an angle of about 45°. The green molecule is so large that it can only align parallel to the channel. *Middle channel*: Arrangement of noninteracting, large molecules that align themselves along the channel axis due to their size and shape. *Bottom channel*: Molecules with appropriate sizes and shapes can stack with each other, leading to excitonic states.

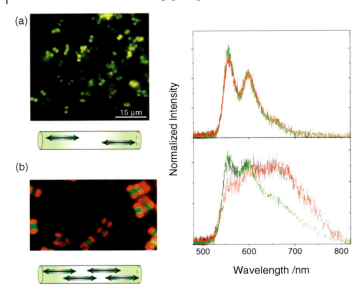

Figure 9.10 Evidence of guest aggregate formation by means of fluorescence microscopy. Fluorescence microscopy images of zeolite L crystals loaded with green emitting perylene dyes. The dye loading increases from 1% (a) to 30% (b). On the right, emission spectra were recorded for single dye-loaded zeolite crystals with the corresponding loading levels. The red emission band observed at higher loadings results from the formation of J-aggregates in the channels.

the channel entrances, while the middle part of the crystal exhibits the typical green emission. This color change is due to increased coupling of the transition dipole moments of the guests at shorter distances imposed by the higher packing. The single-crystal emission spectra given in the left part of Figure 9.10 were recorded at different locations. In both cases, light emitted from the center of the crystal (green curve) shows features that are characteristic of DXP monomers. The spectra recorded in the region of the channel entrances (red curves), however, exhibit a new band at longer wavelengths for higher dye loadings. This band results from the formation of J-aggregates.

The formation of aggregates is advantageous for electron transfer, as the formation of eximers improves the orbital overlap needed for this transport process. Porous materials, such as MOFs, have recently been used for mechanical entrapment of molecular exciplexes [30]. This phenomenon is less useful for FRET applications, as dye aggregation leads to luminescence quenching. The degree of molecular interaction can often be tuned by molecular design, and is especially successful in one-dimensional channel materials. By the addition of bulky substituents that have little effect on the electronic states of the chromophore (such as saturated aliphatic chains or tertiary butyl groups), one can prevent electronic transition dipole moment coupling by increasing the intermolecular distance.

The transport properties in the host–guest materials can thus be controlled and fine-tuned by different types of chemistry. The degree of chromophore–chromophore interaction not only depends on the shape and size of the molecules but also on the cavity geometry of the host. The host–guest interaction is controllable through molecular design of the guest species and proper choice of the host system. Host–chromophore interactions, on the other hand, rely on compatibilities (or lack thereof) between the guests and the host's inner walls. This offers additional tuning options for transport processes by selective modifications of the internal cavity walls, which is especially useful for mesoporous materials.

9.5
Self-Assembly in Solution and on Surfaces

9.5.1
Self-Assembly in Solution

Self-assembly is a process that is not achievable by standard synthetic techniques, in which atoms, molecules, and molecular systems spontaneously arrange themselves into patterns [149]. These newly formed systems either amplify the properties of the individual components by cooperative effects or possess new properties unique to the ensemble, such as luminescence, electrical conductivity or resistance, photoinduced processes, or vectorial movement. We are primarily concerned with the self-assembly processes involving noncovalent interactions such as electrostatic attractions, van der Waals forces, π–π stacking, metal coordination, hydrophobic effects, and hydrogen bonding [150]. Great efforts have been directed to induce and control such processes in the laboratory, therefore, self-assembly has been largely and successfully first explored on the molecular level [151]. More recently, attention has shifted from molecules to nanoobjects [66], nanoparticles [152], rods [63], and plates [153] in order to assemble them into 1D or 2D arrays analogous to what has been done with molecules. The use of nanometer-sized objects can easily bridge the gap between the molecular (Angstrom-sized) and microscopic (micron-sized) worlds. The potential applications of nanoparticles to form assemblies are limited due to the typical shape of nanoparticles, which are generally spherical and without geometrical preference toward site-specific functionalization. Unlike spherical nanoparticles, zeolite L nanocrystals can undergo controlled self-assembly resulting in defined 1D, 2D, or 3D structures, owing to their controlled base and surface functionalization [122]. Other advantages offered by zeolite L are its optical transparency, stiffness, defined morphology, and ease of chemical functionalization.

Recently, De Cola and coworkers reported a way to assemble zeolite L crystals in solution. In this study, formation of one-dimensional zeolite assemblies through coordinative interactions between Zn^{2+} and zeolites functionalized with terpyridine derivative was demonstrated (Figure 9.11) [154]. Since the channel entrances and coat of zeolite L can be functionalized separately, they can self-assemble into rod-like architectures [122]. The use of Zn^{2+} as a "chemical glue" for terpyridine-

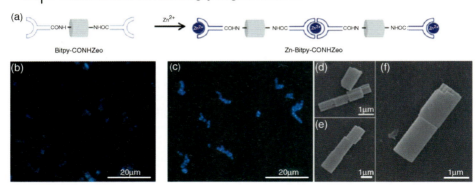

Figure 9.11 Assemblies of zeolite L crystals. (a) Terpyridine-terminated zeolite L crystals were mixed with Zn^{2+} to induce self-assembly. (b) Optical micrograph of unordered zeolite crystals. (c) Optical micrograph after addition of $ZnCl_2$. (d–f) SEM images of the assembled crystals [154].

functionalized zeolite L is desirable since the complexation–decomplexation of Zn^{2+} with terpyridine is a dynamic and reversible process [155]. The complexation is followed by distinctive changes in the photophysical properties (i.e., absorption and emission) and the Zn–terpyridine complexes follow either a 1 : 1 or a 1 : 2 stoichiometry leading to rod-like assemblies.

Furthermore, multicolor systems have been created by using different fluorophores loaded into the zeolite L channels in order to study energy transfer (Figure 9.12). Applications for such systems range from microbarcodes used for bioimaging and tagging purposes to the creation of unidirectional light-harvesting systems [156].

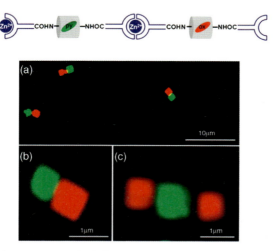

Figure 9.12 Dye-loaded zeolite pair [154].

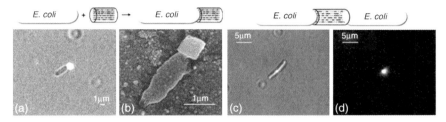

Figure 9.13 1:1 zeolite L/bacterium assembly in PBS buffer solution. (a) White and blue light illumination in an optical microscope. (b) SEM image of the assembly. (c) Self-assembly of two bacteria with functionalized 1 μm zeolite L as the junction. (d) In the same sample, fluorescence of the pyronine-filled zeolite can be observed at the junction, under blue light illumination.

An extension of this study involves the assembly with living systems, or in a more specific context, the assembly of bacteria with zeolite L as connecting abiotic units [157]. Functionalized zeolite L crystals were noncovalently bound to nonpathogenic bacteria (*Escherichia coli*) and these assemblies were visualized by fluorescence spectroscopy (Figure 9.13). Moreover, owing to the particularly defined geometrical arrangement of the zeolite and bacteria, self-organization of two bacteria was accomplished by using the zeolite as bridging element. This assembly is formed by electrostatic interactions between the negatively charged outer cell membrane and the positively charged zeolite L crystal. The positive charges were located only at the entrance of the channels, while the remaining surface retained its overall negative charge. The zeolite–bacterium assembly was able to live under physiological conditions and movement of the living system was not prevented. The organism was able to "swim" in the solution even with a heavy load such as the 1 μm zeolite.

Optically transparent, flexible, and long zeolite fibers can be obtained by taking advantage of ethanol-assisted self-assembly of zeolite nanocrystals [158]. A drop of an ethanol suspension of nanosilicalite was first placed on a glass slide. As the ethanol evaporated, the monodisperse nanocrystals self-assembled into many long cylindrical fibers with a diameter of about 27–33 μm and an average length of 1.5 cm. The fibers were optically transparent both before and after the calcination at 550 °C and show good flexibility when twisted. For the construction of such assemblies, the concentration and reactivity of silanol groups on the zeolite nanocrystals are crucial for hydrogen bond formation and for subsequent cross-linking (Si–O–Si covalent bond) during calcination. Ethanol is advantageous over water as a dispersion medium because ethanol is more volatile and thus easier to remove from the self-assembled structures. Ethanol wets both hydrophilic and hydrophobic surfaces and stabilizes zeolite nanocrystals by partially grafting them with ethyl groups through esterification/hydrogen bond formation [159]. This is important during the self-assembly process, as surface grafting reduces cohesive forces between zeolite nanocrystals, thus preventing premature aggregation.

Self-assembly of nanoparticles into three-dimensional objects like hollow spheres have many applications in chemistry, biotechnology, and materials science [160].

Various methods, including cosurfactant [161], phase separation [162], rapid quench [163], and ultrasonication [164], have been reported for the preparation of hollow spheres with mesoporous shells. Since spheres are symmetric, most of the spherical shaped nanoparticles have been employed for 3D assemblies. However, nanoparticles that are nonspherical have a tendency to orient randomly within the three-dimensional structures. Zeolites are an ideal construction material for the realization of hollow spheres owing to their high thermal stability, large porosity, high shape selectivity, and intrinsic chemical activity. However, the methods used to prepare hollow mesoporous silica spheres cannot be used to fabricate hollow zeolite spheres due to differences in synthetic conditions and formation mechanisms of mesoporous silica and zeolites. Therefore, alternative methods based on layer-by-layer accumulation of zeolite nanocrystals on polystyrene spheres or sonication of nanocrystals with toluene-dispersed water droplets were reported in order to prepare hollow spheres of microporous zeolites [165].

Tang and coworkers reported the fabrication of hollow spheres with differently sized zeolite β-particles through layer-by-layer self-assembly of nanozeolite–polymer multilayers on polystyrene (PS) latex, coupled with removal of the core by calcination [166]. The negatively charged zeolite framework leads to self-assembly of zeolite–polymer multilayers on colloidal templates due to electrostatic attraction between a negatively charged nanozeolite and a positively charged polymer. The pH and ionic strength of the colloidal solution, crystal size of nanozeolites, and size of PS latex templates were also reported as factors that affect the preparation of hollow zeolite spheres.

Yoon and coworkers reported a novel method to align nonspherical cubic zeolite A, octahedral zeolite X, and cylindrical zeolite L on the surface of spherulites during emulsion-templated assembly of nanocrystals into 3D structures (Figure 9.14) [167]. Water droplets dispersed in toluene were used as templates for the assembly of zeolites into microspherulites. Negatively charged zeolites aligned at the water–oil interface with a high degree of ordering. The microspherulites did not show any disintegration even after sonication in water. This indicates the formation of a strong bond between the zeolite nanocrystals. It was proposed that during sonication, condensation reactions occur between surface hydroxyl groups of neighboring zeolite nanocrystals and formed siloxyl links between them.

Yoon and coworkers demonstrated for the first time the self-assembly of an enzyme and a target compound grafted on zeolites. β-glucosidase and D-glucose-

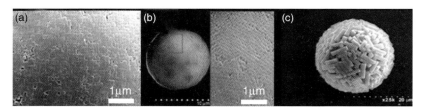

Figure 9.14 SEM images of zeolite assemblies. (a) Aligned zeolite A crystals. (b) Zeolite X spherulite. (c) Zeolite L spherulite [167].

tethered zeolite crystals self-assembled into thin (2–20 μm) and long fibrils (>1 cm). The enzymatic activity of β-glucosidase is also preserved even after the modified zeolites were stored in water for 6 months at room temperature [168]. It was proposed that this fibrillation phenomenon occurs through complexation of β-glucosidase with the D-glucose moieties tethered to the zeolite surfaces followed by crystallization of the enzyme on top of the surface-coating enzyme layer. In the absence of glucose tethered zeolite A crystals, β-glucosidase alone, either in the presence or in the absence of a large amount (100 mM) of plain, unmodified D-glucose, does not spontaneously crystallize into such fibrils. The fibrillation of avidin and D-biotin-tethering zeolite A was also studied by Yoon *et al*. This fibrillation is very similar to the previously described fibrillation of β-glucosidase and D-glucose-tethered zeolite-A [169].

9.5.2
Self-Assembly on Surfaces

Self-assembly strategies for the organization of matter make hierarchical ordering especially attractive from fundamental and applied points of view by avoiding expensive techniques such as photolithography. Assembling zeolite crystals into well-defined macroscopic structures is of particular interest, since zeolites are ideal host materials for supramolecular organization [72].

The generation of zeolite patterns adds a further level of organization, thus extending the ordering from the molecular to the macroscopic scale. Large zeolite crystals (longer than 100 μm) can be aligned by an electric field [170], or by mechanically shaking them into the grooves of a microstructured surface [171].

As demonstrated and calculated by Whitesides and coworkers, strategies employing self-assembly by minimization of interfacial free energy can be applied to position small objects on the micrometric scale [172]. At this scale length, surface–tension interactions should be considered as the tool of choice for manipulation [173, 174]. Such interactions have been exploited by Calzaferri and others to organize zeolite L crystals filled with oriented fluorophores in an hexagonal pattern on polydimethyl-siloxane (PDMS) films obtained through molding of self-assembled, ordered, macroporous polystyrene (PS) films, as shown in Figure 9.15 [175].

With these simple methods based on self-assembling processes, it is possible to hierarchically organize host–guest structures at a macroscopic level, thereby achieving a 2D chromophore arrangement. Another efficient and simple way to obtain such an organization of microobjects to be used for further optoelectronic or biomedical applications is represented by the formation of compact layers of single crystals, also called self-assembled monolayers (SAMs).

9.5.2.1 Zeolite L Monolayers: Synthesis and Characterization
The preparation of zeolite monolayers has been first reported in 1995 [176] and was then optimized by several other groups [56, 57]. Recently, the first example of using high-quality zeolite monolayers for specific chemical reactions such as *in situ* or "ship-in-a-bottle" synthesis has been demonstrated [49].

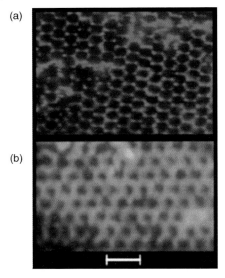

Figure 9.15 Fluorescence microscopy images of (a) Ox^+-and (b) Py^+-loaded zeolite crystals, hexagonally assembled on a patterned PDMS film. Scale bar = 7 (a) and 5 μm (b) [175].

An important step toward adding further functionality to dye–zeolite systems and achieving higher levels of organization is the controlled assembly of zeolite crystals into ordered and oriented structures. In the case of zeolite L, such assembly implies alignment of the unidirectional channels over a large number of crystals. One of the first methods for postsynthetic orientation of zeolite crystals was developed by Caro et al. [170], who observed that crystals of $AlPO_4$-5 and ZSM-5 can be oriented parallel upon application of an electric field of $2-3\,kV\,cm^{-1}$. However, crystals larger than 20 μm and aspect ratios of at least four to five are required for a successful alignment. Large ZSM-5 crystals (several tenths of a micrometer) of uniform size can also be aligned in the grooves of etched silicon wafers, as shown by Scandella et al. [171]. High-quality, closely packed monolayers of zeolite A crystals ranging 2–8 μm in diameter were prepared by sedimentation [56]. This method can be applied to various substrates and extended to other types of zeolites. Covalent bonding between the zeolite (LTA, ZSM-5) and the substrate (usually glass) using a variety of linking molecules was studied by Yoon et al. [56, 57]. This principle was used by the same group to produce micropatterned monolayers of ZSM-5 [58d].

Calzaferri and coworkers have applied these methods to zeolite L, aiming to arrange standing cylindrical crystals. In such an arrangement, the dye-filled channels would run perpendicular to the substrate surface, allowing efficient transport of electronic excitation energy toward the zeolite–substrate interface. The morphology of the crystal determines the degree of orientation since better results are obtained by using disk-shaped zeolites. Further optimization in crystal surface smoothness and size distribution leads to monolayers with even higher packing density [69b]. An underlying principle that has to be considered is that the interaction between the base of the zeolite L crystals and the substrate must be stronger than with the coat and, more important, any interaction between the zeolite crystals. Depending on the target substrate, for example, glass, quartz, gold, or indium tin oxide (ITO), different reagents have to be employed for the preliminary functionalization of both zeolite and substrate surfaces and, consequently, different procedures must be employed for preparing zeolite L monolayers.

Target substrates such as glass or quartz are typically treated with alkoxysilane molecules, acting as covalent linkers toward the zeolite surface (see Figure 9.16). The binding of the crystals onto the activated substrate then proceeds via nucleophilic substitution of the terminal group of the linker to the outer hydroxyl groups on the channel openings of zeolite crystals [40d,49, 67, 177].

A possible mechanism to account for the closely packing phenomenon is surface migration: the initial random attachment of the crystals undergoes continuous cycles of bond breaking/bond making until the zeolites close pack to each other, thus enhancing the bond strength by intercrystal surface hydrogen bond interactions. In order to promote such a mechanism, sonication has been proved to be a better method than reflux [66].

The formation of monolayers on conductive substrates can also be realized without any chemical modification, which would alter the electrochemical properties at the interface. By means of the microcontact printing technique (MCP) [178–181],

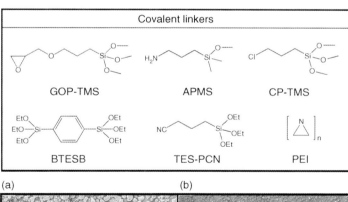

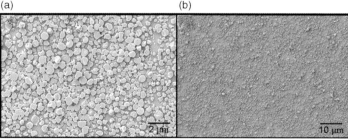

Figure 9.16 Commonly used covalent linkers in the preparation of zeolite monolayers. Examples of self-assembled monolayers on glass substrates, prepared employing (a) chloropropyl-trimethoxysilane (CP-TMS) and (b) polyethyleneimine (PEI).

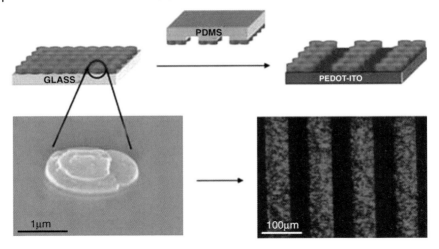

Figure 9.17 Schematic representation of the strategy developed for transferring and patterning zeolite monolayers by means of MCP technique [58f].

well-ordered and uniformly oriented zeolite L monolayers have been obtained on ITO, which is a common substrate used for device fabrication, thus opening the way for further applications in optoelectronics (Figure 9.17) [58f]. Moreover, it has been shown that it is possible to not only apply this method to cylindrical objects by transferring them and maintaining their orientation but also to use the anisotropy of the systems to control the different color emissions of zeolites filled with two different dyes by exciting the materials with polarized light.

As one end of the channels is sealed by the substrate, zeolite monolayers can only be loaded with chromophores from the other side of the channels, thus enabling the synthesis of a unidirectional antenna material [49]. Zeolite monolayers have strongly improved the controlled contact of the stopcocks to the surrounding material, opening a new field of possible applications. A challenging idea is the development of new photonic devices for solar energy conversion and for light-emitting devices. More details on these applications are given in Section 9.6.

9.6
Possible Optical and Biomedical Applications of Nanocontainers

9.6.1
Optical Applications

In view of practical applications, the coupling of zeolite-based functional materials with external devices is of special interest. Organization of dipolar second-order nonlinear optical (NLO) dyes in uniform orientations, for example, represents an essential step toward a successful application of the dyes for second harmonic generation [82]. As demonstrated by Yoon and coworkers, highly ordered systems can be created by inserting a noncentrosymmetric dye into silicalite-1 zeolite films, which fills up the crystals in a preferred orientation [182, 183].

Recently, Calzaferri and coworkers developed a strategy to insert dye-loaded zeolite crystals into a polymer matrix while maintaining transparency in the visible range [52]. The guest molecules are shielded from the polymer matrix by the modified zeolites, while the polymer basically offers them another protective layer. Therefore, moisture- and oxygen-sensitive dye–zeolite materials can be used in such arrangements. These transparent host–guest inorganic–organic hybrid materials offer novel possibilities for the development of optical devices such as lenses, special mirrors, filters, polarizer, grids, optical storage devices, and windows.

The principle of inserting dye-loaded zeolites into polymer matrices has also been extended to zeolite monolayer assemblies. In the former sections, we have seen that anisotropic arrangement of dye molecules is possible within the channel of a zeolite. By assembling the crystals with their c-axis perpendicular to a semiconductor surface, transfer of excitation energy from or to the zeolite–semiconductor interface can occur. Two challenging applications are the development of new photonic devices for solar energy conversion (solar cells) and light emission (LEDs).

9.6.1.1 Hybrid Solar Cells

Over the last few decades, the field of organic solar cell research has developed rapidly [184–186]. The solar power conversion efficiencies have increased from 1% back in the late 1980s to over 3.5% currently [187, 188]. This increase has been triggered by the introduction of new materials, improved engineering, and development of more sophisticated devices [189]. The field of organic solar cells also benefited from the development of light-emitting diodes based on similar technologies. An important feature of active materials in solar cells is efficient light adsorption over the whole solar spectrum. This is a major drawback of silicon for the production of solar cells because silicon is an indirect bandgap material with low absorptivity at wavelengths longer than about 500 nm. Therefore, solar cells based on monocrystalline silicon require a thickness of 200–300 µm in order to be able to absorb enough light. An antenna material that is strongly absorbing in the visible wavelength range and transfers this energy by a radiationless process to the semiconductor appears to be an attractive solution for overcoming this problem [136, 190]. Highly organized zeolite L materials are good candidates for developing such dye-sensitized solar cells or luminescent solar concentrators [49, 191].

In chemical devices, a high degree of supramolecular organization is important for obtaining the desired macroscopic properties [192]. A possibility for achieving this organization is the controlled assembly of the zeolite crystals into oriented structures and the preparation of monodirectional materials. One idea for the realization of a new type of sensitized solar cell is to bridge the absorption gaps of solar cells with dye-loaded zeolites. The inserted dyes absorb light over a broad spectrum and transmit it to the stopcock molecules via radiationless energy transfer mechanism. The stopcock molecules, in turn, will facilitate the transfer of this energy to the active material of the solar cell. The electronic excitation can then produce the electron–hole

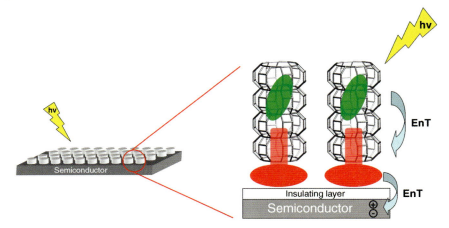

Figure 9.18 Scheme of a sensitized solar cell based on dye-loaded zeolite L antenna systems. The light is absorbed by the monolayer and transported, through the stoppers, to the semiconductor surface. Electron–hole pairs are formed in the semiconductor by energy transfer from the antenna system to the conduction band of the semiconductor.

pairs needed for the cell to function [193]. A possible setup of such a solar cell is schematically presented in Figure 9.18.

It is important that the semiconductor is covered with an insulating layer that makes electron transfer impossible and avoids short circuits in the device. In this approach, the dye–zeolite composite acts as a light-absorbing antenna, whereas the semiconductor is only responsible for the electron–hole pair separation. Therefore, a very thin semiconductor layer on the order of only a few hundred nanometers is sufficient. Furthermore, since the connection between the light-absorbing antenna and the semiconductor layer is realized by a dipole or higher multipole mechanism, and not by electron transfer, there is no need for regeneration of the antenna system.

Energy transfer from dye–zeolite crystals to bulk silicon has recently been demonstrated [194]. The efficiency of energy transfer between antenna and semiconductor was optimized using disk-shaped zeolite L materials. However, the maximum efficiency is limited by the bidirectionality of the antenna system. Since the harvested excitation energy is transferred to acceptor dyes located at both ends of the channels, it can only reach a theoretical maximum of 50%. Monodirectional antenna systems would be the solution to this issue and can be achieved if one side of the channel is selectively blocked.

9.6.1.2 Hybrid Light-Emitting Devices (LEECs and LEDs)

In 2000, the Nobel Prize in chemistry went to Alan Heeger [195], Alan MacDiarmid, and Hideki Shirakawa for their work on conductive polymers [196–199]. Their research enabled the design of polymer light-emitting diodes (PLED). Since then, extensive studies have been conducted in the field of polymer light-emitting diodes to improve their properties [200–203].

Despite the differences in the requirements, of the concepts, and in the kinetics of the processes, it is easy to envisage that mesoporous materials can be used for realizing OLEDs. The use of a zeolite-based supramolecular host–guest system in the emitting layer of such a device could be a viable strategy to stabilize the emitting guest and increase the operation lifetime of the device based on organic compounds and metallic complexes. In this regard, the work of Corma and coworkers has demonstrated a remarkable increase in the stability of polyphenylene vinylene (PPV), an electroluminescent polymer, by encapsulating it in the pores of zeolites X and Y [204]. They have also extended this principle to electrochemiluminescent organometallic complexes and built up a light-emitting electrochemical cell (LEEC) based on Ru $(bpy)_3^{2+}$-loaded zeolite Y [205].

The development of such zeolite-based light-emitting devices is challenging because charge transport between the external electrodes and the internal guests in the zeolite has to be realized. In addition, the construction of submicrometric films of zeolite particles encapsulating an electroluminescent guest is not an easy task.

An interesting approach for hybrid LED fabrication is to use chromophore-loaded zeolite crystals as acceptors for the excitation energy coming from the electroluminescent polymer. This is also expected to increase the stability of the polymer, as it obtains a fast nondecomposing relaxation pathway by transferring excitation energy to the harvesting material. In addition, the final PLED emission color could be easily tuned by varying the chromophores in the host material. The strategy consists of employing dye-loaded zeolite crystals [49a] as part of the hybrid emitting layer that will be interfaced to the polymer layer via stopcock molecules, as illustrated in Figure 9.19 [206]. By using appropriately sized zeolite crystals, with an average dimension of 30–40 nm, a maximum layer thickness of 100 nm would not be exceeded.

Another approach consists of using a semiconductor as energy transfer donor for LEDs [30]. While conventional organic LEDs are excited electrically, limiting the

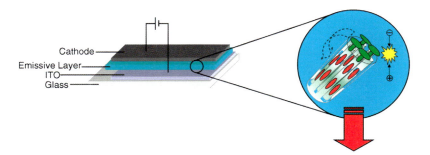

Figure 9.19 A schematic of a potential LED setup containing dye-loaded zeolite crystals in the emissive layer. Stopcock molecules are added to the channel entrances, forming an antenna system that can be embedded in a polymer matrix. Stoppers act as an intermediate between the polymer and the dyes. If the stoppers were based on electroluminescent metal complexes, they would increase the light output efficiency by triplet harvesting [207, 208].

theoretical efficiency considerably [209], the efficiency of an energy transfer LED is only limited by the electron–hole pair recombination, the energy transfer rate, and the fluorescence quantum yield of the dye. The advantage of such a design will then lie on the fact that the excitons formed by charge recombination will not be directly responsible for the emission since the dyes in the zeolite will be excited by energy transfer processes. In such a way, the accumulation of charges will be far from the emitters and a better stability of the emissive material is expected.

9.6.2
Biomedical Applications

Multifunctional nanoscale materials have attracted intense scientific and industrial interest because of their potential applications in optical, electronic, magnetic, biomedical, and many other fields [210]. In biomedical applications, functional nanomaterials have been extensively used for imaging, targeting, drug delivery, gene transfer, and biosensing [211].

Various nanomaterials have been designed and used for diagnostics and cell tracking [212] such as highly luminescent and photostable quantum dots (QD) [213] or less chemically toxic, and still very small (>5 nm) silicon particles and silica nanoparticles [214]. Silica-coated fluorescein– isothiocyanate-conjugated superparamagnetic iron oxide nanoparticles for the noninvasive tracking of stem cells were recently described [215]. Very small silica particles can be used not only for *in vitro* diagnostics [216] but also for *in vivo* applications [217]. Recently, a new generation of near-infrared fluorescent core–shell silica-based nanoparticles (C dots) with hydrodynamic diameters of 3.3–6.0 nm were described. These C dots have improved photophysical characteristics over the parent dye. A neutral organic coating prevents adsorption of serum proteins and facilitates efficient urinary excretion. Carbon particles [218] and different oxide nanomaterials [219] have also been employed for the same purposes. Carbon nanotubes have been used for transporting molecules, like proteins, from outside of a cell into its cytoplasma [220]. In another study, preparation of so-called immuno-carbon nanotubes capable of recognition of target pathogen cells via antibody–antigen interactions in a physiological environment was reported (Figure 9.20) [221]. Phospholipid-based systems such as liposomes [222] and micelles [223] are another type of nanosized materials that can suitably target diseased cells with their payload of imaging agents and are readily taken up by cells. Besides the nanosystems in bulk media, self-assembled monolayers (SAMs) of micrometer- or nanosized materials have been frequently used in biomedicine [224]. For example, self-assembled monolayers of gold nanoparticles have applications as biosensors [225], arrays of magnetic particles are used in microfluidic devices [226], and silica nanoparticle-modified gold electrodes are applied for DNA detection [227]. SAMs of iron oxide magnetic nanoparticles that are modified with organosilane agents are employed as biological assay systems [228]. A recent review by Whitesides and coworkers has elegantly summarized some possible strategies to obtain different surfaces and shows how microfabrication is in close connection with microbiology [229].

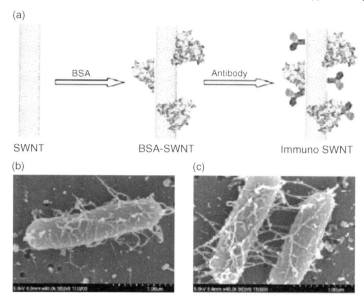

Figure 9.20 (a) Scheme on the formation of immuno SWNT. (b and c) SEM images of immuno SWNT binding to *E. coli* cells [221].

Among the various multifunctional materials, we will focus in this section on inorganic nanocontainers and their biomedical applications. Inorganic nanocontainers provide several functionalities for imaging, targeting, and delivery due to their tunable pore size, porosity, thermal and mechanical properties, biocompatibility, biodegradability, and size compatibility with biological systems [230]. Desirable biomolecules or drugs can be attached to the nanomaterials with appropriate surface modification [231]. Due to their structural characteristics such as channels or hollow interiors, it is possible to use nanocontainers as capsule-like carriers for biomolecules and drugs while protecting their payloads from denaturing agents [232]. Chemical modification of the pore entrances of the nanocontainers allows the targeted delivery of drugs to their desired sites of action and their release in a controlled manner [7]. Large surface areas of these materials can be used for immobilization of (bio)molecules, especially for biosensing [12a], and, finally, their optical properties provide a self-reporting feature that can be monitored *in vivo*, which is important for cell imaging and tracking [233]. However, for biomedical applications, it is important to consider not only the function but also the biocompatibility of a material. In a very recent article, Sailor *et al.* have shown that Si/SiO_2 particles are decomposed in the body of mice and completely excreted after 4 weeks [234]. Silicon exists primarily as a biodegradation product orthosilicic acid (H_4SiO_4), which is the form predominantly absorbed by humans and is naturally found in numerous tissues, including bone, tendons, aorta, liver, and kidney. Silicic acid administered to humans is efficiently excreted from the body through the urine. Furthermore, porous silica-based materials can be decomposed under conditions that do not significantly affect the activity of biomaterials loaded within the templates.

A recent review [235] of porous silicon (PS) gives detailed information of a number of properties that make it an attractive material for controlled drug delivery applications. Their electrochemical synthesis allows one to tailor their pore sizes from the scale of microns to nanometers. Easy modification of porous Si surfaces can be used to control, *in vivo*, the release rate of drug. The optical properties of their photonic structures provide a possible way of monitoring their location *in vivo*. Another interesting paper with respect to porous silicon [236] appeared recently, which reports on the fabrication and applications of PS-based microstructures in biosensing. Different morphologies of PS coated with gold nanoparticles have been described for biomolecular detection. Meso- and macro-PS have also been investigated for protein immobilization and detection using microarray techniques and for DNA biomolecule detection by impedance spectroscopy.

The use of inorganic nanomaterials as nanocontainers has mainly been focused on silica-based materials, such as mesoporous silica nanoparticles, nanorods, nanotubes [232, 237], porous silicon, and zeolites [238]. Examples include mesoporous silica nanomaterials (MSNs) with high surface areas (up to approximately $1500\,m^2\,g^{-1}$), high nanopore volumes (up to approximately $2\,cm^3\,g^{-1}$), and homogeneous nanopore structures, which have been exploited to encapsulate a variety of species, such as proteins, low molecular weight drugs, and nanoparticles [239]. MSNs have been employed as capsules with high, uniformly distributed loadings of catalase and lysozyme [240]. The ability for surface functionalization with various bioactive molecules enhances the nanoparticle accumulation in targeted organs and cells, or mitigates the uptake of the particles by the reticuloendothelial system [241]. The cellular uptake behavior of MSNs depends crucially on its outer surface properties, such as surface functional groups [53] or charges [242], both of which can be easily controlled [243]. Until now, only small molecule ligands, such as folic acid and mannose [244], have been attached to MSNs for the study of their interaction with cells. Although antibodies have been grafted on silica nanobeads [245] and other nanoparticulated systems [246], more specific cell-recognition agents, such as glycoproteins with large molecular size, have not yet been applied to mesoporous silica nanoparticles. Mesoporous silica nanoparticles have been used to deliver cytochrome *c*, a noncell membrane permeable protein [247], to cells and as carriers for various guest molecules such as anticancer drugs [248], genes [249], antibiotics, and nonsteroidal anti-inflammatory drugs [250]. An interesting strategy for controlling the release of drugs from the pores of mesoporous nanospheres and nanorods has been demonstrated (Figure 9.21) [251]. The entry ports of the mesopores were capped with nanoparticles such as quantum dots (CdS) or superparamagnetic iron oxides (Fe_3O_4). In this method, both nanoparticles and pore surfaces were functionalized and covalently linked by a disulfide bond. The cleavage of disulfide bonds by reducing agents allows the release of drugs from the pores.

The mesoporous silica MCM-41 has also been used as an amorphous material to immobilize selected enzymes and for size-selective separation of proteins [252]. Mesoporous materials are excellent supports for enzyme immobilization due to their

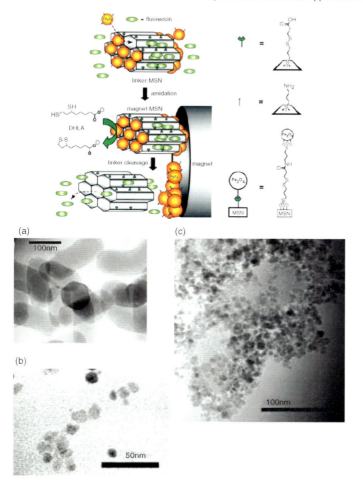

Figure 9.21 Schematic of the stimuli-responsive delivery system. TEM images of (a) linker MSNs, (b) APTS-coated Fe_3O_4 nanoparticles, and (c) Fe_3O_4-capped magnet-MSNs [251].

suitable pore dimensions that match the molecular size of many biomolecules and their large surface area. Both permit relatively high enzyme loading, while allowing the enzymes to retain their activity. Mesoporous silica nanoparticles have also been used as cell markers for magnetic resonance imaging (MRI) and optical imaging [253].

Zeolites are another class of silica-based nanocontainers. Unlike mesoporous silica, zeolite molecular sieves are restricted to a pore size of around 15 Å and, very important, they are crystalline materials. Like mesoporous silica, zeolites are characterized by very large specific surface areas, ordered pore systems, and well-defined pore radius distributions. Synthetic zeolites have two important advantages compared to mesoporous silica. It is relatively easy to control their size and morphology, which depend on the reaction conditions. Their highly selective surface functionalizations, which are described in the previous sections, present another advantage. Recently, natural and synthetic zeolites have attracted widespread attention from scientists of different disciplines, and in particular from biotechnology, for their application in drug delivery, magnetic resonance imaging, and biosensing. Because of their uniform and adjustable surface properties (i.e., surface charges and hydrophilicity/hydrophobicity) and their channel systems, these materials are expected to offer great opportunities for bionanomedical applications.

Zeolites have been investigated for use as carriers for a variety of drugs, such as aspirin, anthelmintics, and antitumor agents [254]. Natural zeolites (e.g., clinoptilolite) [255] have been advocated as oral drug delivery systems. Synthetic zeolites, in particular, zeolite X and a zeolitic product obtained from a cocrystallization of zeolite X and zeolite A, could offer a good means of delivering ketoprofen successfully without adverse effects [256]. Use of zeolite Y as a slow release agent for anthelmintic drugs [257] was found to be more successful in killing adult worms in rodent patients than administration of the pure drug. In addition, zeolite nanocrystals have been applied in the enrichment and identification of low-abundance peptides and proteins [258], as well as for enhancement of transfection efficiencies of pDNA. Proteins adsorbed on the zeolite surface were shown to enter the endosomal pathway after phagocytosis and could be cleaved by the endosomal proteases [259]. Zeolite-based silicalite nanoparticles can enhance the transfection efficiencies generated by polyethylene imine–plasmid DNA (PEI–pDNA) complexes via a sedimentation mechanism, as well as that of pDNA alone when surface functionalized with amine groups [260]. Some types of zeolites may coimmobilize enzymes and mediators for preparing biosensors [261]. Gd^{3+}-doped zeolite nanoparticles have proven to be good contrast agents for magnetic resonance imaging in medical diagnosis [262]. Zeolite Y has also been employed in similar applications [263]. Zeolites are also used as antibacterial agents [264], especially when silver ions are incorporated into these materials by ion exchange or in the adjuvant treatment of cancer [265].

Recently, we have reported on the use of zeolite L nanocrystals for diagnostics, imaging, and targeting. Zeolite L nanocrystals containing hydrophobic fluorophores are developed as new labels for biosensor systems. The external surface of the particle was modified with biomolecules such as antibodies that can be cross-linked in order to stabilize them on the zeolite surface. A microarray fluorescent sandwich immunoassay based on such nanoparticulate labels showed high sensitivity in thyroid stimulation hormone (TSH) assay [266]. The zeolite L nanocrystal channels contain a strongly emitting molecular dye that can be used as a fluorescent label for optical imaging.

In another application, the surface of zeolite L nanocrystals was modified with a gadolinium complex, introducing a probe for magnetic resonance imaging (Figure 9.22) [54a]. In order to bind the paramagnetic ions strongly to its surface, the dye-filled zeolite was functionalized on the entire surface by chelating ligands. The combination of the fluorescent dyes in the channels and surface coverage with a paramagnetic species leads to a composite material that combines optical and magnetic properties for dual imaging.

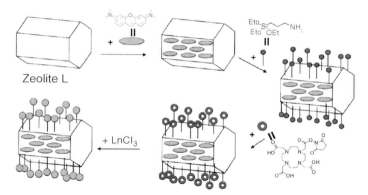

Figure 9.22 Schematic presentation of synthesis of Ln-DOTA functionalized zeolite L crystals [54a].

Another interesting example is the use of zeolite L for targeting, labeling, and photoinactivation of pathogenic and antibiotic-resistant bacteria. A highly green-luminescent dye was inserted into the channels of zeolite L nanocrystals for imaging and to label the cells. The outer surface was functionalized with a photosensitizer (a silicon phthalocyanine) that forms toxic singlet oxygen upon red light irradiation and with additional amino groups for binding to living microorganisms (Figure 9.23) [267].

Finally, we have recently described the radiolabeling of nanometer-sized zeolite L crystals that were injected in rats for *in vivo* experiments (Figure 9.24) [268]. Biocompatible zeolite L crystals have been filled with the radiometal $^{111}In^{3+}$ and closed using a tailor-made stopper molecule. The surface of the zeolites has been functionalized with different chemical groups in order to investigate the different biodistributions depending on the nature of the functionalizations.

It is also worthwhile to provide some selected examples of the use of other known nanocontainers in biomedicine. Cai *et al.* used nickel-embedded magnetic carbon nanotubes that align with the magnetic flux when exposed to a magnetic field, generating a perpendicular hit on the cell membrane [269]. The method denoted as "nanotube spearing" allowed successful gene delivery into primary B cells and neurons. Lanthanide-based MOFs have been used as multimodal cellular probes [270] and drug storage and release has been evaluated for some Tb^{3+}-, Fe^{2+}-, and Cr^{3+}-based MOFs [271]. Biocompatible capsules have been prepared from various polymer pairs and investigated for cellular uptake [272]. Targeting and uptake of polymer capsules biofunctionalized with antibodies that bind specifically to tumor-associated markers on colorectal cancer cells has been reported [273]. Multifunctional lipid/quantum dot hybrid nanocontainers were described for controlled targeting of live cells.

To conclude, the focus of medical science is currently on improving the therapeutic efficacy and specific targeting of nano delivery systems. Some types of hybrid nanomaterials can provide two or more capabilities for imaging, targeting, and drug

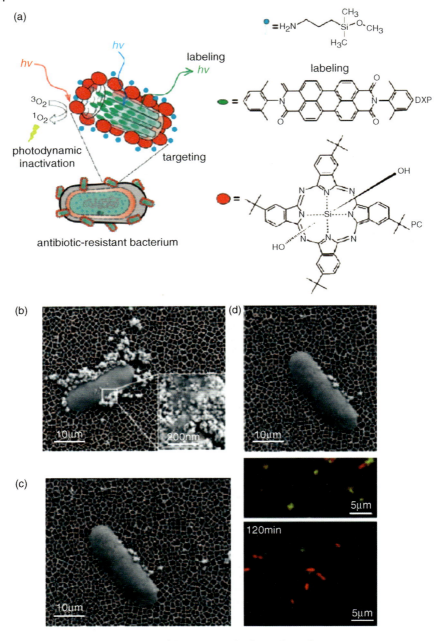

Figure 9.23 (a) Schematic view of the multifunctional nanomaterial. (b)-(d) SEM images of bacteria treated with the amino functionalized material. The bottom right part of the figure shows fluorescence microscopy images (λ_{exe} = 470–490nm) of E. coli cells during photodynamic treatment in PBS, red emission arises from dead bacteria [267].

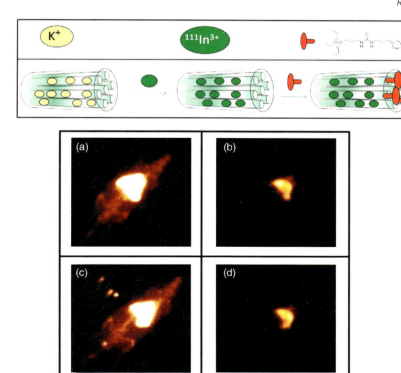

Figure 9.24 Radiolabeling of zeolite L nanocrystals and static scintigraphic images of the rats. Images (a) and (b) show the zeolite L injected into rat 1 in high and low contrast, respectively. Images (c) and (d) show the surface-modified (PEGylated) zeolite L injected into rat 2 in high and low contrast, respectively.

delivery. Taking advantage of recent progress, it may be possible to construct multifunctional nanocontainers that not only provide controlled delivery, selective targeting, imaging, and direction control but also biocompatibility besides alleviating cytotoxicity.

References

1 (a) Schmid, G. (2003) *Nanoparticles: From Theory to Application*, Wiley-VCH Verlag GmbH, Weinheim; (b) Rotello, V.E. (2003) *Nanoparticles: Building Blocks for Nanotechnology*, Nanostructure Science and Technology, Springer, Berlin; (c) Wang, Z. and Lu, Y. (2009) Functional DNA directed assembly of nanomaterials for biosensing. *J. Mater. Chem.*, **19**, 1788–1798; (d) Corchero, J.L. and Villaverde, A. (2009) Biomedical applications of distally controlled magnetic nanoparticles. *Trends Biotechnol.*, **27** (8), 468–476; (e) Chopra, N., Gavalas, V.G., Hinds, B.J., and Bachas, L.G. (2007) Functional one-dimensional nanomaterials: applications in nanoscale biosensors. *Anal. Lett.*, **40**,

2067–2096; (f) Guo, J., Ping, Q., Jiang, G., Huang, L., and Tong, Y. (2003) Chitosan-coated liposomes: characterization and interaction with leuprolide. *Int. J. Pharm.*, **260**, 167–173.

2 (a) Blanazs, A., Armes, S.P., and Ryan, A.J. (2009) Self-assembled block copolymer aggregates: from micelles to vesicles and their biological applications. *Macromol. Rapid Commun.*, **30**, 267-L 277; (b) Pouton, C.W. and Porter, C.J.H. (2008) Formulation of lipid-based delivery systems for oral administration: materials, methods and strategies. *Adv. Drug Deliv. Rev.*, **17**, 625–637.

3 Rijcken, C.J.F., Soga, O., Hennink, W.E., and van Nostrum, C.F. (2007) Triggered destabilization of polymeric micelles and vesicles by changing polymers polarity: an attractive tool for drug delivery. *J. Control. Release*, **120**, 131–148.

4 Andresen, T.L., Jensen, S.S., and Jorgensen, K. (2005) Advanced strategies in liposomal cancer therapy: problems and prospects of active and tumor specific drug release. *Prog. Lipid Res.*, **44**, 69–97.

5 Yaghi, O.M., O'Keeffe, M., Ockwig, N.W., Chae, H.K., Eddaoudi, M., and Kim, J. (2003) Reticular synthesis and the design of new materials. *Nature*, **423**, 705–714.

6 (a) Lu, F., Gu, L., Meziani, M.J., Wang, X., Luo, P.G., Veca, L.M., Cao, L., and Sun, Y.P. (2009) Advances in bioapplications of carbon nanotubes. *Adv. Mater.*, **21**, 139–152; (b) Duclaux, L., Bantignies, J.L., Alvarez, L., Almairac, R., and Sauvajol, J.L. (2006) Structure of doped single wall carbon nanotubes. *Annu. Rev. Nano Res.*, **1**, 215–254.

7 Slowing, I.I., Vivero-Escoto, J.L., Wu, C.-W., and Lin, V.S.-Y. (2008) Mesoporous silica nanoparticles as controlled release drug delivery and gene transfection carriers. *Adv. Drug. Deliv. Rev.*, **60**, 1278–1288.

8 (a) Buck, S.M., Koo, Y.-E.L., Park, E., Xu, H., Philbert, M.A., Brasuel, M.A., and Kopelman, R. (2004) Optochemical nanosensor PEBBLEs: photonic explorers for bioanalysis with biologically localized embedding. *Curr. Opin. Chem. Biol.*, **8**, 540–546; (b) Sukhorukov, G.B., Rogach, A.L., Zebli, B., Liedl, T., Skirtach, A.G., Kçhler, K., Antipov, A.A., Gaponik, N., Susha, A.S., Winterhalter, M., and Parak, W.J. (2005) Nanoengineered polymer capsules: tools for detection, controlled delivery, and site-specific manipulation. *Small*, **1**, 194–200.

9 Soussan, E., Cassel, S., Blanzat, M., and Rico-Lattes, I. (2009) Drug delivery by soft matter: matrix and vesicular carriers. *Angew. Chem., Int. Ed.*, **48**, 274–288.

10 Banerjee, R., Phan, A., Wang, B., Knobler, C., Furukawa, H., O'Keeffe, M., and Yaghi, O.M. (2008) High-throughput synthesis of zeolitic imidazolate frameworks and application to CO_2 capture. *Science*, **319**, 939–943.

11 Cao, Q. and Rogers, J.A. (2009) Ultrathin films of single-walled carbon nanotubes for electronics and sensors: a review of fundamental and applied aspects. *Adv. Mater.*, **21**, 29–53.

12 (a) Slowing, I.I., Trewyn, B.G., Giri, S., and Li, V.S.-Y. (2007) Mesoporous silica nanoparticles for drug delivery and biosensing applications. *Adv. Funct. Mater.*, **17**, 1225–1236; (b) Slowing, I., Trewyn, B.G., and Lin, V.S.Y. (2006) Effect of surface functionalization of MCM-41-type mesoporous silica nanoparticles on the endocytosis by human cancer cells. *J. Am. Chem. Soc.*, **128**, 14792–14793.

13 Lin, Y., Taylor, S., Li, H., Fernando, K.A.S., Qu, L., Wang, W., Gu, L., Zhou, B., and Sun, Y.-P. (2004) Advances toward bioapplications of carbon nanotubes. *J. Mater. Chem.*, **14**, 527–541.

14 Demas, J.N., DeGraff, B.A., and Coleman, P.B. (1999) Oxygen sensors based on luminescence quenching. *Anal. Chem.*, **71**, 793A–800A.

15 Greathouse, J.A. and Allendorf, M.D. (2006) The interaction of water with MOF-5 simulated by molecular dynamics. *J. Am. Chem. Soc.*, **128**, 10678–10679.

16 Tanabe, K.K., Wang, Z., and Cohen, S.M. (2008) Systematic functionalization of a metal-organic framework via a postsynthetic modification approach. *J. Am. Chem. Soc.*, **130**, 8508–8517.

17 (a) Trewyn, B.G., Slowing, I.I., Giri, S., Chen, H.-T., and Lin, V.S.-Y. (2007) Synthesis and functionalization of a mesoporous silica nanoparticle based on the sol–gel process and applications in controlled release. *Acc. Chem. Res.*, **40**, 846–853; (b) Barbé, C., Bartlett, J., Kong, L., Finnie, K., Lin, H.Q., Larkin, M., Calleja, S., Bush, A., and Calleja, G. (2004) Silica particles: a novel drug delivery system. *Adv. Mater.*, **16**, 1959–1966.

18 Moller, K. and Bein, T. (1998) Inclusion chemistry in periodic mesoporous hosts. *Chem. Mater.*, **10**, 2950–2963.

19 (a) Khashab, N.M., Trabolsi, A., Lau, Y.A., Ambrogio, M.W., Friedman, D.C., Khatib, H.A., Zink, J.I., and Stoddart, J.F. (2009) Redox- and pH-controlled mechanized nanoparticles. *Eur. J. Org. Chem.*, 1669–1673; (b) Angelos, S., Yang, Y.W., Patel, K., Stoddart, J.F., and Zink, J.I. (2008) pH-Responsive supramolecular nanovalves based on cucurbit[6]uril pseudorotaxanes. *Angew. Chem., Int. Ed.*, **47**, 2222–2226.

20 Fu, Q., Rao, G.V.R., Ista, L.K., Wu, Y., Andrzejewski, B.P., Sklar, L.A., Ward, T.L., and López, G.P. (2003) Control of molecular transport through stimuli-responsive ordered mesoporous materials. *Adv. Mater.*, **15**, 1262–1266.

21 (a) Montera, R., Vivero-Escoto, J., Slowing, I.I., Garrone, E., Onida, B., and Lin, V.S.Y. (2009) Cell-induced intracellular controlled release of membrane impermeable cysteine from a mesoporous silica nanoparticle-based drug delivery system. *Chem. Commun.*, 3219–3221; (b) Zhao, Y., Trewyn, B.G., Slowing, I.I., and Lin, V.S.Y. (2009) Mesoporous silica nanoparticle-based double drug delivery system for glucose-responsive controlled release of insulin and cyclic AMP. *J. Am. Chem. Soc.*, **131**, 8398–8400.

22 (a) Vivero-Escoto, J.L., Slowing, I.I., Wu, C.Y., and Lin, V.S.Y. (2009) Photoinduced intracellular controlled release drug delivery in human cells by gold-capped mesoporous silica nanosphere. *J. Am. Chem. Soc.*, **131**, 3462–3463; (b) Aznar, E., Marcos, M.D., Martinez-Mañez, R., Sancenón, F., Soto, J., Amorós, P., and Guillem, C. (2009) pH- and photo-switched release of guest molecules from mesoporous silica supports. *J. Am. Chem. Soc.*, **131**, 6833–6843.

23 Aznar, E., Coll, C., Marcos, M.D., Martinez-Mañez, R., Sancenón, F., Soto, J., Amorós, P., Cano, J., and Ruiz, E. (2009) Borate-driven gatelike scaffolding using mesoporous materials functionalised with saccharides. *Chem. Eur. J.*, **15**, 6877–6888.

24 (a) Auerbach, S.M., Carrado, K.A., and Dutta, P.K. (eds) (2003) *Handbook of Zeolite Science and Technology*, Marcel Dekker, New York; (b) Breck, D.W. (1974) *Zeolite Molecular Sieves*, John Wiley & Sons, Inc., New York.

25 Sherman, J.D. (1999) Synthetic zeolites and other microporous oxide molecular sieves. *Proc. Natl. Acad. Sci. USA*, **96**, 3471–3478.

26 (a) Bärlocher, C., Meier, W.M., and Olsen, D.H. (2001) *Atlas of Zeolite Framework Types*, 5th edn, Elsevier, Amsterdam, p. 19; (b) Ohsuna, T., Slater, B., Gao, F., Yu, J., Sakamoto, Y., Zhu, G., Terasaki, O., Vaughan, D.E.W., Qiu, S., and Catlow, C.R. (2004) Fine structures of zeolite-linde-L (LTL): surface structures, growth unit and defects. *Chem. Eur. J.*, **10**, 5031–5040.

27 Lehn, J.M. (1995) *Supramolecular Chemistry*, Wiley-VCH Verlag GmbH, Weinheim.

28 Calzaferri, G. and Devaux, A. (2010) Manipulation of energy transfer processes within the channels of zeolite L, in *Supramolecular Effects in Photochemical and Photophysical Processes* (eds V. Ramamurthy and Y. Inoue), John Wiley & Sons, Inc., Hoboken, NJ.

29 Minkowski, C. and Calzaferri, G. (2005) Förster type energy transfer along a specified axis. *Angew. Chem., Int. Ed.*, **44**, 5325–5329.

30 Klosterman, J.K., Iwamura, M., Tahara, T., and Fujita, M. (2009) Energy transfer in a mechanically trapped exciplex. *J. Am. Chem. Soc.*, **131**, 9478–9479.

31 (a) Slowing, I.I., Vivero-Escoto, J.L., Wu, C.-W., and Lin, V.S.-Y. (2008) Mesoporous

silica nanoparticles as controlled release drug delivery and gene transfection carriers. *Adv. Drug Deliv. Rev.*, **60**, 1278–1288; (b) De Jong, W.H. and Borm, P.J.A. (2008) Drug delivery and nanoparticles: applications and hazards. *Int. J. Nanomedicine*, **3**, 133–149.

32 Yu, J. and Xu, R. (2006) Insight into the construction of open-framework aluminophosphates. *Chem. Soc. Rev.*, **35**, 593–604.

33 Müller, M., Devaux, A., De Cola, L., and Fischer, R.A. (2009) Highly emissive metal-organic framework composites by host–guest chemistry. *Photobiol. Sci.*, **9** 846–853.

34 (a) Schulz-Ekloff, G., Wöhrle, D., van Duffel, B., and Schoonheydt, R.A. (2002) Chromophores in porous silicas and minerals: preparation and optical properties. *Microporous Mesoporous Mater.*, **51**, 91–138; (b) Corma, A. and García, H. (2004) Supramolecular host–guest systems in zeolites prepared by ship-in-a-bottle synthesis. *Eur. J. Inorg. Chem.*, **6**, 1143–1164; (c) Ramamurthy, V. (1991) *Photochemistry in Organized and Constrained Media*, Wiley-VCH Verlag GmbH, New York.

35 (a) Inagaki, S., Guan, S., Ohsuna, T., and Terasaki, O. (2002) An ordered mesoporous organosilica hybrid material with a crystal-like wall structure. *Nature*, **416**, 304–307; (b) Inagaki, S., Ohtani, O., Goto, Y., Okamoto, K., Ikai, M., Yamanaka, K.-I., Tani, T., and Okada, T. (2009) Light harvesting by a periodic mesoporous organosilica chromophore. *Angew. Chem., Int. Ed.*, **48**, 4042–4046; (c) Goto, Y., Mizoshita, N., Ohtani, O., Okada, T., Shimada, T., Tani, T., and Inagaki, S. (2008) Synthesis of mesoporous aromatic silica thin films and their optical properties. *Chem. Mater.*, **20**, 4495–4498.

36 Mauro, M., Schuermann, K., Prétôt, R., Hafner, A., Mercandelli, P., Sironi, A., and De Cola, L. (2010) Complex Iridium (III) salts: luminescent porous crystalline materials. *Angew. Chem., Int. Ed.*, **49** (7), 1222–1226.

37 Calzaferri, G. and Lutkouskaya, K. (2008) Mimicking the antenna system of green plants. *Photochem. Photobiol. Sci.*, **7**, 879–910.

38 Brühwiler, D., Calzaferri, G., Torres, T., Ramm, J.-H., Gartmann, N., Dieu, L.-Q., López-Duarte, I., and Martínez-Días, M. (2009) Nanochannels for supramolecular organization of luminescent guests. *J. Mater. Chem.*, **19**, 8040–8067.

39 Valtchev, V., Mintova, S., and Tsapatsis, M. (2009) Ordered porous solids, in *Recent Advances and Prospects*, Elsevier.

40 (a) Li, H., Wang, Y., Zhang, W., Liu, B., and Calzaferri, G. (2007) Fabrication of oriented zeolite L monolayers employing luminescent perylenediimide-bridged silsesquioxane precursor as the covalent linker. *Chem. Commun.*, **27**, 2853–2854; (b) Wang, Y., Li, H., Liu, B., Gan, Q., Dong, Q., Calzaferri, G., and Sun, Z. (2008) Fabrication of oriented zeolite L monolayer via covalent molecular linkers. *J. Solid State Chem.*, **181**, 2469–2472; (c) Calzaferri, G., Huber, S., Devaux, A., Zabala Ruiz, A., Li, H., Bossart, O., and Dieu, L.-Q. (2006) Light-harvesting host–guest antenna materials for photonic devices. *Proc. SPIE*, **6192**, 619216–619221.

41 (a) Turro, N.J. (2000) From boiling stones to smart crystals: supramolecular and magnetic isotope control of radical–radical reactions in zeolites. *Acc. Chem. Res.*, **33**, 637–646; (b) Chrétien, M.N. (2007) Supramolecular photochemistry in zeolites: from catalysts to sunscreens. *Pure Appl. Chem.*, **79**, 1–20.

42 Bein, T. (2007) Introduction to zeolite chemistry, in *Studies in Surface Science and Catalysis*, 3rd edn, vol. 168 (eds J. Cejka, H. van Bekkum, A. Corma, and F. Schüth), Elsevier, Amsterdam, pp. 611–658.

43 Calzaferri, G., Li, H., and Brühwiler, D. (2008) Dye-modified nanochannel materials for photoelectronic and optical devices. *Chem. Eur. J.*, **14**, 7442–7449.

44 Bein, T. (2007) Tuning functionality and morphology of periodic mesoporous materials. *Stud. Surf. Sci. Catal.*, **170**, 41–53.

45 Albuquerque, R.Q. and Calzaferri, G. (2007) Proton activity inside the channels of zeolite L. *Chem. Eur. J.*, **13**, 8939–8952.

46 (a) Fisher, K.A., Huddersman, K.D., and Taylor, M.J. (2003) Comparison of micro- and mesoporous inorganic materials in the uptake and release of the drug model fluorescein and its analogues. *Chem. Eur. J.*, **9**, 5873–5878; (b) Pearce1, M.E., Mai, H.Q., Lee, N., Larsen, S.C., and Salem, A.K. (2008) Silicalite nanoparticles that promote transgene expression. *Nanotechnology*, **19**, 175103–175110; (c) Uglea, C.V., Albu, I., Vatajanu, A., Croitoru, M., Antoniu, S., Panaitescu, L., and Ottenbrite, R.M. (1994) Drug delivery systems based on inorganic materials: I. Synthesis and characterization of a zeolite–cyclophosphamide system. *J. Biomater. Sci. Polym. Ed.*, **6**, 633–637.

47 Calzaferri, G., Maas, H., Pauchard, M., Pfenniger, M., Megelski, S., and Devaux, A. (2002) Supramolecularly organized luminescent dye molecules in the channels of zeolite L, in *Advances in Photochemistry*, vol. 27 (eds D.C. Neckers, G. von Bünau, and W.S. Jenks), Wiley–Interscience, pp. 1–50.

48 Albuquerque, R.Q., Popovic, Z., De Cola, L., and Calzaferri, G. (2006) Luminescence quenching by O_2 of a Ru^{2+} complex attached to zeolite L. *ChemPhysChem*, **7**, 1050–1053.

49 (a) Zabala Ruiz, A., Li, H., and Calzaferri, G. (2006) Organizing supramolecular functional dye-zeolite crystals. *Angew. Chem., Int. Ed.*, **45**, 5282–5287; (b) Leiggener, C. and Calzaferri, G. (2004) Monolayers of zeolite A containing luminescent silver sulfide clusters. *Chemphyschem*, **5**, 1593–1596; (c) Calzaferri, G., Pauchard, M., Maas, H., Huber, S., Khatyr, A., and Schaafsma, T. (2002) Photonic antenna systems for light harvesting, transport and trapping. *J. Mater. Chem.*, **12**, 1–13.

50 (a) Angelos, S., Choi, E., Vögtle, F., De Cola, L., and Zink, J.I. (2007) Photo-driven expulsion of molecules from mesostructured silica nanoparticles. *J. Phys. Chem. C*, **111**, 6589–6592; (b) Du, L., Liao, S., Khatib, H.A., Stoddart, J.F., and Zink, J.I. (2009) Controlled-access hollow mechanized silica nanocontainers. *J. Am. Chem. Soc.*, **131**, 15136–15142.

51 Burns, A.A., Vider, J., Ow, H., Herz, E., Penate-Medina, O., Baumgart, M., Larson, S.M., Wiesner, U., and Bradbury, M. (2009) Fluorescent silica nanoparticles with efficient urinary excretion for nanomedicine. *Nano Lett.*, **18**, 442–448.

52 Suárez, S., Devaux, A., Bañnuelos, J., Bossart, O., Kunzmann, A., and Calzaferri, G. (2007) Transparent zeolite–polymer hybrid materials with tunable properties. *Adv. Funct. Mater.*, **17**, 2298.

53 Rosenholm, J.M., Meinander, A., Peuhu, E., Niemi, R., Eriksson, J.E., Sahlgren, C., and Lindén, M. (2009) Targeted intracellular delivery of hydrophobic agents using mesoporous hybrid silica nanoparticles as carrier systems. *Nano Lett.*, **3**, 197–206.

54 (a) Tsotsalas, M., Busby, M., Gianolio, E., Aime, S., and De Cola, L. (2008) Functionalized nanocontainers as dual magnetic and optical probes for molecular imaging applications. *Chem. Mater.*, **20**, 5888–5893; (b) Balkus, K.J., Jr., Bresinska, I., Kowalak, S., and Young, S.W. (1991) The application of molecular sieves as magnetic resonance imaging contrast agents. *MRS Symp. Proc.*, **233**, 225–230.

55 (a) Yan, Y. and Bein, T. (1992) Molecular recognition on acoustic-wave devices: sorption in chemically anchored zeolite monolayers. *J. Phys. Chem.*, **96**, 9387–9393; (b) Bein, T. (1996) Synthesis and applications of molecular sieve layers and membranes. *Chem. Mater.*, **8**, 1636–1653; (c) Pradhan, A.R., Macnaughtan, M.A., and Raftery, D. (2000) Zeolite-coated optical microfibers for intrazeolite photocatalysis studied by *in situ* solid-state NMR. *J. Am. Chem. Soc.*, **122**, 404–405.

56 Lainé , P., Seifert, R., Giovanoli, R., and Calzaferri, G. (1997) Convenient preparation of close-packed monograin layers of pure zeolite A microcrystals. *New J. Chem.*, **21**, 453–460.

57 (a) Choi, S.Y., Lee, Y.-J., Park, Y.S., Ha, K., and Yoon, K.B. (2000) Monolayer

assembly of zeolite crystals on glass with fullerene as the covalent linker. *J. Am. Chem. Soc.*, **122**, 5201–5209; (b) Kulak, A., Lee, Y.-J., Park, Y.S., and Yoon, K.B. (2000) Orientation-controlled monolayer assembly of zeolite crystals on glass and mica by covalent linkage of surface-bound epoxide and amine groups. *Angew. Chem., Int. Ed.*, **39**, 950–953; (c) Chun, Y.S., Ha, K., Lee, Y.-J., Lee, J.S., Kim, H.S., Park, Y.S., and Yoon, K.B. (2002) Diisocyanates as novel molecular binders for monolayer assembly of zeolite crystals on glass. *Chem. Commun.*, **17**, 1846–1847.

58 (a) Ha, K., Lee, Y.-J., Jung, D.Y., Lee, J.H., and Yoon, K.B. (2000) Micropatterning of oriented zeolite monolayers on glass by covalent linkage. *Adv. Mater.*, **12**, 1614–1617; (b) Choi, J., Ghosh, S., Lai, Z., and Tsapatsis, M. (2006) Uniformly a-oriented MFI zeolite films by secondary growth. *Angew. Chem., Int. Ed.*, **45**, 1154–1158; (c) Hashimoto, S., Samata, K., Shoji, T., Taira, N., Tomita, T., and Matsuo, S. (2009) Preparation of large-scale 2D zeolite crystal array structures to achieve optical functionality. *Microporous Mesoporous Mater.*, **117**, 220–227; (d) Lee, J.S., Ha, K., Lee, Y.-J., and Yoon, K.B. (2009) Effect of method on monolayer assembly of zeolite microcrystals on glass with molecular linkages. *Top. Catal.*, **52**, 119–139; (e) Cucinotta, F., Popović, Z., Weiss, E.A., Whitesides, G.M., and De Cola, L. (2009) Microcontact transfer printing of zeolite monolayers. *Adv. Mater.*, **21**, 1142–1145.

59 Bashouti, M., Salalha, W., Brumer, M., Zussman, E., and Lifshitz, E. (2006) Alignment of colloidal CdS nanowires embedded in polymer nanofibers by electrospinning. *ChemPhysChem*, **7**, 102–106.

60 Huo, S.-J., Xue, X.-K., Li, Q.-X., Xu, S.-F., and Cai, W.-B. (2006) Seeded-growth approach to fabrication of silver nanoparticle films on silicon for electrochemical ATR surface-enhanced IR absorption spectroscopy. *J. Phys. Chem. B*, **110**, 25721–25728.

61 Dai, L., Patil, A., Gong, X., Guo, Z., Liu, L., Liu, Y., and Zhu, D. (2003) Aligned nanotubes. *ChemPhysChem*, **4**, 1150–1169.

62 Ozin, G.A. and Arsenault, A.C. (2005) *Nanochemistry: A Chemical Approach to Nanomaterials*, Royal Society of Chemistry, London.

63 Hurst, S.J., Payne, E.K., Qin, L.D., and Mirkin, C.A. (2006) Multisegmented one-dimensional nanorods prepared by hard-template synthetic methods. *Angew. Chem., Int. Ed.*, **45**, 2672–2692.

64 Bein, T. (2005) Zeolitic host–guest interactions and building blocks for the self-assembly of complex materials. *MRS Bull.*, **30**, 713–720.

65 Gouzinis, A. and Tsapatsis, M. (1998) On the preferred orientation and microstructural manipulation of molecular sieve films prepared by secondary growth. *Chem. Mater.*, **10**, 2497–2504.

66 Yoon, K.B. (2007) Organization of zeolite microcrystals for production of functional materials. *Acc. Chem. Res.*, **40**, 29–40.

67 Lee, J.S., Lim, H., Ha, K., Cheong, H., and Yoon, K.B. (2006) Facile monolayer assembly of fluorophore-containing zeolite rods in uniform orientations for anisotropic photoluminescence. *Angew. Chem., Int. Ed.*, **45**, 5288–5292.

68 Calzaferri, G., Huber, S., Mass, H., and Minkowski, C. (2003) Host–guest antenna materials. *Angew. Chem., Int. Ed.*, **42**, 3732–3758.

69 (a) Megelski, S. and Calzaferri, G. (2001) Tuning the size and shape of zeolite L-based inorganic–organic host–guest composites for optical antenna systems. *Adv. Funct. Mater.*, **11**, 277–286; (b) Zabala Ruiz, A., Brühwiler, D., Ban, T., and Calzaferri, G. (2005) Synthesis of zeolite L. Tuning size and morphology. *Monatshefte für Chemie*, **136**, 77–89; (c) Zabala Ruiz, A., Brühwiler, D., Dieu, L.-Q., and Calzaferri, G. (2008) in *Materials Syntheses: A Practical Guide* (eds U. Schubert, N. Hüsing, and R. Laine), Springer, New York.

70 (a) Ernst, S. and Weitkamp, J. (1994) Synthesis of large-pore aluminosilicates. *J. Catal. Today*, **94**, 27–60; (b) Lee, Y.-J., Lee, J.S., and Yoon, K.B. (2005) Synthesis

of long zeolite-L crystals with flat facets. *Microporous Mesoporous Mater.*, **80**, 237–246; (c) Ban, T., Saito, H., Naito, M., Ohya, Y., and Takahashi, Y. (2007) Synthesis of zeolite L crystals with different shapes. *J. Porous Mater.*, **14**, 119–126; (d) Brent, R. and Anderson, M.W. (2008) Fundamental crystal growth mechanism in zeolite L revealed by atomic force microscopy. *Angew. Chem., Int. Ed.*, **47**, 5327–5330.

71 Larlus, O. and Valtchev, V.P. (2004) Crystal morphology control of LTL-type zeolite crystals. *Chem. Mater.*, **16**, 3381–3389.

72 Brühwiler, D. and Calzaferri, G. (2004) Molecular sieves as host materials for supramolecular organization. *Microporous Mesoporous Mater.*, **72**, 1–23.

73 Gfeller, N., Megelski, S., and Calzaferri, G. (1998) Transfer of electronic excitation energy between dye molecules in the channels of zeolite L. *J. Phys. Chem. B*, **102**, 2433–2436.

74 Maas, H. and Calzaferri, G. (2002) Trapping energy from and injecting energy into dye-zeolite nanoantennae. *Angew. Chem., Int. Ed.*, **41**, 2284–2288.

75 (a) Mortazavi, M., Yoon, H., and Teng, C. (1993) Optical power handling properties of polymeric nonlinear optical waveguides. *J. Appl. Phys.*, **74**, 4871; (b) Cha, M., Torruellas, W.E., Stegeman, G.I., Horsthuis, W.H.G., Möhlmann, G.R., and Meth, J. (1994) Two photon absorption of di-alkyl-amino-nitro-stilbene side chain polymer. *Appl. Phys. Lett.*, **65**, 2648; (c) Zhang, Q., Canva, M., and Stegeman, G. (1998) Wavelength dependence of 4-dimethylamino-4′-nitrostilbene polymer thin film photodegradation. *Appl. Phys. Lett.*, **73**, 912.

76 Kalogeras, I.M., Neagu, E.R., Vassilikou-Dova, A., and Neagu, R.M. (2002) Physical and chemical effects of the β-relaxation of PMMA in rhodamine 6G + PMMA + SiO_2 matrices. *Mat. Res. Innovat.*, **6**, 198.

77 Suib, S.L. (1993) Zeolitic and layered materials. *Chem. Rev.*, **93**, 803.

78 Ramamurthy, V., Eaton, D.F., and Caspar, J.V. (1992) Photochemical and photophysical studies of organic molecules included within zeolites. *Acc. Chem. Res.*, **25**, 299.

79 Turro, N.J. (1986) Photochemistry of organic molecule in microscopic reactors. *Pure Appl. Chem.*, **58**, 1219.

80 Laeri, F., Schüth, F., Simon, U., and Wark, M. (2003) *Host–Guest-Systems Based on Nanoporous Crystals*, Wiley-VCH Verlag GmbH, Weinheim.

81 Wöhrle, D. and Schulz-Ekloff, G. (1994) Molecular sieves encapsulated organic dyes and metal chelates. *Adv. Mater.*, **6**, 875.

82 Caro, J., Marlow, F., and Wübbenhorst, M. (1994) Chromophore–zeolite composites: the organizing role of molecular sieves. *Adv. Mater.*, **6**, 413.

83 Ramamurthy, V. (2000) Controlling photochemical reactions via confinement: zeolites. *J. Photochem. Photobiol. C*, **1**, 145.

84 Ackley, M.W., Rege, S.U., and Saxena, H. (2003) Application of natural zeolites in the purification and separation of gases. *Microporous Mesoporous Mater.*, **61**, 25.

85 Davis, R.J. (2003) New perspectives on basic zeolites as catalysts and catalyst supported. *J. Catal.*, **216**, 396.

86 Okamoto, M. and Sakamoto, E. (2001) Application of natural zeolites to purify polluted river water, *Stud. Surf. Sci. Catal.*, **135**, 5237–5245.

87 Schulz-Ekloff, G. (1991) Zeolite-hosted metals and semiconductors as advanced materials. *Stud. Surf. Sci. Catal.*, **69**, 65.

88 Martin-Palma, R.J., Hernandez-Velez, M., Diaz, I., Villavicencio-Garcia, H., Garcia-Poza, M.M., Martinez-Duart, J.M., and Perez-Pariente, J. (2001) Optical properties of semiconductors clusters grown into mordenite and MCM-41 matrices. *Mater. Sci. Eng. C*, **15**, 163.

89 De Bruin, M., De Vos, D.E., and Jacobs, P.A. (2002) Chemoselective hydrogen transfer reduction of unsaturated ketones to allylic alcohols with solid Zr and Hf catalysts. *Adv. Synth. Catal.*, **344**, 1120.

90 Gessner, F., Olea, A., Lobaugh, J.H., Johnston, L.J., and Scaiano, J.C. (1989) Intrazeolite photochemistry. 5. Use of zeolites in the control of photostationary

ratios in sensitized *cis–trans* isomerizations. *J. Org. Chem.*, **54**, 259.

91 Gfeller, N., Megelski, S., and Calzaferri, G. (1999) Fast energy migration in pyronine-loaded zeolite L microcrystals. *J. Phys. Chem. B*, **103**, 1250.

92 Calzaferri, G., Lanz, M., and Li, J. (1995) Methyl viologen–zeolite electrodes: intrazeolite charge transfer. *J. Chem. Soc. Chem. Commun.*, 1313.

93 Yoon, K.B., Park, Y.S., and Kochi, J.K. (1996) Interfacial electron transfer to the zeolite-encapsulated methylviologen acceptor from various carbonylmanganate donors. Shape selectivity of cations in mediating electron conduction through the zeolite framework. *J. Am. Chem. Soc.*, **118**, 12710.

94 Vietze, U., Krauss, O., Laeri, F., Ihnlein, G., Schüth, F., Limburg, B., and Abraham, M. (1998) Zeolite-dye microlasers. *Phys. Rev. Lett.*, **81**, 4628.

95 Cox, S.D., Gier, T.E., Stucky, G.D., and Bierlein, J. (1998) Inclusion tuning of nonlinear optical materials: switching the SHG of *p*-nitroaniline and 2-methyl-*p*-nitroaniline with molecular sieve hosts. *J. Am. Chem. Soc.*, **110**, 2986.

96 Werner, L., Caro, J., Finger, G., and Kornatowski, J. (1992) Optical second harmonic generation (SHG) on *p*-nitroaniline in large crystals of AlPO$_4$-5 and ZSM-5. *Zeolites*, **12**, 658.

97 Ehrl, M., Deeg, F.W., Bräuchle, C., Franke, O., Sobbi, A., Schulz-Ekloff, G., and Wöhrle, D. (1994) High-temperature non-photochemical hole-burning of phthalocyanine–zinc derivatives embedded in a hydrated AlPO$_4$-5 molecular sieve. *J. Phys. Chem.*, **98**, 47.

98 Liu, X. and Thomas, J.K. (1994) Photophysical properties of pyrene in zeolites: adsorption and distribution of pyrene molecules on the surface of zeolite L and mordenite. *Chem. Mater.*, **6**, 2303.

99 Hashimoto, S., Ikuta, S., Asahi, T., and Masuhara, H. (1998) Fluorescence spectroscopy studies of anthracene adsorbed into zeolites: from the detection of cation-π interaction to the observation of dimers and crystals. *Langmuir*, **14**, 4284.

100 Corrent, S., Hahn, P., Pohlers, G., Connolly, T.J., Scaiano, J.C., Fornés, V., and García, H. (1998) Intrazeolite photochemistry. 22. Acid–base properties of coumarin 6. Characterization in solution, the solid state, and incorporated into supramolecular systems. *J. Phys. Chem. B*, **102**, 5852.

101 Pauchard, M., Devaux, A., and Calzaferri, G. (2000) Dye-loaded zeolite L sandwiches as artificial antenna systems for light transport. *Chem. Eur. J.*, **6**, 3456.

102 Hashimoto, S., Hagiri, M., Matsubara, N., and Tobita, S. (2001) Photophysical studies of neutral aromatic species confined in zeolite L: comparison with cationic dyes. *Phys. Chem. Chem. Phys.*, **3**, 5043.

103 (a) Calzaferri, G. and Gfeller, N. (1992) Thionine in the cage of zeolite L. *J. Phys. Chem.*, **96**, 3428; (b) Ramamurthy, V., Sanderson, D.R., and Eaton, D.F. (1993) Control of dye assembly within zeolites: role of water. *J. Am. Chem. Soc.*, **115**, 10438; (c) Yoon, K.B., Huh, T.J., Corbin, D.R., and Kochi, J.K. (1993) Shape-selective assemblage of charge-transfer complexes within channel-type zeolites. *J. Phys. Chem.*, **97**, 6492; (d) Ganesan, V. and Ramaraj, R. (2001) Spectral properties of proflavin in zeolite-L and zeolite-Y. *J. Lumin.*, **92**, 167.

104 (a) Ganschow, M., Schulz-Ekloff, G., Wark, M., Wendschuh-Josties, M., and Wöhrle, D. (2001) Microwave-assisted preparation of uniform pure and dye-loaded AlPO$_4$-5 crystals with different morphologies for use as microlaser systems. *J. Mater. Chem.*, **11**, 1823; (b) Seebacher, C., Hellriegel, C., Bräuchle, C., Ganschow, M., and Wöhrle, D. (2003) Orientational behavior of single molecules in molecular sieves: a study of oxazine dyes in AlPO$_4$-5 crystals. *J. Phys. Chem. B*, **107**, 5445.

105 (a) Meyer, G., Wöhrle, D., Mohl, M., and Schulz-Ekloff, G. (1984) Synthesis of faujasite supported phthalocyanines of cobalt, nickel and copper. *Zeolites*, **4**, 30; (b) Herron, N., Stucky, G.D., and Tolman, C.A. (1985) The reactivity of tetracarbonylnickel encapsulated in zeolite X. A case history of intrazeolite

coordination chemistry. *Inorg. Chim. Acta*, **100**, 135; (c) Herron, N. (1986) A cobalt oxygen carrier in zeolite Y. A molecular "ship in a bottle". *Inorg. Chem.*, **25**, 4714.

106 Calzaferri, G., Huber, S., Maas, H., and Minkowski, C. (2003) Host–guest antenna materials. *Angew. Chem., Int. Ed.*, **42**, 3732.

107 Maas, H., Huber, S., Khatyr, A., Pfenniger, M., Meyer, M., and Calzaferri, G. (2003) *Molecular and Supramolecular Photochemistry*, vol. 9 (eds V. Ramamurthy and K.S. Schanze), Marcel Dekker, Inc., New York, pp. 309–351.

108 Karge, H.G., Beyer, H.K., and Borbely, G. (1988) Solid-state ion exchange in zeolites: Part II. Alkaline earth chlorides/mordenite. *Catal. Today*, **3**, 41.

109 Maas, H., Khatyr, A., and Calzaferri, G. (2003) Phenoxazine dyes in zeolite L, synthesis and properties. *Microporous Mesoporous Mater.*, **65**, 233–242.

110 Cox, B.G., Schneider, H., and Stroka, J. (1978) Kinetics of alkali metal complex formation with cryptands in methanol. *J. Am. Chem. Soc.*, **100**, 4746.

111 Dietrich, B., Lehn, J.-M., and Sauvage, J.P. (1973) Cryptates: control over bivalent/monovalent cation selectivity. *J. Chem. Soc. Chem. Commun.*, **1**, 15.

112 Ramamurthy, V., Lakshminarasimhan, P., Grey, C.P., and Johnston, L.J. (1998) Energy transfer, proton transfer and electron transfer reactions within zeolites. *J. Chem. Soc. Chem. Commun.*, 2411.

113 Thomas, K.J., Sunoj, R.B., Chandrasekhar, J., and Ramamurthy, V. (2000) Cation-π-interaction promoted aggregation of aromatic molecules and energy transfer within Y zeolites. *Langmuir*, **16**, 4912.

114 Sendor, D. and Kynast, U. (2002) Efficient red-emitting hybrid materials based on zeolites. *Adv. Mater.*, **14**, 1570.

115 Lezhnina, M.M. and Kynast, U.H. (2004) Luminescence from tungstate functionalized sodalites involving NIR states. *J. Alloys Compd.*, **380**, 55.

116 (a) Jüstel, T., Wiechert, D.U., Lau, C., Sendor, D., and Kynast, U. (2001) Optically functional zeolites: evaluation of UV and VUV stimulated photoluminescence properties of Ce^{3+}- and Tb^{3+}-doped zeolite X. *Adv. Funct. Mater.*, **11**, 105; (b) Lezhnina, M.M., Kätker, H., and Kynast, U.H. (2005) Rare-earth ions in porous matrices. *Phys. Solid State*, **47**, 1423.

117 Kondo, R. (1965) The synthesis and crystallography of a group of new compounds belonging to the hauyne type structure. *J. Ceram. Soc. Jpn.*, **73**, 1.

118 (a) Borgmann, C., Sauer, J., Jüstel, T., Kynast, U.H., and Schüth, F. (1999) Efficiently emitting rare-earth sodalites by phase transformation of zeolite X and by direct synthesis. *Adv. Mater.*, **11**, 45; (b) Esmeria, J.M., Jr., Ishii, H., Sato, M., and Ito, H. (1995) Efficient continuous-wave lasing operation of $Nd:KGd(WO_4)_2$ at 1.067 µm with diode and Ti:sapphire laser pumping. *Opt. Lett.*, **20**, 1538; (c) Rico, M., Pujol, M.C., Diaz, F., and Zaldo, C. (2001) Green up-conversion of Er^{3+} $KGd(WO_4)_2$ crystals. Effects of sample orientation and erbium concentration. *Appl. Phys. B Lasers Opt.*, **72**, 157; (d) Lezhnina, M.M., Laeri, F., Benmouhadi, L., and Kynast, U.H. (2006) Efficient near-infrared emission from sodalite derivatives. *Adv. Mater.*, **18**, 280–283.

119 Devaux, A., Popovic, Z., Bossart, O., De Cola, L., Kunzmann, A., and Calzaferri, G. (2006) Solubilization of dye-loaded zeolite L nanocrystals. *Microporous Mesoporous Mater.*, **90**, 69–72.

120 (a) Ganschow, M., Wark, M., Wöhrle, D., and Schulz-Ekloff, G. (2000) Anchoring of functional dye molecules in MCM-41 by microwave-assisted hydrothermal cocondensation. *Angew. Chem., Int. Ed.*, **39**, 160; (b) Rohlfing, Y., Wöhrle, D., Wark, M., Schulz-Ekloff, G., Rathousky, J., and Zukal, A. (2000) Covalent attachment of dye molecules to the inner surface of MCM-41. *Stud. Surf. Sci. Catal.*, **129**, 295; (c) Ban, T., Brühwiler, D., and Calzaferri, G. (2004) Selective modification of the channel entrances of zeolite L with triethoxysilylated coumarin. *J. Phys. Chem. B*, **108**, 16348.

121 (a) Khatyr, A., Maas, H., and Calzaferri, G. (2002) Synthesis of new molecules

containing head, spacer, and label moieties. *J. Org. Chem.*, **67**, 6705; (b) Bossart, O., De Cola, L., Welter, S., and Calzaferri, G. (2004) Injecting electronic excitation energy into an artificial antenna system through an Ru^{2+} complex. *Chem. Eur. J.*, **10**, 5771.

122. (a) Huber, S. and Calzaferri, G. (2004) Sequential functionalization of the channel entrances of zeolite L crystals. *Angew. Chem., Int. Ed.*, **43**, 6738; (b) Li, H., Devaux, A., Popovic, Z., De Cola, L., and Calzaferri, G. (2006) Carboxyester functionalized dye-zeolite L host–guest materials. *Microporous Mesoporous Mater.*, **95**, 112–117.

123. Busby, M., Kerschbaumer, H., Calzaferri, G., and De Cola, L. (2008) Orthogonally bifunctional fluorescent zeolite L microcrystals. *Adv. Mater.*, **20**, 1614–1618.

124. Guerrero-Martinez, A., Fibikar, S., Pastoriza-Santos, I., Liz-Marzán, L.M., and De Cola, L. (2009) Microcontainers with fluorescent anisotropic zeolite L cores and isotropic silica shells. *Angew. Chem., Int. Ed.*, **48**, 1266–1270.

125. (a) Férey, G. (2008) Hybrid porous solids: past, present, future. *Chem. Soc. Rev.*, **37**, 191; (b) Li, H., Eddaoudi, M., O'Keefe, M., and Yaghi, O.M. (1999) Design and synthesis of an exceptionally stable and highly porous metal–organic framework. *Nature*, **402**, 276; (c) Kitagawa, S. and Matsuda, R. (2007) Chemistry of coordination space of porous coordination polymers. *Coord. Chem. Rev.*, **251**, 2490; (d) Tranchemontagne, D.J., Ni, Z., ÓKeeffe, M., and Yaghi, O.M. (2008) Reticular chemistry of metal–organic polyhedra. *Angew. Chem., Int. Ed.*, **47**, 5136–5147.

126. (a) Férey, G., Latroche, M., Serre, C., Millange, F., Loiseau, T., and Percheron-Guégan, A. (2003) Hydrogen adsorption in the nanoporous metal-benzenedicarboxylate M(OH)(O$_2$C-C$_6$H$_4$-CO$_2$) (M=Al^{3+}, Cr^{3+}), MIL-53. *Chem. Commun.*, 2976; (b) Chen, B., Ockwig, N.W., Millward, A.R., Contreras, D.S., and Yaghi, O.M. (2005) High H$_2$ adsorption in a microporous metal–organic framework with open metal sites. *Angew. Chem., Int. Ed.*, **44**, 4745; (c) Rowsell, J.L.C., Spencer, E.C., Eckert, J., Howard, J.A.K., and Yaghi, O.M. (2005) Gas adsorption sites in a large-pore metal–organic framework. *Science*, **309**, 1350; (d) Latroche, M., Surblé, S., Serre, C., Mellot-Draznieks, C., Llewellyn, P.L., Lee, J.-H., Chang, J.-S., Jhung, S.H., and Férey, G. (2006) Hydrogen storage in the giant-pore metal–organic frameworks MIL-100 and MIL-101. *Angew. Chem., Int. Ed.*, **45**, 8227.

127. (a) Kitaura, R., Seki, K., Akiyama, G., and Kitagawa, S. (2003) Porous coordination-polymer crystals with gated channels specific for supercritical gases. *Angew. Chem., Int. Ed.*, **42**, 428; (b) Ma, S., Sun, D., Wang, X.-S., and Zhou, H.-C. (2007) A mesh-adjustable molecular sieve for general use in gas separation. *Angew. Chem., Int. Ed.*, **46**, 2458.

128. (a) Qiu, L.-G., Xie, A.-J., and Zhang, L.-D. (2005) Encapsulation of catalysts in supramolecular porous frameworks: size- and shape-selective catalytic oxidation of phenols. *Adv. Mater.*, **17**, 689; (b) Zou, R.-Q., Sakurai, H., and Zu, Q. (2006) Preparation, adsorption properties, and catalytic activity of 3D porous metal–organic frameworks composed of cubic building blocks and alkali-metal ions. *Angew. Chem., Int. Ed.*, **45**, 2542; (c) Horcajada, P., Surblé, S., Serre, C., Hong, D.-Y., Seo, Y.-K., Chang, J.-S., Grenéche, J.-M., Margiolaki, I., and Férey, G. (2007) Synthesis and catalytic properties of MIL-100(Fe), an iron(III) carboxylate with large pores. *Chem. Commun.*, 2820.

129. (a) Horcajada, P., Serre, C., Vallet-Regi, M., Sebban, M., Taulelle, F., and Férey, G. (2006) Metal–organic frameworks as efficient materials for drug delivery. *Angew. Chem., Int. Ed.*, **45**, 5974; (b) Horcajada, P., Serre, C., Maurin, G., Ramsahye, N.A., Balas, F., Vallet-Regi, M., Sebban, M., Taulelle, F., and Férey, G. (2008) Flexible porous metal–organic frameworks for a controlled drug delivery. *J. Am. Chem. Soc.*, **130**, 6774.

130. Chen, B., Yang, Y., Zapata, F., Lin, G., Qian, G., and Lobkovsky, E.B. (2007) Luminescent open metal sites within a

130 metal–organic framework for sensing small molecules. *Adv. Mater.*, **19**, 1693.

131 (a) De Lill, D.T., Gunning, N.S., and Cahill, C.L. (2005) Toward templated metal–organic frameworks: synthesis, structures, thermal properties, and luminescence of three novel lanthanide-adipate frameworks. *Inorg. Chem.*, **44**, 258; (b) Harbuzaru, B.V., Corma, A., Rey, F., Atienzar, P., Jordá, J.L., García, H., Ananias, D., Carlos, L.D., and Rocha, J. (2008) Metal–organic nanoporous structures with anisotropic photoluminescence and magnetic properties and their use as sensors. *Angew. Chem., Int. Ed.*, **47**, 1080.

132 Maspoch, D., Ruiz-Molina, D., and Veciana, J. (2004) Magnetic nanoporous coordination polymers. *J. Mater. Chem.*, **14**, 2713.

133 Férey, G., Millange, F., Morcrette, M., Serre, C., Doublet, M.-L., Grenéche, J.-M., and Tarascon, J.-M. (2007) Mixed-valence Li/Fe-based metal–organic frameworks with both reversible redox and sorption properties. *Angew. Chem., Int. Ed.*, **46**, 3259.

134 (a) Yaghi, O.M., Li, G., and Li, H. (1995) Selective binding and removal of guests in a microporous metal–organic framework. *Nature*, **378**, 703–706; (b) Hong, D.-Y., Hwang, Y.K., Serre, C., Férey, G., and Chang, J.-S. (2009) Porous chromium terephthalate MIL-101 with coordinatively unsaturated sites: surface functionalization, encapsulation, sorption and catalysis. *Adv. Funct. Mater.*, **19**, 1537–1552.

135 Millward, A.R. and Yaghi, O.M. (2005) Metal–organic frameworks with exceptionally high capacity for storage of carbon dioxide at room temperature. *J. Am. Chem. Soc.*, **127**, 17998–17999.

136 (a) Förster, Th. (1948) Zwischenmolekulare Energiewanderung und Fluoreszenz. *Ann. Physik Chemie B*, **21**, 55–75; (b) Förster, Th. (1951) *Fluoreszenz Organischer Verbindungen*, Vandenboeck & Ruprecht, Göttingen; (c) Förster, Th. (1960) Excitation transfer, in *Comparative Effects of Radiation* (ed. M. Barton), John Wiley & Sons, Inc., New York, pp. 300–319.

137 Lutkouskaya, K. and Calzaferri, G. (2006) Transfer of electronic excitation energy between randomly mixed dye molecules in the channels of zeolite L. *J. Phys. Chem. B*, **110**, 5633–5638.

138 (a) Pfenniger, M. and Calzaferri, G. (2000) Intrazeolite diffusion kinetics of dye molecules in the nanochannels of zeolite L, monitored by energy transfer. *ChemPhysChem*, **4**, 211–217; (b) Vohra, V., Devaux, A., Dieu, L.-Q., Scavia, G., Catellani, M., Calzaferri, G., and Botta, C. (2009) Energy transfer in fluorescent nanofibers embedding dye loaded zeolites L crystals. *Adv. Mater.*, **21**, 1146–1150.

139 Hennessy, B., Megelski, S., Marcolli, C., Shklover, V., Bärlocher, C., and Calzaferri, G. (1999) Characterization of methylviologen in the channels of zeolite L. *J. Phys. Chem. B*, **103**, 3340–3351.

140 Yoon, K.B. (1993) Electron- and charge-transfer reactions within zeolites. *Chem. Rev.*, **93**, 321–339.

141 (a) Walcarius, A. (2008) Electroanalytical applications of microporous zeolites and mesoporous (organo)silicas: recent trends. *Electroanalysis*, **20**, 711–738; (b) Kim, Y.I., Keller, S.W., Krueger, J.S., Saupe, G.B., Yonemoto, E.H., and Mallouk, T.E. (1997) Photochemical charge transfer and hydrogen evolution mediated by oxide semiconductor particles in zeolite-based molecular assemblies. *J. Phys. Chem. B*, **101**, 2491–2500; (c) Yonemoto, E.H., Kim, Y.I., Schmehl, R.H., Wallin, J.O., Shoulders, B.A., Richardson, B.R., Haw, J.F., and Mallouk, T.E. (1994) Photoinduced electron transfer reactions in zeolite-based donor–acceptor and donor–donor–acceptor diads and triads. *J. Am. Chem. Soc.*, **116**, 10557–10563.

142 Stein, A., Melde, B.J., and Schroden, R.C. (2000) Hybrid inorganic–organic mesoporous silicates: nanoscopic reactors coming of age. *Adv. Mater.*, **12**, 1403–1419.

143 (a) Mercier, L. and Pinnavaia, T.J. (1997) Access in mesoporous materials: advantages of a uniform pore structure in the design of a heavy metal ion adsorbent for environmental remediation. *Adv.*

Mater., **9**, 500–503; (b) Walcarius, A., Etienne, M., and Lebeau, B. (2003) Rate of access to the binding sites in organically modified silicates. 2. Ordered mesoporous silicas grafted with amine or thiol groups. *Chem. Mater.*, **15**, 2161–2173.

144 (a) Holmlin, R.E., Haag, R., Chabinyc, M.L., Ismagilov, R.F., Cohen, A.E., Terfort, A., Rampi, M.A., and Whitesides, G.M. (2001) Electron transport through thin organic films in metal–insulator–metal junctions based on self-assembled monolayers. *J. Am. Chem. Soc.*, **123**, 5075–5085; (b) Chiechi, R.C., Weiss, E.A., Dickey, M.D., and Whitesides, G.M. (2008) Eutectic gallium–indium (EGaIn): a moldable liquid metal for electrical characterization of self-assembled monolayers. *Angew. Chem., Int. Ed.*, **47**, 142–144.

145 Walcarius, A., Rozanska, S., Bessière, J., and Wang, J. (1999) Screen-printed zeolite-modified carbon electrodes. *Analyst*, **124**, 1185–1190.

146 (a) Rolison, D.R. (1994) The intersection of electrochemistry with zeolite science. *Stud. Surf. Sci. Catal.*, **85**, 543–586; (b) Rolison, D.R. (1990) Zeolite-modified electrodes and electrode-modified zeolites. *Chem. Rev.*, **90**, 867–878.

147 Evmiridis, N.P., Demertzis, M.A., and Vlessidis, A.G. (1991) Effect of treatment of synthetic zeolite–polymer membranes on their electrochemical-potential response characteristics. *Fresenius J. Anal. Chem.*, **340**, 145–152.

148 Busby, M., Blum, C., Tibben, M., Fibikar, S., Calzaferri, G., Subramaniam, V., and De Cola, L. (2008) Time, space, and spectrally resolved studies on J-aggregate interactions in zeolite L nanochannels. *J. Am. Chem. Soc.*, **130**, 10970–10976.

149 (a) Huck, W.T.S. (ed.) (2005) *Nanoscale Assembly*, Springer, New York, p. 217. (b) Whitesides, G.M. and Grzybowski, B. (2002) Self-assembly at all scales. *Science*, **295**, 2418–2421; (c) Whitesides, G.M. and Boncheva, M. (2002) Beyond molecules: self-assembly of mesoscopic and macroscopic components. *Proc. Natl. Acad. Sci. USA*, **99**, 4769–4774.

150 (a) Lehn, J.M. (2002) Toward self-organization and complex matter. *Science*, **295**, 2400–2403; (b) Lehn, J.M. (1988) Supramolecular chemistry: molecules, supermolecules, and molecular functional units. *Angew. Chem., Int. Ed.*, **100**, Nobel lecture, 91–116.

151 (a) Whitesides, G.M., Kriebel, J.K., and Love, J.C. (2005) Molecular engineering of surfaces using self-assembled monolayers. *Sci. Prog.*, **88**, 17–48; (b) Mrksich, M. and Whitesides, G.M. (1995) Patterning self-assembled monolayers using microcontact printing: a new technology for biosensors. *Trends Biotechnol.*, **13**, 228–235; (c) Whitesides, G.M., Mathias, J.P., and Seto, C.T. (1991) Molecular self-assembly and nanochemistry: a chemical strategy for the synthesis of nanostructures. *Science*, **254**, 1312–1319.

152 Daniel, M.C. and Astruc, D. (2004) Gold nanoparticles: assembly, supramolecular chemistry, quantum-size-related properties, and applications toward biology, catalysis, and nanotechnology. *Chem. Rev.*, **104**, 293–346.

153 Clark, T.D., Tien, J., Duffy, D.C., Paul, K.E., and Whitesides, G.M. (2001) Self-assembly of 10-µm-sized objects into ordered three-dimensional arrays. *J. Am. Chem. Soc.*, **123**, 7677–7682.

154 (a) Popovic, Z., Busby, M., Huber, S., Calzaferri, G., and De Cola, L. (2007) Assembling micro crystals through cooperative coordinative interactions. *Angew. Chem., Int. Ed.*, **46**, 8898–8902; (b) Busby, M., De Cola, L., Kottas, G.S., and Popovic, Z. (2007) Assembling photo- and electroresponsive molecules and nano-objects. *MRS Bull.*, **32**, 556–560.

155 Yu, S.C., Kwok, C.C., Chan, W.K., and Che, C.M. (2003) Self-assembled electroluminescent polymers derived from terpyridine-based moieties. *Adv. Mater.*, **15**, 1643–1647.

156 Han, M.Y., Gao, X.H., Su, J.Z., and Nie, S. (2001) Quantum-dot-tagged microbeads for multiplexed optical coding of biomolecules. *Nat. Biotechnol.*, **19**, 631–635.

157 Popović, Z., Otter, M., Calzaferri, G., and De Cola, L. (2007) Self-assembling living systems with functional nanomaterials. *Angew. Chem., Int. Ed.*, **46**, 6188–6191.

158 Huang, L., Wang, Z., Sun, J., Miao, L., Li, Q., Yan, Y., and Zhao, D. (2000) Fabrication of ordered porous structures by self-assembly of zeolite nanocrystals. *J. Am. Chem. Soc.*, **122**, 3530–3531.

159 Kawai, T. and Tsutsumi, K. (1998) Reactivity of silanol groups on zeolite surfaces. *Colloid Polym. Sci.*, **276**, 992–998.

160 Caruso, F. (2000) Hollow capsule processing through colloidal templating and self-assembly. *Chem. Eur. J.*, **6**, 413–419.

161 (a) Lin, H.P., Mou, C.Y., Liu, S.B., and Tang, C.Y. (2001) Hollow spheres of MCM-41 aluminosilicate with pinholes. *Chem. Commun.*, **19**, 1970–1971; (b) Lin, H.P., Cheng, Y.R., and Mou, C.Y. (1998) Hierarchical order in hollow spheres of mesoporous silicates. *Chem. Mater.*, **10**, 3772–3776.

162 Schacht, S., Huo, Q., Voigt-Martin, I.G., Stucky, G.D., and Schuth, F. (1996) Oil–water interface templating of mesoporous macroscale structures. *Science*, **273**, 768–771.

163 Fowler, C.E., Khushalani, D., and Mann, S. (2001) Interfacial synthesis of hollow microspheres of mesostructured silica. *Chem. Commun.*, **19**, 2028–2029.

164 Prouzet, E., Cot, F., Boissiere, C., Kooyman, P.J., and Larbot, A.J. (2002) Nanometric hollow spheres made of MSU-X-type mesoporous silica. *J. Mater. Chem.*, **12**, 1553–1556.

165 Rhodes, K.H., Davis, S.A., Caruso, F., Zhang, B., and Mann, S. (2000) Hierarchical assembly of zeolite nanoparticles into ordered macroporous monoliths using core–shell building blocks. *Chem. Mater.*, **12**, 2832–2834.

166 Wang, X.D., Yang, W.L., Tang, Y., Wang, Y.J., Fu, S.K., and Gao, Z. (2000) Fabrication of hollow zeolite spheres. *Chem. Commun.*, **21**, 2161–2162.

167 Kulak, A., Lee, Y.-J., Park, S., Yong, H.S., Kim, G., Lee, S., and Yoon, K.B. (2002) Anionic surfactants as nanotools for the alignment of non-spherical zeolite nanocrystals. *Adv. Mater.*, **14**, 526–529.

168 Lee, G.S., Lee, Y.J., Choi, S.Y., Park, Y.S., and Yoon, K.B. (2000) Self-assembly of β-glucosidase and D-glucose-tethering zeolite crystals into fibrous aggregates. *J. Am. Chem. Soc.*, **122**, 12151–12157.

169 Um, S.H., Lee, G.S., Lee, Y.J., Koo, K.K., Lee, C., and Yoon, K.B. (2002) Self-assembly of avidin and D-biotin-tethering zeolite microcrystals into fibrous aggregates. *Langmuir*, **18**, 4455–4459.

170 Caro, J., Finger, G., Kornatowski, J., Richter-Mendau, J., Werner, L., and Zibrowius, B. (1992) Aligned molecular sieve crystals. *Adv. Mater.*, **4**, 273.

171 Scandella, L., Binder, G., Gobrecht, J., and Jansen, J.C. (1996) Alignment of single-crystal zeolites by means of microstructured surfaces. *Adv. Mater.*, **8**, 137.

172 Bowden, N., Terfort, A., Carbeck, J., and Whitesides, G.M. (1997) Self-assembly of mesoscale objects into ordered two-dimensional arrays. *Science*, **276**, 233.

173 Gu, Z., Chen, Y., and Gracias, D.H. (2004) Surface tension driven self-assembly of bundles and networks of 200 nm diameter rods using a polymerizable adhesive. *Langmuir*, **20**, 11308.

174 Srinivasan, U., Liepmann, D., and Howe, R.T. (2001) Microstructure to substrate self-assembly using capillary forces. *J. Microelectromech. Syst.*, **10**, 17.

175 Yunus, S., Spano, F., Patrinoiu, G., Bolognesi, A., Botta, C., Brühwiler, D., Ruiz, A.Z., and Calzaferri, G. (2006) Hexagonal network organization of dye-loaded zeolite L crystals by surface-tension-driven autoassembly. *Adv. Funct. Mater.*, **16**, 2213–2217.

176 Li, J., Pfanner, K., and Calzaferri, G. (1995) Silver-zeolite-modified electrodes: an intrazeolite electron transport mechanism. *J. Phys. Chem.*, **99**, 2119.

177 Ha, K., Park, J.S., Oh, K.S., Zhou, Y.S., Chun, Y.S., Lee, Y.J., and Yoon, K.B. (2004) Aligned monolayer assembly of zeolite crystals on platinum, gold and indium–tin oxide surfaces with molecular linkages. *Microporous Mesoporous Mater.*, **72**, 91–98.

178 Kumar, A., Biebuyck, H.A., and Whitesides, G.M. (1994) Patterning self-assembled monolayers: applications in materials science. *Langmuir*, **10**, 1498.

179 Xia, Y. and Whitesides, G.M. (1998) Soft lithography. *Angew. Chem., Int. Ed.*, **37**, 550.

180 Ahn, J.H., Kim, H.S., Lee, K.J., Jeon, S., Kang, S.J., Sun, Y., Nuzzo, R.G., and Rogers, J.A. (2006) Heterogeneous three-dimensional electronics by use of printed semiconductor nanomaterials. *Science*, **314**, 1754.

181 Meitl, M.A., Zhu, Z.T., Kumar, V., Lee, K.J., Feng, X., Huang, Y.Y., Adesida, I., Nuzzo, R.G., and Rogers, J.A. (2006) Transfer printing by kinetic control of adhesion to an elastomeric stamp. *Nat. Mater.*, **5**, 33.

182 Kim, H.S., Lee, S.M., Ha, K., Jung, C., Lee, Y.-J., Chun, Y.S., Kim, D., Rhee, B.K., and Yoon, K.B. (2004) Aligned inclusion of hemicyanine dyes into silica zeolite films for second harmonic generation. *J. Am. Chem. Soc.*, **126**, 673–682.

183 Kim, H.S., Sohn, K.W., Jeon, Y., Min, H., Kim, D., and Yoon, K.B. (2007) Aligned inclusion of *n*-propionic acid tethering hemicyanine into silica zeolite film for second harmonic generation. *Adv. Mater.*, **19**, 260–263.

184 Chamberlain, G.A. (1983) Organic solar cells: a review. *Solar Cells*, **8**, 47.

185 Wöhrle, D. and Meissner, D. (1991) Organic solar cells. *Adv. Mater.*, **3**, 129.

186 Brabec, C.J., Sariciftci, N.S., and Hummelen, J.C. (2001) Plastic solar cells. *Adv. Funct. Mater.*, **11**, 15.

187 Peumans, P., Yakimov, A., and Forrest, S.R. (2003) Small molecular weight thin-film photodetectors and solar cells. *J. Appl. Phys.*, **93**, 3693.

188 Shaheen, S.E., Brabec, C.J., Sariciftci, N.S., Padinger, F., Fromherz, T., and Hummelen, J.C. (2001) 2.5% Efficient organic plastic solar cells. *Appl. Phys. Lett.*, **78**, 841.

189 Hoppe, H. and Sariciftci, N.S. (2004) Organic solar cells: an overview. *J. Mater. Res.*, **19**, 1924.

190 Jang, S., Newton, M.D., and Silbey, R.J. (2004) Multichromophoric Förster resonance energy transfer. *Phys. Rev. Lett.*, **92**, 218301.

191 Batchelder, J.S., Zewail, A.H., and Cole, T. (1979) Luminescent solar concentrators. 1: Theory of operation and techniques for performance evaluation. *Appl. Opt.*, **18**, 3090.

192 Hoeben, F.J.M., Schenning, A.P.H.J., and Meijer, E.W. (2005) Energy-transfer efficiency in stacked oligo(*p*-phenylene vinylene)s: pronounced effects of order. *ChemPhysChem*, **6**, 2337.

193 Dexter, D.L. (1979) Two ideas on energy transfer phenomena: ion-pair effects involving the OH stretching mode, and sensitization of photovoltaic cells. *J. Lumin.*, **18/19**, 779.

194 Huber, S. and Calzaferri, G. (2004) Energy transfer from dye-zeolite L antenna crystals to bulk silicon. *ChemPhysChem*, **5**, 239.

195 Heeger, A.J., Kivelson, S., Schrieffer, J.R., and Su, W.P. (1988) Solitons in conducting polymers. *Rev. Mod. Phys.*, **60**, 781.

196 Shirakawa, H., Louis, E.J., MacDiarmid, A.G., Chiang, C.K., and Heeger, A.J. (1977) Synthesis of electrically conductive organic polymers: halogen derivatives of polyacetylene, $(CH)_x$. *J. Chem. Soc. Chem. Commun.*, 578.

197 Ito, T., Shirakawa, H., and Ikeda, S. (1974) Simultaneous polymerization and formation of polyacetylene film on the surface of concentrated soluble Ziegler-type catalyst solution. *J. Polym. Sci. A Polym. Chem.*, **12**, 11.

198 Chiang, C.K., Fincher, C.R., Park, Y.W., Heeger, A.J., Shirakawa, H., Louis, E.J., Gau, S.C., and MacDiarmid, A.G. (1977) Electrical-conductivity in doped polyacetylene. *Phys. Rev. Lett.*, **39**, 1098.

199 Chiang, C.K., Druy, M.A., Gau, S.C., Heeger, A.J., Louis, E.J., MacDiarmid, A.G., Park, Y.W., and Shirakawa, H. (1978) Synthesis of highly conductive films of derivatives of polyacetylene, $(CH)_x$. *J. Am. Chem. Soc.*, **100**, 1013.

200 Burroughes, J.H., Bradley, D.D.C., Brown, A.R., Marks, R.N., Mackay, K., Friend, R.H., Burn, P.L., and Holmes, A.B. (1990) Light-emitting diodes based

on conjugated polymers. *Nature*, **347**, 539.
201 Greenham, N.C., Moratti, S.C., Bradley, D.D.C., Friend, R.H., and Holmes, A.B. (1993) Efficient light-emitting diodes based on polymers with high electron affinities. *Nature*, **365**, 628.
202 Vestweber, H., Greiner, A., Lemmer, U., Mahrt, R.F., Richert, R., Heitz, W., and Bässler, H. (1992) Progress towards processible materials for light-emitting devices using poly(*p*-phenylphenylenevinylene). *Adv. Mater.*, **4**, 661.
203 Shinar, J. (2004) *Organic Light-Emitting Devices: A Survey*, Springer, New York.
204 (a) Alvaro, M., Corma, A., Ferrer, B., Galletero, M.S., García, H., and Peris, E. (2004) Increasing the stability of electroluminescent phenylenevinylene polymers by encapsulation in nanoporous inorganic materials. *Chem. Mater.*, **16**, 2142–2147; (b) Alvaro, M., Cabeza, J.F., Corma, A., García, H., and Peris, E. (2007) Electrochemiluminescence of zeolite-encapsulated poly(*p*-phenylenevinylene). *J. Am. Chem. Soc.*, **129**, 8074.
205 Alvaro, M., Cabeza, J.F., Fabuel, D., Corma, A., and García, H. (2007) Electrochemiluminescent cells based on zeolite-encapsulated host–guest systems: encapsulated ruthenium tris-bipyridyl. *Chem. Eur. J.*, **13**, 3733–3738.
206 Calzaferri, G., Pauchard, M., Maas, H., Huber, S., Khatyr, A., and Schaafsma, T. (2002) Photonic antenna system for light harvesting, transport and trapping. *J. Mater. Chem.*, **12**, 1.
207 Yersin, H. (2004) Triplet emitters for OLED applications. Mechanisms of exciton trapping and control of emission properties. *Top. Curr. Chem.*, **241**, 1.
208 Yersin, H. (2008) *Highly Efficient OLEDs with Phosphorescent Materials*, Wiley-VCH Verlag GmbH, Weinheim.
209 Cleave, V., Yahioglu, G., Le Barny, P., Friend, R.H., and Tessler, N. (1999) Harvesting singlet and triplet energy in polymer LEDs. *Adv. Mater.*, **11**, 285.
210 (a) Hu, J., Odom, T.W., and Lieber, C.M. (1999) Chemistry and physics in one dimension: synthesis and properties of nanowires and nanotubes. *Acc. Chem. Res.*, **32**, 435–445; (b) Alivisatos, A.P. (1996) Semiconductor clusters, nanocrystals, and quantum dots. *Science*, **271**, 933–937.
211 (a) Jiang, W., Kim, B.Y.S., Rutka, J.T., and Chan, W.C.W. (2008) Nanoparticle-mediated cellular response is size-dependent. *Nat. Nanotechnol.*, **3**, 145–150; (b) Peer, D., Karp, J.M., Hong, S., Farokhzad, O.C., Margalit, R., and Langer, R. (2007) Nanocarriers as an emerging platform for cancer therapy. *Nat. Nanotechnol.*, **2**, 751–760; (c) Duncan, R. (2006) Polymer conjugates as anticancer nanomedicines. *Nat. Rev. Cancer*, **6**, 688–701; (d) Gao, X., Cui, Y., Levenson, R.M., Chung, L.W.K., and Nie, S. (2004) In *vivo* cancer targeting and imaging with semiconductor quantum dots. *Nat. Biotechnol.*, **22**, 969–976; (e) Michalet, X., Pinaud, F.F., Bentolila, L.A., Tsay, J.M., Doose, S., Li, J.J., Sundaresan, G., Wu, A.M., Gambhir, S.S., and Weiss, S. (2005) Quantum dots for live cells, *in vivo* imaging, and diagnostics. *Science*, **307**, 538–544; (f) Cauda, V., Onida, B., Platschek, B., Muehlstein, L., and Bein, T. (2009) Large antibiotic molecule diffusion in confined mesoporous silica with controlled morphology. *J. Mater. Chem.*, **18**, 5888–5899; (g) Cauda, V., Schlossbauer, A., Kecht, J., Zuerner, A., and Bein, T. (2009) Multiple core–shell functionalized colloidal mesoporous silica nanoparticles. *J. Am. Chem. Soc.*, **131**, 11361–11370.
212 (a) Lee, J.H., Huh, Y.M., Jun, Y., Seo, J., Jang, J., Song, H.T., Kim, S., Cho, E.J., Yoon, H.G., Suh, J.S., and Cheon, J. (2007) Artificially engineered magnetic nanoparticles for ultra-sensitive molecular imaging. *Nat. Med.*, **13**, 95–99; (b) Torchilin, V.P. (2005) Recent advances with liposomes as pharmaceutical carriers. *Nat. Rev. Drug Discov.*, **4**, 145–160.
213 (a) Knopp, D., Tang, D., and Niessner, R. (2009) Bioanalytical applications of biomolecule-functionalized nanometer-sized doped silica particles. *Anal. Chim. Acta*, **647** (1), 14–30; (b) Chan, W.C. and Nie, S. (1998) Quantum dot bioconjugates for ultrasensitive

nonisotopic detection. *Science*, **281**, 2016–2018; (c) Bruchez, M., Jr., Moronne, M., Gin, P., Weiss, S., and Alivisatos, A.P. (1998) Semiconductor nanocrystals as fluorescent biological labels. *Science*, **281**, 2013–2016.

214 (a) Petrou, P.S., Kakabakos, S.E., and Misiakos, K. (2009) Silicon optocouplers for biosensing. *Int. J. Nanotechnol.*, **6** (1/2), 4–17; (b) Tallury, P., Payton, K., and Santra, S. (2008) Silica-based multimodal/multifunctional nanoparticles for bioimaging and biosensing applications. *Nanomedicine*, **3** (4), 579–592.

215 Lu, H., Hsiao, Y., Yao, J.K., Chung, M., Lin, T.H., Wu, Y.S., Hsu, S.H., Liu, S.C., Mou, H.M., Yang, C.Y., Huang, C.S., and Chen, D.M. (2007) Bifunctional magnetic silica nanoparticles for highly efficient human stem cell labelling. *Nano Lett.*, **7**, 149–154.

216 (a) Zanarini, S., Rampazzo, E., Ciana, L.D., Marcaccio, M., Marzocchi, E., Montalti, M., Paolucci, F., and Prodi, L. (2009) Ru(bpy)3 covalently doped silica nanoparticles as multicenter tunable structures for electrochemiluminescence amplification. *J. Am. Chem. Soc.*, **131**, 2260–2267; (b) Piao, Y., Burns, A., Kim, J., Wiesner, U., and Hyeon, T. (2008) Platinum lead nanostructures: formation, phase behavior, and electrocatalytic properties. *Adv. Funct. Mater.*, **18**, 2745–2753.

217 Burns, A.A., Vider, J., Ow, H., Herz, E., Medina, O.P., Baumgart, M., Larson, S.M., Wiesner, U., and Bradbury, M. (2009) Fluorescent silica nanoparticles with efficient urinary excretion for nanomedicine. *Nano Lett.*, **9**, 442–448.

218 Seo, W.S., Lee, J.H., Sun, X., Suzuki, Y., Mann, D., Liu, Z., Terashima, M., Yang, P.C., McConnell, M.V., Nishimura, D.G., and Dai, H. (2006) FeCo/graphitic-shell nanocrystals as advanced magnetic-resonance-imaging and near-infrared agents. *Nat. Mater.*, **5**, 971–976.

219 Heymer, A., Haddad, D., Weber, M., Gbureck, U., Jakob, P.M., Eulert, J., and Nöth, U. (2008) Iron oxide labelling of human mesenchymal stem cells in collagen hydrogels for articular cartilage repair. *Biomaterials*, **10**, 1473–1483.

220 Cellot, G., Cilia, E., Cipollone, S., Rancic, V., Sucapane, A., Giordani, S., Gambazzi, L., Markram, H., Grandolfo, M., Scaini, D., Gelain, F., Casalis, L., Prato, M., Giugliano, M., and Ballerini, L. (2009) Carbon nanotubes might improve neuronal performance by favouring electrical shortcuts. *Nat. Nanotechnol.*, **4**, 126–133.

221 Elkin, T., Jiang, X., Taylor, S., Lin, Y., Gu, L., Yang, H., Brown, J., Collins, S., and Sun, Y.P. (2005) Immuno-carbon nanotubes and recognition of pathogens. *Chembiochem*, **6**, 640–643.

222 (a) Mulder, W.J.M., Douma, K., Koning, G.A., van Zandvoort, M.A., Lutgens, E., Daemen, M.J., Nicolay, K., and Strijkers, G.J. (2006) Liposome-enhanced MRI of neointimal lesions in the ApoE-KO mouse. *Magn. Reson. Med.*, **55**, 1170–1174; (b) Mulder, W.J.M., Strijkers, G.J., Griffioen, A.W., van Bloois, L., Molema, G., Storm, G., Koning, G.A., and Nicolay, K. (2004) Immuno-carbon nanotubes and recognition of pathogens. *Bioconjug. Chem.*, **15**, 799–806.

223 Accardo, A., Tesauro, D., Roscigno, P., Gianolio, E., Paduano, L., D'Errico, G., Pedone, C., and Morelli, G. (2004) Physicochemical properties of mixed micellar aggregates containing CCK peptides and Gd complexes designed as tumor specific contrast agents in MRI. *J. Am. Chem. Soc.*, **126**, 3097–3107.

224 (a) Pattani, V.P., Li, C., Desai, T.A., and Vu, T.Q. (2008) Microcontact printing of quantum dot bioconjugate arrays for localized capture and detection of biomolecules. *Biomed. Microdevices*, **10**, 367–374; (b) Khang, D.Y., Xiao, J., Kocabas, C., MacLaren, S., Banks, T., Jiang, H., Huang, Y.Y., and Rogers, J.A. (2008) Molecular scale buckling mechanics in individual aligned single-wall carbon nanotubes on elastomeric substrates. *Nano Lett.*, **7**, 124–130; (c) Jiang, X., Bruzewicz, D.A., Thant, M.M., and Whitesides, G.M. (2004) Palladium as a substrate for self-assembled monolayers used in

biotechnology. *Anal. Chem.*, **76**, 6116–6121; (d) Houseman, B.T., Gawalt, E.S., and Mrksich, M. (2003) Maleimide-functionalized self-assembled monolayers for the preparation of peptide and carbohydrate biochips. *Langmuir*, **19**, 1522–1531.

225 Veiseh, M., Zareie, M.H., and Zhang, M.Q. (2002) Highly selective protein patterning on gold–silicon substrates for biosensor applications. *Langmuir*, **18**, 6671–6678.

226 (a) Lyles, B.F., Terrot, M.S., Hammond, P.T., and Gast, A.P. (2004) Directed patterned adsorption of magnetic beads on polyelectrolyte multilayers on glass. *Langmuir*, **20**, 3028–3021; (b) Pregibon, D.C., Toner, M., and Doyle, P.S. (2006) Magnetically and biologically active bead-patterned hydrogels. *Langmuir*, **22**, 5122–5128.

227 Zhang, D., Chen, Y., Chen, H.-Y., and Xia, X.H. (2004) Silica-nanoparticle-based interface for the enhanced immobilization and sequence-specific detection of DNA. *Anal. Bioanal. Chem.*, **379**, 1025–1030.

228 Osaka, T., Matsunaga, T., Nakanishi, T., Arakaki, A., Niwa, D., and Iida, H. (2006) Synthesis of magnetic nanoparticles and their application to bioassays. *Anal. Bioanal. Chem.*, **384**, 593–600.

229 Weibel, D.B., DiLuzio, W.R., and Whitesides, G.M. (2007) Microfabrication meets microbiology. *Nat. Rev. Microbiol.*, **5**, 209–218.

230 (a) Wadhwa, S., Paliwal, R., Paliwal, S.R., and Vyas, S.P. (2009) Nanocarriers in ocular drug delivery: an update review. *Curr. Pharm. Des.*, **15** (23), 2724–2750; (b) Wang, J. (2009) Biomolecule-functionalized nanowires: from nanosensors to nanocarriers. *ChemPhysChem*, **10** (11), 1748–1755; (c) Vijayaraghavalu, S., Raghavan, D., and Labhasetwar, V. (2007) Nanoparticles for delivery of chemotherapeutic agents to tumors. *Curr. Opin. Investig. Drugs*, **8**, 477–484; (d) Jain, K.K. (2005) Nanotechnology-based drug delivery for cancer. *Technol. Cancer Res. Treat.*, **4**, 407–416; (e) Graff, A., Benito, S.M., Verbert, C., and Meier, W. (2004) Polymer nanocontainers. *Nanobiotechnology: Concepts, Applications and Perspectives* (eds C.M. Niemeyer and C.A. Mirkin), Wiley-VCH Verlag GmbH, pp. 168–184.

231 (a) Rámila, V.R., del Real, A.P.P., and Pérez-Pariente, J. (2001) A new property of MCM-41: drug delivery system. *Chem. Mater.*, **13**, 308–311; (b) Ozin, G. (1992) Nanochemistry: synthesis in diminishing dimensions. *Adv. Mater.*, **4**, 612–649.

232 Trewyn, B.G., Giri, S., Slowing, I.I., and Lin, V.S.Y. (2007) Mesoporous silica nanoparticle based controlled release, drug delivery, and biosensor systems. *Chem. Commun.*, **31**, 3236–3245.

233 Liong, M., Lu, J., Kovochich, M., Xia, T., Ruehm, S.G., Nel, A.E., Tamanoi, F., and Zink, J.I. (2008) Multifunctional inorganic nanoparticles for imaging, targeting, and drug delivery. *ACS Nano*, **2** (5), 889–896.

234 Park, J.-H., Gu, L., von Maltzahn, G., Ruoslahti, E., Bhatia, S.N., and Sailor, M.J. (2009) Biodegradable luminescent porous silicon nanoparticles for *in vivo* applications. *Nat. Mater.*, **8**, 331–336.

235 Anglin, E.J., Cheng, L., Freeman, W.R., and Sailor, M.J. (2008) Porous silicon in drug delivery devices and materials. *Adv. Drug. Deliv. Rev.*, **60** (11), 1266–1277.

236 Kleps, I., Miu, M., Simion, M., Ignat, T., Bragaru, A., Craciunoiu, F., and Danila, M. (2009) Study of the micro- and nanostructured silicon for biosensing and medical applications. *J. Biomed. Nanotechnol.*, **5** (3), 300–309.

237 (a) Wang, S. (2009) Ordered mesoporous materials for drug delivery. *Microporous Mesoporous Mater.*, **117** (1–2), 1–9; (b) Wang, Q., Shantz, D.F., Wang, Q., and Shantz, D.F. (2008) Ordered mesoporous silica-based inorganic nanocomposites. *J. Solid State Chem.*, **181** (7), 1659–1669; (c) Son, S.J., Reichel, J., He, B., Schuchman, M., and Lee, S.B. (2005) Magnetic nanotubes for magnetic-field-assisted bioseparation, biointeraction, and drug delivery. *J. Am. Chem. Soc.*, **127**, 7316–7317; (d) Chen, C.C., Liu, Y.C., Wu, C.H., Yeh, C.C., Su, M.T., and Wu, Y.C. (2005) Preparation of fluorescent silica nanotubes and their application in gene delivery. *Adv. Mater.*, **17**, 404–407.

238 (a) Jane, A., Dronov, R., Hodges, A., and Voelcker, N.H. (2009) Porous silicon biosensors on the advance. *Trends Biotechnol.*, **27**, 230–239; (b) Zhang, Y., Ren, N., and Tang, Y. (2009) in *Ordered Porous Solids* (eds V. Valtchev, S. Mintova, and M. Tsapatsis), Elsevier, p. 441; (c) Egeblad, K., Christensen, C.H., Kustova, M., and Christensen, C.H. (2008) Tailoring the porosity of hierarchical zeolites by carbon-templating, *Stud. Surf. Sci. Catal.*, **174**, 285–288. (d) Schmeltzer, J.M. and Buriak, J.M. (2004) Recent Developments in the Chemistry and Chemical Applications of Porous Silicon, in *The Chemistry of Nanomaterials: Synthesis, Properties and Applications*, (eds C.N.R. Rao, A. Müller, A. K. Cheetham), Wiley-VCH, Weinheim, p. 518–550.

239 (a) Angelatos, A.S., Johnston, A.P.R., Wang, Y., and Caruso, F. (2007) Probing the permeability of polyelectrolyte multilayer capsules via a molecular beacon approach. *Langmuir*, **23**, 4554–4562; (b) Itoh, Y., Matsusaki, M., Kida, T., and Akashi, M. (2006) Enzyme-responsive release of encapsulated proteins from biodegradable hollow capsules. *Biomacromolecules*, **7**, 2715.

240 (a) Wang, Y. and Caruso, F. (2005) Mesoporous silica spheres as supports for enzyme immobilization and encapsulation. *Chem. Mater.*, **17**, 953–961; (b) Yu, A., Wang, Y., Barlow, B., and Caruso, F. (2005) Mesoporous silica particles as templates for preparing enzyme-loaded biocompatible microcapsules. *Adv. Mater.*, **17**, 1737–1741.

241 Gabizon, A., Shmeeda, H., and Barenholz, Y. (2003) Pharmacokinetics of pegylated liposomal doxorubicin: review of animal and human studies. *Clin. Pharmacokinet.*, **42**, 419–436.

242 Chung, T.H., Wu, S.H., Yao, M., Lu, C.W., Lin, Y.S., Hung, Y., Mou, C.Y., Chen, Y.C., and Huang, D.M. (2007) The effect of surface charge on the uptake and biological function of mesoporous silica nanoparticles in 3T3-L1 cells and human mesenchymal stem cells. *Biomaterials*, **28**, 2959–2966.

243 Gu, J.L., Fan, W., Shimojima, A., and Okubo, T. (2007) Organic–inorganic mesoporous nanocarriers integrated with biogenic ligands. *Small*, **3**, 1740–1744.

244 Brevet, D., Bobo, M.G., Raehm, L., Richeter, S., Hocine, O., Amro, K., Loock, B., Couleaud, P., Frochot, C., Morere, A., Maillard, P., Garcia, M., and Durand, J.O. (2009) Mannose-targeted mesoporous silica nanoparticles for photodynamic therapy. *Chem. Commun.*, 1475–1477.

245 Bottini, M., Cerignoli, F., Mills, D.M., D'Annibale, F., Leone, M., Rosato, N., Magrini, A., Pellecchia, M., Bergamaschi, A., and Mustelin, T. (2007) Luminescent silica nanobeads: characterization and evaluation as efficient cytoplasmatic transporters for T-lymphocytes. *J. Am. Chem. Soc.*, **129**, 7814–7823.

246 (a) Sun, B.F., Ranganathana, B., and Feng, S.S. (2008) Multifunctional poly(D,L-lactide-*co*-glycolide)/montmorillonite (PLGA/MMT) nanoparticles decorated by trastuzumab for targeted chemotherapy of breast cancer. *Biomaterials.*, **29**, 475–486; (b) Wuang, S.C., Neoh, K.G., Kang, E.T., Pack, D.W., and Leckband, D.E. (2008) HER-2-mediated endocytosis of magnetic nanospheres and the implications in cell targeting and particle magnetization. *Biomaterials*, **29**, 2270–2279.

247 Slowing, I.I., Trewyn, B.G., and Lin, V.S.Y. (2007) Mesoporous silica nanoparticles for intracellular delivery of membrane-impermeable proteins. *J. Am. Chem. Soc.*, **129**, 8845–8849.

248 Lu, J., Liong, M., Zink, J.I., and Tamanoi, F. (2007) Mesoporous silica nanoparticles as a delivery system for hydrophobic anticancer drugs. *Small*, **3**, 1341–1346.

249 Torney, F., Trewyn, B.G., Lin, V.S.Y., and Wang, K. (2007) Mesoporous silica nanoparticles deliver DNA and chemicals into plants. *Nat. Nanotechnol.*, **2**, 295–300.

250 Mellaerts, R., Jammaer, J., Van Speybroeck, M., Hong, C., Van Humbeeck, J., Augustijns, P., Van den Mooter, G., and Martens, J.P. (2008) Physical state of poorly water soluble therapeutic molecules loaded into SBA-15 ordered mesoporous silica carriers: a case study with itraconazole and ibuprofen. *Langmuir*, **24**, 8651–8659.

251 (a) Giri, S., Trewyn, B.G., Stellmaker, M.P., and Lin, V.S.Y. (2005) Stimuli-responsive controlled-release delivery system based on mesoporous silica nanorods capped with magnetic nanoparticles. *Angew. Chem., Int. Ed.*, **44**, 5038–5044; (b) Lai, C.Y., Trewyn, B.G., Jeftinija, D.M., Jeftinija, K., Xu, S., Jeftinija, S., and Lin, V.S.Y. (2003) A mesoporous silica nanosphere-based carrier system with chemically removable CdS nanoparticle caps for stimuli-responsive controlled release of neurotransmitters and drug molecules. *J. Am. Chem. Soc.*, **125**, 4451–4459.

252 Yiu, H.H.P. and Wright, P.A. (2004) Nanoporous materials as supports for enzyme immobilization. *Ser. Chem. Eng.*, **4**, 849–872.

253 (a) Taylor, K.M.L., Kim, J.S., Rieter, W.J., An, H., Lin, W.L., and Lin, W.B. (2008) Mesoporous silica nanospheres as highly efficient MRI contrast agents. *J. Am. Chem. Soc.*, **130**, 2154–2155; (b) Kim, J., Kim, H.S., Lee, N., Kim, T., Kim, H., Yu, T., Song, I.C., Moon, W.K., and Hyeon, T. (2008) Multifunctional uniform nanoparticles composed of a magnetite nanocrystal core and a mesoporous silica shell for magnetic resonance and fluorescence imaging and for drug delivery. *Angew. Chem., Int. Ed.*, **47**, 8438–8441.

254 (a) Lam, A. and Rivera, A. (2001) Channel model for the theoretical study of aspirin adsorption on clinoptilolite. Water influence. *Stud. Surf. Sci. Catal.*, **135**, 5304–5308; (b) Uglea, C.V., Albu, I., Vatajanu, A., Croitoru, M., Antoniu, S., Panaitesc, L., and Ottenbrite, M. (1994) Drug delivery systems based on inorganic materials: I. Synthesis and characterization of a zeolite–cyclophosphamide system. *J. Biomater. Sci. Polym. Ed.*, **6**, 633–637.

255 Rivera, A., Farias, T., Ruiz-Salvador, A.R., and de Menorval, L.C. (2003) Preliminary characterization of drug support systems based on natural clinoptilolite. *Microporous Mesoporous Mater.*, **61**, 249–259.

256 Rimoli, M.G., Rabaioli, M.R., Melisi, D., Curcio, A., Mondello, S., Mirabelli, R., and Abignente, E.S. (2008) Synthetic zeolites as a new tool for drug delivery. *J. Biomed. Mater. Res. A*, **87** (1), 156–164.

257 Dyer, A., Morgan, S., Wells, P., and Williams, C. (2000) The use of zeolites as slow release anthelmintic carriers. *J. Helminthol.*, **74** (2), 137–141.

258 (a) Zhang, Y., Wang, X., Shan, W., Wu, B., Fan, H., Yu, X., Tang, Y., and Yang, P. (2005) Enrichment of low-abundance peptides and proteins on zeolite nanocrystals for direct MALDI-TOF MS analysis. *Angew. Chem., Int. Ed.*, **44**, 615–617; (b) Zhang, Y., Yu, X., Wang, X., Shan, W., Yang, P., and Tang, Y. (2004) Zeolite nanoparticles with immobilized metal ions: isolation and MALDI-TOF-MS/MS identification of phosphopeptides. *Chem. Commun.*, 2882–2883.

259 Dahm, A. and Eriksson, H. (2004) Ultra-stable zeolites: a tool for in-cell chemistry. *J. Biotechnol.*, **111** (5), 279–290.

260 Pearce, M.E., Mai1, H.Q., Lee, N., Larsen, S.C., and Salem, A.K. (2008) Silicalite nanoparticles that promote transgene expression. *Nanotechnology*, **19**, 175103–175110.

261 Liu, B., Yan, F., Kong, J., and Deng, J. (1999) A reagentless amperometric biosensor based on the coimmobilization of horseradish peroxidase and methylene green in a modified zeolite matrix. *Anal. Chim. Acta*, **386**, 31–39.

262 Iglesias, C.P., Elst, L.V., Zhou, W., Muller, R.N., Geraldes, C.F.G.C., Maschmeyer, T., and Peters, J.A. (2002) Zeolite GdNaY nanoparticles with very high relaxivity for application as contrast agents in magnetic resonance imaging. *Chem. Eur. J.*, **8**, 5121–5131.

263 Balkus, K.J. and Shi, J. (1997) Synthesis of hexagonal Y type zeolites in the presence of Gd(III) complexes of 18-crown-6. *Microporous Mater.*, **11**, 325–333.

264 Hotta, M., Nakayima, H., Yamamoto, K., and Aono, M. (1998) Antibacterial temporary filling materials: the effect of adding various ratios of Ag–Zn–zeolite. *J. Oral. Rehabil.*, **25**, 485–489.

265 (a) Zarkovic, N., Zarkovic, K., Kralj, M., Borovic, S., Sabolovic, S., Poljak-Blazi, M., Cipak, A., and Pavelic, K. (2003) Anticancer and antioxidative effects of micronized zeolite linoptilolite. *Anticancer Res.*, **23**, 1589–1595; (b) Pavelic, K., Hadzija, M., Bedrica, L., Pavelic, J., Dikic, I., Katic, M., Kralj, M., Bosnar, M.H., Kapitanovic, S., and Colic, M. (2001) Natural zeolite clinoptilolite: new adjuvant in anticancer therapy. *J. Mol. Med.*, **78**, 708–720.

266 Li, Z., Luppi, G., Geigerb, A., Josel, H.P., and De Cola, L. (2010) Submitted.

267 Strassert, C.A., Otter, M., Albuquerque, R.Q., Höne, A., Vida, Y., Maier, B., and De Cola, L. (2010) Photoactive hybrid nanomaterial for targeting, labeling, and killing antibiotic-resistant bacteria. *Angew. Chem., Int. Ed.*, **48**, 7928–7931.

268 Tsotsalas, M.M., Kopka, K., Luppi, G., Wagner, S., Law, M., Schäfers, M., and De Cola, L. (2010) Encapsulating ^{111}In in nanocontainers for scintigraphic imaging: synthesis, characterization and in vivo biodistribution. *ACS Nano*, **4**, 342–348.

269 Cai, D., Mataraza, J.M., Qin, Z.H., Huang, Z., Huang, J., Chiles, T.C., Carnahan, D., Kempa, K., and Ren, Z. (2005) Highly efficient molecular delivery into mammalian cells using carbon nanotube spearing. *Nat. Methods*, **2**, 449–454.

270 Rieter, W.J., Taylor, K.M.L., An, H.Y., Lin, W.L., and Lin, W.B. (2006) Nanoscale metal–organic frameworks as potential multimodal contrast enhancing agents. *J. Am. Chem. Soc.*, **128**, 9024–9025.

271 (a) Horcajada, P., Serre, C., Maurin, G., Ramsahye, N.A., Balas, F., Vallet-Regi, M., Sebban, M., Taulelle, F., and Ferey, G. (2008) Flexible porous metal–organic frameworks for a controlled drug delivery. *J. Am. Chem. Soc.*, **130**, 6774–6780; (b) Rieter, W.J., Pott, K.M., Taylor, K.M.L., and Lin, W.B. (2008) Nanoscale coordination polymers for platinum-based anticancer drug delivery. *J. Am. Chem. Soc.*, **130**, 11584–11585; (c) Horcajada, P., Serre, C., Vallet-Regi, M., Sebban, M., Taulelle, F., and Ferey, G. (2006) Metal–organic frameworks as efficient materials for drug delivery. *Angew. Chem., Int. Ed.*, **45**, 5974–5978.

272 Ai, H., Pink, J.J., Shuai, X., Boothman, D.A., and Gao, J. (2005) Interactions between self-assembled polyelectrolyte shells and tumor cells. *J. Biomed. Mater. Res. A*, **73A**, 303–312.

273 Cortez, C., Tomaskovic-Crook, E., Johnston, A.P.R., Radt, B., Cody, S.H., Scott, A.M., Nice, E.C., Heath, J.K., and Caruso, F. (2006) Targeting and uptake of multilayered particles to colorectal cancer cells. *Adv. Mater.*, **18**, 1998–2003.

10
Soft Lithography for Patterning Self-Assembling Systems
Xuexin Duan, David N. Reinhoudt, and Jurriaan Huskens

10.1
Introduction

Nanoscience is an important, central theme in fundamental research. It is an emerging science of objects that are intermediate in size between the largest molecules and the smallest structures that can be fabricated by current lithography techniques; that is, the science of objects with the smallest dimensions ranging from a few nanometers to less than 100 nm [1–3]. Objects and structures with these dimensions exhibit peculiar and interesting characteristics [4–7]. In addition to their use in the fundamental research field, nanostructures are central to the development of nanotechnologies. In almost all applications of nanostructures, fabrication represents the first and one of the most significant challenges. Two main lines of nanofabrication strategies have evolved: bottom-up and top-down methods.

Bottom-up methods build highly ordered nanostructures from smaller elementary components. The most efficient and applied method is the self-assembly of molecules, surfactants, colloids, block copolymers, and so on. The key idea in self-assembly is that the final structure is close to or at thermodynamic equilibrium, and it thus tends to form spontaneously and is prone to error correction. Therefore, it can provide routes to achieve structures with greater order than what can be reached in nonself-assembling systems and thus mainly offer new lithographic possibilities [8–13].

Self-assembly provides simple and low-cost processes to make large-area periodic nanostructures. However, for bulk production of nanostructures, self-assembly has drawbacks, such as the lack of long-range order of the structures, as in most cases they are isotropic and random. In contrast, "top-down" methods, consisting of a series of lithographic approaches, that are based on traditional photolithography [14, 15] offer arbitrary geometrical designs and superior nanometer-level precision, accuracy, and registration. The combination of "bottom-up" self-assembly with "top-down" patterned templates can provide new opportunities for fundamental studies of self-

Functional Supramolecular Architectures. Edited by Paolo Samorì and Franco Cacialli
Copyright © 2011 WILEY-VCH Verlag GmbH & Co. KGaA, Weinheim
ISBN: 978-3-527-32611-2

assembly behavior in confined environments, as well as a source of innovation for alternative nanofabrication methods. Among them, soft lithography [16–21], a series of fabrication techniques that replicate structures on a master by conformal contact with a substrate to pattern self-assembling systems, is most attractive since it allows the patterning of a wide range of materials and material precursors in a relatively cheap and facile way. As such, soft lithography has particularly found its way into the research environment, where rapid prototyping and materials' versatility are crucial and access to high-end photolithography is limited.

This chapter gives an overview of patterning of self-assembling systems by soft lithography. It starts with a brief introduction of various self-assembling systems and two main soft lithography techniques, microcontact printing (μCP) and micromolding in capillaries (MIMIC). Thereafter, recent developments in the use of μCP and MIMIC for patterning self-assembling systems will be presented. The main focus is on advances in resolution.

10.2
Self-Assembling Systems

Self-assembly is the spontaneous organization of molecules or objects into stable, well-defined structures by noncovalent forces. It has been the focus of vast research efforts in the past four decades. These efforts have produced a solid foundation of understanding of the physics and chemistry of self-organizing processes [22].

Self-assembly starts with the smallest units – molecular assembly. In molecular self-assembly, multiple, weak reversible interactions based on, for example, hydrogen bonding and van der Waals forces, drive individual molecular subunits to assemble into stable aggregates. The assembled structure typically represents a thermodynamic minimum that results from equilibration of these interactions. There are hundreds of examples of systems that are based on the self-assembly of molecules, such as liquid crystals, micelles, and lipid membranes. Although many of the studies in self-assembly have focused on molecular components, the domain of self-assembly is not limited to the molecular level but extends to structural organizations on various length scales. Energy minimization remains the key motivation for self-assembly at these scales. Self-assembly of larger components, for example, nanoparticles, bimolecules, block polymers, and so on, shows a lot of promise and is extremely important in the emerging fields of nanostructured materials, bioinspired templating, DNA computing, and microelectronics [23].

In general, self-assembly can be classified into (i) nontemplated self-assembly, where individual components interact to produce a larger structure without the assistance of external forces or spatial constraints, and (ii) templated self-assembly, where individual components interact with each other guided by an external force or within spatial constraints. In this chapter, we will focus on the recent advances in templated self-assembly, in particular, examples patterned by soft lithography, and evaluate their application in the nanofabrication of 1D, 2D, and 3D materials.

10.3
Soft Lithography

Photolithography is exclusively used for the fabrication of microelectronic devices [14, 15]. However, photolithography has limitations when targeting the patterning of self-assembling systems: (i) it requires expensive instruments and facilities with high capital investment; (ii) the pattern resolution is limited by optical diffraction, and obtaining submicron patterns requires extensive technology upgrading; (iii) it is not suitable for patterning all types of polymers as only photosensitive resist materials (photoresists) can be directly patterned; and (iv) it requires harsh processing conditions such as exposure to UV-radiation and chemical etching, and therefore, it is not suitable for patterning sensitive materials. Some of the limitations of photolithography, particularly the resolution limit, can be overcome by using next-generation lithography, for example, deep UV and extreme UV photolithography [24, 25], soft X-ray lithography [26], electron-beam writing [27], and ion-beam lithography [28]. However, all these state-of-the-art lithographic techniques have high costs and low accessibility. There is a large demand for low-cost and large-area manufacturing techniques, both for real applications in actual devices and, in particular, for rapid prototyping in research environments. Soft lithography, which is a collective term for a number of nonphotolithographic techniques, fulfills these requirements, to some extent, by removing the need for clean-room facilities. Furthermore, it allows direct patterning of a wide range of materials.

Soft lithography encompasses a family of techniques that employ a soft mold or stamp made by replica molding (REM) a soft polymer using a hard master. Varying the way in which the molds are used results in different techniques: microcontact printing [29], MIMIC [30, 31], REM [32], microtransfer molding (μTM) [33], and solvent-assisted micromolding (SAMIM) [34]. This chapter will focus on the first two methods since they offer the possibilities of patterning self-assembling systems in solid state or from their solutions. Emphasis is put on new developments that provide access to difficult length scales, that is, between a few tens and hundreds of nanometers.

10.3.1
Microcontact Printing

Whitesides *et al.* invented the μCP process in the beginning of 1990s. The general procedure of the μCP is remarkably simple (Figure 10.1). It works by the creation of a flexible, polymeric stamp with patterned reliefs (typically made from poly(dimethylsiloxane), PDMS) and by dipping the stamp in an alkanethiol "ink." Once the stamp has been inked and dried, it is then briefly pressed against a gold (or other thiol-compatible) substrate via conformal contact, and the alkanethiol molecules are transferred from the polymer to the substrate and self-assemble into a self-assembled monolayer (SAM) in the contact areas. The bare gold regions can either be etched, used directly, or backfilled with a different adsorbate. In the latter way, a binary-component SAM is formed that can act as a template for the selective deposition of specific molecules (e.g., proteins).

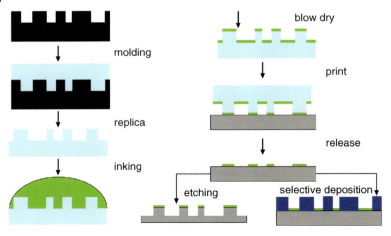

Figure 10.1 Schematic representation of the microcontact printing process.

The μCP as a method of patterning SAMs has grown in popularity due to the ease of fabrication of the printing tools, relatively high spatial resolution of the features produced, and large printing capacities [20, 35–37]. The low costs and simplicity of the technique have inspired the interest in creating smaller patterns with higher edge resolution and in broadening the versatility of the technique. Although initially mainly used as a method for patterning self-assembled alkanethiol monolayers on gold surfaces [37–39], μCP has been extended to alkylsilanes on silicon oxide [40], and this has resulted in numerous biotechnology applications, such as the patterned growth of a variety of cells and the fabrication of microarrays for biosensor purposes. The range of ink molecules has been extended from alkylsilanes and alkanethiols to various particles and organic molecules with higher molecular weights, ranging from Langmuir–Blodgett films [41] to DNA [42, 43] and proteins [44]. Nowadays, μCP has been recognized as a remarkable surface patterning technique and has triggered enormous interest not only from the surface science community but also from engineers and biologists.

10.3.2
Micromolding Injection in Capillaries

Micromolding in capillaries is a simple and versatile soft-lithographic method that forms complex microstructures on both planar and curved surfaces. It was introduced by Whitesides and coworkers in 1995 [31]. MIMIC, its scheme is shown in Figure 10.2, can be considered a precursor of micro- and nanofluidics, and it can be used to pattern many soluble materials.

In a typical MIMIC scheme, a soft elastomeric mold usually made from PDMS with parallel protrusions is placed in contact with a smooth surface, so that the grooves form channels (capillaries). When a solution is poured at the open end of the channels, the liquid spontaneously fills the channels under the effect of capillary pressure. As the solution volume gradually shrinks because of solvent evaporation,

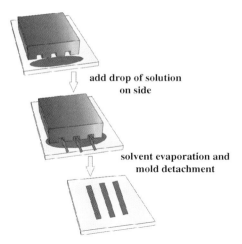

Figure 10.2 Schematic representation of the micromolding in capillaries process.

the capillary forces drive the formation of a meniscus under the roof of the stamp channels. After the complete evaporation of the solvent, the stamp is gently removed leaving the patterns on the surface.

Depending on the concentration of the solution, two kinds of patterns can be obtained. In the high-concentration regime, if the solution reaches supersaturation when the microchannel is still full of solution, the pattern replicates the size of the microchannel. In the dilute regime, when the solution reaches supersaturation with most of the solvent having evaporated and the volume of the residual solution not enough to fill the channel completely, the solution tends to accumulate on the boundaries of the channel, giving rise to some defects in the microstripes or to split lines.

The capability of MIMIC has been demonstrated by the fabrication of patterned structures and devices from a variety of functional materials at the micrometer and nanometer length scales. Although MIMIC was initially used without solvent [30], that is, using a prepolymer instead of a solution, the most recent applications employ solutions of functional molecules, for which the only requirement is that the solvent does not swell the polymeric stamp.

Capillarity is the main driving force in MIMIC for filling the channels. During the solvent evaporation, the self-organization of the solute comes into play when the solution reaches supersaturation, so that spatially organized nanodots, wires, or crystallites can be fabricated. It proved that MIMIC is an excellent tool both for patterning and for studying self-assembling systems.

10.3.3
Main Limitations of Soft Lithography

The key element in soft lithography is the use of a patterned elastomeric stamp that mediates intimate contact between the elastomeric stamp and the substrate. The definition of this intimate or "conformal" contact in high-resolution printing goes beyond the contact between the asperities of two flat, hard surfaces. Adhesion forces mediate this elastic adaptation, and even without the application of external pressure, an elastomer can spontaneously compensate for some degree of substrate roughness, depending on the materials properties [35].

The elastomer stamp most commonly used in soft lithography is PDMS. This is favored because it is commercially available, resistant to many types of chemicals and pH environments, optically transparent, and nontoxic. PDMS stamps can be reused many times without any noticeable degradation. The material cures under moderate conditions and is easily removed from surfaces, making it amenable to patterning complex structures on delicate or nonplanar surfaces.

The use of a soft polymer is also at the origin of the main problems of soft lithography. Deformation of the stamp during stamp removal from the template and during the contacting of the substrate limits the resolution of patterning. Such deformations are illustrated in Figure 10.3. The height of the features divided by their lateral dimensions defines the aspect ratio of a pattern. When the aspect ratio is high, buckling and lateral collapse of the PDMS features can occur, while at low aspect ratios roof collapse is possible [45]. Any deformation of the stamp will affect the printed pattern and decrease reproducibility. These phenomena are enhanced when the sizes of the corrugations reach the submicron or nanoscale.

Besides the stamp deformation, there are some other major drawbacks when using PDMS as the stamp or mold in soft lithography. For example, the patterns can be contaminated with unpolymerized low molecular weight siloxanes cotransferred from the stamp and thus decrease the pattern quality [46–49]. Swelling and distortion of the stamp may occur with certain solvents, especially in the case of nonpolar solvents, and may result in patterns with increased sizes and pattern defects [50–52].

These problems have limited the size of the patterns that can be attained by soft lithography to the > 500 nm scale. In recent years, efforts have been made to shrink

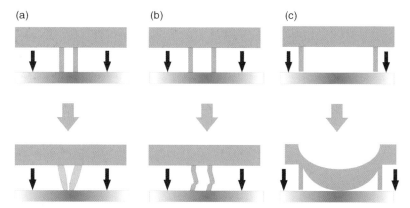

Figure 10.3 The most commonly observed stamp deformations: pairing (a), buckling (b), and roof collapse (c).

the size of the patterns to the nanoscale. To overcome the obstacles described above, optimization of stamp materials and improvements of the patterning conditions are crucial. Furthermore, new patterning strategies combined with self-assembling systems have made patterning with nanoscale dimensions possible through microcontact printing or micromolding in capillaries.

10.4
Contact Printing of SAMs with High Resolution

Self-assembled monolayers constitute an important class of self-assembling systems. SAMs are ordered molecular assemblies formed by the adsorption of an active surfactant onto a solid surface. The order in these two-dimensional systems is produced by a spontaneous chemisorption process at the interface, as the system approaches equilibrium [53]. SAMs have shown potential applications in the control of wettability, biocompatibility, and corrosion resistance of the surfaces of a wide range of materials [54, 55]. Among all SAM systems, the assembly of sulfur-containing molecules, such as alkanethiols and dialkyldisulfides, on noble metal surfaces (especially on gold) have been the most widely studied. The assembly is held together both by the bonds between the sulfur head groups and the gold surface and by the van der Waals interactions between the neighboring hydrocarbon chains. The surface chemistry can be entirely defined by the nature of the functional group at the end of the long-chain alkanethiols used. The facile formation of these self-assembled monolayers has opened a general route to molecular-level control over surface order and chemistry.

Patterning SAMs offers the possibility to form SAMs in predetermined spatial distributions and has thus created applications for nanotechnology [53, 56]. Patterning SAMs in the plane of the surface has been achieved by a wide variety of techniques [57]. Among them, microcontact printing, which relies on the contact of a rubber stamp inked with alkanethiols with a substrate, offers a most interesting combination of convenience and new capabilities. Contact times are typically on the order of seconds, but despite these short contact times, high-quality SAMs are formed that differ very little from the crystalline SAMs formed from solution over hours as demonstrated by scanning tunneling microscopy (STM) studies [58].

Although as simple as µCP is, it has been very difficult to print SAMs with sub-500-nm lateral resolution. Besides the stamp deformation problem, which is mentioned in Section 10.3.3, another important issue is the ink diffusion problem. In µCP, the successful transfer of molecules to the substrate is typically achieved by ink diffusion of absorbed molecules from the PDMS bulk to the surface. However, an excess of ink results in an uncontrolled transfer and spreading of the ink molecules. Delamarche *et al.* have studied in detail the ink transfer mechanism during the microcontact printing process [59]. They proved that ink transfer to the noncontact areas can happen because of ink diffusion through the stamp, ink spreading, surface diffusion, and vapor-phase deposition. All these issues limit the lateral resolution of the contact printing of SAMs. Here, examples of recent advances in microcontact printing to achieve high resolution will be reviewed.

10.4.1
Improvements of the Stamp

10.4.1.1 Hard PDMS and Composite Stamps

In order to solve stamp deformation with the conventional (Sylgard 184 system) PDMS, the IBM group has identified a commercially available Si-based elastomer that has a relatively high compression modulus [60]. This material (hard-PDMS or h-PDMS) is prepared from trimethylsiloxy-terminated vinylmethylsiloxane-dimethylsiloxane (VDT-731; Gelest) and methylhydrosiloxane-dimethylsiloxane (HMS-301; Gelest) copolymers. The h-PDMS system has cross-linkers that have relatively short lengths compared to those in the Sylgard 184 system; thus, it has a relatively high modulus (≈ 9 MPa). With this "hard" PDMS, the contact printing of structures down to 80 nm was achieved. Choi and Rogers [61, 62] designed a new stiffer PDMS stamp that incorporates a rigid urethane methacrylate cross-linker into the PDMS polymeric network. It has a modulus of 4 MPa that is higher than conventional Sylgard 184 (≈ 2 MPa) but lower than h-PDMS. It performs better in printing in many respects than the commercially available materials. In addition, the physical toughness and modulus of this material can be adjusted by controlling the cross-linking density. Moreover, its photocurability allows the elements to be patterned by exposure to light through a mask. Suh et al. [63] developed a PDMS stamp with corrugations that were reinforced by chemical vapor deposition-induced polymerization of poly(p-xylylene) on the side walls of the structure. It also proved to be a good candidate for the fabrication of deformation-proof stamps.

Drawbacks associated with the relatively brittle nature of h-PDMS and the thermal curing requirements of the material remain. In order to overcome the brittle behavior of h-PDMS, the IBM group also tested several composite stamp designs that used a rigid support (glass or quartz) and a thin h-PDMS top layer [60]. These designs included multilayer stamps (a thin layer of h-PDMS on a slab of Sylgard 184 system attached to a quartz plate) and two-layer stamps (thin layers of h-PDMS or Sylgard 184 molded to glass foils) [60, 64]. A stamp, consisting of a thin (30 µm) layer of h-PDMS supported on a flexible glass foil (100 µm thick), conformed the best to uneven substrate topologies with minimal distortion over large areas. Whitesides and coworkers [65] modified these fabricated composite stamps using a Sylgard 184 slab to support a thin, stiff h-PDMS top layer. The smallest features that have been replicated using such composite stamps were 50 nm.

It is important to note that by increasing the stiffness of the elastomeric material, higher resolution patterns can be produced; however, stiffer materials decrease the conformability of the stamp or mold, thereby reducing the contact between the stamp and the substrate and causing defects in the patterns. In addition, stiffer materials limit the versatility of the patterning technique since these cannot be used with nonplanar substrates. Therefore, an acceptable balance must be achieved between the conformability and the stiffness of the material in order to produce pattern reproducibly.

10.4.2
Other Stamp Materials

Apart from PDMS, an enormous variety of other polymeric elastomers is available that derive their elastomeric character from, for instance, self-aggregation of thermoreversible, nanosized structures within the bulk of the polymers. Typically, these elastomers are classified as thermoplastic elastomers. The properties of these elastomers with respect to stiffness can be tuned, to a large extent, by a proper selection of their chemical structure, composition, and processing conditions while preserving their intrinsic toughness. Csucs *et al.* [66] reported the use of polyolefin plastomers (POPs) for the printing of protein and block copolymer patterns. It was shown that in the submicrometer range (submicrometer structures with micrometer separations), a much higher printing quality is achievable with the POPs compared to regular PDMS. This fact is probably due to the higher bulk modulus of the POP stamps. Trimbach *et al.* [67] studied two commercially available thermoplastic block copolymer elastomers (poly(styrene-*bl*-butadiene-*bl*-styrene)) and (poly(styrene-*bl*-ethylene-*co*-butylene-*bl*-styrene)) with high stiffness as stamp materials for microcontact printing. They showed that the thermoplastic elastomers possess a high modulus and toughness in comparison to PDMS, and consequently the stamp deformation during printing is decreased. A UV-curable stamp material, a poly(urethane acylate) based on a functionalized prepolymer with acrylate groups for cross-linking and different monomeric modulators, was developed by Lee's group [68]. By varying the modulator, the mechanical properties of the stamp could be tuned. A series of materials based on perfluoropolyether (PFPE) has also been developed as stamp materials for high-resolution printing [69–71]. The fluoropolymer, which is liquid at room temperature, can be cross-linked under UV light to yield elastomers with an extremely low surface energy (≈ 12 mN m^{-1}). A major advantage of PFPE-based materials is that they are solvent-resistant and chemically robust and therefore swell much less than PDMS when exposed to most organic solvents. This property expands the range of materials that can be patterned effectively. Also, unlike PDMS, PFPEs eliminate the surface functionalization step that is often required to avoid adhesion to oxides (e.g., SiO_2 on Si wafers) during the casting and curing steps used to make the patterning elements. These PFPE-based materials allow the replication of sub-100-nm sized features with no indications of limits when going to even smaller sizes.

10.4.3
Flat Stamps

The mechanical issues are a direct consequence of the inclusion of topographical voids as the transport barriers. In principle, a flat stamp can solve many or all stamp stability issues. Geissler *et al.* [72] first used a planar PDMS stamp to print chemical patterns on a substrate. Delamarche *et al.* [73] have shown that flat PDMS stamps can be patterned by a combination of surface oxidation in an oxygen plasma using a mask

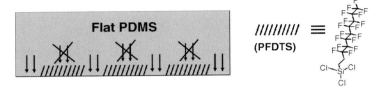

Figure 10.4 Schematic representation of printing thiols with flat stamps: the 1H,1H,2H,2H-perfluorodecyltrichlorosilane-covered area acts as an ink barrier, while the noncovered PDMS transfers the thiols to the gold substrate.

and subsequent stabilization of the hydrophilized areas by reaction with a poly(ethylene oxide) silane. These stamps have been used for the selective deposition and subsequent patterned transfer of proteins from the stamp surface. Later on, they developed a method by using flat stamps to pick up proteins from a hard nanotemplate, thus creating sub-100-nm protein patterns, and called this approach subtractive lithography [74].

In our group, we have introduced the concept of flat, chemically patterned stamps for the μCP of regular thiol inks [75]. Stamp functionalization was achieved by local oxidation of a flat piece of PDMS through a mask, followed by adsorption of a fluorinated silane, 1H,1H,2H,2H-perfluorodecyl-trichlorosilane (PFDTS). It was found that this silane forms densely packed SAMs on oxidized PDMS that constitutes an effective barrier to prevent ink transfer while the rest of the surface allows the diffusion of ink molecules from the bulk of the PDMS to the gold substrates (Figure 10.4). The further development of this technique led to the grafting of 3-aminopropyl-triethoxysilane in the nonfluorinated areas, allowing the efficient printing of polar inks with these chemically patterned flat stamps [76].

The flat stamp design effectively solves the stamp stability issues, and submicrometer-sized features were successfully printed using these chemically patterned flat stamps. However, the main difficulty is the fabrication of the chemical barriers on the stamp's surface as the pattern size is limited by the mask used for selective oxidation. Other methods have been used to fabricate chemical patterns on flat PDMS surfaces, targeting higher resolutions. Zheng et al. recently reported the use of dip-pen nanolithography (DPN) to directly write sub-100-nm chemical patterns on flat PDMS [77]. The limitations of this method are the longer writing time and the difficulties in writing on the flexible substrates. Meanwhile, our group demonstrated that nanoimprint lithography (NIL) can be used to fabricate polymer patterns on flat PDMS [78]. The polymer pattern, produced by thermal NIL followed by residual layer removal, acts as a local mask for the oxidation of the uncovered regions of the PDMS surface. The chemical patterns were subsequently formed by gas-phase evaporation of a fluorinated silane. After removal of the imprint polymer, these stamps were used to transfer alkanethiols as inks to a gold substrate by μCP. Sub-100-nm gold lines were successfully replicated by these stamps.

10.4.4
New Ink Materials

Among all the limitations of μCP, the diffusion of the ink molecules during and after the printing process is a significant problem in getting high-resolution patterns. When feature sizes are smaller than 500 nm, ink diffusion compromises the final resolution. The use of a heavy ink allows the diffusion zone to be reduced to 100 nm for immersion inking and less than 50 nm for contact inking of the stamp [35]. Bass and Lichtenberger [79] showed that a higher molecular weight alkanethiol, such as octadecanethiol or eicosanethiol, diffuses less on a gold surface, compared to hexadecanethiol, and exhibits concomitantly better printing results. Longer chain thiols (longer than eicosanethiol) tend to show more disordered layers on gold, thus leading to poorer etch performance.

Other ways of improving the resolution of the microcontact printed patterns include the use of inks heavier than alkanethiols (Figure 10.5). Liebau et al. [80] investigated some heavy molecular weight thioether derivatives as inks in microcontact printing. Poly(amidoamine) (PAMAM) dendrimers have been used as inks to be transferred to silicon [81], palladium [82], or gold substrates [83]. A sub-100-nm resolution has been achieved by these dendrimer inks. In these studies, the ink molecules did not show diffusion on the patterned surface because of their high molecular weight, thus leading to a more faithful transfer of the inks.

However, heavy inks also have their own limitations, such as longer printing times, less ordered monolayers, and the tendency to crystallize at the surface of the stamp leading to contamination problems. In order to achieve high-resolution printing, there must be a balance between the molecular weights of the inks, the printing time, the inking time, the amount of the inks, and other parameters. Recently, Balmer et al.

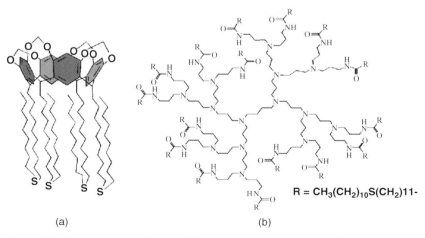

Figure 10.5 Some examples of heavyweight inks: a calix[4]arene thioether derivative (a) and thioether-modified dendrimers (b).

studied in detail the diffusion behavior of several alkanethiols in PDMS stamps [84]. Their results showed that the ink transport through the PDMS stamps follows Fick's laws of diffusion. The diffusion coefficient was also calculated for three different alkanethiols. These studies constitute the basis of future optimizations of the printing conditions.

10.4.5
Alternative µCP Strategies

10.4.5.1 High-Speed µCP

After a series of studies on understanding ink transport and diffusion mechanisms [59, 84], a breakthrough in the timescales involved in stamp contact and monolayer formation was reported in a recent study on high-speed microcontact printing by the IBM group [85]. In a careful analysis of the mechanics of the printing procedure and a numerical diffusion simulation of the ink transfer, the group concluded that µCP can be performed up to three orders of magnitude faster than previously reported. For example, a printing time of 3 ms using a concentration of 16.6 mM of hexadecanethiol ink is sufficient to create the same quality pattern replication in printed and etched gold patterns. These recent results demonstrate that there is a well-defined processing window, in which the combination of ink concentration and contact time yield perfect SAMs (as judged from etch resistance, thus also indicating the formation of monolayers with crystalline order) but where surface diffusion, diffusion through the vapor phase, ink depletion, and stamp distortion are all avoided (Figure 10.6). This major improvement illustrates the possibilities of scaling down the lateral resolution of µCP and indicates that µCP could become a commercially attractive micro(nano)fabrication technology.

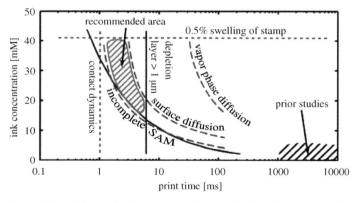

Figure 10.6 High-speed µCP process window as a function of ink concentration and contact time. Reprinted with permission from Ref. [85]. Copyright 2006, the American Chemical Society.

10.4.6
Catalytic Printing

In catalytic printing, the PDMS stamp used can be made catalytically active either by chemical surface modification or by UV treatment in order to induce a chemical reaction with the substrate upon contact. In such a catalytic approach, no ink is needed and therefore the lateral resolution-reducing effect of surface spreading is efficiently eliminated.

In our group, surface-oxidized PDMS stamps have been used to hydrolyze silylether-protected SAMs in the areas of contact, thus forming patterned SAMs. Because there was no ink flow from the stamp to the surface, a high edge resolution (below 60 nm) was obtained [86]. A similar study used plasma-oxidized flat PDMS to promote coupling between amino-terminated SAMs and N-protected amino acids under nanoscale confinement by contact printing [87].

Park *et al.* [88] have developed a printing method to directly transfer the contact surface of the PDMS stamp to a substrate via a UV light-induced surface bonding between the stamp and the substrate. First, a patterned PDMS stamp was prepared that was activated by exposure to UV/ozone and brought into contact with the substrate. Because of the UV/ozone treatment, an irreversible adhesion reaction occurred between the PDMS and the substrate. After the release of the stamp, PDMS patterns were formed on the substrate. The PDMS patterns can be used not only as resists for selective wet etching but also as templates to selectively deposit TiO_2 thin films.

Toone and coworkers used piperidine-functionalized poly(urethane acrylate) stamps to promote the catalytic cleavage of the 9-fluorenylmethoxycarbonyl amino-protecting group [89]. With this inkless method, submicrometer patterns were created by selective deprotection of (9H-fluoren-9-yl)methyl-11-mercaptoundecylcarbamate SAMs on gold. Later on, they extended the use of catalytic µCP to biochemical substrates by immobilizing exonuclease enzymes on biocompatible poly(acrylamide) stamps and created patterned DNA both on glass and on gold surfaces [90].

10.5
Soft Lithography to Pattern Assemblies of Nanoparticles

Nanoparticles (NPs) are most versatile tools for the construction of new and advanced tunable materials [91]. Their intriguing properties make them useful in a wide range of applications in optics, chemical sensors, data storage, and so on [92–94]. A prerequisite for future applications using NPs as functional entities is control over their positions and arrangement on a surface. Doing so with conventional microfabrication techniques is difficult, and it is often time-consuming and inefficient. Soft lithography, in contrast, offers the possibilities to pattern NPs in a relatively easy way. Patterning using imprint lithography has been reviewed before [95]. Here, examples of recent approaches to pattern NPs through soft lithography will be reviewed.

10.5.1
Patterning by Contact Printing

10.5.1.1 Nanoparticles as Inks

When directly using NPs as ink in μCP, the key to successful printing is to tune the surface energy of the particles on the surface. The inking method, the printing conditions, and the chemical functions of the particle surface and the substrate surface are all important. The interaction of the NPs with the targeting surface should be stronger than with the stamp surface.

The earliest example is from the Whitesides group [96]. Micropatterns of palladium NPs were used as inks first absorbed onto a PDMS surface and then stamped onto amino-functionalized substrates. The NPs were transferred to the substrate due to the affinity for the amino surface. The patterned NPs were then used for the selective electroless deposition of copper.

In order to get better ordering of the NPs patterns, Santhanam and Andres used the Langmuir–Blodgett (LB) technique to ink the PDMS surface with NPs [97]. Dense and hexagonally packed monolayers of NPs were assembled first on the air–water interface and transferred to the PDMS stamp surface via an LB setup. Subsequently, the monolayer of densely packed NPs was printed to another substrate. Multilayers of NPs were prepared by repeating the printing process in a layer-by-layer (LBL) scheme, in which subsequent particle layers may be made up of the same or different types of particles. The same method has been applied to print monolayers of magnetic FePt NPs to different substrates to form micrometer-size circles, lines, and squares [98].

In other studies, the PDMS surface is tailored to fine-tune the interactions between the NPs and the stamp surface. Fuchs and coworkers reported that the distribution of the NPs on a structured stamp surface can be controlled by the gas flow rate during the inking process as well as the type and scale of the patterns on the stamp. CdTe NPs stabilized with thioglycolic acid (TGA) were patterned on SiO_2/Si surfaces [99]. Bulovic and coworkers demonstrated a contact printing method for depositing patterned quantum dot (QD) monolayer films that are formed by spin-casting QDs on chemically functionalized PDMS stamps [100]. The chemical functionalization of the PDMS surface was achieved by a coating of a chemical vapor deposited parylene-C layer. Parylene-C is an aromatic polymer; therefore, its surface is optimal for minimizing the surface energy of the QD monolayer, thus facilitating the transfer of QDs to other substrates. Gigli and coworkers proposed another μCP approach to pattern QDs, using the SU-8 photoresist as a protective layer for the PDMS. They used this technique to fabricate a multilayer, hybrid, white-light emitting diode. The advantage of using SU-8 instead of parylene-C is the possibility of deposition by employing a very easy and low-cost spin-coating process [101, 102].

Yang's group first report combined lift-up and μCP techniques to pattern NPs (Figure 10.7) [103]. In their approach, a PDMS stamp with patterned features was brought into contact with a NPs film deposited on a silicon substrate. After the sample was heated and the PDMS stamp was carefully peeled away, a single layer of close-packed particles was transferred to the surface of the PDMS stamp and the corresponding pattern was formed on the colloidal crystal film surface. The NP-

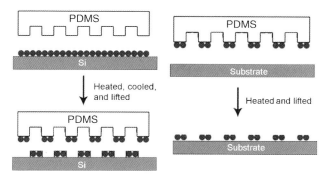

Figure 10.7 Schematic illustration of lift-up combined with microcontact printing to transfer NPs.

inked PDMS stamp was then brought into contact with a thin film of polymer, usually poly(vinyl alcohol) (PVA). After the sample was heated above the T_g of PVA, the PDMS stamp was peeled off, and the NP film was transferred onto the substrate. Later on, the group extended this technique to pattern nonclose-packed arrays of NPs based on the solvent-swelling and mechanical deformation properties of PDMS [104]. With this approach, the lattice structures of the printed 2D particles arrays can be adjusted.

In order to increase the amount of NPs inked onto the stamp surface, our group reported porous stamps structures that are fabricated by one-step phase separation micromolding. With the pore structures functioning as ink reservoirs, multiple printing steps of NPs were achieved without reinking of the stamps [105].

Another method of modifying the interactions between the NPs and the surface is mixing NPs with other functional materials. Wang *et al.* reported the contact printing of a nanocomposite of a polymer poly(styrene-alt-maleic anhydride), (HSMA), and inorganic NPs (hydrolyzed TiO_2) from a dispersion of TiO_2 NPs in a HSMA solution [106]. In the last step, calcination of the composites removed the HSMA polymer and resulted in a nanostructure of TiO_2 NPs. Bittner and coworkers reported the µCP of CdS/dendrimer nanocomposites on a hydroxy-terminated silicon surface [107]. Dendrimers were used as hosts for CdS NPs and facilitated the adsorption of the NPs to the surface via electrostatic forces, hydrogen bonds, and/or van der Waals interactions.

In our group, we exploited host–guest interactions between dendritic guest molecules and β-cyclodextrin (CD)-functionalized NPs for the formation of organic/metal NPs multilayers on a PDMS stamp (Figure 10.8) [108]. The multilayer stacks were transferred to a complementary host surface, while no material remained on the protruding areas of the PDMS stamp. These multilayers showed a well-defined thickness control of 2 nm per bilayer.

10.5.1.2 Convective Assembly

The IBM group introduced the concept of convective assembly via an integrated top-down method, called self-assembly, transfer, and integration (SATI) of NPs with high

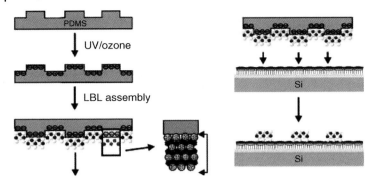

Figure 10.8 The preparation of a multilayered supramolecular nanostructure on PDMS and transfer printing on a CD SAM.

placement accuracy (Figure 10.9) [109, 110]. In their method, top-down lithography was used to create a template and to attach and pattern NPs by means of physical confinement. The height of the template was smaller than the size of the NPs in order to have the NPs protruding from the pattern. The substrate was treated before hand with a fluoroalkyl SAM to minimize the adhesion between the surface and the NPs. A flat PDMS stamp was used to pick up the patterned NPs to subsequently deposit them on a Si surface, having a higher adhesion with the NPs than PDMS. By using this method, the IBM group has shown the printing of a 60-nm Au NPs array with single-particle resolution [111].

In our group, we have combined the concept of convective assembly with supramolecular chemistry to pattern 3D NP structures. β-Cyclodextrin-functionalized NPs were first assembled on a patterned PDMS stamp through convective assembly, then adamantyl dendrimers were used as supramolecular glue to chemically bond neighboring NPs together to form stable and ordered 3D hybrid NP structures. These dendrimer-infiltrated NPs could be transfer-printed [112]. Later by using the same method, the 3D hybrid NPs structures were transferred onto

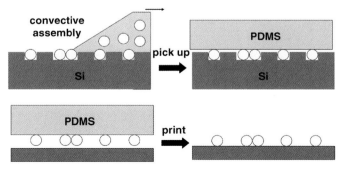

Figure 10.9 Scheme of convective assembly of NPs, pickup by a PDMS stamp, and subsequent transfer printing.

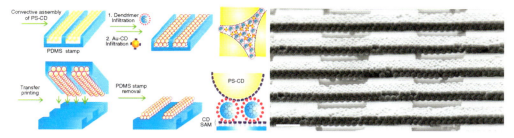

Figure 10.10 Preparation of hybrid particle bridges on a topographically patterned substrate (*left*) and SEM image of the free-standing hybrid particle structures (*right*). Scale bar: 2 μm.

topographically patterned substrates via host–guest interactions. Freestanding particle bridges were formed, and the mechanical robustness and rigidity of the particle bridges could be controlled by manipulating the layer-by-layer cycles of supramolecular glues of gold nanoparticles and dendrimers (Figure 10.10) [112, 113].

10.5.1.3 Templated Assembly

In the templated assembly of NPs, organic molecules are first printed onto substrates, and these function as linkers to immobilize NPs via different interactions. Hammond and coworkers reported the self-organization of SiO_2 and polystyrene (PS) NPs on a μCP-patterned substrate [114]. First, carboxylate-functionalized thiols were printed on a gold surface through μCP and then polyelectrolyte multilayers were selectively deposited onto the printing areas. Multicomponent NPs were assembled on the polyelectrolyte surface driven by electrostatic and hydrophobic interactions. The surface charge density was modulated by pH, ionic strength, and effective surface charge of the polyelectrolyte. Kang and coworkers reported the LBL assembly of QDs on microcontact-printed carboxylate-functionalized SAMs [115]. The LBL assembly was achieved by alternating adsorption of COOH-QDs and 2-mercaptoethanesulfonic acid as the assembly partner. Combinatorially selected peptides and peptide–organic conjugates were used as linkers with controlled structural and organizational conformations to attach quantum dots at contact-printed SAMs [116]. This work establishes a framework for investigating the luminescence properties of surface-immobilized hybrid nanostructures where both the QD–metal distance and the surface attachment density can be monitored and controlled by μCP.

10.5.2
Patterning by Micromolding in Capillaries

MIMIC is an effective method to pattern or process NPs from their solutions. Whitesides and coworkers first studied MIMIC to pattern monolayers and multilayers of microspheres [117]. The procedure can fabricate highly ordered 2D and 3D arrays of microspheres by self-assembly. The mechanism of crystallization of latex particles in capillaries involves nucleation due to capillary attractive forces between the microspheres and growth due to evaporation and influx of suspension to compensate for the loss of solvent. Han and coworkers reported a modified MIMIC approach to fabricate aggregates composed of monodisperse silica microspheres [118, 119]. Two different kinds of contact modes, namely, conformal contact and nonconformal contact, between the PDMS mold and the underlying prepatterned substrate, can be controlled during the micromolding, which result in the formation of different aggregates including wood pile, discoid, and rectangular clusters under the influence of template confinement and capillary forces. Recently, monodisperse Si NPs were also patterned using the MIMIC approach [120, 121]. Large area (cm) and ordered NP arrays were successfully fabricated. However, the pattern size is limited by the micron-size mold.

Processing or patterning of nanoparticles from their solutions is interesting not only for fundamental research but also for promising real applications. Blumel *et al.* reported the use of MIMIC for patterning silver NPs followed by thermal annealing to fabricate silver source/drain electrodes in well-performing bottom–gate/bottom–contact organic field-effect transistors (OFETs) with poly(3-hexylthiophene) as the active layer material [122]. The transistors they fabricated have performances comparable to corresponding devices based on gold electrodes. Yu *et al.* applied MIMIC to fabricate stripes of rare earth ion-doped $LaPO_4$ nanocrystals in a sol–gel process [123]. This class of NPs is particularly important because they are employed in modern lighting and displays, such as fluorescent lamps, cathode-ray tubes, field emission displays, and plasma display panels. In their work, they proved the processability of NPs by MIMIC, controlling the pattern morphology, tuning the quantity of material, and the annealing conditions. Recently, Cavallini *et al.* used MIMIC to pattern magnetic NPs with submicrometer periodic features and a vertical resolution of a monolayer [124]. They demonstrated that the morphology of the patterned NPs can be controlled simply by controlling the solution concentration. Exploiting confinement and competing interactions between the adsorbate and the substrate, they fabricated continuous or split stripes composed of Fe_3O_4 NPs. Working in a dilute regime, they reached a spatial resolution of a few tens of nanoclusters, depositing a single monolayer of NPs. This approach represents a remarkable example of an integrated top-down/bottom-up process.

10.5.3
Patterning by Soft Lithography with Solvent Mediation

There are a few other approaches to pattern NPs from their solutions that exploit the self-organization of NPs with the spatial control provided by the PDMS stamp features. Cavallini *et al.* used a technique they termed "lithographically controlled wetting" (LCW) [125, 126]. By placing a stamp in contact with a thin liquid film, the capillary forces drive the liquid to distribute only under the protrusions of the stamp. As the solvent evaporates, the deposited solute can form nanostripes on a substrate, and by controlling the concentration and the stamp–substrate distance, the pattern size is controlled. More recently, Cheng *et al.* used PDMS stamps to control the shape and location of microdroplets of a solution containing NPs. The pressure exerted on the PDMS stamp and its geometry control the dewetting dynamics of the solution and allows further control of the local nucleation and growth of superlattices [127].

10.6
Soft Lithography Pattern Supramolecular Assembly

Supramolecular interactions play a pivotal role in biology and are being extensively used for other nonbiological applications as well [128]. Supramolecular interactions are directional, specific, and reversible that allow fine-tuning of the adsorption/desorption processes at receptors, which is not feasible for conventional routes of immobilization of molecules, assemblies, and particles on surfaces [129]. The combination of soft lithography with supramolecular host–guest interactions has led to the stable positioning and directed assembly of (bio)molecules.

10.6.1
Affinity Contact Printing

The IBM group reported affinity contact printing (Figure 10.11) [130]. In their work, the surface of a PDMS stamp was functionalized with antimouse IgG that selectively captured ^{125}I-labeled mouse IgG from a crude biological sample. After rinsing to remove unbound molecules from the stamp surface, the stamp was brought into contact with a solid surface. Because of the stronger interaction between the surface and proteins, the captured molecules were transferred to the solid surface. The same group also demonstrated that protein microarrays can be fabricated by using affinity contact printing [131]. Proteins were selectively picked

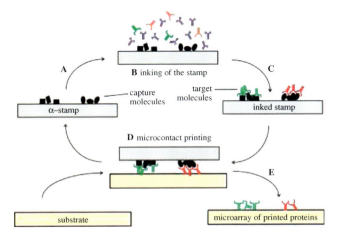

Figure 10.11 Affinity contact printing.

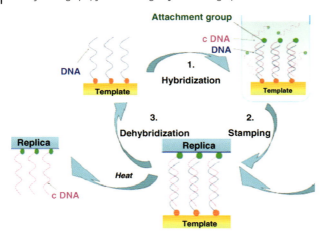

Figure 10.12 Schematic illustration of the supramolecular nanostamping process.

up from crude biological solutions and then printed on substrates by stamps functionalized with reactive groups that bound the proteins from complex mixtures and aided the transfer of these biomolecules onto the chosen substrates. Recently, Yang and coworkers reported a method of transferring complementary target DNA from an aqueous solution onto a solid surface by using the concept of affinity microcontact printing [132].

10.6.2
Supramolecular Nanostamping

Stellacci's group developed a stamping technique they called supramolecular nanostamping (SuNS) [133–138]. The most interesting of this technique is its ability to simultaneously print both spatial and chemical information. It combines chemical reactions, which contact printing can induce, with DNA molecular recognition (see Figure 10.12). The replication of DNA features is achieved through a hybridization–contact–dehybridization cycle. Features made of different single-stranded DNAs can be replicated from a master on a secondary surface in just three steps. First, the master is immersed in a solution of DNA strands complementary to the ones present on its surface, modified – at their distal position – with a group that can form a chemical bond with another surface (hybridization). Second, another substrate is placed onto the hybridized master (contact). Finally, the two substrates are separated by thermally induced melting of the DNA double helices (dehybridization), achieving a copy of the original pattern. SuNS can reproduce DNA patterns with high resolution (40 nm) [137]. Crooks' group reported a similar approach, in which the transfer is based on the affinity between biotin and streptavidin and the separation is obtained through mechanical forces [139–142]. The limitation here is that DNA molecules need to be labeled with biotin before they can be transferred.

10.6.3
Molecular Printboards

Our group has been exploiting supramolecular interactions focusing mainly on multivalency on surfaces [129]. Molecular printboards [143], which are self-assembled monolayers functionalized with β-cyclodextrin as receptor groups, were used as substrates to immobilize molecules through specific and directional supramolecular interactions. A typical procedure for fabricating and patterning guest molecules on molecular printboards on gold and silicon oxide surface was reported by Auletta et al. [144]. A densely packed, well-ordered β-CD SAM was prepared, and guest-functionalized dendrimers, fluorescently labeled molecules, and polymers were used as inks for supramolecular μCP through host–guest interactions (Figure 10.13). Comparison with adsorption onto OH-functionalized SAMs showed that the assembly on the printboard was governed by specific, multivalent host–guest interactions. The guest molecules were found exclusively in the areas of preceding contact between the microcontact printing stamp and the substrate, even after extensive rinsing with water or salt solutions. Only rinsing with 10 mM CD, in order to induce competition for binding the adamantyl guest sites, led to a noticeable desorption. Very similar results were obtained using the adamantyl-functionalized PPI dendrimers [145].

Patterning the adamantyl dendrimers on the printboard on silicon oxide provided one of the first cases of the use of two orthogonal interaction motifs for the formation of more complex architectures [146]. The application of a solution of a negatively

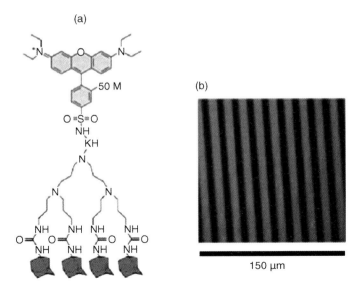

Figure 10.13 Structure of a lissamine–rhodamine-functionalized dendritic guest molecule (a) and confocal microscopy image after μCP of this guest on a CD-terminated SAM on SiO_2 (b).

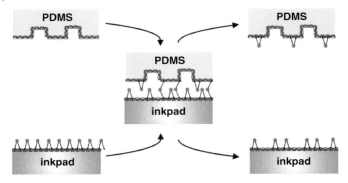

Figure 10.14 Contact inking with a controlled amount of ink by contacting a CD-modified PDMS stamp with a flat β-CD printboard on glass (ink pad) with a preadsorbed ink monolayer and subsequent transfer.

charged fluorescent dye to a substrate patterned by μCP with the adamantyl dendrimers led to localization of the dyes in the dendrimer-printed areas only. This two-step procedure therefore succumbed to an architecture where the dendrimers were bound by multivalent host–guest interactions, whereas the dyes were immobilized inside the dendrimer cores by electrostatic interactions.

By combining adamantyl dendrimers and β-CD-functionalized gold particles, various patterning strategies have been developed to create 3D hybrid nanostructures on a printboard by μCP using an LBL approach [108]. By using the nickel(II) complex of a heterodivalent orthogonal adamantyl nitrilotriacetate linker, hexahistidine (His_6)-tagged proteins were successfully patterned on the printboard through μCP in a specific, stable, multivalent manner [147].

Recently, CD-functionalized PDMS stamps were fabricated and successfully used in supramolecular contact printing of specific guest molecules on printboards [148]. It shows the possibility to selectively recognize ink molecules and to tune the amount of ink molecules transferred. The CD-covered stamps showed a highly selective recognition ability when guest-functionalized molecules were captured by self-assembly from ink mixtures. Uniform, equilibrium-controlled host–guest ink transfer was achieved upon conformal contact between two CD-covered surfaces. A supramolecular ink pad has also been developed by using a solid β-CD printboard, onto which an ink monolayer was assembled, specifically serving as an ink pad. Control over the amount of transferred ink molecules can be exercised by tuning the coverage of the ink monolayer on the ink pad (Figure 10.14).

10.7
Concluding Remarks

Self-assembly of molecular building blocks and nanoparticles by noncovalent interactions is a parallel process that usually occurs in very short times, creating discrete nanostructures and extended 2D and 3D materials. However, free assembly lacks a long-range order, precision, accuracy, and registration.

With the invention of soft lithography, which brought micro(nano)fabrication within the reach of every chemistry, physics, and biology laboratory, without any significant investment in equipment or clean room-type infrastructure, researchers have the chance to control the position, order, and orientation of self-assembling systems with a large freedom over chemical functionalities that are difficult to be afforded by other lithography techniques. It is the aim of this chapter to provide an overview of recent advances in patterning self-assembling systems by soft lithography. The selected examples focus on molding high-resolution features in elastomers that are mechanically stable, the need for specific surface chemistry or template design to adsorb molecules, nanoparticles, and biomolecules from solution onto selected areas of the stamp surface, and new strategies to transfer self-assembly to the target surface.

Even though the major force for downscaling surface patterning to the sub-100-nm regime is driven by the semiconductor industry, soft lithography has also found its particular application in the realm of the life sciences, for example, in cell and tissue engineering, sensing, and (bio)analysis. The life science focus is covered in some recent reviews on soft lithography for bioapplications [149–152]. Soft lithography also contributes to fundamental research, by creating the possibility for making chemically patterned SAMs, the elemental basics of chemically functionalized nanoparticles, and by contributing to the understanding of supramolecular and biological interactions. The self-assembling behavior in nanoconfined conditions is getting more interesting when the patterning resolution is approaching the sub-100-nm range. As such, there are plenty of reasons to expect important breakthroughs in the forthcoming years as the field will be deeply investigated.

References

1 Verberg, R., Pooley, C.M., Yeomans, J.M., and Balazs, A.C. (2004) *Phys. Rev. Lett.*, **93**, 184501.
2 McAlpine, M.C., Friedman, R.S., and Lieber, C.M. (2005) *Proceed. IEEE*, **93**, 1357.
3 Dowling, A. (2004) *Nanosciences and Nanotechnologies: Opportunities and Uncertainties*, The Royal Society and The Royal Academy of Engineering, London.
4 Kastner, M.A. (1993) *Phys. Today*, **46**, 24.
5 Likharev, K.K. and Claeson, T. (1992) *Sci. Amer.*, **266**, 80.
6 Reed, M.A. (1993) *Sci. Amer.*, **268**, 118.
7 Vijayakrishnan, V., Chainani, A., Sarma, D.D., and Rao, C.N.R. (1992) *J. Phys. Chem.*, **96**, 8679.
8 Lehn, J.-M. (1990) *Angew. Chem.*, **102**, 1347.
9 Ostuni, E., Yan, L., and Whitesides, G.M. (1999) *Coll. Surf. B*, **15**, 3.
10 Ruiz, A., Valsesia, A., Bretagnol, F., Colpo, P., and Rossi, F. (2007) *Nanotechnology*, **18**, 505306.
11 Bita, I., Yang, J.K.W., Jung, Y.S., Ross, C.A., Thomas, E.L., and Berggren, K.K. (2008) *Science*, **321**, 939.
12 Segalman, R.A., Yokoyama, H., and Kramer, E.J. (2001) *Adv. Mater.*, **13**, 1152.
13 Shi, J.X., Wang, Z., He, P.S., and Yang, H.Y. (2006) *Chin. J. Chem. Phys.*, **19**, 527.
14 Okazaki, T.I.S. (2000) *Nature*, **406**, 1027.
15 Heier, J., Genzer, J., Kramer, E.J., Bates, F.S., Walheim, S., and Krausch, G. (1999) *J. Chem. Phys.*, **111**, 11101.

16. Brehmer, M., Conrad, L., and Funk, L. (2003) *J. Disp. Sci. Tech.*, **24**, 291.
17. Kane, R.S., Takayama, S., Ostuni, E., Ingber, D.E., and Whitesides, G.M. (1999) *Biomaterials*, **20**, 2363.
18. Whitesides, G.M., Ostuni, E., Takayama, S., Jiang, X.Y., and Ingber, D.E. (2001) *Annu. Rev. Bio. Engin.*, **3**, 335.
19. Xia, Y.N. and Whitesides, G.M. (1998) *Ann. Rev. Mater. Sci.*, **28**, 153.
20. Xia, Y.N. and Whitesides, G.M. (1998) *Angew. Chem., Int. Ed. Eng.*, **37**, 551.
21. Xia, Y.N., Rogers, J.A., Paul, K.E., and Whitesides, G.M. (1999) *Chem. Rev.*, **99**, 1823.
22. Barth, J.V., Costantini, G., and Kern, K. (2005) *Nature*, **437**, 671.
23. Whitesides, G.M. and Grzybowski, B. (2002) *Science*, **295**, 2418.
24. Holmes, S.J., Mitchell, P.H., and Hakey, M.C. (1997) *IBM J. Res. Dev.*, **41**, 7.
25. Stulen, R.H. and Sweeney, D.W. (1999) *IEEE J. Quantum Electron.*, **35**, 694.
26. Silverman, J.P. (1997) *J. Vac. Sci. Technol. B*, **15**, 2117.
27. McCord, M.A. (1997) *J. Vac. Sci. Technol. B*, **15**, 2125.
28. Tandon, U.S. (1992) *Vacuum*, **43**, 241.
29. Kumar, A. and Whitesides, G.M. (1993) *Appl. Phys. Lett.*, **63**, 2002.
30. Xia, Y.N., Kim, E., and Whitesides, G.M. (1996) *Chem. Mater.*, **8**, 1558.
31. Kim, E., Xia, Y.N., and Whitesides, G.M. (1995) *Nature*, **376**, 581.
32. Xia, Y.N. and Whitesides, G.M. (1997) *Langmuir*, **13**, 2059.
33. Zhao, X.M., Stoddart, A., Smith, S.P., Kim, E., Xia, Y., Prentiss, M., and Whitesides, G.M. (1996) *Adv. Mater.*, **8**, 420.
34. Kim, E., Xia, Y.N., Zhao, X.M., and Whitesides, G.M. (1997) *Adv. Mater.*, **9**, 651.
35. Michel, B., Bernard, A., Bietsch, A., Delamarche, E., Geissler, M., Juncker, D., Kind, H., Renault, J.P., Rothuizen, H., Schmid, H., Schmidt-Winkel, P., Stutz, R., and Wolf, H. (2001) *IBM J. Res. Dev.*, **45**, 697.
36. Biebuyck, H.A., Larsen, N.B., Delamarche, E., and Michel, B. (1997) *IBM J. Res. Dev.*, **41**, 159.
37. Xia, Y.N. and Whitesides, G.M. (1995) *J. Am. Chem. Soc.*, **117**, 3274.
38. Deng, L., Mrksich, M., and Whitesides, G.M. (1996) *J. Am. Chem. Soc.*, **118**, 5136.
39. Mrksich, M., Chen, C.S., Xia, Y.N., Dike, L.E., Ingber, D.E., and Whitesides, G.M. (1996) *Proc. Natl. Acad. Sci. U S A*, **93**, 10775.
40. Xia, Y.N., Mrksich, M., Kim, E., and Whitesides, G.M. (1995) *J. Am. Chem. Soc.*, **117**, 9576.
41. Guo, Q.J., Teng, X.W., Rahman, S., and Yang, H. (2003) *J. Am. Chem. Soc.*, **125**, 630.
42. Xu, C., Taylor, P., Ersoz, M., Fletcher, P.D.I., and Paunov, V.N. (2003) *J. Mater. Chem.*, **13**, 3044.
43. Lange, S.A., Benes, V., Kern, D.P., Horber, J.K.H., and Bernard, A. (2004) *Anal. Chem.*, **76**, 1641.
44. Schmalenberg, K.E., Buettner, H.M., and Uhrich, K.E. (2004) *Biomaterials*, **25**, 1851.
45. Rogers, J.A., Paul, K.E., Jackman, R.J., and Whitesides, G.M. (1998) *J. Vac. Sci. Technol. B*, **16**, 59.
46. Delamarche, E., Schmid, H., Michel, B., and Biebuyck, H. (1997) *Adv. Mater.*, **9**, 741.
47. Bessueille, F., Pla-Roca, M., Mills, C.A., Martinez, E., Samitier, J., and Errachid, A. (2005) *Langmuir*, **21**, 12060.
48. Bohm, I., Lampert, A., Buck, M., Eisert, F., and Grunze, M. (1999) *Appl. Surf. Sci.*, **141**, 237.
49. Glasmastar, K., Gold, J., Andersson, A.S., Sutherland, D.S., and Kasemo, B. (2003) *Langmuir*, **19**, 5475.
50. Rogers, J.A., Jackman, R.J., Whitesides, G.M., Wagener, J.L., and Vengsarkar, A.M. (1997) *Appl. Phys. Lett.*, **70**, 7.
51. Graham, D.J., Price, D.D., and Ratner, B.D. (2002) *Langmuir*, **18**, 1518.
52. Lee, J.N., Park, C., and Whitesides, G.M. (2003) *Anal. Chem.*, **75**, 6544.
53. Ulman, A. (1996) *Chem. Rev.*, **96**, 1533.
54. Milner, S.T. (1991) *Science*, **251**, 905.
55. Bal, M., Ursache, A., Touminen, M.T., Goldbach, J.T., and Russell, T.P. (2002) *Appl. Phys. Lett.*, **81**, 3479.
56. Whitesides, G.M., Mathias, J.P., and Seto, C.T. (1991) *Science*, **254**, 1312.
57. Smith, R.K., Lewis, P.A., and Weiss, P.S. (2004) *Prog. Surf. Sci.*, **75**, 1.

58 Larsen, N.B., Biebuyck, H., Delamarche, E., and Michel, B. (1997) *J. Am. Chem. Soc.*, **119**, 3017.

59 Delamarche, E., Schmid, H., Bietsch, A., Larsen, N.B., Rothuizen, H., Michel, B., and Biebuyck, H. (1998) *J. Phys. Chem. B*, **102**, 3324.

60 Schmid, H. and Michel, B. (2000) *Macromolecules*, **33**, 3042.

61 Choi, K.M. (2005) *J. Phys. Chem. B*, **109**, 21525.

62 Choi, K.M. and Rogers, J.A. (2003) *J. Am. Chem. Soc.*, **125**, 4060.

63 Suh, K.Y., Langer, R., and Lahann, J. (2003) *Appl. Phys. Lett.*, **83**, 4250.

64 Burgin, T., Choong, V.E., and Maracas, G. (2000) *Langmuir*, **16**, 5371.

65 Odom, T.W., Love, J.C., Wolfe, D.B., Paul, K.E., and Whitesides, G.M. (2002) *Langmuir*, **18**, 5314.

66 Csucs, G., Kunzler, T., Feldman, K., Robin, F., and Spencer, N.D. (2003) *Langmuir*, **19**, 6104.

67 Trimbach, D., Feldman, K., Spencer, N.D., Broer, D.J., and Bastiaansen, C.W.M. (2003) *Langmuir*, **19**, 10957.

68 Yoo, P.J., Choi, S.J., Kim, J.H., Suh, D., Baek, S.J., Kim, T.W., and Lee, H.H. (2004) *Chem. Mater.*, **16**, 5000.

69 Rolland, J.P., Hagberg, E.C., Denison, G.M., Carter, K.R., and De Simone, J.M. (2004) *Angew. Chem. Int. Ed. Eng.*, **43**, 5796.

70 Hampton, M.J., Williams, S.S., Zhou, Z., Nunes, J., Ko, D.H., Templeton, J.L., Samulski, E.T., and DeSimone, J.M. (2008) *Adv. Mater.*, **20**, 2667.

71 Truong, T.T., Lin, R.S., Jeon, S., Lee, H.H., Maria, J., Gaur, A., Hua, F., Meinel, I., and Rogers, J.A. (2007) *Langmuir*, **23**, 2898.

72 Geissler, M., Bernard, A., Bietsch, A., Schmid, H., Michel, B., and Delamarche, E. (2000) *J. Am. Chem. Soc.*, **122**, 6303.

73 Delamarche, E., Donzel, C., Kamounah, F.S., Wolf, H., Geissler, M., Stutz, R., Schmidt-Winkel, P., Michel, B., Mathieu, H.J., and Schaumburg, K. (2003) *Langmuir*, **19**, 8749.

74 Coyer, S.R., Garcia, A.J., and Delamarche, E. (2007) *Angew. Chem. Int. Ed. Eng.*, **46**, 6837.

75 Sharpe, R.B.A., Burdinski, D., Huskens, J., Zandvliet, H.J.W., Reinhoudt, D.N., and Poelsema, B. (2005) *J. Am. Chem. Soc.*, **127**, 10344.

76 Duan, X.X., Sadhu, V.B., Perl, A., Peter, M., Reinhoudt, D.N., and Huskens, J. (2008) *Langmuir*, **24**, 3621.

77 Zheng, Z., Jang, J.W., Zheng, G., and Mirkin, C.A. (2008) *Angew. Chem. Int. Ed. Eng.*, **47**, 9951.

78 Duan, X.X., Zhao, Y.P., Perl, A., Berenschot, E., Reinhoudt, D.N., and Huskens, J. (2009) *Adv. Mater.*, **21**, 2798.

79 Bass, R.B. and Lichtenberger, A.W. (2004) *Appl. Surf. Sci.*, **226**, 335.

80 Liebau, M., Huskens, J., and Reinhoudt, D.N. (2001) *Adv. Funct. Mater.*, **11**, 147.

81 Li, H.W., Kang, D.J., Blamire, M.G., and Huck, W.T.S. (2002) *Nano Lett.*, **2**, 347.

82 Jang, S.G., Choi, D.G., Kim, S., Jeong, J.H., Lee, E.S., and Yang, S.M. (2006) *Langmuir*, **22**, 3326.

83 Perl, A., Peter, M., Ravoo, B.J., Reinhoudt, D.N., and Huskens, J. (2006) *Langmuir*, **22**, 7568.

84 Balmer, T.E., Schmid, H., Stutz, R., Delamarche, E., Michel, B., Spencer, N.D., and Wolf, H. (2005) *Langmuir*, **21**, 622.

85 Helmuth, J.A., Schmid, H., Stutz, R., Stemmer, A., and Wolf, H. (2006) *J. Am. Chem. Soc.*, **128**, 9296.

86 Li, X.M., Peter, M., Huskens, J., and Reinhoudt, D.N. (2003) *Nano Lett.*, **3**, 1449.

87 Sullivan, T.P., van Poll, M.L., Dankers, P.Y.W., and Huck, W.T.S. (2004) *Angew. Chem. Int. Ed. Eng.*, **43**, 4190.

88 Park, K.S., Seo, E.K., Do, Y.R., Kim, K., and Sung, M.M. (2006) *J. Am. Chem. Soc.*, **128**, 858.

89 Alexander, R.L.C., Shestopalov, A., and Toone, EricJ. (2007) *J. Am. Chem. Soc.*, **129**, 13818.

90 Snyder, P.W., Johannes, M.S., Vogen, B.N., Clark, R.L., and Toone, E.J. (2007) *J. Org. Chem.*, **72**, 7459.

91 Alivisatos, A.P. (1996) *Science*, **271**, 933.

92 Colvin, V.L. (2001) *MRS Bull.*, **26**, 637.

93 Holtz, J.H. and Asher, S.A. (1997) *Nature*, **389**, 829.

94 Gourevich, I., Pham, H., Jonkman, J.E.N., and Kumacheva, E. (2004) *Chem. Mater.*, **16**, 1472.

95 Maury, P.A., Reinhoudt, D.N., and Huskens, J. (2008) *Curr. Opin. Colloid Interf. Sci.*, **13**, 74.

96 Hidber, P.C., Helbig, W., Kim, E., and Whitesides, G.M. (1996) *Langmuir*, **12**, 1375.

97 Santhanam, V. and Andres, R.P. (2004) *Nano Lett.*, **4**, 41.

98 Patel, N., Bhandari, R., Shakesheff, K.M., Cannizzaro, S.M., Davies, M.C., Langer, R., Roberts, C.J., Tendler, S.J.B., and Williams, P.M. (2000) *J. Biomat. Sci. Polym. Ed.*, **11**, 319.

99 Wu, X.C., Lenhert, S., Chi, L.F., and Fuchs, H. (2006) *Langmuir*, **22**, 7807.

100 Kim, L., Anikeeva, P.O., Coe-Sullivan, S.A., Steckel, J.S., Bawendi, M.G., and Bulovic, V. (2008) *Nano Lett.*, **8**, 4513.

101 Rizzo, A., Mazzeo, M., Biasiucci, M., Cingolani, R., and Gigli, G. (2008) *Small*, **4**, 2143.

102 Rizzo, A., Mazzeo, M., Palumbo, M., Lerario, G., D'Amone, S., Cingolani, R., and Gigli, G. (2008) *Adv. Mater.*, **20**, 1886.

103 Yan, X., Yao, J.M., Lu, G.A., Chen, X., Zhang, K., and Yang, B. (2004) *J. Am. Chem. Soc.*, **126**, 10510.

104 Yan, X., Yao, J.M., Lu, G., Li, X., Zhang, J.H., Han, K., and Yang, B. (2005) *J. Am. Chem. Soc.*, **127**, 7688.

105 Xu, H., Ling, X.Y., van Bennekom, J., Duan, X., Ludden, M.J.W., Reinhoudt, D.N., Wessling, M., Lammertink, R.G.H., and Huskens, J. (2009) *J. Am. Chem. Soc.*, **131**, 797.

106 Wang, M.T., Braun, H.G., and Meyer, E. (2002) *Chem. Mater.*, **14**, 4812.

107 Wu, X.C., Bittner, A.M., and Kern, K. (2004) *Adv. Mater.*, **16**, 413.

108 Crespo-Biel, O., Dordi, B., Maury, P., Peter, M., Reinhoudt, D.N., and Huskens, J. (2006) *Chem. Mater.*, **18**, 2545.

109 Malaquin, L., Kraus, T., Schmid, H., Delamarche, E., and Wolf, H. (2007) *Langmuir*, **23**, 11513.

110 Kraus, T., Malaquin, L., Delamarche, E., Schmid, H., Spencer, N.D., and Wolf, H. (2005) *Adv. Mater.*, **17**, 2438.

111 Kraus, T., Malaquin, L., Schmid, H., Riess, W., Spencer, N.D., and Wolf, H. (2007) *Nat. Nanotechnol.*, **2**, 570.

112 Ling, X.Y., Phang, I.Y., Reinhoudt, D.N., Vancso, G.J., and Huskens, J. (2009) *ACS Appl. Mater. Interf.*, **1**, 960.

113 Ling, X.Y., Phang, I.Y., Schönherr, H., Reinhoudt, D.N., Vancso, G.J., and Huskens, J. (2009) *Small*, **5**, 1428.

114 Lee, I.S., Hammond, P.T., and Rubner, M.F. (2003) *Chem. Mater.*, **15**, 4583.

115 Zhou, D.J., Bruckbauer, A., Abell, C., Klenerman, D., and Kang, D.J. (2005) *Adv. Mater.*, **17**, 1243.

116 Zin, M.T., Munro, A.M., Gungormus, M., Wong, N.Y., Ma, H., Tamerler, C., Ginger, D.S., Sarikaya, M., and Jen, A.K.Y. (2007) *J. Mater. Chem.*, **17**, 866.

117 Kim, E., Xia, Y.N., and Whitesides, G.M. (1996) *Adv. Mater.*, **8**, 245.

118 Huang, W.H., Li, J.A., Xue, L.J., Xing, R.B., Luan, S.F., Luo, C.X., Liu, L.B., and Han, Y.C. (2006) *Coll. Surf. A*, **278**, 144.

119 Huang, W.H., Li, J., Luo, C.X., Zhang, J.L., Luan, S.F., and Han, Y.C. (2006) *Coll. Surf. A*, **273**, 43.

120 Singh, A., Malek, C.K., and Kulkarni, S.K. (2008) *Smart Mater. Struct.*, **17**, 069901.

121 Yamauchi, Y., Imasu, J., Kuroda, Y., Kuroda, K., and Sakka, Y. (2009) *J. Mater. Chem.*, **19**, 1964.

122 Blumel, A., Klug, A., Eder, S., Scherf, U., Moderegger, E., and List, E.J.W. (2007) *Org. Electron.*, **8**, 389.

123 Yu, M., Lin, J., Fu, J., Zhang, H.J., and Han, Y.C. (2003) *J. Mater. Chem.*, **13**, 1413.

124 Cavallini, M., Bystrenova, E., Timko, M., Koneracka, M., Zavisova, V., and Kopcansky, P. (2008) *J. Phys. Condens. Matter*, **20**, 204144.

125 Cavallini, M. and Biscarini, F. (2003) *Nano Lett.*, **3**, 1269.

126 Cavallini, M., Bergenti, I., Milita, S., Ruani, G., Salitros, I., Qu, Z.R., Chandrasekar, R., and Ruben, M. (2008) *Angew. Chem. Int. Ed. Eng.*, **47**, 8596.

127 Cheng, W.L., Park, N.Y., Walter, M.T., Hartman, M.R., and Luo, D. (2008) *Nat. Nanotechnol.*, **3**, 682.

128 Reinhoudt, D.N. and Crego-Calama, M. (2002) *Science*, **295**, 2403.

129 Ludden, M.J.W., Reinhoudt, D.N., and Huskens, J. (2006) *Chem. Soc. Rev.*, **35**, 1122.

130 Bernard, A., Fitzli, D., Sonderegger, P., Delamarche, E., Michel, B., Bosshard, H.R., and Biebuyck, H. (2001) *Nat. Biotechnol.*, **19**, 866.

131 Renault, J.P., Bernard, A., Juncker, D., Michel, B., Bosshard, H.R., and Delamarche, E. (2002) *Angew. Chem. Int. Ed. Eng.*, **41**, 2320.

132 Tan, H., Huang, S., and Yang, K.L. (2007) *Langmuir*, **23**, 8607.

133 Thevenet, S., Chen, H.Y., Lahann, J., and Stellacci, F. (2007) *Adv. Mater.*, **19**, 4333.

134 Yu, A.A. and Stellacci, F. (2007) *Adv. Mater.*, **19**, 4338.

135 Akbulut, O., Jung, J.M., Bennett, R.D., Hu, Y., Jung, H.T., Cohen, R.E., Mayes, A.M., and Stellacci, F. (2007) *Nano Lett.*, **7**, 3493.

136 Yu, A.A. and Stellacci, F. (2006) *J. Mater. Chem.*, **16**, 2868.

137 Yu, A.A., Savas, T.A., Taylor, G.S., Guiseppe-Elie, A., Smith, H.I., and Stellacci, F. (2005) *Nano Lett.*, **5**, 1061.

138 Yu, A.A., Savas, T., Cabrini, S., diFabrizio, E., Smith, H.I., and Stellacci, F. (2005) *J. Am. Chem. Soc.*, **127**, 16774.

139 Lin, H.H., Sun, L., and Crooks, R.M. (2005) *J. Am. Chem. Soc.*, **127**, 11210.

140 Kim, J.H. and Crooks, R.M. (2007) *Anal. Chem.*, **79**, 8994.

141 Kim, J. and Crooks, R.M. (2007) *Anal. Chem.*, **79**, 7267.

142 Lin, H.H., Kim, J., Sun, L., and Crooks, R.M. (2006) *J. Am. Chem. Soc.*, **128**, 3268.

143 Huskens, J., Deij, M.A., and Reinhoudt, D.N. (2002) *Angew. Chem. Int. Ed. Eng.*, **41**, 4467.

144 Auletta, T., Dordi, B., Mulder, A., Sartori, A., Onclin, S., Bruinink, C.M., Peter, M., Nijhuis, C.A., Beijleveld, H., Schönherr, H., Vancso, G.J., Casnati, A., Ungaro, R., Ravoo, B.J., Huskens, J., and Reinhoudt, D.N. (2004) *Angew. Chem. Int. Ed. Eng.*, **43**, 369.

145 Bruinink, C.M., Nijhuis, C.A., Peter, M., Dordi, B., Crespo-Biel, O., Auletta, T., Mulder, A., Schönherr, H., Vancso, G.J., Huskens, J., and Reinhoudt, D.N. (2005) *Chem. Eur. J.*, **11**, 3988.

146 Onclin, S., Huskens, J., Ravoo, B.J., and Reinhoudt, D.N. (2005) *Small*, **1**, 852.

147 Ludden, M.L.W., Mulder, A., Schulze, K., Subramaniam, V., Tampe, R., and Huskens, J. (2008) *Chem. Eur. J.*, **14**, 2044.

148 Sadhu, V.B., Perl, A., Duan, X.X., Reinhoudt, D.N., and Huskens, J. (2009) *Soft Matter*, **5**, 1198.

149 Barbulovic-Nad, I., Lucente, M., Sun, Y., Zhang, M.J., Wheeler, A.R., and Bussmann, M. (2006) *Crit. Rev. Biotechnol.*, **26**, 237.

150 Blattler, T., Huwiler, C., Ochsner, M., Stadler, B., Solak, H., Voros, J., and Grandin, H.M. (2006) *J. Nanosci. Nanotechnol.*, **6**, 2237.

151 Falconnet, D., Csucs, G., Grandin, H.M., and Textor, M. (2006) *Biomaterials*, **27**, 3044.

152 Sia, S.K. and Whitesides, G.M. (2003) *Electrophoresis*, **24**, 3563.

11
Colloidal Self-Assembly of Semiconducting Polymer Nanospheres: A Novel Route to Functional Architectures for Organic Electronic Devices

Evelin Fisslthaler and Emil J. W. List

11.1
Introduction

Recent trends in nanoscale electronics show a quest for novel materials and concepts, which allow to control the deposition of active material on nano- to mesoscopic scale by means of top-down and bottom-up approaches. Within this scope, conjugated polymers are a particularly promising class of materials, combining the electrical and optical properties of metals and semiconductors while retaining the attractive mechanical properties of polymers. On the basis of these properties, conjugated polymers may be used in a multitude of applications, such as polymer light-emitting diodes (PLEDs) [1] and light sources [2, 3], light-emitting electrochemical cells (LECs) [4], lasers [5], solar cells [6], photodetectors [7], active sensor devices [8], and field-effect transistors [9]. Furthermore, different available solution-based deposition methods make this class suitable for large-scale large-volume fabrication processes offering a route to low-cost and flexible applications.

Nevertheless, one of their greatest benefits simultaneously constrains their applicability for the fabrication of structured devices: commonly used solution-based deposition methods, such as spin casting or ink-jet printing, offer only limited possibilities for patterning. Mostly, these approaches are only capable of yielding patterned devices if they are combined with rather complex lithographic processes. A typical example is the fabrication of polymer displays by ink-jet printing – the individual pixels have to be separated by previously deposited barriers [10, 11].

However, the cheap and controlled fabrication of structures on micrometer and even nanometer scale is an indispensable precondition for the application of conjugated polymers in industrial micro- and nanoelectronics.

Therefore, the endeavor of current research activities is to combine established structuring methods with novel materials and applications in order to push the limits of miniaturization of the aforementioned devices. One such approach will be exemplified in the following sections via the fabrication steps for nanopatterned light-emitting structures for a polymer light-emitting device.

Functional Supramolecular Architectures. Edited by Paolo Samorì and Franco Cacialli
Copyright © 2011 WILEY-VCH Verlag GmbH & Co. KGaA, Weinheim
ISBN: 978-3-527-32611-2

The prototypical conjugated material used was an electroluminescent polymer that was converted into semiconducting polymer nanospheres (SPNs) [12] with a diameter of approximately 100 nm. Since these nanospheres are dispersed in water, they are subject to various forces acting on them. The specific properties of these nanometer-scale environments can be used for a novel bottom-up approach of nanostructure formation that is directly applicable in the course of the device fabrication process. Before going into detail with respect to the fabrication and deposition of devices and device structures using coating, printing, and soft lithographic methods, a brief introduction to the fabrication of micrometer- and nanometer-sized "spherical" structures based on conjugated polymeric material will be given in conjunction with an introduction to the basic driving forces of nanoparticle self-assembly processes.

11.2
Formation of Semiconducting Polymer Nanospheres

Colloid science offers a novel route for the formation of nanometer-sized particles that are fabricated from conjugated polymers: by using the so-called "miniemulsion" process Landfester *et al.* [13–15] pioneered a process where conjugated polymers that are commonly soluble only in organic solvents can be well dispersed in water in the form of defined nanoparticles. Basically, such a miniemulsion is a system of two immiscible liquids, where small droplets with high stability in a continuous phase are created by high shear stress caused by ultrasonication. Here, droplets of a polymer solution are stabilized in water with a suitable surfactant, as depicted in Figure 11.1.

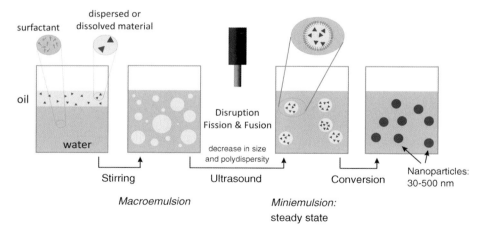

Figure 11.1 Schematic of SPN preparation by a miniemulsion process: Water (with surfactant) and organic solvent (including solved polymer) are mixed and finely dispersed; after evaporation of the organic solvent, the aqueous dispersion is stable for a long time.

After evaporation of the solvent, a stable aqueous dispersion with spherical polymer particles is obtained. The size of the colloids can be controlled by proper choice of surfactant leading to particle sizes in the 30–500 nm range [13]. If the source material is a conjugated polymer, the resulting colloids are called SPNs [12].

The surfactant that is used to stabilize the droplets of organic solvent is characterized by a property that is called amphiphilicity: it possesses both a hydrophobic ("tail") and a hydrophilic ("head") end group. If necessary, the "head" can be positively or negatively charged [16], which can be utilized for directed adsorption of colloids on charged surfaces.

If not mentioned differently, the SPNs described in this chapter were fabricated from methyl-substituted ladder-type poly(*para*-phenylene) (Me-LPPP) [17], which is a commonly used blue-green light-emitting electroluminescent polymer. The diameter of the spheres was in the range of 100 nm, and they were dispersed in water with the aid of Lutensol AT50 (hexadecyl-modified poly(ethylene oxide), $(C_{16}H_{33})(EO)_{50}$) [16], which is a nonionic surfactant. Both decomposition and glass transition temperature of Me-LPPP are beyond 300 °C [12], which is advantageous for processing the SPNs at elevated temperatures without risking deformation of the nanospheres.

11.3
Driving Forces behind Nanoparticle Self-Assembly Processes

On submicrometer scales, the familiar balance of fundamental powers is altered – some forces become negligible, while others start to dominate at that scale.

In order to get micrometer- or nanosized particles that are dispersed in a fluid to assemble in a predetermined pattern, it is necessary to employ suitable forces that are able to affect these particles. Thus, these forces must, on the one hand, be strong enough to overcome Brownian motion. On the other hand, they must be localized on a feature that is fabricated specifically to determine the pattern, for example, by lithographic processes. In addition, they must be able to retain the particles in the pattern after the solvent has evaporated.

Therefore, a short overview of forces that act on nanoparticles is given:

- *Gravitational/buoyancy forces*: For particles with diameters of only few micrometers or smaller that are dispersed in water, these forces can be neglected [18] since they are at most in the range of thermal energy. Actually, such forces are so weak compared to the driving forces that template-assisted self-assembly processes even work upside down [19].
- *Electrostatic forces*: A possible method to assemble nanoparticles is the directed adsorption of charged particles via electrostatic forces on patterned oppositely charged self-assembled monolayers (SAMs) [20], patterns of stored charges in poly(methylmethacrylate) [21], or on patterned polyelectrolyte multilayer surfaces [22]. However, this requires both the fabrication of nanoparticles with charged surfaces (e.g., by using a suitable anionic or cationic surfactant [13]) and the methods for patterning the template surface. This often involves the

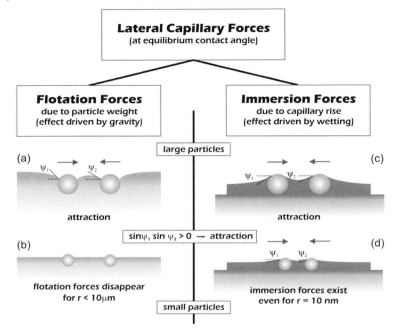

Figure 11.2 Comparison between flotation (a, b) and immersion (c, d) capillary interaction between two identical particles and the resulting attractive forces (ψ_1 and ψ_2 are meniscus slope angles) (adopted from Ref. [23]).

application of intermediate layers, which can be disadvantageous in case the formed structure is intended for device formation.
- *Capillary forces*: Particles that are dispersed in fluids are subject to capillary interaction. These lateral capillary forces occur when the attachment of particles to a fluid-phase boundary is accompanied by perturbations in the interfacial shape, that is, it is caused by the deformation of the surface of the liquid layer by the particles [23]. As depicted in Figure 11.2, there are two different types of capillary forces that act on nanoparticles, depending on whether the particles are floating freely (*flotation capillary force*) or if the liquid level is so low that the particles are already confined between the surface of the liquid and the substrate (*immersion capillary force*):

–*Flotation capillary force:* Such forces cause an attractive interaction between two particles floating freely on the surface of a liquid layer (see Figure 11.2). The meniscus of the liquid deforms to decrease the gravitational potential energy of the two particles when they approach each other. Thus, the source of the force is the particle weight. For two identical particles, the resulting energy is approximately given by [24]

$$\Delta W_F \approx \gamma q^4 r^6 \ln(qL), \quad qL \ll 1 \tag{11.1}$$

where γ is the liquid–vapor interfacial energy, $q^{-1} = [\gamma/g\varrho]^{1/2}$ is the capillary length, with ϱ the fluid density and g the acceleration due to gravity, r is the radius of the spheres, and L is the distance between the particles. Interaction due to flotation capillary force is rather negligible for small particles due to the r^6 dependence – if the particles are smaller than approximately 10 µm, they do not interact because they simply do not deform the surface of the fluid. Therefore, flotation capillary forces can be neglected when analyzing nanoparticle interaction.

–*Immersion capillary force:* As soon as the level of the liquid layer falls below the diameter of the particles, and they start to protrude from the surface of the fluid, the layer is deformed due to wetting effects on the surface of the sphere. The force that results from this deformation is called *immersion capillary force* and can be approximated by [18]

$$\Delta W_I \approx \gamma r^2 \ln(qL) \tag{11.2}$$

Since immersion capillary force depends on r^2, it remains large with respect to kT even for very small particles. Thus, even particles as small as 1 nm experience immersion capillary interaction because they deform the surface of the liquid by wetting. Besides, it is particularly long range: even at $L = 1000\ r$, ΔW is considerably larger than the thermal energy kT [25].

The physical origin of lateral capillary forces is the deformation of the liquid surface by capillary rise of the liquid along the surface of the particles [26]. The larger the interfacial deformation created by the particles, the stronger the capillary interaction between them. Depending on the particle properties, the resulting forces can be either attractive or repulsive: similar particles attract each other (see Figure 11.2), whereas heavy and light particles are pushed apart by flotation forces and hydrophilic and hydrophobic particles by immersion forces, respectively [23].

Lateral capillary forces also give rise to attractive forces between a spherical particle and a wall [27], leading to an assembly of dispersed particles in cavities, corners, and trenches of template structures. Thus, the immersion capillary forces among spherical nanoparticles that are immersed in a liquid layer as well as between single particles and topographic features can be utilized for the formation of all kinds of nanostructures. Wide-area 2D [28] and 3D [29] crystals (see left and central part of Figure 11.3) and complex clusters that are defined by a template surface [30] (as depicted in the right-hand side image in Figure 11.3) can be formed, respectively.

Attractive capillary forces between the particles also cause tightening of the assembled structures as long as the fluid has not fully evaporated – the nanospheres attempt to approach each other to form closely packed monolayers or clusters (depending on the amount of particles in the dispersion and the topography of the substrate). These structures can also be put to various uses: two-dimensional arrays of spherical nanoparticles are applied as shadow masks in an approach called "nanosphere lithography" [31] or for the formation of photonic crystals that are made from polymers [32].

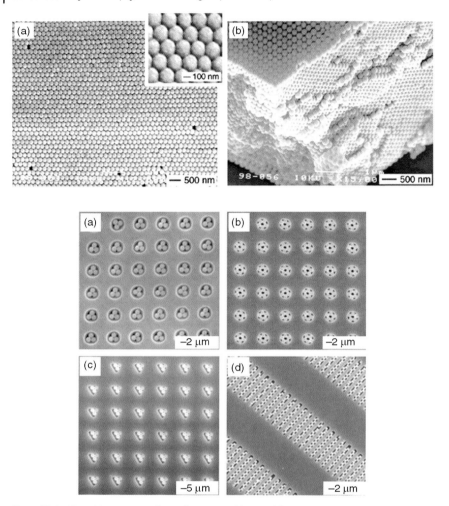

Figure 11.3 Top: SEM images of crystalline arrays fabricated from 220 nm polystyrene beads [29]; outer and bottom: aggregates assembled from polystyrene beads by templating against arrays of holes with different geometries [30].

In brief, apart from electrostatic interaction, immersion capillary forces are by far the strongest forces that act on nanoparticles. Provided that a convenient template is used, nanosized spherical colloids can be arranged in a multitude of different assemblies. It will be demonstrated in the following section that these self-assembly processes are not limited to deposition based on slow-going sedimentation processes. Quite on the contrary, these self-assembly events happen so fast that they can even be employed for fast and high-throughput fabrication processes such as ink-jet printing.

11.4 Deposition Methods for Aqueous Dispersions of Semiconducting Polymer Nanospheres

In order to exploit the full potential of SPNs for the formation of functional architectures, close attention must be paid to the choice of deposition method for the nanocolloid dispersion. If chosen carefully, some approaches are perfectly suitable for the direct formation of polymeric structures on scales that depend on the processing parameters of the chosen method.

The most frequently used solution-based deposition method that is used for conjugated polymers is *spin casting*: a drop of solution containing the material for film formation (e.g., polymer dissolved in a suitable organic solvent) is placed on a substrate that is subsequently rotated at high speed (Figure 11.4, left image). Resulting film thicknesses range from a few to several hundreds of nanometers and depend on the solid content of the fluid, the spin parameters, and the interfacial tension between the substrate and the fluid.

SPNs can be deposited by spin casting, too. In this case, the film thickness depends only on the particle diameter. The thinnest layer possible is a closed monolayer of nanospheres. If the coverage is less dense, a device formed from this material may have short circuits. However, capillary interaction among the particles leads to 2D aggregates, which is beneficial for the formation of closed layers [28].

It is obvious that spin casting may be used only for the formation of homogeneous layers with no specific lateral structure. Fabrication of laterally patterned layers requires alternative approaches as described in the following.

Ink-jet printing, on the other hand, uses a piezoelectric actuator to drive a small drop of fluid from a glass nozzle with a small orifice [33] (usually in the range of tens of micrometers; Figure 11.4, right image). Continuous films can be formed by interlacing of the resulting spots, while the use of a multinozzle printhead and appropriate substrate stage control can result in arbitrary patterns ranging from single spots (minimum diameters are a few micrometers) to large-area patterns. Nowadays, feature sizes of down to 30 μm can be easily obtained by an industrial-scale process delivering droplet volumes in the range of 2–40 pl [34].

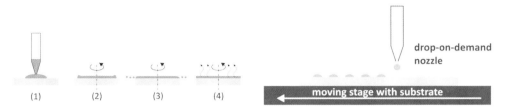

Figure 11.4 Schematics of different solution-based deposition methods for SPN dispersions (*left*: various steps of the spin casting process: (1) deposition; (2) turn on – spreading of fluid; (3) spin off – homogeneous distribution of fluid; and (4) evaporation); *right*: ink-jet printing process with a moving substrate stage).

Fixing the printed ink on the surface and/or allowing a homogeneous spreading on the surface is a key prerequisite for controlling the selected and structured deposition process. In graphical ink-jet applications, this is achieved either by using solvent-based inks fixed by soaking the ink into the substrate or a specific coating or by using UV light-initiated cross-linking of a reactive ink. Alternatively, thermal curing processes of the ink are used [35], which, however, requires a detailed understanding of the evaporation process of the ink-jet-printed droplets and the ongoing mass transport phenomena happening upon drying of the droplet.

The processes of spreading of the drop on the sample surface and evaporation of the solvent are strongly intertwined – especially when they are as small as droplets used for ink-jet printing, which are usually in the range of a few tens of picoliters in volume and a few tens of micrometers in diameter.

This is of particular interest if particles are dispersed in the liquid since the movement of the particles within the droplet is induced by flow processes that are governed by the drying process. After reaching the equilibrium state, the contact line, that is, the boundary line of the drop, is usually "pinned" [36]. This pinning of the edge is not only governed by the contact angle formed between the surface, droplet, and air interface but is also strongly influenced by surface roughness and physical and chemical heterogeneities [37]. Upon drying, the liquid that is removed from this outer area by evaporation is then replenished by a flow of liquid from the interior of the drop. This leads to a migration of the dispersed material within the drop toward the contact line, resulting in the formation of a distinctive rim. This characteristic is also known as the "coffee stain effect" [38, 39].

The undesired formation of ring stain deposits in ink-jet-printed droplets can be deliberately suppressed. For example, this can be done by mixing high-boiling and low-boiling solvents to establish a solvent composition gradient: the composition at the contact line will shift toward a higher fraction of high-boiling solvent than in the bulk due to the increased rate of evaporation at the edge. Therefore, the rate of evaporation at the contact line decreases and a surface tension gradient is established. A flow will be induced from regions of low surface tension to regions of high surface tension. This effect is called "Marangoni flow" – the fluid flows back toward the center of the droplet [40]. Aside from approaches via varying solvent composition, sophisticated surface design can influence the particle transport processes, leading to more homogeneous distribution of the printed particles within the droplet.

An alternative deposition method that is capable of producing patterns from polymer nanodispersions is "*micromolding in capillaries*" (MIMIC) [41]. It is based on the spontaneous filling of a network of small channels by a suitable liquid containing the material (or a precursor thereof) that shall be deposited in the given pattern in a dissolved or dispersed form. The setup is designed especially to allow easy disassembly without destroying the newly formed structure: An elastomeric stamp with a suitable channel structure, commonly fabricated from poly(dimethylsiloxane) (PDMS), [42] is cast from a master structure featuring the desired relief pattern (see Figure 11.5a and b). The geometry of the channel structure of the stamp is mostly in the range of micrometers for channel height and width, and can exceed several centimeters in length. The master structures are usually formed by a

11.4 Deposition Methods for Aqueous Dispersions of Semiconducting Polymer Nanospheres

Figure 11.5 Schematic of the "micromolding in capillaries" process. (a) The elastomeric stamp is cast from a rigid master structure and (b) carefully detached after curing. Then, the stamp is put onto a flat substrate and a drop of fluid is placed next to the inlet, whereupon the capillary network is slowly filled with the liquid (c). After the solvent has evaporated, the stamp is removed, leaving behind the replicated structure (d).

photolithographic process. However, all kinds of suitable structures can be used for stamp casting, for example, diffraction gratings or the surface of a compact disc.

The stamp is then brought in conformal contact with a substrate. Due to the nature of the PDMS, the stamp is slightly adherent, thus providing a leak-proof network of channels on the substrate. A drop of fluid – the solution containing the material that is to be deposited in a solved or dispersed form – of ample volume is placed close to the ingress notch of the PDMS stamp (Figure 11.5c). After the solvent has evaporated, the deposited material can, if necessary, be cured (e.g., by heat or UV light) before the PDMS stamp is carefully peeled off the sample. The low reactivity of the PDMS allows easy separation from both the master and the replicated structure without damage (Figure 11.5d). Since there are no remnants on the stamp, it can be reused many times without any significant loss of quality.

The success of the MIMIC process is mainly governed by the surface properties of the materials forming the channel, and the surface tension and the viscosity of the liquid that should fill it. Thus, the drag that pulls the fluid into the capillary is primarily caused by the wetting properties of the liquid in relation to the capillary surface.

However, the MIMIC approach is limited due to some inherent problems that are related to geometry of the channel system: both the filling rate and the stability of the PDMS channel structure depend on the dimensions of the channel since the rate of liquid flow is determined by both the cross section and the length of the channel [43]. Furthermore, if the channel is disproportionally low it is likely to collapse; hence, some structures are difficult to replicate via MIMIC (e.g., channel structures that are very wide but low).

MIMIC can be used to form a wide variety of microstructures from polymers and dispersed particles. It can replace and complement photolithography in certain cases since it is a simple procedure requiring only one pattern transfer step. Besides, a stamp formed once from a given master can be used numerous times, and it can feature three-dimensional structures (by using an appropriate master), while photolithography is limited to the layer dimensions imposed by the deposition of the resist by spin casting.

A wide range of "MIMICed" microstructures, for example, electrodes for organic transistors formed from carbon [44] or silver nanoparticles [45] and polymeric waveguides [46], have been developed in recent years.

11.5
Organic Electronic Devices from Semiconducting Polymer Nanospheres

Semiconducting polymer nanospheres can be formed from almost all types of conjugated polymers. Therefore, their implementation in organic electronic devices is merely limited by the more elaborate deposition and the various device architectures. In the following sections, several different applications of SPNs in devices are described in more detail. Particular attention is paid to the prospects of formation of micro- and nanostructures, which is the most distinguishing feature of these nanoparticles compared to the bulk polymer.

The most common device fabricated from SPNs is the polymer light-emitting device. Basically, a PLED formed from SPNs can be built in the conventional vertically stacked multilayer OLED configuration [47] ("sandwich structure") that is assembled on an indium tin oxide (ITO) covered glass substrate (as depicted in Figure 11.6).

The most striking differences between a polymer layer fabricated from bulk material and one fabricated from SPNs, respectively, are the properties of the individual solvents (organic solvent versus water) and the variability of the resulting layer thickness. While the thickness of a thin film that is spin cast from a polymer solution can be varied from a few nanometers to several micrometers, the thinnest layer possible with SPN dispersion is a monolayer of nanospheres. Therefore, the minimum layer thickness is defined by the diameter of the spheres. This fact must also be observed for the choice of polymeric material and particle diameter during SPN preparation: materials with a small Stokes shift can show strong self-absorption effects of the high-energy part of the electroluminescence spectrum in "thick" films [48, 49].

A fundamental point in the assembly of an SPN-based PLED is that the polymer layer is free of pinholes in order to avoid the creation of short circuits. This is why self-assembly processes are also relevant for the formation of monolayers: capillary interaction among the spherical nanoparticles gives rise to a contraction process of the SPNs leading to closely packed layers (as depicted in left image in Figure 11.3).

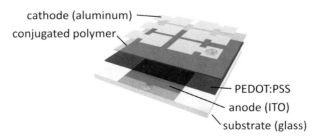

Figure 11.6 Schematic of the multilayer assembly of a conventional PLED: a transparent anode is covered with a thin layer of conjugated polymer and a metal cathode. Additional intermediate layers can be added to enhance device performance.

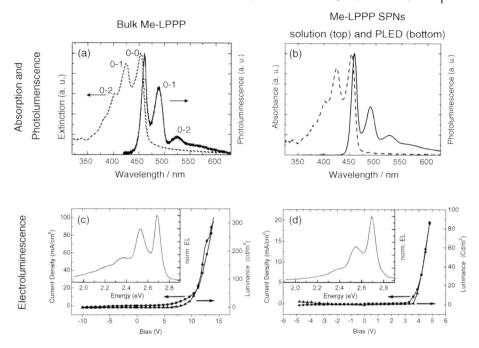

Figure 11.7 Comparison of photophysical and electroluminescent properties of Me-LPPP bulk (*left*) and SPNs (*right*). Both absorption/photoluminescence ((a) from Ref. [17]); (b)) and electroluminescence (inset in (c) and (d), both from Ref. [50]) spectra do not show significant differences. The onset of current and luminescence in the characteristic curves (*bottom*) is even decreased for SPN PLED compared to the device built from bulk Me-LPPP.

A critical point and main precondition for the practical application of SPNs in organic electronics is that the complex fabrication process must not degrade the conjugated structure. Figure 11.7 shows a comparison of the absorbance and photoluminescence (PL) spectra and of the electroluminescence (EL) spectra of PLEDs fabricated from both bulk [17] and SPN Me-LPPP [50]. Apart from a slight decrease in the onset of the monolayer SPN PLED compared to the onset of the bulk PLED (compare Figure 11.7c and d), no significant alteration is observable.

Piok *et al.* were able to show that the miniemulsion process does not impair either the photophysical [51] or the electroluminescent [50] properties of this conjugated polymer, which leads to the conclusion that SPNs are an excellent alternative to common polymers as active materials in organic electronic devices.

This has already been shown for a variety of devices. The majority of these SPN devices feature homogeneous functional layers, and will be described in the following sections, while results discussed in another section show first patterned PLEDs and structured prototypes that make use of the fact that nanospheres are susceptible to capillary forces and can thereby be arranged in patterns by template-assisted fluidic self-assembly processes. This concept may be a first step toward nanoscale organic electronics.

11.5.1
Organic Electronic Devices from SPNs with Homogeneous Functional Layers

The beneficial properties of SPNs have already been applied in a variety of devices fabricated via conventional deposition methods that lead to a homogeneous particle distribution. Since nanospheres offer a lot of possibilities beyond their potential for nanostructure formation, their application has been reported for monolayer [50] and multilayer PLEDs [52], light-emitting electrochemical cells [53], and solar cells [54, 55].

Spin-cast closed layer PLEDs, such as the mono- and multilayer PLEDs presented by Piok *et al.*, seem to profit from the special geometry of the layers created by closely packed spheres, and feature improved optoelectronic characteristics [50, 52]. The solar cells fabricated via spin-casting SPN dispersions by Kietzke *et al.* gained considerable interest due to the fact that they offer two novel approaches to control the dimension of phase separation in polymer blend layers – either by mixing two different types of SPN dispersions before deposition or by creating "mixed" SPNs by emulsifying a two-component polymer blend solution [54, 55].

Finally, Mauthner *et al.* [53] could show that SPN dispersions are exceptionally suitable for ink-jet printing procedures since the water-based dispersions are both easy to handle during the printing process and more environment-friendly than conventional polymer solutions that are based on organic solvents. In addition, they feature very good printing properties due to the absence of the "bead-on-a-string" effect [56].

All these devices rely on the fact that the layers are closed and free of short circuits, which is aided by the aforementioned self-contraction effect induced by the capillary forces acting among the nanospheres when they are deposited in a thin layer of dispersion.

Furthermore, all types of devices gain additional perspectives from the fact that the dispersion is water based, which offers new alternatives concerning polymer multilayer formation and device architecture.

11.5.2
Functional Micro- and Nanopatterns from SPNs for Organic Electronic Devices

In order to form patterned organic electronic devices such as PLEDs from SPNs, precautionary measures must be taken to ensure the devices are free from pinholes and therefore free from short circuits. Although the aqueous SPN dispersions are quite suitable for deposition methods capable of pattern formation, such as ink-jet printing, it is obvious that it is not possible to print isolated areas of polymeric material without covering the rest of the device area that is sandwiched between the electrodes. This is why ink-jet printing is commonly used in combination with barrier structures patterned either by photolithography prior to the printing step [10, 11] or by a previous ink-jet process forming interlevel via holes in an intermediate layer ("via hole" approach) [57].

Water-based SPN dispersions offer the unsurpassed advantage of being capable of forming polymer multilayer structures by consecutive solution-based deposition steps. This is why SPNs can be deposited on top of another polymer layer without redissolving the subjacent layer. By choosing materials with appropriate glass transition temperatures (T_g) for the polymer nanospheres and the polymer matrix layer, it is possible to embed the SPNs in the polymer layer by a heating step following the nanoparticle deposition: The temperature for the embedding step has to be above the glass transition temperatures of the matrix layer but low enough to conserve the spherical shape of the particle. If the dimensions of both SPN diameter and polymer layer are correctly chosen (i.e., the layer has to be a little bit thinner than the diameter of the nanospheres) and contact to both bottom and top electrodes of the device are given, a device, for example, a PLED with a nanosized light-emitting spot can be formed.

This is exemplified in more detail in Figure 11.8, which illustrates the architecture of a PLED that is formed by SPNs that are embedded in an inert polystyrene (PS) matrix layer.

This device is composed of a glass substrate that is covered with a transparent ITO anode, an intermediate layer of poly(3,4-ethylenedioxythiophene) poly(styrene sulfonate) (PEDOT:PSS), a PS layer as a nonemitting, insulating matrix layer to embed the SPNs, SPNs fabricated from an electroluminescent polymer (Me-LPPP), and a metal-top electrode. Typically, the SPN diameter for such a device is in the range of 100 nm.

Despite the fact that the device is a patterned PLED based on nanoparticles, the basic device layout is still a common "sandwich structure" [47] that is assembled via conventional processing steps with the addition of the SPN deposition and embedment steps.

However, the intermediate polymer matrix layer is not only useful for the prevention of short circuits in the device but, with proper pretreatment, it can also be used to trigger self-assembly processes of the SPNs, finally leading to the formation of nanostructures that are formed by the assembled particles. This is achieved by a procedure that can be described as template-assisted self-assembly [29, 30, 58], which is a process that can be applied to entrap micro- or nanoparticles that are dispersed in a fluid in recesses of a surface. The particles are deposited as

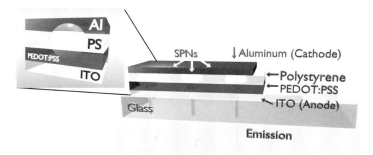

Figure 11.8 Polymer multilayer assembly for structured polymer light-emitting devices from SPNs. The SPNs are embedded into the insulating polystyrene matrix layer and touch both PEDOT:PSS layer and aluminum top electrode.

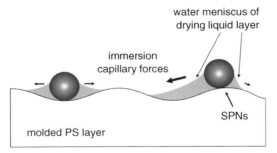

Figure 11.9 Schematic of the template-assisted self-assembly process: Due to immersion capillary forces acting on the spheres, they finally assemble in the grooves of the template while the drop dries.

dispersion; while the carrier fluid dries, they assemble in the given template structure: As soon as the thickness of the liquid film is so low that the SPNs are no longer floating freely but protrude from the surface of the liquid layer, they can get physically confined to the template, and the attractive capillary interaction between the sphere and the substrate rises considerably [23]. As depicted in Figure 11.9, the meniscus of the liquid surrounding a sphere resting on a "hillside" is deformed due to the undulated substrate topography, and the force resulting from the varying contact angle [18–20] drags the sphere into one of the channels.

Similar behavior has been reported for 2D crystal self-assembly processes [28]. In order to provide a suitable template for this self-assembly process, the polymer matrix layer was molded by imprinting a channel structure via a process that is called "soft embossing" [5, 59]: A PDMS stamp fabricated from a suitable master (for the samples presented here, the channel width was 277 nm and the channel depth was 25 nm) was used for the embossing procedure. Polystyrene was dissolved in chloroform and deposited on cleaned glass substrates by spin casting, resulting in a layer thickness of 80 nm. Then, the PDMS stamp was deposited on the PS film and loaded with a weight of 80 g, and both were put in an oven at 120 °C for 2 h (about 25 °C above the T_g of PS). After cooling down, the stamp was carefully peeled off the PS. This procedure is depicted in Figure 11.10a and b. The 3D representation of the atomic force microscopy (AFM) height image in Figure 11.10c shows the result of the successful embossing process, while Figure 11.10d is a section analysis of the same AFM image showing the neat reproduction of the approximately 25 nm deep channels.

Possible minimum structure dimensions are solely limited by the diameter of the particles and by the resolution of the technique that is used to form the pattern of the recessed template. This self-assembly process is triggered due to immersion capillary forces acting between nanoparticles and molded template surface.

The introduction of a template-assisted fluidic assembly process allows the fabrication of nanostructured devices such as PLEDs from aqueous nanoparticle dispersions [58]. However, in order to boost functional nanostructures toward large-scale production processes, potential applications have to be introduced. Another possible utilization that requires controlled arrangement of different components on

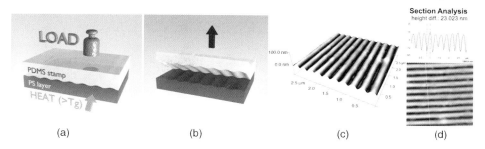

Figure 11.10 Schematic of the soft embossing process: the PDMS stamp is deposited on the PS layer, weighted down, and heated (a); after cooling down, the stamp is peeled off (b), resulting in a PS surface molded according to the master channel structure. (c) and (d) show a 3D representation and a section of an AFM height image of such a PS surface.

a nanometer scale, combined with fast and cheap large-scale production methods, is in organic solar cells. The production and patterning technique for devices presented here is a further step toward controlled structuring of two-phase systems. The choice of appropriate combinations of organic materials for the embedding layer and the SPNs is mostly limited by the necessity of compatible T_g. The major benefit of this approach is the possibility to implement the formation of nanostructures in the device as an additional step in the conventional device preparation procedure; all deposition processes are solution-based, and the imprinting step is an easy procedure that is also applicable for large areas.

Due to the rapid advancements in all involved scientific fields, a significant potential of process optimization is already pending. This will provide the necessary enhancement to the production process for nanostructured devices fabricated from SPNs since a significant reduction of processing time is still necessary to make this approach feasible for industrial-scale fabrication processes. This could, for example, be realized by further optimization of the embossing process, as shown recently for nanostructures in polystyrene that were produced by a soft embossing process that lasted for merely 200 s [60]. Careful adjustment of the process parameters of the SPN embedding procedure might lead to a shorter device fabrication time, too [61].

11.5.2.1 Functional Micro- and Nanopatterns from SPNs Formed by Ink-Jet Printing and Template-Assisted Self-Assembly

As already described, various deposition methods are feasible for SPN device fabrication. However, for nanostructure formation via template-assisted fluidic self-assembly, the SPNs need to be deposited on top of the polymer matrix layer. Due to the high hydrophobicity of these polymeric materials, spin casting of the nanosphere dispersion is not viable, since the resulting particle distribution is sparse and inhomogeneous – the biggest part is just flung off the sample.

Consequently, alternative deposition methods have to be used for patterned device fabrication.

Ink-jet printing an aqueous dispersion results in comparable and reliably reproducible patterns with properties that can be controlled via the printing parameters.

Moreover, ink-jet printing is a well-established technique that is qualified for high-throughput production processes on an industrial scale. Up to now, the demanding task of using this advantageous method for the direct, lithography-free fabrication of patterned PLEDs could not be solved in a satisfactory way due to the fact that both light-emitting polymer and circumjacent material were deposited from related solvents (usually organic ones), which easily leads to pinholes and, therefore, short circuits in the device. By using SPNs, pinhole-free polymer multilayers can be formed by ink-jet printing the dispersion on top of any polymer film. Arbitrary patterns can be composed from individual droplets, thus micrometer-sized structures can be created easily with this procedure [49]. Besides, it is by far faster than the sedimentation-based processes usually used for template-assisted self-assembly [30] since the drying time for an ink-jet-printed droplet of aqueous dispersion is in the range of 1 s [62].

As illustrated in Figure 11.11, individual ink-jet-printed droplets feature a rim that is formed by the coffee stain effect described above [38, 39], clearly visible in the fluorescence micrograph on the left-hand side of the figure. However, when such a structure is used in a micrometrer-sized PLED, emission from this ring is suppressed by the higher layer thickness in these areas. The effect is clearly visible in the images on the right-hand side of the figure: the main image shows an ink-jet-printed PLED with a pixel diameter of 20 μm under operation in complete darkness, whereas the inset shows the same spot of the device with a little bit of background illumination. It is obvious that there is no electroluminescence originating from the rim. As mentioned above, rim formation can also be deliberately suppressed by interlacing procedures or combination of high-boiling and low-boiling solvents [40].

The aqueous Me-LPPP SPN dispersion that was printed had a solid content of 0.09 wt%, and the particle diameter was about 100 nm. For the matrix layer, we chose polystyrene (bulk $T_g = 95\ °C$) [63], which is convenient for both soft embossing and embedding of the SPNs. PS is insulating and nonfluorescent and therefore provides proper contrast both in Kelvin probe force microscopy (KPFM) and in fluorescence microscopy (FM) analyses, and in SPN PLEDs. An 80 nm thick PS layer was spin cast onto a glass substrate and soft embossed at 120 °C according to the procedure depicted in Figure 11.10. The aqueous SPN dispersion was ink-jet printed by using a single nozzle drop-on-demand printer system (nozzle orifice 50 μm) [64] with a

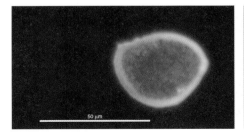 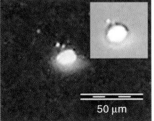

Figure 11.11 Fluorescence microscope image of an ink-jet-printed droplet of SPN dispersion (on PS; *left*); ink-jet-printed 20 μm spot on a patterned SPN-PS PLED (*right*: scale bars in both images are 50 μm).

heatable, semiautomatic substrate stage, producing drops of approximately 20 µm on the substrate. The substrate temperature during the printing process was approximately 30 °C. Prior to investigation and further processing, the samples were heated at 140 °C for 12 h [49].

As mentioned, the PS layer was embossed with a channel structure (25 nm deep and 277 nm wide) by the procedure described above. This results, on the one hand, in a very homogeneous distribution of the SPNs within the droplet, which is beneficial for the fabrication of uniform micrometer-sized PLEDs (see images in Figure 11.11). On the other hand, it leads to the formation of nanostructures by triggering a self-assembly process of the nanospheres within the template structures due to capillary forces acting on the particles. This procedure can be used to form one- or two-dimensional nanostructures [30, 58] across large areas for applications in nanopatterned organic electronic devices. As illustrated in Figure 11.12, the nanoparticles are dispersed in a droplet of water when they are printed onto the substrate. The assembly process does not start immediately: It is triggered as soon as the liquid level drops to the range of the nanoparticle diameter so as to give immersion capillary forces reason to act on the particles. It takes at least 10^{-6} s to drag such a colloidal nanoparticle from a peak into the nearest channel [19]. This is why the result of this nanostructuring process by template-assisted assembly in ink-jet-printed drops depends to a great extent on the substrate temperature. If the liquid evaporates too fast during this stage of the drying process, the self-assembly procedure cannot be successful.

Another consequence is that the nanoparticles start to move closer to each other due to attractive capillary forces among them, leading to the formation of tight particle agglomerations (see Figure 11.3). Consequently, the SPNs assemble in a pearl necklace-like structure in the grooves of the polymer template layer.

The result of this assembly process is shown in the two 3D representations of AFM height images in Figure 11.13: On the left-hand side, the embossed structure of the polymeric matrix layer is still clearly visible, and the SPNs are neatly arranged within the channels. Although the molded recesses seem to be rather shallow (25 nm deep and 277 nm wide in a sinuous pattern), the self-assembly process is very thorough,

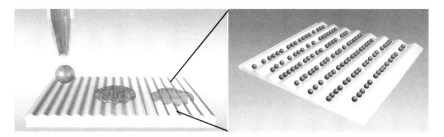

Figure 11.12 Schematic of the fabrication process for functional nanostructures from assembled SPNs: After embossing of the polymeric template layer, the dispersion containing the SPNs is ink-jet printed onto this surface (*left*). Due to capillary forces acting on the spheres, they finally assemble in the grooves of the template while the drop dries (*right*).

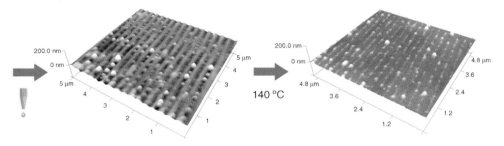

Figure 11.13 AFM images of the results of the fabrication steps depicted above, and a subsequent heating step: 3D AFM image of the SPNs that have assembled in the grooves during ink-jet printing (*left*). Via heating at 140 °C, the linearly assembled SPNs (diameter: 100 nm) are embedded deeply into the 80 nm PS layer (*right*).

and hardly any SPNs that are not arranged at the bottom of the trenches can be found. The right image in Figure 11.13 shows the result of the subsequent heating step at 140 °C: The SPNs are deeply embedded within the polymer matrix layer without losing their spherical shape. Only their tips protruding from the PS film are visible. The successful completion of this processing step is a prerequisite for the fabrication of organic electronic devices from SPNs since the nanospheres must touch both top and bottom electrode.

Figure 11.14 depicts an AFM height image of assembled SPNs (left) with a corresponding KPFM image (right). The contrast in the KPFM image is generated by the difference in surface potential of the two materials. Therefore, the right part of Figure 11.14 illustrates the conjugated polymer (Me-LPPP SPNs, bright) enclosed by the insulating PS matrix. Again, the embossed structure is closely reproduced; almost all particles contribute to the pattern, whereas the surface of the PS layer is flattened again due to the heating procedure necessary for the embedding step.

This technique can be directly used to fabricate organic electronic devices with individual nanospheres as functional spots arranged in a defined pattern.

Patterned ink-jet-printed PLEDs that were fabricated in accordance with the procedure depicted here could already be demonstrated [49]. Thus, by realizing the formation of functional polymeric micro- and nanostructures via an ink-jet printing

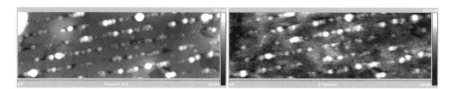

Figure 11.14 AFM topography image (*left*; height scale: 50 nm) and KPFM image (*right*; scale: 100 mV) of ink-jet-printed semiconducting polymer nanospheres (diameter: approximately 100 nm) assembled on and subsequently embedded by annealing into an 80 nm polystyrene layer. The KPFM image shows the contrast in the conjugated polymer (Me-LPPP SPNs, bright) enclosed by the insulating PS matrix. Horizontal scan size is 4 μm for both images.

procedure that is combined with a template-assisted self-assembly process of semiconducting polymer nanospheres, the industrial-scale fabrication of nanopatterned organic electronic devices could become an attainable target within the near future.

11.5.2.2 Functional Micro- and Nanopatterns from SPNs Formed by Micromolding in Capillaries and Template-Assisted Fluidic Self-Assembly

The use of aqueous nanoparticles dispersions instead of polymer solutions can have certain benefits for MIMIC processes that can be used to extend the possibilities of the approach.

First of all, the particles are dispersed in water. This might seem counterproductive for MIMIC since water has a high contact angle on a lot of materials that hinders the filling of the capillary system and slows down the filling rate. But these handicaps can be overcome – a lot of substrate surfaces can be treated to make them more hydrophilic, resulting in lower contact angles. Besides, the addition of suitable surfactants to the aqueous dispersions is of avail, too. However, water is a solvent that does not affect the stamp, in contrast to a number of organic solvents that cause the stamp to swell and detach from the substrate, and it has a low viscosity. Furthermore, it can be used to deposit structures on top of a polymer layer since it does not dissolve the subjacent polymeric material. Finally, the dispersed particles can be used to form structures of high complexity by exploiting both the channel geometry and the disposition of spherical particles to form complex and crystalline structures under the influence of capillary forces.

As illustrated in Figure 11.15, the proportions of the particle diameter–channel dimension relation and the solid content of the dispersion govern the result of the filling of the capillary network. For particles that are only slightly smaller than the channel, long chains of single beads are formed that are contracted even further by immersion capillary forces between the particles (see Figure 11.15a). If particles made of suitable polymeric materials are used, they can even be fused by heating above glass transition temperature, so that the chains can subsequently be separated from the substrate. This was shown for polystyrene microspheres by Li *et al.* [65]. If

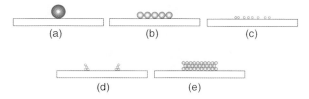

Figure 11.15 Sketch of the different possibilities of particle sedimentation as a result of MIMIC with a dispersion carrying micro- or nanoparticles: (a) single line of spherical microparticles with diameter in the range of channel dimensions; (b) close coverage of the channel bottom; (c) loose coverage of the channel bottom by small particles (sphere diameter "small" or "large" in reference to dimensions of the capillary); (d) corner filling of nanoparticles due to imbibition effects and drying dynamics; and (e) close packing formed by crystallization of larger or quasicrystallization of small particles.

the nanoparticles are smaller, they are carried along through the channel as it is slowly filled by dispersion. Depending on the properties of the liquid (e.g., viscosity) and on the channel geometry, the capillary is either partly or completely filled in propagation direction. When the fluid reservoir is depleted, the dispersion is no longer replenished, and starts to dry in the capillary. The result of this drying process depends on the properties and solid content of the fluid and on the geometry and properties of the inner surfaces of the channel: The bottom of the capillary can be either fully (Figure 11.15b) or partly covered (Figure 11.15c). Due to leading edges caused by imbibing effects [66, 67], parts of the capillary can show sedimentation only in the corners (Figure 11.15d). This effect can be superimposed and fortified by the drying of the liquid inside the channel: the water level first starts to sink and then it retreats toward the corners, carrying along the dispersed particles. Another particular case is shown in Figure 11.15e and in Figure 11.16: The spherical nanoparticles start to form closely packed crystalline structures.

As depicted in detail in Figure 11.16, it is possible to form crystals from spherical nano- and microparticles by MIMIC, provided suitable structures and dispersions are used: As shown by Kim *et al.* for polystyrene nanospheres, the particles are transported along the channel and are deposited at the end of the capillary when the solvent evaporates (see left side of Figure 11.16 and image Figure 11.16c on the right-hand side) [68]. The constant inflow of the dispersion from the reservoir at the channel entrance transports more particles toward the aggregation site, where they are closely stacked due to capillary forces acting among them. The newly formed crystal grows inward from the channel exit, slowly filling the capillary network

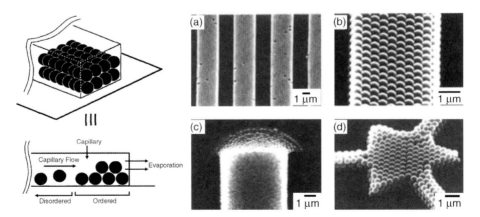

Figure 11.16 Crystallization of spherical polystyrene nanoparticles by MIMIC of dispersions: schematic of the crystallization process close to the end of the capillary – aggregation due to evaporation and attractive capillary forces; the influx of the dispersion transports the spheres to the ordered array (*left*); SEM images of crystallized arrays of nanospheres (*right*; (b) features a close-up of (a); (c) shows a quasicrystalline structure of small nanospheres; (d) demonstrates the formation of complex shapes) (reproduced from Ref. [68]).

(Figure 11.16a and b). As demonstrated in the picture of the triangular shape in Figure 11.16d, this works even for complex structures.

This example nicely demonstrates the way capillary forces govern the processes of capillary filling and nanoparticle assembly within the microcapillary. Apparently, it is an astonishingly simple method to form structures of high complexity by assembling nanospheres in geometrically confined spaces.

If assembled on a suitable substrate and prepared with careful choice of processing parameters, SPN structures formed by MIMIC could also be used for device formation. As for the ink-jet-printed patterned devices, specially designed device architectures, for example, PLED architectures with intermediate polymer layers to prevent pinholes and, thus, short circuits, are necessary. By using this auxiliary polymer layer, MIMIC can be employed to form functional nanostructures assembled from SPNs in the same way as shown above for the ink-jet printing procedure – the same forces employed for the ink-jet-printed nanostructures described above can be applied to manipulate the SPNs that are carried along the PDMS channel. Again, a subjacent polymer layer can be patterned by soft embossing, and nanoparticles have to respond to the forces acting: As soon as the carrier fluid starts to evaporate and the liquid level is so low that the nanospheres protrude from the surface, they get dragged into the recesses of the template by immersion capillary forces and assemble at the lowest point of the trench. This is exactly the same mechanism that causes the SPNs to assemble when they are deposited by ink-jet printing. As described above for the SPN self-assembly procedure by ink-jet printing, a channel structure was molded into an 80 nm thick polystyrene layer by soft embossing. Then, the PDMS stamp with the MIMIC channel structure was put down on top of this substrate; the propagation direction was parallel to the embossed recesses. Figure 11.17a illustrates the drop of nanoparticle dispersion that was placed next to the channel entrance. As indicated by the dark arrow in Figure 11.17b, the fluid advances along the channel as long as it is replenished by the fluid reservoir situated next to the channel entrance. During the final phase of the drying process, the spheres are pulled into the grooves and arrange according to the embossed pattern.

Figure 11.17 (a) MIMIC procedure carried out with nanoparticle dispersion: the colloids are carried along in the fluid that slowly fills the capillary. (b) Schematic of the self-assembly process of SPNs (displayed in white) that are forced into the recesses of a molded template surface by immersion capillary forces: First, the carrier liquid propagates along the channel (dark arrow), but when the fluid reservoir is depleted it dries, moving toward the corners of the rectangular capillary.

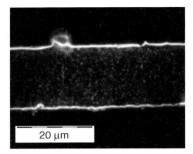

Figure 11.18 Fluorescence microscope image of a MIMIC SPN microstructure fabricated on a PS substrate that was patterned by soft embossing with a shallow trench structure (277 nm wide, 25 nm deep). The (former) channel area is homogeneously covered with Me-LPPP SPNs (diameter 100 nm) that are arranged inside the nanotrenches (scale bar: 20 μm).

An exemplary result of such a procedure can be seen in the fluorescence micrograph in Figure 11.18: The (former) channel bottom area is homogeneously covered with SPNs (diameter 100 nm). Channel width in the image below is about 18 μm, whereas channel height of the PDMS stamp is 1 μm. The agglomerations at the boundaries are probably caused by SPNs that are carried toward the corner of the channel due to the contact angle of the aqueous dispersion with the PDMS channel sidewalls. As for the ink-jet-printed structures with the same solid content (0.09 wt%; Me-LPPP SPNs with a diameter of ~100 nm), this rim is several hundred nanometers high and will therefore not contribute to EL emission if this structure is to be integrated into a PLED assembly.

As for the ink-jet-printed self-assembled SPN nanostructures shown above, the nanospheres assemble in neat, pearl chain-like formations. This is depicted in Figure 11.19 in two 3D representations of AFM height images that show a close-up of the channel region illustrated above after a heating step that flattened the surface of the PS layer and embedded the SPNs therein.

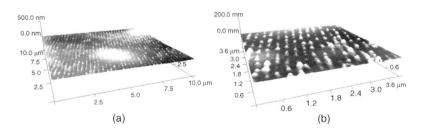

Figure 11.19 3D representation of AFM height images (scan sizes: 10 μm (a) and 4 μm (b)) that show Me-LPPP SPNs deposited by MIMIC that are assembled in a trench structure molded into the PS layer by soft embossing. The PS surface has been flattened by a heating step, and the SPNs are therefore embedded into the PS layer.

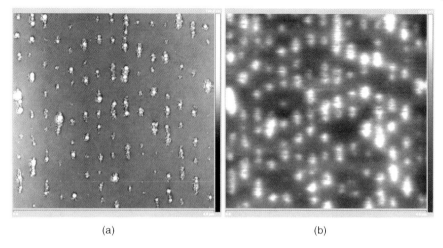

(a)   (b)

Figure 11.20 AFM height (a) and KPFM image (b) showing the contrast between the polystyrene (flat, dark) and the semiconducting polymer nanospheres assembled in lines (elevated, bright). A heating step has flattened the PS surface and embedded the SPNs (scan size is 4 μm).

In order to be able to explicitly distinguish between the insulating polystyrene and the semiconducting polymer nanospheres, AFM height and KPFM images of the MIMIC'ed structures were captured (see Figure 11.20). The contrast in the latter is generated by the difference in surface potential between the two materials – the right image therefore shows the semiconducting material (bright areas) versus the insulating polystyrene (dark areas). This clearly corresponds to the elevated areas in the height image (Figure 11.20a), which are the tops of the nanospheres protruding from the PS surface.

As for all AFM images shown in this chapter, a heating step has been performed. This flattens the undulated PS surface and embeds the SPNs into the PS layer.

Concluding, it could be demonstrated that MIMIC can be used to form both micrometer- and nanometer-sized functional polymer structures on – and, after a heating step also within – polymeric intermediate layers, which is an assembly that can be directly incorporated into organic electronic devices. As for the preparation of such structures via ink-jet printing, template-assisted self-assembly processes of nanoparticles in aqueous dispersions can be used to form nanostructures during the solution-based deposition process.

11.6 Conclusions

The development of novel methods for structuring and processing of layers of conjugated polymers is one of the key elements in the promotion of organic electronics. Newly introduced approaches have to overcome the disadvantages of solution-based deposition methods that are commonly used for the fabrication of organic electronic devices – they must enable the formation of pinhole free patterns on the micro- and even nanometer scale. Simultaneously, they should be applicable in high-throughput production processes and they must provide the opportunity to employ these patterned layers in an organic electronic device.

The methods reviewed in this chapter summarize the recent progress in a number of new approaches for the formation of micro- and nanostructured organic electronic devices. It is based on the application of a novel, very versatile type of material called semiconducting polymer nanospheres. These polymeric nanoparticles are dispersed in water and can therefore be easily used for the formation of polymer multilayer structures by various solution-based deposition methods. Since the electronic properties are not harmed by the conversion step, this material can be used for the same devices as the pristine bulk polymer.

By an appropriate choice of fabrication procedure, structured polymer films with feature sizes as small as several tens of micrometers can be produced. Furthermore, by annealing these particles can be embedded into the subjacent layer made of a second polymer, and the patterned films are thereby "ready for operation" in devices such as PLEDs and solar cells.

However, the use of spherical polymer nanoparticles offers another major benefit: If the dispersion is deposited via a suitable method, the SPNs can be arranged in patterns by various template-assisted fluidic self-assembly methods. The nature of the template depends on the type of force that is used to guide the particles (e.g., electrostatic or capillary), while the minimum feature size of the resulting structure is determined by both the particle diameter and the process of template production. In the method presented here, a topographic template – a polymer layer that is molded by a procedure called "soft embossing" – is used to assemble SPNs with a diameter of about 100 nm in channels with an interval of about 300 nm by immersion capillary forces.

Since this nanostructuring approach can be combined with the device fabrication methods described above, it is directly applicable for the formation of organic electronic devices that are patterned on a nanometer scale. This patterning method is demonstrated for two different solution-based deposition methods: ink-jet printing and micromolding in capillaries. Both lead to comparable nanopatterning results; however, ink-jet printing is particularly well suited since it is a well-established technique that can be applied in industrial-scale production. The high deposition speed does not interfere with the self-assembly process. The variety of possible template designs and materials is large, as is the possible choice of materials for the fabrication of the semiconducting polymer nanospheres. This makes this method remarkably versatile and thus applicable for different types of devices.

The approach demonstrated here might therefore offer an answer to one of the most pressing questions in modern organic electronics – a novel path to micro- and nanopatterned organic electronic devices by combining a bottom-up assembly technique with a rather cheap and simple template fabrication method and a fast deposition process.

Acknowledgment

Part of this work was performed within the RPC "ISOTEC" project funded by the Austrian NANO initiative (project DevAna 0706).

References

1 Burroughes, J.H., Bradley, D.D.C., Brown, A.R., Marks, R.N., Mackay, K., Friend, R.H., Burns, P.L., and Holmes, A.B. (1990) Light-emitting diodes based on conjugated polymers. *Nature*, **347**, 539–541.

2 Duggal, A.R., Shiang, J.J., Heller, C.M., and Foust, D.F. (2002) Organic light-emitting devices for illumination quality white light. *Appl. Phys. Lett.*, **80**, 3470–3472.

3 Kappaun, S., Eder, S., Sax, S., Saf, R., Mereiter, K., List, E.J.W., and Slugovc, C. (2006) WPLEDs prepared from main-chain fluorene–iridium(III)polymers. *J. Mater. Chem.*, **16**, 4389–3492.

4 Pei, Q., Yu, G., Zhang, C., Yang, Y., and Heeger, A.J. (1995) Polymer light-emitting electrochemical cells. *Science*, **269**, 1086–1088.

5 Gaal, M., Gadermaier, C., Plank, H., Moderegger, E., Pogantsch, A., Leising, G., and List, E.J.W. (2003) Imprinted conjugated polymer laser. *Adv. Mater.*, **15**, 1165–1167.

6 Brabec, C.J., Sariciftci, N.S., and Hummelen, J.C. (2001) Plastic solar cells. *Adv. Funct. Mater.*, **11**, 15–26.

7 Schilinsky, P., Waldauf, C., Hauch, J., and Brabec, C.J. (2004) Polymer photovoltaic detectors: progress and recent developments. *Thin Solid Films*, **451–452**, 105–108.

8 Sax, S., Fisslthaler, E., Kappaun, S., Konrad, C., Waich, K., Mayr, T., Slugovc, C., Klimant, I., and List, E.J.W. (2009) SensLED: an electro-optical active probe for oxygen determination. *Adv. Mater.*, **21**, 3483–3487.

9 Garnier, F., Hajlaoui, R., Yassar, A., and Srivastava, P. (1994) All-polymer field effect transistor realized by printing techniques. *Science*, **265**, 1684–1686.

10 Shimoda, T., Kanbe, S., Kobayashi, H., Seki, S., Kiguchi, H., Yudasaka, I., Kimura, M., Miyashita, S., Friend, R.H., Burroughes, J.H., and Towns, C.R. (1999) Multicolor pixel patterning of light-emitting polymers by ink-jet printing. *SID 99 DIGEST*, 376–379.

11 Forrest, S.R. (2004) The path to ubiquitous and low-cost organic electronic appliances on plastic. *Nature*, **428**, 911–918.

12 Landfester, K., Montenegro, R., Scherf, U., Güntner, R., Asawapirom, U., Patil, S., Neher, D., and Kietzke, T. (2002) Semiconducting polymer nanospheres in aqueous dispersion prepared by a miniemulsion process. *Adv. Mater.*, **14**, 651–655.

13 Landfester, K. (2001) Polyreactions in miniemulsions. *Macromol. Rapid Commun.*, **22**, 896–936.

14 Landfester, K. (2001) The generation of nanoparticles in miniemulsions. *Adv. Mater.*, **13**, 765–768.

15 Landfester, K. and Antonietti, M. (2004) Miniemulsions for the convenient synthesis of organic and inorganic nanoparticles and "single molecule" applications in materials chemistry, in *Colloids and Colloid Assemblies* (ed. F. Caruso), Wiley-VCH Verlag GmbH, Weinheim, pp. 175–215.

16 Landfester, K., Bechthold, N., Tiarks, F., and Antonietti, M. (1999) Miniemulsion polymerization with cationic and nonionic surfactants: a very efficient use of surfactants for heterophase polymerization. *Macromolecules*, **32**, 2679–2683.

17 Scherf, U. (1999) Ladder-type materials. *J. Mater. Chem.*, **9**, 1853–1864.

18 Liddle, J.A., Cui, Y., and Alivisatos, P. (2004) Lithographically directed self-assembly of nanostructures. *J. Vac. Sci. Technol. B*, **22** (6), 3409–3414.

19 Mathur, A., Brown, A.-D., and Erlebacher, J. (2006) Self-ordering of colloidal particles on shallow nanoscale surface corrugations. *Langmuir*, **22**, 582–589.

20 Aizenberg, J., Braun, P.V., and Wiltzius, P. (2000) Patterned colloidal deposition controlled by electrostatic and capillary forces. *Phys. Rev. Lett.*, **84**, 2997–3000.

21 Jacobs, H.O., Campbell, S.A., and Steward, M.G. (2002) Approaching nanoxerography: the use of electrostatic forces to position nanoparticles with a 100nm scale resolution. *Adv. Mater.*, **14**, 1553–1557.

22 Zheng, H., Lee, I., Rubner, M.F., and Hammond, P.T. (2002) Two component particle arrays on patterned polyelectrolyte multilayer templates. *Adv. Mater.*, **14**, 569–572.

23 Kralchevsky, P.A. and Nagayama, K. (2000) Capillary interactions between particles bound to interfaces, liquid films and biomembranes. *Adv. Colloid Interf. Sci.*, **85**, 145–192.

24 Paunov, V.N., Kralchevsky, P.A., Denkov, N.D., and Nagayama, K. (1993) Lateral capillary forces between floating submillimeter particles. *J. Colloid Interf. Sci.*, **157**, 100–112.

25 Kralchevsky, P.A., Paunov, V.N., Ivanov, I.B., and Nagayama, K. (1992) Capillary meniscus interaction between colloidal particles attached to a liquid-fluid interface. *J. Colloid Interf. Sci.*, **151**, 79–94.

26 Kralchevsky, P.A. and Nagayama, K. (1994) Capillary forces between colloidal particles. *Langmuir*, **10**, 23–36.

27 Paunov, V.N., Kralchevsky, P.A., Denkov, N.D., Ivanov, I.B., and Nagayama, K. (1992) Capillary meniscus interaction between a microparticle and a wall. *Coll. Surf.*, **67**, 119–138.

28 Denkov, N.D., Velev, O.D., Kralchevsky, P.A., Ivanov, I.B., Yoshimura, H., and Nagayama, K. (1992) Mechanism of formation of two-dimensional crystals from latex particles on substrates. *Langmuir*, **8**, 3183–3190.

29 Xia, Y., Gates, B., Yin, Y., and Lu, Y. (2000) Monodispersed colloidal spheres: old materials with new applications. *Adv. Mater.*, **12**, 693–713.

30 Xia, Y., Yin, Y., Lu, Y., and McLellan, J. (2003) Template-assisted self-assembly of spherical colloids into complex and controllable structures. *Adv. Funct. Mater.*, **13**, 907–918.

31 Hulteen, J.C., Treichel, D.A., Smith, M.T., Duval, M.L., Jensen, T.R., and Van Duyne, R.P. (1999) Nanosphere lithography: size-tunable silver nanoparticle and surface cluster arrays. *J. Phys. Chem. B*, **103**, 3854–3863.

32 Deutsch, M., Vlasov, Y.A., and Norris, D.J. (2000) Conjugated-polymer photonic crystals. *Adv. Mater.*, **12**, 1176–1180.

33 Zoltan, S.L. (1974) Pulse Drop Ejction System, US Patent No. 3,857,049.

34 Yoshioka, Y. and Jabbour, G.E. (2005) Nonlithographic patterning: application of inkjet printing in organic based devices, in *Nanolithography and Patterning Techniques in Microelectronics* (ed. D. Bucknall), Woodhead Publishing Limited, Cambridge, pp. 349–372.

35 Lim, T., Han, S., Chung, J., Chung, J.T., Ko, S., and Grigoropoulos, C.P. (2009) Experimental study on spreading and evaporation of inkjet printed pico-liter droplet on a heated substrate. *Int. J. Heat Mass Transf.*, **52**, 431–441.

36 Dong, H., Carr, W.W., and Bucknall, D.G. (2007) Temporally resolved inkjet drop impaction on surfaces. *AICHE J.*, **53**, 2606–2017.

37 Popov, Y.O. (2005) Evaporative deposition patterns: spatial dimensions of the deposit. *Phys. Rev. E*, **71**, 036313-1–036313-17.

38 Deegan, R.D., Bakajin, O., Dupont, T.F., Huber, G., Nagel, S.R., and Witten, T.A. (1997) Capillary flow as the cause of ring stains from dried liquid drops. *Nature*, **389**, 827–829.

39 Deegan, R.D., Bakajin, O., Dupont, T.F., Huber, G., Nagel, S.R., and Witten, T.A. (2000) Contact line deposits in an evaporating drop. *Phys. Rev. E.*, **62**, 756–765.

40 de Gans, B.-J. and Schubert, U.S. (2004) Inkjet printing of well-defined polymer dots and arrays. *Langmuir*, **20**, 7789–7793.

41 Kim, E., Xia, Y., and Whitesides, G.M. (1995) Polymer microstructures formed by moulding in capillaries. *Nature*, **376**, 581–584.

42 e.g. "Sylgard 184" by Dow Corning.

43 Kim, E., Xia, Y., and Whitesides, G.M. (1996) Micromolding in capillaries: applications in material science. *J. Am. Chem. Soc.*, **118**, 5722–5731.

44 Rogers, J.A., Bao, Z., and Raju, V.R. (1998) Nonphotolithographic fabrication of organic transistors with micron feature sizes. *Appl. Phys. Lett.*, **72**, 2716–2718.

45 Blümel, A., Klug, A., Eder, S., Scherf, U., Moderegger, E., and List, E.J.W. (2007) Micromolding in capillaries and microtransfer printing of silver nanoparticles as soft-lithographic approach for the fabrication of source/drain electrodes in organic field-effect transistors. *Org. Electron.*, **8**, 389–395.

46 Zhao, X.-M., Stoddart, A., Smith, S.P., Kim, E., Xia, Y., Prentiss, M., and Whitesides, G.M. (1996) Fabrication of single-mode polymeric waveguides using micromolding in capillaries. *Adv. Mater.*, **8**, 420–424.

47 Tang, C.W. and VanSlyke, S.A. (1987) Organic electroluminescent diodes. *Appl. Phys. Lett.*, **51**, 913–915.

48 Wenzl, F.P., Pachler, P., List, E.J.W., Somitsch, D., Knoll, P., Patil, S., Guentner, R., Scherf, U., and Leising, G. (2002) Self-absorption effects in a LEC with low Stokes shift. *Physica E*, **13**, 1251–1254.

49 Fisslthaler, E., Sax, S., Scherf, U., Mauthner, G., Moderegger, E., Landfester, K., and List, E.J.W. (2008) Inkjet printed polymer light emitting devices fabricated by thermal embedding of semiconducting polymer nanospheres in an inert matrix. *Appl. Phys. Lett.*, **92**, 183305. (1–3).

50 Piok, T., Gamerith, S., Gadermaier, C., Plank, H., Wenzl, F.P., Patil, S., Montenegro, R., Kietzke, T., Nehrer, D., Scherf, U., Landfester, K., and List, E.J.W. (2003) Organic light-emitting devices fabricated from semiconducting nanospheres. *Adv. Mater.*, **15**, 800–804.

51 Piok, T., Gadermaier, C., Wenzl, F.P., Patil, S., Montenegro, R., Landfester, K., Lanzani, G., Cerullo, G., Scherf, U., and List, E.J.W. (2004) The photophysics of organic semiconducting nanospheres: a comprehensive study. *Chem. Phys. Lett.*, **389**, 7–13.

52 Piok, T., Plank, H., Mauthner, G., Gamerith, S., Gadermaier, C., Wenzl, F.P., Patil, S., Montenegro, R., Bouguettaya, M., Reynolds, J.R., Scherf, U., Landfester, K., and List, E.J.W. (2005) Solution processed conjugated polymer multilayer structures of light emitting devices. *Jpn. J. Appl. Phys.*, **44**, 479–484.

53 Mauthner, G., Landfester, K., Köck, A., Brückl, H., Kast, M., Stepper, C., and List, E.J.W. (2008) Inkjet printed surface cell light-emitting devices from a water-based polymer dispersion. *Org. Electron.*, **9**, 164–174.

54 Kietzke, T., Nehrer, D., Landfester, K., Montenegro, R., Güntner, R., and Scherf, U. (2003) Novel approaches to polymer blends based on polymer nanoparticles. *Nat. Mat.*, **2**, 408–412.

55 Kietzke, T., Nehrer, D., Kumke, M., Montenegro, R., Landfester, K., and Scherf, U. (2004) A nanoparticle approach to control the phase separation in polyfluorene photovoltaic devices. *Macromolecules*, **37**, 4882–4890.

56 Mun, R.P., Byars, J.A., and Boger, D.V. (1998) The effect of polymer concentration and molecular weight on the breakup of laminar capillary jets. *J. Nonnewtonian Fluid Mech.*, **74**, 285–297.

57 Xia, Y. and Friend, R.H. (2007) Nonlithographic patterning through inkjet printing via holes. *Appl. Phys. Lett.*, **90**, 253513. (1–3).

58 Fisslthaler, E., Blümel, A., Landfester, K., Scherf, U., and List, E.J.W. (2008) Printing functional nanostructures: a novel route towards nanostructuring of organic electronic devices via soft embossing, inkjet printing and colloidal self assembly of semiconducting polymer nanospheres. *Soft Matter*, **4**, 2448–2453.

59 Xia, Y. and Whitesides, G.M. (1998) Soft lithography. *Angew. Chem. Int. Ed. Eng.*, **37**, 550–575.

60 Barbero, D.R., Saifullah, M.S.M., Hoffmann, P., Mathieu, H.J., Anderson, D., Jones, G.A.C.,

Welland, M.E., and Steiner, U. (2007) High resolution nanoimprinting with a robust and reusable polymer mold. *Adv. Funct. Mater.*, **17**, 2419–2425.

61 Teichroeb, J.H. and Forrest, J.A. (2003) Direct imaging of nanoparticle embedding into PS films. *Mater. Res. Soc. Symp. Proc.*, **734**, B3.2. 75-80.

62 Sele, C.W., von Werne, T., Friend, R.H., and Sirringhaus, H. (2005) Lithography-free, self-aligned inkjet printing with sub-hundred-nanometer resolution. *Adv. Mater.*, **17**, 997–1001.

63 Sigma Aldrich, average M_w ∼350 000.

64 www.microfab.com.

65 Li, H., Zhu, J.-H., Shi, J.-X., and He, P.-S. (2006) Ordered arrangement of micro-beads in single line by micromolding in capillaries. *Chin. J. Chem. Phys.*, **19**, 352–354.

66 Kim, E. and Whitesides, G.M. (1997) Imbibition and flow of wetting liquids in noncircular capillaries. *J. Phys. Chem. B*, **101**, 855–863.

67 Dong, M. and Chatzis, I. (2004) An experimental investigation of retention of liquids in corners of a square capillary. *J. Colloid Interf. Sci.*, **273**, 306–312.

68 Kim, E., Xia, Y., and Whitesides, G.M. (1996) Two- and three-dimensional crystallization of polymeric microspheres by micromolding in capillaries. *Adv. Mater.*, **8**, 245–247.

12
Photolithographic Patterning of Organic Electronic Materials

John DeFranco, Alex Zakhidov, Jin-Kyun Lee, Priscilla Taylor, Hon Hang Fong, Margarita Chatzichristidi, Ha Soo Hwang, Christopher Ober, and George Malliaras

12.1
Introduction

Over the course of the development of the field of organic electronics [1], patterning has played an increasingly important role. Early work was focused on material properties, such as charge transport [2], charge injection, luminous efficiency, and film growth [3]. All these parameters are still very important, especially in the new materials that are continually being developed. A new emphasis has been placed, however, on device architecture and manufacturability on greater-than-lab scales. The move from single organic light emitting diode (OLED) devices [4] to pixelated displays, from single, large organic thin-film transistors [5] (OTFT) to circuits and display backplanes [6], and from small organic photovoltaic devices [7] (OPV) to large-area, fault-tolerant architectures has necessitated advances in patterning numerous layers, with both organic and inorganic material compositions.

While a variety of methods for patterning these organic materials have been developed, there has been a steadily growing interest in using methods that are already well established in the semiconductor manufacturing industry, namely, photolithography. Several reviews have covered organic patterning methods previously [8, 9]. Although some have mentioned methods of photolithography applied to organic electronics as a side note, none to date has attempted a focused review. This chapter will cover some of the requirements and popular methods of patterning organic electronic materials, giving special attention to the use of photolithography in patterning a range of organic electronic devices.

12.1.1
Patterning Requirements in Organic Electronics

OFTFs use patterning in many steps of fabrication. The most obvious and important use of patterning is in defining source and drain contacts and the transistor channel. Channel length (L) and width (W), or more specifically the ratio of the two, helps

Functional Supramolecular Architectures. Edited by Paolo Samorì and Franco Cacialli
Copyright © 2011 WILEY-VCH Verlag GmbH & Co. KGaA, Weinheim
ISBN: 978-3-527-32611-2

determine the drain current of a transistor. The better the patterning technique can decrease L while maintaining a reasonable width (given the space requirements) the better performing the transistor will be. The natural limit of this scaling is set by the contact resistance since, ohmic contacts might become injection limited at small channel lengths [10]. However, once a proper metal/organic semiconductor interface is chosen this length may be as small as 30 nm [11].

The source–drain electrodes are not the only layer requiring patterning in organic TFTs. Preventing leakage between source and drain through ungated material, or even between devices, requires that the active layer is patterned and material is removed from regions far from the channel. Patterning the organic semiconductor typically leads to the enhancement of on–off ratio, as parasitic current is minimized when no gate voltage is applied.

The need for patterning in OLEDs for use in displays and signage is primarily to define the individual pixels and subpixels. The resolution requirements for display pixels vary widely, depending on the screen size, resolution, and subpixel architecture. Television displays might have a pixel pitch of 80 μm, with subpixels around one third of that. OLED microdisplays [12] often have ~12 μm pixel pitches in order to get full-screen resolution into a sub- 1-inch diagonal display (Figure 12.1). Displays meant for signage naturally have much large pixels than either of these.

The need for patterning is greatly reduced in the case of lighting and photovoltaic devices. Such devices are meant to cover large areas and do not require pixelation. Some sort of segmentation might be required, however, in order to give

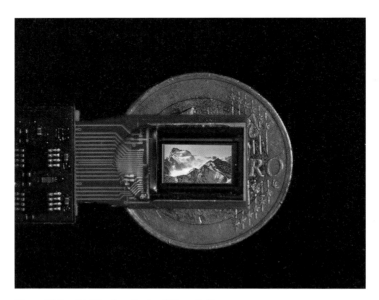

Figure 12.1 OLED microdisplay (Microoled).

the systems fault tolerance for defects in the film and to tune the input/output characteristics.

There are a variety of organic electronics-based sensors [13], from OLED-based devices that utilize the absorption spectra of target molecules to conducting polymer-based sensors, which detect changes in doping level from an analyte in solution [14]. Since the architectures are similar to other organic electronic devices, the pattern needed can be much the same. The levels of integration with other elements, such as microfluidic channels, can make patterning decisions more crucial [15].

12.1.2
Common Methods of Patterning Organic Electronic Materials

12.1.2.1 Solution-Deposited Materials

One of the advantages of solution-processed materials, whether polymer or small molecule, is the ability to use drop-on-demand delivery to allow simultaneous deposition and patterning of active layers. Ink-jet printing (IJP) of organic TFTs has been demonstrated where the source–drain electrodes and active region have been patterned or where all layers, including the gate and gate dielectric, are printed with ink-jet [16].

Ink-jet printing is also of importance to the display industry [17]. All the necessary layers needed for an OLED stack can be printed, making it possible to produce three-color active matrix displays using only ink-jet, and in fact some companies such as Cambridge Display Technologies are pursuing that route commercially. For ink-jet printing to work, specially formulated inks must be made from the various solution-processable polymer and small-molecule electronic materials available. It takes a lot of effort to produce inks that have the proper rheological properties to print and still have good performance as organic semiconductors. Lifetimes of printable OLED materials are still lagging behind vacuum-deposited same-color equivalents, but are catching up, and some printable small molecules are there already.

A wide variety of printing and coating methods that have analogues in graphic arts printing have been modified for use with organic electronic materials [18]. The speed at which the roll-to-roll printers pattern inks is no doubt the biggest motivator in using printing for electronics manufacturing. Techniques such as gravure printing [19], for example, can coat at $1\,\text{m}\,\text{s}^{-1}$ or more. High throughput would undoubtedly lead to low-cost devices provided the functionality remains adequate.

Laser patterning can be used with organic materials, subtractively as an ablation tool [20–22], as a means of changing the chemical properties of the deposited film or additively. The second method can take different forms. Abdou *et al.* [23] used a laser to cross-link regions of a P3HT film, rendering it insoluble to the chloroform/acetone solvent mixture used to remove the unexposed regions. The film was then doped with nitrosonium tetrafluoroborate or ferric chloride, creating conducting lines. The exposed region of organic material may also be rendered nonconducting through laser-induced photobleaching [24], for transistor isolation. Additive laser patterning, where material is moved from a donor film to the substrate through local laser heating, is currently a top candidate for large OLED displays. The laser-induced thermal imaging (LITI) process consists of laminating a solution-coatable transfer sheet to a substrate, imaging the patterns with a laser to transfer the material, and then peeling off the sheet from the substrate. Multiple layers can be simultaneously transferred, and the materials can be polymers or small molecules. Blanchet et al. [25] used this technique to pattern conducting polymers for organic TFTs (Figure 12.2). Other similar techniques that do not require direct lamination include radiation-induced sublimation transfer (RIST), which uses an 810-nm laser to locally heat the material, and laser-induced local transfer (LIST), in which an IR absorber is heated with a 1064-nm laser.

Physical pattern transfer from a stamp is another major area of micropatterning that may be applied to organic electronic patterning. Using a hard mask to transfer a pattern into resist or active layer is known as nanoimprint lithography [26]. The mask can be any hard material – though usually silicon or glass – that is patterned with lithographic methods, including photolithography and ebeam lithography. The latter allows resolutions less than 100 nm that are faithfully reproduced in the underlying film, making this technique one of the highest resolution possible. Kao et al. applied this method [27] to make OLED devices, demonstrating not high resolution but rather reproducible, fairly large-area patterning.

The more common approach to stamping in organic electronics is the use of soft elastomeric stamps to transfer materials onto a substrate (an excellent overview can

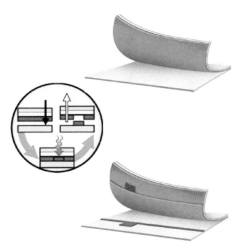

Figure 12.2 Illustration of the printing process. The two flexible films, a multilayer donor and a receiver, are held together by vacuum. The laser beam is focused on a thin absorbing layer that converts light into heat, an optional ejection layer placed directly underneath, and a DNNSA–PANI/SWNT conducting layer coated on top. The heat generated at the metal interface decomposes the surrounding organic material creating a gas bubble that while expanding propels the conducting layer to the receiver. After imaging is completed, the donor and receiver films are separated (from Ref. [25]).

be found in Ref. [8]). Inorganic materials [28], resist [29–31], and organic materials may be patterned with these techniques. Examples of the latter include patterning light-emitting materials [32, 33] and photonic crystals for making organic lasers [34]. Another important use is in depositing electrodes onto an elastomeric stamp and then laminating the stamp on organic devices, thus avoiding damage from metal deposition [35], and make small, high-performance transistors [36]. It is also possible to laminate two organic layers together, to form interfaces that would otherwise be difficult or impossible to fabricate [37].

12.1.2.2 Vapor-Deposited Materials

The simplest and lowest cost method to pattern vapor-deposited small molecules on a small scale is the use of a shadow mask. The method involves etching holes in a thin piece of metal, which is used to selectively block the passage of molecules during the deposition process. Shadow masking is used extensively in industry to make OLED displays for cell phones on substrates up to $\sim$1 m (gen 4.5). It may be difficult to scale the method much beyond this, however, because of the difficulty of working with large, thin sheets of metal. In the first place, sagging becomes an issue at these larger sizes, unless substrates are held vertically (which most now are). It is also a problem that dust can break off and contaminate the substrate if the mask is allowed to flex too much. Furthermore, handling and cleaning of the masks can be expensive and cumbersome at larger sizes. Shadow masking does have the benefit of producing the highest performing OLED devices because all deposition can be done while in vacuum and no chemicals touch the organic materials before encapsulation. Shadow masks are widely acknowledged as a stopgap technology that works now but will need to be replaced if the industry is to reach larger substrates and production volumes. In principle, deposition through shadow mask (also sometimes referred to as stencil lithography) can lead to very high resolution and give features down to a 10-nm scale [38]; however, commercial application so far is limited to low-resolution features due to problems mentioned above.

Organic vapor jet printing (OVJP) [39] was developed in order to bring the deposit-on-demand nature of ink-jet printing to materials that are normally deposited in vacuum. The technique is a printing version of a deposition technique called organic vapor-phase deposition [40] that use an inert carrier gas to create a steady, laminar flow of the deposited material (Figure 12.3). Besides enabling the patterning of small molecules with high resolutions, the carrier gas deposition method gives greater control over deposition conditions, including rate, morphology, and materials' interfaces, than sublimation. Improvements to materials performance with this method have been reported [41].

12.1.3
Photolithographic Patterning Techniques

Photolithography is the nearly ubiquitous method of patterning thin films that uses a layer that can be patterned by selective exposure of UV light, generally through a preformed photomask with a pattern on it made of chrome, and then developed,

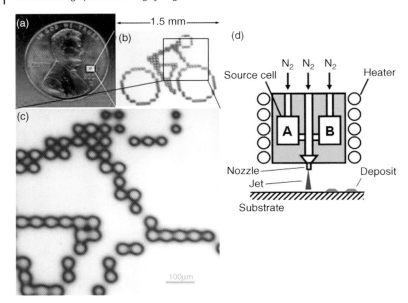

Figure 12.3 (a)–(c) A 24 × 32 pixel bitmap image of a cyclist figure printed by OVJP on silicon using Alq3, with $a = 20$ lm, $L = 100$ lm, $s = (20 \pm 10)$ lm, a dwell time of 2 s above each pixel location, and <0.2 s time interval for translation between each pixel. The coin is shown for size comparison. (d) Schematic of the OVJP apparatus, shown with two of the five source cells used in our experiments, a center dilution channel, and a modular collimating nozzle, all heated from the outside. A hot inert carrier gas enters the apparatus, picks up the organic vapor, and ejects the gas mixture through the nozzle. The collimated vapor jet impinges on a cooled substrate where the organic molecules selectively physisorb, forming a well-defined deposit (from Ref. [39]).

leaving either the exposed material (negative tone resist) or unexposed material (positive tone resist) behind to control the next processing step. Once patterned, resist can be used to shield material beneath it from an etchant, so that only uncovered materials are etched (subtractive patterning). Alternatively, the resist can be used as a contact mask to pattern a deposited layer (additive patterning). Figure 12.4 shows both processes step by step. Subtractive etching is far more common in silicon industry since fast, selective etchants are available for most materials and liftoff itself can leave residue. On the other hand, certain materials, platinum, for example, are usually patterned with a liftoff method.

There are potential advantages that photolithography may bring to organic electronics, which have driven the research in this area. It is an extremely high-resolution process, capable of producing features <32 nm (current manufacturing), which is improving almost yearly [42]. Through the development of large-scale LCD manufacturing [43], large-area substrates (>2 m on a side) can be processed and since it is a parallel process (many features exposed at one time), the time to pattern a large substrate is less than a minute, making throughput relatively high (though clearly not as high as the high-speed printing techniques). A key advantage is in overlay, where

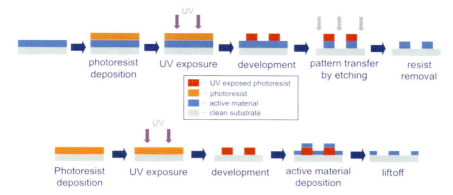

Figure 12.4 *Top*: Subtractive process where photoresist is spun on top of substrate, patterned with UV light and development steps, and then acts as a mask for an etch step. *Bottom*: Liftoff process where resist is patterned on the substrate, material is deposited on top of the resist layer, and the resist is removed leaving material where the resist had been previously removed.

one layer can be precisely and accurately (<0.1 μm) placed on another layer, a difficult task with many alternative techniques. Factors such as relative cost and performance compared to other competing patterning techniques will determine the extent to which photolithography may be adopted in the field. Sheats [44] has done an analysis of the costs associated with a variety of organic electronics manufacturing techniques, and further study of the issue is needed as all these methods improve over time.

One reason that photolithography is not widely used is the general difficulty of working with solvents and developers that tend to damage the active materials being patterned. There are many varieties of photoresist available, but diazonaphthoquinone-sensitized phenol formaldehyde resin (DNQ-novolak) is a very common formulation used for mercury arc lamp sourced photolith tools (wavelengths between 436 nm g-line and 365 nm i-line). DNQ-novolak is typically deposited from propylene glycol monomethyl ether acetate (PGMEA), a generally mild solvent that is nevertheless capable of damaging organic thin films. Most of the process damage, however, comes from the tetramethyl ammonium hydroxide (TMAH) developer. The developer is water based and a strong caustic, both very bad for semiconducting organic films. Water alone has been studied extensively as the cause of degradation in OLEDs [45, 46], OTFTs [47–49], and OPVs. The resist is removed after processing is over using stronger organic solvents than the one used for deposition; acetone is common. Most organic electronic materials do not survive the process (they dissolve, crack, delaminate, etc.), or perform very poorly afterward.

Photolithography has been used in device fabrication for patterning certain layers in OTFT devices, specifically metal for source–drain electrodes for bottom contact transistors. All the processing in that case takes place before the active organic material is introduced, thus avoiding solvent damage. There are some limitations to making devices this way; for instance, mobility and charge injection for bottom contact devices tend to suffer either because of lack of a clean interface or due to nonideal film formation conditions [50]. In certain cases, this can be overcome in clever ways [51]. But photolithography has not been applied as widely to patterning the active layer or subsequent layers.

12.2
Photolithographic Methods for Patterning Organic Materials

12.2.1
Hybrid Techniques

Photolithography has been used in combination with other techniques in order to mitigate their shortcomings, whether these are reproducibility or resolution and registration. These hybrid techniques usually use photolithography early in the process to introduce prepatterns, avoiding any possible chemical damage during the later deposition process.

Shadow masking has limitations, as was mentioned earlier, in high-resolution patterning (subpixels less than ~30 µm over large areas) and registration. This is a problem for high-resolution patterning of cathode layers on top of OLEDs for passive matrix displays. One way to improve this is to include an additional layer of patterned photoresist to act as an integrated mask [52].

Another hybrid technique involves the use of photolithographically patterned polymer films to modify the wetting properties of the substrate surface to enhance ink-jet printing. In order to overcome deficiencies in the achievable resolution of ink-jet printing, surface energy patterning was used to form channels for polymer TFT electrodes. Sirringhaus *et al.* used photolithographically patterned polyimide along with ink-jet printing of PEDOT:PSS to form OTFT channels with 5 µm channel length [53] (see Figure 12.5), far below what ordinary ink-jet printing would allow (>20 µm). Further improvements have pushed that length to 250 nm [54], and below [55], although stamping, and not lithography, was used in these cases.

12.2.2
Organic Materials Directly Processed with UV Light

Taking advantage of the chemical tunability of organic materials, direct patterning by exposure to UV has been used by many to create devices. The advantage of this method is that no other layers are needed to transfer the pattern into the material, potentially saving on materials costs.

Photoinduced cross-linking was used by Müller *et al.* [56] to make oxetane-functionalized light-emitting spirobifluorene-*co*-fluorene polymers that behave like negative photoresists. The specific mechanism involved a photoacid generator (PAG), specifically (4-[(2-hydroxytetradecyl)oxyl]phenyl)phenyliodonium hexafluorantimonate, that produces a proton when activated by UV light. The oxetane ring of the precursor is opened and this initiates the chain polymerization, cross-linking the material. Full-color displays [57] were demonstrated with this method, as shown in

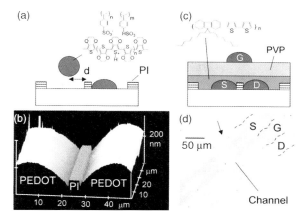

Figure 12.5 (a) Schematic diagram of high-resolution ink-jet printing on a prepatterned substrate. (b) AFM showing accurate alignment of ink-jet-printed PEDOT/PSS source and drain electrodes separated by a repelling polyimide (PI) line with $L = 5\,\mu m$. (c) Schematic diagram of the top-gate IJP TFT configuration with an F8T2 semiconducting layer (S, source; D, drain; and G, gate). (d) Optical micrograph of an IJP TFT ($L = 5\,\mu m$). The image was taken under crossed polarizers so that the TFT channel appears bright blue because of the uniaxial monodomain alignment of the F8T2 polymer on top of rubbed polyimide. Unpolarized background illumination is used to make the contrast in the remaining areas visible, where the F8T2 film is in an isotropic multidomain configuration. The arrow indicates pronounced roughness of the unconfined PEDOT boundary (from Ref. [50])

Figure 12.6. Similarly, hole transport layers (HTL) cross-linked with oxetane side-groups were developed by Bacher *et al.* [58], using bis(diarylamine) as the HTL.

Solomeshch *et al.* [59] extended this concept to use cross-linkable triarylamine photopolymers as binders for higher efficiency light-emitting materials, in this case PPV derivatives. Triarylamine was chosen because it was chemically compatible with the light-emitting polymer and had a higher bandgap to prevent quenching and charge trapping.

Another class of materials, liquid crystalline OLED materials, show similar cross-linking abilities and were studied by the Kelly group [60] and used to make a three-color display [61].

Another group to successfully utilize this concept was that of Afzali *et al.* [62] who fabricated photopatternable pentacene precursor for OTFT application.

The conducting polymer PEDOT has been used with a variety of patterning techniques, including photolithography. Selectively cross-linking PEDOT using a photoinitiator and UV light can turn PEDOT:PSS into a kind of negative tone photoresist, as was demonstrated by Touwslager *et al.* [63] in the fabrication of organic code generators. This is because the film is nearly insoluble once cast as a thin film. The common form of PEDOT:PSS, which is spin-coated from a water solution, is actually only a suspension. It can become delaminated during the development stage, however, and the film integrity may be compromised during a standard photolithographic process. Using cross-linking agents, the PEDOT layers, and in

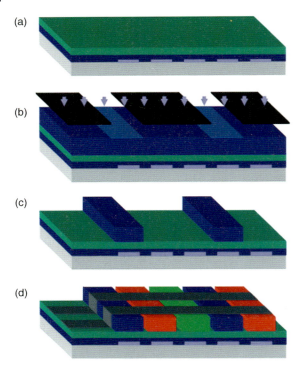

Figure 12.6 Schematic illustration of the direct lithography process. (a) On top of an ITO line structure, homogenous layers of PEDOT: PSS and a cross-linkable hole transport material are spin-coated. (b) Subsequently, a solution of the blue light-emitting polymer is deposited and exposed to UV light through an aligned shadow mask. (c) After a soft-curing step, the noncross-linked parts are dissolved in an organic solvent. Repeating this procedure for green and red results in parallel stripes of blue-, green-, and red-emitting polymers. (d) The device is completed by evaporating a metal cathode through a shadow mask (from Ref. [56]).

some cases the layers placed on top, can be made impervious to delamination during processing [64] (Figure 12.7).

12.2.3
Materials Choice or Modification

Certain organic electronic materials, some with structural modifications, can be used with standard photolithographic processes. Polymeric materials are usually more robust to chemical damage from photolithography.

An example of successful patterning using photolithography was demonstrated by Huang et al. [65]. The negative tone photoresist (resist is left in UV-exposed areas) SU-8, stacks of polymer OLED materials are selectively removed through a liftoff process during development of the resist, along with the unexposed resist (Figure 12.8). Since the developer used is PGMEA, a relatively mild organic solvent,

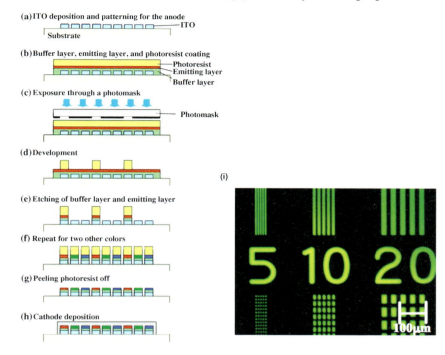

Figure 12.7 (a) Polymer OLEDs fabricated with photolithography. (b) Resolution achievable with the technique (from Ref. [64]).

little damage occurs and multiple layers can be patterned in this way. A downside to this method is that the SU-8 remains under the stack, which can diminish the optical transmission (reported at 80% in the visible spectrum) and block electrical connection to the substrate.

Careful choice of an active material can allow direct patterning with more common types of photoresist. In a very early example [66], metal is patterned on top of poly(2,5-dialkoxy-*p*-phenylene-vinylene) (PDAOPV), a PPV derivative. It was found in that experiment that the most damaging step was baking the light-emitting polymer, not dipping it in chemicals. Regioregular polythiophene (P3HT) has been successfully patterned directly with standard photoresist chemistry, using both liftoff [67] method and etching [68].

12.2.4
Patterning with an Interlayer

Perhaps the most popular way to avoid contact between the harsh chemicals used in photolithography and the active layer of an organic device is to use an inert, protective interlayer. The properties of such a layer must include a benign deposition process, low etch rates in the solvents used in photolithography, low permeability to those same solvents, and the ability to be selectively etched in another

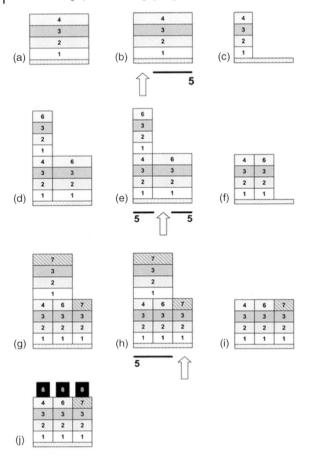

Figure 12.8 Schematic diagram representing the procedure for patterning multiple devices, incorporating different component materials, on a single substrate, for example, the red, green, and blue pixels of a color display. See main text for details. The numeric labels correspond to the following materials: (1) SU8 resist; (2) Baytron P; (3) Baytron P VP AI 4083; (4) the blue light-emitting polyfluorene; (5) a photomask; (6) the green light-emitting polyfluorene; (7) the red light-emitting polyfluorene; and (8) aluminum-capped LiF (from Ref. [65]).

way, usually by oxygen plasma RIE. Removal of the material after patterning is usually optional, as the protective layer may double as an encapsulant in many cases.

Parylene is often used in this kind of process. It is an inert polymer deposited through a room-temperature chemical vapor deposition process. The material itself is nearly impervious to chemical attack, showing no appreciable etch rate in most organic solvents or even strong acids. The conformal coating property of the film allows it to effectively encapsulate the entire substrate while further processing occurs, preventing lower layers from getting exposed to solvents and aqueous developers. It etches easily in oxygen plasma, which allows a facile patterning of lower organic layers with minimal undercutting.

Kymissis et al. [69] demonstrated subtractively patterned pentacene transistors using parylene and oxygen plasma. DeFranco et al. [70] showed a similar technique on a wider range of organic materials, using parylene to pattern electrodes made of PEDOT:PSS with narrow channels. Features as small as 1 μm were patterned on SiO_2 substrates.

Polyvinyl alcohol (PVA) is a spin-coatable water-soluble material that is also commonly used for this kind of process [71]. Zhou et al. [72] used photosensitized PVA to pattern pentacene TFTs for an active matrix flexible display. Other organic circuits have been made in a similar fashion [73].

Inorganic oxide layers have been used as hard etch masks as well, with the additional benefit of serving as dielectric layers for top-gated TFTs. Li et al. [74] used silicon nitride (SiN) deposited with plasma-enhanced chemical vapor deposition (PECVD) as a protective hard mask and gate dielectric on top of P3HT transistors. Chang et al. [75] demonstrated that atomic layer deposition (ALD) deposited aluminum oxide allowed photolithographic patterning of MEH-PPV-based OLEDs. In both cases, the layers were left intact and used for another purpose, which is necessary in such a scheme because of the inability to later remove the layer without damaging the rest of the device.

Metal can also be quite an effective encapsulant [76], and in the case of OLEDs, this metal layer can double as a top electrode. Tian et al. [77] used a thick metal layer (>1 μm) to protect a vacuum-deposited OLED from process conditions, allowing three side-by-side OLED pixels to be patterned even though harsh chemical etchants were used, including hydrofluoric acid. Lamprecht et al. [78] used a thinner (150 nm aluminum) metal barrier to protect an OLED/OPV stack from a photoresist-masked subtractive etch.

Materials that can be effectively removed from the substrate without the use of solvents can be used to lift off an organic material to pattern it. In this way, the technique shares some features with stamping; controlling the surface energy between different materials so that sacrificial layer releases from the substrate but the patterned organic material remains stuck to the surface. DeFranco et al. [70] demonstrated that the parylene layer shows poor adhesion to the surface of an oxide-coated silicon wafer and so can be peeled from that surface to allow mechanical dry liftoff of a variety of polymers and small molecules (Figure 12.9). Other materials besides parylene can also be used for this purpose; for example, Chang and Sirringhaus showed a similar delamination method with polyimide [79]. Patterning of polyimide does not require oxygen plasma etching since it dissolves in the developer used on the top lying photoresist. That developer, however, might damage lower layers, potentially limiting multilayer patterning. This of course is a limitation of most interlayer liftoff techniques because they seek to avoid damage by using all the damaging chemicals before the organic layer is deposited. If other layers are there first, then this strategy does not work as well.

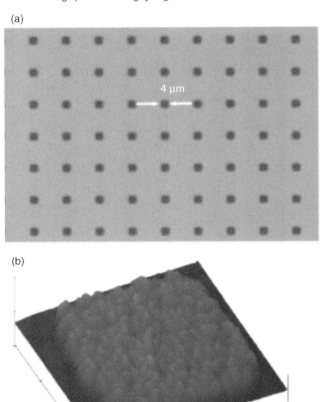

Figure 12.9 Four-micrometer pentacene squares patterned with parylene liftoff technique (from Ref. [70]).

12.2.5
Modification of the Resist Chemistry

One last method of protecting the organic layers from chemical damage from resist processing is to change the chemicals used in the photolithography process. Two recent examples of this approach are resist materials processed with supercritical fluids and those processed with highly fluorous solvents. In order to be completely benign to the layers being patterned, all the harmful chemical steps must be eliminated, including deposition, development, and stripping (lift off), since each contributes to damage. Once this is accomplished, a truly simple method of patterning is achieved so that organic electronic materials can be treated in much the same way as any inorganic material.

Super critical carbon dioxide (scCO$_2$) is a phase of carbon dioxide raised to a pressure and temperature past its critical temperature and pressure ($T_c = 31.1\,°C$; $P_c = 73.8\,bar$). The scCO$_2$ has a high diffusivity and zero surface tension (which is important for both high aspect ratio features and extremely small features), and has proven to be an excellent solvent in dry lithography for high-resolution nanofabrication. Another very important aspect is that scCO$_2$ is a nontoxic, noninflammable, environment-friendly fluid, which is proposed as a key driver in the new field of "green" chemistry [80]. In recent years, scCO$_2$ is being considered as a low-cost substitute solvent in microelectronics processing such as wafer cleaning, spin coating, etching, and resist stripping, in view of its physical and chemical advantages [81]. Besides these general advantages, the fact that scCO$_2$ is a poor solvent for most ionic, high molecular weight, and low vapor pressure organic compounds makes it an extremely promising and nondestructive development medium for the vast majority of materials utilized in organic electronics. Essential to this patterning approach is the ability to develop the photoresist without harming the active organic material. On the basis of the high affinity of scCO$_2$ for fluorocarbons, it has been demonstrated that appropriate statistical copolymers with high fluorine content are effective negative photoresists for development with scCO$_2$. Another important advantage of the fluorinated polymers is their hydrophobicity and oleophobicity, that is, they repel and/or resist both water and most organic solvents. Moreover, for those types of materials one can use a high fluorine content solvent (trifluorotoluene, hydrofluoroethers, etc.) with essentially the same properties.

Similar to scCO$_2$, highly fluorinated liquids, in general, do not interact with nonfluorinated materials. It is thus anticipated that these fluids will extend the options for solvent orthogonality. Among the various fluorinated solvents, hydrofluoroethers (HFEs) are particularly attractive because they are environment-friendly, nontoxic, nonflammable, and nonozone-depleting solvents [82].

To establish HFEs as universal processing media, a possible deterioration of the performance of organic electronic devices in these solvents was investigated by Zakhidov et al. [83]. Poly-3-hexylthiophene (P3HT), a prototypical conjugated polymer soluble in nonpolar organic solvents, did not show any performance variation in terms of field-effect mobility after extensive exposure to HFEs. The benign nature of HFEs was further demonstrated in an experiment in which OLEDs prepared with poly(dioctylfluorene) and [Ru(bpy)$_3$]$^{2+}$(PF$_6^-$)$_2$ were operated in boiling HFE. Even boiling for an hour in that solvent did not cause any substantial change in device brightness and efficiency. In addition, optical and atomic force microscopy was conducted on a variety of polymeric electronic materials before and after immersion in boiling HFE-7100. No significant change in morphology, no cracking, or no delamination was observed, confirming the orthogonality of HFEs even under extreme conditions. Figure 12.10 shows a light-emitting device lit up in a boiling beaker of the HFE solvent.

Lee and coworkers [84] then synthesized a fluorinated solvent-compatible photoresist system, based on a chemically amplified resorcinarene molecular glass resist.

Figure 12.10 Photograph of a $[Ru(bpy)_3]^{2+}(PF_6)_2$ electroluminescent device operating in boiling HFE-7100 (from Ref. [83]).

The resist was used to pattern polymer and small-molecule organic semiconductors through both additive and subtractive methods. Multilayer patterning was also accomplished using this method (as is shown in Figure 12.11), through successive liftoff steps of polyfluorene and ionic complex $[Ru(bpy)_3]^{2+}(PF_6^-)_2$. Resist was also used to create micropatterned high-voltage solar battery consisting of an array of 300 solar cells in series, with a period of 50 μm and a yield of open-circuit voltage of 90 V [85]. A polymeric, nonchemically amplified resist was subsequently made by Taylor *et al.* [86] and used to pattern PEDOT:PSS for small-channel OTFTs.

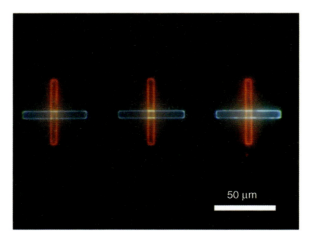

Figure 12.11 Fluorescent microscope image of overlaid patterns (feature width 50 μm) of poly(9,9-dioctylfluorene) and $[Ru(bpy)_3]^{2+}(PF_6)_2$ (from Ref. [84]).

The advantage of this approach is its general applicability to a wide range of materials. It is unusual for a patterning method to be used equally with both small molecules and polymers (not to mention inorganic materials such as metals), and this fact opens up many possibilities of hybrid devices.

12.3
Conclusions and General Considerations

Continued improvement in the technology has moved organic electronics closer to a commercial reality. The technique that will play a role in helping make it happen depends on many complex technological and market factors. Photolithography has many potential advantages over rival patterning schemes, not the least of which is that it uses manufacturing equipment that already exists at high volumes in established industries such as LCD manufacturing. The potential reduction in start-up costs could give organic electronics a way to scale up volumes cheaply and to eventually transition to higher throughput alternatives (to lower costs further) or more flexible alternatives (to enable custom electronics manufacturing) once popular acceptance is achieved. The best opportunities for this kind of scenario is in a market where organic electronic devices are seen as clearly advantageous over other technologies, such as in OLED displays, or flexible electronics-based products.

Further advancements in organics research may also be aided by these lithographic techniques. Leveraging the high resolutions to explore smaller and smaller material domains and interfaces is a rich avenue of research. It is especially important to use techniques that allow a wide range of materials to be patterned, which is the case in many of the reviewed methods, in order to broaden the scope of study. Indeed, the goal of all patterning technologies should be to make the processing of materials the easiest part of device design, putting the emphasis back on materials development, where it began.

References

1 Malliaras, G.G. and Friend, R.H. (2005) An organic electronics primer. *Phys. Today*, **58**, 53–58.
2 Coropceanu, V., Cornil, J., da Silva Filho, D.A., Olivier, Y., Silbey, R., and Bredas, J.L. (2007) Charge transport in organic semiconductors. *Chem. Rev.*, **107** (4), 926–952.
3 Ruiz, R., Choudhary, D., Nickel, B., Toccoli, T., Chang, K.C., Mayer, A.C., Clancy, P., Blakely, J.M., Headrick, R.L., Iannotta, S., and Malliaras, G.G. (2004) Pentacene thin film growth. *Chem. Mater.*, **16**, 4497–4508.
4 Tang, C. and Van Slyke, S.A. (1987) Organic electroluminescent diodes. *Appl. Phys. Lett.*, **51** (12), 913.
5 Dimitrakopoulos, C.D. and Malenfant, P.R.L. (2002) Organic thin film transistors for large area electronics. *Adv. Mater.*, **14** (2), 99.
6 Jackson, T.N., Lin, Y.Y., Gundlach, D.J., and Klauk, H. (1998) Organic thin-film transistors for organic light-emitting flat-panel display backplanes. *IEEE J. Sel. Top. Quant. Electron*, **4** (1), 100–104.
7 Tang, C. (1986) Two-layer organic photovoltaic cell. *Appl. Phys. Lett.*, **48**, 183.

8 Menard, E., Meitl, M.A., Sun, Y., Park, J.-U., Shir, D.J.-L., Nam, Y.-S., Jeon, S., and Rogers, J.A. (2007) Micro- and nanopatterning techniques for organic electronic and optoelectronic systems. *Chem. Rev.*, **107**, 1117–1160.

9 Forrest, S.R. (2004) The path to ubiquitous and low-cost organic electronic appliances on plastic. *Nature*, **428** (6986), 911–918.

10 Shen, Y.L., Hosseini, A.R., Wong, M.H., and Malliaras, G.G. (2004) How to make ohmic contacts to organic semiconductors. *Chemphyschem*, **5**, 16–25.

11 Zhang, Y., Petta, J.T., Ralph, D., and Malliaras, G.G. (2003) 30nm channel length pentacene transistors. *Adv. Mater.*, **15**, 1632.

12 Howard, W.E. and Prache, O.F. (2001) Microdisplays based upon organic light-emitting diodes. *IBM J. Res. Dev.*, **45** (1), 115–127.

13 Mabeck, J.T. and Malliaras, G.G. (2006) Chemical and biological sensors based on organic thin-film transistors. *Anal. Bioanal. Chem.*, **384**, 343–353.

14 Berggren, M. and Richter-Dahlfors, A. (2007) Organic bioelectronics. *Adv. Mater.*, **19** (20), 3201–3213.

15 Yang, S.Y., DeFranco, J.A., Sylvester, Y.A., Gobert, T.J., Macaya, D.J., Owens, R.M., and Malliaras, G.G. (2009) Integration of a surface-directed microfluidic system with an organic electrochemical transistor array for multi-analyte biosensors. *Lab Chip*, **9** (5), 704–708.

16 Knobloch, A., Manuelli, A., Bernds, A., and Clemens, W. (2004) Fully printed integrated circuits from solution processable polymers. *J. Appl. Phys.*, **96** (4), 2286–2291.

17 de Gans, B.J., Duineveld, P.C., and Schubert, U.S. (2004) Inkjet printing of polymers: state of the art and future developments. *Adv. Mater.*, **16** (3), 203–213.

18 Krebs, F.C. (2009) Fabrication and processing of polymer solar cells: a review of printing and coating techniques. *Sol. Energ. Mat. Sol. C*, **93**, 394–412.

19 Kapur, N. (2003) A parametric study of direct gravure coating. *Chem. Eng. Sci.*, **58**, 2875–2882.

20 Wong, T.K.S., Gao, S., Hu, X., Liu, H., Chan, Y.C., and Lam, Y.L. (1998) Patterning of poly(3-alkylthiophene) thin films by direct-write ultraviolet laser lithography. *Mater. Sci. Eng. B*, **55**, 71–78.

21 Yagi, I., Tsukagoshi, K., and Aoyagi, Y. (2004) Direct observation of contact and channel resistance in pentacene four-terminal thin-film transistor patterned by laser ablation method. *Appl. Phys. Lett.*, **84**, 813.

22 Schrodner, M., Stohn, R.I., Schultheis, K., Sensfuss, S., and Roth, H.K. (2005) *Org. Elec.*, **6**, 161.

23 Abdou, M.S.A., Zi, W.X., Leung, A.M., and Holdcroft, S. (1992) Laser, direct-write microlithography of soluble polythiophenes. *Synth. Met.*, **52**, 159.

24 Itoh, E., Torres, I., Hayden, C., and Taylor, D.M. (2006) Excimer-laser micropatterned photobleaching as a means of isolating polymer electronic devices. *Synth. Met.*, **156**, 129.

25 Blanchet, G.B., Loo, Y.L., Rogers, J.A., Gao, F., and Fincher, C.R. (2003) Large area, high resolution, dry printing of conducting polymers for organic electronics. *Appl. Phys. Lett.*, **82**, 463.

26 Clavijo Cedeño, C., Seekamp, J., Kam, A.P., Hoffmann, T., Zankovych, S., Sotomayor Torres, C.M., Menozzi, C., Cavallini, M., Murgia, M., and Ruani, G. (2002) Nanoimprint lithography for organic electronics. *Microelectron. Eng.*, **61**, 25–31.

27 Kao, P.C., Chu, S.Y., Chen, T.Y., Zhan, C.Y., Hong, F.C., Chang, C.Y., Hsu, L.C., Liao, W.C., and Hon, M.H. (2005) Fabrication of large-scaled organic light emitting devices on the flexible substrates using low-pressure imprinting lithography. *IEEE Trans. Electron Dev.*, **52**, 1722.

28 Meitl, M.A., Zhu, Z.T., Kumar, V., Lee, K.J., Feng, X., Huang, Y.Y., Adesida, I., Nuzzo, R.G., and Rogers, J.A. (2006) Transfer printing by kinetic control of adhesion to an elastomeric stamp. *Nat. Mater.*, **5**, 33–38.

29 Gates, B.D. and Whitesides, G.M. (2003) Replication of vertical features smaller

than 2nm by soft lithography. *J. Am. Chem. Soc.*, **125** (49), 14986–14987.
30. Bailey, T.C., Johnson, S.C., Sreenivasan, S.V., Ekerdt, J.G., Willson, C.G., and Resnick, D.J. (2002) Step and flash imprint lithography: an efficient nanoscale printing technology. *J. Photopolym. Sci. Technol.*, **15**, 481.
31. Khang, D.Y., Kang, H., Kim, T., and Lee, H.H. (2004) Low-pressure nanoimprint lithography. *Nano Lett.*, **4**, 633.
32. Rogers, J.A., Bao, Z.N., and Dhar, L. (1998) Fabrication of patterned electroluminescent polymers that emit in geometries with feature sizes into the submicron range. *Appl. Phys. Lett.*, **73**, 294.
33. Cheng, X., Hong, Y.T., Kanicki, J., and Guo, L.J. (2002) High-resolution organic polymer light-emitting pixels fabricated by imprinting technique. *J. Vac. Sci. Technol. B*, **20**, 2877.
34. Schueller, O.J.A., Whitesides, G.M., Rogers, J.A., Meier, M., and Dodabalapur, A. (1999) Fabrication of photonic crystal lasers by nanomolding of solgel glasses. *Appl. Opt.*, **38**, 5799.
35. Bernards, D.A., Biegala, T., Samuels, Z.A., Slinker, J.D., Malliaras, G.G., Flores-Torres, S., Abruña, H.D., and Rogers, J.A. (2004) Organic light-emitting devices with laminated top contacts. *Appl. Phys. Lett.*, **84**, 3675.
36. Loo, Y.-L., Someya, T., Baldwin, K.W., Bao, Z., Ho, P., Dodabalapur, A., Katz, H.E., and Rogers, J.A. (2002) Soft, conformable electrical contacts for organic semiconductors: high-resolution plastic circuits by lamination. *Proc. Natl. Acad. Sci. USA*, **99**, 10252.
37. Bernards, D.A., Flores-Torres, S., Abruña, H.D., and Malliaras, G.G. (2006) Observation of electroluminescence and photovoltaic response in ionic junctions. *Science*, **313**, 1416–1419.
38. Champagne, A.R., Couture, A.J., Kuemmeth, F., and Ralpha, D.C. (2003) Nanometer-scale scanning sensors fabricated using stencil lithography. *Appl. Phys. Lett.*, **82**, 1111.
39. Shtein, M., Peumans, P., Benziger, J.B., and Forrest, S.R. Direct mask- and solvent-free printing of molecular organic semiconductors. (2004) *Adv. Mater.*, **16**, 1615.
40. Baldo, M.A., Deutsch, M., Burrows, P., Gossenberger, H., Gerstenberg, M., Ban, V., and Forrest, S.R. (1998) Organic vapor phase deposition. *Adv. Mater.*, **10** (18), 1505–1514.
41. Lunt, R.R., Benziger, J.B., and Forrest, S.R. (2007) Real-time monitoring of organic vapor-phase deposition of molecular thin films using high-pressure reflection high-energy electron diffraction. *Appl. Phys. Lett.*, **90**, 181932.
42. Lai, K., Burns, S., Halle, S., Zhuang, L., Colburn, M., Allen, S., Babcock, C., Baum, Z., Burkhardt, M., Dai, V., Dunn, D., Geiss, E., Haffner, H., Han, G., Lawson, P., Mansfield, S., Meiring, J., Morgenfeld, B., Tabery, C., Zou, Y., Sarma, C., Tsou, L., Yan, W., Zhuang, H., Gil, D., and Medeiros, D. (2008) 32nm logic patterning options with immersion lithography. *Proc. SPIE*, **6924** 69243C.
43. den Boer, W. (2005) *Active Matrix Liquid Crystal Displays: Fundamentals and Applications*, Newnes, Oxford, UK.
44. Sheats, J. (2004) Manufacturing and commercialization issues in organic electronics. *J. Mater. Res.*, **19** (7), 1974–1989.
45. Papadimitrakopoulos, F., Zhang, X.M., and Higginson, K.A. (1998) Chemical and morphological stability of aluminum tris (8-hydroxyquinoline) (Alq(3)): effects in light-emitting devices. *IEEE J. Sel. Top. Quant. Electron.*, **4** (1), 49–57.
46. Slinker, J.D., Kim, J.-S., Flores-Torres, S., Delcamp, J.H., Abruna, H.D., Friend, R.H., and Malliaras, G.G. (2007) In situ identification of a luminescence quencher in an organic light-emitting device. *J. Mater. Chem.*, **17**, 76–81.
47. Sirringhaus, H. (2009) Reliability of organic field-effect transistors. *Adv. Mater.*, **21** (38–39), 3859–3873.
48. Kim, S.H., Yang, H., Yang, S.Y., Hong, K., Choi, D., Yang, C., Chung, D.S., and Park, C.E. (2008) Effect of water in ambient air on hysteresis in pentacene field-effect transistors containing gate dielectrics coated with polymers with

different functional groups. *Org. Elect.*, **9** (5), 673–677.

49 De Angelis, F., Gaspari, M., Procopio, A., Cuda, G., and Di Fabrizio, E. (2009) Direct mass spectrometry investigation on pentacene thin film oxidation upon exposure to air. *Chem. Phys. Lett.*, **468** (4–6), 193–196.

50 Gundlach, D.J., Zhou, L., Nichols, J.A., Jackson, T.N., Necliudov, P.V., and Shur, M.S. (2006) An experimental study of contact effects in organic thin film transistors. *J. Appl. Phys.*, **100**, 024509–024509.

51 Kymissis, I., Dimitrakopoulos, C.D., and Purushothaman, S. (2001) High-performance bottom electrode organic thin-film transistors. *IEEE T. Electron. Dev.*, **48** (6), 1060–1064.

52 Huang, Z.H., Qi, G.J., Zeng, X.T., Lukito, D., and Su, W.M. (2009) Design and fabrication of a novel integrated shadow mask for passive matrix OLED devices. *Thin Solid Films*, **517** (17), 5280–5283.

53 Sirringhaus, H., Kawase, T., Friend, R.H., Shimoda, T., Inbasekaran, M., Wu, W., and Woo, E.P. (2000) High-resolution inkjet printing of all-polymer transistor circuits. *Science*, **290** (5499), 2123–2126.

54 Wang, J.Z., Zheng, Z.H., Li, H.W., Huck, W.T.S., and Sirringhaus, H. (2004) Dewetting of conducting polymer inkjet droplets on patterned surfaces. *Nat. Mater.*, **3** (3), 171–176.

55 Wang, J.Z., Gu, J., Zenhausem, F., and Sirringhaus, H. (2006) Low-cost fabrication of submicron all polymer field effect transistors. *Appl. Phys. Lett.*, **88**, 133502–133503.

56 Müller, C.D., Falcou, A., Reckefuss, N., Rojahn, M., Wiederhirn, V., Rudati, P., Frohne, H., Nuyken, O., Becker, H., and Meerholz, K. (2003) Multi-colour organic light-emitting displays by solution processing. *Nature*, **421** (6925), 829–833.

57 Gather, M.C., Köhnen, A., Falcou, A., Becker, H., and Meerholz, K. (2007) Solution-processed full-color polymer organic light-emitting diode displays fabricated by direct photolithography. *Adv. Funct. Mater.*, **17** (2), 191–200.

58 Bacher, E., Jungermann, S., Rojahn, M., Wiederhirn, V., and Nuyken, O. (2004) Photopatterning of crosslinkable hole-conducting materials for application in organic light-emitting devices. *Macromol. Rapid Commun.*, **25** (12), 1191–1196.

59 Solomeshch, O., Medvedev, V., Mackie, P.R., Cupertino, D., Razin, A., and Tessler, N. (2006) Electronic formulations: photopatterning of luminescent conjugated polymers. *Adv. Funct. Mater.*, **16** (16), 2095–2102.

60 Aldred, M.P., Eastwood, A.J., Kelly, S.M., and Vlachos, P. (2010) Light-emitting fluorene photoreactive liquid crystals for organic electroluminescence. *Chem. Mater.*, **16**, 4928–4936.

61 Aldred, M.P., Contonoret, A.E.A., Farrar, S.R., Kelly, S.M., Mathieson, D., O'Neill, M., Tsoi, W., and Vlachos, P. (2005) A full-color electroluminescent device and patterned photoalignment using light-emitting liquid crystals. *Adv. Mater.*, **17**, 1368–1372.

62 Afzali, A., Dimitrakopoulos, C., and Graham, T. (2003) Photosensitive pentacene precursor: synthesis, photothermal patterning, and application in thin-film transistors. *Adv. Mater.*, **15**, 2066–2069.

63 Touwslager, F.J., Willard, N.P., and de Leeuw, D.M. (2002) I-Line lithography of poly-(3,4-ethylenedioxythiophene) electrodes and application in all-polymer integrated circuits. *Appl. Phys. Lett.*, **81** (24), 4556–4558.

64 Tachikawa, T., Itoh, N., Handa, S., and Aoki, D. (2005) High resolution full color polymer light emitting devices using photolithography. *SID Symp. Dig. Tech. Papers*, **36**, 1280.

65 Huang, J., Xia, R., Kim, Y., Wang, X., Dane, J., Hofmann, O., Mosley, A., De Mello, A.J., De Mello, J.C., and Bradley, D.D.C. (2007) Patterning of organic devices by interlayer lithography. *J. Mater. Chem.*, **17** (11), 1043.

66 Lidzey, D.G., Pate, M.A., Weaver, M.S., Fisher, T.A., and Bradley, D.D.C. (1996) Photoprocessed and micropatterned conjugated polymer LEDs. *Synth. Met.*, **82** (2), 141–148.

67 Chan, J.R., Huang, X.Q., and Song, A.M. (2006) Nondestructive photolithography of conducting polymer structures. *J. Appl. Phys.*, **99** (2), 023710.

68 Balocco, C., Majewski, L.A., and Song, A.M. (2006) Non-destructive patterning of conducting-polymer devices using subtractive photolithography. *Org. Elect.*, **7**, 500–507.

69 Kymissis, I., Dimitrakopoulos, C.D., and Purushothaman, S. (2002) Patterning pentacene organic thin film transistors. *J. Vac. Sci. Technol. B*, **20**, 956–959.

70 DeFranco, J.A., Schmidt, B.S., Lipson, M., and Malliaras, G.G. (2006) Photolithographic patterning of organic electronic materials. *Org. Elect.*, **7**, 22–28.

71 Steudel, S., Myny, K., De Vusser, S., Genoe, J., and Heremans, P. (2006) Patterning of organic thin film transistors by oxygen plasma etch. *Appl. Phys. Lett.*, **89** (18), 183503.

72 Zhou, L.S., Wanga, A., Wu, S.C., Sun, J., Park, S., and Jackson, T.N. (2006) All-organic active matrix flexible display. *Appl. Phys. Lett.*, **88**, 083502–083503.

73 Kane, M.G., Campi, J., Hammond, M.S., Cuomo, F.P., Greening, B., Sheraw, C.D., Nichols, J., Gundlach, D.J., Huang, J.R., and Kuo, C.C. (2000) Analog and digital circuits using organic thin-film transistors on polyester substrates. *IEEE Electron. Device Lett.*, **21** (11), 534–536.

74 Li, F.M., Vygranenko, Y., Koul, S., and Nathan, A. (2006) Photolithographically defined polythiophene organic thin-film transistors. *J. Vac. Sci. Technol. A*, **24** (3), 657–662.

75 Chang, C.-Y., Tsai, F.-Y., Jhuo, S.-J., and Chen, M.-J. (2008) Enhanced OLED performance upon photolithographic patterning by using an atomic-layer-deposited buffer layer. *Org. Elect.*, **9** (5), 667–672.

76 Wu, C.C., Sturm, J.C., Register, R.A., and Thompson, M.E. (1996) Integrated three-color organic light-emitting devices. *Appl. Phys. Lett.*, **69**, 3117.

77 Tian, P.F., Burrows, P.E., and Forrest, S.R. (1997) Photolithographic patterning of vacuum-deposited organic light emitting devices. *Appl. Phys. Lett.*, **71**, 3197.

78 Lamprecht, B., Kraker, E., Weirum, G., Ditlbacher, H., Jakopic, G., Leising, G., and Krenn, J.R. (2008) Organic optoelectronic device fabrication using standard UV photolithography. *Phys. Stat. Sol. (RRL)*, **2** (1), 16–18.

79 Chang, J.-F. and Sirringhaus, H. (2009) Patterning of solution-processed semiconducting polymers in high-mobility thin-film transistors by physical delamination. *Adv. Mater.*, **21** (24), 2530.

80 Poliakoff, M., Fitzpatrick, J.M., Farren, T.R., and Anastas, P.T. (2002) Green chemistry: science and politics of change. *Science*, **297**, 807–810.

81 Jones, C.A., Zweber, A., DeYoung, J.P., McClain, J.B., Carbonell, R., and DeSimone, J.M. (2004) *Crit. Rev. Solid State Mater. Sci.*, **29**, 97.

82 Tsai, W.-T. (2005) *J. Hazard. Mater.*, **119**, 69.

83 Zakhidov, A.A., Lee, J.-K., Fong, H.H., DeFranco, J.A., Chatzichristidi, M., Taylor, P.G., Ober, C.K., and Malliaras, G.G. (2008) Hydrofluoroethers as orthogonal solvents for the chemical processing of organic electronic materials. *Adv. Mater.*, **20** (18), 3481.

84 Lee, J.-K., Chatzichristidi, M., Zakhidov, A.A., Taylor, P.G., DeFranco, J.A., Hwang, H.S., Fong, H.H., Holmes, A.B., Malliaras, G.G., and Ober, C.K. (2008) Acid-sensitive semiperfluoroalkyl resorcinarene: an imaging material for organic electronics. *J. Am. Chem. Soc.*, **130** (35), 11564.

85 Lim, Y.-F., Lee, J.-K., Zakhidov, A.A., DeFranco, J.A., Fong, H.H., Taylor, P.G., Ober, C.K., and Malliaras, G.G. (2009) High voltage polymer solar cell patterned with photolithography. *J. Mater. Chem.*, **19**, 5394–5397.

86 Taylor, P.G., Lee, J.-K., Zakhidov, A.A., Chatzichristidi, M., Fong, H.H., DeFranco, J.A., Malliaras, G.G., and Ober, C.K. (2009) Orthogonal patterning of PEDOT: PSS for organic electronics using hydrofluoroether solvents. *Adv. Mater.*, **21** (22), 2314.

Part Four
Scanning Probe Microscopies

13
Toward Supramolecular Engineering of Functional Nanomaterials: Preprogramming Multicomponent 2D Self-Assembly at Solid–Liquid Interfaces

Carlos-Andres Palma, Artur Ciesielski, Massimo Bonini, and Paolo Samorì

13.1
Introduction

Supramolecular *engineering* comprises the design, synthesis, and self-assembly of well-defined molecular modules into tailor-made architectures. The incorporation of functional units in these molecular modules makes it possible to provide a preprogrammed function to the overall architecture and material, thus paving the way toward its technological application in a myriad of fields [1]. For this purpose, the last decade has witnessed an increasing interest toward the 3D engineering of supramolecular materials [2–6]. The colossal task of achieving full control over self-assembled systems, however, is the ongoing endeavor of the supramolecular scientists and to date only few systems may be preprogrammed to undergo controlled self-assembly in three dimension [7]. In contrast, 2D interfaces provide a simplified platform for supramolecular and crystal engineering, and the field has known an exciting growth [8–11]. These early attempts of 2D crystal engineering are exemplified in the preprogrammed molecular (self-) organization at surfaces and interfaces, a realm that holds *per se* a great potential for the generation of novel 2D nanoscale functional materials and devices [12] or devices with custom-made properties, such as charge injection [13], transport [14, 15], and storage [16].

In this regard, the need for exploring ordered architectures at the molecular scale has made scanning tunneling microscope (STM) [17] a widely employed, though extremely powerful, tool to study supramolecular materials at interfaces with a submolecular resolution [18–20]. STM investigations provide electronic and thus chemical insights into the subnanometer scale [21].

The working principle of STM is the tunneling of electrons from a sharp scanning tip to a substrate. Since the tunneling current is proportional to the electron density of the molecule within an energy range [22], the STM contrast in the so-called constant (tip) height mode will appear brighter within electron-rich aromatic molecules than aliphatic groups. In addition, quantitative measurements on the electronic structure of the molecules can be obtained by means of scanning tunneling spectroscopy (STS), a mode that probes the elastic tunneling changes associated with the local density of states (LDOSs) [23–25]. This method was successfully employed to investigate on isolated molecules various phenomena such as charge transfer [26], rectification [12, 27], and switching [28]. The highest spatial resolution that can be attained by STM imaging allows one to gain detailed information on molecular interactions; thus, it is a crucial tool to assist the design of molecular modules that undergo controlled self-assembly at interfaces at any desired condition (temperature, pressure, and concentration) to form the chosen supramolecular motifs and ultimately functional materials.

Self-assembly at room temperature and atmospheric pressure is highly appealing, especially when working at the solid–liquid interface: in fact, thanks to the possibility of operating under equilibrium conditions and to mimic phenomena occurring in Nature, it represents a privileged playground where functional molecular systems can be elaborated. At such interface, the self-assembly may be thermodynamically driven, making it possible to lay the foundation for 2D crystal engineering [29, 30]. In fact, supramolecular *engineering* relies on the prediction of the thermodynamic state occupied by the grand majority of molecules at a given time in a free energy *hypersurface*, that is, the prediction of the exact architecture resulting from the molecular self-assembly. When a system reaches equilibrium conditions, its state is the result of the difference in free energy between all possible configurations. Unfortunately, the computation of free energy differences is still a major research challenge [31], limiting supramolecular *engineering* to a semiquantitative approach, which takes into account only enthalpic contributions to predict short-range self-assembly [29, 32, 33]. In this semiquantitative approach, the prediction of the most probable 2D pattern relies naturally on the differences in potential energies (including both intermolecular and adsorption energies) per unit area between probable polymorphs [29]. Minimization of the potential energy per unit area translates into polymorphs being dependent on compensation of adsorption energies by intermolecular energies. Alternatively, one can make a system independent of adsorption energies by working in excess surface area, that is, (sub)monolayer regimes, where only intermolecular energies per molecule are considered [29, 34]. A prior knowledge of intermolecular interaction energies is thus of primordial importance, being the base for qualitative supramolecular *engineering*. As such, qualitative supramolecular *engineering* at surfaces relies on the deep understanding of noncovalent interactions, such as *van der Waals* (vdW), *hydrogen bond*, and *metal–ligand*.

Herein, we will discuss the role of the previous intermolecular interactions in the self-assembly of prototype functionalities, *enroute* toward fully integrated functional supramolecular architectures. In Section 13.1, we focus on the exploration of systems engineered exclusively through van der Waals interactions, introducing the basic of adsorption at interfaces. We then present recent examples in 2D crystal engineering using alkyl chain interdigitation as the main building strategy. Further on, we put into perspective the strategies employed in some vdW systems for bottom-up self-assembly, as well as the electronic and logical characterization of some prototype systems. In Section 13.2, we focus mainly on systems engineered through the formation of hydrogen bonds and the influence of multiple hydrogen bonds on the supramolecular entities. Finally, in Section 13.3, we address metal–ligand interactions in the formation of 2D crystals and their capacity to mediate large-area conformational switching.

13.2
van der Waals Interactions

13.2.1
Adsorption

In 1873, Johaness Diderik van der Waals hypothesized the existence of intermolecular forces to formulate his equation of state for molecular gases [35]. van der Waals forces come into play at first instance in the adsorption of molecules on atomically flat surfaces. Atomically flat surfaces that are commonly used for experiments at the solid–liquid interface include highly oriented pyrolitic graphite (HOPG), Au(111), and MoS_2, because of their ease of cleaning, stability in air, and their inert character. However, to properly understand the nature of molecular physisorption, one must first consider the interaction of molecules with the surfaces. For instance, the theoretical values of heat of adsorption for benzene on HOPG may vary between 8.6 and 11.1 kcal mol^{-1} for some popular density functional theory and force field calculations [36] as compared to the experimentally determined values of -11.5 ± 0.6 kcal mol^{-1} from thermal desorption experiments [37]. Interestingly, alkanes have been reported to have a stronger adsorption on HOPG compared to aromatics having the same number of carbon atoms. For example, hexane has a desorption barrier of approximately $\sim$17 kcal mol^{-1} on HOPG [38] and slightly lower, about $\sim$13 kcal mol^{-1}, on Au(111) [39]. Analogous to large aromatic compounds [40], these apparent large exothermic changes are heavily countered by endothermic desolvation phenomena, such that the choice of the solvent heavily influences the overall exothermic adsorption process. Moreover, even in the absence of a solvent to weaken adsorption energies, diffusion at surfaces is fast ($\sim$mm s^{-1}) [41] with hopping events from adsorption sites occurring in the picosecond timescale [42].

From the above diffusion rates, it follows that crystallization at the solid–liquid interface relies not only on the *strong* intermolecular interactions but also on the *directional* nature of such interactions. Again, a careful choice of the solvent is needed to prevent further entropic effects that disfavor crystallization. STM imaging can be in principle performed in almost any solvent, as long as the faradic and capacitive currents are minimized. This can be achieved by insulating the STM tip using the coatings or the low-dielectric media, to allow working in the picoampere regime. Some of the most common solvents utilized when picoampere currents are sought are 1,2,4-trichlorobenzene (TCB), heptanoic acid, 1-phenyloctane, *n*-tetradecane, and 1-octanol, with relative dielectric constants near 2.5 [43]. On the other hand, when working in aqueous or in other high-dielectric media, it is common to use a reference electrode and an insulated tip. This setup is usually termed an electrochemical STM (ECSTM, not to be confused with SECM, scanning electrochemical microscopy) [44, 45].

Order through van der Waals interactions: alkyl chain interdigitation. When dealing with atoms or molecules, vdW forces have their origin on *London dispersion forces* [46, 47], which arise from the attraction of the instantaneous dipoles generated by atomic electron clouds. Exploiting *London's dispersion* forces, stabilization of the

2D self-assembly has been achieved by equipping the molecules of interest with long aliphatic chains; in this way the interdigitation of alkyl chains belonging to adjacent molecules made it possible to form crystalline patterns [48]. This occurs through the entrapping of an alkyl chain in between two adjacent ones with an approximate energy of interaction of ~0.7 kcal mol^{-1} per methylene unit [49]. This value is markedly lower than the energy of adsorption on HOPG, which is about four times larger, as normalized per methylene unit [38, 50]. Such an interdigitation between alkyl chains features a *pseudo-directionality* because the interdigitation approach rests on a fairly preferential sorption of the rod-like alkyl chains along the six identical main crystallographic lattice directions of HOPG (the nonredundant directions having Miller–Bravais indices of {1010}, {0110}, {1100}. As a result, the physisorption is not characterized by a surface commensurability with HOPG *per se*, although the sorption of the methylene hydrogens into the centers of the graphite rings would be preferred [48, 51]. Charra and coworkers [52] have recently made use of this approach to steer the orientation of phenyl vinylenes on HOPG (Figure 13.1a), a versatile approach that can result in patterning of all five 2D Bravais lattices.

Very recently, De Feyter and coworkers [53] have further employed this strategy to realize the first example of a four-component 2D crystal (Figure 13.1b). It has also been possible to observe one of the smallest aromatic molecules, that is, an unsubstituted triphenylene (yellow arrows), immobilized in a 2D crystal. It is worth noting that neither triphenylene nor coronene can be visualized when deposited from solutions on HOPG as single component, because of their high diffusion at room temperature, due to the absence of stabilizing lateral interactions with adjacent molecules. Such result confirms the importance of lateral interactions in forming a 2D crystal at room temperature. Figure 13.1c shows an example of the same strategy employed for the encapsulation of a triangular polyaromatic molecule [54]. Despite the huge adsorption energy that polyaromatics may exhibit, disorder is always evident in the pores, which measure 4.7 nm edge to edge, with a record 7.0 nm edge to edge when using longer alkyl chains ($C_{30}H_{61}$) [55].

De Feyter and coworkers also demonstrated that the interdigitation approach is not flawless [56]. Given the amount of conformational freedom of alkyl chains, almost all patterns relying on alkyl chain interdigitation are affected by polymorphism. Nevertheless, interdigitation represents an elegant strategy to decorate the nanopattern surfaces.

From 2D to 3D architectures: there is plenty of room at the top! Once accurate 2D positioning is mastered, the controlled placement of objects in the third dimension may be envisaged. A fundamental question arises concerning all different intermolecular interactions that can be exploited for 3D control: Is there a minimum boundary in the magnitude of a force needed for effective bottom-up self-assembly?

Liquid crystalline materials are the proof that very weak interactions are sufficient to create complex 3D architectures [57]. By using liquid crystal forming molecules such as alkyl-substituted hexa-*peri*-hexabenzocoronene (HBC) derivatives, it was possible to visualize by STM the supramolecular architectures featuring a high degree of order extended even to the third dimension in the region near to the surface [49]. Similar result has also been obtained on bicomponent assemblies of an

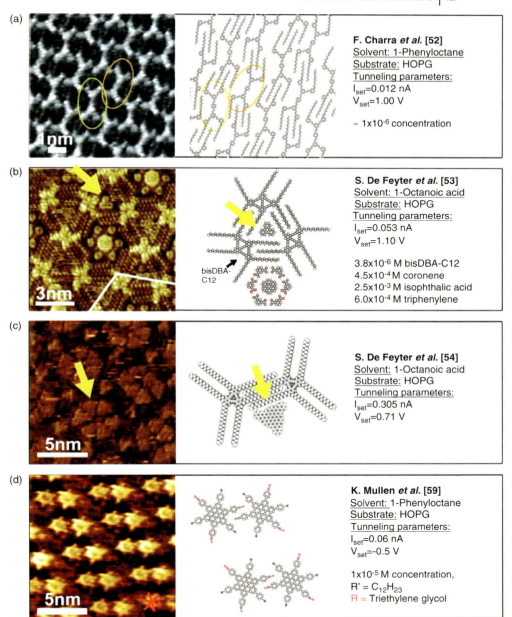

Figure 13.1 (a–c) Two-dimensional crystal engineering through alkyl chain interdigitation. (d) Bottom-layer structure of discotic liquid crystals.

electron donor and acceptor systems [58]. Figure 13.1d reports an STM image of a complex C_3 symmetric hexa-*peri*-hexabenzocoronene [59]. In the bulk, this molecular module forms columnar phase dictated by the π–π stacking [36, 60, 61], with an interplanar distance of 0.35 nm. Despite its attractiveness for charge transport [61, 62], π–π stacking is not the most controlled strategy toward bottom-up construction, as it is poorly directional and features modest selectivity.

Recently, Charra and coworkers have shown that nanopillars consisting of [2.2] paracyclophanes can be consistently positioned onto the 2D platforms, thus forming highly crystalline 3D patterning (Figure 13.2a) [63]. Visualization of such covalently linked double-decker molecules extending into the third dimension was pioneered by Rabe and coworkers on HBC double-deckers [64]. This strategy was later applied with double-deckers made from phtalocyanines complexes [65–67]. These complex modules have been chosen since phtalocyanines are known to feature long-range order in 2D monolayers [68], which can be then used as playground to control the position of such 3D architectures. For instance, Weiss and coworkers [66] have visualized the formation of highly ordered double-decker phtalocyanines in equilibrium with single phtalocyanines (Figure 13.2b). Another sophisticated and versatile approach to access the third dimension was shown by Herges and coworkers [69]. Making use of triazatriangulenium ionic platforms, they introduced a variety of functional units, such as aryl or azobenzene substituents, that extend perpendicularly to the surface into the third dimension (Figure 13.2c). A similar approach can be pursued by using metal–ligand interactions. Feringa and coworkers [70] gave recently a proof-of-principle on how metal–organic interactions can be extended into the third dimension, which may serve for bottom-up 3D construction [2, 71], by complexation of pyridines onto a self-assembled layer of zinc porphyrins. Beyond equilibrium conditions, nonequilibrium organization such as layer-by-layer assembly [72] can also be used in conjunction with van der Waals interactions to control the 3D structure of architectures. In 2004, Fichou and coworkers demonstrated that alkyl chains can be, in a first step, self-assembled at the solid–liquid interface and subsequently used to self-assemble alkyl-substituted hexa-*peri*-hexabenzocoronene in an epitaxial fashion (Figure 13.2d) [73]. Alkyl-substituted HBCs are highly symmetric aromatic disks that have been extensively studied by Rabe and coworkers [49, 58, 74, 75]. This strategy can be further refined by taking advantage of the formation of charge transfer complexes (see later in this chapter) [49, 76, 77].

By and large, control of 2D assembly represents the first step toward the tailoring of 3D architectures for a wide range of applications, for example, spanning from charge and gas storage [2] to optoelectronics [78].

The era of molecular electronics: rectifiers and switches. The beginning of the organic electronics era was marked by the studies on electrical conductivity and luminescence in polymers over 30 years ago [79] and it has bloomed into full maturity during the last few decades. The discovery of carbon nanotubes less than 20 years ago [80] and the dawn of graphene [81] have definitely consolidated the importance of organic electronics. Because of this, interest in the electronic properties of low-dimensional materials is now ubiquitous. In this regard, low-dimensional graphenes [82] like graphene nanoribbons are novel architectures that present important

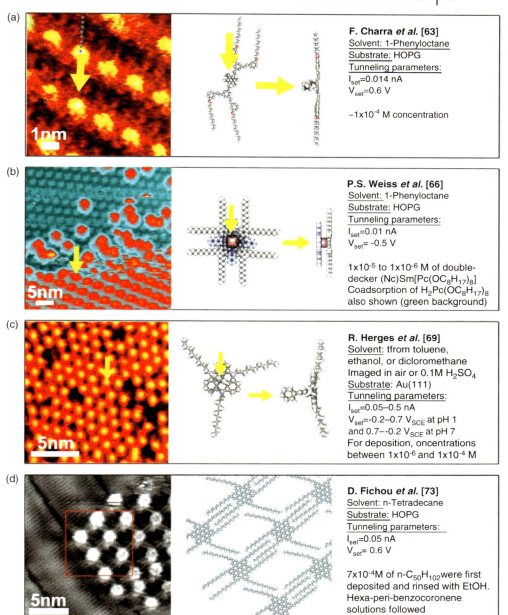

Figure 13.2 (a–c) Three-dimensional architectures at interfaces. (d) A two-layered assembly of polyaromatics on top of hydrocarbons.

electronic properties [83]. Toward this goal, Müllen and coworkers [84] have successfully produced and imaged long (~10 nm) ribbons at the HOPG interface (Figure 13.3a, solubilizing groups are not shown). Accessing such "1D" electronically active objects by synthetic methods opens interesting perspectives in supramolecular electronics.

In semiconducting systems, STS gives qualitative information about the electronic nature of the molecules, and may give quantitative information when the first derivative of the current is analyzed (i.e., the differential tunneling current, dI/dV) [22]. For example, Figure 13.3b features an STM image of a HBC-based architecture in which coronone diimide acceptor has been codeposited [85]. The exact location of the acceptor in the assembly can be revealed by the electron acceptor nature of diimides. The figure displays the I–V (current–voltage) STS characteristics exhibiting clear asymmetric behavior, which can be analyzed considering *orbital-mediated tunneling* (OTS) or, more in general, *resonant tunneling* [23, 24]. When the substrate (being HOPG with a work function of ~4.7 eV or gold with a work function of ~5.1 eV [43]) is positively biased, electrons from the platinum/iridium (Pt/Ir) tip (having a work function of ~5.7 eV [24]) flow to the sample assisted by the orbital aligned between the electrodes, which is normally the lowest unoccupied molecular orbital (LUMO) of an *acceptor* molecule (whose electron affinity varies normally between 1 and 3 eV [43]). This results in a higher current at positive sample biases for the coronene diimide (red curve) than from the bright cores (black curve), proving that indeed coronene diimide is located between the diagonals of the bright cores.

The general host–guest complexation is a common strategy in the supramolecular chemist's toolbox. Host–guest interactions are encompassed behind a broader definition of fundamental interactions, arising from the 1890 seminal lock and key paper by Emil Fischer [86]. In this context, De Feyter and coworkers showed that supramolecular recognition can occur at a surface between a macrocycle and a charged guest, a tropylium ion (Figure 13.3c) [87]. In molecular electronics, host–guest interaction might be mediated by the formation of charge transfer complexes [77] and is of special importance for designing charge separation [88], energy harvesting [89], and rectifying devices [90]. For instance, complexation of the well-known donor–acceptor thiophene and fullerene (C_{60}) pair [91] has been characterized at the solid–liquid interface by Freyland and coworkers [92]. By using a macrocycle decorated with dithiophene units, C_{60} can undergo preferential adsorption onto the dithiophene "host" sites (in Figure 13.3d, the macrocycle-ordered monolayer is only shown, while the illustration shows where C_{60} is adsorbed after deposition).

By using a similar approach on oligothiophene macrocycle building blocks, Mena-Osteritz and Bäuerle [93] went one step further into the electronic characterization of the complex formation. Figure 13.4a shows two STS curves, before and after the adsorption of C_{60} onto the macrocycle, giving evidence of the formation of the complex (Figure 13.4a). Moreover, Lee and coworkers [94] have recently shown that when using pentacene as the electron donor and depositing C_{60} on top of it, a marked rectifying behavior is observed. No apparent contribution of the LUMO of C_{60} to the resonant tunneling at positive sample biases is present (Figure 13.4b, the STM image shows only the first pentacene layer). It is expected that such kind of architectures will

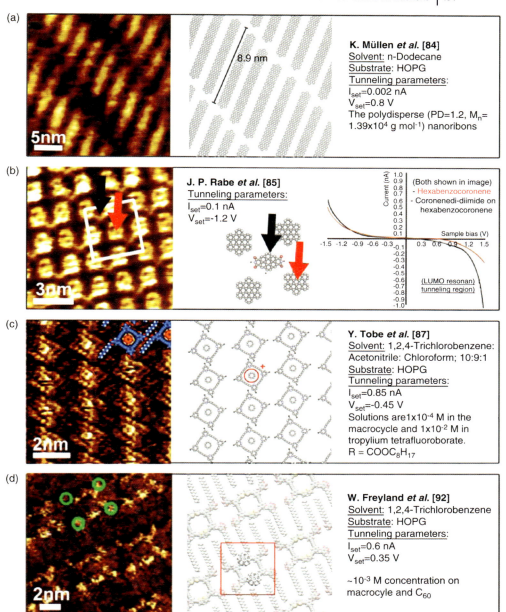

Figure 13.3 (a) Nanographene ribbons at interfaces. (b–d) Host–guest interactions at interfaces.

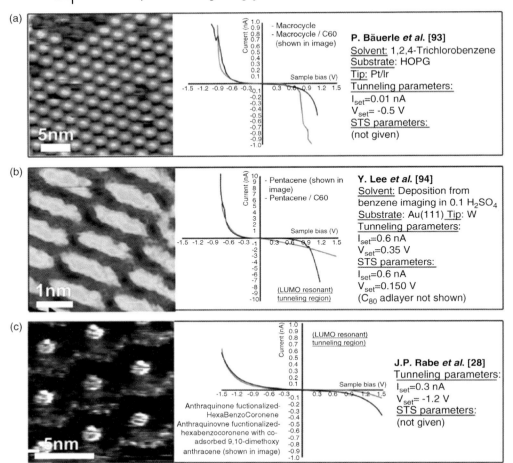

Figure 13.4 (a and b) Scanning tunneling spectroscopic characterization of 2D donor–acceptor complexes. (c) A prototype molecular transistor.

become more and more routinary in the next few years, following bright examples already existing, such as the prototype described by Aviram and Ratner [95] and Coulomb-blockade molecular diodes [12, 14, 90, 96]. Moreover, two-layered architectures are of special interest to the field of nanomechanical transistors (NMTs) [97], where switching of a molecular transistor may be accomplished by mechanical compression of C_{60} using the STM tip apex.

To date, one prototype molecular transistor at the solid–liquid interface has been reported by Rabe and coworkers [28]. In a first step, a molecular dyad [61, 76, 98, 99] consisting of HBC decorated with anthraquinone substituents is used to pattern a surface. When an electron donor (9,10-dimethoxyanthracene) is added to the supernatant solution, a domain where the electron donor forms a complex with anthraquinone appears, corresponding to the STM image in Figure 13.4c. When STS is performed directly on top of the HBC in both domains, a significant difference is observed in the domains where the 9,10-dimethoxyanthracene was coadsorbed, which exactly corresponds to the newly induced surface dipoles.

13.3
Hydrogen–bonding Interactions

Another widely employed noncovalent interaction in supramolecular chemistry is hydrogen bond. This interaction type is unique: it offers a high level of control over the molecular self-assembly process, since it combines reversibility, directionality, specificity, and cooperativity [100–103]. In particular, the association of molecules bearing complementary recognition groups is the most straightforward pathway for preprogramming network self-assembly of multiple components [104, 105], both in solution and in the solid state [106–108]. H-bonding interactions benefit among all from a wide range of interaction energies, depending on the number and position of consecutive strong hydrogen bond donor (D) and acceptor (A) moieties. Jeffrey [109] reported that the energy per single A–D pair in the solution quantified by typical methods like NMR or UV-Vis amounts to 1.0–1.4 kcal mol^{-1} per hydrogen bond. In the case of three parallel H-bonds in ADA–DAD pairs, *ab initio* calculated interaction energies can rise to 18 kcal mol^{-1} [29] and up to 26 kcal mol^{-1} in DDD–AAA pairs [110, 111]. For four consecutive H-bonds, AADD–DDAA (i.e., a tetrahapto moiety) in ureidopyrimidone dimers [112, 113], the association energy gives a record value of 47 kcal mol^{-1}. The computed geometries and harmonic vibrational frequencies of the H-bonded structures, calculated using *ab initio* methods, are in a good agreement with experimental data [113–116]. However, energetic values must be taken with caution, since hydrogen bonds highly depend on the molecular environment, which may completely differ within a supramolecular architecture. Therefore, the presented values may be regarded as upper limits of the stabilization due to the hydrogen bond itself. In the following paragraphs, we will focus on the advantages and disadvantages of the use of different successive hydrogen bond motifs for the formation of stable networks.

13.3.1
Weak and Single Hydrogen Bonds

Although strong complementary hydrogen bonds may be the motif of choice for engineering architectures [29, 34, 117], weak hydrogen bonds can readily direct the formation of 2D crystalline structures [118, 119]. For instance, the self-assembly of oligopyridine derivatives at the HOPG-1,2,4-trichlorobenzene (TCB) reveals a lattice held together by eight weak H-bonding interactions per molecule (Figure 13.5a) [118]. However, such weak intermolecular interactions will hardly rule over van der Waals and adsorption energies (which steer close-packed molecular tiling) making pattern engineering with weak hydrogen bonds a challenging topic. Astonishingly, the use of single hydrogen bonds proved to be robust enough for the

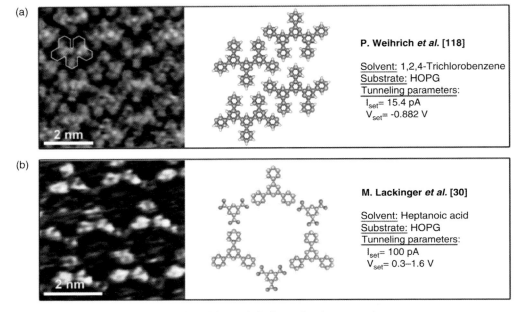

Figure 13.5 (a) Weak and (b) single hydrogen bonds in 2D architectures.

engineering of 2D porous networks. The bicomponent network of 1,3,5-tris(4-pyridyl)-2,4,6-triazine (TPT) and trimesic acid (TMA) with a sixfold symmetry has been described by Lackinger et al. [30]. Figure 13.5b reports a representative STM image of the monolayer. Each molecule can be clearly identified based on their size. Single TPT molecules appear as a triangle with a central depression. This threefold symmetry of TPT in the STM contrast agrees with the symmetry of the molecular structure. Furthermore, based on the STM image, undefined structures are physisorbed in the cavities. This might stem from a transient adsorption of guests (TMA, TPT, or heptanoic acid molecules).

13.3.2
Carboxylic Acids as Dihapto Hydrogen–bonding Moieties: Chirality and Polymorphism

The hydrogen bond form between two adjacent carboxylic acids is the most widely used dihapto H-bond type for preprogramming 2D networks [10]. Many groups [120–122] have reported systems relying on such a dihapto recognition at the solid–solution interface. Among them, Lackinger et al. [120] showed the formation of linear polymer arrays by the deposition of an isophthalic acid (1,3-benzene-dicarboxylic acid) solution in heptanoic acid onto HOPG surface. In the STM image (Figure 13.6a), two types of molecules with different contrasts are identified. Each unit cell contains two molecules differing in their azimuthal orientation and adsorption site. The self-assembly is ruled to a great extent by double H-bonding

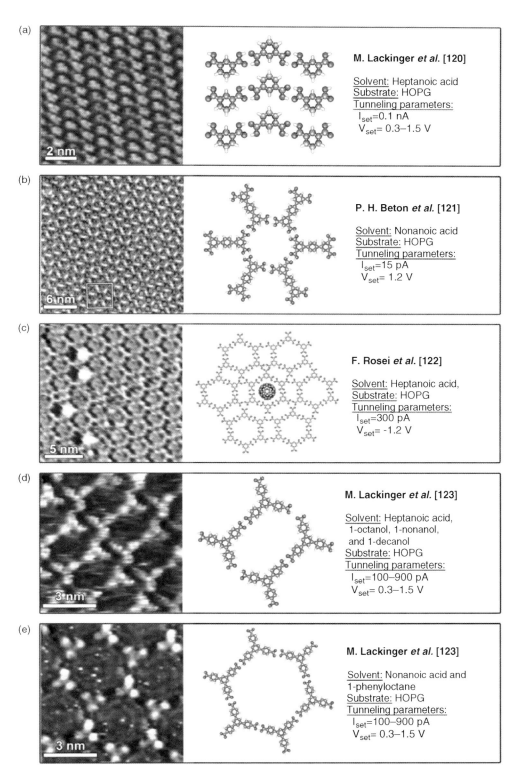

Figure 13.6 Carboxylic acids in 2D crystal engineering.

interactions that are established between consecutive molecules, arranged in *zigzag* chains within an oblique Bravais lattice, featuring an angle of 108°. Such a deviation from the ideal angle of 120° was explained invoking interchain and adsorbate–substrate interactions.

By using different molecules bearing multiple carboxylic acids, other polymorphs (i.e., other monolayer domain arrangements from the same molecular building blocks) may be formed [120–122]. One among such 2D patterns is the workbench for recent studies on random tiling and topological defects at the solid–liquid interface [121]. Beton and coworkers have shown that terphenyl-3,5,3′,5′-tetracarboxylic acids (TPTC) arrange without translational symmetry, but featuring a hexagonal superstructure (Figure 13.6b). The star-shaped structures (highlighted in the square) represent one among five possible arrangements of TPTC molecules.

Polymorphism when using carboxylic acids may also arise from the possibility of trimerization of carboxylic acids on surfaces. For instance, pure TMA regions based on trimerization of the carboxyl moieties may form two polymorphs: the chicken wire and the flower structure (Figure 13.6c). These TMA phases can further host "guest" C_{60} molecules within their pores, whereas the TMA/alcohol bicomponent network does not offer any stable adsorption site for the C_{60} molecules [122].

Interestingly, the selectivity of one polymorph among others may be modulated by the choice of the solvent. Lackinger and coworkers described this phenomenon in the 2D pattern formation of 1,3,5-benzenetribenzoic acid (BTB) [123]. The evolution of different structures for different solvents has been discussed focusing on adsorption rates and stabilization of polar units. In more detail, BTB monolayers at the HOPG–saturated solution interface consist of two different nondensely packed polymorphic architectures. These two structures differ in their H-bonding pattern and their particular molecular packing density. Depending on the ability of the solvent to hydrogen bond the solute molecules, either an oblique or a hexagonal lattice was observed, with the former structure exhibiting rectangular pores and the latter exhibiting circular pores. Representative STM images of both structures are reported in Figure 13.6d and e, respectively. The less dense BTB polymorph is a sixfold chicken wire structure with circular cavities (approximately 2.8 nm wide), while the other structure exhibits an oblique unit cell and a higher surface density. The remarkably large cavity size of the chicken wire structure is the result of combination of the hydrogen bonds between carboxylic groups and the low conformational flexibility of BTB.

In some cases, the solvent can participate actively in the self-assembled pattern, by coadsorbing to form bicomponent networks. A recent work [124] has shown that controlled codeposition of the solvent molecules with melamine can be achieved. Coadsorption phenomena have also been previously reported in systems consisting of TMA and alcohols [122, 125]. The process involves the self-assembly of TMA with alcohols of varying length, which coadsorb on HOPG to form linear patterns. The mixed TMA/alcohol supramolecular structure consists of a hydrogen-bonded network with alternating hydrophobic and hydrophilic regions. The incorporation of the alcohols allows a precise control over the periodicity of the nanostructures.

13.3.3
Combined Dihapto Hydrogen–bond and van der Waals Interactions

AD–DA motifs other than carboxylic acids are ideal for preprogramming the formation of networks, due to their vast geometrical diversity. Such (double) hydrogen bond moieties might be complemented by introducing additional van der Waals interactions, which can be obtained through alkyl chain functionalization. For instance, Ziener and coworkers [126] made use of lactim/lactam moieties with alkoxy-substituted phthalhydrazide molecules, which allows the formation of an architecture that is directed by both van der Waals and H-bond interactions. In this case, the self-assembled patterns were found to depend on the solvent employed. The long-chain phthalhydrazide in TCB self-assembles in a stripe-like structure, whereas monolayers formed from 1-chloronaphthalene solutions form trimeric star-like units that arrange in a hexagonal lattice. Figure 13.7a shows an inverted contrast STM image, therefore, the molecules appear darker than HOPG surface. The hexagons are built up from symmetrical three-arm stars, implying that the chirality of a single hexagon emerges from the self-assembly of achiral (prochiral) molecules. Each star consists of three molecules of the phthalhydrazide derivative, each trimer giving rise to six hydrogen bonds. This arrangement generates voids in the centers of the hexagons, as these voids are too small to accommodate an additional trimer. The molecules of 1-chloronaphthalene are able to stabilize the whole assembly by filling the voids, due to the extended conjugation characterizing the system as compared to TCB, thereby compensating the loss of enthalpy per unit area due to the lower surface coverage from TCB.

The two N–H hydrogen bonds between diaminotriazine units have also been used extensively to form hexagonal motifs at the solid–liquid interface. This approach has enabled De Feyter and coworkers to form cyclic hexameric rosette structures, reminiscent of six-bladed windmills, featuring a 2D chiral character [127]. The rosettes were found to be ordered in rows and form a hexagonal 2D crystal lattice. In Figure 13.7b, an STM image of a small area with submolecular resolution is reported, showing in detail the 2D crystal structure. The blades of the "windmill" appear as bright rods and correspond to the conjugated backbone of OPV, while the alkyl chains are located in the darker areas between the bright blades. Two out of three alkyl chains, which are lying along the direction of one of the main graphite crystallographic axes, can be visualized. Their interdigitation contributes to the stabilization of the 2D crystal structure. Although the location of the third alkyl chain is not clear as it cannot be viewed, it can reasonably be considered to be back folded into the supernatant solution [76]. The diaminotriazine moieties at the opposite end of each molecule are pointing toward the center of the rosettes, suggesting that their formation occurs through self-complementary hydrogen bonds between adjacent triazine moieties. Molecular chirality is transferred to the nanostructures, which in turn form chiral 2D crystalline patterns that belong to the P6 space group. This is the first example of 2D chirality belonging to this space group. Further, not only the 2D pattern is chiral but also the chirality of the 2D crystalline structures is expressed by the relative orientation of the rosettes with respect to the main crystallographic axes of graphite.

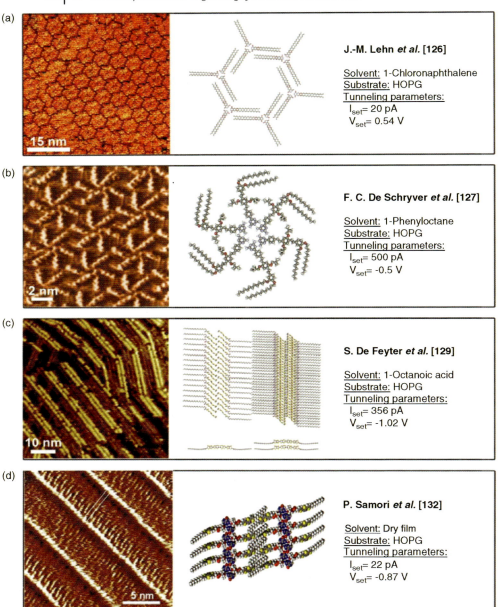

Figure 13.7 Hydrogen-bonded (a) trimers, (b) hexamers, and (c) and (d) 1D polymers.

Amide groups may also be used for the formation of interlamellar hydrogen bonds to the point of using peptides for hydrogen bond recognition [128]. This is especially important when a specific arrangement of the lamella tilting angle (with respect to graphite) is desired. Of particular interest is the use of interlamella spacing dictated by amide units to control the π–π stacking. Using this strategy, TTF (tetrathiafulvalene) derivatives exposing an amido group in the side chain were found to be physisorbed on HOPG from solutions in 1-octanoic acid in a head-to-head and tail-to-tail motif (Figure 13.7c), with lamellar structures stabilized by intermolecular hydrogen bonds [129]. In the case of a monoalkylated TTF derivative, multilayer structures are formed at the interface. The single amide functionality is then sufficient to force π–π stacking of the TTF moieties via intermolecular hydrogen bonding and to overcome the intrinsic tendency of TTF to adsorb flat on the graphite substrate. These results are in line with those obtained by using ureido groups [130, 131] or nucleosides [132] to template the self-assembly of oligothiophenes and for scaffolding.

Harmonizing the functionalities of individual moieties in a supramolecular network represents a versatile approach for developing well-defined polymeric architectures with preprogrammed conformations and tailored properties. In this frame, we studied lipophilic guanosines [133–136] that are very versatile building blocks: depending on the experimental conditions, they can undergo different self-assembly pathways, leading to the formation of either H-bonded ribbons or quartet-based columnar structures. Given the possibility to functionalize the guanosines in the side chains, they appear as ideal building blocks for the fabrication of complex architectures with a controlled high rigidity, thus paving the way toward their future use for scaffolding, that is, to locate functional units in preprogrammed positions. We have shown that guanosine derivatives can form H-bonded ribbon-like or quartet-based supramolecular architectures in solution, depending on the conditions [132]. SPM characterization showed that such molecules can self-assemble from a solution in an apolar solvent into ordered crystalline architectures on surfaces (Figure 13.7d). This self-assembly is governed by the formation of H-bonds between guanosines that dictates the spatial localization of oligothiophenes, ultimately forming 1D conjugated arrays that may be employed as prototypes of supramolecular nanowires.

13.3.4
Trihapto Hydrogen–bonding Moieties

Our group has recently demonstrated the controlled formation of bicomponent porous networks based upon three consecutive H-bonding [29, 34]. The hydrogen-bonded networks are formed by modulating the length of diimide derivative modules, capable of forming three parallel hydrogen bonds with complementary modules, like melamine or diacetyl diaminopyridines [9, 117]. The use of melamine as a cornerstone was pioneered by the Whitesides group [137] and demonstrated to self-assemble with perylene-3,4,9,10-tetracarboxylic diimide (PTCDI) into hexagonal porous networks using UHV conditions by Beton and coworkers [138]. We have extended this approach to a wide variety of diimide modules at the TCB–HOPG interface (Figure 13.8a) [117, 139]. The porous network forms favorably at low concentrations in the solution (1–10 μM), whereas coexistence of the phases formed by the pure components takes place at concentrations ranging 20–50 μM. It is interesting to note that when using flexible linkers, a wide polygon distribution is found. Figure 13.8a shows a pentagon bridging two of the ideal hexagonal structures.

Such robust porous networks may also be combined with chemisorbed self-assembled monolayers (SAMs) to functionalize substrates. This approach offers considerable design flexibility, with the network providing an effective pathway to pattern the substrate and the SAM allowing the surface functionalization with submolecular resolution. The assembly of melamine and PTCDI in dimethylformamide allows the formation of a bicomponent supramolecular network

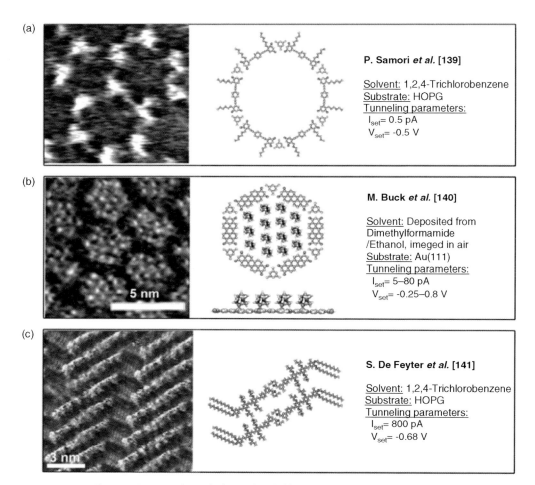

Figure 13.8 (a) Trihapto hydrogen-bonded bicomponent nanoporous network. (b) Nanoporous network acting as surface template for thiols. (c) Tetrahapto hydrogen-bonded dimers.

(Figure 13.8b) [140]. Upon addition of adamantine thiol on top of melamine/PTCDI network, the generation of hybrid structures was achieved. Figure 13.6b demonstrates that the supramolecular network serves as a general template for a range of thiolated molecules.

13.3.5
Tetrahapto Hydrogen–bonding Moieties

Although no 2D networks connected solely by quadruple hydrogen bond have been constructed, much work has been done in two-dimensional pattern formations. A good example of two-dimensional pattern formation driven by H-bonding has been described by Meijer *et al.* by the physisorption of chiral phenylene vinylene derivatives at the solid–liquid interface [141]. Quadruple hydrogen bonding dominates intermolecular interactions, leading to the formation of dimers on the surface. Figure 13.8c shows an STM height image obtained at the TCB–HOPG interface, where the π-conjugated segments appear as bright regions and the alkyl tails are located in the dark parts. The long bright rods as observed in Figure 13.8c often appear as two individual bright rods, separated by a dark trough, corresponding to the location of the hydrogen-bonded moieties. The conjugated segments as well as the aliphatic side chains are lying with their long axes parallel to the basal plane of the graphite substrate. The dimer formation indicates hydrogen bonding between two OPV molecules, which based upon the ratio between their length and width, correspond to all-*trans* vinylene bonds. The space between two dimers in a lamella is sufficient to accommodate a face-on orientation of the conjugated parts and fully extended chiral (*S*)-2-methylbutoxy side chains. The face-on orientation allows the overlap between the orbitals of OPV and those of graphite, resulting in a maximum enthalpy gain. At the same time, the intermolecular interactions between the lateral aliphatic dodecyloxy chains contribute to the stabilization of the lamellar structure of the 2D assembly. Therefore, the orientation of the monomers within the dimers, the orientation of the dimers within the lamellae, and the propagation direction of the lamellae with respect to the symmetry of the substrate are the result of the molecular design, including the chirality of the OPV molecules.

13.4
Metal–Ligand Interactions

Supramolecular architectures combining organic moieties with transition metal ions have been widely employed during the last decades [142–147]. The interest in such assemblies arises from both their robustness and their electronic features, leading to the development of applications in various research fields [148]. For instance, monomolecular transistors based on terpyridine–cobalt–terpyridine complexes [149], as well as dynamic chemical devices undergoing reversible extension/contraction through a pH-triggered complexation of Pb ions have been reported [150]. Recently, the use of self-assembly processes to generate hybrid molecular arrays with predefined structures on surfaces has been explored by combining metal centers and organic units [151]. STM has been used to study various self-assembled systems such as metal–organic 2D coordination networks [152, 153], discrete complexes on surfaces [154–156], or more recently 1D coordination networks based on tectons bridging two coordination poles with different denticities [157].

In 2005, Stang and coworkers reported an STM study on the self-assembled supramolecular metallacyclic rectangle cyclobis[(1,8-bis(*trans*-Pt(PEt$_3$)$_2$)anthracene) (1,4′-bis(4-ethynylpyridyl)benzene](PF$_6$)$_4$ [158], the dimensions of a single rectangle being 3.1 nm × 1.2 nm. The monolayers were studied on both HOPG and Au(111) surfaces to investigate the effect of substrate on their structure. The molecules self-organize into well-ordered 2D patterns on both surfaces, and high-resolution STM results clearly reveal a rectangle on both HOPG (Figure 13.9a) and Au(111) surfaces. The symmetry and molecular orientation adopted by the molecules depend on the substrate. The long edge of the rectangles sits on the HOPG surface, while it lays flat

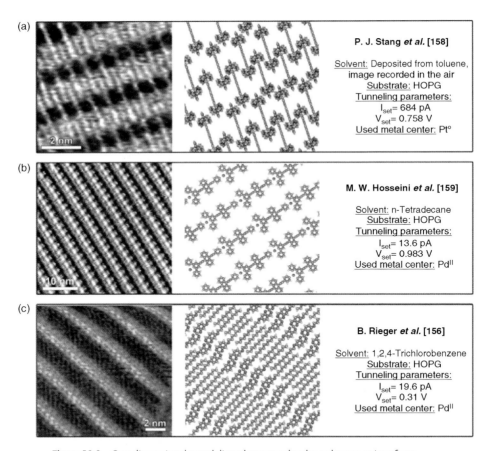

Figure 13.9 One-dimensional metal–ligand supramolecular polymers at interfaces.

on the Au(111) surface, showing that the molecular self-organization of the supramolecular metallacyclic rectangle can be tuned by the appropriate choice of substrate.

In a recent project, the controlled formation of linear coordination suprapolymers at the solid–liquid interface using palladium(II) ions was demonstrated [159]. The metal–organic framework was formed by using a rigid anthracene derivative bearing two units (pyridine and terpyridine) at the extremities, and palladium(II) or cobalt(I) ions as metal centers. One-dimensional arrays were obtained only when using Pd(II), whereas 2D arrays were observed when $CoCl_2 \times H_2O$ salt was added. Because of the dicationic nature of palladium, its interaction with the neutral anthracene derivative generates a charged square–planar complex formed by the palladium center and one terpyridine and one pyridine unit belonging to consecutive anthracene derivatives. Figure 13.9b shows an STM image of the 1D network obtained by self-assembly of the ligand with Pd(II) on HOPG. The spots observed along the rows can be attributed to the metal centers, consistently with their high electron density. The lateral correlation between the 1D networks forming the crystalline layer suggests a cooperative process that takes place during their assembly on graphite. Depending on the overall charge of the network and thus on the nature of the tectons used, it is possible to obtain either 1D or 2D arrays that expose metal centers in preprogrammed positions, with a resolution of few Ångströms.

STM investigations of a series of neutral salicylaldehyde- and aldemine-derived M (II) complexes with various positions and lengths of the alkoxy substituents as well as exchangeable functional groups have been recently reported by Rieger and coworkers [156]. Figure 13.9c shows an STM image of bis(salicylaldehydato)palladium (II) complex at the TCB/HOPG interface. The images allow a detailed analysis of the 2D array, revealing metal–metal distances of 6.7 ± 0.2 Å within the lamellae. The arrangement of the complex cores suggests attractive interactions of the carbonyl protons with both the phenolate and carbonyl oxygen atoms of adjacent complexes. Applying typical geometric parameters to the 2D structure allows one to evaluate the distance between these atoms to be about 2.6 ± 0.2 Å, which is in agreement with the typical length of weak inter- and intramolecular hydrogen bonds formed by aldehyde or aldimine protons with adjacent oxygen atoms. On the basis of these findings, the model of the surface structure of Pd(II) complex was established.

In 2002, De Schryver and coworkers reported an STM investigation of *in situ* complexation of $Pd(OAc)_2$ by a monolayer of a bipyridine derivative at the HOPG surface [160]. The formed monolayer can be then used as a template to build nanostructures. After the successful imaging of the 5,5′-dinonadecyl-2,2′-bipyridine monolayers at the HOPG/1-phenyloctane interface, a drop of $Pd(OAc)_2$ in 1-phenyloctane was applied. A spontaneous change of the monolayer pattern was observed. Figure 13.10a shows a representative STM image of 5,5′-dinonadecyl-2,2′-bipyridine monolayer physisorbed at the HOPG/air interface after the addition of $Pd(OAc)_2$ in 1-heptanol and drying for two days under ambient conditions.

The same year J.-M. Lehn and coworkers reported the synthesis and characterization of bis(terpyridine)-derived ligands that can form [2 × 2] grid-like complexes [161, 162]. Additional pyridine substituents on these ligands do not interfere with the complexation process. STM investigations showed that the physisorption of

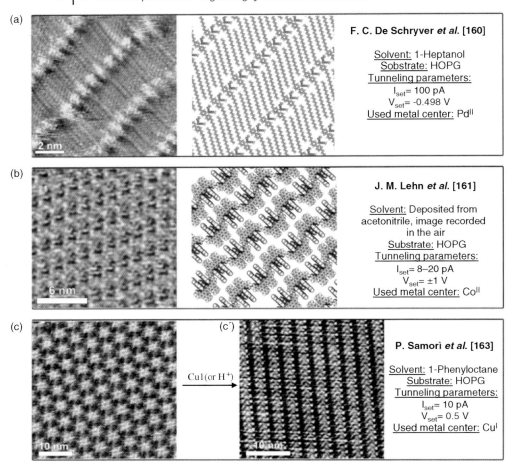

Figure 13.10 (a and b) Metal–organic complexes packed at interfaces. (c) Dynamic conformational switching of a 2D pattern through complexation with copper or protonation.

a pure ligand on HOPG led to the formation of highly ordered structures stabilized by additional intermolecular C—H··· N hydrogen bonds, partially through the extra pyridines. Similar adsorption experiments with one of the corresponding [2 × 2] Co (II) grid-type complexes on HOPG led to a well-organized structure with interdigitation of the extra pyridine moieties (Figure 13.10b).

Our group has recently demonstrated a two-dimensional ordering of molecules adsorbed on surfaces at the solid–liquid interface that are capable to undergo large conformational changes upon the application of an external chemical stimulus, which can be either a change of pH or the addition of metal ions [163]. As a model system, we have chosen a 2,6-bis(1-aryl-1,2,3-triazol-4-yl)pyridine (BTP) derivative, which incorporates a tridentate coordination site consisting of two triazole moieties bridged by a central pyridine ring. The BTP core is decorated with three alkoxy side chains providing enhanced solubility as well as improved propensity to physisorption on HOPG at the solid–liquid interface. Due to favorable electrostatic interactions, the "kinked" *anti,anti* conformation (Figure 13.10c) of the BTP core dominates in solution at neutral pH, whereas the repulsive interactions between the lone pair of the nitrogen atoms destabilize the alternative "extended" *syn,syn* conformation (Figure 13.10ć). This repulsive interaction can be switched into an attractive one by the addition of metal ions or acid to the solution, causing either metalation or protonation of the BTP core followed by a large structural change from the "kinked" to its corresponding "extended" conformation.

13.5
Conclusions and Outlook

In summary, we have given extensive details on how single molecules, decorated with molecular recognition units, are able to self-organize into 2D supramolecular crystalline materials. Supramolecular materials, that is, materials exploiting the use of reversible and dynamic noncovalent interactions, have the intrinsic ability to recognize and exchange their constituents; they are constitutionally dynamic materials and may in principle select their constituents in response to external stimuli. Increasing their complexity is one of the biggest challenges in materials science, leading to the elaboration of smart materials as well as *meta*-materials. Both smart and *meta*-materials are defined by their ability to transform energy, that is, the transformation of external stimuli into the desired smart or *meta*-property (photons into electrons, heat into electrons, electrons into displacement, and electrons into algorithms), making supramolecular materials excellent candidates for such colossal tasks.

We have also showed that how less resistive the path for the generation of complex supramolecular materials is, by thermodynamics, self-assembly, and their complexity and multifunctionality, depends on *engineering*, that is, the accurate positioning of single molecular functionalities in space. Following this line of thought, we have shown here the early attempts of *preprogramming* self-assembly as a predecessor of 2D supramolecular *engineering*, laying the foundations for an atomistic 3D bottom-up assembly. Two-dimensional supramolecular *engineering* is a science standing both at the margin of soft matter and solid crystals, as it remains an interfacial science of molecular dimensionality. The control of the amplitude of the crystallinity in two dimension is one of the first objectives of future scanning tunneling microscopy studies. Of much interest is also the characterization of 2D self-assembly at curved interfaces, much like micelles, membranes, nanotubes, and rough interfaces, which are representative objects of the curved space that we inhabit.

Finally, as we venture into Nature's art of supramolecular *engineering*, we face two major challenges: the implementation of new synthetic strategies for increasingly complex molecular building blocks and the development of an accurate atomistic modeling of full molecular ensembles. Continuous and progressive discoveries in the field of supramolecular chemistry along with realistic dynamic atomistic models will allow the understanding and implementation of the rules that govern the formation of nanostructures and provoke the extraordinary properties of functionalized (nano)materials.

Acknowledgments

This work was supported by the EC Marie-Curie RTNs PRAIRIES (MRTN-CT-2006-035810), ITN-SUPERIOR (PITN-GA-2009-238177), the EC FP7 ONE-P large-scale project no. 212311, the NanoSci-E+ project SENSORS, and the International Center for Frontier Research in Chemistry (FRC, Strasbourg).

References

1. Lehn, J.-M. (1990) Perspectives in supramolecular chemistry: from molecular recognition towards molecular information processing and self-organization. *Angew. Chem., Int. Ed.*, **29** (11), 1304–1319.
2. Yaghi, O.M., O'Keeffe, M., Ockwig, N.W., Chae, H.K., Eddaoudi, M., and Kim, J. (2003) Reticular synthesis and the design of new materials. *Nature*, **423** (6941), 705–714.
3. Mann, S. (2009) Self-assembly and transformation of hybrid nano-objects and nanostructures under equilibrium and non-equilibrium conditions. *Nat. Mater.*, **8**, 781–792.
4. Braga, D., Grepioni, F., and Desiraju, G.R. (1998) Crystal engineering and organometallic architecture. *Chem. Rev.*, **98** (4), 1375–1406.
5. Palmer, L.C., Velichko, Y.S., Olvera de la Cruz, M., and Stupp, S.I. (2007) Supramolecular self-assembly codes for functional structures. *Phil. Trans. R. Soc. A*, **365** (1855), 1417–1433.
6. Hosseini, M.W. (2005) Molecular tectonics: from simple tectons to complex molecular networks. *Acc. Chem. Res.*, **38** (4), 313–323.
7. Winfree, E., Liu, F.R., Wenzler, L.A., and Seeman, N.C. (1998) Design and self-assembly of two-dimensional DNA crystals. *Nature*, **394** (6693), 539–544.
8. Elemans, J.A.A.W., Lei, S.B., and De Feyter, S. (2009) Molecular and supramolecular networks on surfaces: from two-dimensional crystal engineering to reactivity. *Angew. Chem., Int. Ed.*, **48** (40), 7298–7332.
9. Bonifazi, D., Mohnani, S., and Llanes-Pallas, A. (2009) Supramolecular chemistry at interfaces: molecular recognition on nanopatterned porous surfaces. *Chem. Eur. J.*, **15** (29), 7004–7025.
10. Lackinger, M. and Heckl, W.M. (2009) Carboxylic acids: versatile building blocks and mediators for two-dimensional supramolecular self-assembly. *Langmuir*, **25** (19), 11307–11321.
11. Palermo, V. and Samorì, P. (2007) Molecular self-assembly across multiple length scales. *Angew. Chem., Int. Ed.*, **46** (24), 4428–4432.
12. Joachim, C., Gimzewski, J.K., and Aviram, A. (2000) Electronics using hybrid-molecular and mono-molecular devices. *Nature*, **408** (6812), 541–548.
13. Hamadani, B.H., Corley, D.A., Ciszek, J.W., Tour, J.M., and Natelson, D. (2006) Controlling charge injection in organic field-effect transistors using self-assembled monolayers. *Nano Lett.*, **6** (6), 1303–1306.
14. Metzger, R.M. (2008) Unimolecular electronics. *J. Mater. Chem.*, **18** (37), 4364–4396.
15. Reed, M.A., Zhou, C., Muller, C.J., Burgin, T.P., and Tour, J.M. (1997)

Conductance of a molecular junction. *Science*, **278** (5336), 252–254.

16 Arico, A.S., Bruce, P., Scrosati, B., Tarascon, J.-M., and van Schalkwijk, W. (2005) Nanostructured materials for advanced energy conversion and storage devices. *Nat. Mater.*, **4** (5), 366–377.

17 Binnig, G., Rohrer, H., Gerber, C., and Weibel, E. (1982) Tunneling through a controllable vacuum gap. *Appl. Phys. Lett.*, **40** (2), 178–180.

18 Elemans, J.A.A.W. and De Feyter, S. (2009) Structure and function revealed with submolecular resolution at the liquid–solid interface. *Soft Matter*, **5** (4), 721–735.

19 Cyr, D.M., Venkataraman, B., and Flynn, G.W. (1996) STM investigations of organic molecules physisorbed at the liquid–solid interface. *Chem. Mater.*, **8** (8), 1600–1615.

20 Yang, Y. and Wang, C. (2009) Hierarchical construction of self-assembled low-dimensional molecular architectures observed by using scanning tunneling microscopy. *Chem. Soc. Rev.*, **38** (9), 2576–2589.

21 Piot, L., Bonifazi, D., and Samorì, P. (2007) Organic reactivity in confined spaces under scanning tunneling microscopy control: tailoring the nanoscale world. *Adv. Funct. Mater.*, **17** (18), 3689–3693.

22 Tersoff, J. and Hamann, D.R. (1985) Theory of the scanning tunneling microscope. *Phys. Rev. B*, **31** (2), 805–813.

23 Hipps, K.W. and Scudiero, L. (2005) Electron tunneling, a quantum probe for the quantum world of nanotechnology. *J. Chem. Educ.*, **82** (5), 704.

24 Hipps, K.W. and Vij, D.R. (2006) *Handbook of Applied Solid State Spectroscopy*, Springer.

25 Samorì, P. and Rabe, J.P. (2002) Scanning probe microscopy explorations on conjugated (macro)molecular architectures for molecular electronics. *J. Phys. Condens. Matter*, **14** (42), 9955–9973.

26 Jäckel, F., Perera, U.G.E., Iancu, V., Braun, K.F., Koch, N., Rabe, J.P., and Hla, S.W. (2008) Investigating molecular charge transfer complexes with a low temperature scanning tunneling microscope. *Phys. Rev. Lett.*, **100** (12), 126102–126104.

27 Jäckel, F., Wang, Z., Watson, M.D., Müllen, K., and Rabe, J.P. (2004) Nanoscale array of inversely biased molecular rectifiers. *Chem. Phys. Lett.*, **387** (4–6), 372–376.

28 Jäckel, F., Watson, M.D., Müllen, K., and Rabe, J.P. (2004) Prototypical single-molecule chemical-field-effect transistor with nanometer-sized gates. *Phys. Rev. Lett.*, **92** (18), 188303.

29 Palma, C.A., Bjork, J., Bonini, M., Dyer, M.S., Llanes-Pallas, A., Bonifazi, D., Persson, M., and Samorì, P. (2009) Tailoring bicomponent supramolecular nanoporous networks: phase segregation, polymorphism, and glasses at the solid–liquid interface. *J. Am. Chem. Soc.*, **131** (36), 13062–13071.

30 Kampschulte, L., Werblowsky, T.L., Kishore, R.S.K., Schmittel, M., Heckl, W.M., and Lackinger, M. (2008) Thermodynamical equilibrium of binary supramolecular networks at the liquid–solid interface. *J. Am. Chem. Soc.*, **130** (26), 8502–8507.

31 Meirovitch, H. (2007) Recent developments in methodologies for calculating the entropy and free energy of biological systems by computer simulation. *Curr. Opin. Struct. Biol.*, **17** (2), 181–186.

32 Silly, F., Weber, U.K., Shaw, A.Q., Burlakov, V.M., Castell, M.R., Briggs, G.A.D., and Pettifor, D.G. (2008) Deriving molecular bonding from a macromolecular self-assembly using kinetic Monte Carlo simulations. *Phys. Rev. B*, **77** (20), 201408–201414.

33 Palma, C.-A., Bonini, M., Breiner, T., and Samorì, P. (2009) Supramolecular crystal engineering at the solid–liquid interface from first principles: toward unraveling the thermodynamics of 2D self-assembly. *Adv. Mater.*, **21** (13), 1383–1386.

34 Palma, C.A., Bonini, M., Llanes-Pallas, A., Breiner, T., Prato, M., Bonifazi, D., and Samorì, P. (2008) Pre-programmed

bicomponent porous networks at the solid–liquid interface: the low concentration regime. *Chem. Commun.*, (42), 5289–5291.

35 van der Waals, J.D. (1920) The equation of state for gases and liquids. Nobel lecture.

36 Björk, J., Hanke, F., Palma, C.-A., Samorì, P., Karplus, M., Cecchini, M., and Persson, M. (2009) in press.

37 Zacharia, R., Ulbricht, H., and Hertel, T. (2004) Interlayer cohesive energy of graphite from thermal desorption of polyaromatic hydrocarbons. *Phys. Rev. B*, **69** (15), 155406.

38 Gellman, A.J. and Paserba, K.R. (2002) Kinetics and mechanism of oligomer desorption from surfaces: *n*-alkanes on graphite. *J. Phys. Chem. B*, **106** (51), 13231–13241.

39 Wetterer, S.M., Lavrich, D.J., Cummings, T., Bernasek, S.L., and Scoles, G. (1998) Energetics and kinetics of the physisorption of hydrocarbons on Au (111). *J. Phys. Chem. B*, **102** (46), 9266–9275.

40 Kastler, M., Pisula, W., Wasserfallen, D., Pakula, T., and Müllen, K. (2005) Influence of alkyl substituents on the solution- and surface-organization of hexa-*peri*-hexabenzocoronenes. *J. Am. Chem. Soc.*, **127** (12), 4286–4296.

41 Zhdanov, V.P. (1991) Arrhenius parameters for rate processes on solid surfaces. *Surf. Sci. Report*, **12** (5), 185–242.

42 Barth, J.V. (2007) Molecular architectonic on metal surfaces. *Annu. Rev. Phys. Chem.*, **58** (1), 375–407.

43 Lide, D.R. (ed.) (2007) *Handbook of Chemistry and Physics*, 87 edn, CRC Press.

44 Wiesendanger, R., Guntherodt, H.J., and Siegnhaler, H. (1992) *Scanning Tunneling Microscopy II*, Springer, Basel.

45 Itaya, K. (1998) *In situ* scanning tunneling microscopy in electrolyte solutions. *Prog. Surf. Sci.*, **58** (3), 121–247.

46 Margenau, H. (1939) van der Waals forces. *Rev. Mod. Phys.*, **11** (1), 1–35.

47 Israelachvili, J. (1992) *Intermolecular & Surface Forces*, Academic Press, London.

48 Rabe, J.P. and Buchholz, S. (1991) Commensurability and mobility in two-dimensional molecular patterns on graphite. *Science*, **253** (5018), 424–427.

49 Samorì, P., Fechtenkötter, A., Jäckel, F., Böhme, T., Müllen, K., and Rabe, J.P. (2001) Supramolecular staircase via self-assembly of disklike molecules at the solid–liquid interface. *J. Am. Chem. Soc.*, **123** (46), 11462–11467.

50 Yin, S., Wang, C., Qiu, X., Xu, B., and Bai, C. (2001) Theoretical study of the effects of intermolecular interactions in self-assembled long-chain alkanes adsorbed on graphite surface. *Surf. Interface Anal.*, **32** (1), 248–252.

51 Ilan, B., Florio, G.M., Hybertsen, M.S., Berne, B.J., and Flynn, G.W. (2008) Scanning tunneling microscopy images of alkane derivatives on graphite: role of electronic effects. *Nano Lett.*, **8** (10), 3160–3165.

52 Bleger, D., Kreher, D., Mathevet, F., Attias, A.J., Schull, G., Huard, A., Douillard, L., Fiorini-Debuischert, C., and Charra, F. (2007) Surface noncovalent bonding for rational design of hierarchical molecular self-assemblies. *Angew. Chem., Int. Ed.*, **46** (39), 7404–7407.

53 Adisoejoso, J., Tahara, K., Okuhata, S., Lei, S., Tobe, Y., and De Feyter, S. (2009) Two-dimensional crystal engineering: a four-component architecture at a liquid–solid interface. *Angew. Chem., Int. Ed.*, **48** (40), 7267.

54 Lei, S., Tahara, K., Feng, X., Furukawa, S., De Schryver, F.C., Müllen, K., Tobe, Y., and De Feyter, S. (2008) Molecular clusters in two-dimensional surface-confined nanoporous molecular networks: structure, rigidity, and dynamics. *J. Am. Chem. Soc.*, **130** (22), 7119–7129.

55 Tahara, K., Lei, S., Mossinger, D., Kozuma, H., Inukai, K., Van der Auweraer, M., De Schryver, F.C., Hoger, S., Tobe, Y., and De Feyter, S. (2008) Giant molecular spoked wheels in giant voids: two-dimensional molecular self-assembly goes big. *Chem. Commun.*, (33), 3897–3899.

56 Lei, S., Tahara, K., De Schryver, F.C., Van der Auweraer, M., Tobe, Y., and De Feyter, S. (2008) One building block, two

different supramolecular surface-confined patterns: concentration in control at the solid–liquid interface. *Angew. Chem., Int. Ed.*, **120** (16), 3006–3010.
57 De Gennes, P.G. and Prost, J. (1993) *The Physics of Liquid Crystals*, 2nd edn, Oxford Science Publication, Paris.
58 Samorì, P., Severin, N., Simpson, C.D., Müllen, K., and Rabe, J.P. (2002) Epitaxial composite layers of electron donors and acceptors from very large polycyclic aromatic hydrocarbons. *J. Am. Chem. Soc.*, **124** (32), 9454–9457.
59 Feng, X., Pisula, W., Kudernac, T., Wu, D., Zhi, L., De Feyter, S., and Mullen, K. (2009) Controlled self-assembly of C_3-symmetric hexa-*peri*-hexabenzocoronenes with alternating hydrophilic and hydrophobic substituents in solution, in the bulk, and on a surface. *J. Am. Chem. Soc.*, **131** (12), 4439–4448.
60 Grimme, S. (2008) Do special noncovalent π–π stacking interactions really exist? *Angew. Chem., Int. Ed.*, **47** (18), 3430–3434.
61 Mativetsky, J.M., Kastler, M., Savage, R.C., Gentilini, D., Palma, M., Pisula, W., Müllen, K., and Samorì, P. (2009) Self-assembly of a donor–acceptor dyad across multiple length scales: functional architectures for organic electronics. *Adv. Funct. Mater.*, **19** (15), 2486–2494.
62 Feng, X., Marcon, V., Pisula, W., Hansen, M.R., Kirkpatrick, J., Grozema, F., Andrienko, D., Kremer, K., and Müllen, K. (2009) Towards high charge-carrier mobilities by rational design of the shape and periphery of discotics. *Nat. Mater.*, **8** (5), 421–426.
63 Bleger, D., Kreher, D., Mathevet, F., Attias, A.J., Arfaoui, I., Metge, G., Douillard, L., Fiorini-Debuisschert, C., and Charra, F. (2008) Periodic positioning of multilayered [2.2] paracyclophane-based nanopillars. *Angew. Chem., Int. Ed.*, **47** (44), 8412–8415.
64 Watson, M.D., Jäckel, F., Severin, N., Rabe, J.P., and Müllen, K. (2004) A hexa-*peri*-hexabenzocoronene cyclophane: an addition to the toolbox for molecular electronics. *J. Am. Chem. Soc.*, **126** (5), 1402–1407.
65 Klymchenko, A.S., Sleven, J., Binnemans, K., and De Feyter, S. (2006) Two-dimensional self-assembly and phase behavior of an alkoxylated sandwich-type bisphthalocyanine and its phthalocyanine analogues at the liquid–solid interface. *Langmuir*, **22** (2), 723–728.
66 Takami, T., Arnold, D.P., Fuchs, A.V., Will, G.D., Goh, R., Waclawik, E.R., Bell, J.M., Weiss, P.S., Sugiura, K., Liu, W., and Jiang, J. (2006) Two-dimensional crystal growth and stacking of bis (phthalocyaninato) rare earth sandwich complexes at the 1-phenyloctane/graphite interface. *J. Phys. Chem. B*, **110** (4), 1661–1664.
67 Gomez-Segura, J., Diez-Perez, I., Ishikawa, N., Nakano, M., Veciana, J., and Ruiz-Molina, D. (2006) 2-D self-assembly of the bis(phthalocyaninato)terbium(iii) single-molecule magnet studied by scanning tunnelling microscopy. *Chem. Commun.*, (27), 2866–2868.
68 Qiu, X., Wang, C., Zeng, Q., Xu, B., Yin, S., Wang, H., Xu, S., and Bai, C. (2000) Alkane-assisted adsorption and assembly of phthalocyanines and porphyrins. *J. Am. Chem. Soc.*, **122** (23), 5550–5556.
69 Baisch, B., Raffa, D., Jung, U., Magnussen, O.M., Nicolas, C., Lacour, J., Kubitschke, J., and Herges, R. (2009) Mounting freestanding molecular functions onto surfaces: the platform approach. *J. Am. Chem. Soc.*, **131** (2), 442–443.
70 Visser, J., Katsonis, N., Vicario, J., and Feringa, B.L. (2009) Two-dimensional molecular patterning by surface-enhanced Zn–porphyrin coordination. *Langmuir*, **25** (10), 5980–5985.
71 Shekhah, O., Wang, H., Paradinas, M., Ocal, C., Schupbach, B., Terfort, A., Zacher, D., Fischer, R.A., and Woll, C. (2009) Controlling interpenetration in metal–organic frameworks by liquid-phase epitaxy. *Nat. Mater.*, **8** (6), 481–484.
72 Decher, G. (1997) Fuzzy nanoassemblies: toward layered polymeric multicomposites. *Science*, **277** (5330), 1232–1237.

73 Piot, L., Marchenko, A., Wu, J., Müllen, K., and Fichou, D. (2005) Structural evolution of hexa-*peri*-hexabenzocoronene adlayers in heteroepitaxy on *n*-pentacontane template monolayers. *J. Am. Chem. Soc.*, **127** (46), 16245–16250.

74 Stabel, A., Herwig, P., Müllen, K., and Rabe, J.P. (1995) Diode-like current–voltage curves for a single molecule-tunneling spectroscopy with submolecular resolution of an alkylated, pericondensed hexabenzocoronene. *Angew. Chem., Int. Ed.*, **34** (15), 1609–1611.

75 Ito, S., Wehmeier, M., Brand, J.D., Kübel, C., Epsch, R., Rabe, J.P., and Müllen, K. (2000) Synthesis and self-assembly of functionalized hexa-*peri*-hexabenzocoronenes. *Chem. Eur. J.*, **6** (23), 4327–4342.

76 Samorì, P., Yin, X.M., Tchebotareva, N., Wang, Z.H., Pakula, T., Jäckel, F., Watson, M.D., Venturini, A., Müllen, K., and Rabe, J.P. (2004) Self-assembly of electron donor–acceptor dyads into ordered architectures in two and three dimensions: surface patterning and columnar "double cables". *J. Am. Chem. Soc.*, **126** (11), 3567–3575.

77 May, V. and Kühn, O. (2004) *Charge and Energy Transfer Dynamics in Molecular Systems*, Wiley-VCH Verlag GmbH.

78 Liscio, A., De Luca, G., Nolde, F., Palermo, V., Müllen, K., and Samorì, P. (2008) Photovoltaic charge generation visualized at the nanoscale: a proof of principle. *J. Am. Chem. Soc.*, **130** (3), 780–781.

79 Chiang, C.K., Fincher, C.R., Park, Y.W., Heeger, A.J., Shirakawa, H., Louis, E.J., Gau, S.C., and MacDiarmid, A.G. (1977) Electrical conductivity in doped polyacetylene. *Phys. Rev. Lett.*, **39** (17), 1098.

80 Iijima, S. (1991) Helical microtubules of graphitic carbon. *Nature*, **354** (6348), 56–58.

81 Geim, A.K. and Novoselov, K.S. (2007) The rise of graphene. *Nat. Mater.*, **6** (3), 183–191.

82 Samorì, P., Simpson, C.D., Müllen, M., and Rabe, J.P. (2002) Ordered monolayers of nanographitic sheets processed from solutions via oxidative cyclodehydrogenation. *Langmuir*, **18** (11), 4183–4185.

83 Son, Y.-W., Cohen, M.L., and Louie, S.G. (2006) Half-metallic graphene nanoribbons. *Nature*, **444** (7117), 347–349.

84 Yang, X., Dou, X., Rouhanipour, A., Zhi, L., Rader, H.J., and Müllen, K. (2008) Two-dimensional graphene nanoribbons. *J. Am. Chem. Soc.*, **130** (13), 4216–4217.

85 Jackel, F., Watson, M.D., Mullen, K., and Rabe, J.P. (2006) Tunneling through nanographene stacks. *Phys. Rev. B*, **73** (4), 1–6.

86 Kunz, H. (2002) Emil Fischer: unequalled classicist, master of organic chemistry research, and inspired trailblazer of biological chemistry. *Angew. Chem., Int. Ed.*, **41** (23), 4439–4451.

87 Tahara, K., Lei, S., Mamdouh, W., Yamaguchi, Y., Ichikawa, T., Uji-I, H., Sonoda, M., Hirose, K., De Schryver, F.C., De Feyter, S., and Tobe, Y. (2008) Site-selective guest inclusion in molecular networks of butadiyne-bridged pyridino and benzeno square macrocycles on a surface. *J. Am. Chem. Soc.*, **130** (21), 6666.

88 Wasielewski, M.R. (2006) Energy, charge, and spin transport in molecules and self-assembled nanostructures inspired by photosynthesis. *J. Org. Chem.*, **71** (14), 5051–5066.

89 Balaban, T.S. (2005) Tailoring porphyrins and chlorins for self-assembly in biomimetic artificial antenna systems. *Acc. Chem. Res.*, **38** (8), 612–623.

90 Mujica, V., Kemp, M., Roitberg, A., and Ratner, M.A. (1996) Current–voltage characteristics of molecular wires: eigenvalue staircase, Coulomb blockade, and rectification. *J. Chem. Phys.*, **104** (18), 7296–7305.

91 Smilowitz, L., Sariciftci, N.S., Wu, R., Gettinger, C., Heeger, A.J., and Wudl, F. (1993) Photoexcitation spectroscopy of conducting-polymer–C_{60} composites: photoinduced electron transfer. *Phys. Rev. B*, **47** (20), 13835.

92 Pan, G.-B., Cheng, X.-H., Hoger, S., and Freyland, W. (2006) 2D supramolecular

structures of a shape-persistent macrocycle and Co-deposition with fullerene on HOPG. *J. Am. Chem. Soc.*, **128** (13), 4218–4219.

93 Mena-Osteritz, E. and Bäuerle, P. (2006) Complexation of C_{60} on a cyclothiophene monolayer template. *Adv. Mater.*, **18** (4), 447–451.

94 Yang, Y.C., Chang, C.H., and Lee, Y.L. (2007) Complexation of fullerenes on a pentacene-modified Au(111) surface. *Chem. Mater.*, **19** (25), 6126–6130.

95 Aviram, A. and Ratner, M.A. (1974) Molecular rectifiers. *Chem. Phys. Lett.*, **29** (2), 277–283.

96 Dorogi, M., Gomez, J., Osifchin, R., Andres, R.P., and Reifenberger, R. (1995) Room-temperature Coulomb blockade from a self-assembled molecular nanostructure. *Phys. Rev. B*, **52** (12), 9071.

97 Joachim, C., Gimzewski, J.K., and Tang, H. (1998) Physical principles of the single-C_{60} transistor effect. *Phys. Rev. B*, **58** (24), 16407.

98 Tchebotareva, N., Yin, X., Watson, M.D., Samorì, P., Rabe, J.P., and Müllen, K. (2003) Ordered architectures of a soluble hexa-*peri*-hexabenzocoronene–pyrene dyad: thermotropic bulk properties and nanoscale phase segregation at surfaces. *J. Am. Chem. Soc.*, **125** (32), 9734–9739.

99 Surin, M. and Samorì, P. (2007) Multicomponent monolayer architectures at the solid–liquid interface: towards controlled space-confined properties and reactivity of functional building blocks. *Small*, **3** (2), 190–194.

100 Greef, T.F.A. and Meijer, E.W. (2008) Materials science: supramolecular polymers. *Nature*, **453** (7192), 171–173.

101 Brunsveld, L., Folmer, B.J.B., Meijer, E.W., and Sijbesma, R.P. (2001) Supramolecular polymers. *Chem. Rev.*, **101** (12), 4071–4097.

102 Sherrington, D.C. and Taskinen, K.A. (2001) Self-assembly in synthetic macromolecular systems via multiple hydrogen bonding interactions. *Chem. Soc. Rev.*, **30** (2), 83–93.

103 Ciesielski, A., Schaeffer, G., Petitjean, A., Lehn, J.M., and Samorì, P. (2009) STM insight into hydrogen-bonded bicomponent 1D supramolecular polymers with controlled geometries at the liquid–solid interface. *Angew. Chem., Int. Ed.*, **48** (11), 2039–2043.

104 Fouquey, C., Lehn, J.-M., and Levelut, A.-M. (1990) Molecular recognition directed self-assembly of supramolecular liquid crystalline polymers from complementary chiral components. *Adv. Mater.*, **2** (5), 254–257.

105 Lehn, J.-M. (1995) *Supramolecular Chemistry: Concepts and Perspectives*, Wiley-VCH Verlag GmbH, New York.

106 Subramanian, S. and Zaworotko, M.J. (1994) Exploitation of the hydrogen-bond: recent developments in the context of crystal engineering. *Coord. Chem. Rev.*, **137**, 357–401.

107 Fan, E., Vicent, C., Geib, S.J., and Hamilton, A.D. (1994) Molecular recognition in the solid-state: hydrogen-bonding control of molecular aggregation. *Chem. Mater.*, **6** (8), 1113–1117.

108 Macdonald, J.C. and Whitesides, G.M. (1994) Solid-state structures of hydrogen-bonded tapes based on cyclic secondary diamides. *Chem. Rev.*, **94** (8), 2383–2420.

109 Jeffrey, G.A. (1997) *An Introduction to Hydrogen Bonding*, Oxford University Press.

110 Murray, T.J. and Zimmerman, S.C. (1992) New triply hydrogen-bonded complexes with highly variable stabilities. *J. Am. Chem. Soc.*, **114** (10), 4010–4011.

111 Djurdjevic, S., Leigh, D.A., McNab, H., Parsons, S., Teobaldi, G., and Zerbetto, F. (2007) Extremely strong and readily accessible AAA–DDD triple hydrogen bond complexes. *J. Am. Chem. Soc.*, **129** (3), 476–477.

112 Söntjens, S.H.M., Sijbesma, R.P., van Genderen, M.H.P., and Meijer, E.W. (2000) Stability and lifetime of quadruply hydrogen bonded 2-ureido-4[1*H*]-pyrimidinone dimers. *J. Am. Chem. Soc.*, **122** (31), 7487–7493.

113 Lukin, O. and Leszczynski, J. (2002) Rationalizing the strength of hydrogen-bonded complexes. *Ab initio* HF and DFT studies. *J. Phys. Chem. A*, **106** (29), 6775–6782.

114 Mebel, A.M., Morokuma, K., and Lin, M.C. (1995) Modification of the Gaussian-2 theoretical-model: the use of coupled-cluster energies, density-functional geometries, and frequencies. *J. Chem. Phys.*, **103** (17), 7414–7421.

115 Gu, J.D. and Leszczynski, J. (2000) A remarkable alteration in the bonding pattern: an HF and DFT study of the interactions between the metal cations and the Hoogsteen hydrogen-bonded G-tetrad. *J. Phys. Chem. A*, **104** (26), 6308–6313.

116 Gu, J.D. and Leszczynski, J. (2000) Structures and properties of the planar G·C·G·C tetrads: *ab initio* HF and DFT studies. *J. Phys. Chem. A*, **104** (31), 7353–7358.

117 Piot, L., Palma, C.-A., Llanes-Pallas, A., Prato, M., Szekrényes, Z., Kamarás, K., Bonifazi, D., and Samorì, P. (2009) Selective formation of bi-component arrays through H-bonding of multivalent molecular modules. *Adv. Funct. Mater.*, **19** (8), 1207–1214.

118 Meier, C., Ziener, U., Landfester, K., and Weihrich, P. (2005) Weak hydrogen bonds as a structural motif for two-dimensional assemblies of oligopyridines on highly oriented pyrolytic graphite: an STM investigation. *J. Phys. Chem. B*, **109** (44), 21015–21027.

119 Mu, Z., Shu, L., Fuchs, H., Mayor, M., and Chi, L. (2008) Two dimensional chiral networks emerging from the Aryl–F · · · H hydrogen-bond-driven self-assembly of partially fluorinated rigid molecular structures. *J. Am. Chem. Soc.*, **130** (33), 10840–10841.

120 Lackinger, M., Griessl, S., Markert, T., Jamitzky, F., and Heckl, W.M. (2004) Self-assembly of benzene-dicarboxylic acid isomers at the liquid solid interface: steric aspects of hydrogen bonding. *J. Phys. Chem. B*, **108** (36), 13652–13655.

121 Blunt, M.O., Russell, J.C., Gimenez-Lopez, M.D., Garrahan, J.P., Lin, X., Schroder, M., Champness, N.R., and Beton, P.H. (2008) Random tiling and topological defects in a two-dimensional molecular network. *Science*, **322** (5904), 1077–1081.

122 MacLeod, J.M., Ivasenko, O., Perepichka, D.F., and Rosei, F. (2007) Stabilization of exotic minority phases in a multicomponent self-assembled molecular network. *Nanotechnology*, **18** (42), 424031.

123 Kampschulte, L., Lackinger, M., Maier, A.K., Kishore, R.S.K., Griessl, S., Schmittel, M., and Heckl, W.M. (2006) Solvent induced polymorphism in supramolecular 1,3,5-benzenetribenzoic acid monolayers. *J. Phys. Chem. B*, **110** (22), 10829–10836.

124 Walch, H., Maier, A.K., Heckl, W.M., and Lackinger, M. (2009) Isotopological supramolecular networks from melamine and fatty acids. *J. Phys. Chem. C*, **113** (3), 1014–1019.

125 Nath, K.G., Ivasenko, O., Miwa, J.A., Dang, H., Wuest, J.D., Nanci, A., Perepichka, D.F., and Rosei, F. (2006) Rational modulation of the periodicity in linear hydrogen-bonded assemblies of trimesic acid on surfaces. *J. Am. Chem. Soc.*, **128** (13), 4212–4213.

126 Mourran, A., Ziener, U., Möller, M., Suarez, M., and Lehn, J.M. (2006) Homo- and heteroassemblies of lactim/lactam recognition patterns on highly ordered pyrolytic graphite: an STM investigation. *Langmuir*, **22** (18), 7579–7586.

127 Miura, A., Jonkheijm, P., De Feyter, S., Schenning, A.P.H.J., Meijer, E.W., and De Schryver, F.C. (2005) 2D self-assembly of oligo(*p*-phenylene vinylene) derivatives: from dimers to chiral rosettes. *Small*, **1** (1), 131–137.

128 Matmour, R., De Cat, I., George, S.J., Adriaens, W., Leclère, P., Bomans, P.H.H., Sommerdijk, N.A.J.M., Gielen, J.C., Christianen, P.C.M., Heldens, J.T., van Hest, J.C.M., Löwik, D.W.P.M., De Feyter, S., Meijer, E.W., and Schenning, A.P.H.J. (2008) Oligo (*p*-phenylenevinylene)–peptide conjugates: synthesis and self-assembly in solution and at the solid–liquid interface. *J. Am. Chem. Soc.*, **130** (44), 14576–14583.

129 Lei, S., Puigmarti-Luis, J., Minoia, A., Van der Auweraer, M., Rovira, C., Lazzaroni, R., Amabilino, D.B., and De Feyter, S. (2008) Bottom-up assembly of high

density molecular nanowire cross junctions at a solid/liquid interface. *Chem. Commun.*, (6), 703–705.

130 Gesquiere, A., De Feyter, S., De Schryver, F.C., Schoonbeek, F., van Esch, J., Kellogg, R.M., and Feringa, B.L. (2001) Supramolecular π-stacked assemblies of bis(urea)-substituted thiophene derivatives and their electronic properties probed with scanning tunneling microscopy and scanning tunneling spectroscopy. *Nano Lett.*, **1** (4), 201–206.

131 Gesquiere, A., Abdel-Mottaleb, M.M.S., De Feyter, S., De Schryver, F.C., Schoonbeek, F., van Esch, J., Kellogg, R.M., Feringa, B.L., Calderone, A., Lazzaroni, R., and Bredas, J.L. (2000) Molecular organization of bis-urea substituted thiophene derivatives at the liquid/solid interface studied by scanning tunneling microscopy. *Langmuir*, **16** (26), 10385–10391.

132 Spada, G.P., Lena, S., Masiero, S., Pieraccini, S., Surin, M., and Samorì, P. (2008) Guanosine-based hydrogen-bonded scaffolds: controlling the assembly of oligothiophenes. *Adv. Mater.*, **20** (12), 2433–2439.

133 Gottarelli, G., Masiero, S., Mezzina, E., Pieraccini, S., Rabe, J.P., Samorì, P., and Spada, G.P. (2000) The self-assembly of lipophilic guanosine derivatives in solution and on solid surfaces. *Chem. Eur. J.*, **6** (17), 3242–3248.

134 Giorgi, T., Lena, S., Mariani, P., Cremonini, M.A., Masiero, S., Pieraccini, S., Rabe, J.P., Samorì, P., Spada, G.P., and Gottarelli, G. (2003) Supramolecular helices via self-assembly of 8-oxoguanosines. *J. Am. Chem. Soc.*, **125** (48), 14741–14749.

135 Pieraccini, S., Masiero, S., Pandoli, O., Samorì, P., and Spada, G.P. (2006) Reversible interconversion between a supramolecular polymer and a discrete octameric species from a guanosine derivative by dynamic cation binding and release. *Org. Lett.*, **8** (14), 3125–3128.

136 Lena, S., Brancolini, G., Gottarelli, G., Mariani, P., Masiero, S., Venturini, A., Palermo, V., Pandoli, O., Pieraccini, S., Samorì, P., and Spada, G.P. (2007) Self-assembly of an alkylated guanosine derivative into ordered supramolecular nanoribbons in solution and on solid surfaces. *Chem. Eur. J.*, **13** (13), 3757–3764.

137 Seto, C.T. and Whitesides, G.M. (1990) Self-assembly based on the cyanuric acid–melamine lattice. *J. Am. Chem. Soc.*, **112** (17), 6409–6411.

138 Theobald, J.A., Oxtoby, N.S., Phillips, M.A., Champness, N.R., and Beton, P.H. (2003) Controlling molecular deposition and layer structure with supramolecular surface assemblies. *Nature*, **424** (6952), 1029–1031.

139 Llanes-Pallas, A., Palma, C.A., Piot, L., Belbakra, A., Listorti, A., Prato, M., Samorì, P., Armaroli, N., and Bonifazi, D. (2009) Engineering of supramolecular H-bonded nanopolygons via self-assembly of programmed molecular modules. *J. Am. Chem. Soc.*, **131** (2), 509–520.

140 Madueno, R., Raisanen, M.T., Silien, C., and Buck, M. (2008) Functionalizing hydrogen-bonded surface networks with self-assembled monolayers. *Nature*, **454** (7204), 618–621.

141 Gesquiere, A., Jonkheijm, P., Hoeben, F.J.M., Schenning, A.P.H.J., De Feyter, S., De Schryver, F.C., and Meijer, E.W. (2004) 2D-structures of quadruple hydrogen bonded oligo(p-phenylenevinylene)s on graphite: self-assembly behavior and expression of chirality. *Nano Lett.*, **4** (7), 1175–1179.

142 Batten, S.R. and Robson, R. (1998) Interpenetrating nets: ordered, periodic entanglement. *Angew. Chem., Int. Ed.*, **37** (11), 1460–1494.

143 Blake, A.J., Champness, N.R., Hubberstey, P., Li, W.S., Withersby, M.A., and Schroder, M. (1999) Inorganic crystal engineering using self-assembly of tailored building-blocks. *Coord. Chem. Rev.*, **183**, 117–138.

144 Moulton, B. and Zaworotko, M.J. (2001) From molecules to crystal engineering: supramolecular isomerism and polymorphism in network solids. *Chem. Rev.*, **101** (6), 1629–1658.

145 Janiak, C. (2003) Engineering coordination polymers towards

applications. *Dalton Trans.*, (14), 2781–2804.
146 Kitagawa, S., Kitaura, R., and Noro, S. (2004) Functional porous coordination polymers. *Angew. Chem., Int. Ed.*, **43** (18), 2334–2375.
147 Ferey, G., Mellot-Draznieks, C., Serre, C., and Millange, F. (2005) Crystallized frameworks with giant pores: Are there limits to the possible? *Acc. Chem. Res.*, **38** (4), 217–225.
148 Thompson, A.M.W.C. (1997) The synthesis of 2,2′:6′,2″-terpyridine ligands: versatile building blocks for supramolecular chemistry. *Coord. Chem. Rev.*, **160**, 1–52.
149 Liang, W.J., Shores, M.P., Bockrath, M., Long, J.R., and Park, H. (2002) Kondo resonance in a single-molecule transistor. *Nature*, **417** (6890), 725–729.
150 Barboiu, M. and Lehn, J.M. (2002) Dynamic chemical devices: modulation of contraction/extension molecular motion by coupled-ion binding/pH change-induced structural switching. *Proc. Natl. Acad. Sci. USA*, **99** (8), 5201–5206.
151 Kurth, D.G., Severin, N., and Rabe, J.P. (2002) Perfectly straight nanostructures of metallosupramolecular coordination–polyelectrolyte amphiphile complexes on graphite. *Angew. Chem., Int. Ed.*, **41** (19), 3681–3683.
152 Stepanow, S., Lin, N., Barth, J.V., and Kern, K. (2006) Surface-template assembly of two-dimensional metal–organic coordination networks. *J. Phys. Chem. B*, **110** (46), 23472–23477.
153 Stepanow, S., Lingenfelder, M., Dmitriev, A., Spillmann, H., Delvigne, E., Lin, N., Deng, X.B., Cai, C.Z., Barth, J.V., and Kern, K. (2004) Steering molecular organization and host–guest interactions using two-dimensional nanoporous coordination systems. *Nat. Mater.*, **3** (4), 229–233.
154 Kikkawa, Y., Koyama, E., Tsuzuki, S., Fujiwara, K., Miyake, K., Tokuhisa, H., and Kanesato, M. (2007) Odd–even effect and metal induced structural convergence in self-assembled monolayers of bipyridine derivatives. *Chem. Commun.*, (13), 1343–1345.
155 Newkome, G.R., Wang, P.S., Moorefield, C.N., Cho, T.J., Mohapatra, P.P., Li, S.N., Hwang, S.H., Lukoyanova, O., Echegoyen, L., Palagallo, J.A., Iancu, V., and Hla, S.W. (2006) Nanoassembly of a fractal polymer: a molecular "Sierpinski hexagonal gasket". *Science*, **312** (5781), 1782–1785.
156 Zell, P., Mogele, F., Ziener, U., and Rieger, B. (2006) Fine-tuning of relative metal–metal distances within highly ordered chiral 2D nanopatterns. *Chem. Eur. J.*, **12** (14), 3847–3857.
157 Ciesielski, A., Piot, L., Samorì, P., Jouaiti, A., and Hosseini, M.W. (2009) Molecular tectonics at the solid/liquid interface: controlling the nanoscale geometry, directionality, and packing of 1D coordination networks on graphite surfaces. *Adv. Mater.*, **21** (10–11), 1131–1136.
158 Gong, J.R., Wan, L.J., Yuan, Q.H., Bai, C.L., Jude, H., and Stang, P.J. (2005) Mesoscopic self-organization of a self-assembled supramolecular rectangle on highly oriented pyrolytic graphite and Au(111) surfaces. *Proc. Natl. Acad. Sci. USA*, **102** (4), 971–974.
159 Surin, M., Samorì, P., Jouaiti, A., Kyritsakas, N., and Hosseini, M.W. (2007) Molecular tectonics on surfaces: bottom-up fabrication of 1D coordination networks that form 1D and 2D arrays on graphite. *Angew. Chem., Int. Ed.*, **46** (1–2), 245–249.
160 Abdel-Mottaleb, M.M.S., Schuurmans, N., De Feyter, S., Van Esch, J., Feringa, B.L., and De Schryver, F.C. (2002) Submolecular visualisation of palladium acetate complexation with a bipyridine derivative at a graphite surface. *Chem. Commun.*, (17), 1894–1895.
161 Ziener, U., Lehn, J.M., Mourran, A., and Möller, M. (2002) Supramolecular assemblies of a bis(terpyridine) ligand and of its [2×2] grid-type Zn-II and Co-II complexes on highly ordered pyrolytic graphite. *Chem. Eur. J.*, **8** (4), 951–957.

162 Pace, G., Stefankiewicz, A., Harrowfield, J., Lehn, J.M., and Sarnori, P. (2009) Self-assembly of alkoxy-substituted bis(hydrazone)-based organic ligands and of a metallosupramolecular grid on graphite. *ChemPhysChem*, **10** (4), 699–705.

163 Piot, L., Meudtner, R.M., El Malah, T., Hecht, S., and Samorì, P. (2009) Modulating large-area self-assembly at the solid–liquid interface by pH-mediated conformational switching. *Chem. Eur. J.*, **15** (19), 4788–4792.

14
STM Characterization of Supramolecular Materials with Potential for Organic Electronics and Nanotechnology

Kevin R. Moonoosawmy, Jennifer M. MacLeod, and Federico Rosei

14.1
Introduction

Nanotechnology is a broad field of research that encompasses fundamental and applied science and relies on the controlled fabrication of materials at the nanoscale [1]. The quest for faster and smaller devices, in accordance with Moore's empirical law [2], has prompted great interest in cost-effective design and implementation of functional nanostructured materials [3–5]. One of the fastest emerging areas in nanotechnology is that of organic electronics, which includes field-effect transistors (FETs), light-emitting devices (LEDs), chemical sensors, and photovoltaic (PV) cells, that is, devices that make use of thin layers of organic molecules as active material [3, 6–9]. These organic layers are generally formed by self-assembly, that is, long-range order is driven by noncovalent interactions such as van der Waals forces and hydrogen bonding. This type of approach, usually referred to as "bottom-up," has a considerable advantage over top-down techniques (e.g., lithography) since the smallest feature size that can be defined through self-assembly depends only on the size of the molecular building blocks [10].

Supramolecular chemistry [11–13] has been largely inspired by biological systems in nature [14, 15]. Supramolecular materials are generated by the assembly of organic building blocks, which can be thought of as elementary components of complex architectures. One of the advantages of using organic molecules as building blocks is that they can be tailor-made, by an assortment of synthetic routes, to tune their properties and processability for specific applications [16–21]. Pivotal to the organization of molecules is the interplay between intermolecular and interfacial interactions. In some cases, the latter can lead to two-dimensional supramolecular assemblies or thin films with properties that are different from their bulk counterparts [4, 5].

Understanding the fundamental processes that drive nanoscale assembly is essential to harnessing the full potential of supramolecular structures for organic electronics and nanotechnology. This chapter highlights how high-resolution scanning tunneling microscopy (STM), as well as some of its specialized counterpart techniques, can address fundamental questions regarding the adsorption of organic

---

Functional Supramolecular Architectures. Edited by Paolo Samorì and Franco Cacialli
Copyright © 2011 WILEY-VCH Verlag GmbH & Co. KGaA, Weinheim
ISBN: 978-3-527-32611-2

molecules at surfaces. We will not provide a detailed overview here of the mechanisms for supramolecular self-assembly; instead, we direct the reader to recent reviews that provide excellent descriptions of different aspects of H-bonded self-assembled molecular networks (SAMNs) [22, 23] and of metal–organic coordination networks (MOCNs) [24–30]. Selected examples are used to cover numerous aspects of molecule surface interactions. Finally, we conclude by presenting a summary and broad perspective of future directions.

14.2
Characterization Using STM and Related Technologies

In the 28 years since its invention, STM has emerged as one of the most powerful and ubiquitous surface characterization techniques. This is in no small part due to the widespread adoption of STM both as a tool and as a technical challenge; while commercial production of high-quality STMs has made the technique commonplace in research labs, a community of scientists interested in pushing the boundaries of STM is continually modifying and improving different instruments in one way or another. In this section, we will provide both an overview of the basic capabilities of STM and a brief survey of some of the cutting-edge modifications being implemented in specialized instruments. In both cases, we will emphasize the relevance of STM to the study of supramolecular systems.

14.2.1
The Capabilities of STM

14.2.1.1 High-Resolution Imaging

STM is predominantly used to provide nanoscale images of conducting and semiconducting surfaces. With a sufficiently sharp tip, the STM can routinely provide submolecular resolution of isolated and supramolecularly associated molecules on a surface. Although the features in STM images simultaneously reflect both topographic and electronic contributions, with some knowledge of the electronic structure of the substrate and overlying molecules it is possible to extract structural information from an image [31, 32], or through comparison with simple molecular models [33]. More sophisticated molecular modeling approaches such as density functional theory (DFT) can take the substrate into account, allowing the generation of simulated STM images [34, 35], which can then be directly compared with experimental data [36–45].

Most STM instruments are capable of imaging areas from a few nanometers to a few microns. While large-scale images can lack submolecular, or even molecular, resolution, they can provide important information about the size, ordering, and interaction of domains in molecular networks [46–50]. Large-scale images can also be invaluable for identifying structured regions of molecules in cases where conditions do not allow surface-wide self-assembly or where multiple phases coexist [51–53]. These types of localized phenomena are especially important for systems to be used in molecular electronics, where the performance of devices is known to depend critically on the ordering of the first few molecular layers.

14.2.1.2 Electronic Characterization

The appearance of features in STM images can depend very strongly on the selected bias voltage, which dictates the energy levels accessible to the tunneling electrons. With carefully selected bias voltages, important characteristics such as the molecular orbitals can be investigated [54]. In addition, by simply changing the polarity of the bias voltage, STM can access both filled and empty states in the sample surface; with other surface science techniques, states on either side of the Fermi level usually need to be probed by complementary methods (e.g., photoemission and inverse photoemission).

By using the STM outside of the standard imaging mode, additional electronic characterization can be performed. Many types of spectroscopies are possible, but the most common involve measuring current/voltage characteristics or probing the local density of states (LDOS) by measuring $(dI/dV)/(I/V)$ [55, 56]. This type of characterization can be applied to any system amenable to STM; one notable study of this type provided early elucidation of the relationship between wrapping angle, size, and electronic properties in carbon nanotubes [57].

Even subtler techniques can be used to obtain information about surface states from the STM. The standing wave patterns formed by electron scattering from surface defects or steps can be analyzed to reveal momentum–space data such as the surface Fermi contour or the dispersion relations for surface states [58, 59]. Recently, the same technique has been extended to investigate the surface states associated with small supramolecular islands of 3,4,9,10-perylenetetracarboxylic acid dianhydride (PTCDA) on Ag(111); the electron standing wave patterns formed from scattering at the edges of the islands allowed Temirov *et al.* to measure the electron effective mass and the dispersion for a PTCDA-related surface state [60].

14.2.1.3 Manipulation and Synthesis with the STM

One of the most striking capabilities of the STM is its potential to directly manipulate matter at the atomic scale. There are two primary modes in which the STM can be used to this end: acting as a nanomanipulator, that is, a device that can reposition adatoms [61] or molecules [62, 63] on a surface, or acting as a localized electron source, for example, to liberate adsorbates [64, 65] or initiate bond dissociation or formation in underlying molecules [66]. In Section 14.7 of this chapter, we discuss some examples of chemical reactions initiated with the STM.

The interaction of molecules and surface structures with the STM tip can lead to new possibilities for using the STM for electronic characterization. For example, measurements of molecular conductivity have been carried out on thiol-terminated molecules that spontaneously form a molecular wire spanning the tunneling gap between a gold tip and substrate [67]. Lafferentz *et al.* have very recently exploited the ability of STM to pick up molecules from the surface to make measurements of the conductivity of surface-synthesized polyfluorine chains on the Au(111) surface [68]. By attaching one end of a polymer chain to the STM tip (Figure 14.1a), and measuring

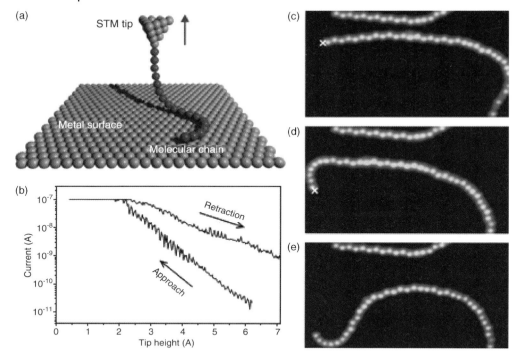

Figure 14.1 Measuring polymer conductivity directly using the STM. (a) A diagram of the experimental geometry: the polymer is bonded to the tip at one end, and the tip is retracted to measure the current flow along the length of the suspended polymer. (b) A typical plot of the change in current along the polymer for tip retraction and approach. (c–e) A polymer that is repositioned after manipulation by attachment of the tip at the end marked with X. Reproduced from Ref. [55]. Originally Figure 2 in Ref. [68].

the current as a function of height for a constant bias voltage as the tip is retracted away from the surface, the authors were able to measure the conductance of the polymer wires.

14.2.2
Modified STM Systems for Specialized Probing of Electronic, Magnetic, and Photonic Properties

14.2.2.1 Two- and Four-Probe STM Systems

Even though the STM is capable of a broad range of electrical characterizations, it is not possible to use a standard STM to directly measure in-plane conductivity. This type of measurement is essential both for determining the suitability of surface architectures for use in devices and for conducting fundamental studies of the relationship between molecular adsorption geometry and charge carrier mobility. For example, Song et al. recently used nanoscale four-probe measurements of cobalt phthalocyanine deposited on the Si(111)-$\sqrt{3} \times \sqrt{3}$ surface to show that the molecular geometry (standing-up versus flat) has profound implications for the conductivity [69]. To make conductivity measurements on nanoscale surface systems, it is necessary to make contact with the surface via two (or, better, four) sharp probe tips that are spaced from one another by only tens or hundreds of nanometers, which presents a rather formidable technical challenge. The development of technologies to meet this challenge is a relatively new field [70]. Although monolithic probe systems, in which the probes are fabricated on a chip or cantilever [71–73], are perhaps the most elegant solution to making conductance measurements, the use of multitip STM instruments confers at least two major benefits: the ability to obtain high-resolution images of the area being characterized and the ability to use tunneling, rather than point contact, to inject current into the sample. Design and instrument descriptions have been published together with initial experimental results for a number of multiprobe STM systems (e.g., Refs [74–77]), and commercial options have emerged on the market, although the systems have not yet been extensively used to characterize molecular systems.

14.2.2.2 Spin-Polarized STM

Notable recent reports of magnetically structured molecular overlayer systems [67–69] have created excitement around the idea of directing the assembly of molecular systems so as to magnetically structure a surface. A specialized mode of STM, known as spin-polarized STM (SP-STM), offers the ability to image a surface with spin sensitivity at the subnanometer scale [70]. SP-STM uses the tip as a source of spin-polarized electrons, a requirement that has been addressed through optically pumped GaAs tips [71], ferromagnetic [72] or antiferromagnetic [73, 74] tips, or nonmagnetic tips that are coated with a magnetic material [75]. The tunneling of spin-polarized electrons is then dictated by what is known as the "spin valve" effect, which implies that the resistance across the tunnel gap will increase when the spin polarization of the tunneling electrons is opposite to that of the electrons in the substrate. Thus, information about the magnetic structure of the surface can be extracted from an analysis of the tunneling current.

We expect to see an increasing number of reports of SP-STM being used for the characterization of molecular systems as this nascent field expands.

14.2.2.3 Light-Emission STM

By adding a photon detector to an STM, it becomes an instrument capable of doing spatially resolved optical spectroscopy. Measuring the photon yield concurrently with the normal STM parameters means that luminescent processes initiated by the tunneling electrons can be correlated with the bias voltage and position of the STM tip. In an early demonstration of the technique, Berndt *et al.* reported enhanced photon emission confined to ~ 4 Å regions centered on C_{60} molecules on a Au(110) surface [78]. The mechanism for photon emission has been the subject of some debate; some studies have found it to depend on plasmon emission from the underlying metal substrate [79–81], whereas others suggest that it results from the direct injection of hot electrons into molecular states [82, 83]. These types of fundamental arguments suggest that further understanding may be necessary before light-emission STM becomes a widely used characterization technique for surface molecular systems.

14.3
Molecular Systems with Applications in Electronics

A broad library of organic semiconductors (OSCs) comprising poly(phenylenes), oligo-thiophenes, acenes, phthalocyanine, and their derivatives have been used in organic electronics [84–87]. Synthetic approaches can be exploited to tune the properties of OSCs. Investigations on these molecular systems have enabled important relationships to be established between the molecular structure and the self-assembly of thin films on surfaces [88], where the morphology of the first layer plays an important role in device performance. STM has been used to probe molecule–molecule interactions and molecule–substrate interactions. These studies provide a deeper insight into the supramolecular architectures, paving the way for a synergistic approach toward a rational design of organic electronics.

14.3.1
Thiophenes

14.3.1.1 Thiophene Layers

Thiophene-based molecular systems constitute one of the most studied OSCs, with potential applications in thin film transistors, photovoltaic solar cells, light-emitting diodes, light modulators, and photochromic switches [89, 90]. The self-assembly of these prototype materials has been widely investigated to develop thin films for the aforementioned applications. Conjugated oligothiophenes (particularly all α-linked thiophene rings denoted by α-nT, where n defines the number of thiophene unit in the oligomeric chain) have been used as well-defined model compounds and starting monomers for polythiophenes. The oligothiophenes are relatively more amenable to front-end processing than their parent polymers since a high degree of ordering can be achieved via solution or vacuum deposition.

The physical properties of thiophene-based materials can be chemically controlled by substituent effects and length. These, in turn, drastically influence the ordering of the films on a surface, which is crucial for charge carrier mobility [85]. STM has been used as a tool to probe the interfacial interactions of these molecules on a conductive surface. Fukunaga et al. [91] studied the 2D organization of monothiophene with long alkyl chains in the β positions at the liquid/highly oriented pyrolytic graphite (HOPG) interface. The thiophene rings oriented diagonally on the surface, lying head to head, form a lamellar structure, with the alignment of the alkyl chains promoted by the surface. Similarly, superstructures of a homologous series of oligothiophenes with alkyl groups in the β-position was imaged by Bäuerle et al. [92] at the liquid/HOPG interface. They observed that the separation of the lamellae is dictated by the length of the oligothiophenes.

The position of substituents bestows regioregularity upon the molecule and this can alter its packing arrangement. Stabel and Rabe [93] addressed the difference in intermolecular spacing by probing regioregular and nonregioregular 2,5-didodecyl-sexithiophene, which physisorbed at the liquid/HOPG interface to form highly ordered monolayers with the alkyl chains and the molecular backbone oriented parallel to the HOPG surface. The effect of increasing the length of alkyl substituents on quarterthiophenes was extensively studied at the 1,2,4 trichlorobenzene/HOPG interface. Longer alkyl chains such as hexyl and dodecyl moieties lead to a long-range order with a lamellae-type orientation, and in the case of dodecyl substituents a row separation of approximately 20 Å was observed [94]. This contrasted the herringbone structure observed when propyl substituents were used; the intermolecular interactions were too weak to promote ordered self-assembly.

Polar substituents such as iodine atoms [95] and formyl groups [96], introduced in the α-position of an oligothiophene, have been studied to explore their influence on the self-assembly process. Symmetrically substituted didodecylquarterthiophenes (both parent molecule and α-ω diiodo-substituted molecule) produce similar lamellar structures with comparable unit cells, whereas the monoiodo derivative forms a dimer along the lamella. This compensates for the dipole moments and the interaction results in a large separation between the molecules. Hydrogen bonding was found to control the spatial arrangement between bis-urea-substituted thiophene derivatives containing one, two, and three thiophene rings [97]. It was suggested that the thiophene rings within the observed lamella structure were tilted with respect to the surface with partially overlapping π-systems.

The influence of the substrate on the organization of thiophene derivatives can be investigated by comparing adsorption on Ag(111) [98], Cu(110) [99], Au (111) [100, 101], and Cu–O templates [102]. Cycloalkane endcapped quinquethiophene formed highly ordered monolayers on Ag(111) under ultrahigh vacuum (UHV) conditions. The molecules are organized in parallel rows lying flat on the substrate with the axis of the molecule nearly orthogonal to the direction of the rows. Yang *et al.* [101] compared the adsorption behavior of asymmetric and symmetric thiophene derivatives on both Au(111) and HOPG. They found that both symmetric and asymmetric thiophenes formed highly ordered wave-like and quasihexagonal adlayers, respectively, on HOPG. However, a random adsorption of the two molecules was observed on Au(111) that highlighted the influence of the substrate on the self-assembly of those molecular systems. A low dosage of 3-thiophene carboxylate molecules on Cu(110) resulted in the molecules preferentially aligned along the [110] direction with the molecular plane parallel to the surface forming a $c(4 \times 8)$ periodicity [99]. At higher dosage, a $p(2 \times 1)$ periodicity was observed, where steric repulsions caused the thiophene rings to twist out of the plane of the carboxylate group. Cicoira *et al.* [102] found that at low coverage α-quinquethiophene preferentially adsorbed on the pristine Cu(110) stripes of a nanotemplated Cu–O surface. The substrate is generated by controlled oxidation of Cu(110) surface to generate rows of oxidized Cu and pristine Cu(110). The influence of the symmetry bestowed by low Miller index surfaces has been extensively investigated and will be discussed in Section 14.3.2.

14.3.1.2 Host–Guest Systems Containing Thiophenes and Fullerenes

Macrocycle Hosts with Fullerene Guests Thiophene macrocycles are well-defined conjugated systems without the perturbing end effects of their linear counterparts [103]. An STM investigation of cyclo[n]thiophene revealed well-ordered monolayers on the HOPG surface [104]. The shape-persistent molecule adopted a hexagonal structure with the molecule exitaxially oriented with respect to the underlying HOPG substrate in a nonplanar conformation with completely stretched alkyl chains. The porous network formed from the ring-shaped cyclo[12]thiophene is amenable to host–guest molecules such as C_{60} fullerenes, which were deposited from solution on the thiophene template HOPG structure. The thiophene macrocycles are p-type and the fullerenes are n-type organic semiconductors that, when assembled, form a donor–acceptor [104, 105] system poised for solar energy conversion and molecular electronic device applications [106]. At low coverage, the C_{60} fullerenes were found to adsorb both at the rim and inside the cavity of the macrocycle. The adsorption at the rim was found to be more stable due to enhanced interactions between the host and the guest molecules [105, 107]. Higher coverage led to a periodic arrangement of C_{60} with a spacing of 23 Å, which is much larger than the spacing of 10 Å observed during adsorption of only fullerenes on a surface. The spacing is dictated by the interaction with the macrocyclic thiophene network, rather than short-range intermolecular fullerene interactions.

Terthienobenzenetricarboxylic Acid Networks Although the early studies of fullerene guests in networks of thiophene-containing macrocycles broke new ground in the study of thiophene/fullerene systems, the complicated synthesis of the large macrocycles might hinder these systems from being adopted in production-scale organic electronics. The first study of a host/guest system comprising a self-assembled porous network of thiophene-containing molecules populated with fullerene guests was recently published by MacLeod et al., who used the C_3-symmetric trimesic acid (TMA) analogue terthienobenzenetricarboxylic acid (TTBTA) as their molecular building block [108]. Like TMA, TTBTA self-assembles into a porous chicken wire structure at the solution–solid interface. Adding C_{60} guest molecules to the solution populates the pores with fullerenes. The authors observe that domains form with identical fullerene fillings in each pore, varying from a single fullerene to the theoretical filling limit of three fullerene molecules per pore. The fullerenes also exhibit identical arrangements within the pores, as shown in Figure 14.2. The authors attribute this effect to both the symmetry-breaking influence of the underlying HOPG substrate and the electronic interaction due to fullerene charge transfer.

14.3.2
Influence of Symmetry on Adsorption of Rubrene

In this section, we elaborate on the impact of different facets of copper on the self-assembly of rubrene. Rubrene (5,6,11,12-tetraphenyltetracene, $C_{42}H_{28}$) is a molecule

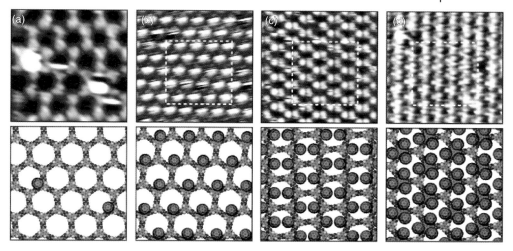

Figure 14.2 C$_{60}$ fullerene guests in a porous TTBTA chicken wire network. Although the fullerene guests appear to be only loosely confined at low coverages (a), at higher coverages they form ordered domains where the number and position of fullerenes within each cavity are identical; (b), (c), and (d) show occupancies of one, two, and three guests per cavity, respectively. The ordering is likely motivated by charge transfer between the fullerenes and the thiophene within the TTBTA. Reproduced from Ref. [108]. Originally Figure 7 in Ref. [108].

with exceptional electronic properties that is being widely used in organic electronic devices such as LEDs [109–111], FETs [112–119], and recently PV cells [120–122]. It is a benchmark material for single-crystal organic FETs since it has one of the highest hole field mobilities ($\leq 15\,\text{cm}^2\,\text{V}^{-1}\,\text{s}^{-1}$) [113, 116, 117] ever reported for organic semiconductors. The previously debated molecular structure of rubrene [123, 124] has recently been established by X-ray crystallography [125]. The molecule is nonplanar and flexible; it is made up of a tetracene backbone with two pairs of phenyl substituents attached to either side (Figure 14.3a). It maintains a twisted chiral geometry in the gas phase to relieve the strain caused by the intermolecular steric hindrance of the phenyl groups.

Copper is often used as an electrode material in the fabrication of organic electronic devices. The interaction of organic molecules with the interfaces of a device is primordial to understanding their adsorption and self-assembly, especially when nanoscale ordering of the thin film is required for device optimization. We describe here a series of experiments in which the adsorption of rubrene was investigated on three low-index copper facets under UHV conditions [126, 127]. Imaging by STM revealed a highly ordered herringbone-like pattern after adsorption of rubrene onto Cu(100) [126] with a molecular packing density of 0.55 molecules nm^{-2} as shown in Figure 14.3b. The individual molecules are observed as bright elongated lobes ascribed to the phenyl substituents that are separated by a dark tetracene backbone. The interaction between the metal surface and the π-orbitals of

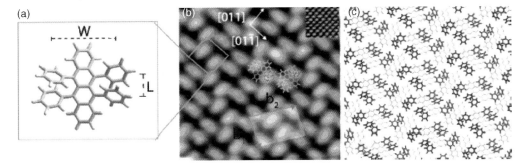

Figure 14.3 (a) Rubrene on Cu(100) where the structure is defined with length (L) and width (W). (b) A high-resolution STM image of the herringbone-like pattern with inset representing the underlying surface and an overlayer model of a unit cell is depicted. (c) A model of the STM image in (b). STM parameters: 8×8 nm^2, $V_b = 2.05$ V, and $I_t = 0.36$ nA and 2×2 nm^2, $V_b = -0.32$ V, and $I_t = 0.93$ nA. Reproduced from Ref. [126], Figure 3.

the molecule enforces the adsorption geometry observed. The fourfold geometry of the surface directs the assembly of rubrene along the [001] and [00-1] directions. Miwa et al. [126] also observed intriguing features on the surface described as "pinwheels" that are formed at domain boundaries.

Two highly ordered patterns referred to as "box" and "zigzag" were observed upon self-assembly of rubrene on Cu(110). The box structure is shown in Figure 14.4a. The high-resolution image (Figure 14.4c) revealed that the molecules parallel to the [-110] reside on top of the row while their tetracene backbone reside in a perpendicular orientation in between the 110 rows. The zigzag (Figure 14.4b and d) structure is formed with the backbone of the molecule aligned along [-110]. The twofold geometry that the surface imparts is consistent with the pattern formation. The molecular packing densities are 0.60 and 0.45 molecules nm^{-2} for the box and zigzag structure, respectively [126]. The close packing of the box structure enhances the network stability while the zigzag structure is stabilized by the commensurability of the tetracene backbone with the [-110] direction. Using DFT calculations, the authors confirmed that the molecule retains its chirality after adsorption on the surface.

The self-assembly of rubrene on Cu(111) is distinctly different from the other two low-index copper facets Cu(110) and Cu(100). Figure 14.5 shows the two structures (dimer and trimers) observed by STM under UHV conditions [127]. The packing density was found to be 0.61 molecules nm^{-2} for both structures. The backbones of rubrene molecules in the dimer structure are aligned along the [10-1] direction with the molecules forming rows along the [01-1] direction as depicted in Figure 14.5a. Six domain orientations are possible because of the combination of the dimer structure and the threefold symmetry of the surface. Four domain orientations were imaged by Miwa et al. [127] who rationalized that the backbone of the molecule is oriented along the high symmetry directions of the surface: [10-1], [01-1], or [-110]. The trimer structure (Figure 14.5b) is composed of three molecules oriented along the high-symmetry surface directions. The structure was also found to be chiral in nature

Figure 14.4 Polymorphs of rubrene on Cu (110). (a) Large STM images (70 × 70 nm^2) of the box structure ($V_b = -1.97$ V, $I_t = 0.34$ nA) and (b) the zigzag structure ($V_b = -1.58$ V, $I_t = 0.42$ nA). (c) High-resolution images (7 × 7 nm^2) of the box structure ($V_b = -0.53$ V, $I_t = 0.34$ nA) and (d) the zigzag structure ($V_b = -1.90$ V, $I_t = 0.30$ nA). Inset in (c) represents a 2 × 2 nm^2 Cu(110) with parameters $V_b = -0.37$ V, $I_t = 1.28$ nA. Reprinted from Ref. [126], Figure 7a–d.

much like the chiral adsorption of rubrene on Au(111), which was suggested to trigger the formation of supramolecular chiral structures [128]. These examples illustrate the balance between van der Waals forces and substrate-mediated interactions, and highlight the need to dissect and understand all the intricate interactions so as to optimize the design of electronic devices.

14.4
Optically Active Molecules

A chiral center (e.g., carbon atom) contains four different functional groups whose spatial arrangement confers on it the ability to rotate a plane of polarized light either to the left or to the right, thence giving rise to its left-handedness or right-handedness.

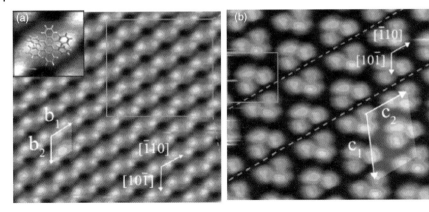

Figure 14.5 (a) STM image of the dimer structure (11.5 × 11.5 nm^2; $V_b = 2.05$ V, $I_t = 0.34$ nA) with an inset showing a molecule (1.6 × 11.5 nm^2) with a superimposed model. (b) An image of the trimer structure (11.5 × 11.5 nm^2, $V_b = -2.20$ V, and $I_t = 0.38$ nA). Reproduced from Ref. [126]. Originally Figures 2a and 4b in Ref. [127].

Therefore, a chiral moiety has a nonsuperimposable mirror image. Besides this difference, most enantiomers (chiral objects) can have similar properties, and the detection and separation of opposite enantiomers is not trivial. The functionality bestowed by the chiral center is of great scientific interest in areas such as pharmacological drug design of stereospecific species [129, 130], stereospecific interactions of DNA base pair essential toward understanding the origins of life [131, 132] and in the field of materials research (e.g., liquid crystal displays [133]).

The high resolution achieved by STM provides a unique opportunity to gain deeper insight into these systems. Imaging can be performed in either constant current mode or constant height mode; in the first case, the current is held constant and the change in tip height is monitored as the tip is rastered across the surface, and in the latter case the height is held constant and the corresponding variation in current generates an image with different contrast that is ascribed to different tunneling current probability across the surface. The observed contrast varies either due to the topography of the surface [134] or due to its electronic structure. For example, the overlap between the orbital of the molecule and the substrate can give rise to a peculiar contrast, such as in the case of aromatic molecules, which appear brighter on certain surfaces [135].

This research area has stimulated an exhaustive list of studies. The direct determination of chirality of individual molecules has been achieved thanks to the use of STM [136, 137]. On the basis of their adsorption conformation, the chirality of the molecules can be determined. Chiral surface structures are formed not only by enantiopure molecules but also by racemic mixtures [138] and achiral molecules [139, 140]. High Miller index surfaces can also induce chirality in adsorbed molecules [141]. Here, we focus on selected technologically relevant examples to highlight the impact of STM as a characterization technique that has greatly improved the body of knowledge on supramolecular assembly of chiral functional materials.

14.4.1
Obliqueness of Chiral Unit Cell and its Angular Mismatch

Liquid crystals have two important aspects of symmetry: (i) the chirality of the building block can be systematically designed and (ii) in turn it directs its chirality toward the formation of helical columns in the mesophase (3D structure) [142]. Thus, understanding the formation of the first 2D layers of these chiral liquid crystals on a surface is critical to predicting its columnar growth in 3D. Organic molecules such as (*RS*)-4-(1-(methylheptyl)oxy)phenyl 4-formamidobenzoate (Figure 14.6) have been studied using STM to gain better control over their self-assembly [143]. The formamide moiety is separated from the alkyl chain by a phenyl benzoate group, which is used to determine the orientation of the molecule at the 1-heptanol/graphite interface. From the STM images in Figure 14.6a–c, De Feyter *et al.* [143] were able to determine that chirality is observed at two levels. The unit cell of the chiral structure is oblique and the overlayer forms an angle with respect to the graphite lattice. The packing of the racemate revealed an intriguing arrangement based on the bright contrast of the phenyl benzoate groups. The lamellae formed were found to be nearly collinear with the lattice of the substrate and the lamellae of both enantiomers (R and S) were situated in a nonperiodic fashion. This implies that the chirality did not impose a preferential angle with graphite during adsorption and that the flexibility of the alkyl chain reinforced the interaction of the molecules via interdigitation.

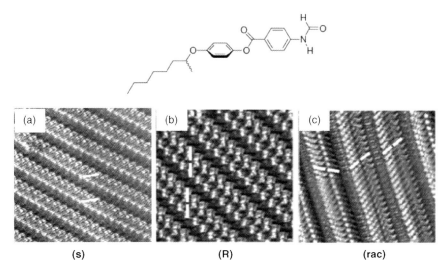

Figure 14.6 STM images of 4-(1-(methylheptyl)oxy)phenyl 4-formamidobenzoate at the 1-heptanol/graphite interface showing (a) the S enantiomer, (b) the R enantiomer, and (c) the racemate with the bar indicating the orientation of the phenyl benzoate moiety. STM parameters for (a) and (b) 16.1 × 16.1 nm^2, I_t = 1.0 nA, and V_b = 0.51 V and 0.73 V, respectively. (c) 13.6 × 13.6 nm^2, I^t = 1.0 nA, and V_b = 0.54 V. Adapted from Ref. [143], Figure 3.

14.4.2
Height Difference Revealed by Contrast

The study of the effect of the number of chiral centers on the self-assembly of symmetric porphyrin derivatives was performed by Linares et al. [144]. Most porphyrin derivatives form racemic domains or achiral overlayers as they are achiral [145, 146]. However, the presence of four identical chiral centers on 5,10,15,20-tetra[4-(R,R,R,R)-2-N-octadecylamidoethyloxiphenyl]porphyrin leads to the formation of single-handed domains at the heptanol/graphite interface as observed by STM [144]. Polycyclic aromatic hydrocarbons (PAHs) are disk-shaped molecules used as liquid crystals (often termed discotic liquid crystals) [4]. These molecules have a rigid core surrounded by flexible chains that promote the formation of a columnar mesophase. Derivatives of hexa-peri-hexabenzocoronenes (HBC) are an important class of PAHs where the self-assembly of the first monolayer is crucial to understanding the subsequent columnar growth.

STM images of (S)-phenylene-alkyl-substituted hexa-peri-hexabenzocoronenes (HBC-PhC$_8$) revealed an intriguing architecture after self-assembly at the solution (1,2,4-trichlorobenzene)/HOPG interface [134]. Samori and coworkers observed a hexagonal 2D packing and a periodic motif distribution of the contrast across the plane of the surface as shown in Figure 14.7. The molecules with similar contrast assume a hexagonal $\sqrt{3} \times \sqrt{3}R30°$ superstructure.

By analyzing the molecular contrast, the authors were able to deduce the arrangement of the molecules on the surface. The STM tunneling current varies exponentially with the tip–sample separation, conferring contrast on images acquired in constant current mode. A change of 1 Å is known to result in about one order of

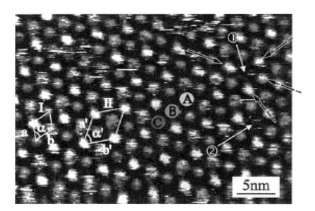

Figure 14.7 Three types of molecular contrast are observed in the STM current image of HBC-PhC$_8$ at the solution–HOPG interface. The black arrows point to type A molecules and the white arrows to two vacancies observed. The unit cell of structure I: $a = 2.55 \pm 0.10$ nm; $b = 2.44 \pm 0.10$ nm; $\alpha = 58 \pm 3°$; and that of the superstructure II: $a' = 4.28 \pm 0.17$ nm; $b' = 4.28 \pm 0.17$ nm; $\alpha' = 58 \pm 3°$. The STM parameters are $V_b = 1$ V and average $I_t = 50$ pA. Reproduced from Ref. [134], Figure 4.

magnitude change in the current [147]. The authors extracted an estimate of the mean variation in tunneling current across multiple line profiles on the surface to quantify the different contrast observed. The latter is attributed to three types of neighboring molecules (A, B, and C) that are packed at different heights (Figure 14.7). There is a combination of intramolecular, intermolecular, and interfacial interactions. The aromatic core of the HBC-PhC$_8$ molecule loses its planarity due to the presence of the phenylene groups and the chiral center in the alkyl moiety. This causes the molecules to assume different conformations that station them at different positions above the surface forming a staircase architecture, thereby minimizing intermolecular steric hindrance.

14.4.3
STM Reveals the Nature of First Layer Growth

Chirality has also been observed in other organic systems with potential electronic applications. π-Conjugated molecules such as oligothiophenes and their derivatives are commonly used to study their interfacial adsorption on numerous substrates [148–151]. These studies are motivated by the quest for a deeper insight into the formation of the first monolayer that acts as a stencil for the growth of subsequent layers [152]. We will focus on α-sexithiophene (6T), which is composed of six thiophene groups linked at the *para* positions. Upon adsorption at room temperature, well-ordered rows of 6T are observed on the reconstructed Au(111) surface, as seen in Figure 14.8a [153]. Contrast due to the underlying herringbone structure is also observed. The rows form a small angle (∼7°) with respect to the high-symmetry direction [11-2]. Individual molecules are observed as rod-like objects with six maxima, which are assigned to the thiophene groups, giving rise to an oblique unit

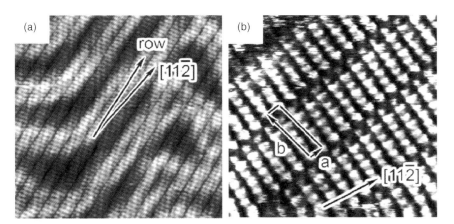

Figure 14.8 STM images of α-sexithiophene (6T) on Au(111) taken at (a) 38 × 38 nm^2, $I_t = 0.3$ nA, and $V_b = -1.5$ V and (b) 10 × 10 nm^2, $I_t = 0.6$ nA, $V_b = 0.18$. Adapted from Ref. [153], Figure 2(d and e).

cell. Kiel et al. [153] observed that the molecules were perpendicular to the propagation of the row and there is also a misalignment between adjacent rows (Figure 14.8b). These are indications of chirality at the surface. The 6T molecule has an all-*trans* conformation and because it has an even number of thiophenes it can be either right- or left-handed depending on the orientation of the sulfur groups.

14.5
Magnetic Systems

Supramolecular chemistry can be used to address the problem of creating organized nanoscale magnetic systems with potential electronic applications. Spintronic devices, quantum computation, and data storage are examples of applications where local control of the magnetic moment is required [154–159]. Data storage densities are rapidly expanding with the need of storing more information in smaller scales [160]. However, decreasing the size of magnetic systems is not trivial since a decrease in the size of ferromagnetic domains also leads to a decrease in the magnetic anisotropy energy, which means that in small domains thermal fluctuations can indiscriminately flip the magnetization direction.

14.5.1
The Spin–Electron Interaction

The magnetic moment of a transition metal can interact with the sea of fermions presented by a metallic substrate; there is a coherent exchange of spin between the localized state of the metal center and the delocalized electrons from the substrate. This spin-related phenomenon is known as the Kondo effect [161]. This effect has been experimentally observed on single atoms and molecules by STM [162–166]. STM and scanning tunneling spectroscopy (STS) studies performed by Iancu et al. [167] using a porphyrin molecule 5,10,15,20-tetrakis-(4-bromophenyl)-porphyrin-Co (TBrPP-Co) on Cu(111) revealed the 2D assembly of this metal–organic complex composed of planar and saddle molecules. They manipulated the planar ribbon (Figure 14.9a) structure formed with the STM tip to generate a small hexagonal clustered assembly of seven molecules (Figure 14.9b) from which single porphyrin units were sequentially removed until only one was left. By measuring the energy dependence of the LDOS around the Fermi level, the authors were able to identify the change in Kondo resonance (dip observed in STS data) from the differential conductance (dI/dV) tunneling spectra (Figure 14.9c). The Kondo temperature can be estimated within experimental uncertainty from the spectrum [167]. Their systematic study revealed that the influence of the number of neighbors on the local spin–electron interaction coupling strength depended on the hybridization of the assembly and distortion due to the packing of the molecules on the surface. In a separate study [168], the authors showed that removal of peripheral atom(s) from a porphyrin unit, using voltage pulses, also resulted in alteration of the spin–electron coupling via a switching mechanism.

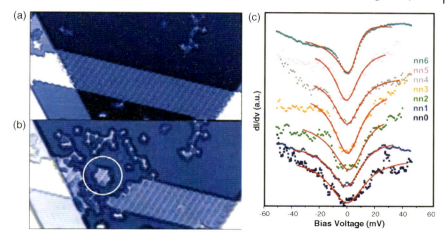

Figure 14.9 (a) Ribbons of planar porphyrin units are manipulated by the STM tip into (b) a hexagonal assembly. The dI/dV spectra measured above the central molecule for different nearest neighbors. STM parameters 28×50 nm, $V_b = 1$ V, and $I_t = 0.2$ nA. Reproduced from Ref. [167], Figures 2a and b and 3b.

14.5.2
Metal–Organic Coordination Networks

The magnetic properties of the transition metal in a metal–organic network can be tuned by the specific interaction with the organic linkers and the underlying substrate. Numerous STM studies have systematically probed MOCNs formed from organic molecules connected to transition metal centers amenable to multiple coordination. Each organic molecule is bound to a metal center via metal–ligand-type complexation [169] that makes MOCNs more robust than H-bonded networks. Organic molecules with extended π-systems are used to encourage adsorption parallel to the surface and a wide variety of chelating functionalities (e.g., carboxylate, pyridine, hydroxyl, and carbonitrile groups) are accessible via coordination chemistry. The symmetry of the metal centers and the organization of the organic ligands drive the formation of the supramolecular structure by way of directional bonds. Supramolecular organization provides an opportunity to precisely control the assembled architecture at the nanoscale.

Lingenfelder *et al.* [170] used STM imaging to systematically study the interaction of benzenepolycarboxylic acid species with iron (Fe) atoms on Cu(100) surface. The nonmagnetic surface is used to generate architectures of regularly spaced Fe spins. A variety of 2D networks are formed depending on the relative concentrations of metal atoms and ligand molecules. After deposition of low Fe concentrations (0.3 atoms per linker molecule), the STM data revealed a 6×6 array with square lattice of 15 Å. Each of the four linkers (terephthalate, TPA, $C_6H_4(COO)_2$) chelated one metal atom via one carboxylate group. The linkers are also stabilized by intermolecular H-bonds. By increasing the ratio of Fe atoms to linker molecules, a ladder structure was observed [171] when using a trimellate linker. STM characterization revealed that a further increase in this ratio to two Fe atoms per linker molecule (using 4,1′,4″,1-terphenyl-1,4″-dicarboxylic acid (TDA), which contains one more benzene ring than TPA) resulted in a rectangular network with well-ordered nanocavities [172]. As exemplified by TPA, the symmetry and length of the linker are very important. TMA has threefold symmetry [173] and cannot assemble to square arrays as compared to the linear TPA [170], and the length of the linker confers different periodicity (TPA [170] versus TDA [172]). Replacing the Cu(100) substrate with Cu(110) results in linear metal–organic chains bestowed by the 1D anisotropy of the surface [174].

14.5.3
Controlling Magnetic Anisotropy of Supramolecular MOCN

Gambardella *et al.* [175] recently showed that ordered high-spin mononuclear centers of Fe can be generated from lateral coordination of TPA molecules on Cu(100). STM characterization revealed the previously reported 2D network [170] as shown in Figure 14.10a. Fe atoms on Cu(100) exhibit a Kondo temperature of ∼55 K [166] despite the considerable overlap of the Fe 3d and 4s states with the Cu substrate. The Fe atoms adsorbed on Cu(110) are chemically active with resultant coordinatively unsaturated sites bridged by four lateral carboxylate ligands. The authors elegantly made use of oxygen, which adsorbed on top of the Fe centers. A change in contrast was observed in the STM image (Figure 14.10b) due to a selective uptake of O_2. Other techniques such as DFT calculations and X-ray magnetic circular dichroism were used to confirm that the spin on the Fe atoms changed upon exposure to oxygen. There are challenges to the design of these structures such as mobility of metal atoms on surfaces and alloying from intermixing processes at the surface. This makes the formation of supramolecular architectures nontrivial, and as such MOCNs are not considered a self-assembly *per se* but rather a synergistic organization of the building

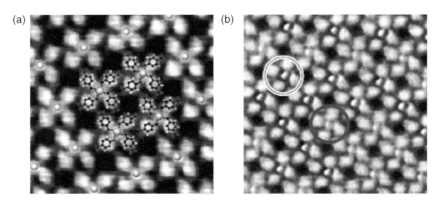

Figure 14.10 (a) STM image of the planar MOCN network formed from TPA/Fe on Cu(100) with the TPA structure overlaid along with gray Fe atoms; (b) following selective uptake of O_2 on the surface, the contrast of the Fe atoms changes. Image parameters 62×62 Å^2 and 84×84 Å^2. Reprinted from Ref. [175].

14.6
STM Characterization and Biological Surface Science

The nanoscale modification of surfaces is important not only in device fabrication but also in the realm of biological processes. STM can be used to probe the adsorption of biomolecules on conducting surfaces, the interfacial interactions, and the nanostructure on a surface for either enhanced biocompatibility or biosensing. In the following section, we illustrate these concepts with selected examples from the recent literature.

14.6.1
Probing the Fundamentals of Life

STM can be used to probe the interfacial interactions of biomolecules with a surface at the nanoscale, which is a length scale salient to biological processes such as cell signaling and recognition [176–178]. This has inspired a vast variety of research geared both toward fundamental understanding of the recognition capabilities and interactions of the smaller cellular building blocks such as biomolecules and toward nanotechnological applications such as biosensors [179–183] and biocompatibility of implants [30, 177, 184–188].

Self-assembly of biomolecules such as deoxyribonucleic acid (DNA) and ribonucleic acid (RNA) is of fundamental interest for understanding the processes that may be related to the appearance of life [189]. Genetic information is encoded in the DNA cells. The DNA cell consists of two strands (an unbranched polymer) that are composed of four different bases: adenine (A), thymine (T), guanine (G), and cytosine (C). The two strands are held together via hydrogen bonding of complementary base pairs in each strand, which results in a helical structure. High-resolution STM imaging has been used to investigate the adsorption of the separate bases on various surfaces to understand the molecule–molecule and molecule–substrate interactions [51, 52, 139, 189–211].

UHV STM provides the opportunity to study interactions of single DNA base in a clean and controlled environment with subatomic resolution [200]. Using a variable temperature (VT)-STM, Otero *et al.* [212] were able to identify the type of networks formed by the adsorption of guanine, adenine, and cytosine at room temperature (RT) on Au(111) (Figure 14.11). The molecules easily diffuse on the surface due to weak interactions with the inert substrate and self-assemble into networks defined by hydrogen bonding and the functional moieties present on the bases.

Guanine (G) deposited on inert Au(111) substrates at room temperature self-assembles into well-ordered islands of irregular shapes. The Au(111) surface appears to interact minimally with guanine. The G networks observed describe an almost square symmetry despite the threefold symmetry of the surface. The unit cell is

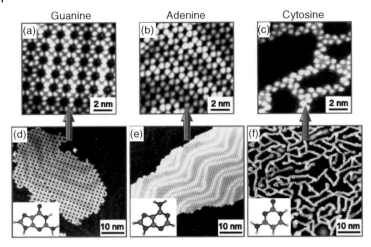

Figure 14.11 High-resolution STM images (constant current mode) of (a) guanine, (b) adenine, and (c) cytosine deposited on Au (111) at RT. The bigger area scan is shown in (d–f), respectively. STM parameters: $I_t = -0.3$–0.7 nA and $V_b = -800$–1500 mV. Reproduced from Ref. [212], Figure 1.

composed of four G molecules (the guanine quartet), with the individual G molecules appearing as triangular protrusions (Figure 14.11a), from which the chirality of the structure can be deduced [207]. Otero *et al.* were able to determine that cooperative hydrogen bonding contributed to the stabilization of the observed quartet [207]. Cytosine (C) deposited on Au(111) resulted in the formation of a random network where three elementary motifs were identified. The supramolecular network is defined by zigzag filaments, also described as ribbon-like structures (Figure 14.11b), and five- and six-membered rings, where most of the C molecules are held to only two neighbors through hydrogen bonding [203]. Kelly *et al.* [199] reported the coexistence of two structures upon self-assembly of adenine (A) on Au(111). A well-ordered 2D assembly consisting of two types of honeycomb network was observed that depended not only on the orientation of the A molecule but also on its interaction with three neighbors via hydrogen bonding (Figure 14.11c). A flat-lying adsorption geometry is observed where the aromatic rings of the bases are parallel to the surface of the substrate. The herringbone reconstruction is still visible as modulation of contrast in the adsorbed structures [199].

Following these studies, the next investigation undertaken by Besenbacher's group was to probe complementary systems such as coadsorption of G and C [210] at the liquid–solid interface on HOPG graphite. The adsorption of this complementary system was compared to noncomplementary systems such as A and C on Au(111) in UHV. The UHV-STM provides a clean environment to study the possible interactions of nucleobase recognition. Cytosine (C) was initially deposited on Au(111), which produced the previously observed random zigzag (ribbon-like) filament and ring network (Figure 14.12a and d). Codeposition of the complementary base G, on top of the partially C-covered surface, produced a binary nucleobase mixture with an increase in the number of fivefold rings as seen in Figure 14.12b. These structures

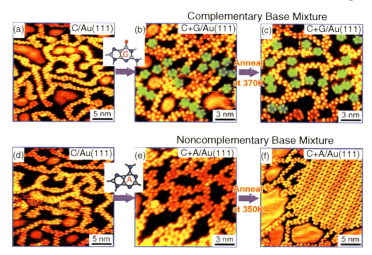

Figure 14.12 Sequential deposition of complementary bases (a–c) Cysteine + Guanine (C + G) and noncomplementary ones (d–f) Cysteine–Adenine (C + A). In (a) and (d), the initial amount of C deposited at RT is kept constant. Deposition of G on C/Au(111) in (b) resulted in the formation of fivefold rings (green), which are not observed in (e) after deposition of A on C/Au(111). The complementary C + G mixture maintains disorder after heating (c), while noncomplementary C + A mixture separates out into islands of A and zigzag branches of C as seen in (f). Reproduced from Ref. [212], Figure 2.

were not observed when the noncomplementary base A was used instead (Figure 14.12e). Subsequent annealing of the binary mixtures at a temperature below the desorption temperature of both bases led to drastic differences. The complementary bases G + C maintained a similar structure at 373 K (Figure 14.12c) and only the C molecules desorbed when heated at 400 K. In contrast, the noncomplementary base A + C began to segregate at 353 K leading to islands of A molecules with zigzags surrounding them (Figure 14.12f). Otero *et al.* [212] also varied the amount of material deposited and observed similar structures. Analysis of their high-resolution STM images allowed the authors to determine that the fivefold ring formation of the complementary bases G + C is made up of one G molecule surrounded by four C molecules. These STM studies provided both molecular recognition and an investigation of the thermal stabilities of these base pairs, allowing deeper insight into the interactions involved. These elements can be crucial to understanding the fundamentals of biological systems.

14.6.2
Bionanotechnological Applications

Biomedical devices such as implants are routinely administered to millions of people worldwide. These implants are employed in many areas of medicine such as orthopedics, dentistry, and cardiology, among others [213]. For these applications, metals have been adopted as the best biologically compatible materials. The commonly used metals are stainless steels, CrCoMo alloys, tantalum and titanium, and related alloys [5, 184–187, 214–220]. The choice of these metals depends on their minimal immune adverse reactions and high resistance to corrosion, which can influence their mechanical strength when load bearing of the implant is required [221, 222]. The surface chemistry and topography are crucial to biological reactions at the implant's interface; the implant's successful integration depends on the local interactions between the material and the cells [176, 223]. Intensive research has focused on controlling and characterizing morphological features as well as surface modifications to improve biocompatibility.

The self-assembly of molecular building blocks to form monolayers on a surface has provided an elegant method for surface modifications. Self-assembled monolayers (SAMs), in particular thiols on gold, have been widely studied and have been thoroughly reviewed elsewhere [22, 224]. SAMs have been typically employed to alter surface hydrophobicity, adhesion properties, and chemical functionality [219, 225, 226]. On metal oxide surfaces used as implants, thiol-based SAMs are not suitable as they incorporate undesirable organics [227]. On the other hand, phosphoric acid-terminated alkyl chains can be used to provide a strong chemical bond toward surface oxides [228–231].

Clair and coworkers [219] recently used STM, among other techniques, to characterize the surface modification of titanium, which is commonly used for implants. Both chemically etched and smooth surfaces were probed before and after coating with dodecylphosphoric acid (DPPA). Individual molecules could be resolved, and the high-resolution images revealed a cauliflower-like structure, with islands 4 nm in diameter. The authors found that nanostructuring the surface greatly improved the contact angle, and the porous nature of the surface promoted more efficient molecular packing [219]. This study highlights the need to better control and characterize surfaces at the nanoscale. Adsorbed proteins are only a few nanometers in size, and cell signaling, growth, and interactions all occur at the nanoscale [176, 177, 232].

14.7
Using the STM to Initiate Chemical Reactions

In addition to the types of characterization already highlighted in this chapter, STM can be used to initiate reactions in molecules adsorbed on a surface. In this section, we provide a brief overview of some notable examples of STM-induced surface chemistry.

14.7.1
Molecular Dissociation

Less than 10 years after the invention of STM, Dujardin *et al.* published an account of STM-induced dissociation of decaborane ($B_{10}H_{14}$) on Si(111)-7 × 7 [233]. The authors dissociated the molecules by scanning with a bias voltage greater than

4 V; subsequent imaging at lower bias voltages revealed that initially intact molecules had been displaced, dissociated, or fragmented. Similarly, Martel et al. were able to initiate the dissociation of O_2 on the same surface by scanning with a bias voltage greater than 6 V [234]. In later studies, such as the work described by Stipe et al. [235], the STM tip was carefully positioned directly over a molecule, and its dissociation was induced through the application of a controlled voltage pulse. In this manner, Stipe et al. were able to measure the rate of dissociation as a function of current for O_2 on Pt (111) at different bias voltages.

The controlled dissociation of surface-adsorbed molecules by STM has provided a new avenue for investigating surface reactions. In particular, inelastic electron tunneling spectroscopy (IETS) can be used to identify reaction products via their vibrational characteristics. This allows a complete analysis of the progression of the molecular dissociation in single- or multistep reactions. Gaudioso et al. used this technique to investigate the decomposition of ethylene (C_2H_4) on Ni(110) at low temperature and found that the molecule could be successively dehydrogenated to acetylene (C_2H_2) and then to bare carbon fragments through STM voltage pulses [236]. In a similar experiment, the same group later found that ethylene on Cu(001) at low temperature could be decomposed first to ethynyl (C_2H) and then to dicarbon (CC) via C—H bond breaking induced with voltage pulses [237]. Kim et al. also induced the dissociation of a C—H bond, this time in trans-2-butene molecules on the Pd(110) at low temperature, creating 1,3-butadiene [238]. These studies represent just a small sampling of the wealth of literature describing these processes. However, even this brief overview illustrates how the comparison of decomposition in these types of experiments not only leads to a better understanding of molecule/substrate interactions but also provides information about how to control surface chemistry with STM.

14.7.2
Molecular Synthesis

In addition to breaking bonds to induce molecular decomposition, the tunneling electrons from the STM tip can be used to initiate bond formation. Lee and Ho provided an elegant account of this phenomenon in 1999, describing experiments in which they were able to bond CO molecules to Fe atoms on a low-temperature Ag (110) surface, forming Fe(CO) by initiating a single bond and $Fe(CO)_2$ by initiating two bonds [66]. Figure 14.13 shows STM images illustrating the process: the authors first move a CO molecule to a Fe atom (a) and initiate bond formation via tunneling electrons, forming FeCO (b). If the same process is repeated with a second CO molecule (c), a $Fe(CO)_2$ molecule can be formed (d).

More recent reports of molecular synthesis via STM include Hla's demonstration of an STM-facilitated Ulmann reaction, wherein two iodobenzene molecules were dehalogenated via tunneling electrons, and the resulting phenyls were moved into proximity to one another and covalently bonded into biphenyl with another STM voltage pulse [239]. Repp et al. have also shown that it is possible to use the STM to covalently bond a gold atom to a pentacene molecule; the spatial precision

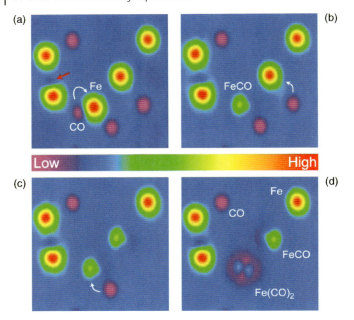

Figure 14.13 STM-induced bond formation. By using the tip to manipulate a CO molecule on a Fe atom (white arrow, a) and subsequently applying a voltage pulse with the STM tip, a FeCO molecule can be formed (b). By moving another CO molecule (white arrow, c) to FeCO, and applying a voltage pulse, a $Fe(CO)_2$ molecule can be formed (d). STM image parameters: $V_b = 70$ mV bias, $I_t = 0.1$ nA, $T = 13$ K, and Area $= 6.3 \times 6.3$ nm^2. Reproduced from Ref. [66], Figure 2.

of this technique makes it possible to form various isomers of the metal–organic complex [240].

14.7.3
Inducing Reactions Over a Larger Spatial Extent

The chain-reaction synthesis of polydiacetylene represents a special case of synthesis by STM. In this particular reaction, a voltage pulse from the STM tip is used to initiate a self-propagating reaction where diacetylene moieties are converted to polydiacetylene (PDA), creating one-dimensional polymers that span a spatial extent much larger than the area directly beneath the STM tip [241–245]. We mention this phenomenon only briefly here since we have recently published a much broader review of the topic [25].

The spatial extent of STM-induced reactions can also be extended by using lateral carrier transport via the surface states associated with the underlying substrate. Working on various metal substrates, Maksymovych and coworkers used the STM tip to inject electrons into a substrate–surface state (or resonance), and subsequently observed the dissociation of isolated CH_3SSCH_3 molecules up to 50 nm away [246].

Nouchi *et al.* were able to induce the polymerization of C_{60} in the same manner: after injecting electrons or holes into a C_{60} thin film, they observed the formation of a ring of polymerized molecules surrounding the injection point [247].

14.8
Conclusions and Perspective

The study of soft matter is one of the important frontiers in advanced materials, both from a fundamental point of view and for a number of applications, including implantable biomaterials, drug delivery, chemical and biological sensors, and organic electronic and photovoltaic devices. Organic building blocks can be used to coat a substrate, thereby conferring specific functionalities; alternatively, several to hundreds of layers can be grown and the resulting thin film can be used as an active material for sensing, solar energy conversion, or transporting charge carriers in an electronic device. Such alternatives to the inorganic materials traditionally in use are advantageous because bottom-up fabrication processes are often inexpensive, and the synthesis of the building blocks allows a great flexibility (using advanced synthetic strategies) to tailor specific properties. Such bottom-up approaches aim at the spontaneous formation of a desired architecture either on a surface or in a thin film form. They are increasingly used in research environments, yet are challenging to implement in developing technologies because of the somewhat limited, and sometimes even poor, control of feature sizes and their distributions. There are both advantages and disadvantages to the fact that soft matter structures are usually held together by noncovalent bonds, such as hydrogen bonds and van der Waals forces. The latter can promote the formation of long-range ordered structures and can even correct local defects through self-repair mechanisms, yet are not strong enough to form electronically and mechanically robust structures.

The aim of this chapter was to provide a broad overview of the use of STM in investigating the early stages of the growth of organic building blocks at solid surfaces. This is an emerging area at the boundary of modern surface science and physical chemistry. This approach has developed only over the last two decades thanks to the advent of scanning probe microscopy, which allows the imaging of surfaces and interfaces with atomic resolution and offers other exciting features such as local electronic and vibrational spectroscopy, manipulation of single atoms and molecules, diffusion studies of individual adsorbates, and the selective rupture and formation of chemical bonds. The studies reported herein take advantage of self-assembly processes, which occur under equilibrium conditions and use noncovalent bonds to form two-dimensional supramolecular structures that result from a balance between intermolecular forces and molecule–substrate interactions. In addition to the intermolecular interactions mentioned earlier, more intricate processes can involve hierarchical architectures, which typically use nanoscale patterns at surfaces to guide the formation of subsequent patterns by providing nanoscale cues. These porous networks show promise with respect to the controlled growth of ordered arrays with alternating donor–acceptor molecules that would be very useful for solar energy conversion.

The impact of the STM in studying such model systems cannot be overemphasized, as described in the selected examples that we presented in this chapter. Important phenomena that have recently been reported include the observation of chirality at surfaces, the formation of magnetically structured MOCNs, and the direct visualization of the preferential bonding of complimentary DNA base pairs. Furthermore, the STM itself can be an integral part of the experiment, as is the case in light emission induced by electron injection or chemical synthesis induced via tunneling electrons. These are just a few of the approaches that STM provides for studying and interacting with nanoscale systems; the more in-depth examples provided throughout this chapter highlight the ways in which STM has established itself as an indispensable tool in the pursuit of new technologies incorporating 2D supramolecular architectures.

List of Acronyms

6T	α-sexithiophene
A	adenine
C	cytosine
DFT	density functional theory
DNA	deoxyribonucleic acid
DPPA	dodecylphosphoric acid
FET	field-effect transistor
G	guanine
HBC	hexa-*peri*-hexabenzocoronenes
HOPG	highly oriented pyrolytic graphite
IETS	inelastic electron tunneling spectroscopy
LDOS	local density of states
LED	light-emitting device
MOCN	metal–organic coordination network
OSC	organic semiconductor
PAH	polycyclic aromatic hydrocarbon
PTCDA	3,4,9,10-perylenetetracarboxylic-acid-dianhydride
PV	photovoltaic
RNA	ribonucleic acid
RT	room temperature
SAM	self-assembled monolayer
SAMN	self-assembled molecular network
SP-STM	spin-polarized scanning tunneling microscopy/microscope
STM	scanning tunneling microscopy/microscope
STS	scanning tunneling spectroscopy
T	thymine
TBrPP-Co	5, 10, 15, 20-Tetrakis-(4-bromophenyl)-porphyrin-Co

TDA	4,1′,4″,1-terphenyl-1,4″-dicarboxylic acid
TMA	trimesic acid
TPA	terephthalate
TTBTA	terthienobenzenetricarboxylic acid
UHV	ultrahigh vacuum
VT	variable temperature

References

1. Foster, L.E. (2009) *Supramolecular Chemistry*, 2nd edn (eds J.W. Steed and J.L. Atwood), John Wiley & Sons, Ltd., p. 900.
2. Moore, G.E. (1965) *Electronics*, **38**, 114–117.
3. Forrest, S.R. (2004) *Nature*, **428**, 911–918.
4. Moriarty, P. (2001) *Rep. Prog. Phys.*, **64**, 297–381.
5. Rosei, F. (2004) *J. Phys. Condens. Matter*, **16**, S1373–S1436.
6. Dimitrakopoulos, C.D. and Malenfant, P.R.L. (2002) *Adv. Mater.*, **14**, 99–117.
7. Hoeben, F.J.M., Jonkheijm, P., Meijer, E.W., and Schenning, A. (2005) *Chem. Rev.*, **105**, 1491–1546.
8. Martin, R.E. and Diederich, F. (1999) *Angew. Chem. Int. Ed. Eng.*, **38**, 1350–1377.
9. Gunes, S., Neugebauer, H., and Sariciftci, N.S. (2007) *Chem. Rev.*, **107**, 1324–1338.
10. Xia, Y. and Whitesides, G.M. (1998) *Angew. Chem., Int. Ed. Eng.*, **37**, 550–575.
11. Lehn, J.M. (2002) *Science*, **295**, 2400–2403.
12. Steed, J.W. and Atwood, J.L. (2009) *Supramolecular Chemistry*, 2nd edn, John Wiley & Sons Ltd., Chichester.
13. Whitesides, G.M., Mathias, J.P., and Seto, C.T. (1991) *Science*, **254**, 1312–1319.
14. Sanchez, C., Arribart, H., and Giraud Guille, M.M. (2005) *Nat. Mater.*, **4**, 277–288.
15. Sarikaya, M., Tamerler, C., Jen, A.K.Y., Schulten, K., and Baneyx, F. (2003) *Nat. Mater.*, **2**, 577–585.
16. Hwang, I.-W., Kamada, T., Ahn, T.K., Ko, D.M., Nakamura, T., Tsuda, A., Osuka, A., and Kim, D. (2004) *J. Am. Chem. Soc.*, **126**, 16187–16198.
17. Grill, L., Dyer, M., Lafferentz, L., Persson, M., Peters, M.V., and Hecht, S. (2007) *Nat. Nanotechnol.*, **2**, 687–691.
18. Brusso, J.L., Hirst, O.D., Dadvand, A., Ganesan, S., Cicoira, F., Robertson, C.M., Oakley, R.T., Rosei, F., and Perepichka, D.F. (2008) *Chem. Mater.*, **20**, 2484–2494.
19. Cicoira, F., Santato, C., Dadvand, A., Harnagea, C., Pignolet, A., Bellutti, P., Xiang, Z., Rosei, F., Meng, H., and Perepichka, D.F. (2008) *J. Mater. Chem.*, **18**, 158–161.
20. Dadvand, A., Cicoira, F., Chernichenko, K.Y., Balenkova, E.S., Osuna, R.M., Rosei, F., Nenajdenko, V.G., and Perepichka, D.F. (2008) *Chem. Comm.*, 5354–5356.
21. Eremtchenko, M., Schaefer, J.A., and Tautz, F.S. (2003) *Nature*, **425**, 602–605.
22. Dubois, L.H. and Nuzzo, R.G. (1992) *Annu. Rev. Phys. Chem.*, **43**, 437–463.
23. Ulman, A. (1996) *Chem. Rev.*, **96**, 1533–1554.
24. Barth, J.V. (2009) *Surf. Sci.*, **603**, 1533–1541.
25. MacLeod, J.M. and Rosei, F. (2010) *Comprehensive Nanoscience and Technology* (eds D. Andrews, G. Scholes and G. Wiedderecht), Elsevier, Amsterdam.
26. Eddaoudi, M., Moler, D.B., Li, H.L., Chen, B.L., Reineke, T.M., O'Keeffe, M., and Yaghi, O.M. (2001) *Acc. Chem. Res.*, **34**, 319–330.
27. James, S.L. (2003) *Chem. Soc. Rev.*, **32**, 276–288.
28. Yaghi, O.M., O'Keeffe, M., Ockwig, N.W., Chae, H.K., Eddaoudi, M., and Kim, J. (2003) *Nature*, **423**, 705–714.

29 Lin, N., Stepanow, S., Ruben, M., and Barth, J.V. (2009) *Templates in Chemistry III, Topics in Current Chemistry*, vol. 287, Springer-Verlag, Berlin, pp. 1–44.
30 Cicoira, F., Santato, C., and Rosei, F. (2008) In: *STM and AFM Studies on (Bio) Molecular Systems, Topics in Current Chemistry*, vol. 285, Springer-Verlag, Berlin, pp. 203–267.
31 Frommer, J. (1992) *Angew. Chem. Int. Ed. Eng.*, **31**, 1298–1328.
32 Claypool, C.L., Faglioni, F., Goddard, W.A., Gray, H.B., Lewis, N.S., and Marcus, R.A. (1997) *J. Phys. Chem. B*, **101**, 5978–5995.
33 Jung, T.A., Schlittler, R.R., and Gimzewski, J.K. (1997) *Nature*, **386**, 696–698.
34 Tersoff, J. and Hamann, D.R. (1983) *Phys. Rev. Lett.*, **50**, 1998–2001.
35 Sautet, P. and Joachim, C. (1991) *Chem. Phys. Lett.*, **185**, 23–30.
36 Bevan, K.H., Zahid, F., Kienle, D., and Guo, H. (2007) *Phys. Rev. B*, **76**, 045325.
37 He, Y., Tilocca, A., Dulub, O., Selloni, A., and Diebold, U. (2009) *Nat. Mater.*, **8**, 585–589.
38 Hofer, W.A., Fisher, A.J., Lopinski, G.P., and Wolkow, R.A. (2001) *Surf. Sci.*, **482–485**, 1181–1185.
39 Li, B., Zeng, C., Li, Q., Wang, B., Yuan, L., Wang, H., Yang, J., Hou, J.G., and Zhu, Q. (2003) *J. Phys. Chem. B*, **107**, 972–984.
40 Maksymovych, P., Sorescu, D.C., and Yates, J.T. (2006) *J. Phys. Chem. B*, **110**, 21161–21167.
41 Orzali, T., Forrer, D., Sambi, M., Vittadini, A., Casarin, M., and Tondello, E. (2007) *J. Phys. Chem C*, **112**, 378–390.
42 Rogers, B.L., Shapter, J.G., and Ford, M.J. (2004) *Surf. Sci.*, **548**, 29–40.
43 Rose, M.K., Mitsui, T., Dunphy, J., Borg, A., Ogletree, D.F., Salmeron, M., and Sautet, P. (2002) *Surf. Sci.*, **512**, 48–60.
44 Schofield, S.R., Saraireh, S.A., Smith, P.V., Radny, M.W., and King, B.V. (2007) *J. Am. Chem. Soc.*, **129**, 11402–11407.
45 Wang, F., Sorescu, D.C., and Jordan, K.D. (2002) *J. Phys. Chem. B*, **106**, 1316–1321.
46 Kunitake, M., Akiba, U., Batina, N., and Itaya, K. (1997) *Langmuir*, **13**, 1607–1615.
47 Hibino, M., Sumi, A., and Hatta, I. (1996) *Thin Solid Films*, **273**, 272–278.
48 Lackinger, M., Griessl, S., Kampschulte, L., Jamitzky, F., and Heckl, W.M. (2005) *Small*, **1**, 532–539.
49 Wakayama, Y., Hill, J.P., and Ariga, K. (2007) *Surf. Sci.*, **601**, 3984–3987.
50 Tao, C.G., Liu, Q., Riddick, B.S., Cullen, W.G., Reutt-Robey, J., Weeks, J.D., and Williams, E.D. (2008) *Proc. Natl. Acad. Sci. USA*, **105**, 16418–16425.
51 Mamdouh, W., Dong, M.D., Xu, S.L., Rauls, E., and Besenbacher, F. (2006) *J. Am. Chem. Soc.*, **128**, 13305–13311.
52 Mamdouh, W., Dong, M.D., Kelly, R.E.A., Kantorovich, L.N., and Besenbacher, F. (2007) *J. Phys. Chem. B*, **111**, 12048–12052.
53 MacLeod, J.M., Ivasenko, O., Perepichka, D.F., and Rosei, F. (2007) *Nanotechnology*, **18**, 424031.
54 Bocquet, M.L. and Sautet, P. (1996) *Surf. Sci.*, **360**, 128–136.
55 Stroscio, J.A., Feenstra, R.M., and Fein, A.P. (1986) *Phys. Rev. Lett.*, **57**, 2579–2582.
56 Feenstra, R.M. and Stroscio, J.A. (1987) *J. Vac. Sci. Technol., B*, **5**, 923–929.
57 Wildoer, J.W.G., Venema, L.C., Rinzler, A.G., Smalley, R.E., and Dekker, C. (1998) *Nature*, **391**, 59–62.
58 Petersen, L., Sprunger, P.T., Hofmann, P., Laegsgaard, E., Briner, B.G., Doering, M., Rust, H.P., Bradshaw, A.M., Besenbacher, F., and Plummer, E.W. (1998) *Phys. Rev. B*, **57**, R6858–R6861.
59 Sato, N., Takeda, S., Nagao, T., and Hasegawa, S. (1999) *Phys. Rev. B*, **59**, 2035.
60 Temirov, R., Soubatch, S., Luican, A., and Tautz, F.S. (2006) *Nature*, **444**, 350–353.
61 Crommie, M.F., Lutz, C.P., and Eigler, D.M. (1993) *Science*, **262**, 218–220.
62 Meyer, G., Neu, B., and Rieder, K.H. (1995) *Appl. Phys. A*, **60**, 343–345.

63 Jung, T.A., Schlittler, R.R., Gimzewski, J.K., Tang, H., and Joachim, C. (1996) *Science*, **271**, 181–184.

64 Foley, E.T., Kam, A.F., Lyding, J.W., and Avouris, P. (1998) *Phys. Rev. Lett.*, **80**, 1336–1339.

65 Avouris, P., Walkup, R.E., Rossi, A.R., Akpati, H.C., Nordlander, P., Shen, T.C., Abeln, G.C., and Lyding, J.W. (1996) *Surf. Sci. Rep.*, **363**, 368–377.

66 Lee, H.J. and Ho, W. (1999) *Science*, **286**, 1719–1722.

67 Haiss, W., Nichols, R.J., van Zalinge, H., Higgins, S.J., Bethell, D., and Schiffrin, D.J. (2004) *Phys. Chem. Chem. Phys.*, **6**, 4330–4337.

68 Lafferentz, L., Ample, F., Yu, H., Hecht, S., Joachim, C., and Grill, L. (2009) *Science*, **323**, 1193–1197.

69 Song, F., Wells, J.W., Handrup, K., Li, Z.S., Bao, S.N., Schulte, K., Ahola-Tuomi, M., Mayor, L.C., Swarbrick, J.C., Perkins, E.W., Gammelgaard, L., and Hofmann, P. (2009) *Nat. Nanotechnol.*, **4**, 373–376.

70 Hofmann, P. and Wells, J.W. (2009) *J. Phys. Condens. Matter*, **21**, 21.

71 Boggild, P., Hansen, T.M., Kuhn, O., Grey, F., Junno, T., and Montelius, L. (2000) *Rev. Sci. Instrum.*, **71**, 2781–2783.

72 Petersen, C.L., Hansen, T.M., Boggild, P., Boisen, A., Hansen, O., Hassenkam, T., and Grey, F. (2002) *Sens. Actuators A*, **96**, 53–58.

73 Lin, R., Boggild, P., and Hansen, O. (2004) *J. Appl. Phys.*, **96**, 2895–2900.

74 Tsukamoto, S., Siu, B., and Nakagiri, N. (1991) *Rev. Sci. Instrum.*, **62**, 1767–1771.

75 Grube, H., Harrison, B.C., Jia, J.F., and Boland, J.J. (2001) *Rev. Sci. Instrum.*, **72**, 4388–4392.

76 Guise, O., Marbach, H., Yates, J.T., Jung, M.C., Levy, J., and Ahner, J. (2005) *Rev. Sci. Instrum.*, **76**, 8.

77 Kim, T.H., Wang, Z.H., Wendelken, J.F., Weitering, H.H., Li, W.Z., and Li, A.P. (2007) *Rev. Sci. Instrum.*, **78**, 7.

78 Berndt, R., Gaisch, R., Gimzewski, J.K., Reihl, B., Schlittler, R.R., Schneider, W.D., and Tschudy, M. (1993) *Science*, **262**, 1425–1427.

79 Hoffmann, G., Libioulle, L., and Berndt, R. (2002) *Phys. Rev. B*, **65**, 4.

80 Liu, H.W., Ie, Y., Nishitani, R., Aso, Y., and Iwasaki, H. (2007) *Phys. Rev. B*, **75**, 5.

81 Zhang, Y., Tao, X., Gao, H.Y., Dong, Z.C., Hou, J.G., and Okamoto, T. (2009) *Phys. Rev. B*, **79**, 5.

82 Dong, Z.C., Guo, X.L., Trifonov, A.S., Dorozhkin, P.S., Miki, K., Kimura, K., Yokoyama, S., and Mashiko, S. (2004) *Phys. Rev. Lett.*, **92**, 4.

83 Qiu, X.H., Nazin, G.V., and Ho, W. (2003) *Science*, **299**, 542–546.

84 Gundlach, D.J., Lin, Y.-Y., Jackson, T.N., and Schlom, D.G. (1997) *Appl. Phys. Lett.*, **71**, 3853–3855.

85 Garnier, F., Horowitz, G., Peng, X.Z., and Fichou, D. (1991) *Synth. Met.*, **45**, 163–171.

86 Anthony, J.E. (2006) *Chem. Rev.*, **106**, 5028–5048.

87 Bao, Z., Lovinger, A.J., and Dodabalapur, A. (1996) *Appl. Phys. Lett.*, **69**, 3066–3068.

88 Facchetti, A. (2007) *Mater. Today*, **10**, 28–37.

89 Fichou, D. (1999) *Handbook of Oligo- and Polythiophenes*, Wiley-VCH Verlag GmbH, Weinheim.

90 Perepichka, I.F., Perepichka, D.F., Meng, H., and Wudl, F. (2005) *Adv. Mater.*, **17**, 2281–2305.

91 Fukunaga, T., Harada, K., Takashima, W., and Kaneto, K. (1997) *Jpn. J. Appl. Phys.*, **36**, 4466.

92 Bäuerle, P., Fischer, T., Bidlingmeier, B., Rabe, J.P., and Stabel, A. (1995) *Angew. Chem. Int. Ed. Eng.*, **34**, 303–307.

93 Stabel, A. and Rabe, J.P. (1994) *Synth. Met.*, **67**, 47–53.

94 Kirschbaum, T., Azumi, R., Mena-Osteritz, E., and Bäuerle, P. (1999) *New J. Chem.*, **23**, 241–250.

95 Abdel-Mottaleb, M.M.S., Gotz, G., Kilickiran, P., Bauerle, P., and Mena-Osteritz, E. (2005) *Langmuir*, **22**, 1443–1448.

96 Müller, H., Petersen, J., Strohmaier, R., Gompf, B., Eisenmenger, W., Vollmer, M.S., and Effenberger, F. (1996) *Adv. Mater.*, **8**, 733–737.

97 Gesquiere, A., Abdel-Mottaleb, M.M.S., De Feyter, S., De Schryver, F.C., Schoonbeek, F., van Esch, J., Kellogg,

R.M., Feringa, B.L., Calderone, A., Lazzaroni, R., and Bredas, J.L. (2000) *Langmuir*, **16**, 10385–10391.

98 Soukopp, A., Glöckler, K., Bäuerle, P., Sokolowski, M., and Umbach, E. (1996) *Adv. Mater.*, **8**, 902–906.

99 Frederick, B.G., Chen, Q., Barlow, S.M., Condon, N.G., Leibsle, F.M., and Richardson, N.V. (1996) *Surf. Sci.*, **352–354**, 238–247.

100 Noh, J., Ito, E., Nakajima, K., Kim, J., Lee, H., and Hara, M. (2002) *J. Phys. Chem. B*, **106**, 7139–7141.

101 Yang, Z.-Y., Zhang, H.-M., Yan, C.-J., Li, S.-S., Yan, H.-J., Song, W.-G., and Wan, L.-J. (2007) *Proc. Natl. Acad. Sci. USA*, **104**, 3707–3712.

102 Cicoira, F., Miwa, J.A., Melucci, M., Barbarella, G., and Rosei, F. (2006) *Small*, **2**, 1366–1371.

103 Bäuerle, P. (1992) *Adv. Mater.*, **4**, 102–107.

104 Krömer, J., Rios-Carreras, I., Fuhrmann, G., Musch, C., Wunderlin, M., Debaerdemaeker, T., Mena-Osteritz, E., and Bäuerle, P. (2000) *Angew. Chem.*, **39**, 3481–3486.

105 Mena-Osteritz, E. and Bauerle, P. (2006) *Adv. Mater.*, **18**, 447–451.

106 Otsubo, T., Aso, Y., and Takimiya, K. (2002) *J. Mater. Chem.*, **12**, 2565–2575.

107 Pan, G.B., Cheng, X.H., Hoger, S., and Freyland, W. (2006) *J. Am. Chem. Soc.*, **128**, 4218–4219.

108 MacLeod, J.M., Ivasenko, O., Fu, C., Taerum, T., Rosei, F., and Perepichka, D.F. (2009) *J. Am. Chem. Soc.*, **131** (46), 16844–16850.

109 Aziz, H. and Popovic, Z.D. (2002) *Appl. Phys. Lett.*, **80**, 2180–2182.

110 Li, G. and Shinar, J. (2003) *Appl. Phys. Lett.*, **83**, 5359–5361.

111 Sakamoto, G., Adachi, C., Koyama, T., Taniguchi, Y., Merritt, C.D., Murata, H., and Kafafi, Z.H. (1999) *Appl. Phys. Lett.*, **75**, 766–768.

112 Podzorov, V., Menard, E., Borissov, A., Kiryukhin, V., Rogers, J.A., and Gershenson, M.E. (2004) *Phys. Rev. Lett.*, **93**, 086602.

113 Takeya, J., Yamagishi, M., Tominari, Y., Hirahara, R., Nakazawa, Y., Nishikawa, T., Kawase, T., Shimoda, T., and Ogawa, S. (2007) *Appl. Phys. Lett.*, **90**, 102120–102123.

114 Briseno, A.L., Mannsfeld, S.C.B., Ling, M.M., Liu, S., Tseng, R.J., Reese, C., Roberts, M.E., Yang, Y., Wudl, F., and Bao, Z. (2006) *Nature*, **444**, 913–917.

115 Takahashi, T., Takenobu, T., Takeya, J., and Iwasa, Y. (2006) *Appl. Phys. Lett.*, **88**, 033505–33513.

116 Zeis, R., Besnard, C., Siegrist, T., Schlockermann, C., Chi, X., and Kloc, C. (2005) *Chem. Mater.*, **18**, 244–248.

117 Sundar, V.C., Zaumseil, J., Podzorov, V., Menard, E., Willett, R.L., Someya, T., Gershenson, M.E., and Rogers, J.A. (2004) *Science*, **303**, 1644–1646.

118 Stassen, A.F., de Boer, R.W.I., Iosad, N.N., and Morpurgo, A.F. (2004) *Appl. Phys. Lett.*, **85**, 3899–3901.

119 Podzorov, V., Sysoev, S.E., Loginova, E., Pudalov, V.M., and Gershenson, M.E. (2003) *Appl. Phys. Lett.*, **83**, 3504–3506.

120 Chan, M.Y., Lai, S.L., Fung, M.K., Lee, C.S., and Lee, S.T. (2007) *Appl. Phys. Lett.*, **90**, 023504–023513.

121 Taima, T., Sakai, J., Yamanari, T., and Saito, K. (2006) *Jpn. J. Appl. Phys.*, **45**, L995.

122 Pandey, A.K. and Nunzi, J.-M. (2007) *Appl. Phys. Lett.*, **90**, 263508–263513.

123 Bergmann, E. and Herlinger, E. (1936) *J. Chem. Phys.*, **4**, 532–534.

124 Schonberg, A. (2002) *J. Am. Chem. Soc.*, **58**, 182–182.

125 Jurchescu, O.D., Meetsma, A., and Palstra, T.T.M. (2006) *Acta Crystallogr. B*, **62**, 330–334.

126 Miwa, J.A., Cicoira, F., Bedwani, S., Lipton-Duffin, J., Perepichka, D.F., Rochefort, A., and Rosei, F. (2008) *J. Phys. Chem. C*, **112**, 10214–10221.

127 Miwa, J.A., Cicoira, F., Lipton-Duffin, J., Perepichka, D.F., Santato, C., and Rosei, F. (2008) *Nanotechnol.*, **19**, 424021.

128 Pivetta, M., Blüm, M.-C., Patthey, F., and Schneider, W.-D. (2008) *Angew. Chem. Int. Ed. Eng.*, **47**, 1076–1079.

129 Zhang, J., Albelda, M.T., Liu, Y., and Canary, J.W. (2005) *Chirality*, **17**, 404–420.

130 Barlow, S.M. and Raval, R. (2003) *Surf. Sci. Rep.*, **50**, 201–341.

131 Hazen, R.M. and Sholl, D.S. (2003) *Nat. Mater.*, **2**, 367–374.
132 Bentley, R. (2005) *Chem. Soc. Rev.*, **34**, 609–624.
133 Solladié, G. and Zimmermann, R.G. (1984) *Angew. Chem. Int. Ed. Eng.*, **23**, 348–362.
134 Samori, P., Fechtenkotter, A., Jackel, F., Bohme, T., Mullen, K., and Rabe, J.P. (2001) *J. Am. Chem. Soc.*, **123**, 11462–11467.
135 Lazzaroni, R., Calderone, A., Bredas, J.L., and Rabe, J.P. (1997) *J. Chem. Phys.*, **107**, 99–105.
136 Lopinski, G.P., Moffatt, D.J., Wayner, D.D.M., and Wolkow, R.A. (1998) *Nature*, **392**, 909–911.
137 Fang, H., Giancarlo, L.C., and Flynn, G.W. (1998) *J. Phys. Chem. B*, **102**, 7311–7315.
138 Mamdouh, W., Uji-i, H., Gesquiere, A., De Feyter, S., Amabilino, D.B., Abdel-Mottaleb, M.M.S., Veciana, J., and De Schryver, F.C. (2004) *Langmuir*, **20**, 9628–9635.
139 Sowerby, S., Heckl, W., and Petersen, G. (1996) *J. Mol. Evol.*, **43**, 419–424.
140 Charra, F. and Cousty, J. (1998) *Phys. Rev. Lett.*, **80**, 1682.
141 McFadden, C.F., Cremer, P.S., and Gellman, A.J. (1996) *Langmuir*, **12**, 2483–2487.
142 Goodby, J.W. (1991) *J. Mater. Chem.*, **1**, 307–318.
143 Feyter, S.D., Gesquière, A., Wurst, K., Amabilino, D.B., Veciana, J., and Schryver, F.C.D. (2001) *Angew. Chem. Int. Ed. Eng.*, **40**, 3217–3220.
144 Linares, M., Iavicoli, P., Psychogyiopoulou, K., Beljonne, D., De Feyter, S., Amabilino, D.B., and Lazzaroni, R. (2008) *Langmuir*, **24**, 9566–9574.
145 Otsuki, J., Nagamine, E., Kondo, T., Iwasaki, K., Asakawa, M., and Miyake, K. (2005) *J. Am. Chem. Soc.*, **127**, 10400–10405.
146 Wang, H.N., Wang, C., Zeng, Q.D., Xu, S.D., Yin, S.X., Xu, B., and Bai, C.L. (2001) *Surf. Interface Anal.*, **32**, 266–270.
147 Tersoff, J. and Hamann, D.R. (1985) *Phys. Rev. B*, **31**, 805–813.
148 Kiguchi, M., Yoshikawa, G., and Saiki, K. (2003) *J. Appl. Phys.*, **94**, 4866–4870.
149 Yoshikawa, G., Kiguchi, M., Ikeda, S., and Saiki, K. (2004) *Surf. Sci.*, **559**, 77–84.
150 Umbach, E., Sokolowski, M., and Fink, R. (1996) *Appl. Phys. A*, **63**, 565–576.
151 Santato, C., Cicoira, F., and Rosei, F. (2009) *Handbook of Thiophene-Based Materials: Applications in Organic Electronics and Photonics*, vol. 2 (eds I.F. Perepichka and D.F. Perepichka), John Wiley & Sons Ltd., Chichester, pp. 517–548.
152 Prato, S., Floreano, L., Cvetko, D., Renzi, V.D., Morgante, A., Modesti, S., Biscarini, F., Zamboni, R., and Taliani, C. (1999) *J. Phys. Chem. B*, **103**, 7788–7795.
153 Kiel, M., Duncker, K., Hagendorf, C., and Widdra, W. (2007) *Phys. Rev. B*, **75**, 195439.
154 Allwood, D.A., Xiong, G., Faulkner, C.C., Atkinson, D., Petit, D., and Cowburn, R.P. (2005) *Science*, **309**, 1688–1692.
155 Hanson, R., Kouwenhoven, L.P., Petta, J.R., Tarucha, S., and Vandersypen, L.M.K. (2007) *Rev. Mod. Phys.*, **79**, 1217–1265.
156 Wolf, S.A., Awschalom, D.D., Buhrman, R.A., Daughton, J.M., von Molnar, S., Roukes, M.L., Chtchelkanova, A.Y., and Treger, D.M. (2001) *Science*, **294**, 1488–1495.
157 Steane, A. (1998) *Rep. Prog. Phys.*, **61**, 117–173.
158 Yoffe, A.D. (2001) *Adv. Phys.*, **50**, 1–208.
159 O'Grady, K. and Laidler, H. (1999) *J. Magn. Magn. Mater.*, **200**, 616–633.
160 Thompson, D.A. and Best, J.S. (2000) *IBM J. Res. Dev.*, **44**, 311.
161 Kondo, J. (1964) *Prog. Theor. Phys.*, **32**, 37.
162 Manoharan, H.C., Lutz, C.P., and Eigler, D.M. (2000) *Nature*, **403**, 512–515.
163 Li, J., Schneider, W.-D., Berndt, R., and Delley, B. (2893) *Phys. Rev. Lett.*, **80**, 1998.
164 Heinrich, A.J., Gupta, J.A., Lutz, C.P., and Eigler, D.M. (2004) *Science*, **306**, 466–469.
165 Madhavan, V., Chen, W., Jamneala, T., Crommie, M.F., and Wingreen, N.S. (1998) *Science*, **280**, 567–569.

166 Wahl, P., Diekhöner, L., Wittich, G., Vitali, L., Schneider, M.A., and Kern, K. (2005) *Phys. Rev. Lett.*, **95**, 166601.

167 Iancu, V., Deshpande, A., and Hla, S.-W. (2006) *Phys. Rev. Lett.*, **97**, 266603.

168 Iancu, V., Deshpande, A., and Hla, S.-W. (2006) *Nano. Lett.*, **6**, 820–823.

169 Lin, N., Payer, D., Dmitriev, A., Strunskus, T., Woll, C., Barth, J.V., and Kern, K. (2005) *Angew. Chem. Int. Ed. Eng.*, **44**, 1488–1491.

170 Lingenfelder, M.A., Spillmann, H., Dmitriev, A., Stepanow, S., Lin, N., Barth, J.V., and Kern, K. (2004) *Chem. Eur. J.*, **10**, 1913–1919.

171 Dmitriev, A., Spillmann, H., Lin, N., Barth, J.V., and Kern, K. (2003) *Angew. Chem. Int. Ed. Eng.*, **42**, 2670–2673.

172 Stepanow, S., Lingenfelder, M., Dmitriev, A., Spillmann, H., Delvigne, E., Lin, N., Deng, X.B., Cai, C.Z., Barth, J.V., and Kern, K. (2004) *Nat. Mater.*, **3**, 229–233.

173 Messina, P., Dmitriev, A., Lin, N., Spillmann, H., Abel, M., Barth, J.V., and Kern, K. (2002) *J. Am. Chem. Soc.*, **124**, 14000–14001.

174 Classen, T., Fratesi, G., Costantini, G., Fabris, S., Stadler, F.L., Kim, C., de Gironcoli, S., Baroni, S., and Kern, K. (2005) *Angew. Chem. Int. Ed. Eng.*, **44**, 6142–6145.

175 Gambardella, P., Stepanow, S., Dmitriev, A., Honolka, J., de Groot, F.M.F., Lingenfelder, M., Gupta, S.S., Sarma, D.D., Bencok, P., Stanescu, S., Clair, S., Pons, S., Lin, N., Seitsonen, A.P., Brune, H., Barth, J.V., and Kern, K. (2009.) *Nat. Mater.*, **8**, 189–193.

176 Puleo, D.A. and Nanci, A. (1999) *Biomaterials*, **20**, 2311–2321.

177 Kasemo, B. (2002) *Surf. Sci.*, **500**, 656–677.

178 Rosei, F., Schunack, M., Naitoh, Y., Jiang, P., Gourdon, A., Laegsgaard, E., Stensgaard, I., Joachim, C., and Besenbacher, F. (2003) *Prog. Surf. Sci.*, **71**, 95–146.

179 Prime, K.L. and Whitesides, G.M. (1993) *J. Am. Chem. Soc.*, **115**, 10714–10721.

180 Nelson, B.P., Grimsrud, T.E., Liles, M.R., Goodman, R.M., and Corn, R.M. (2000) *Anal. Chem.*, **73**, 1–7.

181 Storri, S., Santoni, T., Minunni, M., and Mascini, M. (1998) *Biosens. Bioelectron.*, **13**, 347–357.

182 Ostroff, R.M., Hopkins, D., Haeberli, A.B., Baouchi, W., and Polisky, B. (1999) *Clin. Chem.*, **45**, 1659–1664.

183 Cicoira, F. and Rosei, F. (2006) *Surf. Sci.*, **600**, 1–5.

184 Variola, F., Vetrone, F., Richert, L., Jedrzejowski, P., Yi, J.-H., Zalzal, S., Clair, S., Sarkissian, A., Perepichka, D.F., Wuest, J.D., Rosei, F., and Nanci, A. (2009) *Small*, **5**, 996–1006.

185 Variola, F., Yi, J.H., Richert, L., Wuest, J.D., Rosei, F., and Nanci, A. (2008) *Biomaterials*, **29**, 1285–1298.

186 Vetrone, F., Variola, F., de Oliveira, P.T., Zalzal, S.F., Yi, J.H., Sam, J., Bombonato-Prado, K.F., Sarkissian, A., Perepichka, D.F., Wuest, J.D., Rosei, F., and Nanci, A. (2009) *Nano. Lett.*, **9**, 659–665.

187 Yi, J.H., Bernard, C., Variola, F., Zalzal, S.F., Wuest, J.D., Rosei, F., and Nanci, A. (2006) *Surf. Sci.*, **600**, 4613–4621.

188 Castner, D.G. and Ratner, B.D. (2002) *Surf. Sci.*, **500**, 28–60.

189 Sowerby, S.J. and Heckl, W.M. (1998) *Orig. Life Evol. B*, **28**, 283–310.

190 Andersen, E.S., Dong, M., Nielsen, M.M., Jahn, K., Subramani, R., Mamdouh, W., Golas, M.M., Sander, B., Stark, H., Oliveira, C.L.P., Pedersen, J.S., Birkedal, V., Besenbacher, F., Gothelf, K.V., and Kjems, J. (2009) *Nature*, **459**, U73–U75.

191 Chen, Q., Frankel, D.J., and Richardson, N.V. (2002) *Langmuir*, **18**, 3219–3225.

192 Chen, Q. and Richardson, N.V. (2003) *Nat. Mater.*, **2**, 324–328.

193 Furukawa, M., Tanaka, H., and Kawai, T. (1997) *Surf. Sci.*, **392**, L33–L39.

194 Furukawa, M., Tanaka, H., and Kawai, T. (2000) *Surf. Sci.*, **445**, 1–10.

195 Furukawa, M., Tanaka, H., and Kawai, T. (2001) *J. Chem. Phys.*, **115**, 3419–3423.

196 Furukawa, M., Tanaka, H., Sugiura, K.-I., Sakata, Y., and Kawai, T. (2000) *Surf. Sci.*, **445**, L58–L63.

197 Kawai, T., Tanaka, H., and Nakagawa, T. (1997) *Surf. Sci.*, **386**, 124–136.

198 Kelly, R.E.A., Lukas, M., Kantorovich, L.N., Otero, R., Xu, W., Mura, M., Laegsgaard, E., Stensgaard, I., and Besenbacher, F. (2008) *J. Chem. Phys.*, **129**, 184707.

199 Kelly, R.E.A., Xu, W., Lukas, M., Otero, R., Mura, M., Lee, Y.J., Laegsgaard, E., Stensgaard, I., Kantorovich, L.N., and Besenbacher, F. (2008) *Small*, **4**, 1494–1500.

200 Kuhnle, A., Linderoth, T.R., Hammer, B., and Besenbacher, F. (2002) *Nature*, **415**, 891–893.

201 Mamdouh, W., Kelly, R.E.A., Dong, M.D., Jacobsen, M.F., Ferapontova, E.E., Kantorovich, L.N., Gothelf, K.V., and Besenbacher, F. (2009) *J. Phys. Chem. B*, **113**, 8675–8681.

202 Mamdouh, W., Kelly, R.E.A., Dong, M.D., Kantorovich, L.N., and Besenbacher, F. (2008) *J. Am. Chem. Soc.*, **130**, 695–702.

203 Otero, R., Lukas, M., Kelly, R.E.A., Xu, W., Laegsgaard, E., Stensgaard, I., Kantorovich, L.N., and Besenbacher, F. (2008) *Science*, **319**, 312–315.

204 Otero, R., Naitoh, Y., Rosei, F., Jiang, P., Thostrup, P., Gourdon, A., Laegsgaard, E., Stensgaard, I., Joachim, C., and Besenbacher, F. (2004) *Angew. Chem. Int. Ed. Eng.*, **43**, 2092–2095.

205 Otero, R., Rosei, F., and Besenbacher, F. (2006) *Annu. Rev. Phys. Chem.*, **57**, 497–525.

206 Otero, R., Rosei, F., Naitoh, Y., Jiang, P., Thostrup, P., Gourdon, A., Laegsgaard, E., Stensgaard, I., Joachim, C., and Besenbacher, F. (2004) *Nano. Lett.*, **4**, 75–78.

207 Otero, R., Schock, M., Molina, L.M., Laegsgaard, E., Stensgaard, I., Hammer, B., and Besenbacher, F. (2005) *Angew. Chem. Int. Ed. Eng.*, **44**, 2270–2275.

208 Tao, N.J., DeRose, J.A., and Lindsay, S.M. (2002) *J. Phys. Chem.*, **97**, 910–919.

209 Tao, N.J. and Shi, Z. (2002) *J. Phys. Chem.*, **98**, 1464–1471.

210 Xu, S.L., Dong, M.D., Rauls, E., Otero, R., Linderoth, T.R., and Besenbacher, F. (2006) *Nano. Lett.*, **6**, 1434–1438.

211 Yokoyarna, K. and Taira, S. (2005) In: *Immobilisation of DNA on Chips II, Topics in Current Chemistry*, vol. 261, Springer, pp. 91–112.

212 Otero, R., Xu, W., Lukas, M., Kelly, Ross E.A., Lægsgaard, E., Stensgaard, I., Kjems, J., Kantorovich, LevN., and Besenbacher, F. (2008) *Angew. Chem. Int. Ed. Eng.*, **47**, 9673–9676.

213 Ratner, B.D., Hoffman, A.S., Schoen, F.J., and Lemons, J.E. (2004) *Biomaterials Science: An Introduction to Materials in Medicine*, Elsevier Academic Press, San Diego, USA.

214 Brown, S.A. and Lemmons, J.E. (1996) *Medical Applications of Titanium and Its Alloys: The Material and Biological Issues*, ASTM International, West Conshohocken, PA, USA.

215 Mani, G., Feldman, M.D., Patel, D., and Agrawal, C.M. (2007) *Biomaterials*, **28**, 1689–1710.

216 Bhat, S.V. (2002) *Biomaterials*, Springer, London.

217 Black, J. (1988) In: *Orthopedic Biomaterials in Research and Practice*, Churchill, Livingstone, New York.

218 Levine, B.R., Sporer, S., Poggie, R.A., Della Valle, C.J., and Jacobs, J.J. (2006) *Biomaterials*, **27**, 4671–4681.

219 Clair, S., Variola, F., Kondratenko, M., Jedrzejowski, P., Nanci, A., Rosei, F., and Perepichka, D.F. (2008) *J. Chem. Phys.*, **128**, 144705.

220 Richert, L., Vetrone, F., Yi, J.H., Zalzal, S.F., Wuest, J.D., Rosei, F., and Nanci, A. (2008) *Adv. Mater.*, **20**, 1488–1492.

221 Black, J. (2006) *Biological Performance of Materials: Fundamentals of Biocompatibility*, 4th edn, CRC Press, Boca Raton, USA.

222 Wooley, P.H. and Schwarz, E.M. (2004) *Gene. Ther.*, **11**, 402–407.

223 Liu, H. and Webster, T.J. (2007) *Biomaterials*, **28**, 354–369.

224 Poirier, G.E. (1997) *Chem. Rev.*, **97**, 1117–1128.

225 Schreiber, F. (2000) *Prog. Surf. Sci.*, **65**, 151–257.

226 Love, J.C., Estroff, L.A., Kriebel, J.K., Nuzzo, R.G., and Whitesides, G.M. (2005) *Chem. Rev.*, **105**, 1103–1169.
227 Kanta, A., Sedev, R., and Ralston, J. (2006) *Colloids Surf. A*, **291**, 51–58.
228 Nilsing, M., Lunell, S., Persson, P., and Ojamäe, L. (2005) *Surf. Sci.*, **582**, 49–60.
229 Hofer, R., Textor, M., and Spencer, N.D. (2001) *Langmuir*, **17**, 4014–4020.
230 Tosatti, S., Michel, R., Textor, M., and Spencer, N.D. (2002) *Langmuir*, **18**, 3537–3548.
231 Zwahlen, M., Tosatti, S., Textor, M., and Hähner, G. (2002) *Langmuir*, **18**, 3957–3962.
232 Stevens, M.M. and George, J.H. (2005) *Science*, **310**, 1135–1138.
233 Dujardin, G., Walkup, R.E., and Avouris, P. (1992) *Science*, **255**, 1232–1235.
234 Martel, R., Avouris, P., and Lyo, I.W. (1996) *Science*, **272**, 385–388.
235 Stipe, B.C., Rezaei, M.A., Ho, W., Gao, S., Persson, M., and Lundqvist, B.I. (1997) *Phys. Rev. Lett.*, **78**, 4410–4413.
236 Gaudioso, J., Lee, H.J., and Ho, W. (1999) *J. Am. Chem. Soc.*, **121**, 8479–8485.
237 Lauhon, L.J. and Ho, W. (2000) *Phys. Rev. Lett.*, **84**, 1527–1530.
238 Kim, Y., Komeda, T., and Kawai, M. (2002) *Phys. Rev. Lett.*, **89**, 4.
239 Hla, S.-W., Bartels, L., Meyer, G., and Rieder, K.-H. (2000) *Phys. Rev. Lett.*, **85**, 2777.
240 Repp, J., Meyer, G., Paavilainen, S., Olsson, F.E., and Persson, M. (2006) *Science*, **312**, 1196–1199.
241 Okawa, Y. and Aono, M. (2001) *Nature*, **409**, 683–684.
242 Sullivan, S.P., Schmeders, A., Mbugua, S.K., and Beebe, T.P. (2005) *Langmuir*, **21**, 1322–1327.
243 Takajo, D., Okawa, Y., Hasegawa, T., and Aono, M. (2007) *Langmuir*, **23**, 5247–5250.
244 Nishio, S., I-i, D., Matsuda, H., Yoshidome, M., Uji-i, H., and Fukumura, H. (2005) *Jpn. J. Appl. Phys.*, **44**, 5417–5420.
245 Miura, A., De Feyter, S., Abdel-Mottaleb, M.M.S., Gesquiere, A., Grim, P.C.M., Moessner, G., Sieffert, M., Klapper, M., Mullen, K., and De Schryver, F.C. (2003) *Langmuir*, **19**, 6474–6482.
246 Maksymovych, P., Dougherty, D.B., Zhu, X.Y., and Yates, J.T. (2007) *Phys. Rev. Lett.*, **99**, 016101.
247 Nouchi, R., Masunari, K., Ohta, T., Kubozono, Y., and Iwasa, Y. (2006) *Phys. Rev. Lett.*, **97**, 196101.

15
Scanning Probe Microscopy Insights into Supramolecular π-Conjugated Nanostructures for Optoelectronic Devices

Mathieu Surin, Gwenaëlle Derue, Simon Desbief, Olivier Douhéret, Pascal Viville, Roberto Lazzaroni, and Philippe Leclère

15.1
Introduction: SPM Techniques for the Nanoscale Characterization of Organic Thin Films

Scanning probe microscopies (SPMs) are recent, powerful experimental techniques for the characterization of surface microstructures and properties with atomic resolution. The story began with the advent of scanning tunneling microscopy (STM) [1], which was introduced in 1982 for studying conductive surfaces, and atomic force microscopy (AFM, developed in 1986) [2], which can be used for morphological analysis on any type of surface. In polymer science, AFM studies began at the end of the 1980s, with the pioneering works of Hansma and coworkers [3], Reneker and coworkers [4], and Cantow and coworkers [5]. Besides, the seminal STM work of Rabe on organic monolayers (alkanes) physisorbed on graphite [6] has since led to the study of a plethora of organic assemblies in 2D at the liquid/solid interface (see recent reviews [7–10] and related chapters in this book). STM is nevertheless restricted to one or two organic monolayers adsorbed on conductive surface, while the AFM technique makes it possible to probe all types of materials, which is an obvious advantage for organic, polymeric, and biological surfaces. Over the past 15 years, AFM has provided a wealth of exciting results in all fields of polymer science and more specifically in the field of organic π-conjugated supramolecular structures and materials, which is the subject of this chapter.

The principle of AFM, also called scanning force microscopy (SFM), is very similar to profilometry, where a hard tip is scanned across the surface and its vertical movements resulting from the tip–sample interactions are monitored. As a result of the miniature size of the AFM tip, which is mounted at the free end of a cantilever-like spring, it is possible to image the corrugation of the sample at the nanometer scale (Figure 15.1). The general scheme of any AFM apparatus includes several major components that allow line-by-line scanning with an ideally atomically sharp tip while monitoring cantilever deflections in vertical and horizontal directions. Precise (within a fraction of a nanometer) 3D displacements of either the sample or the

Functional Supramolecular Architectures. Edited by Paolo Samorì and Franco Cacialli
Copyright © 2011 WILEY-VCH Verlag GmbH & Co. KGaA, Weinheim
ISBN: 978-3-527-32611-2

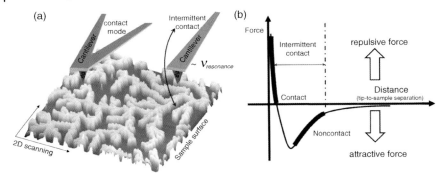

Figure 15.1 (a) Basic schemes of AFM in contact or intermittent modes; (b) force versus tip–surface distance curve, showing the interaction mode regimes.

cantilever toward each other are provided by a tube piezoelement, and the AFM tip deflection is monitored by a detection scheme (e.g., an array of photodiodes, an interferometer setup, or a piezocantilever). The microfabricated probe usually consists in a silicon or silicon nitride cantilever with the integrated (mostly pyramidal) tip of several micrometers in height with a tip apex radius in the range of 5–200 nm. Note that knowledge of the cantilever spring constants is critical for the determination of the absolute values of the vertical and torsional forces acting on the tip.

The basic modes of topographical imaging of any AFM instrument are determined by the scanning type. The first operational mode is called "contact mode." In this regime, the SPM tip is dragged over the sample surface with a constant velocity and under a constant normal load. The tip is in close contact with the surface, that is, the atoms of the tip and the surface are in the repulsive range of van der Waals forces (Figure 15.1b). The AFM measurement can be carried out in two different manners: (i) the height mode, while the cantilever deflection is kept constant by the extending–retracting piezoelement tube, and the adjustments of the vertical extension are recorded. (ii) The deflection mode, while the piezotube extension is constant and the cantilever deflection is recorded. The contact mode operation allows tracking of the surface topography with a high accuracy and can provide a lateral resolution of 0.2–0.3 nm, but it imposes a high local pressure and shear stress on the surface. The small tip–surface contact area (few tens of nm^2) and significant contribution of the capillary forces due to adsorbed water are major features of the SPM operations in humid air. Thus, even very low normal forces (in the nN range) result in a very high normal pressure and shear stresses within the 0.1–100 GPa range. Such high local stress can easily damage soft materials. An alternative for reducing the normal pressure applied to the surface is to operate in a dry atmosphere (nitrogen, argon) that eliminates the capillary forces but affects the easiness of the AFM measurement methods with tedious experimental constraints. To overcome the drawbacks of contact mode (i.e., surface damage problems), dynamical force modes have therefore been introduced, which correspond to an intermittent contact time with the sample surface, minimizing the contact between the tip apex and the sample. A stiff SPM tip oscillates close to its resonance frequency (few hundreds of kHz) in the vicinity of the surface. In such a scheme, weak interactions can significantly change the amplitude of tip oscillations and lead to a phase shift. Thus, images can be recorded with only a very brief intermittent contact over a cycle of oscillation or even with no contact at all. As a result, if not totally removed, shear forces are significantly reduced. The development of dynamic modes can be seen as a breakthrough in scanning probe microscopy as their ability to probe minute forces has opened avenues in imaging soft materials reproducibly and routinely, reaching atomic resolution in ultrahigh vacuum conditions [11].

Two operation modes can be employed for measuring changes in the oscillating behavior when the tip interacts with a surface [12]:

- The amplitude modulation (AM) mode [13, 14], often called "tapping mode," in which the tip–cantilever system is excited at a fixed drive frequency and with fixed amplitude. While the tip is scanning the surface, the acquired topographic image (the "height" image) corresponds to the changes in the piezoactuator height that are necessary to maintain the oscillation amplitude fixed, through an electronic feedback loop. This intermittent mode reduces the typical operational forces by at least one order of magnitude, eliminates the shear stresses, and reduces the time of the tip–surface contact by two orders of magnitude. However, the lateral resolution of this mode is lower than that of contact mode, being about 1 nm for topography. As microscopes using this mode are commercially available and measurements do not require particularly stringent constraints, the AM mode is the most commonly used one. Its first major success was its ability to image polymers without inducing severe damage, in contrast to contact mode [15–17]. As a result, many AM studies are dedicated to very soft samples such as biological systems and polymers [18, 19]. Moreover, this technique can bring new information on the local mechanical properties of heterogeneous materials, by introducing phase detection imaging, that is, images containing the changes in the oscillator phase delay relative to the excitation signal. The phase measurement yields in many cases images reflecting tiny variations of the local mechanical properties of the sample surface. On this basis, it is possible to extract useful information, especially on samples showing compositional heterogeneity at a small scale, as blends of hard and soft materials. The first demonstration of the potential of phase imaging was the elucidation of the phase-separated nanostructures in thin films of block copolymers [15]. The major factors contributing to the phase contrast are still under debate, but they are thought to be a result of viscoelastic response and adhesive forces rather than elastic surface behavior [20–23].
- The frequency modulation (FM) mode, often called non-contact (resonant) mode [24]. The FM mode was first used under ultrahigh vacuum to image semiconductor surfaces. The technique uses a negative resonant frequency shift at constant oscillation amplitude as the error signal to control the distance between the tip and the surface. In doing so, the interaction between the tip and the surface remains attractive so that the tip never "touches" the surface.

The key experimental achievement was to show that the FM mode is able to produce images with atomic resolution [25]. The experimental parameters that depend on the interaction between the tip and the surface are the resonance frequency variation and the damping signal extracted from the energy per cycle supplied to keep the oscillation amplitude constant. In contrast to the AM mode (in which the oscillation amplitude is kept constant), the physical origin of the changes in the oscillating behavior at the resonance frequency is straightforwardly attributed, rendering data interpretation *a priori* simpler. The frequency shift is related only to conservative interactions, and the additional damping signal is governed only by the dissipative interaction. However, the high sensitivity of the FM mode imposes severe conditions on the quality of the sample (only nearly flat surfaces with little contamination), compared to the AM mode.

These dynamic modes have extended the use of AFM to numerous macromolecular systems [26, 27] and led to a better understanding of their properties, such as the folding of polymer chains in single crystals, the local mechanical properties of multiphase polymer films, or the local chemical composition [28, 29]. Moreover, permanent or reversible modifications of polymer surfaces can be done with the AFM tip including the creation of holes [30] or nanorubbing (see below), the formation of textured patterns, the drawing of figures at a submicrometer level by the dip-pen lithography technique [31], the triggering of local chemical reactions, and so on. In addition to developments in the AFM dynamic modes, there has been a great progress in the design of liquid cells and components (e.g., tips) for imaging in aqueous media, where the AFM tip and cantilever are immersed in the liquid droplet [32]. This has allowed to study the biological samples under physiological conditions in many different areas of research, with tip–sample forces down to tens of pN [33]. For example, it has recently been shown by Ido and Yamada that true atomic resolution can be achieved in the noncontact mode not only in ultrahigh vacuum but also in fluid, making it possible to distinguish the major and the minor grooves of DNA.[1)]

Within the two past decades, many new operational modes have been successfully introduced, giving birth to a family of SPM methods for high-resolution characterization of organic and polymer thin films in terms of their electrical properties (e.g., electrostatic force microscopy (EFM), Kelvin probe force microscopy (KPFM), conductive AFM (C-AFM)) [34], magnetic properties (magnetic force microscopy [35]), optical properties (scanning near-field optical microscopy [36]), thermal properties (scanning thermal microscopy [37]), and chemical composition discrimination (chemical force microscopy [38, 39]). All these advances have made SPMs tools of choice for studying self-assembled organic structures arising from specific supramolecular assembly of (macro)molecules, from intermolecular interactions such as hydrogen bonding and/or π–π interactions, or from phase separation in block copolymers or polymer blends [8, 40]. The observed morphologies have been

---

1) Ido, S. and Yamada, H., private communication.

linked with the results of atomistic or coarse-grain molecular modeling, providing physicochemical models of supramolecular assembly.

In this chapter, we will review some recent contributions of SPM to the field of organic electronics, from the morphological studies of novel supramolecular materials to a better understanding of the optoelectronic properties and device performances. The chapter is organized as follows: in Section 15.2, we describe some recent work on controlling the nanoscale ordering of π-conjugated molecules, carried out by taking advantage of self-assembly (Section 15.2.1 for biotemplated assemblies and Section 15.2.2 for nanophase separation in block copolymers) or by nanorubbing thin films with an AFM tip (Section 15.2.3), in line with models of supramolecular assembly. Section 15.3 is dedicated to the studies of supramolecular structures, self-assembled or templated morphologies in organic optoelectronic devices. We focus on the use of SPM for a better understanding of the relationship between the nanomorphology and the performances of light-emitting devices (e.g., for white lighting applications, Section 15.3.1), field-effect transistors (Section 15.3.2), and organic photovoltaic diodes for solar cells applications (Section 15.3.3). In addition, we also report recent progress toward the novel characterization of nanoscale optical or electrical properties based on SPM approaches.

15.2
Controlling the Supramolecular Assembly and Nanoscale Morphology of π-Conjugated (Macro)Molecules

Both molecular conformation and intermolecular interactions have a major influence on the optical and electronic properties of π-conjugated materials, and great effort has been directed toward the study of their photophysical and electronic properties in relation to the supramolecular order and mesoscale morphology in thin films [41–43]. In view of tailoring these properties, chemical and/or physical approaches can be proposed to direct the assembly of a number of conjugated systems (see Figure 15.2). Although there is a vast literature in this field, we describe only a few of these approaches, with particular emphasis on nanostructures obtained by biotemplated self-organization or by phase separation in block copolymers. We also report our recent progress toward aligning chains into well-defined textured patterns by rubbing with an AFM tip.

15.2.1
Biotemplates for Assembling Conjugated Molecules into Supramolecular Nanowires

Many π-conjugated oligomers and polymers can self-assemble into the so-called "fibrillar" morphology, also referred to as nanowires, nanoribbons, or nanowhiskers in the literature. Without being exhaustive, this is the case for phenylene ethynylene oligomers [44], oligo(phenylene vinylene) derivatives [45], some substituted oligothiophenes and regioregular poly(alkyl-thiophene)s [46, 47], linear alkyl-substituted poly(2,7-fluorene)s and poly(2,8-indenofluorene)s [48], and phenylene-based

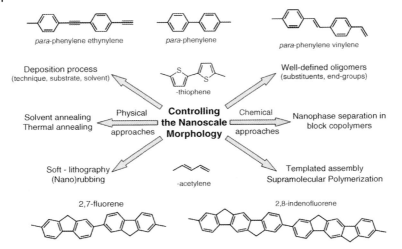

Figure 15.2 A few pathways for controlling the nanoscale morphology in conjugated materials. Some of the most studied π-conjugated structures are shown as dimers (see details in the text).

polymers and dendrimers (see also recent reviews [49, 50]). The driving force behind this type of assembly is π-stacking: usually, for linear, substituted oligomers and polymers, this leads to a packing in which the long axis of the molecules (backbones) is perpendicular to the fibril axis. For short, unsubstituted π-conjugated oligomers (e.g., pentacene, oligothiophenes), the usually observed assembly is a 2D herringbone packing (T-shape π–π interactions) where the long axis of the molecules is almost perpendicular to the surface plane, leading to a terrace-like morphology, as observed with AFM [51, 52]. Specific substituents or end groups can be used to hinder the π-stacking or to constraint the chains in specific conformations. For instance, Yashima and coworkers studied polyacetylene derivatives substituted with phenyl groups by high-resolution AFM and showed well-defined helical structures made of single chains, which aggregate into chiral bundles in thin films [53, 54].

Biotemplated assembly and supramolecular polymerization are new tools for the controlled bottom-up assembly of conjugated structures into nanowires. The concept lies in the binding of a conjugated molecule to a (bio)template, which is able to drive the supramolecular assembly via directional interactions such as hydrogen bonding (H-bonding). Moreover, one can control the assembly by tuning the preparation conditions, such as the pH of aqueous solution and temperature. Biotemplates can be made of peptide oligomers, which can undergo a well-defined folding into α-helix, or parallel or antiparallel β-sheet via H-bonding. Bäuerle and coworkers recently reported on the design of an oligothiophene end capped with oligopeptide-poly (ethylene oxide) conjugates [55], and AFM revealed the formation of well-defined μm-long fibrillar objects (width of about 8 nm, height of 3 nm). Interestingly, by decreasing the pH, the tendency of the peptide oligomer to adopt a β-sheet structure is altered via the protonation of a switch ester segment. At low pH, the AFM images show left-handed helical fibrils, while increasing the pH converts the peptide

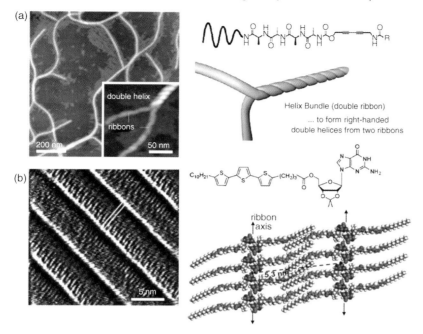

Figure 15.3 (a) AFM height image showing the helical fibrils observed by hierarchical assembly of a diacetylene–oligopeptide conjugate, as schematically displayed on the right (reproduced with permission from Ref. [56]); (b) STM current image showing highly directional ribbons formed by self-assembly of a terthiophene-guanosine derivative, and an atomistic model on the right (reproduced with permission from Ref. [57], copyright Wiley-VCH Verlag GmbH & Co. KGaA).

segments into the native β-sheet form, that is, fibrils without helical structure. Frauenrath and coworkers used another original approach, which consists in linking a diacetylene moiety to an oligopeptide, in order to exploit the H-bonding-driven self-assembly of the latter to build hierarchical structures and then polymerize the diacetylene with UV light (converting neighboring diacetylenes in polydiacetylenes via a topochemical reaction) [56]. Figure 15.3a shows the corresponding morphology, as observed by AFM, displaying double helical fibrils made of ribbons. Each ribbon is presumed to be a double tape, consisting of single parallel β-sheet structures. It is worth pointing out that polymerization with UV light does not disrupt the preformed fibrillar and tape superstructures, as convincingly demonstrated by the authors.

Both oligonucleotides and nucleobases, the building blocks of DNA, can also be used as biotemplates. The single nucleobase approach has been exploited by Spada *et al.*, who designed a guanosine derivative covalently linked to an oligothiophene [57]. The advantage of guan(os)ine-based derivatives relies on the variety of H-bonding patterns they can form depending on the solution conditions. By modifying the nature of the organic solvent (chloroform or acetonitrile) or by adding alkali ions, supramolecular ribbons, helical oligothiophene stacks, or G-quartet assemblies can be obtained. The ribbon-like structures assemble in chloroform, and consist in Hoogsten H-bonding pattern, as observed by STM and AFM [57]. Figure 15.3b shows highly directional ribbons assembled on a graphite surface, as observed in the form of bright parallel lines in the STM current image. Regularly spaced spots are present along these bright lines, and the observed unit cell has an area of 3.07 nm^2 (see white rectangle in Figure 15.3b left). On the basis of the correlation between STM, AFM, and molecular modeling, these lamellae have been attributed to adjacent ribbons arising from H-bonds between guanosines, the brightest lines in the STM images corresponding to the guanine network in the so-called "B-type" ribbon morphology. When the authors dissolved potassium picrate into a chloroform solution of this compound, they observed that guanosines rearrange into an octamer complex, made of a potassium ion sandwiched at the center of two G-quartet assemblies (square planar network of four guanines). Importantly, this process can be reversed by using a complexing agent that selectively binds the potassium cation, composing a switch between the quartet and the ribbon ordered assemblies with two forms of oligothiophene–oligothiophene packing.

Besides the single nucleobase approach, oligonucleotides (i.e., DNA oligomers) are valuable templates for fabricating nanowires and nanogrids thanks to the control in the design of their architecture and the selectivity of the base recognition process [58, 59]. Nowadays, it is possible to control both the length and the base sequence of oligonucleotides from a few bases to about 150 bases via the "base-by-base" solid-phase synthesis. The use of the H-bonding recognition for selectively binding π-conjugated oligomers to nucleobases along a single-stranded oligonucleotide appears therefore as an interesting approach to design monodisperse aggregates. Moreover, one could control the sequence of different oligomers on complementary bases along the single-stranded DNA (ssDNA) [60–62]. Recently, we have used a diaminotriazine unit end capped to π-conjugated oligomers to direct the positioning of these oligomers along the oligonucleotides (see Figure 15.4). The diaminotriazine unit is complementary to bases of DNA, forming three H-bonds with each thymine of the oligonucleotides (while adenine-thymine base pairing consists in only two Watson–Crick H-bonds). The studied systems consist in thymine oligonucleotides of controllable size, from a few thymine units (5, 10, 20, or 40 units) to long ssDNA (prepared by enzymatic synthesis), H-bonded to diaminotriazine-capped π-conjugated oligomers [62, 63]. The studied guest oligomers are naphthalene or oligo(phenylene vinylene) derivatives, see Figure 15.4. The AFM studies performed in aqueous solutions placed in a liquid cell have revealed the adsorption of these complexes on Mg^{2+}-modified mica, which leads to monodisperse globular aggregates (for ssDNA-naphthalene hybrids) or elongated nanowires (for ssDNA-oligo(phenylene vinylene)s hybrids). The thickness of these objects as observed with AFM (around 2.7 nm for DNA-naphthalene and 4.5 nm for DNA- oligo(phenylene vinylene) hybrids) is in agreement with the molecular dynamics simulations and transmission electron microscopy measurements, confirming the adsorption of fully bound supramolecular complexes [62, 63]. This shows that single-stranded DNA constitute promising templates that can serve as "tape measure" for controlled stacking and (helical) alignment of functional organic molecules. Moreover, the base

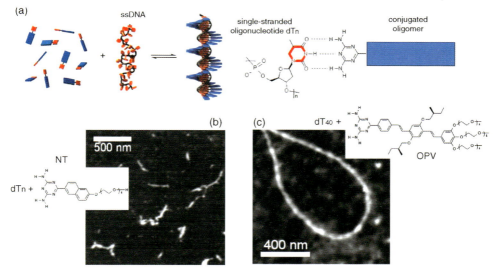

Figure 15.4 (a) Scheme describing the principle of ssDNA-templated self-assembly; (b) and (c) liquid-cell AFM height images of structures obtained from aqueous mixtures deposited on Mg^{2+}-modified mica surfaces of (b) $[T]_{dTn} = 0.25$ mM and $[NT] = 0.5$ mM; (c) $[dT_{40}] = 0.125$ mM and $[OPV] = 0.25$ mM. Reproduced with permission from Ref. [62], copyright Wiley-VCH Verlag GmbH & Co. KGaA.

recognition can be further exploited to make multichromophoric arrays, with a defined position and distance between the different π-conjugated molecules.

15.2.2
Nanophase Separation in Rod–Coil Block Copolymers

An elegant way for obtaining specific nanoscale morphologies is to exploit the spontaneous tendency of polymers of different chemical nature to phase separate. In block copolymers, the different segments are linked by a covalent bond, which leads to nanometer-sized phase-separated domains, in contrast to polymer blends usually giving rise to micrometer-sized domains. The phase diagram of classical block copolymers ("coil–coil") has been thoroughly studied both theoretically and experimentally (see, for instance, Ref. [64]). Usually, four main factors determine the phase behavior: the chemical nature of the comonomers, the molecular architecture, the degree of polymerization, and the composition (expressed as the average volume fraction of a given component). Very recently, there has been a significant interest for "rod–coil" block copolymers, that is, copolymers with a stiffness asymmetry between the blocks, which leads to a phase diagram markedly different from that of a classical (coil–coil) copolymer. The stiffness asymmetry leads to an increase in the Flory–Huggins parameter (χ, linked to the chemical affinity between the two blocks) and to phase-separated domains at chain lengths where the different blocks in coil–coil copolymers are still miscible. The rod segment can be an α-helical polypeptide block, a mesogenic block (e.g., nematic, smectic, and so on), or a π-conjugated segment [65–67]. One can exploit the potential of these compounds to respond to external stimuli, such as temperature (using crystallizable or mesogenic rod blocks) or light (using conjugated rod blocks), to modify and manipulate the supramolecular assembly [68]. In particular, the study of rod–coil copolymers including a π-conjugated sequence constitutes a promising strategy for optoelectronic applications where the construction of well-defined nanometer-sized domains remains a major challenge, for instance, in organic solar cells. Several groups have observed different ordered phase-separated morphologies, such as bilayer morphologies and microporous hollow micelles [69, 70]. Various types of rod–coil short molecules and block copolymers architectures have shown a tendency to form nanoribbon-like structures, similar to the so-called fibrillar structures observed for conjugated homopolymers [71–73]. Figure 15.5 displays a few examples of these structures, from our group on rod–coil polymers and from the Stupp group on dendron–rod–coil molecules. The observed fibrillar structures are in all cases from a few hundred nanometers to a few micrometers long, a few nanometers wide, and arranged either in a web (Figure 15.5a and c) or in a bundle of fibrils (Figure 15.5b and d).

Although a large number of rod–coil molecules have been studied so far, no general picture of the influence of various factors such as molecular size, composition, molecular architecture, and so on on the rod–coil copolymers phase diagram has been firmly established. In this frame, we have studied the microscopic morphology of poly(9,9′-dioctyl-2,7-fluorene)-b-poly(ethylene oxide) copolymers (PDOF-PEO) with various architectures (diblock and triblock copolymers) and with varying volume fraction and segment lengths of the components. Figure 15.6 shows AFM images of thin deposits of PDOF-PEO diblock and triblock copolymers, illustrating the effect of the block ratio on the self-assembly and microscopic morphology. The motivation comes from the fact that polyfluorenes are promising blue light emitters for full-color organic light-emitting diodes, as illustrated in many examples [74, 75]. PEO is a crystallizable nonconjugated flexible (coil) polymer, therefore its crystallization can compete with the tendency of PDOF to form fibrillar constructs by π-stacking, and this can therefore lead to different morphologies with varying luminescence properties [48]. Interestingly, while the deposition originates from a good solvent for both blocks (tetrahydrofuran or chloroform), the fibrillar morphology persists up to an average volume ratio in PEO f_{PEO} of around 0.31 ($f_{PF} = 0.69$), indicating that π-stacking is a stabilizing process even at relatively large contents of nonconjugated PEO. At higher PEO content ($f_{PEO} \sim 0.44$), the crystallization of PEO overcomes the assembly of the PDOF segments, and leads to platelet-like morphologies (Figure 15.6), which are reminiscent of the dendritic lamellae observed for very high PEO content ($f_{PEO} \sim 0.93$). These dendrites (about 6 nm thick) are the signature of the crystalline growth of PEO by diffusion-limited aggregation. This constitutes a nice example of the various nano- to microscopic morphologies attainable by varying the composition of rod–coil block copolymers for controlling the optoelectronic properties of the thin films.

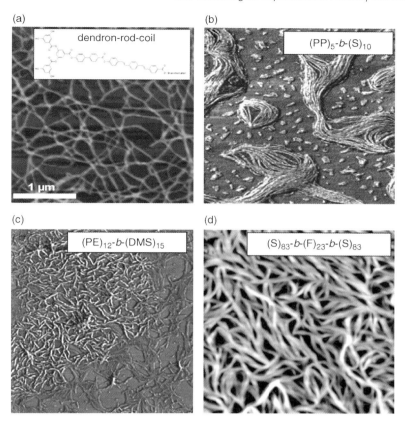

Figure 15.5 AM-AFM images of deposits of different rod–coil molecules with a π-conjugated segment as the rod: (a) a "dendron–rod–coil molecule" (chemical structure in inset) from Stupp et al. (reproduced with permission from Ref. [72], copyright 2004 American Chemical Society); (b) Poly(para-phenylene)-b-poly(styrene) diblock copolymer. Scan size 3.0 μm (reproduced with permission from Ref. [71], copyright Wiley-VCH Verlag GmbH & Co. KGaA.); (c) Poly(para-phenylene ethynylene)-b-poly(dimethylsiloxane) diblock copolymer. Scan size 5.0 μm; (d) poly(styrene)-b-Poly(fluorene)-b-poly(styrene) triblock copolymer. Scan size 5.0 μm. The number of monomeric units is indicated on the images (reproduced with permission from *Progress in Polymer Science* (2003), 28, 55, copyright 2003 Elsevier).

15.2.3
Nanorubbing: a Tool for Orienting π-Conjugated Chains in Thin Films

Since the charge transport properties of π-conjugated materials are intimately related to the long-range order, it is essential to enhance the degree of structural order in conjugated polymer thin films for improving device performance, for example, for field-effect transistor applications [76–78]. Recently, we deposited thin films of regioregular poly(3-hexylthiophene) (RRP3HT) and carried out rubbing with a velvet

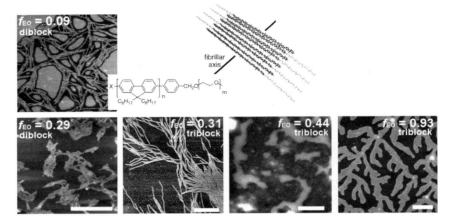

Figure 15.6 AM-AFM images of submonolayer deposits of PDOF-PEO and PEO-PDOF-PEO copolymers. The white scale bars are 500 nm long. The chemical structure and model of supramolecular assembly are shown for the diblock copolymer with $f_{EO}=0.09$ (in triblock copolymers, X stands for the symmetric PEO block). Reproduced with permission from Ref. [73], copyright Wiley-VCH Verlag GmbH & Co. KGaA.

cloth, which led to the orientation of the chains at the polymer surface parallel to the rubbing direction [79]. This type of surface modification can also be achieved at microscopic length scales, with precise control of both size and localization of the rubbed areas [80, 81], by using the tip of an AFM operating in contact mode. In a typical experiment, the film is rubbed along one direction, that is, along the fast scan direction with the AFM tip in contact mode. Scans of $10 \times 10\,\mu m^2$ squares where drawn and afterward the morphology of the rubbed regions was investigated by AM-AFM, which does not damage the polymer films. Figure 15.7a shows the height and phase images of the area across the modified region. The rubbed area is hardly visible in the height image (*left*) but appears more clearly in the phase image (*right*). This phase contrast originates from differences in energy dissipation related to local differences in tip–sample interactions, which indicates that the nanorubbing process induces a modification in the polymer mechanical response. Since the rubbing tends to favor polymer crystallization via chain alignment [82], it is most likely that the change in AM-AFM phase signal over the rubbed areas is due to a local increase in stiffness, as a result of RRP3HT chain ordering (see below). Upon closer examination, we observed that the rubbed area presents a regular pattern made of grooves parallel to the rubbing direction both in the height and in the phase images (Figure 15.7b). From the Fourier transform of the height image (see inset), the period of the pattern is found to be about 40 nm. This value is directly related to the experimental conditions of the rubbing process: since the rubbed $10 \times 10\,\mu m^2$ area was scanned by 256 parallel lines, the lateral shift between two lines is 39 nm, which corresponds to the period of the pattern. Under the conditions used here (i.e., 10 scans of rubbing with a load of 200 nN), the vertical amplitude of the pattern is

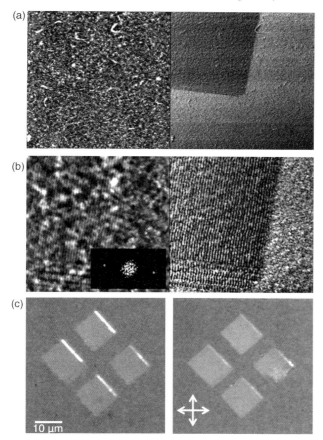

Figure 15.7 (a) $10 \times 10\,\mu m^2$ and (b) $2.5 \times 2.5\,\mu m^2$ AM-AFM height (*left*) and phase (*right*) images of a nanorubbed area on a RRP3HT film. The vertical gray scale is 10 nm and 20° for the height and phase image, respectively. (b) *Inset*: Fourier transform of the height image. (c) Two optical micrographs between crossed polars of RRP3HT films with four $10 \times 10\,\mu m^2$ rubbed squares. The crossed polars are oriented as shown by the white arrows. Reprinted with permission from Ref. [83], copyright 2008 Elsevier.

1.5 ± 0.5 nm. On the basis of these observations, it therefore becomes possible to generate patterns with a well-defined periodicity by selecting the size of the rubbed area and the spatial resolution of the scanning upon rubbing operation.

In order to highlight a possible orientation of the polymer chains within the rubbed zone, the samples were analyzed with optical microscopy between crossed polars. For example, two P3HT films with four rubbed square areas (each of $10 \times 10\,\mu m^2$) are shown in Figure 15.7c. Clearly, all the nanorubbed domains appear with a high birefringence. In addition, there is a complete extinction of these squares when the polars are oriented either parallel or perpendicular to the rubbing direction. These observations clearly testify the orientation of the polymer chains induced by the rubbing process. This shows that the nanorubbing with an AFM tip clearly induces deformation of the surface, in the form of grooves along the fast scan direction, together with a modification of the mechanical response corresponding most probably to shear-induced orientation of crystalline domains in which the RRP3HT chains are oriented along the rubbing direction.

Other physical approaches can be used to align π-conjugated molecules along one direction in a film. Soft lithography methods have been exploited by different groups, as exemplified in other chapters of this book, leading to a variety of templated nanoscale morphologies that can be used to optimize the device performances (e.g., field-effect transistors) [84–86]. The use of an electric field during the film formation also appears as a promising approach in order to align fibrillar structures between electrodes, as recently demonstrated for small molecules forming supramolecular fibers with an intrinsic dipole moment [87].

15.3
Effect of the Nanoscale Morphology on the Optoelectronic Properties and Device Performances

15.3.1
Luminescence Properties of Supramolecular π-Conjugated Structures and Applications for White Light OLED

Luminescent π-conjugated materials are of great interest in the fabrication of flexible organic light-emitting diodes (OLEDs). Although commercial products have appeared recently, research is still ongoing to assess new materials and their interfaces with electrodes or additional charge injection layers. In this view, the SPM characterization of the morphology of interfaces (e.g., between the luminescent organic semiconductor and the electrodes) and of the local-scale electrical properties can provide valuable information on the effects of disorder and defects in order to improve the understanding of the OLED properties [88, 89].

An important SPM tool for characterizing the luminescence properties of the active organic layer with submicrometer resolution is near-field scanning optical microscopy (NSOM, also called SNOM) [36]. This technique enables the simultaneous mapping of the topography and fluorescence by scanning in the near field with an optical fiber ending with a subwavelength aperture. Vanden Bout and coworker have exploited this technique for studying films of conjugated polymers such as poly (9,9'-dihexyl-2,7-fluorene) [90, 91]. They observed the presence of nanoribbon-like structures (about 40 nm wide, see Figure 15.8). Importantly, by mapping the polarized fluorescence at orthogonal polarizations (Figure 15.8c and d), they observed that the fluorescence intensity is higher when collected perpendicular to the ribbon axis. Since the fluorescence is polarized along the polymer backbone, this reveals that the chains are perpendicular to the ribbon axis, in agreement with the models of self-assembly for fluorene-based polymers and copolymers described in Section 15.2.

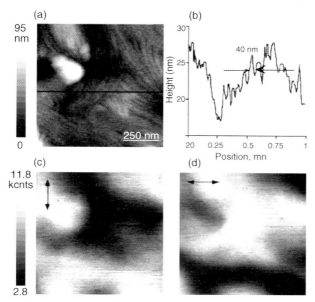

Figure 15.8 1 × 1 μm² NSOM images of PDHF. (a) Topography, (b) topography cross section, (c) horizontal, and (d) vertical polarized NSOM fluorescence. Reprinted with permission from Ref. [90], copyright 2001 American Chemical Society.

In the growing field of solid-state lighting, white light-emitting plastic materials are attractive candidates for flexible light-emitting devices and as backlights for liquid crystalline displays. Common approaches to obtain emission over the whole visible range include energy transfer from donor polymers blended with luminescent acceptor dyes or copolymers containing moieties emitting at different wavelengths. Multicomponent π-conjugated systems based on supramolecular interactions combine the advantages of small molecules with those of polymers and can facilitate energy transfer processes due to a close spatial proximity and controlled orientation between the different species. This concept has been exploited to create white light-emitting organogels in electroluminescent devices.

Owing to its strength, reversibility and directionality, multiple H-bonding is one of the most frequently used principles in supramolecular chemistry. A common example is the self-complementary 2-ureido-4[1H]-pyrimidinone group (UPy) that is able to form noncovalent polymers with high molecular weight [92]. The use of different monomers in the design of well-defined oligomers allows the easy tuning of bandgap over a wide range, resulting in emission colors that span the entire visible spectrum and reach even to the near-infrared region. By combining these approaches (H-bonding and defined cooligomers), several π-conjugated structures end capped with UPy groups have been prepared: blue light-emitting oligofluorenes, green light-emitting oligo(*para*-phenylenevinylene)s, and a red light-emitting perylene diimide to create red–green–blue supramolecular materials [93]. These structures self-assemble by UPy recognition, and the resulting degrees of polymerization were high enough to create white photoluminescent supramolecular polymers both in solution and in films. The solid-state structures, as monitored by AFM, are uniform and the results indicate that the assembly is governed by the combined effects of H-bonding, π-stacking, and microphase separation (due to repulsion of alkyl chains and the aromatic backbones). White light-emissive organogels have been obtained by mixing the different oligomer-based conjugated materials, and AFM images of the blends show very smooth surfaces that closely resemble those of the pure components, ruling out phase separation. These results indicate that supramolecular statistical copolymers are present in these mixed films, and that the emission color can be tuned by partial energy transfer in a modular approach, ultimately giving rise to white light OLEDs (WOLEDs).

Another approach to creating white light-emitting organogels is the design of *polycatenars* [94]. These constitute an intriguing class of liquid crystalline, rigid molecules equipped with flexible wedges at the ends (the "poly" referring to the total number of chains in the wedges). These structures give rise to a rich variety of mesophases, both thermotropic and lyotropic. Recently, the group of Meijer and Schenning designed a series of polycatenars with fluorene-based oligomers as the rigid parts and 3,4,5-tridodecyl benzoic amide wedges at the periphery (Figure 15.9) [94]. For the rigid parts, two fluorene units were linked by a central aromatic moiety that was chosen to cover a wide range of electron accepting character to vary the absorption and emission properties. These moieties are fluorene, quinoxaline, benzothiadiazole, or thienopyrazine. Amide linkers between the chromophores and the wedges were chosen in order to introduce H-bonding as a noncovalent ordering interaction. The formation of organogels (prepared in methylcyclohexane solution) is confirmed by AFM studies (Figure 15.9), showing typical tape-like structures with a constant width and thickness of ∼20 nm and 4.3 nm, respectively. Additional clustering into thicker structures and overlap between crossing fibrils are also observed. The thickness of the ribbons, as measured by

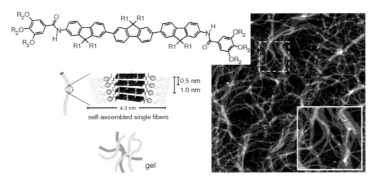

Figure 15.9 *Left*: Structure of the fluorene-based bolaamphiphile oligomer forming an organogel (R_1 and R_2 are alkyl derivatives) and model of self-assembly yielding an organogel. *Right*: $10 \times 10\,\mu m^2$ AM-AFM image of a drop-cast film on glass (the inset image size is $2 \times 2\,\mu m^2$).

AFM, roughly corresponds to the long *d*-spacing (4.74 nm) observed in the wide-angle X-ray scattering (WAXS) analysis of the supramolecular order in the bulk. The combined results from WAXS, molecular modeling, AFM, and optical spectroscopy allow to propose the self-assembly mechanism described in Figure 15.9. H-bonding between the amide functionalities, mutual repulsion of aliphatic chains and the aromatic backbone, and to some extent also π–π interactions induce the molecules to pack on top of each other. Subsequently, these fibrils self-assemble into ribbons by interdigitation of the alkyl chains in their periphery, and, at sufficiently high concentration, these ribbons align and form an organogel due to entanglements between the nanoribbons. In a similar approach, the energy transfer in the organogels was studied by mixing small amounts of acceptor and donor molecule-based polycatenars. By variation of the mixing ratio, energy transfer from the blue light-emitting oligofluorene matrix to the embedded energy acceptors is observed upon excitation of the oligofluorene, leading to stable white light-emitting gels at a specific composition.

15.3.2
Polythiophene Nanostructures for Charge Transport Applications: Rubbing, Annealing, and Doping

Organic field-effect transistors (OFETs) are of interest for fabricating (via deposition from solution) low-cost, flexible electronic devices (see Refs [77, 95] and chapters related to these types of devices in this book). OFETs are also important in materials research; they are used for characterizing the charge transport properties in organic semiconductors and, in particular, the charge carrier mobility, a fundamental parameter for understanding the electronic processes in semiconductors. Poly(thiophene)s have emerged as promising polymeric semiconductors for FET applications since they exhibit relatively high charge carrier mobility (i.e., in the range of 10^{-3}–$10^{-1}\,cm^2\,V^{-1}\,s^{-1}$). Such performances have been reached by improving the regioregularity of the alkyl groups along the poly(thiophene) backbone and increasing the polymer molecular weight [76, 96]. In this section, we focus on different approaches aiming at the improvement of the charge transport properties in RRP3HT thin films.

15.3.2.1 Nanorubbing P3HT Thin Films in OFET Channels
In Section 15.2.3, we have described the use of the AFM tip as a rubbing tool for aligning π-conjugated molecules in defined microscopic patterns. On the basis of this approach, we used nanorubbing for texturing RRP3HT active layers inside the channels of OFETs, with the objective of studying the effect of chain orientation on the transistor properties. We applied the technique to RRP3HT films deposited by dip-coating at very high dipping speeds, about 70 mm/min, which leads to homogeneous, nontextured amorphous layers [79, 83]. Nanorubbing was then carried out at different orientations with respect to the channel direction, and we observed that FETs show different performances depending on the orientation of the alignment process [97]. The AM-AFM images after nanorubbing are shown in Figure 15.10.

15 SPM Insights into Supramolecular π-Conjugated Nanostructures

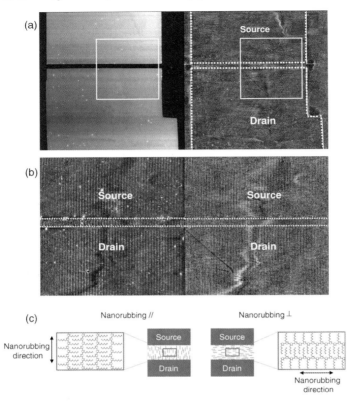

Figure 15.10 (a) 35 × 35 μm² and (b) 15 × 15 μm² AM-AFM height (*left*) and phase (*right*) images of a channel-rubbed RRP3HT FET. Images (b) correspond to squared areas in images (a). The vertical gray scale is 7 nm for height images and 25° for phase images (reproduced with permission from Ref. [83], copyright 2008, Elsevier). (c) Schematic representation of the RRP3HT chain arrangement when the nanorubbing is performed parallel (*left*) and perpendicular (*right*) to the source–drain axis. Reproduced with permission from Ref. [83], copyright 2008, Elsevier.

Figure 15.10a left displays the P3HT channel as the dark horizontal region in between the source and the drain electrodes appearing in gray. The corresponding phase image (Figure 15.10a right) shows no contrast between the channel and the electrodes, consistently with the fact that the whole device surface is covered with RRP3HT. A close-up view of the channel is shown in Figure 15.10b, in which the pattern generated by the rubbing (vertical lines) is clearly observed. Again, polarized optical microscopy indicates that the RRP3HT chains are oriented along the rubbing direction (see Section 15.2.3): the rubbed area is very bright when the rubbing direction is at 45° with respect to the polars, which indicates that the rubbed polymer is birefringent. Consistently, a complete extinction of the rubbed square is observed when the rubbing direction is either parallel or perpendicular to the polarization directions, testifying to the polymer chain orientation within the FET channel [83].

15.3 Effect of the Nanoscale Morphology on the Optoelectronic Properties and Device Performances

To quantify the impact of nanorubbing process on FET characteristics, we determined the linear (hole) mobility μ_{lin} (see Ref. [83]). We observed that μ_{lin} values for the nanorubbed FETs were slightly more than two times higher than for the pristine unrubbed systems (using these preparation conditions, that is, nonannealed devices, see [83]): 6.4×10^{-3} cm^2 V^{-1} s^{-1} and 2.6×10^{-3} cm^2 V^{-1} s^{-1}, respectively. Note that this relatively low charge mobility (compared to continuous fibrillar films) is typical of amorphous RRP3HT films (*vide supra*). After rubbing the RRP3HT channel along the electrode direction (nanorubbing//, see Figure 15.10c), the FETs operate with improved charge carrier mobilities. In contrast, when the nanorubbing is performed perpendicular to the electrode direction (nanorubbing $\perp$), the mobility is lower, approaching 10^{-4} cm^2 V^{-1} s^{-1}. This difference is most probably related to the difference in chain orientation induced by the rubbing. We have shown that rubbing the P3HT films on a macroscopic scale leads to a global orientation of the chains at the surface, with the (100) plane perpendicular to the polymer layer [79]. In other words, the plane of individual polymer molecules is parallel to the surface upon rubbing, in the plane of the polymer film. Let us note that such an arrangement is not the most favorable for charge transport (an "edge-on" organization has been shown to lead to the highest mobilities) [76]. When rubbing is carried out perpendicular to the source–drain axis, the chains are therefore arranged flat on the surface and perpendicular to the transport direction (nanorubbing $\perp$). In such configuration, the charge transport in the source–drain direction implies chain-to-chain hopping over a large distance (because of the presence of the alkyl groups), which is consistent with the low mobility observed. In contrast, when nanorubbing is performed parallel to the source–drain axis (nanorubbing//), the chains are oriented parallel to the transport direction. Intramolecular transport is favored and only a few interchain hopping events are necessary for charges crossing the channel. This is consistent with the twofold increase in mobility with respect to the pristine layer, in which the chain orientation is random. Altogether, this shows that the nanorubbing approach, although being relatively slow, is a valuable technique for improving or tuning the properties of individual thin-film devices.

15.3.2.2 Influence of the Film Morphology on Field-Effect Transistor's Performances

Recently, the correlation between the nanoscopic morphology within RRP3HT films and the performance of OFET devices, in particular the charge carrier mobility, was thoroughly studied by us and others. We have shown that it is possible to improve the charge carrier mobility (hole mobility, as estimated from the saturation regime) of these devices up to a value of about 0.25 cm^2 V^{-1} s^{-1}, with device current on/off ratio on the order of 10^5, by engineering the RRP3HT layer morphology. This can be done (i) by modifying the type of solution-based deposition technique (using spin coating, dip coating, or drop casting); (ii) by changing the nature of the solvent (solvents with different boiling temperatures); (iii) by depositing self-assembled monolayers on the dielectric surface; and (iv) by applying a post-thermal annealing to the device [98, 99]. The FET current–voltage characteristics were examined in the accumulation regime of hole transport (i.e., negative gate–source voltage) since holes are the major charge carriers in oligo- and poly(thiophene)s using this type of device architecture (gold electrodes, silicon oxide dielectric layer). The analysis of the device performances in the light of AFM studies carried out within the FET channel shows that

i. the higher the uniform extent of fibrillar morphology in the FET channel, the higher the charge mobility. The crystalline ordering of RRP3HT into fibrillar structures appears to lead to optimal FET performances. By modifying the solvent and preparation conditions, one can fabricate completely amorphous films into films with very high content of fibrillar order. The latter show charge mobilities (μ, as extracted from the saturation regime) of around 0.1 cm^2 V^{-1} s^{-1}, which is two orders of magnitude higher than for amorphous films. This is attributed to the fibrils acting as "conduits" for the charge carrier transport. Indeed, the length of the fibrils can cover the distance between source and drain electrodes (a few micrometers in this case), providing an efficient pathway for the charges from one electrode to the other (see Figure 15.11). The fibrillar morphology consists in domains of stacked crystalline lamellae with a typical width and thickness of 20–35 nm and a few nanometers, respectively. Each single fibril is made of

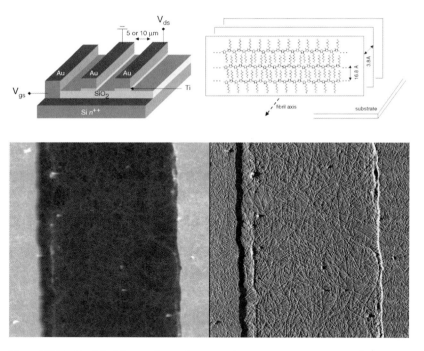

Figure 15.11 *Top left*: Scheme of typical FET architecture used for depositing RRP3HT films. *Top right*: Schematic representation of the RRP3HT packing into the fibrillar structures (reprinted with permission from Ref. [99], copyright 2006 the American Institute of Physics). *Bottom*: 7.5 × 7.5 µm^2 AM-AFM height (*left*) and phase (*right*) across the FET channel, showing a continuous RRP3HT fibrillar film prepared by slow dip coating. The gray stripes in the height image correspond to the source and drain gold electrodes.

π-stacked chains, with the backbone axis perpendicular to the fibril axis (the stacking direction), as suggested by structural studies using X-ray diffraction and AFM measurements [47, 99–101]. Each fibril corresponds to a crystalline lamella, with an interchain distance of 0.38 nm (stacking distance between thiophene backbones) and an interlamellar distance around 1.7 nm due to interdigitation of hexyl groups, as depicted in Figure 15.11 top right.

ii. starting from a fibrillar film (prepared by dip coating), subsequent annealing of the device at the crystallization temperature of RRP3HT (about 150 °C) increases the charge mobility by a factor of 2. This increase is attributed to a higher crystallinity of the film, which is observed by AFM as a lateral extension of the fibrillar structures (up to 30–40 nm). However, when the device is annealed at higher temperature, μ significantly decreases due to a morphological transition from fibrils to grains occurring above 150 °C. This is illustrated in Figure 15.12, showing AFM phase images within the FET channel at different annealing temperatures. The pristine film shows the characteristic fibrillar morphology and this morphology is preserved upon annealing at a temperature below 150 °C. However, the thickness of the fibrils tends to increase, up to 30–40 nm, consistent with the annealing temperature dependence of μ up to 150 °C (larger crystalline domains and interchain contacts, which are likely to improve the charge transport). At higher annealing temperatures, the fibrils appear to break up into smaller "grains" along the axis and then completely disappear as the annealing temperature increases up to 180 °C (see Figure 15.12). Since these grains are very small (diameter of a few tens of nanometers) compared to fibrils (micrometer long), this transition is most probably responsible for the abrupt decrease in μ above 150 °C. Note that this morphological transition around 150 °C can be directly identified by plotting the route mean square (RMS) roughness of the RRP3HT films as a function of the annealing temperature, see Figure 15.12 bottom right. The nonannealed films show a relatively low RMS roughness of $\sim$2.5 nm (almost identical to that of the underlying silicon Si/SiO_2 dielectric substrate). Although films annealed at 150 °C show no change in the roughness, films annealed at 155 °C exhibit a sharp increase in the roughness to $\sim$3.8 nm due to the appearance of the grains. The RMS roughness decreases again at annealing temperatures around 160 °C (relaxation of the film) and remains unchanged up to 240 °C.

This study and others [102–104] illustrate the need of SPM (with complementary techniques) to better understand the electrical properties of organic semiconductors within the OFET channel and at the interface with source and drain electrodes or with the dielectric layer, ultimately providing further clues for improving the device performance.

15.3.2.3 Doping of P3HT Nanofibrils

The doping of poly(thiophene)s and particularly P3HT has been studied for more than 20 years, both on amorphous and on crystalline films, with iodine (I_2) [105, 106], nitrosonium tetrafluoroborate ($NOBF_4$) [107], nitrosonium hexafluorophosphate

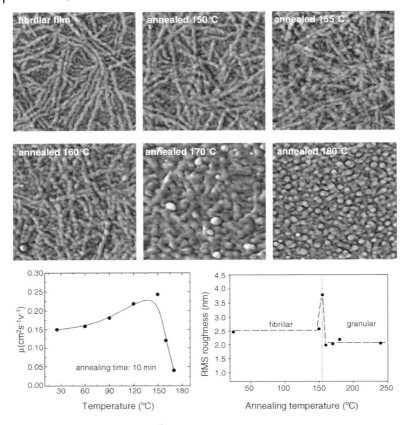

Figure 15.12 *Top*: 800 × 800 nm² AM-AFM phase images within the FET channel of thermally annealed RRP3HT films. *Bottom*: Charge mobility (*left*), as extracted from OFET transfer characteristics, and the RMS roughness (*right*), as measured with AFM, as a function of the annealing temperature. Reprinted with permission from Ref. [98], copyright 2006 American Institute of Physics.

($NOPF_6$) [108–110], and oxygen (O_2) [111], with the charges predominantly present as polarons and bipolarons [112]. The doping process results in a strong electrical conductivity increase. Since the semicrystalline fibrillar structure is known to be favorable for charge transport, the effect of doping the P3HT fibrils on their nanoscale morphology and electrical properties has recently been studied to yield even higher conductivity [113]. After determining the optimal deposition conditions for generating films made of a single layer of close packed fibrils, the doping process is carried out in a liquid phase containing the oxidizing agent: the polymer film is immersed in a $NOPF_6$ acetonitrile solution, which is a nonsolvent of RRP3HT, until a color change occurs (usually from a few seconds up to 1 min, depending on the film thickness), typical of the oxidized polymer. Then, the sample is rinsed with the pure solvent and dried with gentle airflow. The doping chemically proceeds as follows: in acetonitrile, NOPF$_6$ dissociates in NO$^+$ and PF$_6^-$; the NO$^+$ ions remove an electron from the polymer backbone [109], creating positive polarons compensated by PF$_6^-$ counterions.

Figure 15.13 shows $2 \times 2\,\mu m^2$ AFM height images of RRP3HT before and after doping. Clearly, the fibrillar structure is preserved upon doping, even though the doping process occurs in contact with a liquid phase. This observation is crucial since this particular fibrillar structure is very favorable for interchain charge transport. The histogram in Figure 15.13 shows the fibril width distribution before and after the doping process: the average width of 20.2 nm for the undoped fibrils is consistent with the length of RRP3HT chains oriented perpendicular to the fibril axis, while the doped fibrils are wider by almost 4 nm. We can hypothesize that the PF$_6^-$ counterions are in the vicinity of the conjugated backbones, presumably leading to some lateral shift of the polymer chains with respect to each other, which may lead to the observed broadening of the fibrils.

Doped and undoped fibrillar RRP3HT films were deposited on silicon microfabricated patterns with 8 nm thick platinum contacts, with variable channel width

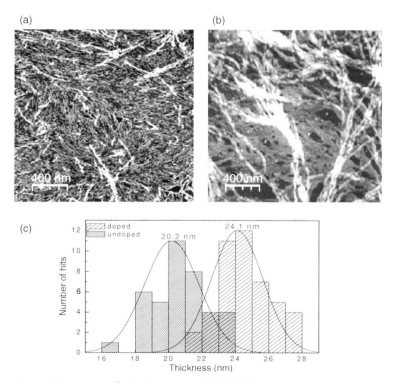

Figure 15.13 $2 \times 2\,\mu m^2$ AM-AFM images of RRP3HT films. (a) Undoped; (b) doped with a 0.5 mg/mL NOPF$_6$ solution in acetonitrile. (c) The fibril width distribution for the doped and undoped films. Reproduced with permission of *European Physics Journal of Applied Physics* from Ref. [113].

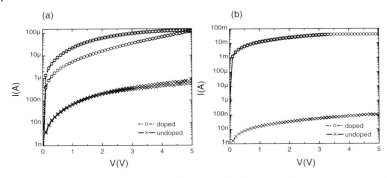

Figure 15.14 Current–Voltage plots of RRP3HT films between 10 μm long electrodes with a 50 nm channel length. (a) For a monolayer of fibrils and (b) for a 1 μm thick film. Reproduced with permission of *European Physics Journal of Applied Physics* from Ref. [113].

(from 3 to 100 μm) and length (from 30 to 1000 nm). The smaller gaps ensure that some fibrils connect the two electrodes. Electrical measurements were carried out in a N_2-filled glove box, with H_2O and O_2 levels maintained below 1 ppm. Figure 15.14 shows typical *I–V* curves recorded on those structures for thin (a monolayer of fibrils) and thick (~1 μm) films (Figure 15.14a and b, respectively). Surprisingly, even a single monolayer film of RRP3HT fibrils becomes conducting upon doping: the doped film has a maximum current value two orders of magnitude higher than the undoped one. This demonstrates that even a single layer of RRP3HT fibrils is continuous enough to effectively conduct charges with high conductivity values, that is, close to the bulk values for conducting polymers. When the film is thicker (1 μm for Figure 15.14b), the maximum current is six orders of magnitude higher than in the undoped film, certainly due to a high number of conduction paths within the film made of a thick web of fibrils. It should be noted that the conductivity of the films significantly decreases with time when exposed to ambient atmosphere, which may be due to dedoping or induced by water and oxygen absorption (see intermediate curve in Figure 15.14a).

15.3.3
SPM Characterization of Organic Bulk Heterojunction Solar Cells

15.3.3.1 A Brief Introduction to Organic Photovoltaics

Among the driving forces sustaining the growing interest in organic electronics, the emerging needs of alternative and renewable energy sources have considerably promoted active research in the field of photovoltaic technologies for solar cell applications. According to NanoMarkets, the sales forecast of materials for both organic photovoltaic diodes (OPVs) and hybrid dye-sensitized solar cells is expected to reach $600 million by 2016 [114].

The basic geometry of an OPV device is presented in Figure 15.15b. It consists in a stack of thin films successively deposited on a transparent substrate, where two electrodes sandwich one or several photoactive organic semiconducting layers. The

15.3 Effect of the Nanoscale Morphology on the Optoelectronic Properties and Device Performances

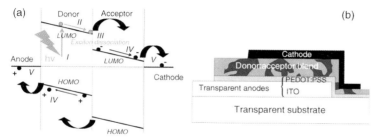

Figure 15.15 (a) Schematic band diagram of an organic photovoltaic device. The absorbing material is, in this case, the donor. (b) Schematic architecture of a bulk heterojunction solar cell. The PEDOT:PSS layer between the ITO anode and the photovoltaic blend serves to smoothen the device and stabilize the contact prior to the deposition of the blend.

principles running the device are described in Figure 15.15a. These imply various mechanisms, each of these to be optimized to yield ultimate performances of the device. First, the photoactive layers should exhibit appropriate absorption of photons within the visible and near-IR spectral range (I). Specific to organic semiconducting materials is the formation of short-lifetime excitons that do not spontaneously dissociate into electron–hole pairs. This dissociation occurs at donor–acceptor interfaces to be reached within the short lifetime of the exciton (II) and where sufficient band offsets between the LUMOs (> 300 meV) is likely to overcome the exciton binding energy (III). The free charges have then to be transported (IV) and collected (V) at their respective electrodes. The total power efficiency of a device is then defined as the product of the efficiency rate of each of these steps (to be all optimized for ultimate performances). A thorough description of OPV devices is available in the books recently published by Krebs [115] and Brabec et al. [116]. Several pathways for optimizing device performances are considered: (i) technological: tandem cells and other types of device architectures; (ii) chemical: synthesis of new effective organic materials to cope with drastic optical properties and electronic transport requirements; (iii) physical: monitoring the nanoorganization of the photoactive films. The latter two approaches are of peculiar importance for achieving low-cost OPVs. The so-called bulk heterorujunction (BH) active layer, one of the most studied OPV strategy nowadays, consists in a blend of the electron donor and acceptor materials forming an interpenetrated network, with the advantage of combining both large interface for exciton splitting and proper pathways for charge transport and collection at the electrodes. High-resolution morphological characterization techniques such as AFM have therefore appeared as imperative tools for further understanding of the nanoscale properties ruling some of the device running mechanisms [117]. In this section, we present some case studies on representative BH OPVs. The high versatility of SPM-based methods has also promoted the application of related high-resolution electrical techniques for studying the electrical and photoelectrical mechanisms at the nanoscale [118–121].

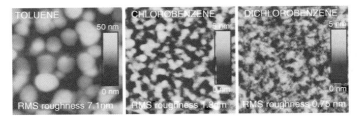

Figure 15.16 $1 \times 1 \mu m^2$ AM-AFM height images of a 1:4 MDMO-PPV:PCBM blend spin coated from different solvents. The RMS roughness of the scan area is indicated on the images. Reproduced with permission from Ref. [123], copyright Wiley-VCH Verlag GmbH & Co. KGaA.

15.3.3.2 Morphological Analysis of Bulk Heterojunction OPVs

The first AFM-based morphological studies have been performed on photovoltaic blends composed of poly(*para*-phenylene-vinylene) (PPV) derivatives as donors and (6,6)-phenyl-C_{61}-butyric acid methyl ester (PCBM) as the acceptor. The blends exhibit spontaneous phase separation upon deposit formation, forming PCBM-rich clusters, dimensions of which are shown to vary with the solution concentration and the solvent nature [122]. Figure 15.16 shows AFM height images of a 1:4 MDMO-PPV:PCBM blend spin-coated from solution in different solvents. The formation of large PCBM clusters is clearly observed using toluene as solvent. The size of these clusters appears to be smaller when using chlorobenzene, while no specific phase separation is observed using dichlorobenzene. Interestingly, the corresponding devices show poor performances except those fabricated from chlorobenzene solutions. Using toluene solution, the phase separation is far too pronounced, resulting not only in reducing the area of donor–acceptor interfaces in the film but also in restricting the dissociation to those excitons solely formed in the near vicinity of the clusters. Using chlorobenzene, however, both larger interface and shorter distance between neighboring clusters contribute to enhance the performances of the device. Conversely, using dichlorobenzene, no phase separation is observed, affecting the formation of pathways for charge transport and collection at the electrode, hence the poor performances observed with the corresponding devices.

AFM characterization of RRP3HT:PCBM blends appears to be more complex because of the ability of both components to crystallize (upon annealing) [124–126]. Figure 15.17 depicts a schematic description of the structural composition of P3HT: PCBM blend prior to and after annealing [127]. Both PCBM and RRP3HT form polycrystalline aggregates for which dimension, geometry, and distribution in the film strongly influence the photovoltaic performances. Even though increasing the content of crystalline RRP3HT domains is greatly desired for enhancing the hole mobility, the formation of PCBM-rich clusters is meanwhile to be avoided. Since the crystallization of RRP3HT is known to occur at lower temperatures than that of PCBM, the annealing temperature is to be set in between these two crystallization temperatures (typically 120–150 °C, depending on other parameters such as the weight ratio and the concentration of the solution [128]).

15.3 Effect of the Nanoscale Morphology on the Optoelectronic Properties and Device Performances

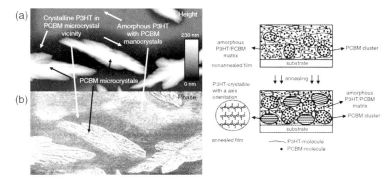

Figure 15.17 *Left*: 10 × 20 μm² AM-AFM (a) height and (b) phase images of a 1:1 RRP3HT:PCBM blend. Post annealing treatment of the film: 120 °C, 30 min. The PCBM-rich microclusters and the highly crystalline P3HT region in the vicinity of the clusters and the surrounding amorphous matrix are indicated with arrows. *Right*: Scheme presenting the structural changes in P3HT:PCBM films upon annealing. Reproduced with permission from Ref. [127], copyright Wiley-VCH Verlag GmbH & Co. KGaA.

Figure 15.17a shows a typical morphology of annealed RRP3HT:PCBM blends, in which micrometer-scale PCBM-rich clusters are formed (irreversibly damaging the device by short-circuiting the electrodes). Interestingly, the corresponding AFM phase image (Figure 15.17b) exhibits contrast variations within the polymer matrix around the clusters. Dark contrast is attributed to crystalline P3HT regions. As expected, RRP3HT crystallization is favored in PCBM-depleted regions, that is, in the vicinity of the PCBM clusters [124]. Transversal morphological studies performed by Kim and coworkers indicate the migration of PCBM molecules to the top of the active layer upon annealing. Better device efficiencies are reported when a transversal gradient in the P3HT:PCBM weight ratio across the film results in a high concentration of donor (acceptor) nearby the anode (cathode) [129]. Moreover, the ability of RRP3HT to self-organize into fibrillar nanostructures (*vide supra*) has also been evidenced in these blends [130]. Although the fibrillar nanostructures have proven to significantly enhance hole mobility (compared to amorphous films), the added-value to photovoltaic devices still remains an open question, with Meerholz and coworker claiming that this type of morphological arrangement might exclude PCBM from the aggregated network [131]. The correlation of photophysical and photovoltaic properties with the morphologies of different films of polyfluorene blended with perylene tetracarboxylic diimide (PTCDI) has been described by Keivanidis *et al*. AFM images of the blends indicate the formation of PTCDI-rich domains upon deposition, and surface smoothening upon thermal annealing is shown to vary depending on the nature of the polyfluorenes forming the blend. Interestingly, blends with PDI aggregating at the anode/blend interface are shown to yield the best performances (F8BT:PTCDI) [132]. Along the same line, a liquid crystalline derivative of PTCDI was blended with a liquid crystalline phthalocyanine, which is a donor compound. Combining the mesogenic properties of the two partners, appropriate deposition conditions, and thermal annealing allows to generate controlled morphologies in thin films, with phase separation at the nanoscale, as revealed with AFM [133].

15.3.3.3 Electrical Characterization of Bulk Heterojunction OPVs by Conductive-AFM

This section deals with high-resolution electrical characterization of photovoltaic blends and related devices using primarily C-AFM [134]. For similar studies by means of electrostatic force microscopy and Kelvin probe force microscopy, the reader may refer to Samorì and coworkers [135–137], Hoppe and coworkers [138, 139], and others [140–144]. The suitability of C-AFM to characterize organic BH systems was first demonstrated by Loos and coworkers in 2005 [145]. Though operating in contact scanning mode, the 2D mapping of the electrical properties has shown to be successful using small tip–sample contact forces, preventing tip-induced topographical damages [124]. The conventional basic AFM setup is augmented with an independent current measurement system including a conductive probe, positive and negative tip–sample bias, and high sensitivity pA-amplifiers. The ability of the system to then detect current variations as low as 100 fA and up to the µA makes C-AFM appropriate to investigate rather resistive semiconductors such as those used in BH organic photovoltaic blends. The electrical spatial resolution of the methods is expected to be intrinsically limited by the contact area between the probe and the sample, thus the radius of curvature of the tip apex. Depending on the type of coating (metallic, boron-doped diamond, etc.), a lateral resolution of 20 nm can be achieved. Figure 15.18 shows simultaneously acquired height and current C-AFM

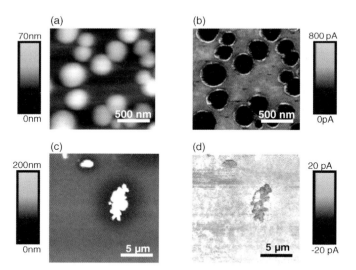

Figure 15.18 *Top*: $2 \times 2\,\mu m^2$ C-AFM height (*left*) and current (*right*) images of a 1:4 MDMO-PPV:PCBM blend spin coated from toluene. The dc tip bias is set to −2.3 V (reproduced with permission from Ref. [146], copyright 2008 Elsevier). *Bottom*: $1 \times 1\,\mu m^2$ C-AFM height (*left*) and current (*right*) images of a 1:1 RRP3HT:PCBM blend spin coated from chlorobenzene (postannealing treatment of the film: 125 °C, 5 min). The dc tip bias is set to −100 mV. Reproduced with permission from Ref. [147], copyright 2006 American Institute of Physics.

images of two representative photovoltaic BH blends (MDMO-PPV:PCBM and RRP3HT:PCBM) [146, 147]. The phase separation is clearly delineated, both in height and current images. Interestingly, in the P3HT:PCBM blend, the current variations in the matrix surrounding the PCBM-rich clusters confirm the differences in the degree of crystallization of RRP3HT within the film.

An additional key characteristic of the method is its ability to perform local current–voltage ($I-V$) profiles in order to provide further insight into the charge transport mechanisms at the nanoscale. Barrier heights, charge carrier mobility, and local power efficiencies are the electrical parameters that can be experimentally determined. Comparing $I-V$ local profiles with those obtained with macroscopic measurements, Ginger and coworkers experimentally evidenced space charge-limited current mechanisms at the nanoscale in pure P3HT films [148]. The difference observed between the macroscopic and nanoscale $I-V$ variations (current offsets) are attributed to the tip–sample contact geometry, the resulting charge injection and transport upon bias modifying the analytical expression of the Mott–Gurney law. The local hole mobility has been experimentally determined and appears to be significantly higher than the values obtained with well-established macroscopic methods. The origin of such discrepancies is still to be explained.

Finally, the spatially resolved photocurrent mapping of photovoltaic BH blends by C-AFM has been investigated by Ginger and coworkers and Marks and coworkers, implementing external sample illumination in the experimental setup [89, 149, 150]. Typical topography and photocurrent C-AFM images of an MDMO-PPV:PCBM blend are presented in Figure 15.19a and b, respectively. The origin of the photocurrent contrast variations observed in the blend is illustrated in Figure 15.19c. The

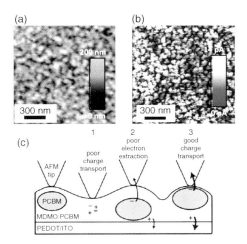

Figure 15.19 $1.5 \times 1.5\,\mu m^2$ C-AFM height (a) and photocurrent (b) images of a 1:4 MDMO-PPV:PCBM film spin coated from chlorobenzene. The photocurrent is recorded at zero dc bias and at an illumination intensity of $5 \times 10^3\,W\,m^{-2}$. (c) Proposed scheme for the observed current variations. Reproduced with permission from Ref. [149], copyright 2007 American Chemical Society.

carriers that are created upon illumination not only require proper mobility for transport across the film but also need electrical pathways to be collected at the respective electrodes. These electrical and morphological requisites are ensured with PCBM-rich domains laying in a near-surface region coated with a thin polymer-rich surrounding shell protruding at the surface of the film (case 3 in Figure 15.19c) [149].

15.4
Conclusions and Perspectives

The high-resolution characterization of π-conjugated materials and devices is generating numerous emerging challenges stimulated by both the continuous development of SPM versatility and the huge diversification of nanostructured organic materials. With today's increasing interest for such materials and the substantial supply expected for novel devices, improving the spatial resolution is also a crucial issue [151]. While AFM measurements are henceforth routinely applied for topographical analysis of polymer-based devices, the contribution of advanced SPM-based electrical characterization methods such as EFM, KPFM, and C-AFM, though very promising, still requires to be further assessed. Not only must the origin of the electrical contrast and the local electrical mechanisms be identified but also should the quantitative measurements of intrinsic parameters (e.g., work function, barrier height, carrier mobility, density of photogenerated charge carriers) be obtained to clearly assess the added value of these techniques. This also implies further modeling and simulation of the tip–sample electrical interactions taking into account the geometry of the probe and the mechanical properties of the studied materials. Although already applied to characterize the electrical properties of self-assembled monolayers and molecular crystals, the use of AFM-based electrical methods, and especially C-AFM, on π-conjugated single molecules has barely been addressed [134, 152–155].

The growing interest in organic materials for electronics and nanotechnology is most likely to urge the contribution of new advanced characterization tools. For instance, SPM tools such as torsion harmonic mode [156], transverse shear microscopy [157], and frequency modulation AFM [158] can bring valuable information on the crystal growth orientation relative to molecular packing (for FET applications) and phase separation in block copolymers and polymer blends (for OPV applications). With a driving force focusing on the analysis and monitoring of both the morphology and the local mechanical, optical, and electrical properties of self-organized nanostructures, SPM-based methods can therefore be expected to play a key role in a near future.

Acknowledgments

The authors express their thanks to Fabio Biscarini, Massimiliano Cavallini, Pascal Damman, Philippe Dubois, Benjamin Grévin, W. Jim Feast, Alan J. Heeger,

Christine Jérôme, Klaus Müllen, Bert Meijer, Paolo Samorì, Gian Piero Spada, and Albert P.H.J. Schenning and their groups for fruitful collaborations. The financial support comes from the Wallonia Region, the European Commission (FSE, FEDER), the National Fund for Scientific Research (FRS-FNRS), and the Belgian Federal Science Policy Office (PAI 6/27). M.S. and P.L. are research associates of the FRS-FNRS.

References

1. Binnig, G., Rohrer, H., Gerber, C., and Weibel, E. (1982) *Appl. Phys. Lett.*, **40**, 178.
2. Binnig, G., Quate, C.F., and Gerber, C. (1986) *Phys. Rev. Lett.*, **12**, 930.
3. Drake, B., Prater, C.B., Weisenhorn, A.L., Gould, S.A., Albrecht, T.R., Quate, C., Cannell, D.S., Hansma, H.G., and Hansma, P.K. (1988) *Science*, **243**, 1586.
4. Patil, R., Kim, S.-J., Smith, E., Reneker, D.H., and Weisenhorn, A.L. (1990) *Polym. Comm.*, **31**, 455.
5. Stocker, W., Bar, G., Kunz, M., Moller, M., Magonov, S.N., and Cantow, H.J. (1991) *Polym. Bull.*, **26**, 215.
6. Buchholz, S. and Rabe, J.P. (1992) *Angew. Chem. Int. Ed. Eng.*, **31**, 189.
7. De Feyter, S., Gesquière, A., Abdel-Mottaleb, M.M., Grim, P.C.M., De Schryver, F., Meiners, C., Sieffert, M., Valiyaveettil, S., and Müllen, K. (2000) *Acc. Chem. Res.*, **33**, 520.
8. Samorì, P. (2005) *Chem. Soc. Rev.*, **34**, 551.
9. De Feyter, S. and De Schryver, F.C. (2005) *J. Phys. Chem. B*, **109**, 4290.
10. Amabilino, D.B., De Feyter, S., Lazzaroni, R., Gomar-Nadal, E., Veciana, J., Rovira, C., Abdel-Mottaleb, M.M., Mamdouh, W., Iavicoli, P., Psychogyiopoulou, K., Linares, M., Minoia, A., Xu, H., and Luis, J.P. (2008) *J. Phys. Condens. Matter*, **20**, 184003–184013.
11. Gross, L., Mohn, F., Moll, N., Liljeroth, P., and Meyer, G. (2009) *Science*, **325**, 1110.
12. Garcia, R. and Perez, R. (2002) *Surf. Sci. Rep.*, **47**, 197.
13. Gleyzes, P., Kuo, K., and Boccara, A.C. (1991) *Appl. Phys. Lett.*, **58**, 2989.
14. Zhong, Q., Inniss, D., Kjoller, K., and Elings, V.B. (1993) *Surf. Sci.*, **290**, L688.
15. Leclère, P., Lazzaroni, R., Brédas, J.L., Yu, J.M., Dubois, P., and Jérôme, R. (1996) *Langmuir*, **12**, 4317.
16. Magonov, S.N., Elings, V., and Wangbo, M.H. (1997) *Surf. Sci.*, **389**, 201.
17. Stocker, W., Beckmann, J., Stadler, R., and Rabe, J.P. (1996) *Macromolecules*, **29**, 7502.
18. Dubourg, F., Couturier, G., Aimé, J.P., Marsaudon, S., Leclère, P., Lazzaroni, R., Salardenne, J., and Boisgard, R. (2001) *Macromol. Symp.*, **167**, 177.
19. Nony, L., Boisgard, R., and Aimé, J.P. (2001) *Biomacromolecules*, **2**, 827.
20. Kopp-Marsaudon, S., Leclère, P., Dubourg, F., Lazzaroni, R., and Aimé, J.P. (2000) *Langmuir*, **16**, 8432.
21. Anczykowski, B., Krüger, D., and Fuchs, H. (1996) *Phys. Rev. B*, **53**, 15485.
22. Boisgard, R., Michel, D., and Aimé, J.P. (1998) *Surf. Sci.*, **401**, 191.
23. Tamayo, J. and Garcia, R. (1997) *Appl. Phys. Lett.*, **71**, 2394.
24. Albrecht, T.R., Grütter, P., Horne, D., and Rugard, D. (1991) *J. Appl. Phys.*, **69**, 668.
25. Sugimoto, Y., Abe, M., Hirayama, S., Oyabu, N., Custance, O., and Morita, S. (2005) *Nat. Mater.*, **4**, 156.
26. Sheiko, S.S. and Möller, M. (2001) *Chem. Rev.*, **101**, 4099.
27. Samorì, P., Surin, M., Palermo, V., Lazzaroni, R., and Leclère, P. (2006) *Phys. Chem. Chem. Phys.*, **8**, 3927.
28. Leclère, P., Cornet, V., Surin, M., Viville, P., Aimé, J.P., and Lazzaroni, R. (2005) *Applications of Scanned Probe Microscopy to Polymers*, ACS Symposium

Series, **897**, American Chemical Society, p. 86.
29 Samorì, P. (ed.) (2006) *Scanning Probe Microscopies: Beyond Imaging*, Wiley-VCH Verlag GmbH, Weinheim.
30 VanLandingham, M., Villarubia, J.S., Guthrie, W.F., and Meyers, G.F. (2001) *Macromol. Symp.*, **167**, 15.
31 Piner, R.D., Zhu, J., Xu, F., Hong, S., and Mirkin, C.A. (1999) *Science*, **283**, 661.
32 Bustamante, C., Rivetti, C., and Keller, D.J. (1997) *Curr. Opin. Struct. Biol.*, **7**, 709.
33 Muller, D.J. and Dufrêne, Y.F. (2008) *Nat. Nanotech.*, **3**, 261.
34 Palermo, V., Liscio, A., Palma, M., Surin, M., Lazzaroni, R., and Samorì, P. (2007) *Chem. Comm.*, 3326.
35 Hartmann, U. (1999) *Annu. Rev. Mater. Res.*, **29**, 53.
36 Dunn, R.C. (1999) *Chem. Rev.*, **99**, 2891.
37 Pollock, H.M. and Ammiche, A. (2001) *J. Phys. D: Appl. Phys.*, **34**, 23.
38 Frisbie, C.D., Rozsynai, L.F., Noy, A., Wrighton, M.S., and Lieber, C.M. (1994) *Science*, **265**, 2071.
39 Noy, A., Vezenov, D.V., and Lieber, C.M. (1997) *Ann. Rev. Mater. Sci.*, **27**, 381.
40 Leclère, P., Surin, M., Brocorens, P., Cavallini, M., Biscarini, F., and Lazzaroni, R. (2006) *Mater. Sci. Eng. R: Rep.*, **55**, 1.
41 Cornil, J., Beljonne, D., Calbert, J.P., and Brédas, J.L. (2001) *Adv. Mater.*, **13**, 1053.
42 Kim, J. (2002) *Pure. Appl. Chem.*, **74**, 2031.
43 Schwartz, B. (2003) *Annu. Rev. Phys. Chem.*, **54**, 141.
44 Samorì, P., Francke, V., Müllen, K., and Rabe, J.P. (1999) *Chem. Eur. J.*, 5, 2312.
45 Jonkheijm, P., van der Schoot, P., Schenning, A.P.H.J., and Meijer, E.W. (2006) *Science*, **313**, 80.
46 Leclère, P., Surin, M., Viville, P., Lazzaroni, R., Kilbinger, A.F.M., Henze, O., Feast, W.J., Cavallini, M., Biscarini, F., Schenning, A.P.H.J., and Meijer, E.W. (2004) *Chem. Mater.*, **16**, 4452.
47 Yang, H.C., Shin, T.J., Yang, L., Cho, K., Ryu, C.Y., and Bao, Z.N. (2005) *Adv. Funct. Mater.*, **15**, 671.
48 Surin, M., Hennebicq, E., Ego, C., Marsitzky, D., Grimsdale, A.C., Müllen, K., Brédas, J.L., Lazzaroni, R., and Leclère, P. (2004) *Chem. Mater.*, **16**, 994.
49 Hoeben, F.J.M., Jonkheijm, P., Meijer, E.W., and Schenning, A.P.H.J. (2005) *Chem. Rev.*, **105**, 1491.
50 Grimsdale, A.C. and Müllen, K. (2005) *Angew. Chem. Int. Ed. Eng.*, **44**, 5592.
51 Biscarini, F., Zamboni, R., Samorì, P., Ostoja, P., and Taliani, C. (1995) *Phys. Rev. B*, **52**, 14868.
52 Ruiz, R., Choudhary, D., Nickel, B., Toccoli, T., Chang, K.C., Mayer, A.C., Clancy, P., Blakely, J.M., Headrick, R.L., Iannotta, S., and Malliaras, G.G. (2004) *Chem. Mater.*, **16**, 4497.
53 Sakurai, S.I., Kuroyanagi, K., Morino, K., Kunitake, M., and Yashima, E. (2003) *Macromolecules*, **36**, 9670.
54 Sakurai, S.I., Okoshi, K., Kumaki, J., and Yashima, E. (2006) *Angew. Chem. Int. Ed. Eng.*, **45**, 1245.
55 Schillinger, E.K., Mena-Osteritz, E., Hentschel, J., Borner, H.G., and Bäuerle, P. (2009) *Adv. Mater.*, **21**, 1562.
56 Jahnke, E., Severin, N., Kreutzkamp, P., Rabe, J.P., and Frauenrath, H. (2008) *Adv. Mater.*, **20**, 409.
57 Spada, G.P., Lena, S., Masiero, S., Pieraccini, S., Surin, M., and Samori, P. (2008) *Adv. Mater.*, **20**, 2433.
58 LaBean, T.H. and Li, H. (2007) *Nano Today*, **2**, 26.
59 Seeman, N.C. (2003) *Nature*, **421**, 427.
60 Iwaura, R., Hoeben, F.J.M., Masuda, M., Schenning, A.P.H.J., Meijer, E.W., and Shimizu, T. (2006) *J. Am. Chem. Soc.*, **128**, 13298.
61 Janssen, P.G.A., Vandenbergh, J., van Dongen, J.L.J., Meijer, E.W., and Schenning, A.P.H.J. (2007) *J. Am. Chem. Soc.*, **129**, 6078.
62 Surin, M., Janssen, P.G.A., Lazzaroni, R., Leclère, P., Meijer, E.W., and Schenning, A.P.H.J. (2009) *Adv. Mater.*, **21**, 1126.
63 Janssen, P.G.A., Jabbari-Farouji, S., Surin, M., Vila, X., Gielen, J.C., de Greef, T.F.A., Vos, M.R.J., Bomans, P.H.H., Sommerdijk, N.A.J.M., Christianen, P.C.M., Leclère, P., Lazzaroni, R., van der

Schoot, P., Meijer, E.W., and Schenning, A.P.H.J. (2009) *J. Am. Chem. Soc.*, **131**, 1222.
64 Bates, F.S. and Fredrickson, G.H. (1999) *Phys. Today*, **52**, 32.
65 Klok, H.-A. and Lecommandoux, S. (2001) *Adv. Mater.*, **13**, 1217.
66 Lee, M., Cho, B.-K., and Zin, W.-C. (2001) *Chem. Rev.*, 101.
67 Zubarev, E.R., Pralle, M.U., Sane, E.D., and Stupp, S.I. (2002) *Adv. Mater.*, **14**, 198.
68 Ruokolainen, J., Makinen, R., Torkkeli, M., Makela, T., Serimaa, R., ten Brinke, G., and Ikkala, O. (1998) *Science*, **280**, 557.
69 Chen, J.T., Thomas, E.L., Ober, C.K., and Mao, G.P. (1996) *Science*, **273**, 343.
70 Jenekhe, S.A. and Chen, X.L. (1999) *Science*, **283**, 372.
71 Leclère, P., Calderone, A., Marsitzky, D., Francke, V., Geerts, Y., Müllen, K., Brédas, J.L., and Lazzaroni, R. (2000) *Adv. Mater.*, **12**, 1042.
72 Messmore, B.W., Hulvat, J.F., Sone, E.D., and Stupp, S.I. (2004) *J. Am. Chem. Soc.*, **126**, 14452.
73 Surin, M., Marsitzky, D., Grimsdale, A.C., Müllen, K., Lazzaroni, R., and Leclère, P. (2004) *Adv. Funct. Mater.*, **14**, 708.
74 Leclerc, M. (2001) *J. Polym. Sci. A: Polym. Chem.*, **39**, 2867.
75 Neher, D. (2001) *Macromol. Rapid Comm.*, **22**, 1366.
76 Sirringhaus, H., Brown, P.J., Friend, R.H., Nielsen, M.M., Bechgaard, K., Langeveld-Voss, B.M.W., Spiering, A.J.H., Janssen, R.A.J., Meijer, E.W., Herwig, P., and de Leeuw, D.M. (1999) *Nature*, **401**, 685.
77 Facchetti, A. (2007) *Mater. Today*, **10**, 28.
78 Sandberg, H.G.O., Frey, G.L., Shkunov, M.N., Sirringhaus, H., Friend, R.H., Nielsen, M.M., and Kumpf, C. (2002) *Langmuir*, **18**, 10176.
79 Derue, G., Coppée, S., Gabriele, S., Surin, M., Geskin, V., Monteverde, F., Leclère, P., Lazzaroni, R., and Damman, P. (2005) *J. Am. Chem. Soc.*, **127**, 8018.
80 Kim, J.H., Yoneya, M., Yamamoto, J., and Yokoyama, H. (2002) *Nanotechnology*, **13**, 133.
81 Kim, J.H., Yoneya, M., and Yokoyama, H. (2002) *Nature*, **420**, 159.
82 Coppée, S., Geskin, V.M., Lazzaroni, R., and Damman, P. (2004) *Macromolecules*, **37**, 244.
83 Derue, G., Serban, D.A., Leclère, P., Melinte, S., Damman, P., and Lazzaroni, R. (2008) *Org. Electron.*, **9**, 821.
84 Cavallini, M., Albonetti, C., and Biscarini, F. (2009) *Adv. Mater.*, **21**, 1043.
85 Cavallini, M., Stoliar, P., Moulin, J.F., Surin, M., Leclère, P., Lazzaroni, R., Breiby, D.W., Andreasen, J.W., Nielsen, M.M., Sonar, P., Grimsdale, A.C., Müllen, K., and Biscarini, F. (2005) *Nano. Letters*, **5**, 2422.
86 Holdcroft, S. (2001) *Adv. Mater.*, **13**, 1753.
87 Sardone, L., Palermo, V., Devaux, E., Credgington, D., De Loos, M., Marletta, G., Cacialli, F., Van Esch, J., and Samorì, P. (2006) *Adv. Mater.*, **18**, 1276.
88 Pingree, L.S.C., Russell, M.T., Scott, B.J., Marks, T.J., and Hersam, M.C. (2007) *Org. Electron.*, **8**, 465.
89 Pingree, L.S.C., Reid, O.G., and Ginger, D.S. (2009) *Adv. Mater.*, **21**, 19.
90 Teetsov, J.A. and Vanden Bout, D.A. (2001) *J. Am. Chem. Soc.*, **123**, 3605.
91 Teetsov, J.A. and Vanden Bout, D.A. (2002) *Langmuir*, **18**, 897.
92 Brunsveld, L., Folmer, B.J., Meijer, E.W., and Sijbesma, R.P. (2001) *Chem. Rev.*, **101**, 4071.
93 Abbel, R., Grenier, C., Pouderoijen, M.J., Stouwdam, J.W., Leclère, P., Sijbesma, R.P., Meijer, E.W., and Schenning, A.P.H.J. (2009) *J. Am. Chem. Soc.*, **131**, 833.
94 Abbel, R., van der Weegen, R., Pisula, W., Surin, M., Leclère, P., Lazzaroni, R., Meijer, E.W., and Schenning, A.P.H.J. (2009) *Chem. Eur. J.*, **15**, 9737.
95 Dimitrakopoulos, C.D. and Malenfant, P.R.L. (2002) *Adv. Mater.*, **14**, 99.
96 Bao, Z., Dodabalapur, A., and Lovinger, A.J. (1996) *Appl. Phys. Lett.*, **69**, 4108.
97 Scheinert, S., Doll, T., Scherer, A., Paasch, G., and Horselmann, I. (2004) *Appl. Phys. Lett.*, **84**, 4427.
98 Cho, S., Lee, K., Yuen, J., Wang, G.M., Moses, D., Heeger, A.J., Surin, M., and Lazzaroni, R. (2006) *J. Appl. Phys.*, **100**, 114503, 1–6.

99 Surin, M., Leclère, P., Lazzaroni, R., Yuen, J.D., Wang, G., Moses, D., Heeger, A.J., Cho, S., and Lee, K. (2006) *J. Appl. Phys.*, **100**, 33712, 1–6.

100 Yang, H.C., Shin, T.J., Yang, L., Bao, Z.N., Ryu, C.Y., and Cho, K.W. (2004) *Abstr. Pap. Am. Chem. Soc.*, **227**, U425.

101 Merlo, J.A. and Frisbie, C.D. (2004) *J. Phys. Chem. B*, **108**, 19169.

102 Chua, L.L., Ho, P.K.H., Sirringhaus, H., and Friend, R.H. (2004) *Adv. Mater.*, **16**, 1609.

103 McCulloch, I., Heeney, M., Chabinyc, M.L., DeLongchamp, D., Kline, R.J., Coelle, M., Duffy, W., Fischer, D., Gundlach, D., Hamadani, B., Hamilton, R., Richter, L., Salleo, A., Shkunov, M., Sporrowe, D., Tierney, S., and Zhong, W. (2009) *Adv. Mater.*, **21**, 1091.

104 Luo, Y., Gustavo, F., Henry, J.Y., Mathevet, F., Lefloch, F., Sanquer, M., Rannou, P., and Grévin, B. (2007) *Adv. Mater.*, **19**, 2267.

105 Prosa, T.J., Winokur, M.J., Moulton, J., Smith, P., and Heeger, A.J. (1995) *Phys. Rev. B*, **51**, 159.

106 Tashiro, K., Kobayashi, M., Kawai, T., and Yoshino, K. (1997) *Polymer*, **38**, 2867.

107 Samitsu, S., Shimomura, T., and Ito, K. (2008) *Thin Solid Films*, **516**, 2478.

108 Kim, Y.H., Spiegel, D., Hotta, S., and Heeger, A.J. (1988) *Phys. Rev. B.*, **38**, 5490.

109 Lazzaroni, R., Logdlund, M., Stafström, S., Salaneck, W.R., and Brédas, J.L. (1990) *J. Chem. Phys.*, **93**, 4433.

110 Logdlund, M., Lazzaroni, R., Stafström, S., Salaneck, W.R., and Brédas, J.L. (1989) *Phys. Rev. Lett.*, **63**, 1841.

111 Rep, D.B.A., Morpurgo, A.F., and Klapwijk, T.M. (2003) *Org. Electron.*, **4**, 201.

112 Nowak, M.J., Rughooputh, S.D.D.V., Hotta, S., and Heeger, A.J. (1987) *Macromolecules*, **20**, 965.

113 Desbief, S., Derue, G., Leclère, P., Lenfant, S., Vuillaume, D., and Lazzaroni, R. (2009) *Eur. Phys. J. Appl. Phys.*, **46**, 12504-1–5.

114 Nolan, R. (2009) Nanomarkets, Vol. Nano-087, Glen Allen.

115 Krebs, F.C. (2008) *Polymer Photovoltaics: A Practical Application*, SPIE, Bellingham.

116 Brabec, C., Dyakonov, V., and Scherf, U. (eds) (2008) *Organic Photovoltaics: Materials, Device Physics and Manufacturing Technologies*, Wiley-VCH Verlag GmbH, Weinheim.

117 Moons, E. (2002) *J. Phys. Condens. Matter*, **14**, 12235.

118 Girard, P. (2001) *Nanotechnology*, **12**, 485.

119 Nonnenmacher, M., Oboyle, M.P., and Wickramasinghe, H.K. (1991) *Appl. Phys. Lett.*, **58**, 2921.

120 Liscio, A., Palermo, V., Gentilini, D., Nolde, F., Müllen, K., and Samorì, P. (2006) *Adv. Funct. Mater.*, **16**, 1407.

121 Olbrich, A., Ebersberger, B., Boit, C., Vancea, J., Hoffmann, H., Altmann, H., Gieres, G., and Wecker, J. (2001) *Appl. Phys. Lett.*, **78**, 2934.

122 Hoppe, H., Niggemann, M., Winder, C., Kraut, J., Hiesgen, R., Hinsch, A., Meissner, D., and Sariciftci, N.S. (2004) *Adv. Funct. Mater.*, **14**, 1005.

123 Douhéret, O., Swinnen, A., Bertho, S., Haeldermans, I., D'Haen, J., D'Olieslaeger, M., Vanderzande, D., and Manca, J.V. (2007) *Prog. Photovolt.*, **15**, 713.

124 Douhéret, O., Swinnen, A., Breselge, M., Van Severen, I., Lutsen, L., Vanderzande, D., and Manca, J. (2007) *Microelectron. Eng.*, **84**, 431.

125 Yang, X.N., Loos, J., Veenstra, S.C., Verhees, W.J.H., Wienk, M.M., Kroon, J.M., Michels, M.A.J., and Janssen, R.A.J. (2005) *Nano. Letters*, **5**, 579.

126 Swinnen, A., Haeldermans, I., vande Ven, M., D'Haen, J., Vanhoyland, G., Aresu, S., D'Olieslaeger, M., and Manca, J. (2006) *Adv. Funct. Mater.*, **16**, 760.

127 Erb, T., Zhokhavets, U., Gobsch, G., Raleva, S., Stuhn, B., Schilinsky, P., Waldauf, C., and Brabec, C.J. (2005) *Adv. Funct. Mater.*, **15**, 1193.

128 Ma, W.L., Yang, C.Y., Gong, X., Lee, K., and Heeger, A.J. (2005) *Adv. Funct. Mater.*, **15**, 1617.

129 Jo, J., Na, S.-I., Kim, S.-S., Lee, T.-W., Chung, Y., Kang, S.-J., Vak, D., and Kim, D.-Y. (2009) *Adv. Funct. Mater.*, **19**, 2398.

130 Berson, S., De Bettignies, R., Bailly, S., and Guillerez, S. (2007) *Adv. Funct. Mater.*, **17**, 1377.

131 Moule, A.J. and Meerholz, K. (2008) *Adv. Mater.*, **20**, 240.
132 Keivanidis, P.E., Howard, I.A., and Friend, R.H. (2008) *Adv. Funct. Mater.*, **18**, 3189.
133 Zucchi, G., Viville, P., Donnio, B., Vlad, A., Melinte, S., Mondeshki, M., Graf, R., Spiess, H.W., Geerts, Y.H., and Lazzaroni, R. (2009) *J. Phys. Chem. B*, **113**, 5448.
134 Kelley, T.W., Granström, E.L., and Frisbie, C.D. (1999) *Adv. Mater.*, **11**, 261.
135 Palermo, V., Palma, M., and Samorì, P. (2006) *Adv. Mater.*, **18**, 145.
136 Palermo, V., Ridolfi, G., Talarico, A.M., Favaretto, L., Barbarella, G., Camaioni, N., and Samorì, P. (2007) *Adv. Funct. Mater.*, **17**, 472.
137 Palermo, V., Otten, M.B.J., Liscio, A., Schwartz, E., de Witte, P.A.J., Castriciano, M.A., Wienk, M.M., Nolde, F., De Luca, G., Cornelissen, J.J.L.M., Janssen, R.A.J., Müllen, K., Rowan, A.E., Nolte, R.J.M., and Samorì, P. (2008) *J. Am. Chem. Soc.*, **130**, 14605.
138 Glatzel, T., Hoppe, H., Sariciftci, N.S., Lux-Steiner, M.C., and Komiyama, M. (2005) *Jpn. J. Appl. Phys.*, **44**, 5370.
139 Hoppe, H., Glatzel, T., Niggemann, M., Hinsch, A., Lux-Steiner, M.C., and Sariciftci, N.S. (2005) *Nano. Letters,* **5**, 269.
140 Burgi, L., Richards, T., Chiesa, M., Friend, R.H., and Sirringhaus, H. (2004) *Synth. Met.*, **146**, 297.
141 Annibale, P., Albonetti, C., Stoliar, P., and Biscarini, F. (2007) *J. Phys. Chem. A*, **111**, 12854.
142 Coffey, D.C. and Ginger, D.S. (2006) *Nature Mater.*, **5**, 735.
143 Lei, C.H., Das, A., Elliott, M., and Macdonald, J.E. (2003) *Appl. Phys. Lett.*, **83**, 482.
144 Maturova, K., Kemerink, M., Wienk, M.M., Charrier, D.S.H., and Janssen, R.A.J. (2009) *Adv. Funct. Mater.*, **19**, 1379.
145 Alexeev, A., Loos, J., and Koetse, M.M. (2006) *Ultramicroscopy,* **106**, 191.
146 Alexeev, A. and Loos, J. (2008) *Org. Electron.*, **9**, 149.
147 Douhéret, O., Lutsen, L., Swinnen, A., Breselge, M., Vandewal, K., Goris, L., and Manca, J. (2006) *Appl. Phys. Lett.*, **89**, 032107.
148 Reid, O.G., Munechika, K., and Ginger, D.S. (2008) *Nano. Letters,* **8**, 1602.
149 Coffey, D.C., Reid, O.G., Rodovsky, D.B., Bartholomew, G.P., and Ginger, D.S. (2007) *Nano. Letters,* **7**, 738.
150 Leever, B.J., Durstock, M.F., Irwin, M.D., Hains, A.W., Marks, T.J., Pingree, L.S.C., and Hersam, M.C. (2008) *Appl. Phys. Lett.*, **92**, 013302.
151 Bonnell, D. (2008) *ACS Nano*, **2**, 1753.
152 Leatherman, G., Durantini, E.N., Gust, D., Moore, T.A., Moore, A.L., Stone, S., Zhou, Z., Rez, P., Liu, Y.Z., and Lindsay, S.M. (1999) *J. Phys. Chem. B*, **103**, 4006.
153 Kelley, T.W. and Frisbie, C.D. (2000) *J. Vac. Sci. Techn. B*, **18**, 632.
154 Loiacono, M.J., Granström, E.L., and Frisbie, C.D. (1998) *J. Phys. Chem. B*, **102**, 1679.
155 Muller, E.M. and Marohn, J.A. (2005) *Adv. Mater.*, **17**, 1410.
156 Sahin, O., Magonov, S., Su, C., Quate, C.F., and Solgaard, O. (2007) *Nat. Nanotech.*, **2**, 507.
157 Kalihari, V., Tadmor, E.B., Haugstad, G., and Frisbie, C.D. (2008) *Adv. Mater.*, **20**, 4033.
158 Grévin, B., Demadrille, R., Linares, M., Lazzaroni, R., and Leclère, P. (2009) *Adv. Mater.*, **21**, 4124.

16
Single-Molecule Organic Electronics: Toward Functional Structures
Simon J. Higgins and Richard J. Nichols

16.1
Introduction

The application of individual molecules or supramolecules as building blocks for electronic devices is the key long-term goal for molecular electronics. Advances in synthetic chemistry mean that one can envisage molecules with "switchable" redox-active, photoactive, or spin-state behavior, designed to work as molecular switches. Indeed, the incorporation of small numbers of Langmuir–Blodgett-aligned switchable rotaxanes into real electronic devices, using fault-tolerant crossbar architectures, has been described [1], and more recently, circuits employing transistors with oligothiophene monolayer channels have been demonstrated [2]. However, reliable integration of *individual* molecules into devices is an even more demanding challenge. In this chapter, we begin by reviewing techniques that have been developed for fabricating metal|molecule|metal junctions down to the single-molecule level and for measuring their electrical properties. We focus mainly on studies performed in ambient conditions, rather than in UHV. We then look at results that have been obtained on molecules in which supramolecular interactions of various kinds play a role in controlling these electrical properties, or in which the molecular components are often found in supramolecular systems.

16.2
Techniques

There are a number of techniques that have been used to trap single molecules between metallic contacts and to measure their electrical properties. These methods commonly use electrical break junctions (BJs), probe microscopy, or nanofabricated electrode gaps [3–13]. Mechanically formed break junctions (MCBJs) are a convenient platform for forming metal|molecule|metal junctions [8–10, 14–17]. MCBJs can be formed by a number of methods, including mechanical cleavage of notched metal wires and by electromigration, to form pairs of metal electrodes separated by

Functional Supramolecular Architectures. Edited by Paolo Samorì and Franco Cacialli
Copyright © 2011 WILEY-VCH Verlag GmbH & Co. KGaA, Weinheim
ISBN: 978-3-527-32611-2

nanometer spacing. In ground-breaking experiments, Reed *et al.* showed that molecules can span the electrode gap produced in a MCBJ [8]. They were able to measure the electrical properties of benzene-1,4-dithiol (BDT) self-assembled monolayers (SAMs) within such a junction, although in the absence of a statistical analysis of the data, they could not unequivocally determine the number of molecules in the junction [8]. Subsequent experiments by other groups showed the technique to be an adaptable method for measuring the electrical properties of metal|molecule|metal junctions, and Weber *et al.* provided evidence for the isolation of single molecules in the junction through the choice of an asymmetric molecular bridge [10]. More recently, electrical conductance through such junctions has been statistically analyzed through the use of conductance histograms (see below) produced from large numbers of electrical measurements made during the opening and closing of the MCBJ [14–16].

Cui *et al.* were the first to provide clear statistical evidence that conductance measurements on *single* molecules can be performed [4]. They used a conducting probe AFM to perform electrical measurements on single alkanedithiol molecules isolated in a monothiol matrix where the top contact to the molecule was achieved with phosphine-protected gold nanoparticles (GNPs). This method is illustrated in Figure 16.1. The monolayer matrix, in which dithiol molecules are embedded at high dilution, is usually formed using an alkanethiol self-assembled monolayer. This monolayer matrix can be formed by coadsorbing the dithiol and monothiol at a given

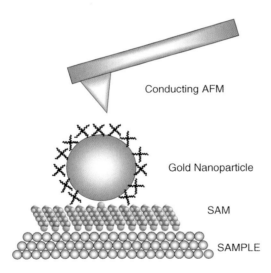

Figure 16.1 An illustration of the formation of molecular junctions using the method developed by Cui *et al.* [4]. The target dithiol molecule is diluted within a matrix of alkanethiols (monolayer matrix isolation). The thiol group of the dithiol on the outer face of the monolayer acts as a site for anchoring gold nanoparticles that then form the top contact of Au|dithiol|Au junction. The major contribution to the current flow is from the dithiol bridge rather than the surrounding monothiol monolayer. Electrical properties of the junction are measured by pressing a conducting AFM (or an STM) tip onto the nanoparticle.

concentration ratio, or by first forming the monothiol SAM and then exposing it to dithiol to achieve insertion of dithiol, generally at defects in the monothiol SAM. The dilution of the monolayer is chosen to be sufficiently high to give a good probability of forming single-molecule junctions (where one dithiol molecule contacts a nanoparticle as illustrated in Figure 16.1). However, since there is a statistical distribution of junctions containing 1,2,3, ... contacted dithiols, many such conducting probe measurements are typically made over many nanoparticles and the results are statistically analyzed to determine the current through a single molecule [4]. From the resulting histograms, the conductance of junctions containing just one dithiol contact can be identified [4]. When analyzing the electrical properties of such junctions, complicating factors such as the contact impedance between tip and nanoparticle and Coulomb blockading have to be taken into account, along with the physical compression of the monolayer by AFM tip contact force [18–20].

The STM has proven to be a very valuable tool for the formation of metal|molecule|metal junctions. In 2003, Xu and Tao developed a method in which molecular junctions were repeatedly formed in STM experiments under organic liquids containing the target molecules [3]. Mechanical contact of the gold STM tip with the gold substrate was established; the tip was then retracted until the metallic contact between tip and substrate cleaved and molecular bridges of the target molecule (either α,ω-dithiol molecules or bipyridine in these original experiments) formed within the gap. They named this method the *in situ* break junction method [3]. The tunneling current versus distance curves obtained during retraction of the STM tip exhibited characteristic current plateaux. These were at integer multiples of a fundamental current value as revealed by statistical analysis using conductance (or current) histograms. This response was associated with the presence of an integer number of molecules in the gap, with the lowest current histogram peak being assigned to a single-molecule junction [3]. In such *in situ* BJ experiments, it is also possible to monitor the conductance due to the cleavage of metallic break junctions (at $\sim G_0 = 77\,\mu S$ and multiples thereof).

In 2003, Haiss *et al.* demonstrated another way of forming single-molecule junctions [6]. By also using an STM, they demonstrated that molecular junctions could be formed from a low coverage phase of flat-lying α,ω-dithiol molecules on a Au (111) substrate by bringing the STM tip close to the substrate, but while avoiding contact between the tip and the sample. The tip was then retracted and the spontaneous formation of molecular bridges between tip and sample was recognized by current plateaux in the current versus distance scans. Haiss *et al.* referred to this method in which molecular bridges form spontaneously in the absence of metallic contact between tip and substrate as the $I(s)$ method (current I and distance s) [6]. Figure 16.2 shows a schematic representation of a metallic junction formed in an *in situ* BJ experiment and a Au|octanedithiol|Au junction that may be formed in either a BJ or an $I(s)$ experiment; experimental determined values for the junction conductance are listed in this figure.

Figure 16.3 shows current–distance scans recorded by the $I(s)$ method for 6-[1'-(6-mercapto-hexyl)-[4,4']bipyridinium]-hexane-1-thiol dication (6V6; illustrated in Figure 16.3) [6, 21]. The black curve shows the exponential decay of the tunneling

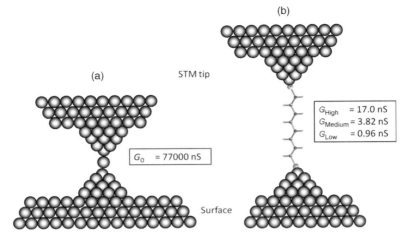

Figure 16.2 A schematic illustration of the formation of junctions in STM-based experiments. (a) A metallic junction that gives a conductance of ∼G_0 (77 μS) for a clean Au atomic junction. (b) A Au|octanedithiol|Au junction such as might be formed by the in situ BJ or $I(s)$ methods. For such molecular junctions, multiple conductance groups have been observed giving G_{high}, G_{medium}, and G_{low} conductance values. The values shown are taken from Ref. [20]. It has been proposed that the differing values relate to differences in the binding between gold surface atoms and the thiol contacting groups.

current in the absence of molecular bridge formation. In contrast, when Au|6V6|Au junctions are formed, a pronounced plateau is observed of height I_w, with a break-off distance $s_{1/2}$ as marked in the figure [6, 21]. Many such $I(s)$ scans are analyzed and histograms constructed of the current plateau (or conductance) values. Later sections

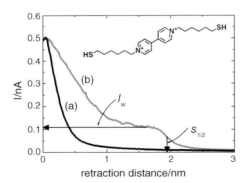

Figure 16.3 Current–distance scans recorded by the $I(s)$ method for 6-[1′-(6-mercapto-hexyl)-[4,4′]bipyridinium]-hexane-1-thiol dication (6V6). The upper (gray) curve shows current flow through a Au|6V6|Au molecular junction. The current plateau values (I_w) are used to build histograms from many such traces, from which single-molecule conductance is determined. The lower curve shows the tunneling decay in the absence of molecule bridge formation (i.e., when no molecule is trapped between the Au STM tip and Au substrate). Adapted with permission from Ref. [21]. Copyright 2004 American Chemical Society.

of this chapter show examples of histograms obtained and used in the analysis of molecular conductance.

A number of other methods have been used for characterizing single molecular electrical properties. In the so-called $I(t)$ technique, molecule conductance is also measured with an STM [5]. However, in contrast to the *in situ* BJ and $I(s)$ methods, a fixed gap separation is employed and the stochastic formation of molecule bridges between a gold STM tip and substrate is recorded as current jumps as molecular bridges form and break. These events are analyzed using a similar histogram representation of the *in situ* BJ and $I(s)$ methods to obtain single-molecule conductance [5]. The method has been particularly useful in determining molecular conductance as a function of the STM tip-to-sample separation [22–24]. In another approach, Dadosh *et al.* have obtained molecular conductance by preassembling in solution pairs of gold nanoparticles bridged by single-dithiol linkers [7]. These dimers were then assembled between metal contact pairs for electrical interrogation. A number of groups have used nanofabricated electrode gaps with sizes tuned to match those of the molecular bridges under study [11, 25–27]. This has involved challenging nanofabrication procedures, but has enabled to perform detailed low-temperature electrical characterizations. Using devices with underlying gate electrodes, molecules could be sequentially charged to different charged states by the imposition of the gate voltage allowing the characterization of different molecular states and their influence on the transmission of the metal(source)|molecule|metal(drain) junction [11, 26].

16.2.1
DNA

DNA and RNA are undoubtedly very versatile in their ability to form supramolecular structures and this has been widely recognized for its potential in nanotechnology, which has included DNA and RNA assemblies used for nanolithography and molecular wires. In terms of molecular electronics, an attractive feature of DNA is its ability to form defined architectures and templates that may lend themselves to the construction of circuits or wired arrays. However, the electrical properties of DNA have been a long-standing issue. A historical survey of the literature shows many conflicting reports, with DNA having been reported to display metal-like conductance [28, 29] and to behave as a semiconductor [30] and even as an insulator [31], although more consensus has been reached in recent years with the advent of new techniques to attach DNA between metallic contacts. A wide variety of methods now exist for examining charge transfer through DNA, ranging from photochemical experiments employing donor–DNA–acceptor systems or similar concepts to direct electrical wiring of single DNA molecules into contact junctions. The photochemical methods typically rely on highly excited radicals, generated by groups either typically located at a terminus of the DNA chain or intercalated between base pairs. Generally, the terminal groups are photoactivated metal complexes or aromatic compounds attached to the DNA, although radicals have also been generated by photoinduced chemical cleavage [32]. Charge transfer along the DNA strands from these photoactivated states is then typically analyzed in terms of distance dependence (for instance, the tunnel factor β); such experiments have demonstrated without doubt the occurrence of long-range electron or hole transfer through DNA double strands [28, 32]. However, electron or hole injection is from an excited state, leaving open the key question as to whether DNA or other oligonucleotides behave like electrically conducting single molecular wires at low "driving forces" or voltages. In this respect, the complete electrical behavior (I–V response) can only be directly obtained with DNA chemically "wired" into an electrical circuit, where the DNA strands are connected between metallic contacts. Particularly over the past 10 years, a number of ingenious methods have been devised to achieve the feat of wiring DNA between metallic contacts.

In their pioneering work, Fink and Schonenberger [29] and Porath et al. [30] showed that DNA can be wired between metallic contacts for electrical transport measurements. The measurements of Fink and Schonenberger [29] and Porath et al. [30] were for ~600 and 10.4 nm length structures, respectively. Experimental challenges to achieve these measurements included the manipulation of DNA "ropes" and the fabrication of metallic nanogap electrodes, respectively. Fink and Schonenberger [29] found simple ohmic current–voltage behavior for ~600 nm long DNA "ropes" (consisting of a bundle of a few DNA molecules), while Porath et al. [30] found that individual 10.4 nm long poly-GC duplexes in vacuum exhibited large bandgap semiconducting behavior. More recently, a number of probe microscopy methods (STM and conducting AFM) have been used to record the electrical properties of relatively short DNA sequences electrically wired between gold contacts. These recent studies will be the focus of this section.

The conductance of DNA sequences has now been measured by a number of single-molecule conductance methods that have been introduced in Section 16.2. These include measurements using the *in situ* BJ method [33, 34], the $I(s)$ method, the $I(t)$ method [35, 36], and adaptations of the Cui method with gold nanoparticle top contacts [37–39]. To reliably measure the conductance of DNA, it is important that good chemical and electronic coupling is achieved between the DNA molecule and the metal contacts; for all these methods, this has been achieved by using thiol linker groups at either end of the target DNA molecule that chemisorbs to the respective gold contacts [33–39]. These methods have all been applied in aqueous buffer solutions or ambient air environments, where it is hoped that denaturation of the sequences will be less of an issue than in UHV.

Using the *in situ* break junction method in aqueous buffer, Xu and coworkers have analyzed the single-molecule conductance of a number of DNA duplexes [33, 34]. Each of the strands forming the double helix was terminated with a -$CH_2CH_2CH_2SH$ group at the 3′-end; these provide the chemical anchoring points to the gold contacts. They compared the conductance of two series of DNA sequences, a 5′-$(GC)_n$-3′-thiol linker with $n = 4, 5, 6,$ and 7 and a 5′-CGCG$(AT)_m$CGCG-3′-thiol linker with $m = 0, 1,$ and 2, where thiol linker signifies the -$CH_2CH_2CH_2$-SH group [33]. This represents an interesting comparison between the electrical properties of GC- and AT-rich domains. In the first instance, they determined that double-stranded (ds) DNA is much more "conductive" than equivalent length alkanes or peptides and that for the ds-DNA sequences measured, the conductance decreases relatively slowly with length. They also found that the directly measured electrical conductance of ds-DNA is indeed dependent on sequence [33]. Interestingly, the GC-rich domains showed an algebraic dependence of the conductance on molecular length (conductance $\propto$ 1/length), while for $(AT)_m$ domains, an exponential dependence on length of the domain was measured with a decay constant of 0.43 Å^{-1} [33]. The latter is characteristic of tunneling or superexchange transport across the A:T domains. The slow algebraic decay for $(GC)_n$ domains points to a different conduction mechanism for this sequence and suggests transport via stacked base pairs. In a subsequent study, also using the *in situ* BJ method, Hihath *et al.* analyzed the influence of single-nucleotide polymorphisms on the conductance of Au|ds-DNA|Au junctions [34]. They employed short sequences with either 11 or 12 base pairs and modified a single base, a single base pair, or two separate bases in the double-stranded sequence. These mismatches could be identified by single-molecule conductance measurements; indeed, it was found that a single base pair mismatch could change the conductance of the sequence by an order of magnitude [34].

A major question concerning the electrical properties of DNA in electrical junctions is whether charge transport is predominantly via the stacked base pairs in double-stranded DNA or whether transport through the backbone is also significant. In this respect, it is interesting to compare the directly measured conductance of single-stranded DNA with that of double-stranded forms. To reliably determine the conductance of single-stranded DNA, it is necessary to have gold-contact binding groups at each end of the single strand. This has been achieved by van Zalinge *et al.* by using thiol linkers at both ends [35]. In this study, the single-stranded oligodeoxynucleotides (ODNs) were derivatized with a C_6 thiol at the 5'-end and a C_3 thiol at the 3'-end. A series of 3, 4, 5, and 7 mer oligodeoxynucleotides containing runs of the single bases adenine (A), guanine (G), cytosine (C), and thymine (T), were analyzed. The single-molecule conductance data for these single-stranded ODNs are shown in Figure 16.4 [35]. As can be seen, there is a dependence of the conductance on the base type, with the oligoguanine strands presenting the highest conductance. The highest conductance of oligoguanine is qualitatively in agreement with the lowest oxidation and corresponding highest HOMO level of guanine giving the most favorable energetics for transport via HOMO-mediated hole transfer [35]. Also marked on Figure 16.4 are conductance values for the double-stranded ODNs $(A)_{15}$-$(T)_{15}$ and $(G)_{15}$-$(C)_{15}$. The conduction of these duplexes (15 base pairs long) is compared to the 3-, 4-, 5-, and 7-base long single-stranded ODNs by extrapolating the single-stranded data. The conductance of the double-stranded $(A)_{15}$-$(T)_{15}$ and $(G)_{15}$-$(C)_{15}$ is orders of magnitude above these extrapolated lines. This shows that the single-stranded ODNs are poorly conductive and that the base stacking in the double-stranded ODNs plays a central role in longer range charge transport through ds-DNA.

As discussed in Section 16.2, the single-molecule conductance may also be determined when the target molecules are embedded within a self-assembled monolayer (monolayer matrix isolation), with the top contact then being provided by chemically contacting to gold nanoparticles through thiol linkage [4]. Such an

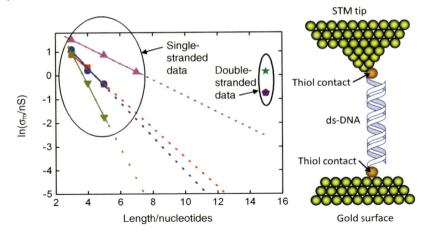

Figure 16.4 Single-molecule conductance data measured with the $I(t)$ method for a selection of single-stranded (ss) and double-stranded (ds) oligodeoxynucleotides (ODNs). For the ss-ODNs, the dependence of the single-molecule conductance of the oligonucleotides on the number of bases is shown: ■ ss-oligo-A, ● ss-oligo-T, ▲ ss-oligo-G and ▼ ss-oligo-C. Conductance values for two selected double strands are also shown: (A)$_{15}$-(T)$_{15}$ (⬟) and (G)$_{15}$-(C)$_{15}$ (★). The dashed lines are extrapolations of the length dependence for the ss-ODN data; these help to compare the shorter single strands with the longer double-stranded ODNs (see text for comparison). The right-hand side shows a schematic illustration of a double strand in the contact junction between the Au STM tip and Au substrate. Adapted from Ref. [35]. Copyright Wiley-VCH Verlag GmbH & Co. KGaA. Reproduced with permission.

approach has been used by Cohen et al. [37, 38] and Ullien et al. [39]. In these studies, the monolayer matrix was provided by single-stranded DNA forming a self-assembled monolayer on the gold substrate. Junctions containing ds-DNA were then formed by exposing these single-stranded monolayers to gold nanoparticles encapsulated with the complementary ss-DNA. In this way Au|ds-DNA|GNP junctions were produced. Using this method, the double-stranded DNA is embedded in a monolayer matrix of single strands, and is hence forced to stand on the surface, reducing the possibility of undesired interactions between the DNA strands and the gold surface; chemical attachment to the contacts is then limited to the thiol groups (-C_3H_6-SH) on the 3′-ends. Using estimates of the surface area occupied by a hybridized ds-DNA molecule, it was concluded that for small nanoparticles (5 nm), the maximal number of ds-DNA under the particle is no more than 2–3 [39]. One may expect this number to increase with nanoparticle size, however, Ullien et al. showed that similar current flowed through Au|ds-DNA|GNP junctions for GNPs of different diameters (5, 10, and 20 nm), which led them to conclude that a similar (and very small) number of double-stranded connections exists for all three nanoparticle sizes [39]. By pressing a conducting AFM tip onto the nanoparticle, they were able to record I–V traces for the Au|ds-DNA|GNP junctions [39]. For a 26 base pair long sequence, a current of 220 nA was recorded at 2 V. Given that the maximal number of ds-DNA molecules in this junction was 2–3, they concluded that each molecule must be capable of carrying at least 70–220 nA. They noted that the large magnitudes of the current imply faster than the previously assumed rates of charge transport through ds-DNA [39].

16.2.2
Base Pair Junctions

The noncovalent interaction of DNA and RNA base pairs is clearly of great significance in molecular biology and in the supramolecular organization of structures such as double-stranded DNA. Recently, the interaction of base pairs has been used in the formation of molecular electrical junctions and this may also be of future significance in electric-based sequencing methods. In a recent interesting development in this direction, Chang et al. [40] have used the $I(t)$ method to measure the conductance of gold junctions spanned by a pairing of nucleoside bases interacting by hydrogen bonding (see Figure 16.5a). Current jumps on a relatively fast timescale and on a slower timescale were observed. The $I(t)$ events with the shorter lifetime were assigned to cleavage of the molecular junction at the hydrogen bonds between the bases (Figure 16.5c), while the longer lifetime events were assigned to a Au–S bond breaking (Figure 16.5b) [40]. Using a conducting probe AFM to perform *in situ* break junction experiments, Tao and coworkers developed a technique for simultaneously measuring the tunneling current and the mechanical force as metal|molecule|metal junctions are formed, extended, and broken [41]. This technique has recently been applied to gold junctions spanned by pairs of H-bonded nucleoside bases [42]. It was found that junctions held together with three H-bonds (e.g., guanine–cytosine interactions) are stiffer than those held together by two (e.g.,

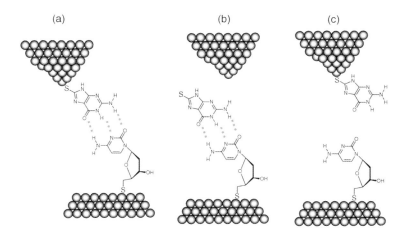

Figure 16.5 Schematic illustration of the measurements performed by Chang et al. using the $I(t)$ technique [40]. The intact molecular wire attached between the STM tip and sample shown in (a) may break either by cleavage of the Au−S bond (b) or by cleavage of the hydrogen bonds between the nucleobases (c).

adenine–thymine interactions). More generally, these studies show how noncovalent interaction such as hydrogen bonding can be used to construct molecular junctions and point toward the future assembly of more complex electrical junctions based on supramolecular chemistry.

16.2.3
Porphyrins

The supramolecular chemistry of porphyrins is very rich and a wide variety of architectures based on metalloporphyrins have been assembled, as recently reviewed by Beletskaya *et al.* [43]. Porphyrin units can be linked together through covalent bonds or noncovalent interactions such as hydrogen bonding or $\pi-\pi$ stacking interactions to give complex molecular assemblies [43]. For example, double-stranded ladder complexes have been formed from Zn–porphyrin oligomers, with the two strands being linked through axial coordination of bridging bipyridine between Zn centers [44]. Metalloporphyrins are an interesting class of compounds for molecular electronics due to their ability to transport charge over extended distances, giving potential for future use as device interconnects. They are also attractive since there is synthetic methodology for preparing porphyrin molecular wires with lengths exceeding 10 nm. Other attractive molecular features include their ability to coordinate a range of metal cations, enabling control over their redox activity for electrochemical switching applications, and also their rich photochemistry that may be exploited in photoswitching applications.

Long-range charge transport through porphyrins has been analyzed by photophysical methods to determine the kinetics of charge transfer between donor and acceptor moieties appended to either end of porphyrin molecular wires. Such measurements have indicated that oligoporphyrins are indeed good conduits for long-range charge transport [45]. Other measurements have shown that enforcing planarity on porphyrin oligomers results in increased conjugation and increased charge carrier mobility [45]. Recently, single-molecule conductance measurements on porphyrins have been made by Sedghi *et al.* using the *I*(*s*) method [46]. The DC electrical response was also recorded and the conductance evaluated as a function of molecular length for covalently linked porphyrin oligomers. The conductance of the homologous series of oligoporphyrins **1a–1c** shown in Figure 16.6 was determined [46]. Figure 16.7a shows a current–distance curve for $N=1$ recorded by the *I*(*s*) method with a gold tip and surface; a histogram of current step values compiled from many such measurements is shown in Figure 16.7b [46]. When such conductance data were analyzed across the homologous series, a remarkably low rate of decay of the conductance with molecular length was observed. An attenuation factor of $\beta = (0.04 \pm 0.006)$ Å^{-1} was obtained from $\ln\sigma_m$ versus distance plots (where σ_m is single-molecule conductance) [46]. By comparison, π-conjugated organic bridges typically give β-values in the range 0.1–0.6 Å^{-1}. Decay constants have also been determined from single-molecule break junction measurements for oligothiophenes, where $\beta = 0.1$ Å^{-1} was reported [47], again attesting to the low β for the oligoporphyrins.

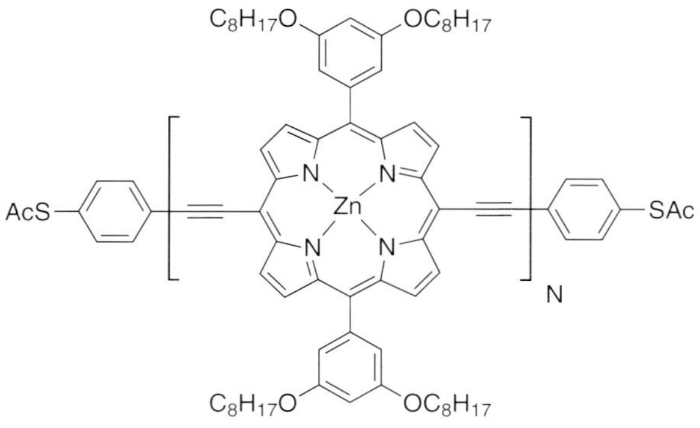

[Structure diagram]

1; N = 1 **(a)**, 2 **(b)**, and 3 **(c)**

Figure 16.6 Oligoporphyrins **1** (a–c) used in the single-molecule conductance study of Ref. [46].

In a recent study, Kiguchi et al. have analyzed the influence of end group position on the formation of single-molecule junctions from porphyrin monomers [48]. The ability of the two compounds, **2** and **3** shown in Figure 16.8, to form junctions was compared. It was found that the positions of the thiol linkers (formed following deprotection of the thioacetate groups in Figure 16.8) had a profound effect on the junction formation process. The junction conductance of Au|**2**|Au was determined to be $5 \times 10^{-4} G_0$ from a peak in conductance histograms obtained by the *in situ* break junction method. However, no peak was apparent for measurements made with **3** [48]. This would indicate that the formation of Au|**3**|Au junctions is inhibited. From surface characterization data, Kiguchi et al. proposed that **3** adsorbed to the gold

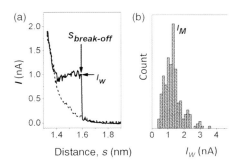

Figure 16.7 (a) $I(s)$ scans recorded by scanning tunneling microscopy under ambient conditions for a low coverage of the porphyrin **1a** (Figure 16.6) on a Au(111) film. Curves in the absence (dotted) and presence (full) of molecular wire formation are shown. (b) A histogram of I_w values measured for the porphyrin monomer with $I_0 = 8$ nA and $V_{bias} = 600$ mV. Reprinted in part with permission from Ref. [46]. Copyright 2007 American Chemical Society.

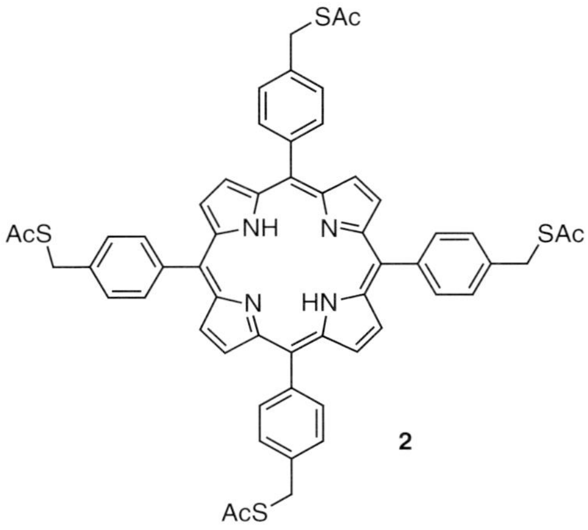

Figure 16.8 Porphyrin wires used by Kiguchi et al. in a study of single-molecule conductance using the *in situ* break junction technique [48]. They conclude that the position of the thiol group has a significant influence on the metal|molecule|metal junction formation process, which is greatly impaired in the case of **3**.

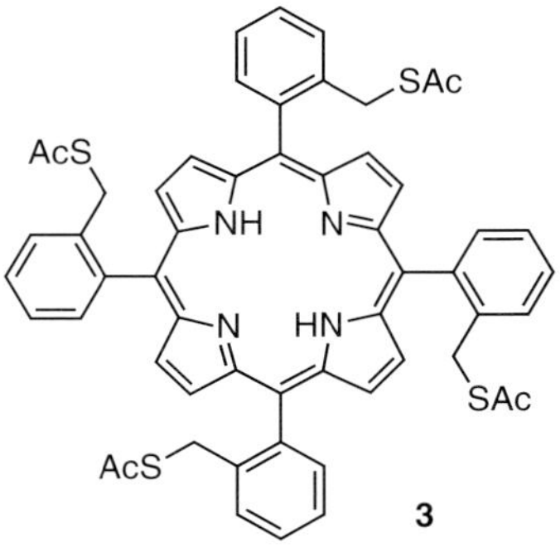

surface through all four thiol linkers and four nitrogens of the porphyrin ring, while **2** adsorbed with the porphyrin ring inclined to the surface (i.e., with much less "chemical attachment" to the surface) [48]. They proposed that this in turn leads to a higher binding energy for **3**, lower surface diffusion, and consequently an impaired ability to form molecular junctions following formation of the metallic break junction.

Scanning tunneling spectroscopy measurements have also been carried out on porphyrins using an STM under electrochemical conditions [49]. Although these experiments were not aimed at quantifying the single-molecule conductance, they showed how tunneling current flow through redox-active porphyrins could be controlled using electrochemical potential and monitored using STM. Figure 16.9 shows the iron–protoporphyrin studied; a free base version was also employed as a control [49]. Ordered 2D arrays of the Fe–protoporphyrin (FePP) compounds mixed with protoporphyrin (PP) were formed on a graphite surface and imaged with STM in an electrochemical environment. By tuning the electrochemical potential to the redox potential for the FePP compound, the tunneling current could be enhanced. This was seen by monitoring the apparent height of individual FePP molecules relative to neighboring PP molecules in the mixed array. At electrochemical potentials far away from the redox potential, the FePP and PP molecules appeared at the same height. On the other hand, at the redox potential for FePP, the apparent height of FePP was greatly enhanced over PP. The increase in apparent height has been attributed to alignment of the Fermi level of the graphite electrode with the lowest unoccupied molecular orbital (LUMO) of FePP. This illustrates how electrochemical potential can be used to control current flow through redox molecules at the single-molecule level.

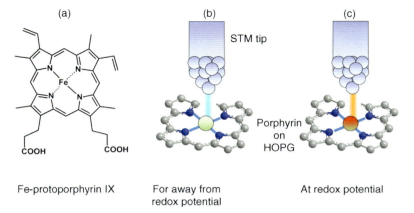

Figure 16.9 The Fe–protoporphyrin compounds employed by Tao in an STM study of tunneling through this molecule [49]. The Fe–protoporphyrin compound was adsorbed in a flat-lying array on the highly ordered pyrolytic graphite (HOPG) surface. Under electrochemical conditions, the redox state of the porphyrin compound could be controlled and used to bring the redox center into resonance in the tunneling gap configuration between STM tip and HOPG surface. (b) Far away from the redox potential, the redox molecule is out of resonance, while (c) at the redox potential, resonant tunneling occurs and an enhanced tunneling current is observed.

16.2.4
Environmental Effects on Junctions

In early metal|molecule|metal junction experiments using STM-based techniques, the focus was on the nature of the contact between the metal contacts and the molecule and upon the nature of the molecule itself, and little attention was paid to the effect of the surrounding medium. In the case of the simplest "model" system, α,ω-alkanedithiols, it was found that the electrical properties of the junctions did not depend on the medium to any significant extent. Chen *et al.* noted that the conductance of a thiol-terminated hepta-aniline oligomer **4** (Figure 16.10) measured under toluene was 0.32 ± 0.03 nS and that the current–bias voltage behavior was linear for tip biases up to -0.5 V [50]. In contrast, in aqueous electrolyte (0.1 M phosphate buffer), at potentials where the oligomer should likewise be in its undoped state, the low-bias conductance was significantly higher (1.8 nS). Moreover, the current versus tip–substrate bias relation showed negative differential resistance characteristics. Oligoanilines are redox active, and in this case, the conductance of the molecule depends on the redox potential, reaching a maximum of 5.3 nS at a potential where the hepta-aniline should be in its partially oxidized (emeraldine salt) redox state, which also coincides with the maximum observed for the conductivity of "bulk" samples of polyaniline [51].

It might be expected that the conductance of an oligoaniline would vary depending on whether it was measured in an aqueous environment, because the conductivity of "bulk" polyaniline is pH as well as redox state dependent. However, we recently found

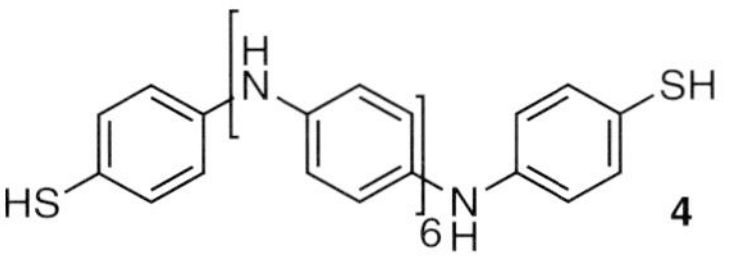

Figure 16.10 Structure of oligoaniline 4.

a much larger solvent effect on the conductance of Au|molecule|Au junctions with oligothiophenes, which are not expected to show pH effects.

Earlier, we had shown that the conductances of junctions involving a range of molecules **5–7** (Figure 16.11) were all similar (0.5–1.5 nS) [52, 53] and much higher than that of $HS(CH_2)_{12}SH$ (0.025 nS) [54], that is, two hexylthiol chains "back-to-back" with no intervening π-conjugated unit. It appears that these molecules act as organic molecular analogues of inorganic double tunneling barrier devices. It was therefore of interest to examine what would happen if the degree of conjugation (and hence, also the length) of the central π-conjugated unit was increased. Oligothiophenes were selected for this purpose, for reasons of synthetic convenience.

The conductance of Au|molecule|Au junctions with **8a–8d** (Figure 16.12a) was measured using the $I(s)$ technique, in ambient air, and was found to increase from **8a** to **8b**; there was little subsequent change for **8c** and **8d** (Figure 16.12b, upper line) [55]. The conductance mechanism is expected to be superexchange for these molecules, since results for longer and more conjugated oligothiophenes, directly connected to Au leads without intervening alkyl "spacers," strongly point to superexchange even for these molecules [47]. Initially, our hypothesis was that the effect of increased conjugation (which might be expected to increase conductance owing to the closer proximity of the frontier orbital energies to the contact Fermi energy) was offset by the increase in the length of the molecules (which would be expected to decrease conductance exponentially). In an effort to test this, the Lambert group at Lancaster carried out transport calculations on this system, using a nonequilibrium Green's function (SMEAGOL [56, 57]) approach. This predicted that in fact conductance should decrease exponentially with the length of the molecules, as indicated in Figure 16.12c (black line) [55].

We realized that the transport calculations assume a vacuum (dry) environment, whereas the experiments were carried out in ambient conditions. Accordingly, we investigated possible effects of atmospheric components (oxygen and water) on

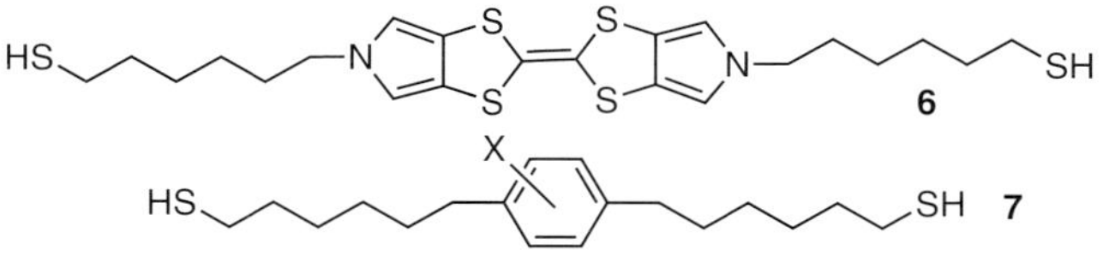

Figure 16.11 Structures of molecules 5–7.

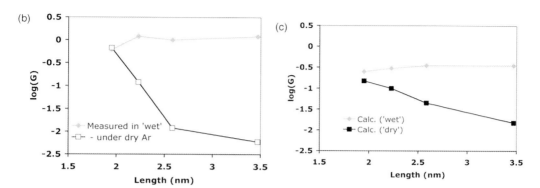

Figure 16.12 Structures of oligothiophenes **8** (a). Experimental (b) and theoretical (c) results for the single-molecule conductance of **8a–8d** in water and in the absence of water. For $n = 3$, the presence or absence of environmental water changes the conductance by two orders of magnitude. Theory reveals that H_2O significantly increases the conductance of the longer molecules (**8c** and **8d**) by interacting with the π-orbitals of the thiophene rings and shifting orbital energies, such that the frontier orbitals move closer to the Au Fermi energy. By cycling between wet and dry argon, the conductance for **8c** could be changed reversibly [55].

junction conductance [55]. While the inclusion of oxygen in transport calculations had almost no effect, water had a dramatic effect. Calculations suggested that water interacts most strongly with the thiophene ring π-system. When such water molecules were included in the model, the transport calculations agreed quite well with the experimental conductance values. Although the interaction of water with the thiophene π-system is quite weak, it is sufficient to shift the transport resonances slightly. This has little effect upon junctions involving the shorter oligothiophenes, because the frontier orbital energies, and hence the transport resonances, are too far from the Au contact Fermi energy for the zero bias conductance to be much affected, but for the more conjugated **8c** and **8d**, the effect is to increase conductance very significantly [55].

Prompted by the results of the calculations, we then attempted to measure the conductances of junctions with **8a–8d** in the absence of water, by using an argon atmosphere chamber. We found that indeed the conductances of **8c** and **8d** were much lower when measured in dry conditions (Figure 16.12b, black line). In fact, the conductance of **8d** under dry Ar could not be determined as it was too low for the current follower of our STM, meaning that it was at least 200 times smaller in the absence of water (the value indicated in Figure 16.12 is an upper limit). The conductance of **8c** was also approximately two orders of magnitude lower in dry

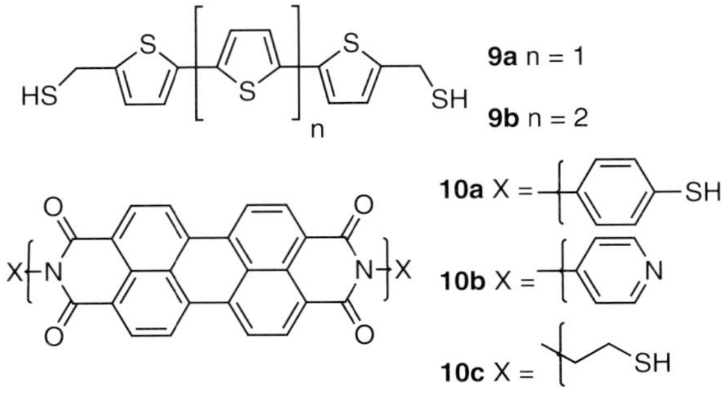

Figure 16.13 Structures of molecules **9** and **10**.

argon than in ambient conditions, or under water-saturated argon (the latter experiment confirming that the presence of oxygen is not a factor in the high conductances) [55].

Xu et al. had earlier studied Au|molecule|Au junctions using the oligothiophenes **9a** and **9b** (Figure 16.13) [58]. The main focus of this paper was the behavior of the conductance of these molecules as a function of electrochemical gating potential, but it is nevertheless interesting to note that the conductance of junctions involving **9b** measured under toluene was $7.5 \times 10^{-5} G_0$ (5.8 nS), but in 0.1 M aqueous $NaClO_4$ (at a potential where the molecule is likewise still in the neutral state), the conductance was $7.7 \times 10^{-4} G_0$ (59 nS; from the histogram in Figure 3 from Ref. [58]), a factor of 10 greater. Although the reason for this observation is not discussed in Ref. [58], we believe that it is related to the presence or absence of water. It is unusual to find a theoretical study that predicts an increase in conductance upon interaction of a molecule with water. More typically, a decrease is predicted. For instance, within the limitations imposed by a rather more crude approximation of the metal contacts, Kula et al. found that water interactions with polar functional groups in oligophenyleneethynylene (OPE) molecules could drastically reduce the conductance of the molecules by causing charge redistributions within the molecule. In this instance, this lowers the off-resonant transition probability [59].

A related case where the electrical properties of a Au|molecule|Au junction depends upon the environment was described by Li et al. [60], and later investigated theoretically by Cao et al. [61]. This work is also concerned with the oxidation state-dependent conductance of redox-active molecules, in this case the perylenetetracarboxylic diimide (PTCDI) derivatives **10a–10c**. It was found that the conductance of junctions involving these molecules increased by up to two orders of magnitude as the molecules were switched electrochemically between their neutral and anionic redox states [60]. The conductance of **10a** in its neutral state in aqueous electrolyte was about $10^{-5} G_0$, or 0.8 nS, and when measured under toluene, it was almost identical (about 1 nS). Therefore, unlike the oligothiophenes, the conductance of this molecule does not appear to be medium dependent. However, the *mechanism* of conductance may be different because whereas under toluene, the conductances of junctions involving **10a** were not temperature dependent, in aqueous electrolyte at potentials where the molecule is in its neutral state, the conductance was temperature dependent. A plot of ln(conductance) versus $1/T$ was linear, with an activation energy of 0.35 eV. These authors [60] suggested that the activation energy corresponded to outer shell solvent reorganization, meaning that in water, a two-step (hopping) mechanism applied at positive potentials. However, in the potential range where the PTCDI moiety is reduced, the temperature dependence disappeared, which is not in accord with a "hopping" model. Cao *et al.* investigated the temperature dependence of the off-state conductance theoretically [61], by combining molecular dynamics simulations of **10a** in water with electron transport calculations. They concluded that the increased conductance at higher temperature could be accounted for by thermal motions of water molecules; even within a one-step tunneling model of conductance, the fact that water solvating the polar carbonyl groups is less ordered and interacts more weakly with the molecule at higher temperature is enough to account for the observed experimental behavior [61].

16.2.5
Influence of π-Stacking on Metal|Molecule|Metal Junctions

The interaction of aromatic molecules via π-stacking is an important factor in determining such issues as crystal packing in the solid state and molecular aggregation in solution, and it plays a key role in biology, most notably in the structure of DNA and in the tertiary structures of proteins. It is perhaps not surprising, therefore, that it can also play a role in metal|molecule|metal junction formation. Using a mechanically controlled break junction technique, Wu *et al.* studied Au|molecule|Au junctions using oligophenyleneethynylene molecules **11–15** (Figure 16.14) [62].

They observed that molecules **11** and **12** gave prominent peaks in their conductance histograms at $(1.2 \pm 0.1) \times 10^{-4} G_0$ (9.2 nS) and $(5.7 \pm 2.4) \times 10^{-5} G_0$ (4.4 nS), respectively. This is as expected, since both molecules have potentially coordinating moieties at each end of the molecule and so should be able to form robust metal|molecule|metal junctions during the dynamic process in which the suspended Au nanocontact is broken in the presence of a solution of the molecules. More

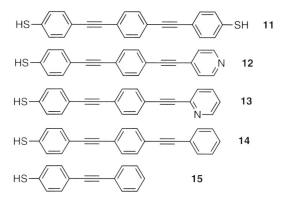

Figure 16.14 Structures of molecules **11–15** from Ref. [62].

remarkable, however, were the results obtained using molecules **13–15**. Molecule **13** has a pyridyl end group, but should be sterically prevented from forming a pyridyl–Au contact, and both **14** and **15** have only one terminal thiol. Nevertheless, molecules **13** and **14** both showed a peak in their conductance histograms at $(6.6 \pm 1.3) \times 10^{-6} G_0$ (0.50 nS) and $(5.9 \pm 2.4) \times 10^{-6} G_0$ (0.45 nS), respectively [62]. This peak was as prominent as those for **11** and **12**, with a similar width at half height. The authors attribute this to the formation of Au|molecule...molecule|Au junctions, in which the two molecules interact by π-stacking. The lower conductance observed for such junctions is expected because of the greater distance between the contacts, and the weaker coupling through the molecules, than for single Au|molecule|Au junctions involving **11** and **12**. Experiments using **15** showed a very broad, ill-defined peak in the conductance histogram, since this molecule is expected to show very much weaker π-stacking interactions as fewer rings from each molecule are capable of overlapping, and so a conductance value could not be assigned in this case [62].

The π-interactions in these junctions have been investigated theoretically by Lin et al. [63]. They used a general Green's function formalism, with two molecules of **14**, each bound to the two semi-infinite gold leads via a triangle of Au atoms, and they chose the center of the HOMO and LUMO of the resulting assembly as the Fermi energy for the transport calculations. Geometry optimization was done at the hybrid density functional B3LYP level with the Lanl2DZ basis set using GAUSSIAN 03. They varied systematically the distance between the two molecules and their relative orientation and found that the transport properties of the junction were very sensitive to both these factors. They concluded that the transport characteristics observed in the experiments could be accounted for if the two molecules were 0.30 nm apart vertically in the junction, with a cofacial stacking arrangement, although the latter seems unlikely given what is known about aromatic π-stacking (*q.v.*).

It is interesting to note that the difference in conductance between **11** and **12** was attributed to weaker coupling between pyridine and Au than between sulfur and Au [62]. However, other workers have observed that the use of asymmetric contacts *in itself* has an adverse effect upon the conductance of Au|molecule|Au junctions. The conductance of such junctions involving molecules $HOOC(CH_2)_nSH$ was significantly lower than those with the symmetrical analogues $HOOC(CH_2)_nCOOH$ and $HS(CH_2)_nSH$ [64].

That the coupling between molecules as a result of π-stacking should be strong enough to allow junctions to form, and to transmit charge moderately well, might seem surprising. Yet support for this comes from an earlier paper by Seferos et al. [65]. They used a magnetically controlled crossed-wire junction technique to measure the electrical properties of molecules **16** and **17** (Figure 16.15). In this technique, a close-packed self-assembled monolayer of the molecule of interest is formed on a 10 mm diameter Au wire. A second 10 mm Au wire is then gently brought into contact crosswise to form the junction, while the tunneling current is monitored. The junctions formed have an area of about 300 nm^2 in this technique, and therefore involve many molecules. Molecule **17** contains a [2.2]paracyclophane unit, in which two benzene rings are constrained to be cofacial and 0.309 nm apart by the two ethylene linkers. Careful electrochemical and surface characterization of monolayers of **16**

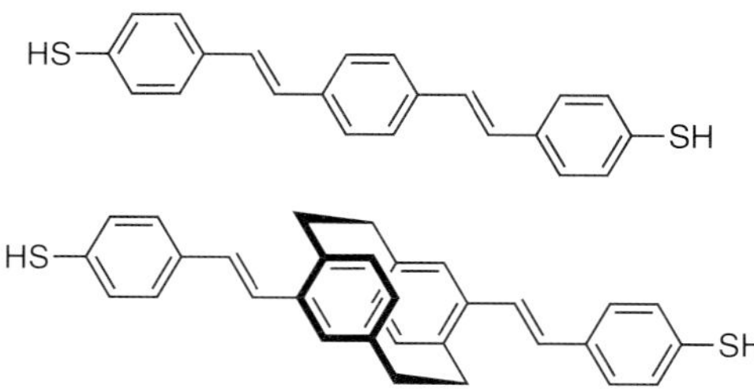

Figure 16.15 Structures of molecules **16** and **17**.

and **17** established that both molecules formed nearly complete monolayers with low defect density, but with approximately twice as many molecules per unit area with **16** than with **17**. Allowing this difference, the authors concluded that the conductance per molecule for junctions involving **16** and **17** was almost identical, in spite of the fact that **16** is fully conjugated and **17** is not [65]. Electronic structure calculations on **16** and **17** with single-terminal Au atoms suggested that the HOMOs of both have significant Au components, and their energies are within 0.5 eV of the Au Fermi energy. In the case of **17**, the π-system HOMO and HOMO-1 are almost isoenergetic, and they span the entire molecule owing to through-space interactions at the cyclophane core (although the HOMO has a node at the cyclophane core).

It is interesting to speculate on the reasons for the differences in findings between these two studies. In the case of molecules **11–15**, the break of conjugation, comparing results for **11** and **14**, causes a drop in conductance by a factor of about 20, whereas the cross-wire results for **16** and **17** suggest similar conductances for these molecules. Apart from the difference in techniques used to fabricate the junctions, in **17** the two π-stacked rings are constrained in a cofacial position by the ethylene spacers. But π-stacking of two molecules of **14** will likely not result in cofacial orientation of the arene rings. The geometrical arrangement of aromatic molecules is essentially governed by electrostatics; net favorable interactions between two arene rings result from π–σ interactions, between the δ^- π and δ^+ σ systems [66]. This will most likely result in a staggered minimum energy conformation for two molecules of **14** in a junction, which may result in a smaller degree of electronic coupling in this case, hence a lower conductance.

A subset of π–π interactions is donor–acceptor interaction. These interactions result in a degree of charge transfer and are characterized by the appearance of intervalence charge transfer bands in the electronic spectra of donor–acceptor complexes. This could be a promising avenue for exploration in single-molecule junction studies, leading ultimately to the prospect of single-molecule electrical sensors. However, to date only one study has been reported, in which charge transfer interactions between monolayers of the hexamethylbenzene donor derivative 1,4-$HSCH_2C_6Me_4CH_2SH$ (**18**) and the strong acceptor tetracyanoethylene (TCNE) were examined on Au(111) [67]. Reflection–absorption infrared spectroscopy (RAIRS), ellipsometry, electrostatic force microscopy, and contact angle measurements supported the formation of a relatively close-packed, upright configuration for monolayers of **18** on Au, in which the molecules were bound through a single sulfur. On treatment of such a monolayer with a solution of TCNE, a charge transfer complex was formed, as judged by changes in the RAIRS spectra, in which the molecules of **18** adopted a "lying down" structure, bound through both thiols, with a complexed TCNE, probably lying flat on top. The TCNE could be removed by treatment of this monolayer with a solution of the stronger donor, trimethyltetrathiafulvalene (Me$_3$TTF), leaving a monolayer consisting solely of **18**, still bound through both thiols. This work predates the publication of the *I(s)* technique, and the monolayers were examined using scanning tunneling spectroscopy (STS), so that statistical evaluation of single-molecule junction electrical properties was not performed. However, it is clear from the STS data that the flat-lying **18**:TCNE charge transfer complex monolayer is much more conductive than the flat-lying free **18**, probably because the charge transfer complex has frontier orbitals closer in energy to the Fermi energy of the contacts.

16.3
Summary and Outlook

Some of the issues that at first puzzled workers in this field, such as apparent multiple conductance values for the same molecule in different studies, appear to be reaching resolution [5, 20, 68, 69]. It appears that the metal–molecule contacts play an important role, and the most-studied example, thiols on gold, are more resistive than at first thought [54]. Work is needed on making more transmissive contacts. For instance, theoretical studies indicate that thiols on palladium might be more transmissive than thiols on gold [70]. Methods for making metal–carbene monolayers that are moderately stable to the ambient environment have been developed [71], and theoretical work suggests that such contacts may be highly transmissive [72], potentially opening the way to organometallic electronics.

It is becoming clear that at least for some structures, supramolecular interactions (H–bonding, π-stacking, donor–acceptor interactions, solvation) are very important in determining the electrical properties of metal|molecule|metal junctions. More work is needed to investigate the range of structures strongly affected by solvent environment. The latter sensitivity is both a problem (for reproducible device fabrication) and an opportunity (e.g., for the development of molecular sensing technologies). There is clearly an interesting role for porphyrins as components in molecular species designed for metal|molecule|metal junction fabrication, given that they appear to be highly transmissive, and that they are key components in the construction of supramolecular systems.

References

1 Green, J.E., Choi, J.W., Boukai, A., Bunimovich, Y., Johnston-Halperin, E., DeIonno, E., Luo, Y., Sheriff, B.A., Xu, K., Shin, Y.S., Tseng, H.R., Stoddart, J.F., and Heath, J.R. (2007) *Nature*, **445**, 414–417.

2 Smits, E.C.P., Mathijssen, S.G.J., van Hal, P.A., Setayesh, S., Geuns, T.C.T., Mutsaers, K., Cantatore, E., Wondergem, H.J., Werzer, O., Resel, R., Kemerink, M., Kirchmeyer, S., Muzafarov, A.M., Ponomarenko, S.A., de Boer, B., Blom, P.W.M., and de Leeuw, D.M. (2008) *Nature*, **455**, 956–959.
3 Xu, B.Q. and Tao, N.J.J. (2003) *Science*, **301**, 1221–1223.
4 Cui, X.D., Primak, A., Zarate, X., Tomfohr, J., Sankey, O.F., Moore, A.L., Moore, T.A., Gust, D., Harris, G., and Lindsay, S.M. (2001) *Science*, **294**, 571–574.
5 Haiss, W., Nichols, R.J., van Zalinge, H., Higgins, S.J., Bethell, D., and Schiffrin, D.J. (2004) *Phys. Chem. Chem. Phys.*, **6**, 4330–4337.
6 Haiss, W., van Zalinge, H., Higgins, S.J., Bethell, D., Höbenreich, H., Schiffrin, D.J., and Nichols, R.J. (2003) *J. Am. Chem. Soc.*, **125**, 15294–15295.
7 Dadosh, T., Gordin, Y., Krahne, R., Khivrich, I., Mahalu, D., Frydman, V., Sperling, J., Yacoby, A., and Bar-Joseph, I. (2005) *Nature*, **436**, 677–680.
8 Reed, M.A., Zhou, C., Muller, C.J., Burgin, T.P., and Tour, J.M. (1997) *Science*, **278**, 252–254.
9 Kergueris, C., Bourgoin, J.P., Palacin, S., Esteve, D., Urbina, C., Magoga, M., and Joachim, C. (1999) *Phys. Rev. B*, **59**, 12505–12513.
10 Weber, H.B., Reichert, J., Weigend, F., Ochs, R., Beckmann, D., Mayor, M., Ahlrichs, R., and von Lohneysen, H. (2002) *Chem. Phys.*, **281**, 113–125.
11 Kubatkin, S., Danilov, A., Hjort, M., Cornil, J., Bredas, J.L., Stuhr-Hansen, N., Hedegard, P., and Bjornholm, T. (2003) *Nature*, **425**, 698–701.
12 Osorio, E.A., O'Neill, K., Stuhr-Hansen, N., Nielsen, O.F., Bjornholm, T., and van der Zant, H.S.J. (2007) *Adv. Mater.*, **19**, 281–285.
13 Venkataraman, L., Klare, J.E., Nuckolls, C., Hybertsen, M.S., and Steigerwald, M.L. (2006) *Nature*, **442**, 904–907.
14 Smit, R.H.M., Noat, Y., Untiedt, C., Lang, N.D., van Hemert, M.C., and van Ruitenbeek, J.M. (2002) *Nature*, **419**, 906–909.
15 Gonzalez, M.T., Wu, S.M., Huber, R., van der Molen, S.J., Schonenberger, C., and Calame, M. (2006) *Nano Lett.*, **6**, 2238–2242.
16 Huber, R., Gonzalez, M.T., Wu, S., Langer, M., Grunder, S., Horhoiu, V., Mayor, M., Bryce, M.R., Wang, C.S., Jitchati, R., Schonenberger, C., and Calame, M. (2008) *J. Am. Chem. Soc.*, **130**, 1080–1084.
17 Champagne, A.R., Pasupathy, A.N., and Ralph, D.C. (2005) *Nano Lett.*, **5**, 305–308.
18 Cui, X.D., Zarate, X., Tomfohr, J., Sankey, O.F., Primak, A., Moore, A.L., Moore, T.A., Gust, D., Harris, G., and Lindsay, S.M. (2002) *Nanotechnology*, **13**, 5–14.
19 Morita, T. and Lindsay, S. (2007) *J. Am. Chem. Soc.*, **129**, 7262.
20 Haiss, W., Martin, S., Leary, E., van Zalinge, H., Higgins, S.J., Bouffier, L., and Nichols, R.J. (2009) *J. Phys. Chem. C*, **113**, 5823–5833.
21 Haiss, W., van Zalinge, H., Höbenreich, H., Bethell, D., Schiffrin, D.J., Higgins, S.J., and Nichols, R.J. (2004) *Langmuir*, **20**, 7694–7702.
22 Haiss, W., Wang, C.S., Grace, I., Batsanov, A.S., Schiffrin, D.J., Higgins, S.J., Bryce, M.R., Lambert, C.J., and Nichols, R.J. (2006) *Nat. Mater.*, **5**, 995–1002.
23 Haiss, W., Wang, C.S., Jitchati, R., Grace, I., Martin, S., Batsanov, A.S., Higgins, S.J., Bryce, M.R., Lambert, C.J., Jensen, P.S., and Nichols, R.J. (2008) *J. Phys. Condens. Matter*, **20**, 374119.
24 Wang, C., Batsanov, A.S., Bryce, M.R., Martin, S., Nichols, R.J., Higgins, S.J., García-Suárez, V., and Lambert, C.J. (2009) *J. Am. Chem. Soc.*, **131**, 15647–15654.
25 Danilov, A., Kubatkin, S., Kafanov, S., Hedegard, P., Stuhr-Hansen, N., Moth-Poulsen, K., and Bjornholm, T. (2008) *Nano Lett.*, **8**, 1–5.
26 Osorio, E.A., O'Neill, K., Wegewijs, M., Stuhr-Hansen, N., Paaske, J., Bjornholm, T., and van der Zant, H.S.J. (2007) *Nano Lett.*, **7**, 3336–3342.
27 Osorio, E.A., Bjornholm, T., Lehn, J.M., Ruben, M., and van der Zant, H.S.J. (2008) *J. Phys. Condens. Matter*, **20**, 374121.
28 Arkin, M.R., Stemp, E.D.A., Holmlin, R.E., Barton, J.K., Hormann, A.,

Olson, E.J.C., and Barbara, P.F. (1996) *Science*, **273**, 475–480.

29. Fink, H.W. and Schonenberger, C. (1999) *Nature*, **398**, 407–410.
30. Porath, D., Bezryadin, A., de Vries, S., and Dekker, C. (2000) *Nature*, **403**, 635–638.
31. Lewis, F.D., Wu, T.F., Zhang, Y.F., Letsinger, R.L., Greenfield, S.R., and Wasielewski, M.R. (1997) *Science*, **277**, 673–676.
32. Meggers, E., Kusch, D., Spichty, M., Wille, U., and Giese, B. (1998) *Angew. Chem., Int. Ed.*, **37**, 460–462.
33. Xu, B.Q., Zhang, P.M., Li, X.L., and Tao, N.J. (2004) *Nano Lett.*, **4**, 1105–1108.
34. Hihath, J., Xu, B.Q., Zhang, P.M., and Tao, N.J. (2005) *Proc. Natl. Acad. Sci. USA*, **102**, 16979–16983.
35. van Zalinge, H., Schiffrin, D.J., Bates, A.D., Haiss, W., Ulstrup, J., and Nichols, R.J. (2006) *ChemPhysChem*, **7**, 94–98.
36. van Zalinge, H., Schiffrin, D.J., Bates, A.D., Starikov, E.B., Wenzel, W., and Nichols, R.J. (2006) *Angew. Chem., Int. Ed.*, **45**, 5499–5502.
37. Cohen, H., Nogues, C., Naaman, R., and Porath, D. (2005) *Proc. Natl. Acad. Sci. USA*, **102**, 11589–11593.
38. Cohen, H., Nogues, C., Ullien, D., Daube, S., Naaman, R., and Porath, D. (2006) *Faraday Discuss.*, **131**, 367–376.
39. Ullien, D., Cohen, H., and Porath, D. (2007) *Nanotechnology*, **18**, 424015.
40. Chang, S.A., He, J., Lin, L.S., Zhang, P.M., Liang, F., Young, M., Huang, S., and Lindsay, S. (2009) *Nanotechnology*, **20**, 185102.
41. Xu, B.Q., Xiao, X.Y., and Tao, N.J. (2003) *J. Am. Chem. Soc.*, **125**, 16164–16165.
42. Chang, S., He, J., Kibel, A., Lee, M., Sankey, O., Zhang, P.M., and Lindsey, S. (2009) *Nat. Nanotechnol.*, **4**, 297–301.
43. Beletskaya, I., Tyurin, V.S., Tsivadze, A.Y., Guilard, R., and Stern, C. (2009) *Chem. Rev.*, **109**, 1659–1713.
44. Grozema, F.C., Houarner-Rassin, C., Prins, P., Siebbeles, L.D.A., and Anderson, H.L. (2007) *J. Am. Chem. Soc.*, **129**, 13370–13371.
45. Winters, M.U., Dahlstedt, E., Blades, H.E., Wilson, C.J., Frampton, M.J., Anderson, H.L., and Albinsson, B. (2007) *J. Am. Chem. Soc.*, **129**, 4291–4297.
46. Sedghi, G., Sawada, K., Esdaile, L.J., Hoffmann, M., Anderson, H.L., Bethell, D., Haiss, W., Higgins, S.J., and Nichols, R.J. (2008) *J. Am. Chem. Soc.*, **130**, 8582–8583.
47. Yamada, R., Kumazawa, H., Noutoshi, T., Tanaka, S., and Tada, H. (2008) *Nano Lett.*, **8**, 1237–1240.
48. Kiguchi, M., Takahashi, T., Kanehara, M., Teranishi, T., and Murakoshi, K. (2009) *J. Phys. Chem. C*, **113**, 9014–9017.
49. Tao, N.J. (1996) *Phys. Rev. Lett.*, **76**, 4066–4069.
50. Chen, F., He, J., Nuckolls, C., Roberts, T., Klare, J.E., and Lindsay, S. (2005) *Nano Lett.*, **5**, 503–506.
51. Brédas, J.L. (1991) *Conjugated Polymers and Related Materials: Proceedings of the 1991 Nobel Symposium in Chemistry*, vol. **NS81**, (eds W.R. Salaneck, I. Lundström, and B. Rånby), Oxford University Press, Oxford, pp. 187–221.
52. Leary, E., Higgins, S.J., van Zalinge, H., Haiss, W., and Nichols, R.J. (2007) *Chem. Commun.*, 3939–3942.
53. Leary, E., Higgins, S.J., van Zalinge, H., Haiss, W., Nichols, R.J., Nygaard, S., Jeppesen, J.O., and Ulstrup, J. (2008) *J. Am. Chem. Soc.*, **130**, 12204–12205.
54. Haiss, W., Martin, S., Scullion, L.E., Bouffier, L., Higgins, S.J., and Nichols, R.J. (2009) *Phys. Chem. Chem. Phys.*, **11**, 10831–10838.
55. Leary, E., Hobenreich, H., Higgins, S.J., van Zalinge, H., Haiss, W., Nichols, R.J., Finch, C.M., Grace, I., Lambert, C.J., McGrath, R., and Smerdon, J. (2009) *Phys. Rev. Lett.*, **102**, 086801.
56. Rocha, A.R., Garcia-Suarez, V.M., Bailey, S.W., Lambert, C.J., Ferrer, J., and Sanvito, S. (2005) *Nat. Mater.*, **4**, 335–339.
57. Rocha, A.R., Garcia-Suarez, V.M., Bailey, S., Lambert, C., Ferrer, J., and Sanvito, S. (2006) *Phys. Rev. B*, **73**, 085414.
58. Xu, B.Q., Li, X.L., Xiao, X.Y., Sakaguchi, H., and Tao, N.J. (2005) *Nano Lett.*, **5**, 1491–1495.
59. Kula, M., Jiang, J., Lu, W., and Luo, Y. (2006) *J. Chem. Phys.*, **125**, 194703.

60 Li, X.L., Hihath, J., Chen, F., Masuda, T., Zang, L., and Tao, N.J. (2007) *J. Am. Chem. Soc.*, **129**, 11535–11542.

61 Cao, H., Jiang, J., Ma, J., and Luo, Y. (2008) *J. Am. Chem. Soc.*, **130**, 6674–6675.

62 Wu, S.M., Gonzalez, M.T., Huber, R., Grunder, S., Mayor, M., Schonenberger, C., and Calame, M. (2008) *Nat. Nanotechnol.*, **3**, 569–574.

63 Lin, L.L., Leng, J.C., Song, X.N., Li, Z.L., Luo, Y., and Wang, C.K. (2009) *J. Phys. Chem. C*, **113**, 14474–14477.

64 Martín, S., Manrique, D.Z., Garcia-Suarez, V.M., Haiss, W., Higgins, S.J., Lambert, C.J., and Nichols, R.J. (2009) *Nanotechnology*, **20**, 125203.

65 Seferos, D.S., Trammell, S.A., Bazan, G.C., and Kushmerick, J.G. (2005) *Proc. Natl. Acad. Sci. USA*, **102**, 8821–8825.

66 Hunter, C.A. and Sanders, J.K.M. (1990) *J. Am. Chem. Soc.*, **112**, 5525–5534.

67 Sundari, B., Kasibhatla, T., Labonté, A.P., Zahid, F., Reifenberger, R.G., Datta, S., and Kubiak, C.P. (2003) *J. Phys. Chem. B.*, **107**, 12378–12382.

68 Li, X.L., He, J., Hihath, J., Xu, B.Q., Lindsay, S.M., and Tao, N.J. (2006) *J. Am. Chem. Soc.*, **128**, 2135–2141.

69 Li, C., Pobelov, I., Wandlowski, T., Bagrets, A., Arnold, A., and Evers, F. (2008) *J. Am. Chem. Soc.*, **130**, 318–326.

70 Seminario, J.M., De la Cruz, C.E., and Derosa, P.A. (2001) *J. Am. Chem. Soc.*, **123**, 5616–5617.

71 Tulevski, G.S., Myers, M.B., Hybertsen, M.S., Steigerwald, M.L., and Nuckolls, C. (2005) *Science*, **309**, 591–594.

72 Ning, J., Qian, Z.K., Li, R., Hou, S.M., Rocha, A.R., and Sanvito, S. (2007) *J. Chem. Phys.*, **126**, 174706.